From Kant to Hilbert:
A Source Book in the Foundations of Mathematics

Volume I

From Kant to Hilbert: A Source Book in the Foundations of Mathematics

Volume I

WILLIAM EWALD
The Law School
University of Pennsylvania

CLARENDON PRESS · OXFORD

Oxford University Press, Walton Street, Oxford, OX2 6DP

Oxford New York
Athens Auckland Bangkok Bombay
Calcutta Cape Town Dar es Salaam Delhi
Florence Hong Kong Istanbul Karachi
Kuala Lumpur Madras Madrid Melbourne
Mexico City Nairobi Paris Singapore
Taipei Tokyo Toronto

and associated companies in
Berlin Ibadan

Oxford is a trade mark of Oxford University Press

Published in the United States
by Oxford University Press Inc., New York

First published 1996
First published in paperback 1999

A catalogue record for this book is available from the British Library

Library of Congress Cataloging in Publication Data
From Kant to Hilbert: readings in the foundations of mathematics/William Ewald.
Includes bibliographical references and indexes

ISBN 0 19 850535 3 (Vol. 1)
ISBN 0 19 850536 1 (Vol. 2)
ISBN 0 19 850537 X (set)

I. Mathematics-Philosophy. I. Ewald, William Bragg, III 1954–
QA8.6.F77 1996 510'.1–dc20 96–1586

Printed in Great Britain by
Bookcraft (Bath) Ltd
Midsomer Norton, Avon.

To my parents

Philosophy probably will always have its mysteries. But these are to be avoided in geometry: and we ought to guard against abating from its strictness and evidence the rather, that an absurd philosophy is the natural product of a vitiated geometry.

—*MacLaurin 1742*, p. 47

CONTENTS

Volume I

Volume II

COPYRIGHT PERMISSIONS

Acknowledgement is due to the holders of copyright for their permission to use or reprint the following materials:

To the Clifford Memorial Library, University of Evansville, Evansville, Indiana, for permission to publish translations of the late correspondence between Cantor and Dedekind.

To Dover Publications, Inc., for permission to reprint the translations by George Bruce Halsted of portions of Poincaré's 'On the nature of mathematical reasoning', 'On the role of intuition and logic in mathematics', and 'Mathematics and logic'. (*Poincaré 1894*, *1900*, *1905b*, *1906a*, and *1906b*.)

To *Fundamenta Mathematicae* for permission to translate Ernst Zermelo's 'Über Grenzzahlen und Mengenbereiche', *Fundamenta Mathematicae*, vol. 14, 339–44 (1930); copyright © 1930 by *Fundamenta Mathematicae*. (*Zermelo 1930*.)

To Kluwer Academic Publishers for permission to reprint the translations by Malcolm F. Lowe of Helmholtz's 'The facts in perception' and 'Numbering and measuring from an epistemological viewpoint', which originally appeared in *Hermann von Helmholtz: Epistemological writings* (ed. Robert Cohen and Marx Wartofsky), copyright © 1977, Reidel publishers. (*Helmholtz 1878b* and *1887*.)

To the Mathematics Association of America for permission to reprint the translation by Arnold Dresden of Bourbaki's 'The architecture of mathematics', *American Mathematical Monthly*, vol. 57, pp. 221–32 (1950); copyright © 1950 by the Mathematics Association of America. (*Bourbaki 1948*.)

To the Mathematisches Institut of the Georg-August Universität, Göttingen, for permission to publish the extract from Hilbert's Göttingen lectures which appears on pp. 943–6.

To the Niedersächsische Staats- und Universitätsbibliothek, Göttingen, for permission to publish the selections from Cantor's late correspondence with Hilbert, and the selection from Dedekind's *Nachlass* which appears on p. 837.

To Springer Verlag for permission to reprint the translation by Gregory H. Moore of 'Five letters on set theory', in Gregory H. Moore, *Zermelo's axiom of choice: its origins, development, and influence*; copyright © 1982 by Springer Verlag. (*Baire* et alii *1905*.)

To Springer Verlag for permission to translate five articles by Hilbert: 'Axiomatic thought', 'The new grounding of mathematics', 'The logical foundations of mathematics', 'Logic and the knowledge of nature', and 'The grounding of elementary number theory'. These articles were published in David Hilbert, *Gesammelte Abhandlungen*, vol. 3, pp. 146–95; copyright © 1935 and 1970 by Springer Verlag. (*Hilbert 1918*, *1922a*, *1923a*, *1930b*, and *1931a*.)

Efforts to locate the current holder of the copyright to the selections from Brouwer were unsuccessful; any information on this point would be appreciated by the editor.

Introduction

The most important advances in the foundations of mathematics have occurred during three periods: classical antiquity, the seventeenth century, and the modern period. In these periods, the great philosophers are also the great mathematicians, and the great problems of mathematics are great problems of philosophy.

The first period gave us Aristotle and Euclid, Plato and Pythagoras, Zeno and Archimedes; irrational numbers, paradoxes of motion, mathematical proofs, axioms of geometry, syllogistic logic—indeed, it furnishes the groundwork for all that was to follow. This period has been intensively studied, and its contours are well known. The principal texts have been translated; commentaries have been written; works such as Heath's *History of Greek mathematics* (*Heath 1921*) provide a comprehensive survey of the main developments.

The second period gave us Galileo and Descartes, Newton and Leibniz; infinitesimals, analytic geometry, the differential and integral calculus, mathematical physics, and a plethora of mathematically inspired systems of metaphysics. This period has been less thoroughly explored. The writings are voluminous, the language is archaic, and the mathematics is complex and unfamiliar. Even the writings of Leibniz are still being published; the writings of lesser figures are often buried in obscure archives. Despite the existence of some superb monographs, the seventeenth century remains the least understood and least accessible of the three periods.

The modern period can usefully be divided into two sub-periods, which overlap both chronologically and in subject-matter. The first period commences with Kant and lasts until Hilbert; the second commences with Frege and continues to the present day. For brevity, the two periods may be referred to as the *nineteenth century* and the *twentieth century*, provided these terms are taken with a pinch of salt.

The present collection is devoted to the first of these two sub-periods. To explain why the span from Kant to Hilbert forms a natural whole, and in what respects it differs from its successor period, it will be helpful to recall the changes that took place in the foundations of mathematics in the last decades of the nineteenth century.

The publication of Frege's *Begriffsschrift* in 1879 is often said to mark the birth of modern logic; and with good reason. Although others—notably De Morgan and Boole—had done important work in logic, their achievements remained on the periphery both of mathematics and of philosophy. But starting in 1879 the logical foundations of mathematics experienced something of a gold-rush. The story is well known, not least because of Jean van Heijenoort's pioneering *Source book in mathematical logic* (*van Heijenoort 1967*).[a] Frege introduces quantification theory, analyses propositions into function and argument rather than subject and predicate, and describes powerful new techniques for studying the foundations of logic and arithmetic. Charles Peirce independently makes comparable advances in logic and the algebra of relations, and his ideas are soon spread to Europe by Schröder. Dedekind produces his set-theoretic analysis of the natural numbers; Cantor discovers non-denumerable multiplicities and develops his theory of transfinite arithmetic. Peano describes a new notation for symbolic logic and uses it to present axiomatic number-theory. Russell discovers the set-theoretic paradoxes, publishes *The principles of mathematics*, and begins his collaboration with Whitehead. Hilbert searches for a consistency proof for arithmetic; Zermelo announces his well-ordering theorem. By 1904, twenty-five years after the *Begriffschrift*, mathematical logic had come into its own, and the new technical discipline of *the foundations of mathematics*, embracing logic, set-theory, formal axiomatics, and the foundations of arithmetic, had been born.

These accomplishments cannot be solely ascribed to Frege's influence (which at the time was slight); nor, indeed, was his discovery of the quantifiers as great an intellectual leap as Cantor's creation of transfinite arithmetic or Dedekind's work on the foundations of number-theory. If mathematical achievement alone were at issue, *Dedekind 1872* or *Cantor 1874* would have as great a claim to inaugurating the new era as *Frege 1879*; certainly the ideas of Cantor and Dedekind have loomed larger in twentieth-century mathematics.

But Frege is important for other reasons as well. In addition to producing the *Begriffschrift*, he wrote about logic and mathematics in a philosophical style that was to become characteristic of the twentieth century. To open his works is to encounter the familiar topics that have had an influence far beyond the philosophy of mathematics: sense and reference, truth and meaning, logical constants, formal languages, proper names, the deficiencies of psychologism, the relation of language to thought—all this at a time when philosophy was dominated by neo-Kantian epistemology. Among his contemporaries, only Peirce wrote on similar topics with similar penetration; and Peirce's philosophical influence was far less then Frege's.

In addition, Frege's interest in the foundations of mathematics is more narrowly focused than that of Cantor and Dedekind. He sees vividly the intrinsic interest, both philosophical and mathematical, of the new logic, and he pursues his foundational investigations primarily for their own sake rather than for their

[a] For an appreciation of van Heijenoort's contribution to the history of logic, see the dedicatory remarks in *Paris Logic Group 1987*, pp. 1–8. The *Source book*, the fruit of a decade's work and a lifetime's thought, is a milestone in its own right; the sagacious choice of documents, the craftsmanship of the introductory notes, and the skill of the translations can hardly be bettered. Special mention should also be made of van Heijenoort's translator, Stefan Bauer-Mengelberg, whose contribution to the foundations of mathematics seems never to have been adequately appreciated. Of the 41 selections in the *Source book*, 31 are translations. Of the 31 translations, Bauer-Mengelberg is responsible, in whole or in part, for 25; and these 25 were, for the most part, the longest and most difficult.

possible applications to other branches of mathematics. Cantor and Dedekind, in contrast, pursue the foundations of mathematics for the light it can shed on problems in real analysis and algebraic number-theory. In this respect, they belong to the nineteenth century, while Frege belongs to the twentieth.

Indeed, it is important at this point to observe that the phrase *the foundations of mathematics* can be used in two distinct senses. In one sense, it is a technical discipline within mathematics—'mathematical logic', if logic is understood to include such matters as set-theory and the foundations of arithmetic. In the other sense, the foundations of mathematics are the concepts, techniques, and structures that are central to mathematical practice—the elements of which mathematics is composed, rather than the groundwork on which it rests. Thus, the concepts of *saturated models* and *primitive recursive functions* are foundational in the logical sense, but not the mathematical; *groups* and *commutative rings* are foundational in the mathematical sense, but not the logical; *sets* and *functions* are foundational in both senses; highly specialized concepts of, say, point-set topology are foundational in neither.

Broadly speaking, there are two important criteria that distinguish the foundations of mathematics of the nineteenth century from those of the twentieth. First, the twentieth century has tended to pursue foundations in the logical sense, while the nineteenth tended to pursue them in the mathematical. Second, the twentieth century approaches the philosophy of mathematics largely *via* logic and the philosophy of language, while the nineteenth century approached it *via* epistemology. Of course there are overlappings and continuities, and these distinctions have blurry edges; but they bring out the difference in flavour between the writings of, say, Helmholtz and those of Carnap.

Frege is an important transitional figure, not only because he introduced many of the philosophical ideas that were to dominate the twentieth century, but also because he represents a narrowing of subject-matter. Apart from some minor contributions to group-theory, Frege's technical research was confined to logical foundations: a pattern that was to be followed by many of the great mathematical logicians of the twentieth century—Russell, Gödel, Herbrand, Turing, Tarski, and their successors. And the twentieth-century philosophy of mathematics has been similarly preoccupied with issues in mathematical logic: the axiom of choice, the logical analysis of number, Church's thesis, the iterative conception of set, Hilbert's programme, the logical paradoxes, the incompleteness theorems, definitions of truth.

The picture in the nineteenth century is very different. Gauss and Bolzano, Riemann, Dedekind, Poincaré are remembered principally for their contributions to the mainstream of mathematics: their outlook is much broader than that of their twentieth-century colleagues. They drew their inspiration from the whole of mathematics—from algebra and geometry, analysis and number-theory, as well as from logic and set-theory. Each of these disciplines produced a crop of discoveries so new and unexpected, so rich in philosophical consequences, that even the mathematicians who made them could scarcely believe their eyes. Cantor, informing Dedekind in 1877 that the points of the unit interval can be correlated one-to-one with the points of a Euclidean space of any dimension, was driven to express his feelings in French: *Je le vois, mais je ne le crois pas*. The comment might stand as a motto for the century.

Throughout the century, the results come thick and fast. Bolzano in 1817 gives his 'purely analytic proof' of the intermediate value theorem, and starts on the path towards a purely arithmetical theory of the real numbers; he also publishes his study of the infinite. Gauss hits upon the idea of non-Euclidean geometry, and paves the way for Riemann by his investigations of curved manifolds; in number-theory, he defends the legitimacy of the complex integers, and describes their geometric representation in the plane.

Now follow a spate of developments in algebra. William Rowan Hamilton widens the number-concept beyond Gauss, introducing *quaternions*—new objects that fail to obey the commutative law of multiplication. Then Hermann Grassmann, in his *Ausdehnungslehre* of 1844, describes vector algebras even more general than those of Hamilton. These results and the investigations of Peacock, Gregory, De Morgan, and Boole put algebra on a new footing. No longer is it assumed that algebraic equations must behave like the familiar operations of elementary arithmetic: the rules obeyed by algebraic equations are to be distinguished from the properties of the objects studied. Boole applies the new algebraic techniques to the analysis of logic and probability; traditional Aristotelian logic is turned into a formal calculus. New algebras with strange new properties crop up everywhere, in almost tropical abundance: octonians, Clifford algebras, linear associative algebras, matrices, vector spaces. Sylvester, calling himself the 'new Adam', provides exotic names for the flora and fauna of the mathematical world—combinants, reciprocants, concomitants, discriminants, zetaic multipliers, plagiographs, skew pantographs, allotrious factors. By 1860, Benjamin Peirce is classifying hundreds of different algebras; the familiar integers now occupy only a small corner of the mathematical zoo. His son, Charles Peirce, studies the algebra of relative terms and the logic of quantification; this work, *via* Schröder, led eventually to the logical investigations of Löwenheim and Skolem.

Developments in geometry are similarly revolutionary. In Göttingen in 1854 Riemann delivers his address on the hypotheses which lie at the foundation of geometry; when this memoir is published in 1868, Clifford suggests almost at once that the new geometry may have physical applications. Helmholtz and Klein rush through the doors opened by Riemann, and develop non-Euclidean geometry into a mature mathematical discipline. Hilbert observes that 'It must be possible to replace in all geometric statements the words *point*, *line*, *plane*, by *table*, *chair*, *mug*'; from this shrewd observation spring not only his investigations into the axioms of geometry, but also, in time, proof theory and model theory.

In analysis, Weierstrass, lecturing from 1859 in Berlin, pursues his project of putting the calculus on a strictly arithmetical foundation. His student Cantor, working on questions raised by Riemann in trigonometric series, discovers the existence of non-denumerable sets of real numbers; he is led to far-reaching studies of the structure of the continuum. Dedekind, too, investigates the set-theoretic foundations of the real numbers and the integers; these investigations are intimately bound up with his creation of modern algebraic number theory. At about the same time, Frege discovers the theory of quantification, and begins his study of the logical foundations of arithmetic. Peano in 1890 discovers space-filling curves, and thereby delivers a blow to settled geometric intuitions.

Not everybody approves of the new trend towards infinitary, set-theoretic mathematics. Kronecker, urging an algorithmic conception of number theory, criticizes Weierstrass, Cantor, Dedekind, and their arithmetical continuum; so do Klein and Poincaré, but from very different points of view. Hilbert, fresh from his investigations of the foundations of geometry, takes up the cause of Cantor's transfinite numbers, and begins to extend his axiomatic method to logic and the foundations of arithmetic.

These developments are not tidy, and the central ideas cut across traditional barriers between subjects. Thus a fundamental issue in the philosophy of algebra—the distinction between manipulating meaningless *symbols* and manipulating *numbers*—is treated by Berkeley in real analysis, by Lambert in geometry, by Gauss in number-theory, by Hamilton, Gregory, and De Morgan in algebra, by Boole in logic, by Pasch, Fano, and Hilbert in geometry, and again by Hilbert in his proof theory.

This complicated network of interconnections constitutes one of the chief reasons for studying the foundations of mathematics of the nineteenth century. More than any other period, the nineteenth century enbraces the entirety of mathematical research, and can serve as a corrective to the narrowness of gaze of twentieth-century philosophy of mathematics. Logic remains an important part of the story; but it here appears in its natural connections to the other branches of mathematics.

Any attempt to assign precise chronological boundaries to this period must be arbitrary. Berkeley's philosophical writings, especially his *Analyst* of 1734, raise many of the epistemological problems that were to trouble nineteenth-century mathematicians. Although Berkeley is not usually thought of as a philosopher of mathematics, he was ahead of his time in his discussions of the foundations of the calculus, geometry, and arithmetic. But Berkeley had almost no influence upon the mathematical thinkers of the nineteenth century: he was remembered principally as an ingenious paradoxer who had denied the existence of matter, and his penetrating criticisms of Newton and Leibniz were forgotten. (Indeed, until the present work Berkeley's *Analyst* has never been reprinted except in complete editions of his writings.)

Kant, in contrast, exerted a powerful influence over the entire period, and set the stage for many of the debates that were to follow. Bolzano, Hamilton, Riemann, Peirce, Helmholtz, Frege, Cantor, Poincaré, Hilbert, and Brouwer all explicitly oriented themselves with respect to his work, whether in agreement or in opposition. His *Critique of pure reason*, the first edition of which was published in 1781, is the most conspicuous feature on the philosophical landscape, and may be taken to mark the beginning of our period.

As for the end, Hilbert stands out as the last great mainstream mathematician to pursue the foundations of mathematics in the nineteenth-century style. His Göttingen lectures from the 1920s take a panoramic view of mathematics as a whole; they embrace mathematical physics and the foundations of geometry as well as set-theory and the philosophy of the infinite. Hilbert treats logic as being of interest in its own right, but also for the contribution it can make to the central questions of mathematics. Something of the spirit of these lectures is conveyed in his last public address, 'Logic and the knowledge of nature' (*Hilbert 1930*). This address—delivered in his and Kant's native city of Königsberg a century and a half after the publication of the *Critique of pure reason*—may be taken to mark the end of our period.

* * * * * * * * * *

The readings in this collection have been selected to present some of the main developments in the foundations of mathematics during the period from Kant to Hilbert.

Several classic works are here translated into English for the first time: Bolzano's *Contributions to a better-grounded presentation of mathematics*, Dedekind's correspondence with Cantor, Cantor's *Grundlagen*, Hilbert's *Axiomatic thought*. These works take their rightful place beside such warhorses as Dedekind's *Was sind und was sollen die Zahlen*? or Riemann's lecture on the foundations of geometry. Numerous subsidiary readings have also been provided to help place these central documents in historical perspective.

In general, an effort has been made to include documents that are either difficult to obtain or that have been unaccountably neglected.[b] Conversely, works that are well known and widely available have been omitted. In particular, there is no reduplication of material from *van Heijenoort 1967* or *Benacerraf and Putnam 1983*, nor from the works of Frege, Russell, or Gödel.

Both mathematical and philosophical writings have been included; most selections are a mixture. It is important to remember that most of the authors in this collection are mathematicians rather than philosophers: their contributions have to be assessed in the light of the mathematics of the age. So some technical material has been included to convey a sense of evolving mathematical styles. The aim here was to choose technical papers that would illustrate changes in style while being as accessible as possible to the non-specialist.

The selections attempt to give adequate representation to foundational work in each of the main branches of mathematics; the hope being to show the interconnections between algebra and geometry, number theory and analysis—and of all of these subjects to logic and set theory.

The selections from the British algebraists in Volume I may need a special

---

[b] In the latter category belong Berkeley's *Analyst*, Lambert's essay on parallel lines, Kant's early writings on geometry, and Charles Peirce's writings on logic. The neglect of Pierce is particularly difficult to explain. His accomplishments in logic were comparable to those of Frege, he wrote on a wide variety of topics central to twentieth-century philosophy, and he did so with uncommon insight; yet today he is virtually forgotten. Even *van Heijenoort 1967*, on most matters an impeccable guide, passes over his discovery of quantification theory in silence.

word of explanation, especially the long selection from Hamilton's *Lectures on quaternions*. Quaternions are no longer important to mathematics, and Hamilton's 'philosophy of pure time' was never important to philosophy. But his essay gives a fine picture of the state of algebra in 1853, and shows him grappling with one of the central foundational questions of nineteenth-century mathematics. The idea of a non-commutative algebra was almost as difficult to accept as the idea of a non-Euclidean geometry; and the idea of an algebra as a purely syntactic calculus took many decades to be absorbed. Hamilton did not achieve a final solution of this problem; but his account of his discovery of quaternions records an important milestone in the history of nineteenth-century mathematics.

This collection cannot pretend to offer a complete picture of the nineteenth-century foundations of mathematics: the subject is too vast for such a thing to be possible. Some important documents had to be excluded because they were too long; others, because they were too technical; yet others, because of difficulties with the copyright. Many important topics receive only cursory treatment; others are not treated at all.

Introductory notes have therefore been provided which attempt to fill gaps and to supply some of the historical background; these notes also make suggestions for further reading. But the material is too voluminous and too diffuse to be happily treated by this expedient. Readers are accordingly urged to use this collection in conjunction with one of the standard histories of the period, such as *Kline 1972*, *Klein 1926–7*, *M. Cantor 1894–1908*, *Bourbaki 1969*, or *Dieudonné 1978*. The selection *Cayley 1883* gives a masterly survey of developments in nineteenth-century mathematics, and may be found a convenient starting-point for readers to whom this period is not familiar.

References appear at the end of Volume I, and an extensive bibliography, incorporating all references from both volumes, appears at the end of Volume II. References are given in the form *Dedekind 1888*; or simply *1888* if the reference to Dedekind is clear from the context. A reference such as (*Peirce 1931–58*, Vol. iii, p. 47) refers to page 47 of volume three of Charles Peirce's *Collected papers*. Benjamin Peirce and Moritz Cantor are designated as *B. Peirce* and *M. Cantor*; Charles Peirce and Georg Cantor are simply *Peirce* and *Cantor*.

No uniform rule has been adopted for dating the selections. In most cases the date is the year of first publication, i.e. the year in which the work first became widely available to the mathematical public. Usually this is the year in which the selection was first printed; sometimes, however, it is the date of a major public address. But not all dates follow this rule.[c] When dating raises

[c] Some selections were written in the author's youth and printed decades after his death—for example, the *Habilitation* addresses of Riemann and Dedekind, which were both delivered in the presence of Gauss in Göttingen in 1854. It would have been incongrous for Dedekind's address, which was first printed in the 1930s, to be assigned a later date than the eight other Dedekind selections; so it appears here as *Dedekind 1854*. On the other hand, Riemann's address was unknown until it was printed in 1868—the same year Helmholtz began independently to write about the foundations of geometry. To have dated it 1854 would have dated it correctly with respect to Dedekind, but would have given a misleading picture of Riemann's relationship to Helmholtz and to the development of non-Euclidean geometry. So it bears the date *Riemann 1868*. Another such posthumously published work is *Gauss 1929*. This selection is so dated because the exact date of composition is unknown.

special problems, full information on dates of composition, of public delivery, and of first printing is given in the Bibliography.

The selections are arranged by author, and, with a few exceptions, are ordered chronologically.[d]

Footnote references have been given in the form in which they originally appeared. Sometimes it is useful to know which edition, translation, or reprinting an author relied upon (or to know that this information is not conveyed by the author's references); for this reason, there seemed little merit in attempting to impose a uniform system of citation. Detailed information on the mathematical works cited can be found in the Bibliography; citations of non-mathematical works (for example Cantor's references to various obscure histories of scholasticism) have not been included in the Bibliography. Asterisks and daggers have often been replaced by numerals; editorial footnotes are lettered rather than numbered. The original numbering of footnotes has, so far as possible, not been disturbed.

These volumes contain 89 selections; of these, 56 are translations; of the translations, 36 appear here for the first time. All new translations are by the editor, except for the translations of *Bolzano 1804*, *1810*, and *1817a* (which are by Stephen Russ), of *Dedekind 1876–7* (which is by David Reed), and of *Zermelo 1930* (which is by Michael Hallett). Hallett also wrote the introductory note to Zermelo; all other notes are by the editor. The editor has also made revisions to the previously existing translations, either to correct errors or to bring the terminology into line with the other translations in this collection; *Dedekind 1872* and *1888* have been heavily revised in this way, as have several of the selections from Poincaré (which were previously only partially translated).

When a translated word or phrase seemed to merit special attention, it has been given in the original language in double square brackets: for example '⟦Mannigfaltigkeit⟧'. Such bracketed words are printed in italics only if they so appeared in the original.

Editorial insertions are always given in double square brackets. Single square brackets are sometimes used for other purposes (for example to indicate deletions or additions between various editions). These uses are always explained in the accompanying note.

German authors sometimes emphasize words by increasing the spacing between letters; these emphases have always been rendered here with italics. It is also a German practice to print proper names in small capitals; this practice has not been preserved in the translations.

In his translations the editor has tried to cleave to the terminology of Stefan Bauer-Mengelberg's translations in *van Heijenoort 1967*; in particular, he has

---

[d] Gauss appears after Bolzano because *Bolzano 1804* continues the discussion of themes raised by Lambert, while *Gauss 1831* leads naturally into the selections from the British algebraists. Hamilton appears after De Morgan and Gregory because his long *Preface to the lectures on quaternions*—the second Hamilton selection—is best read after their writings. Kronecker appears after his juniors Dedekind and Cantor because he is writing in response to their work.

adopted Bauer-Mengelberg's translation of Hilbert's adjective '*inhaltlich*' by the neologism 'contentual'. The editor has also striven not to break up paragraphs or sentences (although in a very few cases some exceptionally long and convoluted German sentences had to be split into two English sentences). Technical vocabulary has deliberately been translated literally. For example, the German word *Fundamentalreihe* has been translated as 'fundamental sequence' rather than as the more familiar 'Cauchy sequence'. 'Fundamental sequence' was good English mathematical usage a hundred years ago, and *Cauchyreihe* is good German mathematical usage today; but to have translated *Fundamentalreihe* as 'Cauchy sequence' would have been anachronistic.

Citations of the works in this collection present a special problem. Many of the works originally appeared in journals or books that are now difficult to obtain. Some have been reprinted in various collected works; others have been printed in more than one forum. Some selections have been translated into other European languages; others have undergone several revised editions. Page numbers in one version do not necessarily provide a clue to page numbers in another. The situation is already immensely confusing; and so it seemed best to attempt to design a context-free system of citation, rather than to add yet another layer of confusion. (This is particularly important as a courtesy to scholars working with the original texts.) The system varies, as it must, from document to document. Sometimes the original pagination is preferred; sometimes section numbers; sometimes paragraph numbers. The preferred form for citations is given in the introductory note at the beginning of each selection.

Readers are urged to regard the translations merely as an introduction to the original texts. Certainly any scholar wishing to work in this area must expect to delve deeply into the primary sources, and to read them in the original languages. This is not just because no translation is perfect (although that is true enough). But, as a glance at the bibliography will show, very little has in fact been translated. For every article in this collection, there exist many related background works, often entire volumes, that have never been translated; and these works must be consulted if the original selection is to be properly understood in its historical setting. At best, a compilation of this kind can serve as a general introduction to the subject; it is no substitute for a library, or for a thorough knowledge of the original documents.[e]

---

[e] This may be an appropriate spot to note some of the important topics that had to be omitted from these two volumes. I should have like to have included some readings from Dirichlet, Riemann, Heaviside, and others on the development of the concept of *function*. Hermann Grassmann's *Ausdehnungslehre* is badly slighted; but this important and influential book does not lend itself to being read in excerpts. Poincaré's work in topology deserves to be represented, as do his late articles on set theory. More extensive selections from the French analysts (notably Borel, Lebesgue, and Baire) would have shed valuable light on the reception of Cantor's ideas; so would appropriate selections from the writings of Schoenflies, Hausdorff, and the early Brouwer. The topic of implicit definitions (raised in *Gergonne 1818* and treated also by Pasch, Fano, Veronese, Peano, and Hilbert) could have formed an instructive counterpoint to the selections on the foundations of algebra, and made it clear that the idea of an uninterpreted formal calculus has roots in geometry as well as in algebra. The topic of constructivism is here treated principally *via* the medium

Acknowledgments. The idea for this collection came from a set of eight seminars given by Daniel Isaacson in Oxford in Hilary Term of 1984. The seminars treated many of the authors represented here, and in particular made a strong case for the superiority of Dedekind's analysis of the natural numbers to Frege's. I am grateful to Isaacson for many suggestions about the readings to be included, and for his constant encouragement; the book would look very different without his influence.

Another participant in the Isaacson seminar, Michael Hallett, was also of great assistance. He gave me the benefit of his detailed knowledge of the Göttingen archives, and guided me towards the *Nachlässe* of Hilbert and Dedekind, as well as towards the Cantor–Hilbert correspondence. He pointed out many documents that needed to be translated, and consented to translate the selection from Zermelo himself. In preparing the introductory notes to the selections from Cantor, I drew heavily on his *Cantorian set-theory and limitation of size*. This book, despite its forbidding title, is one of the most readable and illuminating contributions to the philosophy of mathematics in many years; it is particularly noteworthy for the deft way in which it blends philosophy, mathematics, and history. (In this respect, Hallett's book is very much in the spirit of Isaacson's seminar.) I learned much from studying his work, and hope that his style of scholarship will find many imitators.

While editing this book, I was supported by four institutions: The Queen's College, Oxford; the Philosophisches Seminar of the Georg-August Universität, Göttingen; the Institute for Advanced Study, Princeton; and the Istituto Universitario Europeo, Florence. I am grateful to all four institutions for their support; to the Alexander von Humboldt Stiftung for a fellowship which made my stay in Göttingen possible; and to the Research Foundation of the University of Pennsylvania for assistance with the costs of preparing the manuscript for publication.

I also owe thanks to Stephen Russ for allowing me to publish his translations of Bolzano; to David Reed for translating *Dedekind 1876–7*; and to Angus Bowie for his help with the classical quotations. Ralf Haubrich provided useful advice on Dedekind, as did Paolo Mancosu and Walter van Stigt on Hilbert and Brouwer respectively.

No doubt there are many ways in which this work could be improved; and no doubt it contains many errors. I hope one day to produce a revised and expanded edition. Comments and suggestions will be received with gratitude; they can be communicated either to the Mathematics Editor at Oxford University Press, or to wewald@oyez.law.upenn.edu.

Philadelphia
September 1992

---

of number theory; but it is also important in geometry and early topology. The writings of Weierstrass and his school; Hilbert's unpublished lectures on the foundations of mathematics; Riemann's philosophical speculations; Weyl on the continuum; the chief methodological writings of the mathematical physicists—all of these important topics had to be omitted if the topics represented here were to receive adequate treatment. Anybody who rummages in the titles listed in the Bibliography will be able to extend this list of omissions almost indefinitely.

1
George Berkeley (1685–1753)

Of all treatises written on the subject in the eighteenth century, Berkeley's *Analyst* was the most sustained and penetrating critique of the methodology of the infinitesimal calculus. Despite its early date (1734) and the fact that it was largely ignored by mathematicians, this work foreshadows the foundational research of the nineteenth century, and provides a link between the mathematical preoccupations of the seventeenth and eighteenth centuries and those of the nineteenth.

Broadly speaking, the mathematicians of the seventeenth and eighteenth centuries—and in particular Newton, Leibniz, and Euler—had been more concerned with exploiting and extending the techniques of the differential and integral calculus than with tidying its foundations. The mathematicians of the early nineteenth century, in contrast, bent their efforts towards placing the calculus on an unobjectionable footing—towards chasing away the obscurities that surrounded such notions as *infinitesimal*, *limit*, and *differential*. This project was pursued during the first two-thirds of the nineteenth century by such mathematicians as Gauss, Bolzano, Cauchy, Abel, Fourier, Riemann, and Weierstrass. Their investigations into the foundations of real analysis in turn inspired the later studies by Dedekind, Cantor, Frege, Peano, Peirce, Russell, and Hilbert of set-theory, logic, and the foundations of arithmetic.

Berkeley's *Analyst* prefigures this entire development, and his philosophically-motivated criticisms of the Newtonian mathematics of the eighteenth century raise many issues that will loom large in the selections that follow. These issues may be loosely grouped under four headings.

1. His critique of *infinitesimals* raises not only the issue of justifying the central concepts of the calculus (a principal concern in *MacLaurin 1742*, *D'Alembert 1765a,b*, and *Bolzano 1817a*) but also the more general issue of the legitimacy of the *actual infinite* in mathematics. (This issue is already present in 'Of infinites' *Berkeley 1901* [1707]).) Berkeley's objections to the actual infinite will be encountered again, wearing different masks, in many of the selections that follow: for instance, in the selections from Kronecker, Cantor, Hilbert, Brouwer, and Poincaré.

2. Berkeley, as befits the empiricist philosopher who wrote *An essay towards a new theory of vision* (1709), makes acute comments about the relationship between geometry, human visual perception, and the foundations of the calculus. These themes will reappear in selections from Bolzano; from Riemann and Clifford; and from Helmholtz and Klein.

3. In numerous places Berkeley discusses the problem of the *reference* of

mathematical expressions; declaring, for instance, in *Of infinites*, that ' 'Tis plain to me we ought to use no sign without an idea answering to it'; one of his cardinal criticisms of Newton is that infinitesimals can have no empirical reference. This topic of the reference of mathematical expressions (a topic on which Berkeley occasionally shifted his position) was to be central to the development of algebra during the nineteenth century, and we shall encounter it again in the selections from Lambert, Gauss, Gregory, De Morgan, Hamilton, Boole, Dedekind, and Hilbert.

4. All these mathematical issues—in analysis, geometry, and algebra—are intertwined in Berkeley's thought with more general concerns about *mathematical truth, the rigour* of demonstrations, the *applicability* of mathematics to the empirical world, and the *scope* and *limits* of mathematical knowledge. These topics were not central to the mathematical thought of Berkeley's immediate predecessors and successors, but were to dominate the mathematical research of the nineteenth century; it is remarkable that they should have been explored so early and in such depth in an essentially philosophical treatise. Philosophers more often react to developments in mathematics than anticipate them, and in modern times perhaps only Descartes, Leibniz, and Kant can be said to have equalled Berkeley's insight into the foundations of mathematics .

The connection between Berkeley's mathematical interests and the leading strands of his philosophy was not adventitious: his education at Trinity College, Dublin was in mathematics and logic as well as in philosophy and the classics, and the foundations of mathematics was to be a lifelong preoccupation. His investigations into the philosophy of mathematics commenced with his first published work (in effect, his bachelor's thesis), the *Arithmetica et miscellanea mathematica* (*1707*; probably written 1704); they were continued in his early, unpublished *Philosophical commentaries* (*1707–8*) and *Of infinites* (*1901* [1707])—especially in *Notebook B* of the *Commentaries*, where remarks on optics and on the nature of the soul mingle with observations on algebra, geometry, and the infinitesimal calculus; they play a major though subsidiary role in his *Principles of human knowledge*, *De motu*, and *Alciphron*; and they are once more the centre of attention in his last major philosophical enterprise, the critique of Newton and the Newtonians in *The analyst*.

Berkeley would never have acknowledged a sharp distinction between his studies in metaphysics and his studies in 'natural philosophy', and it is important to remember, even in reading his works on epistemology and immaterialism, that his thought was as heavily affected by Newton and Boyle as by Locke, Bacon, and Malebranche. Indeed, Berkeley's writings show the futility of attempting to draw a sharp boundary between philosophy and mathematics. We have already observed that his philosphical reflections led him to deep criticisms of current mathematical practice. But the influence goes in the other direction as well, and his study of mathematics and the physical sciences affected his general metaphysical position—so much so that the two are often difficult to disentangle. His discussions of infinitesimals are interwoven with arguments about *minima sensibilia* and with his doctrine that, for sensible objects, *esse* is *percipi*; his discussion in the *Principles of human knowledge* of algebra and of geometric reasoning is bound up with his view of language and his rejection of abstract ideas; his critique of the foundations of the calculus is motivated by many of the same epistemological considerations that underlie his critique of the idea of material substance. It is these connections (rather than his powerful but already-known logical criticisms of Newton's reasoning in the *Principia*) that give his philosophy of mathematics its depth and its strength and its present interest.

Despite the penetration of his writings—their wealth of implications both for mathematics and for philosophy—Berkeley had little actual influence on the development of mathematics. He was nobody's inspiration, and indeed to this day is rarely thought of as a philosopher of mathematics at all. The nineteenth century remembered him principally as an ingenious paradoxer, the precursor of Hume who had denied the existence of material substance; his writings on mathematics have largely been forgotten. (Two exceptions are reproduced below: *MacLaurin 1742* and the correspondence between William Rowan Hamilton and Augustus De Morgan.) But even if Berkeley did not himself initiate or influence the great period of nineteenth-century foundational research, he nevertheless glimpsed many of its central themes; and his writings are therefore an appropriate starting-point for these volumes.

The selections below represent the main lines of his thought; but there is a great deal of supplementary material in *Berkeley 1948–57*, especially in Volumes 1 and 4. *Stammler 1922* and *Jesseph 1993* are general studies of Berkeley's philosophy of mathematics; *Boyer 1939*, *Baron 1969*, *Edwards 1979*, and *Grattan-Guinness 1980* give accounts of the mathematical background.

A. *FROM THE* PHILOSOPHICAL COMMENTARIES (*BERKELEY 1707–8)*

The *Philosophical commentaries* (so named by Berkeley's editors in *Berkeley 1948–57*) consist of two notebooks, now in the British Library, each containing some four hundred working notes on philosophical topics. They were written in 1707–8 (roughly the time of Berkeley's election as a Fellow of Trinity College, Dublin) and were first printed in *Berkeley 1871*. The notebooks were physically bound together in the wrong order, so that most of the entries in Notebook B were in fact written before those in Notebook A. The notebooks show him groping towards the doctrines of *An essay towards a new theory of vision* (1709) and *A treatise concerning the principles of human knowledge* (1710). Many of the notes express views that Berkeley later discarded, while others may reflect his reading rather than his own thoughts. For this reason they should be cited with caution.

The individual entries often have symbols beside them in the margin; the precise significance of these symbols is uncertain. The symbol + probably indicates an entry about which Berkeley later changed his mind; the symbol **x** apparently indicates entries that found their way into the *New theory of vision*. Other symbols are explained by Berkeley; for instance, the symbol **Mo.** that appears by entry 769 below indicates that the entry concerns moral philosophy. Berkeley occasionally changed or erased a marginal symbol: another reason for citing the following entries with caution. (A discussion of Berkeley's marginal symbols can be found in the editors' introduction to the *Philosophical commentaries* in *Berkeley 1948–57*, Vol. 1, or more fully in the article on this subject by A.A. Luce in *Hermathena* for 1970.)

The selection below is reprinted from *Berkeley 1948–57*, Vol. 1; emendations in single square brackets are from that edition. References to *Berkeley 1707–8* should be to the paragraph numbers, which first appeared in the edition of 1948. (In the 1948 numeration the paragraphs of the *recto* side of the notebook are numbered consecutively; entries on the facing *verso* side—which seem to reflect Berkeley's later thoughts—are assigned the number of the facing paragraph followed by a small Roman *a*. Thus note 341a is Berkeley's facing-page comment on note 341.)

FROM NOTEBOOK B:

x 341. When a small line upon Paper represents a mile the Mathematicians do not calculate the 1/10000 of the Paper line they Calculate the 1/10000 of the mile 'tis to this the[y] have regard, tis of this the[y] think if they think or have any idea at all. the inch perhaps might represent to their imaginations the mile but ye 1/10000 of the inch can not be made to represent anything it not being imaginable.

x 341a. But the 1/10000 of a mile being somewhat they think the 1/10000 of the inch is somewhat, w^{n} they think of y^{t} they imagine they think on this.

x 351. We need not strain our Imaginations to conceive such little things. Bigger may do as well for intesimals since the integer must be an infinite.

x 352. Evident y^{t} wch has an infinite number of parts must be infinite.

x 353. Qu: whether extension be resoluble into points id ‖sic‖ does not consist of.

x 354. Axiom. No reasoning about things whereof we have no idea. Therefore no reasoning about Infinitesimals.

x 354a. nor can it be objected that we reason about Numbers wch are only words & not ideas, for these Infinitesimals are words, of no use, if not suppos'd to stand for Ideas.

x 355. Much less infinitesimals of infinitesimals &c.

x 356. Axiom. No word to be used without an idea.

+ 368. I'll not admire the mathematicians. tis w^{t} any one of common sense might attain to by repeated acts. I know it by experience, I am but one of common sense, and I etc

+ 372. I see no wit in any of them but Newton. The rest are meer triflers, meer Nihilarians.

+ 375. Mathematicians have some of them good parts, the more is the pity. Had they not been Mathematicians they had been good for nothing. they were such fools they knew not how to employ their parts.

x 395. I can square the circle, &c they cannot, w[ch] goes on the best principles.

FROM NOTEBOOK A:

x 449. If the Disputations of the Schoolmen are blam'd for intricacy triflingness & confusion, yet it must be acknowledg'd that in the main they treated of great & important subjects. If we admire the Method & acuteness of the Math: the length, the subtilty, the exactness of their Demonstrations, we must nevertheless be forced to grant that they are for the most part about trifling subjects & perhaps nothing at all.

x 633. Mem: upon all occasions to use the Utmost Modesty. to Confute the Mathematicians w[th] the utmost civility & respect. not to stile them Nihilarians etc:

x 750. Words (by them meaning all sort of signs) are so necessary that instead of being (w[n] duly us'd or in their own Nature) prejudicial to the Advancement of knowlege, or an hindrance to knowlege that w[th]out them there could in Mathematiques themselves be no demonstration.

751. Mem: To be eternally banishing Metaphisics &c & recalling Men to Common Sense.

x 761. I am better inform'd & shall know more by telling me there are 10000 men than by shewing me them all drawn up. I shall better be able to judge of the Bargain you'd have me make w[n] you tell me how much (i.e. the name of y[e]) mony lies on y[e] Table than by offering & shewing it without naming. In short I regard not the Idea the looks but the names. Hence may appear the Nature of Numbers.

x 762. Children are unacquainted with Numbers till they have made some Progress in language. This could not be if they were Ideas suggested by all the senses.

x 763. Numbers are nothing but Names, never Words.

x 764. Mem: Imaginary roots to unravel that Mystery.

x 765. Ideas of Utility are annexed to Numbers.

x 766. In Arithmetical Problems Men seek not any Idea of Number, they onely seek a Denomination. this is all can be of use to them.

x 767. Take away the signs from Arithmetic & Algebra, & pray w[t] remains?

x 768. These are sciences purely Verbal, & entirely useless but for Practise in Societys of Men. No speculative knowledge, no comparing of Ideas in them.

Mo. 769. Sensual Pleasure is the Summum Bonum. This is the Great Principle of Morality. This once rightly understood all the Doctrines even the severest of the [Gospels] may cleerly be Demonstrated.

x 770. Qu: whether Geometry may not be properly reckon'd among the Mixt Mathematics. Arithmetic and Algebra being the only abstracted pure i.e. entirely Nominal. Geometry being an application of these to Points.

B. OF INFINITES (*BERKELEY 1901* [1707])

The following selection is an undated manuscript in the possession of Trinity College, Dublin; it was probably written in 1707–8. The manuscript lay neglected for many years in the Molyneux archives, and was first printed in 1901. Luce and Jessop, Berkeley's editors, conjecture that it was intended as an address to the Dublin ⟦Philosophical⟧ Society. This Society had been founded in 1683 by William Molyneux, and revived by his son Samuel in or around 1707; most of the documents in the Molyneux archives are in fact learned documents contributed to this society. Whether Berkeley's address was in fact given is unknown. The manuscript deals with infinitesimals, the necessity of having an *idea* to correspond to each *sign*, and the distinction (at least as ancient as Aristotle) between the potential and the completed infinite. The issue of infinity was to preoccupy Berkeley in his later writings, not only because of its mathematical importance (and in particular its centrality in the mathematics of Newton and Leibniz), but also because it poses the following challenge to Berkeley's philosophy: if space is infinitely divisible, then it can be divided beyond the *minimum sensibile*; but then something can exist which cannot be sensed; and this conclusion places in jeopardy the doctrine of *esse est percipi*, which is the very cornerstone of Berkeley's idealist metaphysics and epistemology.

The text below is reprinted from *Berkeley 1948–57*, Vol. 4. References to *Berkeley 1901* [1707] should be to the paragraph numbers, which have been added in this reprinting.

⟦1⟧ THO' some mathematicians of this last age have made prodigious advances, and open'd divers admirable methods of investigation unknown to the ancients, yet something there is in their principles which occasions much controversy & dispute, to the great scandal of the so much celebrated evidence of Geometry. These disputes and scruples, arising from the use that is made of quantitys infinitely small in the above mentioned methods, I am bold to think they might easily be brought to an end, by the sole consideration of one passage in the incomparable Mr. Locke's treatise of *Humane Understanding*, b. 2. ch. 17, sec. 7, where that authour, handling the subject of infinity with that judgement & clearness wch is so peculiar to him, has these remarkable words:

'I guess we cause great confusion in our thoughts when we joyn infinity to any suppos'd idea of quantity the mind can be thought to have, and so discourse or reason about an infinite quantity, *viz.* an infinite space or an infinite duration. For our idea of infinity being as I think an endless growing idea, but the idea of any quantity the mind has being at that time terminated in that idea, to join infinity to it is to adjust a standing measure to a growing bulk; &, therefore, I think 'tis not an insignificant subtilty if I say we are carefully to distinguish between the idea of infinity of space and the idea of space infinite.'

[2] Now if what Mr. Locke says were, *mutatis mutandis*, apply'd to quantitys infinitely small, it would, I doubt not, deliver us from that obscurity & confusion wch perplexes otherwise very great improvements of the Modern Analysis. For he that, with Mr. Locke, shall duly weigh the distinction there is betwixt infinity of space & space infinitely great or small, & consider that we have an idea of the former, but none at all of the later, will hardly go beyond his notions to talk of parts infinitely small or *partes infinitesimae* of finite quantitys, & much less of *infinitesimae infinitesimarum,* and so on. This, nevertheless, is very common with writers of fluxions or the differential calculus, &c. They represent, upon paper, infinitesimals of several orders, as if they had ideas in their minds corresponding to those words or signs, or as if it did not include a contradiction that there should be a line infinitely small & yet another infinitely less than it. 'Tis plain to me we ought to use no sign without an idea answering it; & 'tis as plain that we have no idea of a line infinitely small, nay, 'tis evidently impossible there should be any such thing, for every line, how minute soever, is still divisible into parts less than itself; therefore there can be no such thing as a line *quavis data minor* or infinitely small.

[3] Further, it plainly follows that an infinitesimal even of the first degree is meerly *nothing*, from wt Dr. Wallis, an approv'd mathematician, writes at the 95th proposition of his *Arithmetic of Infinites*,[a] where he makes the asymptotic space included between the 2 asymptotes and the curve of an hyperbola to be in his stile a *series reciproca primanorum*, so that the first term of the series, *viz.*, the asymptote, arises from the division of 1 by 0. Since therefore, unity, *i.e.* any finite line divided by 0, gives the asymptote of an hyperbola, *i.e.* a line infinitely long, it necessarily follows that a finite line divided by an infinite gives 0 in the quotient, *i.e.* that the *pars infinitesima* of a finite line is just nothing. For by the nature of division the dividend divided by the quotient gives the divisor. Now a man speaking of lines infinitely small will hardly be suppos'd to mean nothing by them, and if he understands real finite quantitys he runs into inextricable difficultys.

[4] Let us look a little into the controversy between Mr. Nieuentiit[b] and Mr. Leibnitz. Mr. Nieuentiit allows infinitesimals of the first order to be real quantitys,

---

[a] [Wallis is also mentioned by Berkeley below; see *Berkeley 1734*, §17.]

[b] [Bernard Nieuwentijdt or Nieuwentijt (1654–1718) was a Dutch philosopher and mathematician, and a defender of religion; he is mentioned again in Berkeley's *Siris*, §190. His *Analysis infinitorum, seu curvilineorum proprietates ex polygonorum natura deductae* (Amsterdam, 1695) was one of the first expositions of Leibniz's differential calculus.]

but the *differentiae differentiarum* or infinitesimals of the following orders he takes away making them just so many noughts. This is the same thing as to say the square, cube, or other power of a real positive quantity is equal to nothing; wch is manifestly absurd.

⟦5⟧ Again Mr. Nieuentiit lays down this as a self evident axiom, *viz.*, that betwixt two equal quantitys there can be no difference at all, or, which is the same thing, that their difference is equal to nothing. This truth, how plain soever, Mr. Leibnitz sticks not to deny, asserting that not onely those quantitys are equal which have no difference at all, but also those whose difference is incomparably small. *Quemadmodum* (says he) *si lineae punctum alterius lineae addas quantitatem non auges.*[c] But if lines are infinitely divisible, I ask how there can be any such thing as a point? Or granting there are points, how can it be thought the same thing to add an indivisible point as to add, for instance, the *differentia* of an ordinate, in a parabola, wch is so far from being a point that it is itself divisible into an infinite number of real quantitys, whereof each can be subdivided *in infinitum*, and so on, according to Mr. Leibnitz. These are difficultys those great men have run into, by applying the idea of infinity to particles of extension exceeding small, but real and still divisible.

⟦6⟧ More of this dispute may be seen in the *Acta Eruditorum* for the month of July, A.D. 1695, where, if we may believe the French authour of *Analyse des infiniment petits*, Mr. Leibnitz has sufficiently established & vindicated his principles. Tho' 'tis plain he cares not for having 'em call'd in question, and seems afraid that *nimia scrupulositate arti inveniendi obex ponatur*,[d] as if a man could be too scrupulous in Mathematics, or as if the principles of Geometry ought not to be as incontestable as the consequences drawn from them.

⟦7⟧ There is an argument of Dr. Cheyne's,[e] in the 4th chapter of his *Philosophical Principles of Natural Religion* which seems to make for quantitys infinitely small. His words are as follows:

> 'The whole abstract geometry depends upon the possibility of infinitely great & small quantitys, & the truths discover'd by methods wch depend upon these suppositions are confirm'd by other methods wch have other foundations.'

⟦8⟧ To wch I answer that the supposition of quantitys infinitely small is not essential to the great improvements of the Modern Analysis. For Mr. Leibnitz acknowleges his *Calculus differentialis* might be demonstrated *reductione ad*

---

[c] ⟦'For example, if you add to a line a point of a second line, you do not increase its size'.⟧

[d] ⟦'a barrier would be erected to the art of discovery by too great a concern for accuracy'.⟧

[e] ⟦George Cheyne (1671–1743) was a London physician. In 1702 he published *A new theory of fevers*, a work which attempted to extend the methods of Newtonian celestial mechanics to the human body; Cheyne proposed a quasi-mathematical explication of fevers, based on hydraulics and on a view of the body as a system of pipes and fluids. In 1703 he published a treatise on fluxions, the *Fluxionum methodus inversa*; the work was however riddled with mathematical errors. His *Philosophical principles of natural religion*, referred to by Berkeley, appeared in 1705; it attempts to prove the existence of God from the existence of gravitation. Cheyne later renounced his earlier 'riotous' Newtonian life, moved to Bath, and devoted himself to purely medical writings, producing his *Essay on the gout* in 1720. His name is mentioned by Berkeley in *Berkeley 1707–8a*, §§367, 387, and 459.⟧

absurdum after the manner of the ancients; & Sir Isaac Newton in a late treatise informs us his method of Fluxions can be made out *a priori* without the supposition of quantitys infinitely small.

‖9‖ I can't but take notice of a passage in Mr. Raphson's[f] treatise *De Spatio Reali seu Ente Infinito*, chap. 3, p. 50, where he will have a particle infinitely small to be *quasi extensa*. But wt Mr. Raphson would be thought to mean by *pars continui quasi extensa* I cannot comprehend. I must also crave leave to observe that some modern writers of note make no scruple to talk of a sphere of an infinite radius, or an aequilateral triangle of an infinite side, which notions if thoroughly examin'd may perhaps be found not altogether free from inconsistencys.

‖10‖ Now I am of opinion that all disputes about infinites would cease, & the consideration of quantitys infinitely small no longer perplex Mathematicians, would they but joyn Metaphysics to their Mathematics, and condescend to learn from Mr. Locke what distinction there is betwixt infinity and infinite.

C. LETTER TO SAMUEL MOLYNEUX (*BERKELEY 1709*)

The following letter concerns the relationship between words and ideas in mathematics; Berkeley here denies that words must suggest the ideas they stand for 'at every turn'. He was shortly to revisit this subject in §§19 and 20 of the *Principles of human knowledge*.

The letter was addressed to Samuel Molyneux (1689–1728), to whom Berkeley had dedicated the *Miscellanea mathematica* in 1707. Molyneux (who was four years Berkeley's junior) later became a distinguished astronomer; he was the son of William Molyneux, an astronomer, physicist, and Member of Parliament, whose work on optics, the *Dioptrica nova* (1692), was the basis for Berkeley's *Essay towards a new theory of vision* (1709). (The elder Molyneux had also translated Descartes into English, and posed to Locke the question whether a man blind since birth, on gaining his sight, would be able to distinguish, by sight alone, between a sphere and a cube that he previously knew only by the sense of touch.)

The letter is reprinted from *Berkeley 1948–57*, Vol. 8.

---

[f] ‖Joseph Raphson, FRS, wrote treatises on fluxions and on the nature of space; his *De spatio reali, seu ente infinito conamen mathematico-metaphysicum* (1697) is here referred to by Berkeley Berkeley also mentions his name in *Berkeley 1707–8a*, §§298 and 827, and in his letter to Samuel Johnson of 24 Mar. 1730.‖

Trin. Coll. Dec. 19. 1709

Dr. Molyneux.

You desire to know what a Geometer thinks of when he demonstrates Properties of a Curve formd by a Ray of Light as it passes through the Air. His Imagination or Memory (say You) can affoard him no Idea of it, and as for the rude Idea of this or that Curve which may be suggested to him by Fancy that has no Connexion with his Theorems & Reasonings. I answer first, That in my Opinion he thinks of the Various Density of the Atmosphere the Obliquity of the Incidence and the Nature of Refraction in Air by which the Ray is bent into a Curve, 'tis on these he meditates and from these principles he proceeds to investigate the Nature of the Curve. Secondly. It appears to Me That in Geometricall Reasonings We do not make any Discovery by contemplating the Ideas of the Lines whose Properties are investigated. For Example, In order to discover the Method of drawing Tangents to a Parabola, 'tis true a Figure is drawn on paper & so suggested to your Fancy, but no Matter whether it be of a Parabolic Line or no, the Demonstration proceeds as well tho it be an Hyperbole or the Portion of an Ellipsis, provided that I have regard to the Equation expressing the Nature of a Parabola wherein the Squares of the Ordinates are every where equall to the Rectangles under the Abscissae & Parameters, it being this Equation or the Nature of the Curve thus expressed and not the Idea of it that leads to the Solution‖.‖ Again You tell Me that if, as I think, Words do not at every Turn suggest the respective Ideas they are supposd to stand for it is purely by Chance Our Discourse hangs together, and is found after 2 or 3 hours jingling & permutation of Sounds to agree with our Thoughts. As for what I said of Algebra, You are of Opinion the Illustration will not hold good because there are no Set rules except those of the Syllogisms whereby to range & permute o‖u‖r Words like to the Algebraic Process. In Answer to all which I observe first, That if We put Our Words together any how and at Random then indeed there may be some Grounds for what You say, but if people lay their Words together with Design and according to Rule then there can be no Pretence so far as I can see for your Inference. Secondly. I cannot but dissent from what You say, of there being no Set Rules for the Ranging and Disposition of Words but only the Syllogistic, for to Me it appears That all Grammar & every part Logic contain little else than Rules for Discourse & Ratiocination by Words. And those who do not expresly set themselves to study those Arts do nevertheless learn them insensibly by Custom. I am very sleepy & can say no more but that I am

Yo‖u‖rs &c.
G. BERKELEY.

D. *FROM* A TREATISE CONCERNING THE PRINCIPLES OF HUMAN KNOWLEDGE, PART ONE (*BERKELEY 1710*)

The *Principles of human knowledge, Part One*, Berkeley's major philosophical work, was first printed in 1710 while Berkeley was a Fellow of Trinity College, Dublin. As the title indicates, Berkeley planned to publish a Part Two; however, he lost the manuscript during his travels on the Continent in 1716–20, and never rewrote it. The *Principles* is so well known to students of philosophy that it needs no detailed introduction here; for a modern discussion of the place of the *Principles* in British empiricism, see *Warnock 1982* and the works therein cited. Students of Berkeley's thought and its connections to earlier manifestations of British nominalism are urged to read the perceptive review of his *Works* by C.S. Peirce (*Peirce 1871*). The passages excerpted here discuss abstract ideas and language; the nature of arithmetic and the finite integers; spatial extension and infinite divisibility. The passages have been selected not only for their intrinsic interest, but also for the light they shed on Berkeley's other writings in this collection. In particular, the central role played by Berkeley's attack on abstract ideas should not be overlooked, and the extended discussion of abstract ideas in §§6–25 of the Introduction to the *Principles* ought to be compared with the remarks on language and mathematics in the *Philosophical commentaries* (§§354–6 and 761–70), as well as with the attack upon infinitesimals in *The analyst*.

The text below consists of §§6–25 from the Introduction and of §§118–34 from Part One. The text is based on the second edition of 1734; variants from the first edition are recorded in *Berkeley 1948–57*, Vol. 2. References to *Berkeley 1710* should be to the paragraph numbers, which appeared in the original edition.

---

6 In order to prepare the mind of the reader for the easier conceiving what follows, it is proper to premise somewhat, by way of introduction, concerning the nature and abuse of language. But the unravelling this matter leads me in some measure to anticipate my design, by taking notice of what seems to have had a chief part in rendering speculation intricate and perplexed, and to have occasioned innumerable errors and difficulties in almost all parts of knowledge. And that is the opinion that the mind hath a power of framing *abstract ideas* or notions of things. He who is not a perfect stranger to the writings and disputes of philosophers, must needs acknowledge that no small part of them are spent about abstract ideas. These are in a more especial manner, thought to be the object of those sciences which go by the name of *Logic* and *Metaphysics*, and of all that which passes under the notion of the most abstracted and sublime

learning, in all which one shall scarce find any question handled in such a manner, as does not suppose their existence in the mind, and that it is well acquainted with them.

7 It is agreed on all hands, that the qualities or modes of things do never really exist each of them apart by itself, and separated from all others, but are mixed, as it were, and blended together, several in the same object. But we are told, the mind being able to consider each quality singly, or abstracted from those other qualities with which it is united, does by that means frame to itself abstract ideas. For example, there is perceived by sight an object extended, coloured, and moved: this mixed or compound idea the mind resolving into its simple, constituent parts, and viewing each by itself, exclusive of the rest, does frame the abstract ideas of extension, colour, and motion. Not that it is possible for colour or motion to exist without extension: but only that the mind can frame to itself by *abstraction* the idea of colour exclusive of extension, and of motion exclusive of both colour and extension.

8 Again, the mind having observed that in the particular extensions perceived by sense, there is something common and alike in all, and some other things peculiar, as this or that figure or magnitude, which distinguish them one from another; it considers apart or singles out by itself that which is common, making thereof a most abstract idea of extension, which is neither line, surface, nor solid, nor has any figure or magnitude but is an idea entirely prescinded from all these. So likewise the mind by leaving out of the particular colours perceived by sense, that which distinguishes them one from another, and retaining that only which is common to all, makes an idea of colour in abstract which is neither red, nor blue, nor white, nor any other determinate colour. And in like manner by considering motion abstractedly not only from the body moved, but likewise from the figure it describes, and all particular directions and velocities, the abstract idea of motion is framed; which equally corresponds to all particular motions whatsoever that may be perceived by sense.

9 And as the mind frames to itself abstract ideas of qualities or modes, so does it, by the same precision or mental separation, attain abstract ideas of the more compounded beings, which include several coexistent qualities. For example, the mind having observed that Peter, James, and John, resemble each other, in certain common agreements of shape and other qualities, leaves out of the complex or compounded idea it has of Peter, James, and any other particular man, that which is peculiar to each, retaining only what is common to all; and so makes an abstract idea wherein all the particulars equally partake, abstracting entirely from and cutting off all those circumstances and differences, which might determine it to any particular existence. And after this manner it is said we come by the abstract idea of *man* or, if you please, humanity or human nature; wherein it is true, there is included colour, because there is no man but has some colour, but then it can be neither white, nor black, nor any particular colour; because there is no one particular colour wherein all men partake. So likewise there is included stature, but then it is neither tall stature nor low stature, nor yet middle stature, but something abstracted from all these. And so of the rest. Moreover, there being a great variety of other creatures that partake in some

parts, but not all, of the complex idea of *man*, the mind leaving out those parts which are peculiar to men, and retaining those only which are common to all the living creatures, frameth the idea of *animal*, which abstracts not only from all particular men, but also all birds, beasts, fishes, and insects. The constituent parts of the abstract idea of animal are body, life, sense, and spontaneous motion. By *body* is meant, body without any particular shape or figure, there being no one shape or figure common to all animals, without covering, either of hair or feathers, or scales, &c. nor yet naked: hair, feathers, scales, and nakedness being the distinguishing properties of particular animals, and for that reason left out of the *abstract idea*. Upon the same account the spontaneous motion must be neither walking, nor flying, nor creeping, it is nevertheless a motion, but what that motion is, it is not easy to conceive.

10 Whether others have this wonderful faculty of *abstracting their ideas*, they best can tell: for myself I find indeed I have a faculty of imagining, or representing to myself the ideas of those particular things I have perceived and of variously compounding and dividing them. I can imagine a man with two heads or the upper parts of a man joined to the body of a horse. I can consider the hand, the eye, the nose, each by itself abstracted or separated from the rest of the body. But then whatever hand or eye I imagine, it must have some particular shape and colour. Likewise the idea of man that I frame to myself, must be either of a white, or a black, or a tawny, a straight, or a crooked, a tall, or a low, or a middle-sized man. I cannot by any effort of thought conceive the abstract idea above described. And it is equally impossible for me to form the abstract idea of motion distinct from the body moving, and which is neither swift nor slow, curvilinear nor rectilinear; and the like may be said of all other abstract general ideas whatsoever. To be plain, I own myself able to abstract in one sense, as when I consider some particular parts or qualities separated from others, with which though they are united in some object, yet, it is possible they may really exist without them. But I deny that I can abstract one from another, or conceive separately, those qualities which it is impossible should exist so separated; or that I can frame a general notion by abstracting from particulars in the manner aforesaid. Which two last are the proper acceptations of *abstraction*. And there are grounds to think most men will acknowledge themselves to be in my case. The generality of men which are simple and illiterate never pretend to *abstract notions*. It is said they are difficult and not to be attained without pains and study. We may therefore reasonably conclude that, if such there be, they are confined only to the learned.

11 I proceed to examine what can be alleged in defence of the doctrine of abstraction, and try if I can discover what it is that inclines the men of speculation to embrace an opinion, so remote from common sense as that seems to be. There has been a late deservedly esteemed philosopher,[a] who, no doubt, has given it very much countenance by seeming to think the having abstract general ideas is what puts the widest difference in point of understanding betwixt man and beast. 'The having of general ideas (*saith he*) is that which puts

[a] ‖Locke.‖

a perfect distinction betwixt man and brutes, and is an excellency which the faculties of brutes do by no means attain unto. For it is evident we observe no footsteps in them of making use of general signs for universal ideas; from which we have reason to imagine that they have not the faculty of *abstracting* or making general ideas, since they have no use of words or any other general signs. *And a little after.* Therefore, I think, we may suppose that it is in this that the species of brutes are discriminated from men, and 'tis that proper difference wherein they are wholly separated, and which at last widens to so wide a distance. For if they have any ideas at all, and are not bare machines (as some would have them) we cannot deny them to have some reason. It seems as evident to me that they do some of them in certain instances reason as that they have sense, but it is only in particular ideas, just as they receive them from their senses. They are the best of them tied up within those narrow bounds, and have not (as I think) the faculty to enlarge them by any kind of *abstraction*.' *Essay on Hum. Underst.* B.2. C.11. Sect. 10 and 11. I readily agree with this learned author, that the faculties of brutes can by no means attain to *abstraction*. But then if this be made the distinguishing property of that sort of animals, I fear a great many of those that pass for men must be reckoned into their number. The reason that is here assigned why we have no grounds to think brutes have abstract general ideas, is that we observe in them no use of words or any other general signs; which is built on this supposition, to wit, that the making use of words, implies the having general ideas. From which it follows, that men who use language are able to abstract or generalize their ideas. That this is the sense and arguing of the author will further appear by his answering the question he in another place puts. 'Since all things that exist are only particulars, how come we by general terms?' *His answer is*, 'Words become general by being made the signs of general ideas.' *Essay on Hum. Underst.* B.3. C.3. Sect. 6. But it seems that a word becomes general by being made the sign, not of an abstract general idea but, of several particular ideas, any one of which it indifferently suggests to the mind. For example, when it is said *the change of motion is proportional to the impressed force*, or that *whatever has extension is divisible*; these propositions are to be understood of motion and extension in general, and nevertheless it will not follow that they suggest to my thoughts an idea of motion without a body moved, or any determinate direction and velocity, or that I must conceive an abstract general idea of extension, which is neither line, surface nor solid, neither great nor small, black, white, nor red, nor of any other determinate colour. It is only implied that whatever motion I consider, whether it be swift or slow, perpendicular, horizontal or oblique, or in whatever object, the axiom concerning it holds equally true. As does the other of every particular extension, it matters not whether line, surface or solid, whether of this or that magnitude or figure.

12 By observing how ideas become general, we may the better judge how words are made so. And here it is to be noted that I do not deny absolutely there are general ideas, but only that there are any *abstract general ideas*: for in the passages above quoted, wherein there is mention of general ideas, it is

always supposed that they are formed by *abstraction*, after the manner set forth in Sect. 8 and 9. Now if we will annex a meaning to our words, and speak only of what we can conceive, I believe we shall acknowledge, that an idea, which considered in itself is particular, becomes general, by being made to represent or stand for all other particular ideas of the same sort. To make this plain by an example, suppose a geometrician is demonstrating the method, of cutting a line in two equal parts. He draws, for instance, a black line of an inch in length, this which in itself is a particular line is nevertheless with regard to its signification general, since as it is there used, it represents all particular lines whatsoever; for that what is demonstrated of it, is demonstrated of all lines or, in other words, of a line in general. And is that particular line becomes general, by being made a sign, so the name *line* which taken absolutely is particular, by being a sign is made general. And as the former owes its generality, not to its being the sign of an abstract or general line, but of all particular right lines that may possibly exist, so the latter must be thought to derive its generality from the same cause, namely, the various particular lines which it indifferently denotes.

13 To give the reader a yet clearer view of the nature of abstract ideas, and the uses they are thought necessary to, I shall add one more passage out of the *Essay on Human Understanding*, which is as follows. '*Abstract ideas* are not so obvious or easy to children or the yet unexercised mind as particular ones. If they seem so to grown men, it is only because by constant and familiar use they are made so. For when we nicely reflect upon them, we shall find that general ideas are fictions and contrivances of the mind, that carry difficulty with them, and do not so easily offer themselves, as we are apt to imagine. For example, does it not require some pains and skill to form the general idea of a triangle (which is yet none of the most abstract comprehensive and difficult) for it must be neither oblique nor rectangle, neither equilateral, equicrural, nor scalenon, but *all and none* of these at once. In effect, it is something imperfect that cannot exist, an idea wherein some parts of several different and *inconsistent* ideas are put together. It is true the mind in this imperfect state has need of such ideas, and makes all the haste to them it can, for the conveniency of communication and enlargement of knowledge, to both which it is naturally very much inclined. But yet one has reason to suspect such ideas are marks of our imperfection. At least this is enough to shew that the most abstract and general ideas are not those that the mind is first and most easily acquainted with, nor such as its earliest knowledge is conversant about.' B.4. C.7. Sect. 9. If any man has the faculty of framing in his mind such an idea of a triangle as is here described, it is in vain to pretend to dispute him out of it, nor would I go about it. All I desire is, that the reader would fully and certainly inform himself whether he has such an idea or no. And this, methinks, can be no hard task for anyone to perform. What more easy than for anyone to look a little into his own thoughts, and there try whether he has, or can attain to have, an idea that shall correspond with the description that is here given of the general idea of a triangle, which is, *neither oblique, nor rectangle, equilateral, equicrural, nor scalenon, but all and none of these at once?*

14 Much is here said of the difficulty that abstract ideas carry with them, and the pains and skill requisite to the forming them. And it is on all hands agreed that there is need of great toil and labour of the mind, to emancipate our thoughts from particular objects, and raise them to those sublime speculations that are conversant about abstract ideas. From all which the natural consequence should seem to be, that so difficult a thing as the forming abstract ideas was not necessary for *communication*, which is so easy and familiar to all sorts of men. But we are told, if they seem obvious and easy to grown men, *it is only because by constant and familiar use they are made so*. Now I would fain know at what time it is, men are employed in surmounting that difficulty, and furnishing themselves with those necessary helps for discourse. It cannot be when they are grown up, for then it seems they are not conscious of any such pains-taking; it remains therefore to be the business of their childhood. And surely, the great and multiplied labour of framing abstract notions, will be found a hard task for that tender age. Is it not a hard thing to imagine, that a couple of children cannot prate together, of their sugar-plumbs and rattles and the rest of their little trinkets, till they have first tacked together numberless inconsistencies, and so framed in their minds *abstract general ideas*, and annexed them to every common name they make use of?

15 Nor do I think them a whit more needful for the *enlargement of knowledge* than for *communication*. It is I know a point much insisted on, that all knowledge and demonstration are about universal notions, to which I fully agree: but then it doth not appear to me that those notions are formed by *abstraction* in the manner premised; *universality*, so far as I can comprehend, not consisting in the absolute, positive nature or conception of anything, but in the relation it bears to the particulars signified or represented by it: by virtue whereof it is that things, names, or notions, being in their own nature particular, are rendered *universal*. Thus when I demonstrate any proposition concerning triangles, it is to be supposed that I have in view the universal idea of a triangle; which ought not to be understood as if I could frame an idea of a triangle which was neither equilateral nor scalenon nor equicrural. But only that the particular triangle I consider, whether of this or that sort it matters not, doth equally stand for and represent all rectilinear triangles whatsoever, and is in that sense *universal*. All which seems very plain and not to include any difficulty in it.

16 But here it will be demanded, how we can know any proposition to be true of all particular triangles, except we have first seen it demonstrated of the abstract idea of a triangle which equally agrees to all? For because a property may be demonstrated to agree to some one particular triangle, it will not thence follow that it equally belongs to any other triangle, which in all respects is not the same with it. For example, having demonstrated that the three angles of an isosceles rectangular triangle are equal to two right ones, I cannot therefore conclude this affection agrees to all other triangles, which have neither a right angle, nor two equal sides. It seems therefore that, to be certain this proposition is universally true, we must either make a particular demonstration for every

particular triangle, which is impossible, or once for all demonstrate it of the *abstract idea of a triangle*, in which all the particulars do indifferently partake, and by which they are all equally represented. To which I answer, that though the idea I have in view whilst I make the demonstration, be, for instance, that of an isosceles rectangular triangle, whose sides are of a determinate length, I may nevertheless be certain it extends to all other rectilinear triangles, of what sort or bigness soever. And that, because neither the right angle, nor the equality, nor determinate length of the sides, are at all concerned in the demonstration. It is true, the diagram I have in view includes all these particulars, but then there is not the least mention made of them in the proof of the proposition. It is not said, the three angles are equal to two right ones, because one of them is a right angle, or because the sides comprehending it are of the same length. Which sufficiently shews that the right angle might have been oblique, and the sides unequal, and for all that the demonstration have held good. And for this reason it is, that I conclude that to be true of any obliquangular or scalenon, which I had demonstrated of a particular right-angled, equicrural triangle; and not because I demonstrated the proposition of the abstract idea of a triangle. ‖And here it must be acknowledged that a man may consider a figure merely as triangular, without attending to the particular qualities of the angles, or relations of the sides. So far he may abstract: but this will never prove, that he can frame an abstract general inconsistent idea of a triangle. In like manner we may consider Peter so far forth as man, or so far forth as animal, without framing the forementioned abstract idea, either of man or of animal, in as much as all that is perceived is not considered.‖[b]

17 It were an endless, as well as an useless thing, to trace the Schoolmen, those great masters of abstraction, through all the manifold inextricable labyrinths of error and dispute, which their doctrine of abstract natures and notions seems to have led them into. What bickerings and controversies, and what a learned dust have been raised about those matters, and what mighty advantage hath been from thence derived to mankind, are things at this day too clearly known to need being insisted on. And it had been well if the ill effects of that doctrine were confined to those only who make the most avowed profession of it. When men consider the great pains, industry and parts, that have for so many ages been laid out on the cultivation and advancement of the sciences, and that notwithstanding all this, the far greater part of them remain full of darkness and uncertainty, and disputes that are like never to have an end, and even those that are thought to be supported by the most clear and cogent demonstrations, contain in them paradoxes which are perfectly irreconcilable to the understandings of men, and that taking all together, a small portion of them doth supply any real benefit to mankind, otherwise than by being an innocent diversion and amusement. I say, the consideration of all this is apt to throw them into a despondency, and perfect contempt of all study. But this may

[b]‖Last three sentences of §16 added in second edition.‖

perhaps cease, upon a view of the false principles that have obtained in the world, amongst all which there is none, methinks, hath a more influence over the thoughts of speculative men, than this of abstract general ideas.

18 I come now to consider the source of this prevailing notion, and that seems to me to be language. And surely nothing of less extent than reason itself could have been the source of an opinion so universally received. The truth of this appears as from other reasons, so also from the plain confession of the ablest patrons of abstract ideas, who acknowledge that they are made in order to naming; from which it is a clear consequence, that if there had been no such thing as speech or universal signs, there never had been any thought of abstraction. See B.3. C.6. Sect. 39 *and elsewhere of the Essay on Human Understanding*. Let us therefore examine the manner wherein words have contributed to the origin of that mistake. First then, 'tis thought that every name hath, or ought to have, one only precise and settled signification, which inclines men to think there are certain *abstract, determinate ideas*, which constitute the true and only immediate signification of each general name. And that it is by the mediation of these abstract ideas, that a general name comes to signify any particular thing. Whereas, in truth, there is no such thing as one precise and definite signification annexed to any general name, they all signifying indifferently a great number of particular ideas. All which doth evidently follow from what has been already said, and will clearly appear to anyone by a little reflexion. To this it will be objected, that every name that has a definition, is thereby restrained to one certain signification. For example, a *triangle* is defined to be a *plane surface comprehended by three right lines*; by which that name is limited to denote one certain idea and no other. To which I answer, that in the definition it is not said whether the surface be great or small, black or white, nor whether the sides are long or short, equal or unequal, nor with what angles they are inclined to each other; in all which there may be great variety, and consequently there is no one settled idea which limits the signification of the word *triangle*. 'Tis one thing for to keep a name constantly to the same definition, and another to make it stand everywhere for the same idea: the one is necessary, the other useless and impracticable.

19 But to give a farther account how words came to produce the doctrine of abstract ideas, it must be observed that it is a received opinion, that language has no other end but the communicating our ideas, and that every significant name stands for an idea. This being so, and it being withal certain, that names, which yet are not thought altogether insignificant, do not always mark out particular conceivable ideas, it is straightway concluded that they stand for abstract notions. That there are many names in use amongst speculative men, which do not always suggest to others determinate particular ideas, is what nobody will deny. And a little attention will discover, that it is not necessary (even in the strictest reasonings) significant names which stand for ideas should, every time they are used, excite in the understanding the ideas they are made to stand for: in reading and discoursing, names being for the most part used as letters are in *algebra*, in which though a particular quantity be marked by each letter, yet

to proceed right it is not requisite that in every step each letter suggest to your thoughts, that particular quantity it was appointed to stand for.

20 Besides, the communicating of ideas marked by words is not the chief and only end of language, as is commonly supposed. There are other ends, as the raising of some passion, the exciting to, or deterring from an action, the putting the mind in some particular disposition; to which the former is in many cases barely subservient, and sometimes entirely omitted, when these can be obtained without it, as I think doth not infrequently happen in the familiar use of language. I entreat the reader to reflect with himself, and see if it doth not often happen either in hearing or reading a discourse, that the passions of fear, love, hatred, admiration, disdain, and the like arise, immediately in his mind upon the perception of certain words, without any ideas coming between. At first, indeed, the words might have occasioned ideas that were fit to produce those emotions; but, if I mistake not, it will be found that when language is once grown familiar, the hearing of the sounds or sight of the characters is oft immediately attended with those passions, which at first were wont to be produced by the intervention of ideas, that are now quite omitted. May we not, for example, be affected with the promise of a *good thing*, though we have not an idea of what it is? Or is not the being threatened with danger sufficient to excite a dread, though we think not of any particular evil likely to befall us, nor yet frame to ourselves an idea of danger in abstract? If anyone shall join ever so little reflection of his own to what has been said, I believe it will evidently appear to him, that general names are often used in the propriety of language without the speaker's designing them for marks of ideas in his own, which he would have them raise in the mind of the hearer. Even proper names themselves do not seem always spoken, with a design to bring into our view the ideas of those individuals that are supposed to be marked by them. For example, when a Schoolman tells me *Aristotle hath said it*, all I conceive he means by it, is to dispose me to embrace his opinion with the deference and submission which custom has annexed to that name. And this effect may be so instantly produced in the minds of those who are accustomed to resign their judgment to the authority of that philosopher, as it is impossible any idea either of his person, writings, or reputation should go before. Innumerable examples of this kind may be given, but why should I insist on those things, which everyone's experience will, I doubt not, plentifully suggest unto him?

21 We have, I think, shewn the impossibility of *abstract ideas*. We have considered what has been said for them by their ablest patrons; and endeavoured to shew they are of no use for those ends, to which they are thought necessary. And lastly, we have traced them to the source from whence they flow, which appears to be language. It cannot be denied that words are of excellent use, in that by their means all that stock of knowledge which has been purchased by the joint labours of inquisitive men in all ages and nations, may be drawn into the view and made the possession of one single person. But at the same time it must be owned that most parts of knowledge have been strangely perplexed and darkened by the abuse of words, and general ways of speech

wherein they are delivered. Since therefore words are so apt to impose on the understanding, whatever ideas I consider, I shall endeavour to take bare and naked into my view, keeping out of my thoughts, so far as I am able, those names which long and constant use hath so strictly united with them; from which I may expect to derive the following advantages.

22 First, I shall be sure to get clear of all controversies purely verbal; the springing up of which weeds in almost all the sciences has been a main hindrance to the growth of true and sound knowledge. Secondly, this seems to be a sure way to extricate myself out of that fine and subtle net of *abstract ideas*, which has so miserably perplexed and entangled the minds of men, and that with this peculiar circumstance, that by how much the finer and more curious was the wit of any man, by so much the deeper was he like to be ensnared, and faster held therein. Thirdly, so long as I confine my thoughts to my own ideas divested of words, I do not see how I can easily be mistaken. The objects I consider, I clearly and adequately know. I cannot be deceived in thinking I have an idea which I have not. It is not possible for me to imagine, that any of my own ideas are alike or unlike, that are not truly so. To discern the agreements or disagreements there are between my ideas, to see what ideas are included in any compound idea, and what not, there is nothing more requisite, than an attentive perception of what passes in my own understanding.

23 But the attainment of all these advantages doth presuppose an entire deliverance from the deception of words, which I dare hardly promise myself; so difficult a thing it is to dissolve an union so early begun, and confirmed by so long a habit as that betwixt words and ideas. Which difficulty seems to have been very much increased by the doctrine of *abstraction*. For so long as men thought abstract ideas were annexed to their words, it doth not seem strange that they should use words for ideas: it being found an impracticable thing to lay aside the word, and retain the abstract idea in the mind, which in itself was perfectly inconceivable. This seems to me the principal cause, why those men who have so emphatically recommended to others, the laying aside all use of words in their meditations, and contemplating their bare ideas, have yet failed to perform it themselves. Of late many have been very sensible of the absurd opinions and insignificant disputes, which grow out of the abuse of words. And in order to remedy these evils they advise well, that we attend to the ideas signified, and draw off our attention from the words which signify them. But how good soever this advice may be, they have given others, it is plain they could not have a due regard to it themselves, so long as they thought the only immediate use of words was to signify ideas, and that the immediate signification of every general name was a *determinate*, *abstract idea*.

24 But these being known to be mistakes, a man may with greater ease prevent his being imposed on by words. He that knows he has no other than particular ideas, will not puzzle himself in vain to find out and conceive the abstract idea, annexed to any name. And he that knows names do not always stand for ideas, will spare himself the labour of looking for ideas, where there are none to be had. It were therefore to be wished that everyone would use his utmost

endeavours, to obtain a clear view of the ideas he would consider, separating from them all that dress and encumbrance of words which so much contribute to blind the judgment and divide the attention. In vain do we extend our view into the heavens, and pry into the entrails of the earth, in vain do we consult the writings of learned men, and trace the dark footsteps of antiquity; we need only draw the curtain of words, to behold the fairest tree of knowledge, whose fruit is excellent, and within the reach of our hand.

25 Unless we take care to clear the first principles of knowledge, from the embarras and delusion of words, we may make infinite reasonings upon them to no purpose; we may draw consequences from consequences, and be never the wiser. The farther we go, we shall only lose ourselves the more irrecoverably, and be the deeper entangled in difficulties and mistakes. Whoever therefore designs to read the following sheets, I entreat him to make my words the occasion of his own thinking, and endeavour to attain the same train of thoughts in reading, that I had in writing them. By this means it will be easy for him to discover the truth or falsity of what I say. He will be out of all danger of being deceived by my words, and I do not see how he can be led into an error by considering his own naked, undisguised ideas.

---

118 Hitherto of natural philosophy: we come now to make some inquiry concerning that other great branch of speculative knowledge, to wit, *mathematics*. These, how celebrated soever they may be, for their clearness and certainty of demonstration, which is hardly any where else to be found, cannot nevertheless be supposed altogether free from mistakes; if in their principles there lurks some secret error, which is common to the professors of those sciences with the rest of mankind. Mathematicians, though they deduce their theorems from a great height of evidence, yet their first principles are limited by the consideration of quantity: and they do not ascend into any inquiry concerning those transcendental maxims, which influence all the particular sciences, each part whereof, mathematics not excepted, doth consequently participate of the errors involved in them. That the principles laid down by mathematicians are true, and their way of deduction from those principles clear and incontestable, we do not deny. But we hold, there may be certain erroneous maxims of greater extent than the object of mathematics, and for that reason not expressly mentioned, though tacitly supposed throughout the whole progress of that science; and that the ill effects of those secret unexamined errors are diffused through all the branches thereof. To be plain, we suspect the mathematicians are, as well as other men, concerned in the errors arising from the doctrine of abstract general ideas, and the existence of objects without the mind.

119 *Arithmetic* hath been thought to have for its object abstract ideas of *number*. Of which to understand the properties and mutual habitudes is supposed no mean part of speculative knowledge. The opinion of the pure and

intellectual nature of numbers in abstract, hath made them in esteem with those philosophers, who seem to have affected an uncommon fineness and elevation of thought. It hath set a price on the most trifling numerical speculations which in practice are of no use, but serve only for amusement: and hath therefore so far infected the minds of some, that they have dreamt of mighty *mysteries* involved in numbers, and attempted the explication of natural things by them. But if we inquire into our own thoughts, and consider what hath been premised, we may perhaps entertain a low opinion of those high flights and abstractions, and look on all inquiries about numbers, only as so many *difficiles nugæ*, so far as they are not subservient to practice, and promote the benefit of life.

120 Unity in abstract we have before considered in *Sect.* 13, from which and what hath been said in the Introduction, it plainly follows there is not any such idea. But number being defined a *collection of units*, we may conclude that, if there be no such thing as unity or unit in abstract, there are no ideas of number in abstract denoted by the numerical names and figures. The theories therefore in arithmetic, if they are abstracted from the names and figures, as likewise from all use and practice, as well as from the particular things numbered, can be supposed to have nothing at all for their object. Hence we may see, how entirely the science of numbers is subordinate to practice, and how jejune and trifling it becomes, when considered as a matter of mere speculation.

121 However since there may be some, who, deluded by the specious shew of discovering abstracted verities, waste their time in arithmetical theorems and problems, which have not any use: it will not be amiss, if we more fully consider, and expose the vanity of that pretence; and this will plainly appear, by taking a view of arithmetic in its infancy, and observing what it was that originally put men on the study of that science, and to what scope they directed it. It is natural to think that at first, men, for ease of memory and help of computation, made use of counters, or in writing of single strokes, points or the like, each whereof was made to signify an unit, that is, some one thing of whatever kind they had occasion to reckon. Afterwards they found out the more compendious ways, of making one character stand in place of several strokes, or points. And lastly, the notation of the Arabians or Indians came into use, wherein by the repetition of a few characters or figures, and varying the signification of each figure according to the place it obtains, all numbers may be most aptly expressed: which seems to have been done in imitation of language, so that an exact analogy is observed betwixt the notation by figures and names, the nine simple figures answering the nine first numeral names and places in the former, corresponding to denominations in the latter. And agreeably to those conditions of the simple and local value of figures, were contrived methods of finding from the given figures or marks of the parts, what figures and how placed, are proper to denote the whole or *vice versa*. And having found the sought figures, the same rule or analogy being observed throughout, it is easy to read them into words; and so the number becomes perfectly known. For then the number of any particular things is said to be known, when we know the name or figures (with their due arrangement) that according to the standing analogy belong to

them. For these signs being known, we can by the operations of arithmetic, know the signs of any part of the particular sums signified by them; and thus computing in signs (because of the connection established betwixt them and the distinct multitudes of things, whereof one is taken for an unit), we may be able rightly to sum up, divide, and proportion the things themselves that we intend to number.

122 In *arithmetic* therefore we regard not the *things* but the *signs*, which nevertheless are not regarded for their own sake, but because they direct us how to act with relation to things, and dispose rightly of them. Now agreeably to what we have before observed, of words in general (*Sect.* 19. *Introd.*) it happens here likewise, that abstract ideas are thought to be signified by numeral names or characters, while they do not suggest ideas of particular things to our minds. I shall not at present enter into a more particular dissertation on this subject; but only observe that it is evident from what hath been said, those things which pass for abstract truths and theorems concerning numbers, are, in reality, conversant about no object distinct from particular numerable things, except only names and characters; which originally came to be considered, on no other account but their being *signs*, or capable to represent aptly, whatever particular things men had need to compute. Whence it follows, that to study them for their own sake would be just as wise, and to as good purpose, as if a man, neglecting the true use or original intention and subserviency of language, should spend his time in impertinent criticisms upon words, or reasonings and controversies purely verbal.

123 From numbers we proceed to speak of *extension*, which considered as relative, is the object of geometry. The *infinite* divisibility of *finite* extension, though it is not expressly laid down, either as an axiom or theorem in the elements of that science, yet is throughout the same every where supposed, and thought to have so inseparable and essential a connection with the principles and demonstrations in geometry, that mathematicians never admit it into doubt, or make the least question of it. And as this notion is the source from whence do spring all those amusing geometrical paradoxes, which have such a direct repugnancy to the plain common sense of mankind, and are admitted with so much reluctance into a mind not yet debauched by learning: so is it the principal occasion of all that nice and extreme subtlety, which renders the study of *mathematics* so difficult and tedious. Hence if we can make it appear, that no finite extension contains innumerable parts, or is infinitely divisible, it follows that we shall at once clear the science of geometry from a great number of difficulties and contradictions, which have ever been esteemed a reproach to human reason, and withal make the attainment thereof a business of much less time and pains, than it hitherto hath been.

124 Every particular finite extension, which may possibly be the object of our thought, is an *idea* existing only in the mind, and consequently each part thereof must be perceived. If therefore I cannot perceive innumerable parts in any finite extension that I consider, it is certain they are not contained in it: but it is evident, that I cannot distinguish innumerable parts in any particular line,

surface, or solid, which I either perceive by sense, or figure to my self in my mind: wherefore I conclude they are not contained in it. Nothing can be plainer to me, than that the extensions I have in view are no other than my own ideas, and it is no less plain, that I cannot resolve any one of my ideas into an infinite number of other ideas, that is, that they are not infinitely divisible. If by *finite extension* be meant something distinct from a finite idea, I declare I do not know what that is, and so cannot affirm or deny any thing of it. But if the terms *extension*, *parts*, and the like, are taken in any sense conceivable, that is, for ideas; then to say a finite quantity or extension consists of parts infinite in number, is so manifest a contradiction, that every one at first sight acknowledges it to be so. And it is impossible it should ever gain the assent of any reasonable creature, who is not brought to it by gentle and slow degrees, as a converted Gentile to the belief of *transubstantiation*. Ancient and rooted prejudices do often pass into principles: and those propositions which once obtain the force and credit of a *principle*, are not only themselves, but likewise whatever is deducible from them, thought privileged from all examination. And there is no absurdity so gross, which by this means the mind of man may not be prepared to swallow.

125 He whose understanding is prepossessed with the doctrine of abstract general ideas, may be persuaded, that (whatever be thought of the ideas of sense), extension in *abstract* is infinitely divisible. And one who thinks the objects of sense exist without the mind, will perhaps in virtue thereof be brought to admit, that a line but an inch long may contain innumerable parts really existing, though too small to be discerned. These errors are grafted as well in the minds of *geometricians*, as of other men, and have a like influence on their reasonings; and it were no difficult thing, to shew how the arguments from geometry made use of to support the infinite divisibility of extension, are bottomed on them. At present we shall only observe in general, whence it is that the mathematicians are all so fond and tenacious of this doctrine.

126 It hath been observed in another place, that the theorems and demonstrations in geometry are conversant about universal ideas. *Sect.* 15. *Introd.* Where it is explained in what sense this ought to be understood, to wit, that the particular lines and figures included in the diagram, are supposed to stand for innumerable others of different sizes: or in other words, the geometer considers them abstracting from their magnitude: which doth not imply that he forms an abstract idea, but only that he cares not what the particular magnitude is, whether great or small, but looks on that as a thing indifferent to the demonstration: hence it follows, that a line in the scheme, but an inch long, must be spoken of, as though it contained ten thousand parts, since it is regarded not in it self, but as it is universal; and it is universal only in its signification, whereby it represents innumerable lines greater than it self, in which may be distinguished ten thousand parts or more, though there may not be above an inch in it. After this manner the properties of the lines signified are (by a very usual figure) transferred to the sign, and thence through mistake thought to appertain to it considered in its own nature.

127 Because there is no number of parts so great, but it is possible there may be a line containing more, the inch-line is said to contain parts more than any assignable number; which is true, not of the inch taken absolutely, but only for the things signified by it. But men not retaining that distinction in their thoughts, slide into a belief that the small particular line described on paper contains in it self parts innumerable. There is no such thing as the ten-thousandth part of an *inch*; but there is of a *mile* or *diameter of the earth*, which may be signified by that inch. When therefore I delineate a triangle on paper, and take one side not above an inch, for example, in length to be the *radius*: this I consider as divided into ten thousand or an hundred thousand parts, or more. For though the ten-thousandth part of that line considered in it self, is nothing at all, and consequently may be neglected without any error or inconveniency; yet these described lines being only marks standing for greater quantities, whereof it may be the ten-thousandth part is very considerable, it follows, that to prevent notable errors in practice, the *radius* must be taken of ten thousand parts, or more.

128 From what hath been said the reason is plain why, to the end any theorem may become universal in its use, it is necessary we speak of the lines described on paper, as though they contained parts which really they do not. In doing of which, if we examine the matter throughly, we shall perhaps discover that we cannot conceive an inch it self as consisting of, or being divisible into a thousand parts, but only some other line which is far greater than an inch, and represented by it. And that when we say a line is *infinitely divisible*, we must mean a line which is *infinitely great*. What we have here observed seems to be the chief cause, why to suppose the infinite divisibility of finite extension hath been thought necessary in geometry.

129 The several absurdities and contradictions which flowed from this false principle might, one would think, have been esteemed so many demonstrations against it. But by I know not what *logic*, it is held that proofs *à posteriori* are not to be admitted against propositions relating to infinity. As though it were not impossible even for an infinite mind to reconcile contradictions. Or as if any thing absurd and repugnant could have a necessary connection with truth, or flow from it. But whoever considers the weakness of this pretence, will think it was contrived on purpose to humour the laziness of the mind, which had rather acquiesce in an indolent scepticism, than be at the pains to go through with a severe examination of those principles it hath ever embraced for true.

130 Of late the speculations about infinites have run so high, and grown to such strange notions, as have occasioned no small scruples and disputes among the geometers of the present age. Some there are of great note, who not content with holding that finite lines may be divided into an infinite number of parts, do yet farther maintain, that each of those infinitesimals is it self subdivisible into an infinity of other parts, or infinitesimals of a second order, and so on *ad infinitum*. These, I say, assert there are infinitesimals of infinitesimals of infinitesimals, without ever coming to an end. So that according to them an inch doth not barely contain an infinite number of parts, but an infinity of an infinity

of an infinity *ad infinitum* of parts. Others there be who hold all orders of infinitesimals below the first to be nothing at all, thinking it with good reason absurd, to imagine there is any positive quantity or part of extension, which though multiplied infinitely, can ever equal the smallest given extension. And yet on the other hand it seems no less absurd, to think the square, cube, or other power of a positive real root, should it self be nothing at all; which they who hold infinitesimals of the first order, denying all of the subsequent orders, are obliged to maintain.

131 Have we not therefore reason to conclude, that they are *both* in the wrong, and that there is in effect no such thing as parts infinitely small, or an infinite number of parts contained in any finite quantity? But you will say, that if this doctrine obtains, it will follow the very foundations of geometry are destroyed: and those great men who have raised that science to so astonishing an height, have been all the while building a castle in the air. To this it may be replied, that whatever is useful in geometry and promotes the benefit of human life, doth still remain firm and unshaken on our principles. That science considered as practical, will rather receive advantage than any prejudice from what hath been said. But to set this in a due light, may be the subject of a distinct inquiry. For the rest, though it should follow that some of the more intricate and subtle parts of *speculative mathematics* may be pared off without any prejudice to truth; yet I do not see what damage will be thence derived to mankind. On the contrary, it were highly to be wished, that men of great abilities and obstinate application would draw off their thoughts from those amusements, and employ them in the study of such things as lie nearer the concerns of life, or have a more direct influence on the manners.

132 If it be said that several theorems undoubtedly true, are discovered by methods in which infinitesimals are made use of, which could never have been, if their existence included a contradiction in it. I answer, that upon a thorough examination it will not be found, that in any instance it is necessary to make use of or conceive infinitesimal parts of finite lines, or even quantities less than the *minimum sensibile*: nay, it will be evident this is never done, it being impossible. ‖And whatever mathematicians may think of fluxions or the differential calculus and the like, a little reflection will shew them, that in working by those methods, they do not conceive or imagine lines or surfaces less than what are perceivable to sense. They may, indeed, call those little and almost insensible quantities infinitesimals or infinitesimals of infinitesimals, if they please: but at bottom this is all, they being in truth finite, nor does the solution of problems require the supposing any other. But this will be more clearly made out hereafter.‖[c]

133 By what we have premised, it is plain that very numerous and important errors have taken their rise from those false principles, which were impugned in the foregoing parts of this treatise. And the opposites of those erroneous tenets at the same time appear to be most fruitful principles, from whence do

---

[c] ‖This passage occurs in the first edition only.‖

flow innumerable consequences highly advantageous to true philosophy as well as to religion. Particularly, *matter* or *the absolute existence of corporeal objects*, hath been shewn to be that wherein the most avowed and pernicious enemies of all knowledge, whether human or divine, have ever placed their chief strength and confidence. And surely, if by distinguishing the real existence of unthinking things from their being perceived, and allowing them a subsistence of their own out of the minds of spirits, no one thing is explained in Nature; but on the contrary a great many inexplicable difficulties arise: if the supposition of matter is barely precarious, as not being grounded on so much as one single reason: if its consequences cannot endure the light of examination and free inquiry, but screen themselves under the dark and general pretence of *infinites being incomprehensible*: if withal the removal of this *matter* be not attended with the least evil consequence, if it be not even missed in the world, but every thing as well, nay much easier conceived without it: if lastly, both *sceptics* and *atheists* are for ever silenced upon supposing only spirits and ideas, and this scheme of things is perfectly agreeable both to *reason* and *religion*: methinks we may expect it should be admitted and firmly embraced, though it were proposed only as an *hypothesis*, and the existence of matter had been allowed possible, which yet I think we have evidently demonstrated that it is not.

134 True it is, that in consequence of the foregoing principles, several disputes and speculations, which are esteemed no mean parts of learning, are rejected as useless. But how great a prejudice soever against our notions, this may give to those who have already been deeply engaged, and made large advances in studies of that nature: yet by others, we hope it will not be thought any just ground of dislike to the principles and tenets herein laid down, that they abridge the labour of study, and make human sciences more clear, compendious, and attainable, than they were before.

E. DE MOTU
(*BERKELEY 1721*)

In 1713 Berkeley left Dublin for London, where he published his *Three dialogues between Hylas and Philonous*, a popularized exposition of the doctrines of the *Principles of human knowledge*. For the next eight years he travelled on the Continent, wrote essays in defence of religion, and published little on philosophical topics. His *De motu sive de motus principio & natura, et de causa communicationis motuum*, written on the Continent at the end of this period of Berkeley's life, is his most sustained criticism of the foundations of Newtonian physics. It contains a subtle discussion (§§35–42) of the nature of physical axioms and of the relationship between mathematics and physics; the essay prefigures the critique of Newtonian mathematics in *The analyst*, and shows how Berkeley's general philosophical arguments—in particular his arguments against abstract ideas and material substance—could be applied to the foundations of natural science. For a discussion of Berkeley's philosophy of physics, see *Popper 1953-4*. The translation of the *De motu* is by Arthur Aston Luce, and is reprinted from *Berkeley 1948-57*, Vol. 4. References to *Berkeley 1721* should be to the paragraph numbers, which appeared in the original Luce translation.

OF MOTION

OR

THE PRINCIPLE AND NATURE OF MOTION AND THE CAUSE OF THE COMMUNICATION OF MOTIONS

1 In the pursuit of truth we must beware of being misled by terms which we do not rightly understand. That is the chief point. Almost all philosophers utter the caution; few observe it. Yet it is not so difficult to observe, where sense, experience, and geometrical reasoning obtain, as is especially the case in physics. Laying aside, then, as far as possible, all prejudice, whether rooted in linguistic usage or in philosophical authority, let us fix our gaze on the very nature of things. For no one's authority ought to rank so high as to set a value on his words and terms unless they are found to be based on clear and certain fact.

2 The consideration of motion greatly troubled the minds of the ancient philosophers, giving rise to various exceedingly difficult opinions (not to say absurd) which have almost entirely gone out of fashion, and not being worth a detailed discussion need not delay us long. In works on motion by the more recent and sober thinkers of our age, not a few terms of somewhat abstract and obscure signification are used, such as *solicitation of gravity, urge, dead forces*, etc., terms which darken writings in other respects very learned, and beget opinions at variance with truth and the common sense of men. These terms must be examined with great care, not from a desire to prove other people wrong, but in the interest of truth.

3 *Solicitation* and *effort* or *conation* belong properly to animate beings alone. When they are attributed to other things, they must be taken in a metaphorical sense; but a philosopher should abstain from metaphor. Besides, anyone who has seriously considered the matter will agree that those terms have no clear and distinct meaning apart from all affection of the mind and motion of the body.

4 While we support heavy bodies we feel in ourselves effort, fatigue, and discomfort. We perceive also in heavy bodies falling an accelerated motion towards the centre of the earth; and that is all the senses tell us. By reason,

however, we infer that there is some cause or principle of these phenomena, and that is popularly called *gravity*. But since the cause of the fall of heavy bodies is unseen and unknown, gravity in that usage cannot properly be styled a sensible quality. It is, therefore, an occult quality. But what an occult quality is, or how any quality can act or do anything, we can scarcely conceive—indeed we cannot conceive. And so men would do better to let the occult quality go, and attend only to the sensible effects. Abstract terms (however useful they may be in argument) should be discarded in meditation, and the mind should be fixed on the particular and the concrete, that is, on the things themselves.

5 *Force* likewise is attributed to bodies; and that word is used as if it meant a known quality, and one distinct from motion, figure, and every other sensible thing and also from every affection of the living thing. But examine the matter more carefully and you will agree that such force is nothing but an occult quality. Animal effort and corporeal motion are commonly regarded as symptoms and measures of this occult quality.

6 Obviously then it is idle to lay down gravity or force as the principle of motion; for how could that principle be known more clearly by being styled an occult quality? What is itself occult explains nothing. And I need not say that an unknown acting cause could be more correctly styled substance than quality. Again, *force*, *gravity*, and terms of that sort are more often used in the concrete (and rightly so) so as to connote the body in motion, the effort of resisting, *etc*. But when they are used by philosophers to signify certain natures carved out and abstracted from all these things, natures which are not objects of sense, nor can be grasped by any force of intellect, nor pictured by the imagination, then indeed they breed errors and confusion.

7 About general and abstract terms many men make mistakes; they see their value in argument, but they do not appreciate their purpose. In part the terms have been invented by common habit to abbreviate speech, and in part they have been thought out by philosophers for instructional purposes, not that they are adapted to the natures of things which are in fact singulars and concrete, but they come in useful for handing on received opinions by making the notions or at least the propositions universal.

8 We generally suppose that corporeal force is something easy to conceive. Those, however, who have studied the matter more carefully are of a different opinion, as appears from the strange obscurity of their language when they try to explain it. Torricelli says that force and impetus are abstract and subtle things and quintessences which are included in corporeal substance as in the magic vase of Circe.[1] Leibniz likewise in explaining the nature of force has this: 'Active primitive force which is *ἐντελέχεια ἡ πρώτη* corresponds to the soul or substantial form.' See *Acta Erudit. Lips*. Thus even the greatest men when they give

[1] Matter is nothing else than a magic vase of Circe, which serves as a receptacle of force and of the moments of the impetus. Force and the impetus are such subtle abstractions and such volatile quintessences that they cannot be shut up in any vessel except in the innermost substance of natural solids. See *Academic Lectures*.

way to abstractions are bound to pursue terms which have no certain significance and are mere shadows of scholastic things. Other passages in plenty from the writings of the younger men could be produced which give abundant proof that metaphysical abstractions have not in all quarters given place to mechanical science and experiment, but still make useless trouble for philosophers.

9 From that source derive various absurdities, such as that dictum: 'The force of percussion, however small, is infinitely great'—which indeed supposes that gravity is a certain real quality different from all others, and that gravitation is, as it were, an act of this quality, really distinct from motion. But a very small percussion produces a greater effect than the greatest gravitation without motion. The former gives out some motion indeed, the latter none. Whence it follows that the force of percussion exceeds the force of gravitation by an infinite ratio, *i.e.* is infinitely great. See the experiments of Galileo, and the writings of Torricelli, Borelli, and others on the definite force of percussion.

10 We must, however, admit that no force is immediately felt by itself, nor known or measured otherwise than by its effect; but of a dead force or of simple gravitation in a body at rest, no change taking place, there is no effect; of percussion there is some effect. Since, then, forces are proportional to effects, we may conclude that there is no dead force, but we must not on that account infer that the force of percussion is infinite; for we cannot regard as infinite any positive quantity on the ground that it exceeds by an infinite ratio a zero-quantity or nothing.

11 The force of gravitation is not to be separated from momentum; but there is no momentum without velocity, since it is mass multiplied by velocity; again, velocity cannot be understood without motion, and the same holds therefore of the force of gravitation. Then no force makes itself known except through action, and through action it is measured; but we are not able to separate the action of a body from its motion; therefore as long as a heavy body changes the shape of a piece of lead put under it, or of a cord, so long is it moved; but when it is at rest, it does nothing, or (which is the same thing) it is prevented from acting. In brief, those terms *dead force* and *gravitation* by the aid of metaphysical abstraction are supposed to mean something different from moving, moved, motion, and rest, but, in point of fact, the supposed difference in meaning amounts to nothing at all.

12 If anyone were to say that a weight hung or placed on the cord acts on it, since it prevents it from restoring itself by elastic force, I reply that by parity of reasoning any lower body acts on the higher body which rests on it, since it prevents it from coming down. But for one body to prevent another from existing in that space which *it* occupies cannot be styled the action of that body.

13 We feel at times the pressure of a gravitating body. But that unpleasant sensation arises from the motion of the heavy body communicated to the fibres and nerves of our body and changing their situation, and therefore it ought to be referred to percussion. In these matters we are afflicted by a number of serious prejudices, which should be subdued, or rather entirely exorcised by keen and continued reflection.

14 In order to prove that any quantity is infinite, we have to show that some, finite, homogeneous part is contained in it an infinite number of times. But dead force is to the force of percussion, not as part to the whole, but as the point to the line, according to the very writers who maintain the infinite force of percussion. Much might be added on this matter, but I am afraid of being prolix.

15 By the foregoing principles famous controversies which have greatly exercised the minds of learned men can be solved; for instance, that controversy about the proportion of forces. One side conceding that momenta, motions, and impetus, given the mass, are simply as the velocities, affirms that the forces are as the squares of the velocities. Everyone sees that this opinion supposes that the force of the body is distinguished from momentum, motion, and impetus, and without that supposition it collapses.

16 To make it still clearer that a certain strange confusion has been introduced into the theory of motion by metaphysical abstractions, let us watch the conflict of opinion about force and impetus among famous men. Leibniz confuses impetus with motion. According to Newton impetus is in fact the same as the force of inertia. Borelli asserts that impetus is only the degree of velocity. Some would make impetus and effort different, others identical. Most regard the motive force as proportional to the motion; but a few prefer to suppose some other force besides the motive, to be measured differently, for instance by the squares of the velocities into the masses. But it would be an endless task to follow out this line of thought.

17 *Force, gravity, attraction*, and terms of this sort are useful for reasonings and reckonings about motion and bodies in motion, but not for understanding the simple nature of motion itself or for indicating so many distinct qualities. As for attraction, it was certainly introduced by Newton, not as a true, physical quality, but only as a mathematical hypothesis. Indeed Leibniz when distinguishing elementary effort or solicitation from impetus, admits that those entities are not really found in nature, but have to be formed by abstraction.

18 A similar account must be given of the composition and resolution of any direct forces into any oblique ones by means of the diagonal and sides of the parallelogram. They serve the purpose of mechanical science and reckoning; but to be of service to reckoning and mathematical demonstrations is one thing, to set forth the nature of things is another.

19 Of the moderns many are of the opinion that motion is neither destroyed nor generated anew, but that the quantity of motion remains for ever constant. Aristotle indeed propounded that problem long ago, Does motion come into being and pass away, or is it eternal? *Phys.* Bk. 8. That sensible motion perishes is clear to the senses, but apparently they will have it that the same impetus and effort remains, or the same sum of forces. Borelli affirms that force in percussion is not lessened, but expanded, that even contrary impetus are received and retained in the same body. Likewise Leibniz contends that effort exists everywhere and always in matter, and that it is understood by reason where it is not evident to the senses. But these points, we must admit, are too abstract and obscure, and of much the same sort as substantial forms and entelechies.

20 All those who, to explain the cause and origin of motion, make use of the hylarchic principle, or of a nature's want or appetite, or indeed of a natural instinct, are to be considered as having said something, rather than thought it. And from these they[2] are not far removed who have supposed 'that the parts of the earth are self-moving, or even that spirits are implanted in them like a form' in order to assign the cause of the acceleration of heavy bodies falling. So too with him[3] who said 'that in the body besides solid extension, there must be something posited to serve as starting-point for the consideration of forces'. All these indeed either say nothing particular and determinate, or if there is anything in what they say, it will be as difficult to explain as that very thing it was brought forward to explain.

21 To throw light on nature it is idle to adduce things which are neither evident to the senses, nor intelligible to reason. Let us see then what sense and experience tell us, and reason that rests upon them. There are two supreme classes of things, body and soul. By the help of sense we know the extended thing, solid, mobile, figured, and endowed with other qualities which meet the senses, but the sentient, percipient, thinking thing we know by a certain internal consciousness. Further we see that those things are plainly different from one another, and quite heterogeneous. I speak of things known; for of the unknown it is profitless to speak.

22 All that which we know to which we have given the name *body* contains nothing in itself which could be the principle of motion or its efficient cause; for impenetrability, extension, and figure neither include nor connote any power of producing motion; nay, on the contrary, if we review singly those qualities of body, and whatever other qualities there may be, we shall see that they are all in fact passive and that there is nothing active in them which can in any way be understood as the source and principle of motion. As for gravity we have already shown above that by that term is meant nothing we know, nothing other than the sensible effect, the cause of which we seek. And indeed when we call a body heavy we understand nothing else except that it is borne downwards, and we are not thinking at all about the cause of this sensible effect.

23 And so about body we can boldly state as established fact that it is not the principle of motion. But if anyone maintains that the term *body* covers in its meaning occult quality, virtue, form, and essence, besides solid extension and its modes, we must just leave him to his useless disputation with no ideas behind it, and to his abuse of names which express nothing distinctly. But the sounder philosophical method, it would seem, abstains as far as possible from abstract and general notions (if *notions* is the right term for things which cannot be understood).

24 The contents of the idea of body we know; but what we know in body is agreed not to be the principle of motion. But those who as well maintain something unknown in body of which they have no idea and which they call

[2] Borelli.
[3] Leibniz.

the principle of motion, are in fact simply stating that the principle of motion is unknown, and one would be ashamed to linger long on subtleties of this sort.

25 Besides corporeal things there is the other class, *viz.* thinking things, and that there is in them the power of moving bodies we have learned by personal experience, since our mind at will can stir and stay the movements of our limbs, whatever be the ultimate explanation of the fact. This is certain, that bodies are moved at the will of the mind, and accordingly the mind can be called, correctly enough, a principle of motion, a particular and subordinate principle indeed, and one which itself depends on the first and universal principle.

26 Heavy bodies are borne downwards, although they are not affected by any apparent impulse; but we must not think on that account that the principle of motion is contained in them. Aristotle gives this account of the matter, 'Heavy and light things are not moved by themselves; for that would be a characteristic of life, and they would be able to stop themselves.' All heavy things by one and the same certain and constant law seek the centre of the earth, and we do not observe in them a principle or any faculty of halting that motion, of diminishing it or increasing it except in fixed proportion, or finally of altering it in any way. They behave quite passively. Again, in strict and accurate speech, the same must be said of percussive bodies. Those bodies as long as they are being moved, as also in the very moment of percussion, behave passively, exactly as when they are at rest. Inert body so acts as body moved acts, if the truth be told. Newton recognizes that fact when he says that the force of inertia is the same as impetus. But body, inert and at rest, does nothing; therefore body moved does nothing.

27 Body in fact persists equally in either state, whether of motion or of rest. Its existence is not called its action; nor should its persistence be called its action. Persistence is only continuance in the same way of existing, which cannot properly be called action. Resistance which we experience in stopping a body in motion we falsely imagine to be its action, deluded by empty appearance. For that resistance which we feel is in fact passion in ourselves, and does not prove that body acts, but that we are affected; it is quite certain that we should be affected in the same way, whether that body were to be moved by itself, or impelled by another principle.

28 Action and reaction are said to be in bodies, and that way of speaking suits the purposes of mechanical demonstrations; but we must not on that account suppose that there is some real virtue in them which is the cause or principle of motion. For those terms are to be understood in the same way as the term *attraction*; and just as attraction is only a mathematical hypothesis, and not a physical quality, the same must be understood also about action and reaction, and for the same reason. For in mechanical philosophy the truth and the use of theorems about the mutual attraction of bodies remain firm, as founded solely in the motion of bodies, whether that motion be supposed to be caused by the action of bodies mutually attracting each other, or by the action of some agent different from the bodies, impelling and controlling them. Similarly the traditional formulations of rules and laws of motions, along with

the theorems thence deduced remain unshaken, provided that sensible effects and the reasonings grounded in them are granted, whether we suppose the action itself or the force that causes these effects to be in the body or in the incorporeal agent.

29 Take away from the idea of body extension, solidity, and figure, and nothing will remain. But those qualities are indifferent to motion, nor do they contain anything which could be called the principle of motion. This is clear from our very ideas. If therefore by the term *body* be meant that which we conceive, obviously the principle of motion cannot be sought therein, that is, no part or attribute thereof is the true, efficient cause of the production of motion. But to employ a term, and conceive nothing by it is quite unworthy of a philosopher.

30 A thinking, active thing is given which we experience as the principle of motion in ourselves. This we call *soul*, *mind*, and *spirit*. An extended thing also is given, inert, impenetrable, movable, totally different from the former and constituting a new genus. Anaxagoras, wisest of men, was the first to grasp the great difference between thinking things and extended things, and he asserted that the mind has nothing in common with bodies, as is established from the first book of Aristotle's *De Anima*. Of the moderns Descartes has put the same point most forcibly. What was left clear by him others have rendered involved and difficult by their obscure terms.

31 From what has been said it is clear that those who affirm that active force, action, and the principle of motion are really in bodies are adopting an opinion not based on experience, are supporting it with obscure and general terms, and do not well understand their own meaning. On the contrary those who will have mind to be the principle of motion are advancing an opinion fortified by personal experience, and one approved by the suffrages of the most learned men in every age.

32 Anaxagoras was the first to introduce *nous* to impress motion on inert matter. Aristotle, too, approves that opinion and confirms it in many ways, openly stating that the first mover is immovable, indivisible, and has no magnitude. And he rightly notes that to say that every mover must be movable is the same as to say that every builder must be capable of being built. *Phys*. Bk. 8. Plato, moreover, in the Timaeus records that this corporeal machine, or visible world, is moved and animated by mind which eludes all sense. To-day indeed Cartesian philosophers recognize God as the principle of natural motions. And Newton everywhere frankly intimates that not only did motion originate from God, but that still the mundane system is moved by the same actus. This is agreeable to Holy Scripture; this is approved by the opinion of the schoolmen; for though the Peripatetics tell us that nature is the principle of motion and rest, yet they interpret *natura naturans* to be God. They understand of course that all the bodies of this mundane system are moved by Almighty Mind according to certain and constant reason.

33 But those who attribute a vital principle to bodies are imagining an obscure notion and one ill suited to the facts. For what is meant by being endowed with the vital principle, except to live? And to live, what is it but to move oneself, to

stop, and to change one's state? But the most learned philosophers of this age lay it down for an indubitable principle that every body persists in its own state, whether of rest or of uniform movement in a straight line, except in so far as it is compelled from without to alter that state. The contrary is the case with mind; we feel it as a faculty of altering both our own state and that of other things, and that is properly called vital, and puts a wide distinction between soul and bodies.

34 Modern thinkers consider motion and rest in bodies as two states of existence in either of which every body, without pressure from external force, would naturally remain passive; whence one might gather that the cause of the existence of bodies is also the cause of their motion and rest. For no other cause of the successive existence of the body in different parts of space should be sought, it would seem, than that cause whence is derived the successive existence of the same body in different parts of time. But to treat of the good and great God, creator and preserver of all things, and to show how all things depend on supreme and true being, although it is the most excellent part of human knowledge, is, however, rather the province of first philosophy or metaphysics and theology than of natural philosophy, which to-day is almost entirely confined to experiments and mechanics. And so natural philosophy either presupposes the knowledge of God or borrows it from some superior science. Although it is most true that the investigation of nature everywhere supplies the higher sciences with notable arguments to illustrate and prove the wisdom, the goodness, and the power of God.

35 The imperfect understanding of this situation has caused some to make the mistake of rejecting the mathematical principles of physics on the ground that they do not assign the efficient causes of things. It is not, however, in fact the business of physics or mechanics to establish efficient causes, but only the rules of impulsions or attractions, and, in a word, the laws of motions, and from the established laws to assign the solution, not the efficient cause, of particular phenomena.

36 It will be of great importance to consider what properly a principle is, and how that term is to be understood by philosophers. The true, efficient and conserving cause of all things by supreme right is called their fount and principle. But the principles of experimental philosophy are properly to be called foundations and springs, not of their existence but of our knowledge of corporeal things, both knowledge by sense and knowledge by experience, foundations on which that knowledge rests and springs from which it flows. Similarly in mechanical philosophy those are to be called principles, in which the whole discipline is grounded and contained, those primary laws of motions which have been proved by experiments, elaborated by reason and rendered universal. These laws of motion are conveniently called principles, since from them are derived both general mechanical theorems and particular explanations of the phenomena.

37 A thing can be said to be explained mechanically then indeed when it is reduced to those most simple and universal principles, and shown by accurate

reasoning to be in agreement and connection with them. For once the laws of nature have been found out, then it is the philosopher's task to show that each phenomenon is in constant conformity with those laws, that is, necessarily follows from those principles. In that consist the explanation and solution of phenomena and the assigning their cause, *i.e.* the reason why they take place.

38 The human mind delights in extending and expanding its knowledge; and for this purpose general notions and propositions have to be formed in which particular propositions and cognitions are in some way comprised, which then, and not till then, are believed to be understood. Geometers know this well. In mechanics also notions are premised, *i.e.* definitions and first and general statements about motion from which afterwards by mathematical method conclusions more remote and less general are deduced. And just as by the application of geometrical theorems, the sizes of particular bodies are measured, so also by the application of the universal theorems of mechanics, the movements of any parts of the mundane system, and the phenomena thereon depending, become known and are determined. And that is the sole mark at which the physicist must aim.

39 And just as geometers for the sake of their art make use of many devices which they themselves cannot describe nor find in the nature of things, even so the mechanician makes use of certain abstract and general terms, imagining in bodies force, action, attraction, solicitation, *etc.* which are of first utility for theories and formulations, as also for computations about motion, even if in the truth of things, and in bodies actually existing, they would be looked for in vain, just like the geometers' fictions made by mathematical abstraction.

40 We actually perceive by the aid of the senses nothing except the effects or sensible qualities and corporeal things entirely passive, whether in motion or at rest; and reason and experience advise us that there is nothing active except mind or soul. Whatever else is imagined must be considered to be of a kind with other hypotheses and mathematical abstractions. This ought to be laid to heart; otherwise we are in danger of sliding back into the obscure subtlety of the schoolmen, which for so many ages, like some dread plague, has corrupted philosophy.

41 Mechanical principles and universal laws of motions or of nature, happy discoveries of the last century, treated and applied by aid of geometry, have thrown a remarkable light upon philosophy. But metaphysical principles and real efficient causes of the motion and existence of bodies or of corporeal attributes in no way belong to mechanics or experiment, nor throw light on them, except in so far as by being known beforehand they may serve to define the limits of physics, and in that way to remove imported difficulties and problems.

42 Those who derive the principle of motion from spirits mean by *spirit* either a corporeal thing or an incorporeal; if a corporeal thing, however tenuous, yet the difficulty recurs; if an incorporeal thing, however true it may be, yet it does not properly belong to physics. But if anyone were to extend natural philosophy beyond the limits of experiments and mechanics, so as to

cover a knowledge of incorporeal and inextended things, that broader interpretation of the term permits a discussion of soul, mind, or vital principle. But it will be more convenient to follow the usage which is fairly well accepted, and so to distinguish between the sciences as to confine each to its own bounds; thus the natural philosopher should concern himself entirely with experiments, laws of motions, mechanical principles, and reasonings thence deduced; but if he shall advance views on other matters, let him refer them for acceptance to some superior science. For from the known laws of nature very elegant theories and mechanical devices of practical utility follow; but from the knowledge of the Author of nature Himself by far the most excellent considerations arise, but they are metaphysical, theological, and moral.

43 So far about principles; now we must speak of the nature of motion. Motion, though it is clearly perceived by the senses, has been rendered obscure rather by the learned comments of philosophers than by its own nature. Motion never meets our senses apart from corporeal mass, space, and time. There are indeed those who desire to contemplate motion as a certain simple and abstract idea, and separated from all other things. But that very fine-drawn and subtle idea eludes the keen edge of intellect, as anyone can find for himself by meditation. Hence arise great difficulties about the nature of motion, and definitions far more obscure than the thing they are meant to illustrate. Such are those definitions of Aristotle and the school-men, who say that motion is the act 'of the movable in so far as it is movable, or the act of a being in potentiality in so far as it is in potentiality'. Such is the saying of a famous man[a] of modern times, who asserts that 'there is nothing real in motion except that momentary thing which must be constituted when a force is striving towards a change'. Again, it is agreed that the authors of these and similar definitions had it in mind to explain the abstract nature of motion, apart from every consideration of time and space; but how that abstract quintessence, so to speak, of motion, can be understood I do not see.

44 Not content with this they go further and divide and separate from one another the parts of motion itself, of which parts they try to make distinct ideas, as if of entities in fact distinct. For there are those who distinguish movement from motion, looking on the movement as an instantaneous element in the motion. Moreover, they would have velocity, conation, force, and impetus to be so many things differing in essence, each of which is presented to the intellect through its own abstract idea separated from all the rest. But we need not spend any more time on these discussions if the principles laid down above hold good.

45 Many also define motion by *passage*, forgetting indeed that passage itself cannot be understood without motion, and through motion ought to be defined. So very true is it that definitions throw light on some things, and darkness again on others. And certainly hardly anyone could by defining them make clearer or better known the things we perceive by sense. Enticed by the vain hope of

[a] [The identity of the 'famous man' is uncertain.]

doing so, philosophers have rendered easy things very difficult, and have ensnared their own minds in difficulties which for the most part they themselves produced. From this desire of defining and abstracting many very subtle questions both about motion and other things take their rise. Those useless questions have tortured the minds of men to no purpose; so that Aristotle often actually confesses that motion is 'a certain act difficult to know', and some of the ancients became such pastmasters in trifling as to deny the existence of motion altogether.

46 But one is ashamed to linger on minutiæ of this sort; let it suffice to have indicated the sources of the solutions; but this, too, I must add. The traditional mathematical doctrines of the infinite division of time and space have, from the very nature of the case, introduced paradoxes and thorny theories (as are all those that involve the infinite) into speculations about motion. All such difficulties motion shares with space and time, or rather has taken them over from that source.

47 Too much abstraction, on the one hand, or the division of things truly inseparable, and on the other hand composition or rather confusion of very different things have perplexed the nature of motion. For it has become usual to confuse motion with the efficient cause of motion. Whence it comes about that motion appears, as it were, in two forms, presenting one aspect to the senses, and keeping the other aspect covered in dark night. Thence obscurity, confusion, and various paradoxes of motion take their rise, while what belongs in truth to the cause alone is falsely attributed to the effect.

48 This is the source of the opinion that the same quantity of motion is always conserved; anyone will easily satisfy himself of its falsity unless it be understood of the force and power of the cause, whether that cause be called nature or *nous*, or whatever be the ultimate agent. Aristotle indeed (*Phys*. Bk. 8) when he asks whether motion be generated and destroyed, or is truly present in all things from eternity like life immortal, seems to have understood the vital principle rather than the external effect or change of place.

49 Hence it is that many suspect that motion is not mere passivity in bodies. But if we understand by it that which in the movement of a body is an object to the senses, no one can doubt that it is entirely passive. For what is there in the successive existence of body in different places which could relate to action, or be other than bare, lifeless effect?

50 The Peripatetics who say that motion is the one act of both the mover and the moved do not sufficiently divide cause from effect. Similarly those who imagine effort or conation in motion, or think that the same body at the same time is borne in opposite directions, seem to be the sport of the same confusion of ideas, and the same ambiguity of terms.

51 Diligent attention in grasping the concepts of others and in formulating one's own is of great service in the science of motion as in all other things; and unless there had been a failing in this respect I do not think that matter for dispute could have come from the query, Whether a body is indifferent to motion and to rest, or not. For since experience shows that it is a primary law

of nature that a body persists exactly in 'a state of motion and rest as long as nothing happens from elsewhere to change that state', and on that account it is inferred that the force of inertia is under different aspects either resistance or impetus, in this sense assuredly a body can be called indifferent in its own nature to motion or rest. Of course it is as difficult to induce rest in a moving body as motion in a resting body; but since the body conserves equally either state, why should it not be said to be indifferent to both?

52 The Peripatetics used to distinguish various kinds of motion corresponding to the variety of changes which a thing could undergo. To-day those who discuss motion understand by the term only local motion. But local motion cannot be understood without understanding the meaning of *locus*. Now *locus* is defined by moderns as 'the part of space which a body occupies', whence it is divided into relative and absolute corresponding to space. For they distinguish between absolute or true space and relative or apparent space. That is they postulate space on all sides measureless, immovable, insensible, permeating and containing all bodies, which they call absolute space. But space comprehended or defined by bodies, and therefore an object of sense, is called relative, apparent, vulgar space.

53 And so let us suppose that all bodies were destroyed and brought to nothing. What is left they call absolute space, all relation arising from the situation and distances of bodies being removed together with the bodies. Again, that space is infinite, immovable, indivisible, insensible, without relation and without distinction. That is, all its attributes are privative or negative. It seems therefore to be mere nothing. The only slight difficulty arising is that it is extended, and extension is a positive quality. But what sort of extension, I ask, is that which cannot be divided nor measured, no part of which can be perceived by sense or pictured by the imagination? For nothing enters the imagination which from the nature of the thing cannot be perceived by sense, since indeed the imagination is nothing else than the faculty which represents sensible things either actually existing or at least possible. Pure intellect, too, knows nothing of absolute space. That faculty is concerned only with spiritual and inextended things, such as our minds, their states, passions, virtues, and such like. From absolute space then let us take away now the words of the name, and nothing will remain in sense, imagination, or intellect. Nothing else then is denoted by those words than pure privation or negation, *i.e.* mere nothing.

54 It must be admitted that in this matter we are in the grip of serious prejudices, and to win free we must exert the whole force of our minds. For many, so far from regarding absolute space as nothing, regard it as the only thing (God excepted) which cannot be annihilated; and they lay down that it necessarily exists of its own nature, that it is eternal and uncreate, and is actually a participant in the divine attributes. But in very truth since it is most certain that all things which we designate by names are known by qualities or relations, at least in part (for it would be stupid, to use words to which nothing known, no notion, idea or concept, were attached), let us diligently inquire whether it is possible to form any idea of that pure, real, and absolute space continuing to exist after

the annihilation of all bodies. Such an idea, moreover, when I watch it somewhat more intently, I find to be the purest idea of nothing, if indeed it can be called an idea. This I myself have found on giving the matter my closest attention; this, I think, others will find on doing likewise.

55 We are sometimes deceived by the fact that when we imagine the removal of all other bodies, yet we suppose our own body to remain. On this supposition we imagine the movement of our limbs fully free on every side; but motion without space cannot be conceived. None the less if we consider the matter again we shall find, 1st, relative space conceived defined by the parts of our body; 2nd, a fully free power of moving our limbs obstructed by no obstacle; and besides these two things nothing. It is false to believe that some third thing really exists, *viz.* immense space which confers on us the free power of moving our body; for this purpose the absence of other bodies is sufficient. And we must admit that this absence or privation of bodies is nothing positive.[4]

56 But unless a man has examined these points with a free and keen mind, words and terms avail little. To one who meditates, however, and reflects, it will be manifest, I think, that predications about pure and absolute space can all be predicated about nothing. By this argument the human mind is easily freed from great difficulties, and at the same time from the absurdity of attributing necessary existence to any being except to the good and great God alone.

57 It would be easy to confirm our opinion by arguments drawn, as they say *a posteriori*, by proposing questions about absolute space, *e.g.* Is it substance or accidents? Is it created or uncreated? and showing the absurdities which follow from either answer. But I must be brief. I must not omit, however, to state that Democritus of old supported this opinion with his vote. Aristotle is our authority for the statement, *Phys.* Bk. 1, where he has these words, 'Democritus lays down as principles the solid and the void, of which the one, he says, is as what is, the other as what is not.' That the distinction between absolute and relative space has been used by philosophers of great name, and that on it as on a foundation many fine theorems have been built, may make us scruple to accept the argument, but those are empty scruples, as will appear from what follows.

58 From the foregoing it is clear that we ought not to define the true place of the body as the part of absolute space which the body occupies, and true or absolute motion as the change of true or absolute place; for all place is relative just as all motion is relative. But to make this appear more clearly we must point out that no motion can be understood without some determination or direction, which in turn cannot be understood unless besides the body in motion our own body also, or some other body, be understood to exist at the same time. For *up*, *down*, *left*, and *right* and all places and regions are founded in some relation, and necessarily connote and suppose a body different from the body moved. So that if we suppose the other bodies were annihilated and,

[4] See the arguments against absolute space in my book on *The Principles of Human Knowledge* in the English tongue published ten years ago.

for example, a globe were to exist alone, no motion could be conceived in it; so necessary is it that another body should be given by whose situation the motion should be understood to be determined. The truth of this opinion will be very clearly seen if we shall have carried out thoroughly the supposed annihilation of all bodies, our own and that of others, except that solitary globe.

59 Then let two globes be conceived to exist and nothing corporeal besides them. Let forces then be conceived to be applied in some way; whatever we may understand by the application of forces, a circular motion of the two globes round a common centre cannot be conceived by the imagination. Then let us suppose that the sky of the fixed stars is created; suddenly from the conception of the approach of the globes to different parts of that sky the motion will be conceived. That is to say that since motion is relative in its own nature, it could not be conceived before the correlated bodies were given. Similarly no other relation can be conceived without correlates.

60 As regards circular motion many think that, as motion truly circular increases, the body necessarily tends ever more and more away from its axis. This belief arises from the fact that circular motion can be seen taking its origin, as it were, at every moment from two directions, one along the radius and the other along the tangent, and if in this latter direction only the impetus be increased, then the body in motion will retire from the centre, and its orbit will cease to be circular. But if the forces be increased equally in both directions the motion will remain circular though accelerated—which will not argue an increase in the forces of retirement from the axis, any more than in the forces of approach to it. Therefore we must say that the water forced round in the bucket rises to the sides of the vessel, because when new forces are applied in the direction of the tangent to any particle of water, in the same instant new equal centripetal forces are not applied. From which experiment it in no way follows that absolute circular motion is necessarily recognized by the forces of retirement from the axis of motion. Again, how those terms *corporeal forces* and *conation* are to be understood is more than sufficiently shown in the foregoing discussion.

61 A curve can be considered as consisting of an infinite number of straight lines, though in fact it does not consist of them. That hypothesis is useful in geometry; and just so circular motion can be regarded as arising from an infinite number of rectilinear directions—which supposition is useful in mechanics. Not, however, on that account must it be affirmed that it is impossible that the centre of gravity of each body should exist successively in single points of the circular periphery, no account being taken of any rectilineal direction in the tangent or the radius.

62 We must not omit to point out that the motion of a stone in a sling or of water in a whirled bucket cannot be called truly circular motion as that term is conceived by those who define the true places of bodies by the parts of absolute space, since it is strangely compounded of the motions, not alone of bucket or sling, but also of the daily motion of the earth round her own axis, of her monthly motion round the common centre of gravity of earth and moon,

and of her annual motion round the sun. And on that account each particle of the stone or the water describes a line far removed from circular. Nor in fact does that supposed axifugal conation exist, since it is not concerned with some one axis in relation to absolute space, supposing that such a space exists; accordingly I do not see how that can be called a single conation to which a truly circular motion corresponds as to its proper and adequate effect.

63 No motion can be recognized or measured, unless through sensible things. Since then absolute space in no way affects the senses, it must necessarily be quite useless for the distinguishing of motions. Besides, determination or direction is essential to motion; but that consists in relation. Therefore it is impossible that absolute motion should be conceived.

64 Further, since the motion of the same body may vary with the diversity of relative place, nay actually since a thing can be said in one respect to be in motion and in another respect to be at rest, to determine true motion and true rest, for the removal of ambiguity and for the furtherance of the mechanics of these philosophers who take the wider view of the system of things, it would be enough to bring in, instead of absolute space, relative space as confined to the heavens of the fixed stars, considered as at rest. But motion and rest marked out by such relative space can conveniently be subsituted in place of the absolutes, which cannot be distinguished from them by any mark. For however forces may be impressed, whatever conations there are, let us grant that motion is distinguished by actions exerted on bodies; never, however, will it follow that that space, absolute place, exists, and that change in it is true place.

65 The laws of motions and the effects, and theorems containing the proportions and calculations of the same for the different configurations of the paths, likewise for accelerations and different directions, and for mediums resisting in greater or less degree, all these hold without bringing absolute motion into account. As is plain from this that since according to the principles of those who introduce absolute motion we cannot know by any indication whether the whole frame of things is at rest, or is moved uniformly in a direction, clearly we cannot know the absolute motion of any body.

66 From the foregoing it is clear that the following rules will be of great service in determining the true nature of motion: (1) to distinguish mathematical hypotheses from the natures of things; (2) to beware of abstractions; (3) to consider motion as something sensible, or at least imaginable; and to be content with relative measures. If we do so, all the famous theorems of the mechanical philosophy by which the secrets of nature are unlocked, and by which the system of the world is reduced to human calculation, will remain untouched; and the study of motion will be freed from a thousand minutiæ, subtleties, and abstract ideas. And let these words suffice about the nature of motion.

67 It remains to discuss the cause of the communication of motions. Most people think that the force impressed on the movable body is the cause of motion in it. However, that they do not assign a known cause of motion, and one distinct from the body and the motion, is clear from the preceding argument. It is clear, moreover, that force is not a thing certain and determinate, from the

fact that great men advance very different opinions, even contrary opinions, about it, and yet in their results attain the truth. For Newton says that impressed force consists in action alone, and is the action exerted on the body to change its state, and does not remain after the action. Torricelli contends that a certain heap or aggregate of forces impressed by percussion is received into the mobile body, and there remains and constitutes impetus. Borelli and others say much the same. But although Newton and Torricelli seem to be disagreeing with one another, they each advance consistent views, and the thing is sufficiently well explained by both. For all forces attributed to bodies are mathematical hypotheses just as are attractive forces in planets and sun. But mathematical entities have no stable essence in the nature of things; and they depend on the notion of the definer. Whence the same thing can be explained in different ways.

68 Let us lay down that the new motion in the body struck is conserved either by the natural force by reason of which any body persists in its own uniform state of motion or of rest, or by the impressed force, received (while the percussion lasts) into the body struck, and there remaining; it will be the same in fact, the difference existing only in name. Similarly when the striking movable body loses motion, and the struck body acquires it, it is not worth disputing whether the acquired motion is numerically the same as the motion lost; the discussion would lead into metaphysical and even verbal minutiæ about identity. And so it comes to the same thing whether we say that motion passes from the striker to the struck, or that motion is generated *de novo* in the struck, and is destroyed in the striker. In either case it is understood that one body loses motion, the other acquires it, and besides that, nothing.

69 That the Mind which moves and contains this universal, bodily mass, and is the true efficient cause of motion, is the same cause, properly and strictly speaking, of the communication thereof I would not deny. In physical philosophy, however, we must seek the causes and solutions of phenomena among mechanical principles. Physically, therefore, a thing is explained not by assigning its truly active and incorporeal cause, but by showing its connection with mechanical principles, such as *action and reaction are always opposite and equal*. From such laws as from the source and primary principle, those rules for the communication of motions are drawn, which by the moderns for the great good of the sciences have been already found and demonstrated.

70 I, for my part, will content myself with hinting that that principle could have been set forth in another way. For if the true nature of things, rather than abstract mathematics, be regarded, it will seem more correct to say that in attraction or percussion, the passion of bodies, rather than their action, is equal on both sides. For example, the stone tied by a rope to a horse is dragged towards the horse just as much as the horse towards the stone; for the body in motion impinging on a quiescent body suffers the same change as the quiescent body. And as regards real effect, the striker is just as the struck, and the struck as the striker. And that change on both sides, both in the body of the horse and in the stone, both in the moved and in the resting, is mere passivity. It is not established that there is force, virtue, or bodily action truly and

properly causing such effects. The body in motion impinges on the quiescent body; we speak, however, in terms of action and say that that impels this; and it is correct to do so in mechanics, where mathematical ideas, rather than the true natures of things, are regarded.

71 In physics sense and experience which reach only to apparent effects hold sway; in mechanics the abstract notions of mathematicians are admitted. In first philosophy or metaphysics we are concerned with incorporeal things, with causes, truth, and the existence of things. The physicist studies the series or successions of sensible things, noting by what laws they are connected, and in what order, what precedes as cause, and what follows as effect. And on this method we say that the body in motion is the cause of motion in the other, and impresses motion on it, draws it also or impels it. In this sense second corporeal causes ought to be understood, no account being taken of the actual seat of the forces or of the active powers or of the real cause in which they are. Further, besides body, figure, and motion, even the primary axioms of mechanical science can be called causes or mechanical principles, being regarded as the causes of the consequences.

72 Only by meditation and reasoning can truly active causes be rescued from the surrounding darkness and be to some extent known. To deal with them is the business of first philosophy or metaphysics. Allot to each science its own province; assign its bounds; accurately distinguish the principles and objects belonging to each. Thus it will be possible to treat them with greater ease and clarity.

F. *FROM* ALCIPHRON (*BERKELEY 1732*)

Alciphron, or the minute philosopher was written during Berkeley's brief sojourn in Newport, Rhode Island, and was first printed in London in 1732. The book, in seven dialogues, is principally concerned with defending Christianity against 'those who are called free-thinkers'; but it also contains passages relevant to the philosophy of mathematics. The passage below is taken from the *Seventh dialogue*, and expands upon Berkeley's remarks in his letter to Molyneux and in §§19–20 of the *Principles of human knowledge* about the relationship of *signs* to *ideas*; other remarks on this general theme may be found in §§8–15 of *Alciphron*'s *Fourth dialogue*. Berkeley here denies that every sign must signify an idea, or that the solitary purpose of language is to impart ideas; and in §14 he cites $\sqrt{-1}$ as an example of a symbol which has a use even though it corresponds to no idea. (This precise issue will resurface in the selections from Gauss, DeMorgan, and Rowan Hamilton.)

The text below follows the third edition of 1752; variant readings from the first and second editions are recorded in the critical apparatus to *Berkeley*

1948–57, Vol. 3, from which this selection is reprinted. The passage is taken from the *Seventh dialogue*; references should be to the paragraph numbers, which appeared in the original edition.

---

11. ALCIPHRON. It seems, Euphranor and you would persuade me into an opinion that there is nothing so singularly absurd as we are apt to think in the belief of mysteries; and that a man need not renounce his reason to maintain his religion. But, if this were true, how comes it to pass that, in proportion as men abound in knowledge, they dwindle in faith?

EUPHRANOR. O Alciphron. I have learned from you that there is nothing like going to the bottom of things, and analysing them into their first principles. I shall therefore make an essay of this method, for clearing up the nature of faith: with what success, I shall leave you to determine, for I dare not pronounce myself, on my own judgment, whether it be right or wrong. But thus it seems to me:—The objections made to faith are by no means an effect of knowledge, but proceed rather from an ignorance of what knowledge is; which ignorance may possibly be found even in those who pass for masters of this or that particular branch of knowledge. Science and faith agree in this, that they both imply an assent of the mind: and, as the nature of the first is most clear and evident, it should be first considered in order to cast a light on the other. To trace things from their original, it seems that the human mind, naturally furnished with the ideas of things particular and concrete, and being designed, not for the bare intuition of ideas, but for action and operation about them, and pursuing her own happiness therein, stands in need of certain general rules or theorems to direct her operations in this pursuit; the supplying which want is the true, original, reasonable end of studying the arts and sciences. Now, these rules being general, it follows that they are not to be obtained by the mere consideration of the original ideas, or particular things, but by the means of marks and signs, which, being so far forth universal, become the immediate instruments and materials of science. It is not, therefore, by mere contemplation of particular things, and much less of their abstract general ideas, that the mind makes her progress, but by an apposite choice and skilful management of signs: for instance, force and number, taken in concrete, with their adjuncts, subjects, and signs, are what every one knows; and considered in abstract, so as making precise ideas of themselves, they are what nobody can comprehend. That their abstract nature, therefore, is not the foundation of science is plain: and that barely considering their ideas in concrete is not the method to advance in the respective sciences is what every one that reflects may see; nothing being more evident than that one who can neither write nor read, in common use understands the meaning of numeral words as well as the best philosopher or mathematician.

12. But here lies the difference: the one who understands the notation of numbers, by means thereof is able to express briefly and distinctly all the variety

and degrees of number, and to perform with ease and despatch several arithmetical operations by the help of general rules. Of all which operations as the use in human life is very evident, so it is no less evident that the performing them depends on the aptness of the notation. If we suppose rude mankind without the use of language, it may be presumed they would be ignorant of arithmetic. But the use of names, by the repetition whereof in a certain order they might express endless degrees of number, would be the first step towards that science. The next step would be to devise proper marks of a permanent nature, and visible to the eye, the kind and order whereof must be chose with judgment, and accommodated to the names. Which marking or notation would, in proportion as it was apt and regular, facilitate the invention and application of general rules to assist the mind in reasoning and judging, in extending, recording, and communicating its knowledge about numbers: in which theory and operations, the mind is immediately occupied about the signs or notes, by mediation of which it is directed to act about things, or number in concrete (as the logicians call it), without ever considering the simple, abstract, intellectual, general idea of number. ‖The signs, indeed, do in their use imply relations or proportions of things; but these relations are not abstract general ideas, being founded in particular things, and not making of themselves distinct ideas to the mind, exclusive of the particular ideas and the signs.‖[a] I imagine one need not think much to be convinced that the science of arithmetic, in its rise, operations, rules, and theorems, is altogether conversant about the artificial use of signs, names, and characters. These names and characters are universal, inasmuch as they are signs. The names are referred to things, the characters to names, and both to operation. The names being few, and proceeding by a certain analogy, the characters will be more useful, the simpler they are, and the more aptly they express this analogy. Hence the old notation by letters was more useful than words written at length; and the modern notation by figures, expressing the progression or analogy of the names by their simple places, is much preferable to that, for ease and expedition, as the invention of algebraical symbols is to this, for extensive and general use. As arithmetic and algebra are sciences of great clearness, certainty, and extent, which are immediately conversant about signs, upon the skilful use and management whereof they entirely depend, so a little attention to them may possibly help us to judge of the progress of the mind in other sciences, which, though differing in nature, design, and object, may yet agree in the general methods of proof and inquiry.

13. If I mistake not, all sciences, so far as they are universal and demonstrable by human reason, will be found conversant about signs as their immediate object, though these in the application are referred to things. The reason whereof is not difficult to conceive. For, as the mind is better acquainted with some sort of objects, which are earlier offered to it, strike it more sensibly, or are more

[a] ‖The passage on relations was added in the third edition. Compare §§89 and 101 of the 1734 edition of *Berkeley 1710.*‖

easily comprehended than others, it seems naturally led to substitute those objects for such as are more subtle, fleeting, or difficult to conceive. Nothing, I say, is more natural than to make the things we know a step towards those we do not know; and to explain and represent things less familiar by others which are more so. Now, it is certain we imagine before we reflect, and we perceive by sense before we imagine, and of all our senses the sight is the most clear, distinct, various, agreeable, and comprehensive. Hence it is natural to assist the intellect by imagination, imagination by sense, and other senses by sight. Hence figures, metaphors, and types. We illustrate spiritual things by corporeal; we substitute sounds for thoughts, and written letters for sounds; emblems, symbols, and hieroglyphics, for things too obscure to strike, and too various or too fleeting to be retained. We substitute things imaginable for things intelligible, sensible things for imaginable, smaller things for those that are too great to comprehend easily, and greater things for such as are too small to be discerned distinctly, present things for absent, permanent for perishing, and visible for invisible. Hence the use of models and diagrams. Hence lines are substituted for time, velocity, and other things of very different natures. Hence we speak of spirits in a figurative style, expressing the operations of the mind by allusions and terms borrowed from sensible things, such as *apprehend*, *conceive*, *reflect*, *discourse*, and such-like: and hence those allegories which illustrate things intellectual by visions exhibited to the fancy. Plato, for instance, represents the mind presiding in her vehicle by the driver of a winged chariot, which sometimes moults and droops, and is drawn by two horses, the one good and of a good race, the other of a contrary kind;[b] symbolically expressing the tendency of the mind towards the Divinity, as she soars or is borne aloft by two instincts like wings, the one in the intellect towards truth, the other in the will towards excellence, which instincts moult or are weakened by sensual inclinations; expressing also her alternate elevations and depressions, the struggles between reason and appetite, like horses that go an unequal pace, or draw different ways, embarrassing the soul in her progress to perfection. I am inclined to think the doctrine of signs[c] a point of great importance and general extent, which, if duly considered, would cast no small light upon things, and afford a just and genuine solution of many difficulties.

14. Thus much, upon the whole, may be said of all signs:—that they do not always suggest ideas signified to the mind: that when they suggest ideas, they are not general abstract ideas: that they have other uses besides barely standing for and exhibiting ideas, such as raising proper emotions, producing certain dispositions or habits of mind, and directing our actions in pursuit of that happiness which is the ultimate end and design, the primary spring and motive, that sets rational agents at work: ‖that signs may imply or suggest the relations of

[b] ‖*Phaedrus*, 246A.‖

[c] ‖Compare Locke, *An essay concerning human understanding*, IV, xii, 4, which also uses the expression 'doctrine of signs'.‖

things; which relations, habitudes or proportions, as they cannot be by us understood but by the help of signs, so being thereby expressed and confuted, they direct and enable us to act with regard to things‖:[d] that the true end of speech, reason, science, faith, assent, in all its different degrees, is not merely, or principally, or always, the imparting or acquiring of ideas, but rather something of an active operative nature, tending to a conceived good: which may sometimes be obtained, not only although the ideas marked are not offered to the mind, but even although there should be no possibility of offering or exhibiting any such idea to the mind: for instance, the algebraic mark, which denotes the root of a negative square, hath its use in logistic operations, although it be impossible to form an idea of any such quantity. And what is true of algebraic signs is also true of words or language, modern algebra being in fact a more short, apposite, and artificial sort of language, and it being possible to express by words at length, though less conveniently, all the steps of an algebraical process. And it must be confessed that even the mathematical sciences themselves, which above all others are reckoned the most clear and certain, if they are considered, not as instruments to direct our practice, but as speculations to employ our curiosity, will be found to fall short in many instances of those clear and distinct ideas which, it seems, the minute philosophers of this age, whether knowingly or ignorantly, expect and insist upon in the mysteries of religion.

G. *FROM* NEWTON'S *PRINCIPIA MATHEMATICA* (*NEWTON 1726*)

Berkeley's *Analyst* is a searching and profound critique of the foundations of Newton's calculus, which had been hinted at in the *Principia* (1686), and more fully developed in the *Tractatus de quadratura curvarum* (*Newton 1700*); Newton's ideas on the calculus dated from 1664–8. Before we turn to *The analyst*, let us consider the following passage, which is Newton's lengthiest effort in the *Principia* to justify his methodology; it was the starting-point both for Berkeley's criticisms and for MacLaurin's subsequent defence of Newton (*MacLaurin 1742*).

Newton's proofs in the *Principia* follow the geometric style of the Greek mathematicians (what MacLaurin was to call the 'manner of the ancients'); except that, instead of giving proofs by *reductio ad absurdum*, Newton preferred to use his new method of 'first and last ratios'. Book I, Section I of the *Principia* is entitled, *The method of first and last ratios of quantities, by the help of which we demonstrate the propositions that follow.* Newton's 'first

[d] ‖The passage on relations was added in the third edition. Compare §§89 and 101 of the 1734 edition of *Berkeley 1710*.‖

ratios' (or 'last ratios') are essentially limits of converging, infinite sequences of ratios of finite quantities; and Newton rightly saw that the device of using limits would enable him to dispense with infinitesimals. He did not, however, possess a precise definition of *limit*; nor a theory of when limits in fact exist; nor, in consequence, a fully satisfactory explanation of why his method worked. His remarks in the *Scholium* which follows are somewhat opaque, and rely upon an appeal to one's intuitions about the motion of physical bodies; this passage left his contemporaries confused, and raised questions that were not to be answered until the work of Weierstrass in the middle of the nineteenth century.

The passage is taken from the *Philosophiae naturalis principia mathematica*, Book I, Section I, Lemma XI, *Scholium* (3rd edn 1726) as reprinted in *Newton 1972*; the passage is the conclusion of Section I. The translation is a revision, by William Ewald, of Florian Cajori's revision (*Newton 1934*) of Andrew Motte's 1729 translation of the third edition of the *Principia*. There exist many general accounts of the historical development of the calculus and of the views of Newton and Leibniz on questions of rigour; see, for example, *Boyer 1939*, *Baron 1969*, *Edwards 1979*, *Grattan-Guinness 1980*, and the documents collected in *Struik 1969*.

I state these lemmas to avoid the tediousness of deducing involved demonstrations *ad absurdum*, according to the method of the ancient geometers. For demonstrations are made more concise by the method of indivisibilia. But because the hypothesis of indivisibilia is cruder ‖durior‖ and therefore that method is considered less geometrical, I chose rather to reduce ‖deducere‖ the demonstrations of the following propositions to the first and last sums and ratios of nascent and evanescent quantities, i.e. to the limits of those sums and ratios; and hence to set forth, as briefly as I could, the demonstrations of those limits. For the same thing is accomplished by them as by the method of indivisibilia; and now, those principles having been demonstrated, we may use them more safely. If, therefore, I should hereafter happen to consider quantities as made up of particles, or should treat little curved lines as straight,[a] I do not wish to be understood as meaning indivisibilia, but evanescent divisibilia; not the sums and ratios of determinate parts, but always the limits of sums and ratios; and the force of such demonstrations always depends on the methods laid down in the preceding lemmas.

It may be objected that there is no last proportion of evanescent quantities, because before the quantities have vanished the proportion is not last, while

[a] ‖'si pro rectis usurpavero lineolas curvas'. Newton in Corollary III to Lemma VII had asserted that 'in all our reasoning about last ratios we may freely use any one of those lines [i.e. the curve or its linear approximation] for any other'.‖

afterwards it is nothing. But by the same argument one might argue that a body arriving and stopping at a certain location has no last velocity: for before the body arrives at the location its velocity is not last, while afterwards it has none. And the reply is easy: by last velocity I understand that with which the body is moved, neither before it arrives at the last position and its motion ceases, nor thereafter, but just when it arrives; i.e. the very velocity with which the body arrives at the last location and with which the motion ceases. And similarly, by the last ratio of evanescent quantities is to be understood the ratio of the quantities not before they vanish, nor afterwards, but with which they vanish. Equally, the first ratio of nascent quantities is that with which they begin to be. And the first or last sum is that with which they begin and cease to be (or to be increased or diminished). There is a limit which the velocity at the end of the motion may attain, but not exceed. This is the last velocity. And there is likewise a limit in all quantities and proportions that begin and that cease to be. And since these limits are certain and definite, to determine them is a purely geometrical problem. But anything geometrical may legitimately be used to determine and demonstrate any other thing that is also geometrical.

It may also be maintained that if the last ratios of evanescent quantities are given, their last magnitudes will also be given; hence all quantities will consist of indivisibilia, contrary to what Euclid has demonstrated concerning incommensurables in the Tenth Book of the *Elements*. But this objection rests on a false hypothesis. Those last ratios with which quantities vanish are not truly the ratios of last quantities, but limits towards which the ratios of quantities decreasing without limit always approach ‖appropinquant‖; and to which they approach nearer than by any given difference, but never exceed ‖transgredi‖, nor attain until the quantities diminish *in infinitum*. This is more evident with respect to the infinitely great. If two quantities, whose difference is given, are increased *in infinitum*, their last ratio will be given, namely, the ratio of equality; but this does not mean that the last or greatest quantities themselves, whose ratio that is, will be given. In what follows, therefore, if for the sake of being more easily understood I should happen to refer to quantities as being least, or evanescent, or last, you should beware of thinking that quantities of any determinate magnitude are meant, but rather think of them as always diminishing without limit.

H. THE ANALYST (*BERKELEY 1734*)

The Analyst, or A Discourse Addressed to an Infidel Mathematician, Wherein it is examined whether the object, principles, and inferences of the modern Analysis are more distinctly conceived, or more evidently deduced, than religious Mysteries and points of Faith is Berkeley's most lengthy and sophisticated discussion of the metaphysics of the calculus. At the time, Berkeley was the bishop-elect of Cloyne, and his essay was written with a polemical purpose, namely, to defend revealed religion against physicalist-mathematical scepticism. Berkeley's argument is curious. It is, that mathematicians often reason as badly as theologians: 'He who can digest a second or third fluxion . . . need not, methinks, be squeamish about any point in divinity.' The 'infidel mathematician' of Berkeley's title is not Newton, but most likely the Astronomer Royal, Edmund Halley, a Fellow of The Queen's College, Oxford, who had the reputation of being an atheist; he had written the ode to Newton which prefaced the first edition of the *Principia*.

Berkeley advances several different but interconnected arguments against the mathematicians. The rather obvious logical criticisms with which he begins (§§10–20) are the arguments most frequently cited in histories of mathematics; but although they were more forcefully expressed by Berkeley than by his contemporaries, and although they attracted the attention of MacLaurin, they were not original. The Dutch mathematician Bernard Nieuwentijdt (1654–1718) had already raised similar objections in his public controversy with Leibniz in 1694 and 1695; related problems were also raised by Pierre Varignon in his letter to Leibniz of 28 November 1701, and by Leibniz in his reply of 2 February 1702 (see *Becker 1964*, pp. 162–7). Mathematicians of the day were well aware of the shortcomings in the foundations of the calculus, and many attempted to improve upon the arguments of Newton and Leibniz. Accounts of the various strategies employed and further references can be found in M. Cantor (*1894–1908*, vol. iii, pp. 244–7), Kline (*1972*, pp. 384–5), and Boyer (*1939*). The originality and importance of Berkeley's essay is rather to be sought in the philosophical underpinnings—in his penetrating observations on infinity and the epistemology of mathematics; in his theory of 'contrary errors' (§§21 ff.), where he attempts to account for the success of the calculus in finitary terms; and in his remarks about the importance of language to mathematics.

This last topic arises repeatedly in Berkeley—in the *Philosophical commentaries* (§§354–6 and 761–70), in *Of infinites*, in the *Principles* (§§19 and 122), in *Alciphron* (*Fourth dialogue* and *Seventh dialogue*), and in *The analyst* (§§8, 36, and 37); it is tied to most of the leading themes of his philosophy, as well as to conceptual problems involved in the development of modern algebra. At the time Berkeley wrote, mathematicians had not yet clearly distinguished between manipulating *symbols* and manipulating *mathematical quantities*, and the problem of the reference of algebraic expressions remained a bone of contention until the middle of the nineteenth century. Berkeley's difficult and conflicting remarks on this topic should be compared with the selections below on the growth of abstract algebra; in particular, with the selections from Lambert, Gauss, Boole, Gregory, De Morgan, Hamilton, and Hilbert.

The Analyst gave rise to a flurry of polemical replies, and Berkeley continued the debate in his '*A Defence of Free-thinking in Mathematics, In answer to a pamphlet of Philalethes Cantabrigiensis, intituled Geometry no friend to Infidelity, or a defence of Sir Isaac Newton and the British Mathematicians.*

Also an appendix concerning Mr. Walton's Vindication of the principles of fluxions against the objections contained in the Analyst; wherein it is attempted to put this controversy in such a light as that every reader may be able to judge thereof (first printed in 1735) and his ironically titled '*Reasons for not replying to Mr. Walton's Full answer, in a letter to P.T.P.*' (1735), which contains a lengthy reply to Walton. (Like Shakespeare's 'W.H.', 'P.T.P.' has never been identified.)

These polemical essays on the philosophy of mathematics were Berkeley's last substantial contribution to philosophy. Thereafter, he devoted himself to writing pamphlets about the supposed medicinal virtues of tar water; and although these writings (in particular, the *Siris*) shed a rather pale light on Berkeley's philosophy, they cannot be regarded as major works. Berkeley died in Oxford in 1753, and is buried in Christ Church Cathedral.

References to *Berkeley 1734* should be to the paragraph numbers, which appeared in the original edition.

---

1 Though I am a stranger to your person, yet I am not, Sir, a stranger to the reputation you have acquired in that branch of learning which hath been your peculiar study; nor to the authority that you therefore assume in things foreign to your profession, nor to the abuse that you, and too many more of the like character, are known to make of such undue authority, to the misleading of unwary persons in matters of the highest concernment, and whereof your mathematical knowledge can by no means qualify you to be a competent judge. Equity indeed and good sense would incline one to disregard the judgment of men, in points which they have not considered or examined. But several who make the loudest claim to those qualities do nevertheless the very thing they would seem to despise, clothing themselves in the livery of other men's opinions, and putting on a general deference for the judgment of you, Gentlemen, who are presumed to be of all men the greatest masters of reason, to be most conversant about distinct ideas, and never to take things upon trust, but always clearly to see your way, as men whose constant employment is the deducing truth by the justest inference from the most evident principles. With this bias on their minds, they submit to your decisions where you have no right to decide. And that this is one short way of making Infidels, I am credibly informed.

2 Whereas then it is supposed that you apprehend more distinctly, consider more closely, infer more justly, conclude more accurately than other men, and that you are therefore less religious because more judicious, I shall claim the privilege of a Freethinker; and take the liberty to inquire into the object, principles, and method of demonstration admitted by the mathematicians of the present age, with the same freedom that you presume to treat the principles and mysteries of Religion; to the end that all men may see what right you have to lead, or what encouragement others have to follow you. It hath been an old

remark, that Geometry is an excellent Logic. And it must be owned that when the definitions are clear; when the postulata cannot be refused, nor the axioms denied; when from the distinct contemplation and comparison of figures, their properties are derived, by a perpetual well-connected chain of consequences, the objects being still kept in view, and the attention ever fixed upon them; there is acquired an habit of reasoning, close and exact and methodical: which habit strengthens and sharpens the mind, and being transferred to other subjects is of a general use in the inquiry after truth. But how far this is the case of our geometrical analysts, it may be worth while to consider.

3 The Method of Fluxions is the general key by help whereof the modern mathematicians unlock the secrets of Geometry, and consequently of Nature. And, as it is that which hath enabled them so remarkably to outgo the ancients in discovering theorems and solving problems, the exercise and application thereof is become the main if not sole employment of all those who in this age pass for profound geometers. But whether this method be clear or obscure, consistent or repugnant, demonstrative or precarious, as I shall inquire with the utmost impartiality, so I submit my inquiry to your own judgment, and that of every candid reader. Lines are supposed to be generated[1] by the motion of points, planes by the motion of lines, and solids by the motion of planes. And whereas quantities generated in equal times are greater or lesser according to the greater or lesser velocity wherewith they increase and are generated, a method hath been found to determine quantities from the velocities of their generating motions. And such velocities are called fluxions: and the quantities generated are called flowing quantities. These fluxions are said to be nearly as the increments of the flowing quantities, generated in the least equal particles of time; and to be accurately in the first proportion of the nascent, or in the last of the evanescent increments. Sometimes, instead of velocities, the momentaneous increments or decrements of undetermined flowing quantities are considered, under the appellation of moments.

4 By moments we are not to understand finite particles. These are said not to be moments, but quantities generated from moments, which last are only the nascent principles of finite quantities. It is said that the minutest errors are not to be neglected in mathematics: that the fluxions are celerities, not proportional to the finite increments, though ever so small; but only to the moments or nascent increments, whereof the proportion alone, and not the magnitude, is considered. And of the aforesaid fluxions there be other fluxions, which fluxions of fluxions are called second fluxions. And the fluxions of these second fluxions are called third fluxions: and so on, fourth, fifth, sixth, &c. *ad infinitum*. Now, as our sense is strained and puzzled with the perception of objects extremely minute, even so the imagination, which faculty derives from sense, is very much strained and puzzled to frame clear ideas of the least particles of time, or the least increments generated therein: and much more so to comprehend the

[1] *Introd. ad Quadraturam Curvarum* ‖*Newton 1700*‖.

moments, or those increments of the flowing quantities in *statu nascenti*, in their very first origin or beginning to exist, before they become finite particles. And it seems still more difficult to conceive the abstracted velocities of such nascent imperfect entities. But the velocities of the velocities, the second, third, fourth, and fifth velocities, &c., exceed, if I mistake not, all human understanding. The further the mind analyseth and pursueth these fugitive ideas the more it is lost and bewildered; the objects, at first fleeting and minute, soon vanishing out of sight. Certainly in any sense, a second or third fluxion seems an obscure mystery. The incipient celerity of an incipient celerity, the nascent augment of a nascent augment, *i.e.* of a thing which hath no magnitude: take it in what light you please, the clear conception of it will, if I mistake not, be found impossible; whether it be so or no I appeal to the trial of every thinking reader. And if a second fluxion be inconceivable, what are we to think of third, fourth, fifth fluxions, and so on without end?

5 The foreign mathematicians[a] are supposed by some, even of our own, to proceed in a manner less accurate, perhaps, and geometrical, yet more intelligible. Instead of flowing quantities and their fluxions, they consider the variable finite quantities as increasing or diminishing by the continual addition or subduction of infinitely small quantities. Instead of the velocities wherewith increments are generated, they consider the increments or decrements themselves, which they call differences, and which are supposed to be infinitely small. The difference of a line is an infinitely little line; of a plain an infinitely little plain. They suppose finite quantities to consist of parts infinitely little, and curves to be polygons, whereof the sides are infinitely little, which by the angles they make one with another determine the curvity of the line. Now to conceive a quantity infinitely small, that is, infinitely less than any sensible or imaginable quantity, or than any the least finite magnitude is, I confess, above my capacity. But to conceive a part of such infinitely small quantity that shall be still infinitely less than it, and consequently though multiplied infinitely shall never equal the minutest finite quantity, is, I suspect, an infinite difficulty to any man whatsoever; and will be allowed such by those who candidly say what they think; provided they really think and reflect, and do not take things upon trust.

6 And yet in the *calculus differentialis*, which method serves to all the same intents and ends with that of fluxions, our modern analysts are not content to consider only the differences of finite quantities: they also consider the differences of those differences, and the differences of the differences of the first differences. And so on *ad infinitum*. That is, they consider quantities infinitely less than the least discernible quantity; and others infinitely less than those infinitely small ones; and still others infinitely less than the preceding infinitesimals, and so on without end or limit. Insomuch that we are to admit an infinite succession of infinitesimals, each infinitely less than the foregoing, and infinitely greater than the following. As there are first, second, third,

[a] [Elsewhere Berkeley names Leibniz, Nieuwentijdt, and the Marquis de l'Hospital.]

fourth, fifth, &c. fluxions, so there are differences, first, second, third, fourth, &c., in an infinite progression towards nothing, which you still approach and never arrive at. And (which is most strange) although you should take a million of millions of these infinitesimals, each whereof is supposed infinitely greater than some other real magnitude, and add them to the least given quantity, it shall never be the bigger. For this is one of the modest *postulata* of our modern mathematicians, and is a corner-stone or ground-work of their speculations.

7 All these points, I say, are supposed and believed by certain rigorous exactors of evidence in religion, men who pretend to believe no further than they can see. That men who have been conversant only about clear points should with difficulty admit obscure ones might not seem altogether unaccountable. But he who can digest a second or third fluxion, a second or third difference, need not, methinks, be squeamish about any point in divinity. There is a natural presumption that men's faculties are made alike. It is on this supposition that they attempt to argue and convince one another. What therefore shall appear evidently impossible and repugnant to one may be presumed the same to another. But with what appearance of reason shall any man presume to say that mysteries may not be objects of faith, at the same time that he himself admits such obscure mysteries to be the object of science?

8 It must indeed be acknowledged the modern mathematicians do not consider these points as mysteries, but as clearly conceived and mastered by their comprehensive minds. They scruple not to say that by the help of these new analytics they can penetrate into infinity it self: that they can even extend their views beyond infinity: that their art comprehends not only infinite, but infinite of infinite (as they express it), or an infinity of infinites. But, notwithstanding all these assertions and pretensions, it may be justly questioned whether, as other men in other inquiries are often deceived by words or terms, so they likewise are not wonderfully deceived and deluded by their own peculiar signs, symbols, or species. Nothing is easier than to devise expressions or notations, for fluxions and infinitesimals of the first, second, third, fourth, and subsequent orders, proceeding in the same regular form without end or limit $\dot{x}$. $\ddot{x}$. $\dddot{x}$. $\ddddot{x}$. *&c.* or *dx. ddx. dddx. ddddx. &c.* These expressions indeed are clear and distinct, and the mind finds no difficulty in conceiving them to be continued beyond any assignable bounds. But if we remove the veil and look underneath, if, laying aside the expressions, we set ourselves attentively to consider the things themselves which are supposed to be expressed or marked thereby, we shall discover much emptiness, darkness, and confusion; nay, if I mistake not, direct impossibilities and contradictions. Whether this be the case or no, every thinking reader is intreated to examine and judge for himself.

9 Having considered the object, I proceed to consider the principles of this new analysis by momentums, fluxions, or infinitesimals; wherein if it shall appear that your capital points, upon which the rest are supposed to depend, include error and false reasoning; it will then follow that you, who are at a loss to conduct your selves, cannot with any decency set up for guides to other men. The main point in the method of fluxions is to obtain the fluxion or momentum

of the rectangle or product of two indeterminate quantities. Inasmuch as from thence are derived rules for obtaining the fluxions of all other products and powers; be the coefficients or the indexes what they will, integers or fractions, rational or surd. Now, this fundamental point one would think should be very clearly made out, considering how much is built upon it, and that its influence extends throughout the whole analysis. But let the reader judge. This is given for demonstration.[2] Suppose the product or rectangle AB increased by continual motion: and that the momentaneous increments of the sides A and B are a and b. When the sides A and B were deficient, or lesser by one half of their moments, the rectangle was $\overline{A-\frac{1}{2}a}\times\overline{B-\frac{1}{2}b}$ *i.e.* $AB-\frac{1}{2}aB-\frac{1}{2}bA+\frac{1}{4}ab$. And as soon as the sides A and B are increased by the other two halves of their moments, the rectangle becomes $\overline{A+\frac{1}{2}a}\times\overline{B+\frac{1}{2}b}$ or $AB+\frac{1}{2}aB+\frac{1}{2}bA+\frac{1}{4}ab$. From the latter rectangle subduct the former, and the remaining difference will be $aB + bA$. Therefore the increment of the rectangle generated by the intire increments a and b is $aB + bA$. *Q.E.D.* But it is plain that the direct and true method to obtain the moment or increment of the rectangle AB, is to take the sides as increased by their whole increments, and so multiply them together, $A + a$ by $B + b$, the product whereof $AB + aB + bA + ab$ is the augmented rectangle; whence, if we subduct AB the remainder $aB + bA + ab$ will be the true increment of the rectangle, exceeding that which was obtained by the former illegitimate and indirect method by the quantity ab. And this holds universally be the quantities a and b what they will, big or little, finite or infinitesimal, increments, moments, or velocities. Nor will it avail to say that ab is a quantity exceeding small: since we are told that *in rebus mathematicis errores quam minimi non sunt contemnendi.*[3]

10 Such reasoning as this for demonstration, nothing but the obscurity of the subject could have encouraged or induced the great author of the fluxionary method to put upon his followers, and nothing but an implicit deference to authority could move them to admit. The case indeed is difficult. There can be nothing done till you have got rid of the quantity ab. In order to this the notion of fluxions is shifted: It is placed in various lights: Points which should be clear as first principles are puzzled; and terms which should be steadily used are ambiguous. But notwithstanding all this address and skill the point of getting rid of ab cannot be obtained by legitimate reasoning. If a man, by methods not geometrical or demonstrative, shall have satisfied himself of the usefulness of certain rules; which he afterwards shall propose to his disciples for undoubted truths; which he undertakes to demonstrate in a subtile manner, and by the help of nice and intricate notions; it is not hard to conceive that such his disciples may, to save themselves the trouble of thinking, be inclined to confound the usefulness of a rule with the certainty of a truth, and accept the one for the other; especially if they are men accustomed rather to compute than to think;

---

[2] *Naturalis philosophiæ principia mathematica*, lib. 2. lem. 2.

[3] *Introd. ad Quadraturam Curvarum* ‖'in mathematical matters errors, however small, are not to be contemned‖.

earnest rather to go on fast and far, than solicitous to set out warily and see their way distinctly.

11 The points or mere limits of nascent lines are undoubtedly equal, as having no more magnitude one than another, a limit as such being no quantity. If by a momentum you mean more than the very initial limit, it must be either a finite quantity or an infinitesimal. But all finite quantities are expressly excluded from the notion of a momentum. Therefore the momentum must be an infinitesimal. And, indeed, though much artifice hath been employed to escape or avoid the admission of quantities infinitely small, yet it seems ineffectual. For ought I see, you can admit no quantity as a medium between a finite quantity and nothing, without admitting infinitesimals. An increment generated in a finite particle of time is it self a finite particle; and cannot therefore be a momentum. You must therefore take an infinitesimal part of time wherein to generate your momentum. It is said, the magnitude of moments is not considered: And yet these same moments are supposed to be divided into parts. This is not easy to conceive, no more than it is why we should take quantities less than A and B in order to obtain the increment of AB, of which proceeding it must be owned the final cause or motive is very obvious; but it is not so obvious or easy to explain a just and legitimate reason for it, or shew it to be geometrical.

12 From the foregoing principle, so demonstrated, the general rule for finding the fluxion of any power of a flowing quantity is derived.[4] But, as there seems to have been some inward scruple or consciousness of defect in the foregoing demonstration, and as this finding the fluxion of a given power is a point of primary importance, it hath therefore been judged proper to demonstrate the same in a different manner independent of the foregoing demonstration. But whether this other method be more legitimate and conclusive than the former, I proceed now to examine; and in order thereto shall premise the following lemma. 'If with a view to demonstrate any proposition, a certain point is supposed, by virtue of which certain other points are attained; and such supposed point be it self afterwards destroyed or rejected by a contrary supposition; in that case, all the other points attained thereby, and consequent thereupon, must also be destroyed and rejected, so as from thence forward to be no more supposed or applied in the demonstration.' This is so plain as to need no proof.

13 Now the other method of obtaining a rule to find the fluxion of any power is as follows. Let the quantity x flow uniformly, and be it proposed to find the fluxion of x^n. In the same time that x by flowing becomes $x + o$, the power x^n becomes $\overline{x + o}|^n$, i.e. by the method of infinite series

$$x^n + nox^{n-1} + \frac{nn - n}{2}oox^{n-2} + \&c.,$$

and the increments

[4] *Philosophiæ naturalis principia mathematica*, lib. 2. lem. 2.

$$o \text{ and } nox^{n-1} + \frac{nn-n}{2} oox^{n-2} + \&c.$$

are one to another as

$$1 \text{ to } nx^{n-1} + \frac{nn-n}{2} ox^{n-2} + \&c.$$

Let now the increments vanish, and their last proportion will be 1 to nx^{n-1}. But it should seem that this reasoning is not fair or conclusive. For when it is said, let the increments vanish, *i.e.* let the increments be nothing, or let there be no increments, the former supposition that the increments were something, or that there were increments, is destroyed, and yet a consequence of that supposition, *i.e.* an expression got by virtue thereof, is retained. Which, by the foregoing lemma, is a false way of reasoning. Certainly when we suppose the increments to vanish, we must suppose their proportions, their expressions, and every thing else derived from the supposition of their existence to vanish with them.

14 To make this point plainer, I shall unfold the reasoning, and propose it in a fuller light to your view. It amounts therefore to this, or may in other words be thus expressed. I suppose that the quantity x flows, and by flowing is increased, and its increment I call o, so that by flowing it becomes $x + o$. And as x increaseth, if follows that every power of x is likewise increased in a due proportion. Therefore as x becomes $x + o$, x^n will become $\overline{x + o}|^n$: that is, according to the method of infinite series,

$$x^n + nox^{n-1} + \frac{nn-n}{2} oox^{n-2} + \&c.$$

And if from the two augmented quantities we subduct the root and the power respectively, we shall have remaining the two increments, to wit,

$$o \text{ and } nox^{n-1} + \frac{nn-n}{2} oox^{n-2} + \&c.$$

which increments, being both divided by the common divisor o, yield the quotients

$$1 \text{ and } nx^{n-1} + \frac{nn-n}{2} ox^{n-2} + \&c.$$

which are therefore exponents of the ratio of the increments. Hitherto I have supposed that x flows, that x hath a real increment, that o is something. And I have proceeded all along on that supposition, without which I should not have been able to have made so much as one single step. From that supposition it is that I get at the increment of x^n, that I am able to compare it with the increment of x, and that I find the proportion between the two increments. I now beg leave to make a new supposition contrary to the first, *i.e.* I will suppose that there is no increment of x, or that o is nothing; which second supposition destroys my first, and is inconsistent with it, and therefore with every thing that supposeth it. I do nevertheless beg leave to retain nx^{n-1}, which is an expression obtained in virtue of my first supposition, which necessarily presupposeth such supposition, and which could not be obtained without it: All which seems a most inconsistent way of arguing, and such as would not be allowed of in Divinity.

15 Nothing is plainer than that no just conclusion can be directly drawn from two inconsistent suppositions. You may indeed suppose any thing possible: But afterwards you may not suppose any thing that destroys what you first supposed: or, if you do, you must begin *de novo*. If therefore you suppose that the augments vanish, *i.e.* that there are no augments, you are to begin again and see what follows from such supposition. But nothing will follow to your purpose. You cannot by that means ever arrive at your conclusion, or succeed in what is called by the celebrated author, the investigation of the first or last proportions of nascent and evanescent quantities, by instituting the analysis in finite ones. I repeat it again: You are at liberty to make any possible supposition: And you may destroy one supposition by another: But then you may not retain the consequences, or any part of the consequences, of your first supposition so destroyed. I admit that signs may be made to denote either any thing or nothing: and consequently that in the original notation $x + o$, o might have signified either an increment or nothing. But then which of these soever you make it signify, you must argue consistently with such its signification, and not proceed upon a double meaning: Which to do were a manifest sophism. Whether you argue in symbols or in words the rules of right reason are still the same. Nor can it be supposed you will plead a privilege in mathematics to be exempt from them.

16 If you assume at first a quantity increased by nothing, and in the expression $x + o$, o stands for nothing, upon this supposition as there is no increment of the root, so there will be no increment of the power; and consequently there will be none except the first, of all those members of the series constituting the power of the binomial; you will therefore never come at your expression of a fluxion legitimately by such method. Hence you are driven into the fallacious way of proceeding to a certain point on the supposition of an increment, and then at once shifting your supposition to that of no increment. There may seem great skill in doing this at a certain point or period. Since, if this second supposition had been made before the common division by o, all had vanished at once, and you must have got nothing by your supposition. Whereas, by this artifice of first dividing and then changing your supposition, you retain 1 and nx^{n-1}. But, notwithstanding all this address to cover it, the fallacy is still the same. For, whether it be done sooner or later, when once the second supposition or assumption is made, in the same instant the former assumption and all that you got by it is destroyed, and goes out together. And this is universally true, be the subject what it will, throughout all the branches of human knowledge; in any other of which, I believe, men would hardly admit such a reasoning as this, which in mathematics is accepted for demonstration.

17 It may be not amiss to observe that the method for finding the fluxion of a rectangle of two flowing quantities, as it is set forth in the Treatise of

Quadratures, differs from the above-mentioned taken from the second book of the Principles, and is in effect the same with that used in the *calculus differentialis*.[5] For the supposing a quantity infinitely diminished, and therefore rejecting it, is in effect the rejecting an infinitesimal; and indeed it requires a marvellous sharpness of discernment to be able to distinguish between evanescent increments and infinitesimal differences. It may perhaps be said that the quantity being infinitely diminished becomes nothing, and so nothing is rejected. But according to the received principles it is evident that no geometrical quantity can by any division or subdivision whatsoever be exhausted, or reduced to nothing. Considering the various arts and devices used by the great author of the fluxionary method; in how many lights he placeth his fluxions; and in what different ways he attempts to demonstrate the same point; one would be inclined to think, he was himself suspicious of the justness of his own demonstrations; and that he was not enough pleased with any notion steadily to adhere to it. Thus much at least is plain, that he owned himself satisfied concerning certain points, which nevertheless he could not undertake to demonstrate to others.[6] Whether this satisfaction arose from tentative methods or inductions; which have often been admitted by mathematicians (for instance, by Dr. Wallis in his Arithmetic of Infinites[b]), is what I shall not pretend to determine. But, whatever the case might have been with respect to the author, it appears that his followers have shewn themselves more eager in applying his method, than accurate in examining his principles.

18 It is curious to observe what subtilty and skill this great genius employs to struggle with an insuperable difficulty; and through what labyrinths he endeavours to escape the doctrine of infinitesimals; which as it intrudes upon him whether he will or no, so it is admitted and embraced by others without the least repugnance. Leibniz and his followers in their *calculus differentialis* making no manner of scruple, first to suppose, and secondly to reject, quantities infinitely small; with what clearness in the apprehension and justness in the reasoning, any thinking man, who is not prejudiced in favour of those things, may easily discern. The notion or idea of an *infinitesimal quantity*, as it is an object simply apprehended by the mind, hath been already considered.[7] I shall now only observe as to the method of getting rid of such quantities, that it is done without the least ceremony. As in fluxions the point of first importance, and which paves the way to the rest, is to find the fluxion of a product of two indeterminate quantities, so in the *calculus differentialis* (which method is supposed to have been borrowed from the former with some small alterations) the main point is to obtain the difference of such product. Now the rule for this is got by rejecting the product or rectangle of the differences. And in general it is supposed that no quantity is bigger or lesser for the addition or

---

[5] *Analyse des infiniment petits*, part 1. prop. 2 ‖*L'Hospital 1696*‖.
[6] See letter to Collins, 8 Nov. 1676.
[b] ‖*Wallis 1693*; Wallis is also mentioned in *Berkeley 1901* [1707].‖
[7] Sect. 5 and 6.

subduction of its infinitesimal: and that consequently no error can arise from such rejection of infinitesimals.

19 And yet it should seem that, whatever errors are admitted in the premises, proportional errors ought to be apprehended in the conclusion, be they finite or infinitesimal: and that therefore the *ἀκρίβεια* ‖rigour, exactness‖ of geometry requires nothing should be neglected or rejected. In answer to this you will perhaps say, that the conclusions are accurately true, and that therefore the principles and methods from whence they are derived must be so too. But this inverted way of demonstrating your principles by your conclusions, as it would be peculiar to you gentlemen, so it is contrary to the rules of logic. The truth of the conclusion will not prove either the form or the matter of a syllogism to be true; inasmuch as the illation might have been wrong or the premises false, and the conclusion nevertheless true, though not in virtue of such illation or of such premises. I say that in every other science men prove their conclusions by their principles, and not their principles by the conclusions. But if in yours you should allow your selves this unnatural way of proceeding, the consequence would be that you must take up with Induction, and bid adieu to Demonstration. And if you submit to this, your authority will no longer lead the way in points of Reason and Science.

20 I have no controversy about your conclusions, but only about your logic and method. How you demonstrate? What objects you are conversant with, and whether you conceive them clearly? What principles you proceed upon; how sound they may be; and how you apply them? It must be remembered that I am not concerned about the truth of your theorems, but only about the way of coming at them; whether it be legitimate or illegitimate, clear or obscure, scientific or tentative. To prevent all possibility of your mistaking me, I beg leave to repeat and insist, that I consider the geometrical analyst as a logician, *i.e.* so far forth as he reasons and argues; and his mathematical conclusions, not in themselves, but in their premises; not as true or false, useful or insignificant, but as derived from such principles, and by such inferences. And, forasmuch as it may perhaps seem an unaccountable paradox that mathematicians should deduce true propositions from false principles, be right in the conclusion and yet err in the premises; I shall endeavour particularly to explain why this may come to pass, and shew how error may bring forth truth, though it cannot bring forth science.

21 In order therefore to clear up this point, we will suppose for instance that a tangent is to be drawn to a parabola, and examine the progress of this affair as it is performed by infinitesimal differences. Let AB be a curve, the absciss $AP = x$, the ordinate $PB = y$, the difference of the absciss $PM = dx$, the difference of the ordinate $RN = dy$. Now, by supposing the curve to be a polygon, and consequently BN, the increment or difference of the curve, to be a straight line coincident with the tangent, and the differential triangle BRN to be similar to the triangle TPB, the subtangent PT is found a fourth proportional to $RN : RB : PB$: that is, to $dy : dx : y$. Hence the subtangent will be $\frac{y\,dx}{dy}$. But

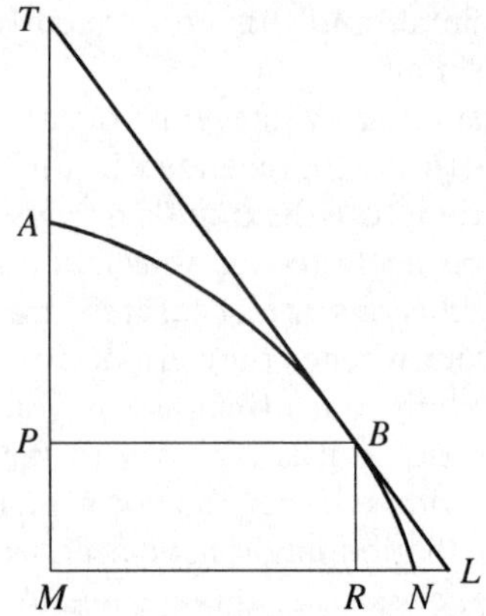

herein there is an error arising from the forementioned false supposition, whence the value of PT comes out greater than the truth: for in reality it is not the triangle RNB but RLB, which is similar to PBT, and therefore (instead of RN) RL should have been the first term of the proportion, *i.e.* $RN + NL$, *i.e.* $dy + z$: whence the true expression for the subtangent should have been $\frac{y\,dx}{dy + z}$. There was therefore an error of defect in making dy the divisor: which error was equal to z, *i.e.* NL the line comprehended between the curve and the tangent. Now by the nature of the curve $yy = px$, supposing p to be the parameter, whence by the rule of differences $2y\,dy = p\,dx$ and $dy = \frac{p\,dx}{2y}$. But if you multiply $y + dy$ by it self, and retain the whole product without rejecting the square of the difference, it will then come out, by substituting the augmented quantities in the equation of the curve, that $dy = \frac{p\,dx}{2y} - \frac{dy\,dy}{2y}$ truly. There was therefore an error of excess in making $dy = \frac{p\,dx}{2y}$, which followed from the erroneous rule of differences. And the measure of this second error is $\frac{dy\,dy}{2y} = z$. Therefore the two errors being equal and contrary destroy each other; the first error of defect being corrected by a second error of excess.

22 If you had committed only one error, you would not have come at a true solution of the problem. But by virtue of a twofold mistake you arrive, though not at science, yet at truth. For science it cannot be called, when you proceed blindfold, and arrive at the truth not knowing how or by what means. To demonstrate that z is equal to $\frac{dy\,dy}{2y}$, let BR or dx be m, and RN or dy be n. By the thirty third proposition of the first book of the Conics of Apollonius,

and from similar triangles, as $2x$ to y so is m to $n + z = \frac{my}{2x}$. Likewise from the nature of the parabola $yy + 2yn + nn = xp + mp$, and $2yn + nn = mp$: wherefore $\frac{2yn + nn}{p} = m$: and because $yy = px$, $\frac{yy}{p}$ will be equal to x. Therefore substituting these values instead of m and x we shall have

$$n + z = \frac{my}{2x} = \frac{2yynp + ynnp}{2yyp}:$$

i.e.

$$n + z = \frac{2yn + nn}{2y}:$$

which being reduced gives

$$z = \frac{nn}{2y} = \frac{dy\,dy}{2y} \quad Q.E.D.$$

23 Now, I observe in the first place, that the conclusion comes out right, not because the rejected square of dy was infinitely small; but because this error was compensated by another contrary and equal error. I observe in the second place, that whatever is rejected, be it ever so small, if it be real and consequently makes a real error in the premises, it will produce a proportional real error in the conclusion. Your theorems therefore cannot be accurately true, nor your problems accurately solved, in virtue of premises which themselves are not accurate; it being a rule in logic that *conclusio sequitur partem debiliorem*. Therefore I observe in the third place, that when the conclusion is evident and the premises obscure, or the conclusion accurate and the premises inaccurate, we may safely pronounce that such conclusion is neither evident nor accurate, in virtue of those obscure inaccurate premises or principles; but in virtue of some other principles which perhaps the demonstrator himself never knew or thought of. I observe in the last place, that in case the differences are supposed finite quantities ever so great, the conclusion will nevertheless come out the same: inasmuch as the rejected quantities are legitimately thrown out, not for their smallness, but for another reason, to wit, because of contrary errors, which, destroying each other do upon the whole cause that nothing is really, though something is apparently, thrown out. And this reason holds equally with respect to quantities finite as well as infinitesimal, great as well as small, a foot or a yard long as well as the minutest increment.

24 For the fuller illustration of this point, I shall consider it in another light, and proceeding in finite quantities to the conclusion, I shall only then make use of one infinitesimal. Suppose the straight line MQ cuts the curve AT in the points R and S. Suppose LR a tangent at the point R, AN the absciss, NR and OS ordinates. Let AN be produced to O, and RP be drawn parallel to NO. Suppose $AN = x$, $NR = y$, $NO = v$, $PS = z$, the subsecant $MN = s$. Let the

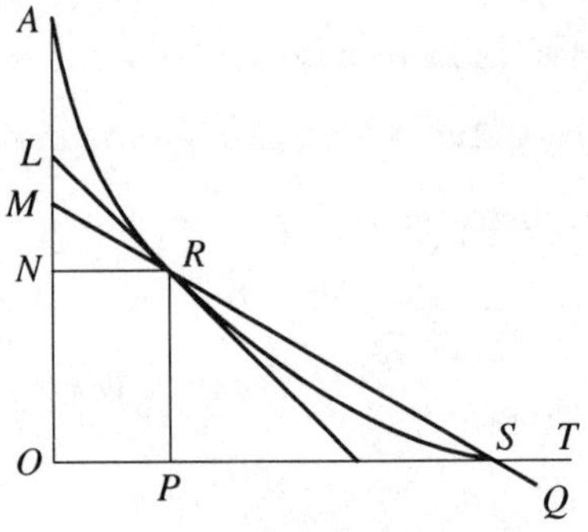

equation $y = xx$ express the nature of the curve: and supposing y and x increased by their finite increments we get

$$y + z = xx + 2xv + vv:$$

whence the former equation being subducted there remains $z = 2xv + vv$. And by reason of similar triangles

$$PS:PR::NR:NM, \text{ i.e. } z:v::y:s=\frac{vy}{z},$$

wherein if for y and z we substitute their values, we get

$$\frac{vxx}{2xv + vv} = s = \frac{xx}{2x + v}.$$

And supposing NO to be infinitely diminished, the subsecant NM will in that case coincide with the subtangent NL, and v as an infinitesimal may be rejected, whence it follows that

$$S = NL = \frac{xx}{2x} = \frac{x}{2}$$

which is the true value of the subtangent. And since this was obtained by one only error, *i.e.* by once rejecting one only infinitesimal, it should seem, contrary to what hath been said, that an infinitesimal quantity or difference may be neglected or thrown away, and the conclusion nevertheless be accurately true, although there was no double mistake, or rectifying of one error by another, as in the first case. But, if this point be thoroughly considered, we shall find there is even here a double mistake, and that one compensates or rectifies the other. For, in the first place, it was supposed that when NO is infinitely diminished or becomes an infinitesimal then the subsecant NM becomes equal to the subtangent NL. But this is a plain mistake; for it is evident that as a secant cannot be a tangent, so a subsecant cannot be a subtangent. Be the difference ever so small, yet still there is a difference. And if NO be infinitely small, there will even then be an infinitely small difference between NM and NL. Therefore

NM or *S* was too little for your supposition (when you supposed it equal to *NL*) and this error was compensated by a second error in throwing out *v*, which last error made *s* bigger than its true value, and in lieu thereof gave the value of the subtangent. This is the true state of the case, however it may be disguised. And to this in reality it amounts, and is at bottom the same thing, if we should pretend to find the subtangent by having first found, from the equation of the curve and similar triangles, a general expression for all subsecants, and then reducing the subtangent under this general rule, by considering it as the subsecant when *v* vanishes or becomes nothing.

25 Upon the whole I observe, *First*, that *v* can never be nothing so long as there is a secant. *Secondly*, that the same line cannot be both tangent and secant. *Thirdly*, that when *v* or *NO*[8] vanisheth, *PS* and *SR* do also vanish, and with them the proportionality of the similar triangles. Consequently the whole expression, which was obtained by means thereof and grounded thereupon, vanisheth when *v* vanisheth. *Fourthly*, that the method for finding secants or the expression of secants, be it ever so general, cannot in common sense extend any further than to all secants whatsoever: and, as it necessarily supposeth similar triangles, it cannot be supposed to take place where there are not similar triangles. *Fifthly*, that the subsecant will always be less than the subtangent, and can never coincide with it; which coincidence to suppose would be absurd; for it would be supposing the same line at the same time to cut and not to cut another given line; which is a manifest contradiction, such as subverts the hypothesis and gives a demonstration of its falsehood. *Sixthly*, if this be not admitted, I demand a reason why any other apagogical demonstration, or demonstration *ad absurdum* should be admitted in geometry rather than this: Or that some real difference be assigned between this and others as such. *Seventhly*, I observe that it is sophistical to suppose *NO* or *RP*, *PS*, and *SR* to be finite real lines in order to form the triangle, *RPS*, in order to obtain proportions by similar triangles; and afterwards to suppose there are no such lines, nor consequently similar triangles, and nevertheless to retain the consequence of the first supposition, after such supposition hath been destroyed by a contrary one. *Eighthly*, that although, in the present case, by inconsistent suppositions truth may be obtained, yet that such truth is not demonstrated: That such method is not conformable to the rules of logic and right reason: That, however useful it may be, it must be considered only as a presumption, as a knack, an art, rather an artifice, but not a scientific demonstration.

26 The doctrine premised may be farther illustrated by the following simple and easy case, wherein I shall proceed by evanescent increments. Suppose $AB = x$, $BC = y$, $BD = o$, and that xx is equal to the area *ABC*: it is proposed to find the ordinate y or *BC*. When x by flowing becomes $x + o$, then xx becomes $xx + 2xo + oo$: And the area *ABC* becomes *ADH*, and the increment of xx will be equal to *BDHC*, the increment of the area, *i.e.* to $BCFD + CFH$. And if we suppose the curvilinear space *CFH* to be qoo, then

---

[8] See the foregoing figure.

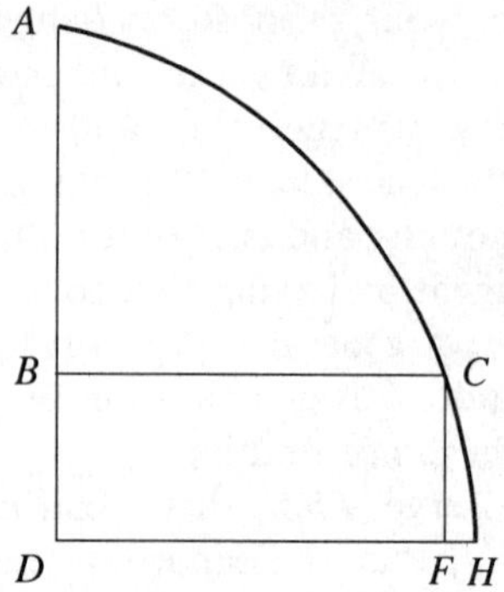

$$2xo + oo = yo + qoo$$

which divided by o gives $2x + o = y + qo$. And, supposing o to vanish, $2x = y$, in which case ACH will be a straight line, and the areas ABC, CFH, triangles. Now with regard to this reasoning, it hath been already remarked,[9] that it is not legitimate or logical to suppose o to vanish, *i.e.* to be nothing, *i.e.* that there is no increment, unless we reject at the same time with the increment it self every consequence of such increment, *i.e.* whatsoever could not be obtained but by supposing such increment. It must nevertheless be acknowledged that the problem is rightly solved, and the conclusion true, to which we are led by this method. It will therefore be asked, how comes it to pass that the throwing out o is attended with no error in the conclusion? I answer, the true reason hereof is plainly this: because q being unit, qo is equal to o: and therefore

$$2x + o - qo = y = 2x,$$

the equal quantities qo and o being destroyed by contrary signs.

27 As on the one hand, it were absurd to get rid of o by saying, Let me contradict my self: Let me subvert my own hypothesis: Let me take it for granted that there is no increment, at the same time that I retain a quantity which I could never have got at but by assuming an increment: So on the other hand it would be equally wrong to imagine that in a geometrical demonstration we may be allowed to admit any error, though ever so small, or that it is possible, in the nature of things, an accurate conclusion should be derived from inaccurate principles. Therefore o cannot be thrown out as an infinitesimal, or upon the principle that infinitesimals may be safely neglected. But only because it is destroyed by an equal quantity with a negative sign, whence $o - qo$ is equal to nothing. And as it is illegitimate to reduce an equation, by subducting from one side a quantity when it is not to be destroyed, or when an equal quantity is not subducted from the other side of the equation: So it must be allowed a very logical and just method of arguing to conclude that if from equals either

[9] Sect. 12 and 13 *supra*.

nothing or equal quantities are subducted, they shall still remain equal. And this is a true reason why no error is at last produced by the rejecting of o. Which therefore must not be ascribed to the doctrine of differences, or infinitesimals, or evanescent quantities, or momentums, or fluxions.

28 Suppose the case to be general, and that x^n is equal to the area ABC, whence by the method of fluxions the ordinate is found nx^{n-1} which we admit for true, and shall inquire how it is arrived at. Now if we are content to come at the conclusion in a summary way, by supposing that the ratio of the fluxions of x and x^n are found[10] to be 1 and nx^{n-1}, and that the ordinate of the area is considered as its fluxion, we shall not so clearly see our way, or perceive how the truth comes out, that method as we have shewed before being obscure and illogical. But if we fairly delineate the area and its increment, and divide the latter into two parts $BCFD$ and CFH,[11] and proceed regularly by equations between the algebraical and geometrical quantities, the reason of the thing will plainly appear. For as x^n is equal to the area ABC, so is the increment of x^n equal to the increment of the area, *i.e.* to $BDHC$; that is to say

$$nox^{n-1} + \frac{nn-n}{2}oox^{n-2} + \&c. = BDFC + CFH.$$

And only the first members on each side of the equation being retained, $nox^{n-1} = BDFC$: And dividing both sides by o or BD, we shall get $nx^{n-1} = BC$. Admitting therefore that the curvilinear space CFH is equal to the rejectaneous quantity

$$\frac{nn-n}{2}oox^{n-2} + \&c.,$$

and that when this is rejected on one side, that is rejected on the other, the reasoning becomes just and the conclusion true. And it is all one whatever magnitude you allow to BD, whether that of an infinitesimal difference or a finite increment ever so great. It is therefore plain that the supposing the rejectaneous algebraical quantity to be an infinitely small or evanescent quantity, and therefore to be neglected, must have produced an error, had it not been for the curvilinear spaces being equal thereto, and at the same time subducted from the other part or side of the equation, agreeably to the axiom, *If from equals you subduct equals, the remainders will be equal.* For those quantities which by the analysts are said to be neglected, or made to vanish, are in reality subducted. If therefore the conclusion be true, it is absolutely necessary that the finite space CFH be equal to the remainder of the increment expressed by

$$\frac{nn-n}{2}oox^{n-2}\ \&c.$$

equal, I say, to the finite remainder of a finite increment.

---

[10] Sect. 13.

[11] See the figure in Sect. 26.

29 Therefore, be the power what you please, there will arise on one side an algebraical expression, on the other a geometrical quantity, each of which naturally divides itself into three members: The algebraical or fluxionary expression, into one which includes neither the expression of the increment of the absciss nor of any power thereof; another which includes the expression of the increment itself; and a third including the expression of the powers of the increment. The geometrical quantity also or whole increased area consists of three parts or members, the first of which is the given area; the second a rectangle under the ordinate and the increment of the absciss; and the third a curvilinear space. And, comparing the homologous or correspondent members on both sides, we find that as the first member of the expression is the expression of the given area, so the second member of the expression will express the rectangle or second member of the geometrical quantity, and the third, containing the powers of the increment, will express the curvilinear space, or third member of the geometrical quantity. This hint may perhaps be further extended, and applied to good purpose, by those who have leisure and curiosity for such matters. The use I make of it is to shew, that the analysis cannot obtain in augments or differences, but it must also obtain in finite quantities, be they ever so great, as was before observed.

30 It seems therefore upon the whole that we may safely pronounce the conclusion cannot be right, if in order thereto any quantity be made to vanish, or be neglected, except that either one error is redressed by another; or that secondly on the same side of an equation equal quantities are destroyed by contrary signs, so that the quantity we mean to reject is first annihilated; or lastly, that from the opposite sides equal quantities are subducted. And therefore to get rid of quantities by the received principles of fluxions or of differences is neither good geometry nor good logic. When the augments vanish, the velocities also vanish. The velocities or fluxions are said to be *primo* and *ultimo*, as the augments nascent and evanescent. Take therefore the *ratio* of the evanescent quantities, it is the same with that of the fluxions. It will therefore answer all intents as well. Why then are fluxions introduced? Is it not to shun or rather to palliate the use of quantities infinitely small? But we have no notion whereby to conceive and measure various degrees of velocity beside space and time; or when the times are given beside space alone. We have even no notion of velocity prescinded from time and space. When therefore a point is supposed to move in given times, we have no notion of greater or lesser velocities, or of proportions between velocities, but only of longer or shorter lines, and of proportions between such lines generated in equal parts of time.

31 A point may be the limit of a line: A line may be the limit of a surface: A moment may terminate time. But how can we conceive a velocity by the help of such limits? It necessarily implies both time and space, and cannot be conceived without them. And if the velocities of nascent and evanescent quantities, *i.e.* abstracted from time and space, may not be comprehended, how can we comprehend and demonstrate their proportions? Or consider their *rationes primæ* and *ultimæ*. For, to consider the proportion or *ratio* of things implies that such things

have magnitude: That such their magnitudes may be measured, and their relations to each other known. But, as there is no measure of velocity except time and space, the proportion of velocities being only compounded of the direct proportion of the spaces, and the reciprocal proportion of the times; doth it not follow that to talk of investigating, obtaining, and considering the proportions of velocities, exclusively of time and space, is to talk unintelligibly?

32 But you will say that, in the use and application of fluxions, men do not overstrain their faculties to a precise conception of the above-mentioned velocities, increments, infinitesimals, or any other such-like ideas of a nature so nice, subtile, and evanescent. And therefore you will perhaps maintain that problems may be solved without those inconceivable suppositions; and, that, consequently, the doctrine of fluxions, as to the practical part, stands clear of all such difficulties. I answer that if in the use or application of this method those difficult and obscure points are not attended to, they are nevertheless supposed. They are the foundations on which the moderns build, the principles on which they proceed, in solving problems and discovering theorems. It is with the method of fluxions as with all other methods, which presuppose their respective principles and are grounded thereon; although the rules may be practised by men who neither attend to, nor perhaps know the principles. In like manner, therefore, as a sailor may practically apply certain rules derived from astronomy and geometry, the principles whereof he doth not understand: And as any ordinary man may solve divers numerical questions, by the vulgar rules and operations of arithmetic, which he performs and applies without knowing the reasons of them: Even so it cannot be denied that you may apply the rules of the fluxionary method: You may compare and reduce particular cases to general forms: You may operate and compute and solve problems thereby, not only without an actual attention to, or an actual knowledge of, the grounds of that method, and the principles whereon it depends, and whence it is deduced, but even without having ever considered or comprehended them.

33 But then it must be remembered that in such case although you may pass for an artist, computist, or analyst, yet you may not be justly esteemed a man of science and demonstration. Nor should any man, in virtue of being conversant in such obscure analytics, imagine his rational faculties to be more improved than those of other men which have been exercised in a different manner and on different subjects; much less erect himself into a judge and an oracle concerning matters that have no sort of connection with or dependence on those species, symbols or signs, in the management whereof he is so conversant and expert. As you, who are a skilful computist or analyst, may not therefore be deemed skilful in anatomy: or *vice versa*, as a man who can dissect with art may, nevertheless, be ignorant in your art of computing: Even so you may both, notwithstanding your peculiar skill in your respective arts, be alike unqualified to decide upon logic, or metaphysics, or ethics, or religion. And this would be true, even admitting that you understood your own principles and could demonstrate them.

34 If it is said that fluxions may be expounded or expressed by finite lines

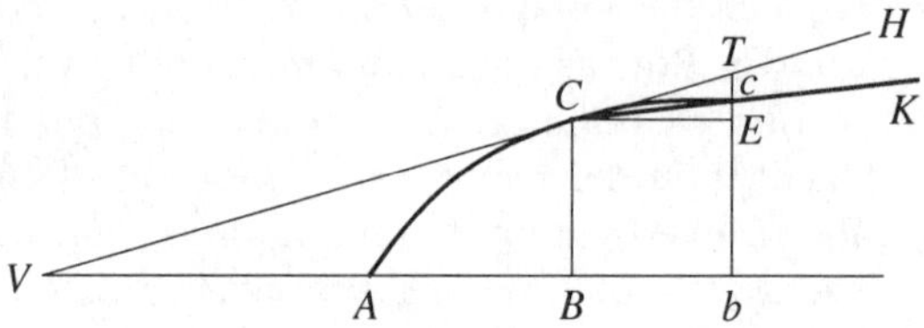

proportional to them: Which finite lines, as they may be distinctly conceived and known and reasoned upon, so they may be substituted for the fluxions, and their mutual relations or proportions be considered as the proportions of fluxions: By which means the doctrine becomes clear and useful. I answer that, if in order to arrive at these finite lines proportional to the fluxions, there be certain steps made use of which are obscure and inconceivable, be those finite lines themselves ever so clearly conceived, it must nevertheless be acknowledged that your proceeding is not clear nor your method scientific. For instance, it is supposed that *AB* being the absciss, *BC* the ordinate, and *VCH* a tangent of the curve *AC*, *Bb* or *CE* the increment of the absciss, *Ec* the increment of the ordinate, which produced meets *VH* in the point *T*, and *Cc* the increment of the curve. The right line *Cc* being produced to *K*, there are formed three small triangles, the rectilinear *CEc*, the mixtilinear *CEc*, and the rectilinear triangle *CET*. It is evident these three triangles are different from each other, the rectilinear *CEc* being less than the mixtilinear *CEc*, whose sides are the three increments above mentioned, and this still less than the triangle *CET*. It is supposed that the ordinate *bc* moves into the place *BC*, so that the point *c* is coincident with the point *C*; and the right line *CK*, and consequently the curve *Cc*, is coincident with the tangent *CH*. In which case the mixtilinear evanescent triangle *CEc* will, in its last form, be similar to the triangle *CET*: and its evanescent sides *CE*, *Ec*, and *Cc*, will be proportional to *CE*, *ET*, and *CT*, the sides of the triangle *CET*. And therefore it is concluded that the fluxions of the lines *AB*, *BC*, and *AC*, being in the last ratio of their evanescent increments, are proportional to the sides of the triangle *CET*, or, which is all one, of the triangle *VBC* similar thereunto.[12] It is particularly remarked and insisted on by the great author, that the points *C* and *c* must not be distant one from another, by any the least interval whatsoever: but that, in order to find the ultimate proportions of the lines *CE*, *Ec*, and *Cc* (*i.e.* the proportions of the fluxions or velocities) expressed by the finite sides of the triangle *VBC*, the points *C* and *c* must be accurately coincident, *i.e.* one and the same. A point therefore is considered as a triangle, or a triangle is supposed to be formed in a point. Which to conceive seems quite impossible. Yet some there are who, though they shrink at all other mysteries, make no difficulty of their own, who strain at a gnat and swallow a camel.

35 I know not whether it be worth while to observe, that possibly some men

---

[12] *Introduct. ad Quad. Curv.*

may hope to operate by symbols and suppositions, in such sort as to avoid the use of fluxions, momentums, and infinitesimals, after the following manner. Suppose x to be an absciss of a curve, and z another absciss of the same curve. Suppose also that the respective areas are xxx and zzz: and that $z - x$ is the increment of the absciss, and $zzz - xxx$ the increment of the area, without considering how great or how small those increments may be. Divide now $zzz - xxx$ by $z - x$, and the quotient will be $zz + zx + xx$: and, supposing that z and x are equal, this same quotient will be $3xx$, which in that case is the ordinate, which therefore may be thus obtained independently of fluxions and infinitesimals. But herein is a direct fallacy: for, in the first place, it is supposed that the abscisses z and x are unequal, without which supposition no one step could have been made; and in the second place, it is supposed they are equal; which is a manifest inconsistency, and amounts to the same thing that hath been before considered.[13] And there is indeed reason to apprehend that all attempts for setting the abstruse and fine geometry on a right foundation, and avoiding the doctrine of velocities, momentums, &c. will be found impracticable, till such time as the object and end of geometry are better understood than hitherto they seem to have been. The great author of the method of fluxions felt this difficulty, and therefore he gave into those nice abstractions and geometrical metaphysics without which he saw nothing could be done on the received principles; and what in the way of demonstration he hath done with them the reader will judge. It must, indeed, be acknowledged that he used fluxions, like the scaffold of a building, as things to be laid aside or got rid of as soon as finite lines were found proportional to them. But then these finite exponents are found by the help of fluxions. Whatever therefore is got by such exponents and proportions is to be ascribed to fluxions: which must therefore be previously understood. And what are these fluxions? The velocities of evanescent increments? And what are these same evanescent increments? They are neither finite quantities, nor quantities infinitely small, nor yet nothing. May we not call them the ghosts of departed quantities?

36 Men too often impose on themselves and others as if they conceived and understood things expressed by signs, when in truth they have no idea, save only of the very signs themselves. And there are some grounds to apprehend that this may be the present case. The velocities of evanescent or nascent quantities are supposed to be expressed, both by finite lines of a determinate magnitude, and by algebraical notes or signs: but I suspect that many who, perhaps never having examined the matter take it for granted, would upon a narrow scrutiny find it impossible to frame any idea or notion whatsoever of those velocities, exclusive of such finite quantities and signs.

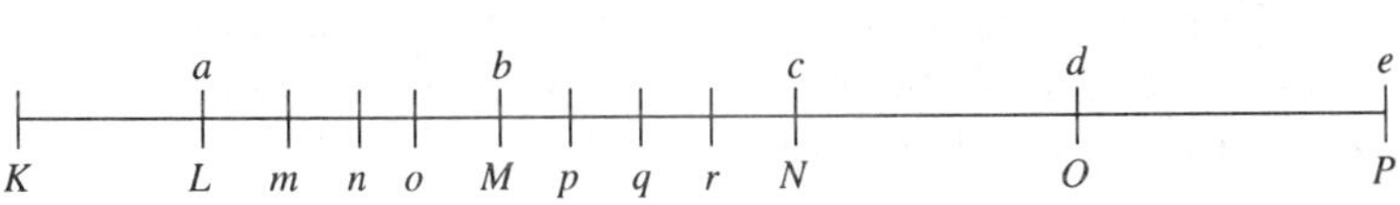

[13] Sect. 15.

Suppose the line *KP* described by the motion of a point continually accelerated, and that in equal particles of time the unequal[c] parts *KL*, *LM*, *MN*, *NO*, &c. are generated. Suppose also that *a*, *b*, *c*, *d*, *e*, &c. denote the velocities of the generating point, at the several periods of the parts or increments so generated. It is easy to observe that these increments are each proportional to the sum of the velocities with which it is described: That, consequently, the several sums of the velocities, generated in equal parts of time, may be set forth by the respective lines *KL*, *LM*, *MN*, &c. generated in the same times: It is likewise an easy matter to say, that the last velocity generated in the first particle of time may be expressed by the symbol *a*, the last in the second by *b*, the last generated in the third by *c*, and so on: that *a* is the velocity of *LM* in *statu nascenti*, and *b*, *c*, *d*, *e*, &c. are the velocities of the increments *MN*, *NO*, *OP*, &c. in their respective nascent estates. You may proceed, and consider these velocities themselves as flowing or increasing quantities, taking the velocities of the velocities, and the velocities of the velocities of the velocities, *i.e.* the first, second, third, &c. velocities *ad infinitum*: which succeeding series of velocities may be thus expressed.

$$a.b - a.c - 2b + a.d - 3c + 3b - a \text{ \&c.}$$

which you may call by the names of first, second, third, fourth fluxions. And for an apter expression you may denote the variable flowing line *KL*, *KM*, *KN*, &c. by the letter x; and the first fluxions by $\dot{x}$, the second by $\ddot{x}$, the third by $\dddot{x}$, and so on *ad infinitum*.

37 Nothing is easier than to assign names, signs, or expressions to these fluxions, and it is not difficult to compute and operate by means of such signs. But it will be found much more difficult to omit the signs and yet retain in our minds the things which we suppose to be signified by them. To consider the exponents, whether geometrical, or algebraical, or fluxionary, is no difficult matter. But to form a precise idea of a third velocity for instance, in itself and by itself, *Hoc opus, hic labor*. Nor indeed is it an easy point to form a clear and distinct idea of any velocity at all, exclusive of and prescinding from all length of time and space; as also from all notes, signs, or symbols whatsoever. This, if I may be allowed to judge of others by myself, is impossible. To me it seems evident that measures and signs are absolutely necessary in order to conceive or reason about velocities; and that consequently, when we think to conceive the velocities simply and in themselves, we are deluded by vain abstractions.

38 It may perhaps be thought by some an easier method of conceiving fluxions to suppose them the velocities wherewith the infinitesimal differences are generated. So that the first fluxions shall be the velocities of the first differences, the second the velocities of the second differences, the third fluxions the velocities of the third differences, and so on *ad infinitum*. But not to mention the insurmountable difficulty of admitting or conceiving infinitesimals, and

[c] [Berkeley's diagram, reproduced here, does not correspond entirely with his description in the text.]

infinitesimals of infinitesimals, &c. it is evident that this notion of fluxions would not consist with the great author's view; who held that the minutest quantity ought not to be neglected, that therefore the doctrine of infinitesimal differences was not to be admitted in geometry, and who plainly appears to have introduced the use of velocities or fluxions, on purpose to exclude or do without them.

39 To others it may possibly seem that we should form a juster idea of fluxions by assuming the finite, unequal, isochronal increments *KL*, *LM*, *MN*, &c. and considering them in *statu nascenti*, also their increments in *statu nascenti*, and the nascent increments of those increments, and so on, supposing the first nascent increments to be proportional to the first fluxions or velocities, the nascent increments of those increments to be proportional to the second fluxions, the third nascent increments to be proportional to the third fluxions, and so onwards. And, as the first fluxions are the velocities of the first nascent increments, so the second fluxions may be conceived to be the velocities of the second nascent increments, rather than the velocities of velocities. By which means the analogy of fluxions may seem better preserved, and the notion rendered more intelligible.

40 And indeed it should seem that in the way of obtaining the second or third fluxion of an equation the given fluxions were considered rather as increments than velocities. But the considering them sometimes in one sense, sometimes in another, one while in themselves, another in their exponents, seems to have occasioned no small share of that confusion and obscurity which is found in the doctrine of fluxions. It may seem therefore that the notion might be still mended, and that instead of fluxions of fluxions, or fluxions of fluxions of fluxions, and instead of second, third, or fourth, &c. fluxions of a given quantity, it might be more consistent and less liable to exception to say, the fluxion of the first nascent increment, *i.e.* the second fluxion; the fluxion of the second nascent increment, *i.e.* the third fluxion; the fluxion of the third nascent increment, *i.e.* the fourth fluxion, which fluxions are conceived respectively proportional, each to the nascent principle of the increment succeeding that whereof it is the fluxion.

41 For the more distinct conception of all which it may be considered that if the finite increment *LM*[14] be divided into the isochronal parts *Lm*, *mn*, *no*, *oM*; and the increment *MN* into the parts *Mp*, *pq*, *qr*, *rN* isochronal to the former; as the whole increments *LM*, *MN* are proportional to the sums of their describing velocities, even so the homologous particles *Lm*, *Mp* are also proportional to the respective accelerated velocities with which they are described. And as the velocity with which *Mp* is generated, exceeds that with which *Lm* was generated, even so the particle *Mp* exceeds the particle *LM*. And in general, as the isochronal velocities describing the particles of *MN* exceed the isochronal velocities describing the particles of *LM*, even so the particles of the former

[14] See the foregoing Scheme in Sect. 36.

exceed the correspondent particles of the latter. And this will hold, be the said particles ever so small. *MN* therefore will exceed *LM* if they are both taken in their nascent states: and that excess will be proportional to the excess of the velocity *b* above the velocity *a*. Hence we may see that this last account of fluxions comes, in the upshot, to the same thing with the first.[15]

42 But notwithstanding what hath been said it must still be acknowledged that the finite particles *Lm* or *Mp*, though taken ever so small, are not proportional to the velocities *a* and *b*; but each to a series of velocities changing every moment, or which is the same thing, to an accelerated velocity, by which it is generated during a certain minute particle of time: That the nascent beginnings or evanescent endings of finite quantities, which are produced in moments or infinitely small parts of time, are alone proportional to given velocities: That therefore, in order to conceive the first fluxions, we must conceive time divided into moments, increments generated in those moments, and velocities proportional to those increments: That, in order to conceive second and third fluxions, we must suppose that the nascent principles or momentaneous increments have themselves also other momentaneous increments, which are proportional to their respective generating velocities: that the velocities of these second momentaneous increments are second fluxions: those of their nascent momentaneous increments third fluxions. And so on *ad infinitum*.

43 By subducting the increment generated in the first moment from that generated in the second, we get the increment of an increment. And by subducting the velocity generating in the first moment from that generating in the second, we get the fluxion of a fluxion. In like manner, by subducting the difference of the velocities generating in the two first moments from the excess of the velocity in the third above that in the second moment, we obtain the third fluxion. And after the same analogy we may proceed to fourth, fifth, sixth fluxions, &c. And if we call the velocities of the first, second, third, fourth moments, *a*, *b*, *c*, *d*, the series of fluxions will be as above, $a.b - a.c - 2b + a.d - 3c + 3b - a$. *ad infinitum*, *i.e.* $\dot{x}.\ \ddot{x}.\ \dddot{x}.\ \ddddot{x}$. *ad infinitum*.

44 Thus fluxions may be considered in sundry lights and shapes, which seem all equally difficult to conceive. And, indeed, as it is impossible to conceive velocity without time or space, without either finite length or finite duration,[16] it must seem above the powers of men to comprehend even the first fluxions. And if the first are incomprehensible what shall we say of the second and third fluxions, &c.? He who can conceive the beginning of a beginning, or the end of an end, somewhat before the first or after the last, may be perhaps sharp-sighted enough to conceive these things. But most men will, I believe, find it impossible to understand them in any sense whatever.

45 One would think that men could not speak too exactly on so nice a subject. And yet, as was before hinted, we may often observe that the exponents of fluxions or notes representing fluxions are confounded with the fluxions

---

[15] Sect. 36.

[16] Sect. 31.

themselves. Is not this the case when, just after the fluxions of flowing quantities were said to be the celerities of their increasing, and the second fluxions to be the mutations of the first fluxions or celerities, we are told that $z''.\ z'.\ z.\ \dot{z}.\ \ddot{z}.\ \dddot{z}.$[17] represents a series of quantities whereof each subsequent quantity is the fluxion of the preceding; and each foregoing is a fluent quantity having the following one for its fluxion?

46 Divers series of quantities and expressions, geometrical and algebraical, may be easily conceived, in lines, in surfaces, in species, to be continued without end or limit. But it will not be found so easy to conceive a series, either of mere velocities or of mere nascent increments, distinct therefrom and corresponding thereunto. Some perhaps may be led to think the author intended a series of ordinates, wherein each ordinate was the fluxion of the preceding and fluent of the following, *i.e.* that the fluxion of one ordinate was itself the ordinate of another curve; and the fluxion of this last ordinate was the ordinate of yet another curve; and so on *ad infinitum*. But who can conceive how the fluxion (whether velocity or nascent increment) of an ordinate should be itself an ordinate? Or more than that each preceding quantity or fluent is related to its subsequent or fluxion, as the area of a curvilinear figure to its ordinate; agreeably to what the author remarks, that each preceding quantity in such series is as the area of a curvilinear figure, whereof the absciss is z, and the ordinate is the following quantity?

47 Upon the whole it appears that the celerities are dismissed, and instead thereof areas and ordinates are introduced. But, however expedient such analogies or such expressions be found for facilitating the modern quadratures, yet we shall not find any light given us thereby into the original real nature of fluxions; or that we are enabled to frame from thence just ideas of fluxions considered in themselves. In all this the general ultimate drift of the author is very clear, but his principles are obscure. But perhaps those theories of the great author are not minutely considered or canvassed by his disciples; who seem eager, as was before hinted, rather to operate than to know, rather to apply his rules and his forms than to understand his principles and enter into his notions. It is nevertheless certain that, in order to follow him in his quadratures, they must find fluents from fluxions; and in order to this, they must know to find fluxions from fluents; and in order to find fluxions, they must first know what fluxions are. Otherwise they proceed without clearness and without science. Thus the direct method precedes the inverse, and the knowledge of the principles is supposed in both. But as for operating according to rules, and by the help of general forms, whereof the original principles and reasons are not understood, this is to be esteemed merely technical. Be the principles therefore ever so abstruse and metaphysical, they must be studied by whoever would comprehend the doctrine of fluxions. Nor can any geometrician have a right to apply the rules of the great author, without first considering his metaphysical

---

[17] *De Quadratura Curvarum*.

notions whence they were derived. These how necessary soever in order to science, which can never be attained without a precise, clear, and accurate conception of the principles, are nevertheless by several carelessly passed over; while the expressions alone are dwelt on and considered and treated with great skill and management, thence to obtain other expressions by methods suspicious and indirect (to say the least) if considered in themselves, however recommended by Induction and Authority; two motives which are acknowledged sufficient to beget a rational faith and moral persuasion, but nothing higher.

48 You may possibly hope to evade the force of all that hath been said, and to screen false principles and inconsistent reasonings, by a general pretence that these objections and remarks are metaphysical. But this is a vain pretence. For the plain sense and truth of what is advanced in the foregoing remarks, I appeal to the understanding of every unprejudiced intelligent reader. To the same I appeal, whether the points remarked upon are not most incomprehensible metaphysics. And metaphysics not of mine, but your own. I would not be understood to infer that your notions are false or vain because they are metaphysical. Nothing is either true or false for that reason. Whether a point be called metaphysical or no avails little. The question is, whether it be clear or obscure, right or wrong, well or ill deduced?

49 Although momentaneous increments, nascent and evanescent quantities, fluxions and infinitesimals of all degrees, are in truth such shadowy entities, so difficult to imagine or conceive distinctly, that (to say the least) they cannot be admitted as principles or objects of clear and accurate science: and although this obscurity and incomprehensibility of your metaphysics had been alone sufficient to allay your pretensions to evidence; yet it hath, if I mistake not, been further shewn, that your inferences are no more just than your conceptions are clear, and that your logics are as exceptionable as your metaphysics. It should seem therefore upon the whole, that your conclusions are not attained by just reasoning from clear principles; consequently, that the employment of modern analysts, however useful in mathematical calculations and constructions, doth not habituate and qualify the mind to apprehend clearly and infer justly; and, consequently, that you have no right, in virtue of such habits, to dictate out of your proper sphere, beyond which your judgment is to pass for no more than that of other men.

50 Of a long time I have suspected that these modern analytics were not scientifical, and gave some hints thereof to the public twenty-five years ago.[d] Since which time, I have been diverted by other occupations, and imagined I might employ myself better than in deducing and laying together my thoughts on so nice a subject. And though of late I have been called upon[e] to make good my suggestions; yet, as the person who made this call doth not appear to think maturely enough to understand either those metaphysics which he

[d] ‖See *Principles of Human Knowledge*, sects. 123–34, reproduced above.‖

[e] ‖Apparently a reference to Andrew Baxter's attack on Berkeley's philosophy (*Baxter c. 1730*). (Baxter's work went through multiple editions.)‖

would refute, or mathematics which he would patronize, I should have spared myself the trouble of writing for his conviction. Nor should I now have troubled you or myself with this address, after so long an intermission of these studies; were it not to prevent, so far as I am able, your imposing on yourself and others in matters of much higher moment and concern. And, to the end that you may more clearly comprehend the force and design of the foregoing remarks, and pursue them still further in your own meditations, I shall subjoin the following Queries.

Qu. 1 Whether the object of geometry be not the proportions of assignable extensions? And whether there be any need of considering quantities either infinitely great or infinitely small?

Qu. 2 Whether the end of geometry be not to measure assignable finite extension? And whether this practical view did not first put men on the study of geometry?

Qu. 3 Whether the mistaking the object and end of geometry hath not created needless difficulties, and wrong pursuits in that science?

Qu. 4 Whether men may properly be said to proceed in a scientific method, without clearly conceiving the object they are conversant about, the end proposed, and the method by which it is pursued?

Qu. 5 Whether it doth not suffice, that every assignable number of parts may be contained in some assignable magnitude? And whether it be not unnecessary, as well as absurd, to suppose that finite extension is infinitely divisible?

Qu. 6 Whether the diagrams in a geometrical demonstration are not to be considered as signs, of all possible finite figures, of all sensible and imaginable extensions or magnitudes of the same kind?

Qu. 7 Whether it be possible to free geometry from insuperable difficulties and absurdities, so long as either the abstract general idea of extension, or absolute external extension be supposed its true object?

Qu. 8 Whether the notions of absolute time, absolute place, and absolute motion be not most abstractedly metaphysical? Whether it be possible for us to measure, compute, or know them?

Qu. 9 Whether mathematicians do not engage themselves in disputes and paradoxes concerning what they neither do nor can conceive? And whether the doctrine of forces be not a sufficient proof of this[18]?

Qu. 10 Whether in geometry it may not suffice to consider assignable finite magnitude, without concerning ourselves with infinity? And whether it would not be righter to measure large polygons having finite sides, instead of curves, than to suppose curves are polygons of infinitesimal sides, a supposition neither true nor conceivable?

Qu. 11 Whether many points which are not readily assented to are not nevertheless true? And whether those in the two following queries may not be of that number?

---

[18] See a Latin treatise *De Motu*, published at London, in the year 1721.

Qu. 12 Whether it be possible that we should have had an idea or notion of extension prior to motion? Or whether, if a man had never perceived motion, he would ever have known or conceived one thing to be distant from another?

Qu. 13 Whether geometrical quantity hath co-existent parts? And whether all quantity be not in a flux as well as time and motion?

Qu. 14 Whether extension can be supposed an attribute of a Being immutable and eternal?

Qu. 15 Whether to decline examining the principles, and unravelling the methods used in mathematics would not shew a bigotry in mathematicians?

Qu. 16 Whether certain maxims do not pass current among analysts which are shocking to good sense? And whether the common assumption that a finite quantity divided by nothing is infinite, be not of this number?

Qu. 17 Whether the considering geometrical diagrams absolutely or in themselves, rather than as representatives of all assignable magnitudes or figures of the same kind, be not a principal cause of the supposing finite extension infinitely divisible; and of all the difficulties and absurdities consequent thereupon?

Qu. 18 Whether from geometrical propositions being general, and the lines in diagrams being therefore general substitutes or representatives, it doth not follow that we may not limit or consider the number of parts into which such particular lines are divisible?

Qu. 19 When it is said or implied, that such a certain line delineated on paper contains more than any assignable number of parts, whether any more in truth ought to be understood, than that it is a sign indifferently representing all finite lines, be they ever so great. In which relative capacity it contains, *i.e.* stands for more than any assignable number of parts? And whether it be not altogether absurd to suppose a finite line, considered in itself or in its own positive nature, should contain an infinite number of parts?

Qu. 20 Whether all arguments for the infinite divisibility of finite extension do not suppose and imply, either general abstract ideas or absolute external extension to be the object of geometry? And, therefore, whether, along with those suppositions, such arguments also do not cease and vanish?

Qu. 21 Whether the supposed infinite divisibility of finite extension hath not been a snare to mathematicians and a thorn in their sides? And whether a quantity infinitely diminished and a quantity infinitely small are not the same thing?

Qu. 22 Whether it be necessary to consider velocities of nascent or evanescent quantities, or moments, or infinitesimals? And whether the introducing of things so inconceivable be not a reproach to mathematics?

Qu. 23 Whether inconsistencies can be truths? Whether points repugnant and absurd are to be admitted upon any subject, or in any science? And whether the use of infinites ought to be allowed as a sufficient pretext and apology for the admitting of such points in geometry?

Qu. 24 Whether a quantity be not properly said to be known, when we know its proportion to given quantities? And whether this proportion can be known but by expressions or exponents, either geometrical, algebraical, or arithmetical? And whether expressions in lines or species can be useful but so far forth as they are reducible to numbers?

Qu. 25 Whether the finding out proper expressions or notations of quantity be not the most general character and tendency of the mathematics? And arithmetical operation that which limits and defines their use?

Qu. 26 Whether mathematicians have sufficiently considered the analogy and use of signs? And how far the specific limited nature of things corresponds thereto?

Qu. 27 Whether because, in stating a general case of pure algebra, we are at full liberty to make a character denote either a positive or a negative quantity, or nothing at all, we may therefore, in a geometrical case, limited by hypotheses and reasonings from particular properties and relations of figures, claim the same licence?

Qu. 28 Whether the shifting of the hypothesis, or (as we may call it) the *fallacia suppositionis* be not a sophism that far and wide infects the modern reasonings, both in the mechanical philosophy and in the abstruse and fine geometry?

Qu. 29 Whether we can form an idea or notion of velocity distinct from and exclusive of its measures, as we can of heat distinct from and exclusive of the degrees on the thermometer by which it is measured? And whether this be not supposed in the reasonings of modern analysts?

Qu. 30 Whether motion can be conceived in a point of space? And if motion cannot, whether velocity can? And if not, whether a first or last velocity can be conceived in a mere limit, either initial or final, of the described space?

Qu. 31 Where there are no increments, whether there can be any *ratio* of increments? Whether nothings can be considered as proportional to real quantities? Or whether to talk of their proportions be not to talk nonsense? Also in what sense we are to understand the proportion of a surface to a line, of an area to an ordinate? And whether species or numbers, though properly expressing quantities which are not homogeneous, may yet be said to express their proportion to each other?

Qu. 32 Whether if all assignable circles may be squared, the circle is not, to all intents and purposes, squared as well as the parabola? Or whether a parabolical area can in fact be measured more accurately than a circular?

Qu. 33 Whether it would not be righter to approximate fairly than to endeavour at accuracy by sophisms?

Qu. 34 Whether it would not be more decent to proceed by trials and inductions, than to pretend to demonstrate by false principles?

Qu. 35 Whether there be not a way of arriving at truth, although the principles are not scientific, nor the reasoning just? And whether such a way ought to be called a knack or a science?

Qu. 36 Whether there can be science of the conclusion where there is not science of the principles? And whether a man can have science of the principles without understanding them? And therefore whether the mathematicians of the present age act like men of science, in taking so much more pains to apply their principles than to understand them?

Qu. 37 Whether the greatest genius wrestling with false principles may not be foiled? And whether accurate quadratures can be obtained without new *postulata* or assumptions? And if not, whether those which are intelligible and consistent ought not to be preferred to the contrary? *See Sect.* 28 and 29.

Qu. 38 Whether tedious calculations in algebra and fluxions be the likeliest method to improve the mind? And whether men's being accustomed to reason altogether about mathematical signs and figures doth not make them at a loss how to reason without them?

Qu. 39 Whether, whatever readiness analysts acquire in stating a problem, or finding apt expressions for mathematical quantities, the same doth necessarily infer a proportionable ability in conceiving and expressing other matters?

Qu. 40 Whether it be not a general case or rule, that one and the same coefficient dividing equal products gives equal quotients? And yet whether such coefficient can be interpreted by o or nothing? Or whether any one will say that if the equation $2 \times o = 5 \times o$, be divided by o, the quotients on both sides are equal? Whether therefore a case may not be general with respect to all quantities and yet not extend to nothings, or include the case of nothing? And whether the bringing nothing under the notion of quantity may not have betrayed men into false reasoning?

Qu. 41 Whether in the most general reasonings about equalities and proportions men may not demonstrate as well as in geometry? Whether in such demonstrations they are not obliged to the same strict reasoning as in geometry? And whether such their reasonings are not deduced from the same axioms with those in geometry? Whether therefore algebra be not as truly a science as geometry?

Qu. 42 Whether men may not reason in species as well as in words? Whether the same rules of logic do not obtain in both cases? And whether we have not a right to expect and demand the same evidence in both?

Qu. 43 Whether an algebraist, fluxionist, geometrician, or demonstrator of any kind can expect indulgence for obscure principles or incorrect reasonings? And whether an algebraical note or species can at the end of a process be interpreted in a sense which could not have been substituted for it at the beginning? Or whether any particular supposition can come under a general case which doth not consist with the reasoning thereof?

Qu. 44 Whether the difference between a mere computer and a man of science be not, that the one computes on principles clearly conceived, and by rules evidently demonstrated, whereas the other doth not?

Qu. 45 Whether, although geometry be a science, and algebra allowed to be a science, and the analytical a most excellent method, in the application nevertheless of the analysis to geometry, men may not have admitted false principles and wrong methods of reasoning?

Qu. 46 Whether, although algebraical reasonings are admitted to be ever so just, when confined to signs or species as general representatives of quantity, you may not nevertheless fall into error, if, when you limit them to stand for particular things, you do not limit your self to reason consistently with the nature of such particular things? And whether such error ought to be imputed to pure algebra?

Qu. 47 Whether the view of modern mathematicians doth not rather seem to be the coming at an expression by artifice, than the coming at science by demonstration?

Qu. 48 Whether there may not be sound metaphysics as well as unsound? Sound as well as unsound logic? And whether the modern analytics may not be brought under one of these denominations, and which?

Qu. 49 Whether there be not really a *philosophia prima*, a certain transcendental science superior to and more extensive than mathematics, which it might behove our modern analysts rather to learn than despise?

Qu. 50 Whether, ever since the recovery of mathematical learning, there have not been perpetual disputes and controversies among the mathematicians? And whether this doth not disparage the evidence of their methods?

Qu. 51 Whether anything but metaphysics and logic can open the eyes of mathematicians and extricate them out of their difficulties?

Qu. 52 Whether, upon the received principles, a quantity can by any division or subdivision, though carried ever so far, be reduced to nothing?

Qu. 53 Whether, if the end of geometry be practice, and this practice be measuring, and we measure only assignable extensions, it will not follow that unlimited approximations compleatly answer the intention of geometry?

Qu. 54 Whether the same things which are now done by infinites may not be done by finite quantities? And whether this would not be a great relief to the imaginations and understandings of mathematical men?

Qu. 55 Whether those philomathematical physicians, anatomists, and dealers in the animal œconomy, who admit the doctrine of fluxions with an implicit faith, can with a good grace insult other men for believing what they do not comprehend?

Qu. 56 Whether the corpuscularian, experimental, and mathematical philosophy, so much cultivated in the last age, hath not too much engrossed men's attention; some part whereof it might have usefully employed?

Qu. 57 Whether from this and other concurring causes the minds of speculative men have not been borne downward, to the debasing and stupifying of the higher faculties? And whether we may not hence account for that prevailing narrowness and bigotry among many who pass for men of science, their incapacity for things moral, intellectual, or theological, their proneness to measure all truths by sense and experience of animal life?

Qu. 58 Whether it be really an effect of thinking, that the same men admire the great author for his fluxions, and deride him for his religion?

Qu. 59 If certain philosophical virtuosi of the present age have no religion, whether it can be said to be for want of faith?

Qu. 60 Whether it be not a juster way of reasoning, to recommend points of faith from their effects, than to demonstrate mathematical principles by their conclusions?

Qu. 61 Whether it be not less exceptionable to admit points above reason than contrary to reason?

Qu. 62 Whether mysteries may not with better right be allowed of in Divine Faith than in Human Science?

Qu. 63 Whether such mathematicians as cry out against mysteries have ever examined their own principles?

Qu. 64 Whether mathematicians, who are so delicate in religious points, are strictly scrupulous in their own science? Whether they do not submit to authority, take things upon trust, and believe points inconceivable? Whether they have not their mysteries, and what is more, their repugnancies and contradictions?

Qu. 65 Whether it might not become men who are puzzled and perplexed about their own principles, to judge warily, candidly and modestly concerning other matters?

Qu. 66 Whether the modern analytics do not furnish a strong *argumentum ad hominem* against the philomathematical infidels of these times?

Qu. 67 Whether it follows from the above-mentioned remarks, that accurate and just reasoning is the peculiar character of the present age? And whether the modern growth of infidelity can be ascribed to a distinction so truly valuable?

FINIS

2
Colin MacLaurin (1698–1746)

Far the ablest reply to Berkeley's *Analyst* was *A treatise of fluxions* (2 vols, 1742) by the Edinburgh mathematician and disciple of Newton, Colin MacLaurin. Until the appearance of Cauchy's *Cours d'analyse* in 1821 (and apart from the researches of Bolzano, which were unknown) this work represented the high point of mathematical rigour in the foundations of the calculus.

MacLaurin had studied mathematics at the University of Glasgow, writing a thesis 'On the power of gravity'. His first important mathematical publication was the *Geometrica organica, sive descriptio linearum curvarum universalis* (1720), which dealt with the properties of conics and plane curves, and which supplied proofs for many theorems which Newton had merely stated. MacLaurin, on Newton's recommendation, was appointed to the chair of mathematics at Edinburgh in 1725; he lectured on Euclid, trigonometry, perspective, the elements of fortification, and Newton's *Principia*. The *Treatise of fluxions* was written as a comprehensive exposition and defence of Newton's methods in the calculus. In the *Preface* MacLaurin explains the motivation of his work:

> A letter published in the year 1734, under the title of *The Analyst*, first gave occasion to the ensuing treatise, and several reasons concurred to induce me to write on this subject at so great length. The author of that piece had represented the method of fluxions as founded on false reasoning, and full of mysteries. His objections seemed to have been occasioned in a great measure by the concise manner in which the elements of this method have been usually described; and their having been so much misunderstood by a person of his abilities appeared to me a sufficient proof that a fuller account of the grounds of them was requisite.
>
> Though there can be no comparison made betwixt the extent or usefulness of the ancient and modern discoveries in geometry, yet it seems to be generally allowed that the ancients took greater care, and were more successful in preserving the character of its evidence entire. This determined me, immediately after that piece came to my hands, (and before I knew any thing of what was intended by others in answer to it,) to attempt to deduce those elements after the manner of the ancients, from a few unexceptionable principles, by demonstrations of the strictest form.

In the passage from the *Treatise of fluxions* reproduced below, MacLaurin gives a concise exposition of the methodology of the Greek geometers, and in particular of Archimedes' technique of using polygons to approximate curved lines and surfaces. This account is particularly valuable as an illustration of the concept of mathematical rigour that prevailed in the middle of the eighteenth century; and indeed, the development that such concepts as *rigour*, *axioms*, and *deduction* underwent in the following century can be seen by comparing MacLaurin's writings with the selections below—for example *Lambert 1786*, *Bolzano 1810*, *Helmholtz 1876*, or *Hilbert 1922a*.

MacLaurin's reliance on the 'manner of the ancient geometricians' was not entirely happy in its consequences for British mathematics, and it led his successors to emulate the mathematical style of Archimedes rather than the new and more fruitful methods of analysis that were being developed on the Continent. In consequence, the development of the calculus in Britain lagged behind the work of the Continental mathematicians until George Peacock, Charles Babbage, and John Frederick William Herschel published their translation of Silvestre François Lacroix's *Elementary treatise on the differential and integral calculus* (*Peacock 1816*), called attention to the achievements of Laplace, and championed the notation of Leibniz over that of Newton.

Although the style of MacLaurin's presentation is taken from the ancients, the content of the principles and axioms described in Book I, Chapter 1 is taken from Newton, and relies heavily on the concept of physical motion (as is clear from pp. 51–9 of the selection below). Indeed, to resume the quotation from the *Preface*:

> In explaining the notion of a fluxion, I have followed Sir Isaac Newton in the first book, imagining that there can be no difficulty in conceiving velocity wherever there is motion; nor do I think that I have departed from his sense in the second book; and in both I have endeavoured to avoid several expressions, which, though convenient, might be liable to exceptions, and, perhaps, occasion disputes. I have always represented fluxions of all orders by finite quantities, the supposition of an infinitely little magnitude being too bold a *postulatum* for such a science as geometry. But, because the method of infinitesimals is much in use, and is valued for its conciseness, I thought it was requisite to account explicitly for the truth, and perfect accuracy, of the conclusions that are derived from it; the rather, that it does not seem to be a very proper reason that is assigned by authors, when they determine what is called the *difference* (but more accurately the fluxion) of a quantity, and tell us, that they reject certain parts of the element because they become infinitely less than the other parts; not only, because a proof of this nature may leave some doubt as to the accuracy of the conclusion; but because it may be demonstrated that those parts ought to be neglected by them at any rate, or that it would be an error to retain them. If an accountant, that pretends to a scrupulous exactness, should tell us that he had neglected certain articles, because he found them to be of small importance; and it should appear that they ought not to have been taken into consideration by him on that occasion, but belong to a different account, we should approve his conclusions as accurate, but not his reason.

In the two volumes of the *Treatise of fluxions* itself, MacLaurin solved a large number of problems in real analysis. He discusses infinite series at length and in a systematic fashion that set a new standard; he gives a geometrical version of the integral test for the convergence of an infinite series, and he gives what is now known as MacLaurin's series for the expansion of a function of x, that is:

$$f(x) = f(0) + xf'(0) + [x^2/2!]f''(0) + [x^3/3!]f'''(0) + \dots$$

In addition, he treats in an original fashion the theory of gravitational attraction, the determination of the maxima and minima of a function, and numerous problems in geometry, statics, and the theory of fluid motion.

In 1745, three years after the appearance of the *Treatise*, MacLaurin (who was also an experimental scientist, with interests in astronomy, mathematical physics, insurance, and cartography) organized the defences of Edinburgh against the Jacobites. Despite his efforts the city fell, and he was forced to take refuge in England; he was soon able to return to Edinburgh, but his health had been ruined by the strains of this adventure, and he died in January 1746 at the age of forty-eight. A biographical sketch with a bibliography of MacLaurin's scientific publications is given by *Scott 1973*; see also *Turnbull 1947* and *1951*. MacLaurin's account of Newton's philosophy, *MacLaurin 1748*, contains further informative comments on the foundations of the calculus. The passages reprinted below should be compared with Newton's remarks on fluxions, reproduced above (*Newton 1726*); for a general account of developments in British mathematics after Newton, see *Cajori 1919* and *1925*.

A. *FROM* A TREATISE OF FLUXIONS (*MACLAURIN 1742*)

The text is taken from the first edition; it comprises pp. 1–12 and 33–50 of the Introduction and the opening pages of Book One, Chapter One. References to *MacLaurin 1742* should be to the original pagination, as given in the margins.

Introduction

GEOMETRY is valued for its extensive usefulness, but has been most admired for its evidence; mathematical demonstration being such as has been always supposed to put an end to dispute, leaving no place for doubt or cavil. It acquired this character by the great care of the old writers, who admitted no principles but a few self-evident truths, and no demonstrations but such as were accurately deduced from them. The science being now vastly enlarged, and applied with success to philosophy and the arts, it is of greater importance than ever that its evidence be preserved perfect. But it has been objected on several occasions, that the modern improvements have been established for the most part upon new and exceptionable maxims, of too abstruse a nature to deserve a place amongst the plain principles of the ancient geometry: And some have proceeded so far as to impute false reasoning to those authors who have contributed most to the late discoveries, and have at the same time been most cautious in their manner of describing them.

In the method of indivisibles, lines were conceived to be made up of points, surfaces of lines, and solids of surfaces; and such suppositions have been emoyed by several ingenious men for proving the old theorems, and discovering new ones, in a brief and easy manner. But as this doctrine was inconsistent with the strict principles of geometry, so it soon appeared that there was some danger of its leading them into false conclusions: Therefore others, in the place of indivisible, substituted infinitely small divisible elements, of which they supposed all magnitudes to be formed; and thus endeavoured to retain, and improve, the advantages that were derived from the former method for the advancement of geometry. After these came to be relished, an infinite scale of infinites and infinitesimals, (ascending and descending always by infinite steps,) was imagined, and proposed to be received into geometry, as of the greatest
1|2 use for penetrating into its abstruse parts. Some have argued for quantities | more than infinite; and others for a kind of quantities that are said to be neither finite nor infinite, but of an intermediate and indeterminate nature.

This way of considering what is called the sublime part of geometry has so far prevailed, that it is generally known by no less a title than the *Science*, *Arithmetic*, or *Geometry of infinites*. These terms imply something lofty, but mysterious; the contemplation of which may be suspected to amaze and perplex, rather than satisfy or enlighten the understanding, in the prosecution of this science; and while it seems greatly to elevate geometry, may possibly lessen its true and real excellency, which chiefly consists in its perspicuity and perfect evidence: For we may be apt to rest in an obscure and imperfect knowledge of so abstruse a doctrine, as better suited to its nature, instead of seeking for that clear and full view we ought to have of geometrical truth; and to this we may ascribe the inclination which has appeared of late for introducing mysteries into a science wherein there ought to be none.

There were some, however, who disliked the making much use of infinites and infinitesimals in geometry. Of this number was Sir ISAAC NEWTON (whose caution was almost as distinguishing a part of his character as his invention) especially after he saw that this liberty was growing to so great a height. In demonstrating the grounds of the method of fluxions* he avoided them, establishing it in a way more agreeable to the strictness of geometry. He considered magnitudes as generated by a flux or motion, and showed how the velocities of the generating motions were to be compared together. There was nothing in this doctrine but what seemed to be natural and agreeable to the ancient geometry. But what he has given us on this subject being very short, his conciseness may be supposed to have given some occasion to the objections which have been raised against his method.

* De quadrat. curvarum.

When the certainty of any part of geometry is brought into question, the most effectual way to set the truth in a full light, | and to prevent disputes, is to 2|3
deduce it from axioms or first principles of unexceptionable evidence, by demonstrations of the strictest kind, after the manner of the antient geometricians. This is our design in the following treatise; wherein we do not propose to alter Sir ISAAC NEWTON's notion of a fluxion, but to explain and demonstrate his method, by deducing it at length from a few self-evident truths, in that strict manner: and, in treating of it, to abstract from all principles and postulates that may require the imagining any other quantities but such as may be easily conceived to have a real existence. We shall not consider any part of space or time as indivisible, or infinitely little; but we shall consider a point as a term or limit of a line, and a moment as a term or limit of time: Nor shall we resolve curve lines, or curvilineal spaces, into rectilineal elements of any kind. In delivering the principles of this method, we apprehend it is better to avoid such suppositions: but after these are demonstrated, short and concise ways of speaking, though less accurate, may be permitted, when there is no hazard of our introducing any uncertainty or obscurity into the science from the use of them, or of involving it in disputes. The method of demonstration, which was invented by the author of fluxions, is accurate and elegant; but we propose to begin with one that is somewhat different; which, being less removed from that of the ancients, may make the transition to his method more easy to beginners, (for whom chiefly this treatise is intended,) and may obviate some objections that have been made to it.

But, before we proceed, it may be of use to consider the steps by which the ancients were able, in several instances, from the mensuration of right-lin'd figures, to judge of such as were bounded by curve lines: for as they did not allow themselves to resolve curvilinear figures into rectilineal elements, it is worth while to examine by what art they could make a transition from the one to the other: And as they were at great pains to finish their demonstrations in the most perfect manner, so by following their example, as much as possible in demonstrating a method so much more general than theirs, we may best guard against exceptions and cavils, and vary less from the old foundations of geometry. | 3|4

They found, that similar triangles are to each other in the duplicate ratio of their homologous sides; and, by resolving similar polygons into similar triangles, the same proposition was extended to these polygons also. But when they came to compare curvilineal figures, that cannot be resolved into rectilineal parts, this method failed. Circles are the only curvilineal plane figures considered in the elements of geometry. If they could have allowed themselves to have considered these as similar polygons of an infinite number of sides, (as some have done who pretend to abridge their demonstrations,) after proving that any similar polygons inscribed in circles are in the duplicate ratio of the diameters, they would have immediately extended this to the circles themselves; and would have considered the second proposition of the twelfth book of the Elements as an easy corollary from the first. But there is ground to think that they would not have admitted a demonstration of this kind. It was a fundamental principle with them, that the difference of any two unequal quantities, by which the greater exceeds the lesser, may be added to itself till it shall exceed any proposed finite quantity of the same kind: and that they founded their propositions concerning curvilineal figures upon this principle in a particular manner, is evident from the demonstrations, and from the express declaration of ARCHIMEDES, who acknowledges it to be the foundation upon which he established his own discoveries*, and cites it as assumed by the ancients in demonstrating all their propositions of this kind. But this principle seems to be inconsistent with the admitting of an infinitely little quantity or difference, which, added to itself any number of times, is never supposed to become equal to any finite quantity whatsoever.

They proceeded therefore in another manner, less direct indeed, but perfectly evident. They found that the inscribed similar polygons, by increasing the number of their sides continually approached to the areas of the circles; so that the decreasing differences betwixt each circle and its inscribed polygon, by still
4|5 further and further divisions of the circular arches which | the sides of the polygons subtend, could become less than any quantity that can be assigned: and that all this while the similar polygons observed the same constant invariable proportion to each other, *viz.* that of the squares of the diameters of the circles. Upon this they founded a demonstration, that the proportion of the circles themselves could be no other than that same invariable ratio of the similar inscribed polygons: of which we shall give a brief abstract, that it may appear in what manner they were able, in this instance, and some others of the same nature, to form a demonstration of the proportions of curvilineal figures, from what they had already discovered of rectilineal ones. And that the general reasoning by which they demonstrated all their theorems of this kind may more easily appear, we shall represent the circles and polygons by right lines, in the same manner as all magnitudes are expressed in the fifth book of the Elements.

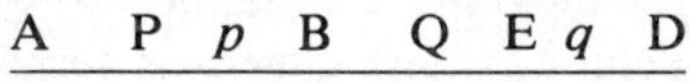

Suppose the right lines AB and AD to represent the two areas of the circles that are compared together; and let AP, AQ represent any two similar polygons inscribed in these circles. By further continual subdivisions of the circular arches which the sides of the polygons subtend, the areas of the polygons increase, and may approach to the circles AB and AD so as to differ from them by less than any assignable measure; the triangle which is subducted from each segment at every new subdivision being always greater than the half the segment. The

*Δείκνυται γὰρ ὅτι πᾶν τμῆμα περιεχόμενον ὑπὸ εὐθείας καὶ ὀρθογωνίου κώνου τομᾶς ἐπίτριτον ἐστί τοῦ τριγώνου, &c. *Archimed. de quadr. parab. ad Dosith.* ‖'For it is here shown that every segment bounded by a straight line and a section of a right-angled cone [a parabola] is four-thirds of the triangle.' The quotation is from Archimedes, *On the quadrature of the parabola*, Dedicatory Epistle; the translation and editorial insertion are by T.L. Heath. In the original the sentence continues 'which has the same base and equal height with the segment'.‖

polygons inscribed in the two circles, as they increase, are ever in the same constant proportion to each other: and this invariable ratio of these polygons must also be the ratio of the circles themselves. For, if it is not, let the ratio of the polygons AP and AQ to each other be, in the first place, the same as the ratio of the circle AB to any magnitude AE less than the circle AD; suppose the subdivisions of the arches of the circle AD to be continued till the difference betwixt the circle and inscribed polygon become less than ED, so that the polygon may be represented by A*q*, greater than AE; and let A*p* represent a polygon inscribed
in the circle AB, similar to the | polygon A*q*. Then, since AP is to AQ as 5|6
AB is to AE by the supposition, and the polygon A*p* is to the similar polygon A*q* as AP is to AQ; it follows, that AB is to AE as A*p* is to A*q*; and that the circle AB being greater than A*p*, a polygon inscribed in it, AE must be greater than A*q*. But A*q* is supposed to be greater than AE; and these being repugnant, it follows, that the polygon AP is not to the polygon AQ as the circle AB is to any magnitude (as AE) less than the circle AD. For the same reason AQ is not to AP as AD is to any magnitude (as AF) less than AB. From which it follows

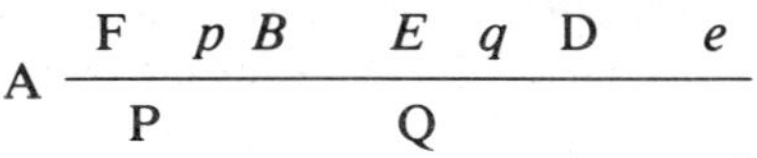

that we cannot suppose AP to be to AQ as AB is to any magnitude A*e* greater than AD; because if we take AF to AB as AD is to A*e*, AF will be less than AB, and AP will be to AQ as AF less than AB to AD; against what has been demonstrated. It follows therefore that AP is not to AQ as AB is to any magnitude greater or less than AD; but that the ratio of the circles AB and AD to each other, must be the same as the invariable ratio of the similar polygons AP and AQ inscribed in them, which is the duplicate of the ratio of their diameters.

In the same manner the ancients have demonstrated, that pyramids of the same height are to each other as their bases, that spheres are as the cubes of their diameters, and that a cone is the third part of a cylinder on the same base and of the same height. In general, it appears from this demonstration, that when two variable quantities, AP and AQ, which always are in an invariable ratio to each other, approach at the same time to two determined quantities, AB and AD, so that they may differ less from them than by any assignable measure, the ratio of these limits AB and AD must be the same as the invariable ratio of the quantities AP and AQ: and this may be considered as the most simple and fundamental proposition in this doctrine, by which we are enabled to
compare curvilineal spaces in some of the more simple cases. | 6|7

This general principle may serve for demonstrating many other propositions,
besides the elementary theorems already mentioned. For example, let ADB be FIG. 1.
a semicircle described on the diameter AB, AEB a semiellipse described on the same right line as its transverse axis; let AFGB be any polygon described in the semicircle; and let FN, GM, perpendicular to AB, meet the semiellipse in H and K, and the axis in N and M. Because any ordinate of the circle is to the ordinate of the ellipse on the same point of the axis as the transverse axis is to the conjugate, it follows, that the triangle ANF is to the triangle ANH, the trapezium

FNMG to the trapezium HNMK, the triangle GMB to the triangle KMB, and the whole polygon AFGB to the polygon AHKB, in the same constant ratio of the transverse to the conjugate axis. Bisect the arch FG in D, and the triangle FDG will be greater than half the segment in which it is inscribed. Let the ordinate DI meet the ellipse in E, and the triangle HEK will be also greater than half the elliptic segment in which it is inscribed; for it is obvious that the right lines GF, KH produced meet the axis in the same point R, and that the tangents at D and E meet it in the same point T; consequently the tangent DT being parallel to the chord FG, the tangent of the ellipse at E is parallel to HK, and the triangle HEK is greater than half the segment HEK. The polygons therefore AFGB, AHKB, by continually bisecting the circular arches, may approach to the areas of the semicircle and semiellipse, so as to differ from them by less than any assignable measure. Hence if the right line AD in the preceeding article A P B Q D represent the area of the semicircle, AB the area of the semiellipse, AQ the polygon inscribed in the semicircle, AP the corresponding polygon inscribed in the semiellipse; it will appear, in the same manner, that the semicircle must be to the semiellipse in the same ratio that is the constant proportion of those inscribed polygons, *viz.* that of the transverse axis of the ellipse to the conjugate axis.

We have given this demonstration a little different from that of ARCHIMEDES in his fifth proposition of *conoids* and *spheroids*, that the same proportion
7|8 might appear to be the ratio of the | area NFDGM in the circle to the area NHEKM in the ellipse. From which it follows, that if C be the common center of the circle and ellipse, the triangles CFN, CHN, and the triangles CGM, CKM being to each other as the transverse axis is to the conjugate, the sector CFDG in the circle must be to the sector CHEK in the ellipse in the same proportion. Let the diameter CE meet its ordinate HK in P, and CP being to CE as CR is to CT, or as CQ is to CD; it appears, that when the ratio of CP to the semidiameter of the ellipse CE is given, then CQ is given, and therefore the sector CFDG is of a determined magnitude; consequently the elliptic sector CHEK, and the segment HEK, must each be of determined invariable magnitude in the same ellipse, when the absciss CP is in a given ratio to the semidiameter CE. The triangle CHK, and the trapezium CH*t*K (formed by the semidiameters CH, CK and the tangents H*t*, K*t*,) are also given in magnitude when this ratio of CP to CE, or of CE to C*t*, is given.

In general, if upon any diameter produced without the ellipse, any number of points be taken, on the same or on different sides of the center, at distances from it that are each in some given ratio to that diameter; and from these points tangents be drawn to the ellipse in any one certain order; the polygon formed by these tangents is always of a given magnitude in a given ellipse, and is equal to a polygon described by a similar construction about a circle, the diameter of which is a mean proportional betwixt the transverse and conjugate axis of the ellipse. The polygon inscribed in the ellipse by joining the points of contact, and the sectors bounded by the semidiameters drawn to these points, are also of given or determined magnitudes; and the Parts of any tangent intercepted betwixt the intersections of the other tangents with it, or betwixt these intersections and the point of contract, are always in the same ratio to each other in the same figure. There is an analogous property of the other conic sections.

When ARCHIMEDES demonstrated, that the area of a circle is equal to a triangle upon a base equal to the circumference of the circle, of a height equal to the radius, it was not by supposing | it to coincide with a circumscribed 8|9
equilateral polygon of an infinite number of sides, but in a more accurate and unexceptionable manner. Let *bd*, the base of the right-angled triangle *abd*, FIG. 2.
be supposed equal to the circumference of the circle ABD, *ab* equal to the radius CA, EFGH any equilateral polygon described about the circle, ABDK a similar polygon inscribed in it, and let CQ perpendicular to AB meet it in Q. As the circumscribed polygon EFGH is greater than the circle, so it is greater than the triangle *abd*; because it is equal to a triangle of a height equal to CA or *ab*, upon a base equal to the perimeter EFGH, which is always greater than *bd* the circumference of the circle. The inscribed polygon is less than the circle; and it is also less than the triangle *abd*, because it is equal to a triangle of a height equal to CQ (which is less than CA or *ab*) upon a base equal to its perimeter ABDK which is less than the circumference of the circle *bd*. Therefore the circle and the triangle *abd* are both constantly limits betwixt the external and internal polygons EFGH, ABDK. Let the arch AB be bisected in L, and the tangent at L meet AE, BE, in M and N; and the angle ELM being right, EM must be greater than LM or AM, the triangle ELM greater than ALM, EMN greater than the sum of the triangles ALM, BLN, and consequently greater than half the space EALB bounded by the tangents EA, EB, and the arch ALB: From which it follows, (by the I.IO. Eucl. the foundation of this method,) that the circumscribed polygon may approach to the circle so as to exceed it by a less quantity than any that can be assigned. The inscribed polygon may also approach to the circle so as that their difference may become less than any assignable quantity, as is shewn in the Elements. Therefore the circle and the triangle *abd*, which are both limits betwixt these polygons, must be equal to each other. For, if the triangle *abd* be not equal to the circle, it must either be greater or less than it. If the triangle *abd* was greater than the circle; then, since the external polygon, by increasing the number of its sides, might be made to approach to the circle so as to exceed it by a quantity less than any difference that can be supposed to be between it and the triangle *abd*, it follows, that the external polygon might become less than that triangle, against | what has been 9|10
demonstrated. If the triangle *abd* was less than the circle, then the inscribed polygon, by being made to approach to the circle, might exceed that triangle: which, by what we have shewn, is also impossible.

In general, let any determined quantity AB be always a limit betwixt two variable quantities AP, AQ, which are supposed to approach continually to it and to each other, so that the difference of either from it may become less than

Fig 1

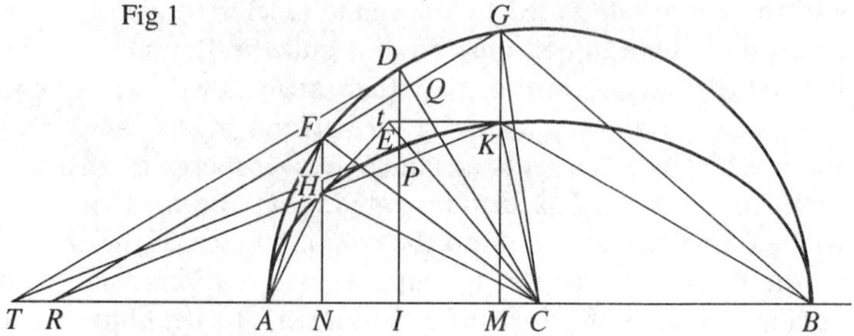

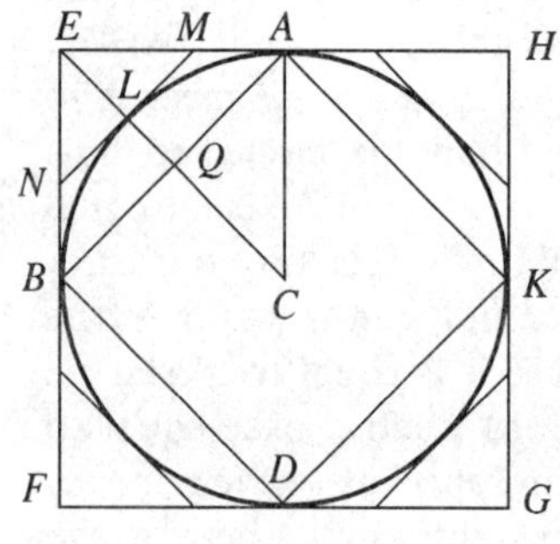

Fig 2

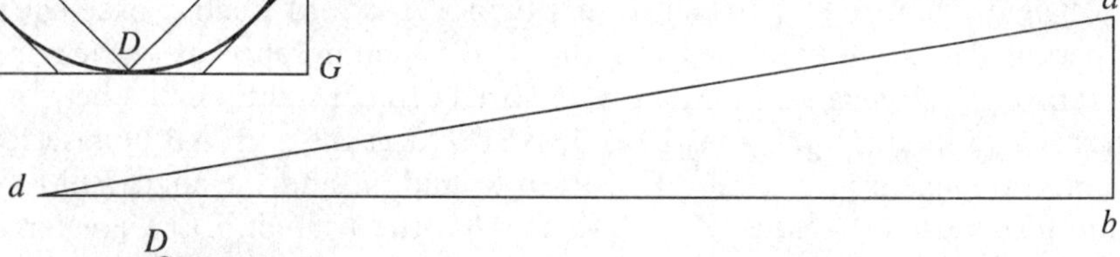

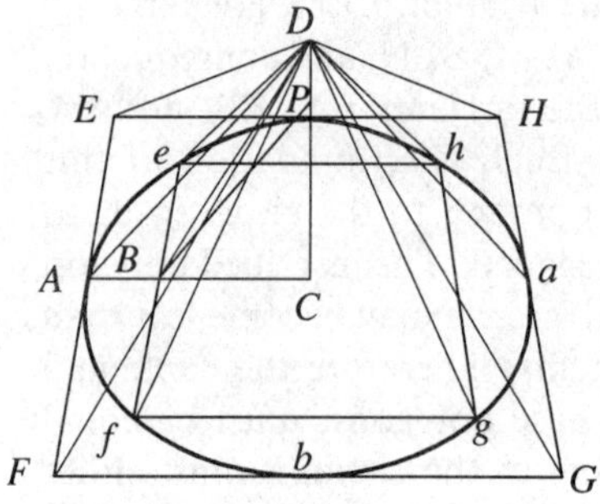

Fig 3

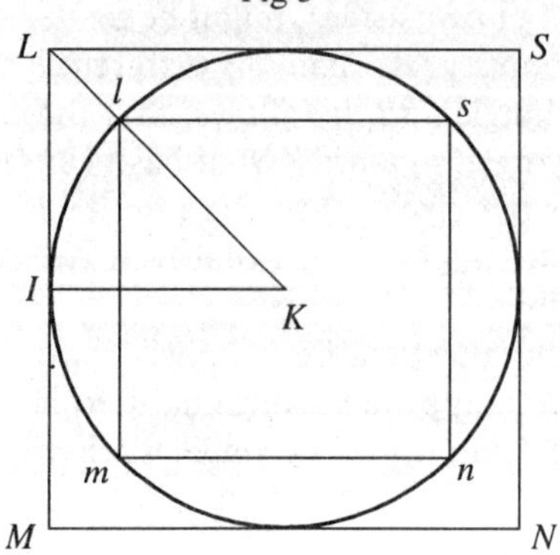

Fig 13

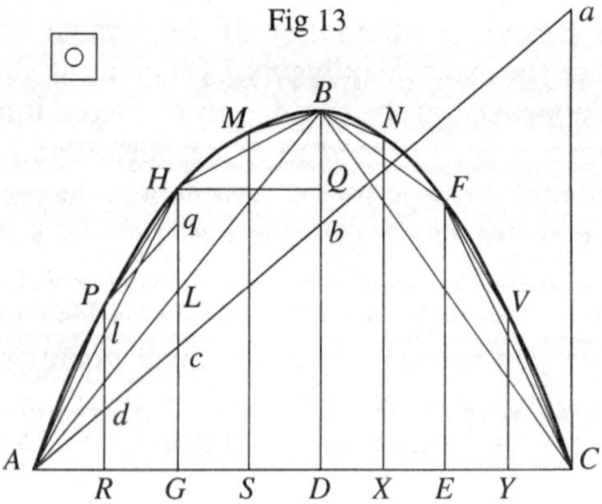

any assignable quantity, or so that the ratio of AQ to AP may become less than any assignable ratio of a greater magnitude to a lesser. Suppose also any other determined quantity *ab* to be always a limit betwixt the quantities *ap* and *aq*;

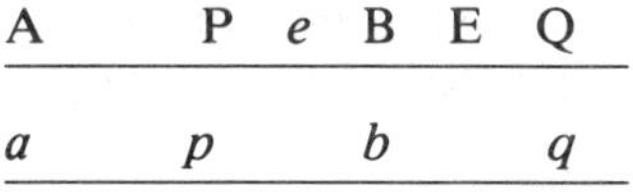

and *aq* being always equal to AQ or less than it, let *ap* be either equal to AP or greater than it. Then shall these limits AB and *ab* be equal to each other. For, if *ab* was equal to any quantity AE greater than AB, then, by supposing the ratio of AQ to AP to become less than that of AE to AB, the ratio of *aq* to AP would become less than the ratio of *ab* to AB: but since *aq* is always greater than *ab*, and AP less than AB, the ratio of *aq* to AP is always greater than the ratio of *ab* to AB; and these being repugnant, it follows, that *ab* is not greater than AB. If *ab* was equal to any quantity *Ae* less than AB, then, by supposing the ratio of AQ to AP to become less than that of AB to *Ae*, the ratio of AQ to *ap* (which, by the supposition, is either equal to AP, or greater than it) would become less than that of AB to *ab*: but since AQ always exceeds AB, and *ap* is always less than *ab*, the ratio of AQ to *ap* must be greater than that of AB to *ab*; and these being repugnant, it follows, that *ab* is not less than AB. These limits therefore AB, *ab* are equal to each other.

In this manner ARCHIMEDES demonstrates the fourteenth proposition of the first book of his admirable treatise concerning the sphere and cylinder. Let FIG. 3.
DA*ha* be an upright cone; D | the vertex, C the center and CA the radius 10|11
of the base. Let KI be a mean proportional betwixt DA the side of the cone and CA the radius of its base; and the convex surface of the cone shall be equal to a circle of the radius KI. Let LMNS be any external equilateral polygon described about this circle, *lmns* a similar polygon inscribed in it; EFGH a similar polygon described about the base of the cone, *efgh* a similar polygon described in it: then the surface of the pyramid DEFGH, which is described about the convex surface of the cone, shall be to the polygon EFGH as DA is to CA, or as the square of KI is to the square of CA, or as the polygon LMNS is to the polygon EFGH; and therefore the surface of the external pyramid DEFGH, without the base, is equal to the polygon LMNS. Let CA, perpendicular to EF, meet *cf* in B; and BR, parallel to DA, meet CD in R: then the polygon *lmns* being to *efgh* as the square of KI is to the square of CA, that is, as DA is to CA, or as BR is to CB; and this being a less ratio than that of DB to CB, which is the same as the ratio of D*efgh*, the surface of the internal pyramid without its base, to the internal polygon *efgh*: it follows, that the surface of the internal pyramid without its base is greater than the polygon *lmns*. The external polygon LMNS is greater than the circle of the radius KI, the internal polygon is less than that circle, and the ratio of the external polygon to the internal may become less than any assignable ratio of a greater quantity to a lesser. The surface of the external pyramid DEFGH, without the base, is greater than the convex surface of the cone; and the surface of the internal pyramid D*efgh*, without the base, is less than it. Therefore by substituting, in the general demonstration of the last article, the circle of the radius KI in place of the

quantity AB, the external polygon LMNS for AQ, the internal polygon *lmns* for AP, the convex surface of the cone for *ab*, the surfaces of the external and internal pyramids without their bases for *aq* and *ap*; it will appear, that the convex surface of the cone and the circle of the radius KI are equal to each other. We have given so full an account of this demonstration the rather that a part of
11|12 it has been sometimes misrepresented. |

It is much in the same manner that he demonstrates the thirty first, fortieth and forty first, or, according to the celebrated Dr. BARROW's numbers, the thirty seventh, forty ninth and fiftieth of the same treatise: which are esteemed by so good a judge, for the usefulness of the propositions, the subtlety of the invention, and the elegance of the demonstration, amongst the most valuable discoveries in geometry.

⟦For the next 21 pages, Maclaurin gives further examples of the Greek style of mathematical reasoning. He continues on page 33:⟧

We have now given a summary account, of the progress that was made by the ancients in measuring and comparing curvilineal figures, and of the method by which they demonstrated all their theorems of this kind. It is often said, that curve lines have been considered by them as polygons of an infinite number of sides. But this principle no where appears in their writings. We never find them resolving any figure, or solid, into infinitely small elements. On the contrary, they seem to avoid such suppositions, as if they judged them unfit to be received into geometry, when it was obvious that their demonstrations might have been sometimes abridged by admitting them. They considered curvilineal areas as the limits of circumscribed or inscribed figures of a more simple kind,
33|34 which approach to these | limits, (by a bisection of lines, or angles, that is continued at pleasure,) so that the difference betwixt them may become less than any given quantity. The inscribed or circumscribed figures were always conceived to be of a magnitude and number that is assignable; and from what had been shewn of these figures, they demonstrated the mensuration, or the proportions, of the curvilineal limits themselves, by arguments *ab absurdo*. They had made frequent use of demonstrations of this kind from the beginning of the Elements; and these are in a particular manner adapted for making a transition from right-lined figures to such as are bounded by curve lines. By admitting them only, they established the more difficult and sublime part of their geometry on the same foundation as the first elements of the science. Nor could they have proposed to themselves a more perfect model.

We have already observed, how solicitous ARCHIMEDES appears to be, that his demonstrations should be found to depend on those principles only that had been universally received before his time. In his treatise of the quadrature of the parabola, he treats of a progression whose terms decrease constantly in the proportion of four to one, which we expressed by the trapezia CD*ba*, DG*cb*,
FIG. 13 GR*dc*, *&c*. But he does not suppose this progression to be continued to infinity,

or mention the sum of an infinite number of terms; though it is manifest, that all which can be understood by those who assign that sum was fully known to him. He appears to have been more fond of preserving to the science all its accuracy and evidence, than of advancing paradoxes; and contents himself with demonstrating this plain property of such a progression. That the sum of the terms continued at pleasure, added to the third part of the last term, amounts always to four thirds of the first term; as the sum of the trapezia C*D*b*a*, DG*cb*, GR*dc*, added to one third part of the last trapezium GR*dc*, amounts always to four thirds of the first trapezium CD*ba*, or to the triangle CA*a*. Nor does he suppose the chords of the curve to be bisected to infinity; so that after an infinite bisection the inscribed polygon might be said to coincide with the parabola. These suppositions had been | new to the Geometricians in his time, 34|35
and such he appears to have carefully avoided.

He has demonstrated many other theorems of this kind from the properties of certain progressions, the terms of which correspond to the circumscribed and inscribed figures: but he never supposed these terms, or figures, to increase or decrease by infinitely small differences, and to become infinite in number; that their sum might be supposed equal to the curvilineal area, or solid. It was sufficient for his purpose, to assign a quantity that is always a limit betwixt the sum of all the terms of the progression, and the same sum without one of the extreme terms; as the area, or solid, is always a limit between the sum of the circumscribed and the sum of the inscribed figures, which sums differ from each other in the extreme figures only. He considered but one decreasing geometrical progression, and shewed how to find in an arithmetical progression the sum of the terms and of their squares only. Of late, other geometrical progressions have been employed successfully for measuring the areas of curves; the sums of the cubes, and of all the other powers, of the terms in an arithmetical progression have served for the same purpose: and each of his discoveries has produced some extensive theory in the modern geometry.

His method has been often represented as very perplexed, and sometimes as hardly intelligible. But this is not a just character of his writings, and the ancients had a different opinion of them*. He finds it necessary indeed to premise several propositions to the demonstration of the principal theorems; and on this account his method has been excepted against as tedious. But the number of steps is not the greatest fault a demonstration may have; nor is this number to be always computed from those that may be proposed in it, but from those that are necessary | to make it full and conclusive. 35|36
Besides, these preliminary propositions are generally valuable on their own account, and render our view of the whole subject more clear and compleat. In his treatise of the sphere and cylinder, for example, by his demonstrating

* Plutarch celebrates the simplicity and plainness with which he treats the most difficult and abstruse questions: *οὐ γὰρ ἔστιν ἐν γεωμετρίᾳ χαλεπωτέρας καὶ βαρυτέρας ὑποθέσεις ἐν ἁπλοτέροις λαβεῖν καὶ καθαρωτέροις στοιχείοις γραφομένας. Plut. in vita Marcelli.* ‖'For it is not possible to find in geometry more complex and profound questions discussed in simpler and clearer terms.'—Plutarch, *Life of Marcellus*, 307D.‖

so fully the mensuration of the surfaces and solids, generated by the internal and external polygons, we not only see how the surface and solid content of the sphere itself is determined, but we acquire a more perfect knowledge of this theory, and of all that relates to it, with a satisfaction that we are sensible is often wanting in the incompleat demonstrations of some other methods.

By so many valuable discoveries demonstrated in so accurate a manner, and by the admirable use he made of his knowledge in the celebrated siege of his native city*, and upon other occasions†, ARCHIMEDES has distinguished himself amongst the Geometricians, and has done the greatest honour to this part of learning. He has not however escaped the censures of some writers, who being unskilful in geometry, and unable to reconcile their own conceits with his demonstrations, have represented him as in an error, and misleading Mathematicians by his authority‖. But though Mathematicians may be grateful, authority has not any place in this science; and no Geometrician ever pretended, from the highest veneration for ARCHIMEDES, Sir ISAAC NEWTON or others,
36|37 to rest on their judgment | in a matter of geometrical demonstration. The pursuit of general and easy methods may have induced some to make use of exceptionable principles; and the vast extent, which the science has of late acquired, may have occasioned their proposing incompleat demonstrations. They may have also sometimes fallen into mistakes: but it will be found difficult to assign one false proposition that has been ever generally received by Geometricians; and it is hardly possible, that accusations of this nature can be more misplaced.

In what ARCHIMEDES had demonstrated of the limits of figures and progressions, there were valuable hints towards a general method of considering curvilineal figures; so as to subject them to mensuration by an exact quadrature, an approximation, or by comparing them with others of a more simple kind. Such methods have been proposed of late in various forms, and upon different principles. The first essays were deduced from a careful attention to his steps*.

---

* How he disconcerted all the efforts of two Roman armies, commanded by the Proconsul Marcellus and by Appius Claudius, in the siege of Syracuse, (till the city being taken by surprise and treachery, an end was put to his life and enquiries at once,) is described at length by Polybius, Livy, Plutarch, etc. He was called *πολυμήχανος* [of many wiles and tricks] *ἑκατόνχειρ* [hundred-handed]; and, according to Plutarch, acquired the reputation of more than human learning. Medals of Syracuse, with figures that are supposed to refer to his discoveries, serve rather to justify his countrymen from the reproach of ingratitude which some have imputed to them, than to do honour to the immortal Archimedes *Paru*[*tæ*] *Sicil. Spanhem. in orat. I. Juliani.* [The editor has not been able to locate the exact source of this reference; however, an edition of the writings of Julian the Apostate was published in 1741, with translations into French by Ezekiel Spanheim and into Latin by Petrus Cunaeus.]

† Diodorus Siculus tells us, (*lib.* 5.) that when Archimedes travelled into Egypt, he invented machines that were of great use to that nation, and procured him an universal reputation.

‖ Decepit illos auctoritas Archimedes, etc. *Hobbes de principiis et ratiocinatione Geometrarum* [*Hobbes 1845*]. The learned Joseph Scaliger and others have also writ against him.

* C'etoit en observant de prês la marche d'Archimede qu'il [*M. de Roberval*] etoit arrivé à cette sublime & merveilleuse science, etc. *Ouvrag. de l' Acad. Royal. 1693.* This is generally acknowledged by the writers of that time.

But, that his method might be more easily extended, its old foundation was abandoned, and suppositions were proposed which he had avoided. It was thought unnecessary to conceive the figures circumscribed or inscribed in the curvilineal area, or solid, as being always assignable and finite; and the precautions of ARCHIMEDES came to be considered as a check upon Geometricians, that served only to retard their progress. Therefore, instead of his assignable finite figures, indivisible or infinitely small elements were substituted; and these being imagined indefinite, or infinite, in number, their sum was supposed to coincide with the curvilineal area, or solid.

It was however with caution that these suppositions were at first employed in geometry by CAVALERIUS, the ingenious author of the method of indivisibles, and by others. He discovered a method, which he found to be of a very
extensive use, | and of an easy application, for measuring or comparing planes 37|38
and solids; and would not deprive the public of so valuable an invention. In preposing it, he strove to avoid* the supposing magnitude to consist of indivisible parts, and to abstract from the contemplation of infinity; but he acknowledged, that there remained some difficulties in this matter which he was not able to resolve. Therefore he subjoined more unexceptionable demonstrations to those he had deduced from his own principles; and the disputes which ensued (the first of any moment that were known between Geometricians) justified his precautions. Afterwards, infinitely small elements were substituted in place of his indivisibles; and various improvements were made in this doctrine. The method of ARCHIMEDES, however, was often kept in view, and frequently appealed to as the surest test of every new invention. The harmony betwixt the conclusions that arose from the old and new methods contributed not a little to the credit which the latter at first acquired; till being more and more relished, they came at length to be generally admitted on their own evidence, and seem'd to merit so favourable a reception, by the great advantages that were derived from them for resolving the most difficult problems, and demonstrating the most general theories, in a brief and easy manner.

But when the principles and strict method of the ancients, which had hitherto preserved the evidence of this science entire, were so far abandoned, it was difficult for the Geometricians to determine where they should stop. After they had indulged themselves in admitting quantities, of various kinds, that were not

---

* Quoad continui compositionem, manifestum est ex præostensis, ad ipsum ex indivisibilibus componendum nos minime cogi: solum enim continua sequi indivisibilium proportionem, & è converso, probare intentum fuit; quod quidem cum utraque positione stare potest. Tandem vero dicta indivisibilium aggregata non ita pertractavimus, ut infinitatis rationem propter infinitas lineas seu plana subire videantur, &c. *Cavalerii Geom. indivis. lib. 7 præf.* ⟦'As for the composition of a continuum, it is clear from what has been shown above that we are not in the least compelled to regard it as actually composed of indivisibilia: for it was only our intention to prove that continua behave analogously to indivisibilia and conversely; which indeed is consistent with either position. Indeed, we have not treated the said aggregates of indivisibilia in such a way that they appear to conform to the nature of infinity on account of their being infinite lines or planes, etc.'—*Cavalieri 1635*, Book 7, preface.⟧

assignable, in supposing such things to be done as could not possibly be effected, (against the constant practice of the ancients,) and had involved themselves in the mazes of infinity; it was not easy for them to avoid perplexity,
38|39 and sometimes error, | or to fix bounds to these liberties when they were once introduced. Curves were not only considered as polygons of an infinite number of infinitely little sides, and their differences deduced from the different angles that were supposed to be formed by these sides; but infinites and infinitesimals were admitted of infinite orders, every operation in geometry and arithmetic applied to them with the same freedom as to finite real quantities, and suppositions of this nature multipled, till the higher parts of geometry (as they were most commonly described) appeared full of mysteries.

From geometry the infinites and infinitesimals passed into philosophy, carrying with them the obscurity and perplexity that cannot fail to accompany them. An actual division, as well as a divisibility of matter *in infinitum*, is admitted by some. Fluids are imagined consisting of infinitely small particles, which are composed themselves of others infinitely less; and this subdivision is supposed to be continued without end. Vortices are proposed, for solving the phænomena of nature, of indefinite or infinite degrees, in imitation of the infinitesimals in geometry; that, when any higher order is found insufficient for this purpose, or attended with an insuperable difficulty, a lower order may preserve so favourite a scheme. Nature is confined in her operations to act by infinitely small steps. Bodies of a perfect hardness are rejected, and the old doctrine of atoms treated as imaginary, because in their actions and collisions they might pass at once from motion to rest, or from rest to motion, in violation of this law. Thus the doctrine of infinites is interwoven with our speculations in geometry and nature. Suppositions, that were proposed at first diffidently, as of use for discovering new theorems in this science with the greater facility, and were suffered only on that account, have been indulged, till it has become crowded with objects of an abstruse nature, which tend to perplex it and the other sciences that have a dependence upon it.

They who have made use of infinites and infinitesimals with the greatest liberty, have not agreed as to the truth and reality they would ascribe to them.
39|40 The celebrated Mr. LEIBNITZ | owns them to be no more than fictions. Others place them on a level with finite quantities, and endeavour to demonstrate their reality, from magnitude's being susceptible of augmentation and diminution without end, from the properties of the progressions of numbers that may be continued at pleasure, and from the infinity which some Geometricians have ascribed to the hyperbolic area. But in these arguments they seem to suppose the infinity which they would demonstrate.

It was a principle of the ancient Geometricians, That any given line may be produced, and its parts subdivided at pleasure: but they never supposed it to be produced, till it should become infinitely great; or to be subdivided, till its parts should become infinitely small. It does not necessarily follow, that, because any given right line may be continued further, it can be produced till it become actually infinite, or that we are able to conceive such a line to be described, so as to admit it in geometry. In general, magnitude is capable of being increased without end; that is, no term or limit can be assigned or supposed beyond which it may not be conceived to be further increased. But from this it cannot be inferred, that we are able to conceive or suppose magnitude to be really infinite*: or, if we | are able to join infinity to any supposed 40|41

* In a late treatise ascribed to a celebrated author, justly esteemed for his various writings, several arguments are proposed, for admitting magnitude actually infinite: not that kind which has no limits, comprehends all, and can receive no addition, which he calls *metaphysical*; but that which he defines to be greater than any finite magnitude, which he distinguishes from the former, and calls *geometrical*. *Puisque la grandeur est susceptible d' augmentation sans fin on la peut concevoir ou supposer augmentée une infinité des fois, c'est-à-dire qu'elle sera devenuë infinie. Et, en effet, il est impossible que la grandeur susceptible d' augmentation sans fin soit dans le même cas que si elle n'en etoit pas susceptible sans fin. Or, si elle ne l'etoit pas, elle demeureroit toujurs finie; donc etant susceptible d' augmentation sans fin, elle peut ne demeurer pas toujours finie, ou, ce qui est le même, devenir infinie.* Elem. de la geom. de l'infini, §83. [*Fontenelle 1727*] Because magnitude is susceptible of augmentation without end, the author concludes, that we may suppose it augmented an infinite number of times. But, by being susceptible of augmentation without end, we understand only, that no magnitude can be assigned or conceived so great but it may be supposed to receive further augmentation, and that a greater than it may still be assigned or conceived. We easily conceive that a finite magnitude may become greater and greater without end, or that no termination | or limit can be assigned of the increase which it may admit: but we do not 40|41 therefore clearly conceive magnitude increased an infinite number of times. Mr. Lock acknowledges, that we easily form an idea of the infinity of number, to the end of whose addition there is no approach: but he distinguishes betwixt this and the idea of an infinite number; and subjoins, that how clear soever our idea of the infinity of number may be, there is nothing more evident than the absurdity of the actual idea of an infinite number.

The latter part of the argument amounts to this: "It is impossible that magnitude being susceptible of augmentation without end, can be in the same case as if it was not susceptible of augmentation without end. But if it was not susceptible of augmentation without end, it would remain always finite. Therefore, since it is susceptible of augmentation without end, the contrary must be allowed; that is, it may not always remain finite, or it may become infinite." The Force of which argument seems to be taken off, by considering, that, if magnitude was not susceptible of augmentation without end, it would not only remain always finite, but there would neccessarily be a term, limit or degree of magnitude which could never be excceded, or there might be a greatest magnitude. And, by allowing that there is no such term or limit, magnitude is not supposed to be in the same case as if it was not susceptible of augmentation without end, though we should refuse that it may become infinite. What is opposite to the supposing magnitude susceptible of augmentation without end, is not the supposing it always finite, (for finite magnitude is capable of being increased without end;) but the supposing it susceptible of no augmentation at all, or of an augmentation that has a limit or end.

The series of numbers, 1, 2, 3, 4, &c. in their natural order, may be continued without end; and it is said, that "we never come nearer the end of the progression, how great soever the number may be to which we arrive; which is a character that cannot belong to a series of a finite number of terms. Therefore this natural series has an infinite number of terms." And it is added, that "though we can go over a finite number of terms only, yet all the terms of this infinite progression are equally real." But if we can conceive this series to have any end, it seems to be evident, that we must approach to this end as we proceed from the beginning towards it; and that, while we advance, the distance of any term from the end must decrease (whether this distance be called finite or infinite) by the same quantity as the distance from any subsequent term decreases, or the distance from the beginning of the series increases. If we cannot conceive the series to have an end, then we can have no idea of its last term. If we suppose this series to be continued to infinity, it would indeed be absurd, after such a supposition, to say that the number of its terms is finite. But, in treating this science strictly, it may perhaps be better to avoid this supposition. For if it is only a finite number of terms we can clearly conceive, how shall we judge of the reality of the rest?

idea of a determinate quantity, and to reason concerning magnitude actually
41|42 infinite, it is not surely with that perspicuity that is required in geometry. | In the same manner, no magnitude can be conceived so small, but a less than

---

or wherein shall we place the reality of those which it is impossible for us to assign? of which two
41|42 kinds are said to be in this same series, | each infinite in number; the first of which are said to be finite, but indeterminable; the latter, actually infinite.

The argument from the infinity of the hyperbolic area is much insisted on. "The hyperbolic area (*Elem. de la geom. de l'infin. pref.*) is as really infinite, as a determined parabolic area is two thirds of the circumscribed parallelogram. It is trifling to say, that the one can be actually described, and the other cannot. Geometry is entirely intellectual, and independent of the actual description and existence of the figures whose properties it discovers. All that is conceived necessary in it has the reality which it supposes in its object. Therefore the infinite which it demonstrates is as real as that which is finite, &c."

And the learned author, after insisting on this subject, concludes, that, "not to receive infinity as it is here represented, with all its necessary consequences, is to reject a geometrical demonstration; and that he who rejects one, ought to reject them all." But though the actual description of the figures which are considered in geometry be not necessary, yet it is requisite that we should be able clearly to conceive that they may exist; and a distinct idea of the manner how they may be supposed to be described or generated is necessary, that they may have a place in this science. Principles that are proposed as of the most extensive use, and as the foundation of all the sublime geometry, ought to be clear and unexceptionable. If this science is entirely intellectual, or if the reality of its objects is to be considered as having a dependence on their being conceived by the mind, it would seem that there must be a difference betwixt the reality of finite assignable lines or numbers, and the reality we can ascribe to infinite lines of numbers, which are not assignable, and cannot be supposed to be produced or generated but in a manner that is allowed to be inconceivable. As for what is said of the parabolic and hyperbolic areas, we can conceive any portion of the parabola to be accurately described, and its area to be determined, though no exact figure of this kind should ever exist. We can also conceive, that the hyperbola and its asymptote may be produced to any assignable distance: but we do not so clearly conceive that they may be produced to a distance greater than what is assignable; and we may well be allowed to hesitate at such a supposition in strict geometry. Any finite space being proposed, the hyperbolic area (terminated by the curve, the asymptote and a given ordinate) will exceed it by producing the curve and asymptote to an assignable distance; and there is no assignable limit in this (as in some other cases) which the area may not surpass in magnitude. Therefore it is said, that this area would be infinite, if the curve and asymptote could be infinitely produced. But no argument for admitting magnitude actually infinite can be deduced from this, which does not more easily appear from hence, that a parallelogram of a given height would be infinite if it could have an infinite base: from which it cannot be inferred that such a base or parallelogram can actually exist. It is often said, that a rectangle of a given height on an imaginary base (as the Analysts speak) is imaginary:
42|43 but we cannot thence infer, that an imaginary line or rectangle can exist. It is not however | our intention to maintain the impossibility of infinite magnitude; but to shew, that such doctrines are not necessary consequences of the received principles of this science, and not very proper to be admitted as the ground work of the high geometry.

As for the hyperbolic areas of a higher kind, which are said to be of a finite magnitude though infinitely produced, the meaning is, that there is a certain finite space which such an area never can equal, though the curve and its asymptote be produced never so far; to which however the area approaches, so that the excess of that finite space above it may become less than any space that may be proposed, by producing the curve and its asymptote to a distance that is assignable:
FIG. 13 As, the sum of the trapezia CD*ba*, DG*cb*, GR*ds*, &c. that are determined by bisecting AF continually, is always less than the triangle AC*a*, but approaches to it so that their difference AR*d* may become less than any given space O: Or, as the sum of the right lines CD, DG, GR, &c. is always less than CA, but approaches to it, so that, by continuing the bisection, the difference AR may become less than any assigned quantity. But we shall have occasion to treat of these afterwards more fully.

In the same treatise (§196.) a proof is offered, to shew, that, in the infinite series of numbers proceeding in their natural order, there are finite numbers whose squares become infinite, which are called indeterminable, and are supposed to occupy the obscure passage from the numbers that are assignable to those that are infinite. A greatest finite square is supposed in this progression,

it may be supposed; but we are not therefore able to conceive a quantity
infinitely small. A given magnitude | may be supposed to be divided into 42|43
any assignable number of parts; but it cannot therefore be conceived to be

---

and represented by *nn*; all that preceed it are finite, and all that follow after it are supposed infinite. The numbers in this progression between *n* and *nn*, being less than *nn*, are finite; but being greater than *n*, their squares are greater than *nn*, and therefore, by the supposition, are infinite. But how can we admit the supposition of a greatest finite square number, such as is here expressed by *nn*? The number *nn*, being finite, is not the next to it in the progression, (which exceeds it by unit only,) also finite. Should we allow, that a finite number becomes infinite by adding unit to it, or even by squaring it, how shall we distinguish finite from infinite? We commonly conceive finite magnitude to be assignable, or to be limited by such as are assignable, and to be susceptible of further augmentation: and therefore infinite magnitude would seem to imply, either that which exceeds all assignable magnitude, or that which cannot admit of any further augmentation; these being directly opposite to what we most clearly conceive of finite magnitude. But neither of these constitute the idea of infinite magnitude, as it must be understood in that treatise. The former is applicable to those numbers which the author calls finite and indeterminable; which, being supposed to produce infinite squares, must therefore exceed all assignable numbers whose squares are assignable and finite. The latter is ascribed to that infinite only which he calls metaphysical, and excludes from geometry. We are at a loss to form a distinct idea even of finite itself as it is here
understood; and it would seem, | that the more art and ingenuity is employed in penetrating into 43|44
the doctrine of infinites, it becomes the more abstruse.

A proof is offered *à posteriori* (§393.) to shew, that there are finite fractions in the series 1, $\frac{1}{2}$, $\frac{1}{3}$, $\frac{1}{4}$, $\frac{1}{5}$, &c. whose squares become infinitely little in the series 1, $\frac{1}{4}$, $\frac{1}{9}$, $\frac{1}{16}$, $\frac{1}{25}$, &c. The sum of the first series corresponds with the area included betwixt the common hyperbola and its asymptote; and is said to be infinite when the series is supposed to be continued to infinity. The sum of the latter series corresponds with the area of an hyperbola of a higher order, and is said to be finite, even when the series is supposed to be continued to infinity; because there is a limit which this sum can never equal, to which however it continually approaches, as we have already described. This being allowed, it is supposed farther, that there is an infinite number of finite terms in the first progression; and it is thence demonstrated, that there are finite fractions in the first series whose squares become infinitely little in the second, thus: "If it should be pretended, that all the finite terms in the first series have their squares finite in the second, there would be an infinite number of finite terms in the second as well as in the first; and the sums of both would be infinite: so that the contrary of an undoubted truth, that is universally received, would be demonstrated." If we could allow that there is an infinite number of finite terms in the first series, this argument might have some weight. But this is a supposition we cannot admit. For the denominator of any fraction in the first series is always equal to the number of terms from the beginning, and must be supposed infinite when the number of terms is supposed infinite; but a fraction that has unit for its numerator, and is supposed to have an infinite number for its denominator, cannot be supposed finite, but infinitely little: so that we cannot suppose an infinite number of terms in the first series to be finite. It is often said in this treatise, that there is an infinite number of finite terms in the natural series 1, 2, 3, 4, 5, &c. continued to infinity. But we are at a loss to conceive how this can be admitted; since, in any such progression, the last or greatest term is always equal to the number of terms from the beginning, and cannot be supposed finite when the number of terms is supposed infinite. There is an assignable limit which the sum of the terms of the second series never amounts to; but there is no assignable limit which the sum of the first series may not surpass, (as we shall shew afterwards;) and the sum of the terms of the first is greater than the sum of the corresponding terms of the second, in a ratio that by continuing the terms may exceed any assignable ratio of a greater magnitude to a lesser: and as this is easily understood and demonstrated, so there is no necessity for having recourse to such abstruse principles in order to account for it.

It is of no use to cite authorities on this subject, but as they may justify us in establishing so noble a part of geometry for avoiding principles that are so much contested. What Aristotle taught of infinite magnitude is well known. Mr. Leibnitz, who for obvious reasons cannot be suspected of any prejudice against the doctrine of infinites, expresses himself thus; *On s'embarasse dans*
les series | des nombres qui vont à l'infini. On conçoit un dernier terme, un nombre infini, ou 44|45
infiniment petit; mais tout cela ne sont que des fictions. Tout nombre est fini et assignable, toute ligne l'est de même. Essai de Theodicée, disc. prelim. §70.

43|44 divided into a number of parts greater than what is assignable. The | parts
of a given line may be supposed to be continually bisected till they become less
than any line that is proposed; and this is sufficient for completing the
44|45 demonstrations of the ancients. | But it is acknowledged by those who have
treated the doctrine of infinites in the fullest manner, that "there is something inconceivable in supposing an infinitely great or infinitely small number or figure to be produced or generated; and that the passage from finite to infinite is obscure and incomprehensible:" and therefore it is better for us, in treating of so strict a science as geometry, to abstract from these suppositions. The abstruse consequences, that have been deduced from them by ingenious men, may the rather induce us to beware of admitting them as necessary principles in this science, and to adhere to its ancient principles.

Mr. LOCK, who wrote his excellent essay, "that we might discover how far the
45|46 powers of the understanding reach, to | what things they are in any degree
proportionate, and where they fail us," observes, "that whilst men talk and dispute of infinite magnitudes, as if they had as compleat and positive ideas of them as they have of the names they use for them, or as they have of a yard, or an hour, or any other determinate quantity, it is no wonder if the incomprehensible nature of the thing they discourse of, or reason about, leads them into perplexities and contradictions; and their minds be overlaid by an object too large and mighty to be surveyed and managed by them." Mathematicians indeed abridge their computations by the supposition of infinites; but when they pretend to treat them on a level with finite quantities, they are sometimes led into such doctrines as verify the observation of this judicious author. To mention an instance or two: The progression of the numbers 1, 2, 3, 4, 5, & c. in their natural order, is supposed to be continued to infinity, till by the continual addition of units an infinite number is produced, which is conceived to be the termination of this series. This infinite number is supposed to be still capable of augmentation and diminution; and yet it is said, "that it is neither increased nor diminished by the addition or subtraction of the same units from which it was supposed to be generated." In a progression of this kind, the number

---

We have subjoined these remarks, at the desire of some persons for whom we have a great regard, to shew why we have not followed an author who has merited so well of Mathematicians, and who on every other occasion has been justly applauded for his clear and distinct way of explaining the abstruse geometry. They who treated of infinites before him proceeded, as he observes, with a timorousness which the contemplation of such an object naturally inspires: *Quand on y etoit arrivé*, (says he) *on s' arrestoit avec une espece d'effroy et de fainte horreur—on regardoit l'infini comme un mistere qu'il falloit respecter, et qu'il n' etoit pas permis d' approfondir*. They stopt when they came to infinity with a sort of holy dread, and respected it as an incomprehensible mystery. He adventures farther, in order to discover the source, and penetrate into the first principles of geometrical truth. Infinity, according to him, is the great trunk from which its various branches are derived, and to which they all lead. In this great pursuit he displays infinite and finite with a freedom that puts us in mind of the ancient Poet and his Gods; whom he represents with the passions of men, and mingles in their battles. We doubt not, that if a full and perfect account of all that is most profound in the high geometry could have been deduced from the doctrine of infinites, it might have been expected from this author: But our ideas of infinites are too obscure and unadequate to answer this end; and there are many things advanced by all those who have applied them with great freedom in geometry, that give ground to a remark like to Mr. de St. Evre-

of terms is always equal to the last or greatest term, and is finite when the last term is finite. If the number of terms be supposed infinite, the last term cannot be finite; and yet it is said, "that in such a progression continued to infinity there is an infinite number of finite terms." It is evident, that no finite number can become infinite by the addition of unit or of any other finite number; and yet "a greatest finite square number is supposed in such a progression, the next to which (though it exceed that finite number by an unit only) is supposed infinite." From these suppositions it is infered, "that in such a progression continued to infinity there are finite numbers whose squares become infinite;" though it seems very evident, that a finite number taken any finite number of times can never produce more than a finite number. We may perceive from these instances, that it is not by founding the higher goemetry on the docrine
of infinites we can propose to | avoid the apparent inconsistencies that have 46|47
been objected to it; and since an excellent author, who has always distinguished himself as a clear and acute writer, has had no better success in establishing it on these principles, it is better for us to avoid them. These suppositions however may be of use, when employed with caution, for abridging computations in the investigation of theorems, or even for proving them where a scrupulous exactness is not required; and we would not be understood to affirm, that the methods of indivisibles and infinitesimals, by which so many uncontested truths have been discovered, are without a foundation. We acknowledge further, that there is something marvellous in the doctrine of infinites, that is apt to please and transport us; and that the method of infinitesimals has been prosecuted of late with an acuteness and subtlety not to be parallelled in any other science. But geometry is best established on clear and plain principles; and these speculations are ever obnoxious to some difficulties. If the accuracy has been always required in this science, in reasoning concerning finite quantities, we apprehend that Geometricians cannot be too scrupulous in admitting or treating infinites, of which our ideas are so imperfect. Philosophy probably will always have its mysteries. But these are to be avoided in geometry: and we ought to guard against abating from its strictness and evidence the rather, that an absurd philosophy is the natural product of a vitiated geometry.

It is just at the same time to acknowledge, that they who first carried geometry beyond its ancient limits, and they who have since enlarged it, have done great service, by describing plainly the methods which they found so advantageous for this purpose, (though they might appear exceptionable in some respects,) that others might proceed with the same facility to improve it. Some of them have been so cautious as to verify their discoveries by demonstrations in the strictest form; and others were able to have done this, had they not chose rather to

mond's; when he observes, that "it is surprising to find the ancient Poets so scrupulous to preserve probability in actions purely human, and so ready to violate it in representing the actions of the Gods." Some have not only admitted infinites and infinitesimals of infinite orders, but have distinguished even nothings into various kinds: and if such liberties continue, it is not easy to foresee
what absurdities may be advanced as discoveries in what is called the sublime geometry. | 46|47

employ their time in extending the science. At first, the variation from the ancient method was not so considerable, but that it was easy to have recourse
47|48 to it, when it should be thought necessary | for the satisfaction of such as required a scrupulous exactness. The Geometricians in the mean time made great improvements. They had the accurate method and examples of ARCHIMEDES before them, by which they might try their discoveries. These served to keep them from error, and the new methods facilitated their progress. Thus their views enlarged; and problems, that appeared at first sight of an insuperable difficulty, were afterwards resolved, and came at length to be despised as too simple and easy. The mensuration of parabolas, hyperbolas, spirals of all the higher orders, and of the famous cycloid, were amongst the earliest productions of this period; some of which seem to have been discovered by several Geometricians almost at the same time. It is not necessary for our purpose to describe more particularly what discoveries were made by TORRICELLI, Mess. de FERMAT and de ROBERVAL, GREGORY à Sto. Vincentio, *&c.* by whom the theorems of ARCHIMEDES were continued, and applied to the mensuration of various figures.

The *Arithmetica infinitorum* of Dr. WALLIS was the fullest treatise of this kind that appeared before the invention of the method of fluxions. ARCHIMEDES had considered the sums of the terms in an arithmetical progression, and of their squares only, (or rather the limits of these sums, described above,) these being sufficient for the mensuration of the figures he had examined. Dr. WALLIS treats this subject in a very general manner, and assigns like limits for the sums of any powers of the terms, whether the exponents be integers or fractions, positive or negative. Having discovered one general theorem that includes all of this kind, he then compounded new progressions from various aggregates of these terms, and enquired into the sums of the powers of these terms, by which he was enabled to measure accurately, or by approximation, the areas of figures without number. But he composed this treatise (as he tells us) before he had examined the writings of ARCHIMEDES, and he proposes his theorems and demonstrations in a less accurate form. He supposes the progressions to be continued to infinity, and investigates, by a kind of induction, the proportion
48|49 of the sum of the powers to the product that would arise by taking the | greatest power as often as there are terms. His demonstrations, and some of his expressions (as when he speaks of quantities more than infinite) have been excepted against. But it was not very difficult to demonstrate the greatest part of his propositions in a stricter method; and this was effected afterwards by himself and others in various instances. He chose to describe plainly a method which he had found very commodious for discovering new theorems; and it must be owned, that this valuable treatise contributed to produce the great improvements which soon followed after. A like apology may be made for others who have promoted this doctrine since his time, but have not given us rigid demonstrations. In general, it must be owned, that if the late discoveries were deduced at length, in the very same method in which the ancients demonstrated their theorems, the life of man could hardly be sufficient for considering them all: so that a general and concise method, equivalent to theirs in

accuracy and evidence, that comprehends innumerable theorems in a few general views, may well be esteemed a valuable invention.

CAVALERIUS was sensible of the difficulties, as well as the advantages that attended his method. He speaks as if he foresaw that it should be afterwards delivered in an unexceptionable form, that might satisfy the most scrupulous Geometrician; and leaves this *Gordian knot*, as he expresses himself, to some *Alexander*. Its form indeed was soon altered, and many improvements were made by the Mathematicians who prosecuted it since his time that deserve to be mentioned with esteem. But the method still remained liable to some exceptions, and was thought to be less perfect than that of the ancients on several grounds.

Sir ISAAC NEWTON accomplished what CAVALERIUS wished for, by inventing the method of Fluxions, and proposing it in a way that admits of strict demonstration, which requires the supposition of no quantities but such as are finite, and easily conceived. The computations in this method are the same as in the method of infinitesimals; but it is founded on accurate principles, agreeable to the ancient geometry. In it, the premisses | and conclusions are 49|50
equally accurate, no quantities are rejected as infinitely small, and no part of a curve is supposed to coincide with a right line. The excellency of this method has not been so fully described, or so generally attended to, as it seems to deserve; and it has been sometimes represented as on a level in all these respects with the method of infinitesimals. The chief design of the following treatise is, to shew its advantages in a clearer and fuller light, and to promote the design of the great inventor, by establishing the higher geometry on plain principles, perfectly consistent with each other and with those of the ancient Geometricians.

The method of demonstration which we make most use of in this treatise, was first suggested to us from a particular attention to Sir ISAAC NEWTON's brief reasoning in that place of his principles of philosophy where he first published the elements of this doctrine. After the greatest part of the following treatise was writ, we had the pleasure to observe, that Geometricians of the first rank had recourse to it long ago on several occasions, as a method of the strictest kind. Mr. de FERMAT, in a letter to GASSENDUS, and Mr. HUYGENS, in his *Horologium oscillatorium*, have employed it for completing the demonstrations of some theorems that were proposed by GALILEUS, and proved by him in a less accurate manner; and Dr. BARROW has demonstrated by it a theorem concerning the tangents of curve lines. The approbation which it appears to have had from so good judges, encouraged us to publish the following treatise; where it is applied for demonstrating the method of Fluxions. The chief pursuit of Geometricians for some time has been to improve their general methods. In proportion as these are valuable, it is important that they be established above all exception: and since they save us so much time and labour, we may allow the more for illustrating these methods themselves. | 50|51

THE ELEMENTS OF THE Method of FLUXIONS,

Demonstrated after the Manner of the

Ancient Geometricians.

BOOK I.

Of the Fluxions of Geometrical Magnitudes.

CHAP. I.

Of the Grounds of this Method.

1. The mathematical sciences treat of the relations of quantities to each other,
and of all their affections that can be subjected to rule or measure. They treat
of the properties of figures that depend on the position and form of the lines
or planes that bound them, as well as those that depend on their magnitude;
of the direction of motion, as well as its velocity; of the composition and resolu-
tion of quantities, and of every thing of this nature that is susceptible of a
51|52 regular determination. We enquire | into the relations of things, rather than
their inward essences, in these sciences. Because we may have a clear conception
of that which is the foundation of a relation, without having a perfect or
adequate idea of the thing it is attributed to*, our ideas of relations are often
clearer and more distinct than of the things to which they belong; and to this
we may ascribe in some measure the peculiar evidence of the mathematics. It
is not necessary that the objects of the speculative parts should be actually
described, or exist without the mind; but it is essential, that their relations
should be clearly conceived, and evidently deduced: and it is useful, that we
should chiefly consider such as correspond with those of external objects, and
may serve to promote our knowledge of nature.

2. In our pursuits after knowledge, we sometimes consider things as they appear to be in themselves, sometimes we judge of them from their causes, and sometimes by their effects. In ordinary enquiries, but especially in philosophy,

---

* Essay concerning the human understanding, book 2. chap. 25. §8.

we employ one or more of these methods according as we find ground for applying them. The two last may be no less satisfactory than the first, when there is a sufficient foundation for them; and by carrying our enquiries to the springs and principles of things, our knowledge of them becomes more perfect, and our views more extensive. In geometry, there are various ways of discovering the affections and relations of magnitudes that correspond to these general methods of enquiry. In the common geometry, we suppose the magnitudes to be already formed, and compare them or their parts, immediately, or by the intervention of others of the same kind, to which they have a relation that is already known. In the doctrine which we propose to explain and demonstrate in this treatise, we have recourse to the genesis of quantities, and either deduce their relations, by comparing the powers which are conceived to generate them; or, by comparing the quantities that are generated, we discover the relations of these powers and of any quantities that are supposed to be represented by them. The power by which magnitudes are conceived to be generated in geometry, is motion; and therefore we must begin with some account of it. | 52|53

3. No quantities are more clearly conceived by us than the limited parts of space and time. They consist indeed always of parts; but of such as are perfectly uniform and similar. Those of space exist together; those of time flow continually: but by motion they become the measures of each other reciprocally. The parts of space are permanent; but being described successively by motion, the space may be conceived to flow as the time. The time is ever perishing; but an image or representation of it is preserved and presented to us at once in the space described by the motion.

4. Time is conceived to flow always in an uniform course, that serves to measure the changes of all things. When the space described by motion flows as the time, so that equal parts of space are described in any equal parts of the time, the motion is uniform; and the velocity is measured by the space that is described in any given time. As this space may be conceived to be greater or less, and to be susceptible of all degrees of assignable magnitude; so may the velocity of the motion by which we suppose the space to be always described in a given time. The velocity of an uniform motion is the same at any term of the time during which it continues. But motion is susceptible of the same variations with other quantities, and the velocity in other instances may increase or decrease while the time increases. In these cases, however, the velocity at any term of the time is accurately measured by the space that would be described in a given time, if the motion was to be continued uniformly from that term.

5. Any space and time being given, a velocity is determined by which that space may be described in that given time: And, conversely, a velocity being given, the space which would be described by it in any given time is also determined. This being evident, it does not seem to be necessary, in pure geometry, to enquire further what is the nature of this power, affection or mode, which is called *Velocity*, and is commonly ascribed to the body that is supposed to move. It seems to be sufficient for our purpose, that while a body is supposed in motion, it must be conceived to have some velocity or other at any term of

53|54 the time during which it moves, and that we can demonstrate accurately | what are the measures of this velocity at any term, in the enquiries that belong to this doctrine, as will appear in the course of this treatise; especially since it is the business of geometry, as we have observed already, to enquire into the measures, rather than unfold the hidden essences of things.

6. But perhaps this explication will not be thought sufficient, and it will be required that we should propose a definition of velocity in form. The excellent Dr. BARROW defines it to be the power by which a certain space may be described in a certain time. Some perhaps may scruple to ascribe power to a body, figure or point in motion. But it is to be observed, that it is of no consequence, in pure geometry, to what the power may be most properly attributed. It is indeed generally allowed, that if a body was to be left to itself from any term of the time of its motion, and was to be affected by no external influence after that term, it would proceed for ever with an uniform motion, describing always a certain space in a given time: and this seems to be a sufficient foundation for ascribing, in common language, the velocity to the body that moves, as a power. It is well known, that what is an effect in one respect, may be considered as a power or cause in another; and we know no cause in common philosophy, but what is itself to be considered as an effect: but this does not hinder us from judging of effects from such causes. However, if any dislike this expression, they may suppose any mover or cause of the motion they please, to which they may ascribe the power, considering the velocity as the action of this power, or as the adequate effect and measure of its exertion, while it is supposed to produce the motion at every term of the time. We have observed already, that the principles of this method are analogous to the general doctrine of powers, or may be considered as a particular application of it. As a power which acts continually and uniformly is measured by the effect that is produced by it in a given time, so the velocity of an uniform motion is measured by the space that is described in a given time. If the action of the power vary, then its exertion at any term of the time is not measured by the effect that is actually produced after that term in a given time, but by the effect that would have been
54|55 produced | if its action had continued uniform from that term: and, in the same manner, the velocity of a variable motion at any given term of time is not to be measured by the space that is actually described after that term in a given time, but by the space that would have been described if the motion had continued uniformly from that term. If the action of a variable power, or the velocity of a variable motion, may not be measured in this manner, they must not be susceptible of any mensuration at all. It will appear afterwards, in the course of this treatise, that the other principles of this method correspond with the plain maxims of the general doctrine of powers that are employed by us on every occasion, and are to be reckoned amongst the most common and evident notions. There are two fundamental principles of this method. The first is, That when the quantities which are generated are always equal to each other, the generating motions must be always equal. The second is the converse of the first, That when the generating motions are always equal to each other, the quantities that are generated in the same time must be always equal. The

first is the foundation of the direct method of fluxions; the second, of the inverse method. But it is obvious, that they may be considered as cases of these two general principles: When the effects produced by two powers are always equal to each other, then (supposing that no other power of any kind affects their operations) these powers must be supposed to act equally at any term of the time; and, conversely, When the actions of two powers are always equal to each other at any term of the time, then the effects produced by them in the same time must be always equal.

7. This method is so well founded, that its rules and operations may be delivered in a way consistent with any general principles that are not repugnant to the most evident notions; though it is impossible for us, in treating of it, to keep to expressions that may appear equally consistent with every scheme of metaphysics. It has been frequently considered in a manner agreeable to the principles of those who suppose quantities to consist of indivisible or infinitely small elements. We are to proceed upon more strict and rigid principles: but it will be hardly possible | for us to avoid always such expressions as may 55|56
have been some time or other matter of dispute amongst philosophers. Their controversies concerning continued and discrete quantity have not been thought to weaken the evidence of the common geometry. Nor can their disputes* concerning motion affect the certainty of this method; since we have occasion in it for no more than the most obvious notions of space, time, motion and velocity, that cannot be said to yield in clearness and evidence to the principles of the common geometry.

8. When we suppose that a body has some velocity or other at any term of the time during which it moves, we do not therefore suppose that there can be any motion in a term, limit or moment of time, or in an indivisible point of space: and as we shall always measure this velocity by the space that would be described by it continued uniformly for some given finite time, it surely will not be said that we pretend to conceive motion or velocity without regard to space and time.

9. But to proceed: When any quantity is proposed, all others of the same kind may be conceived to be generated from it; such as are greater than it, by supposing it to be increased; such as are less, by supposing it to be diminished. In the common arithmetic, integer numbers are conceived to be produced by adding a given quantity or unit to itself continually, and fractions are produced by supposing it to be divided into such parts as by a like addition would generate the given quantity itself. But in geometry, that all degrees of magnitude may be produced, and in such a way as may found a general method of deriving their affections from their genesis, we conceive the quantities to be increased

* De natura motus et recta definitione, de causis ac differentiis, complura subtiliter argutantur Physici; quarum ferè Mathematicis nihil cordi vel curae: sufficere potest his quae communis sensus agnoscit. *Barrow, lect geom*. I. [‘The Natural Philosophers dispute much and with great subtlety about the nature and correct definition of motion, about causes and distinctions: but for this Mathematicians have little taste or time; sufficient for them is what common sense recognizes’ (*Barrow 1670*, I).]

and diminished, or to be wholly generated by motion, or by a continual flux analogous to it. The quantity that is thus generated, is said to flow, and called *a Fluent*.

10. Lines are generated by the motion of points; surfaces, by the motion of lines; solids, by the motion of surfaces; angles, by the rotation of their sides; the flux of time being supposed to be always uniform. The velocity with which
56|57 a line flows, | is the same as that of the point which is supposed to describe or generate it. The velocity with which a surface flows, is the same as the velocity of a given right line, that, by moving parallel to itself, is supposed to generate a rectangle which is always equal to the surface. The velocity with which a solid flows, is the same as the velocity of a given plain surface, that, by moving parallel to itself, is supposed to generate an erect prism or cylinder that is always equal to the solid. The velocity with which an angle flows, is measured by the velocity of a point, that is supposed to describe the arch of a given circle, which always subtends the angle, and measures it. In general, all quantities of the same kind (when we consider their magnitude only, and abstract from their position, figure, and other affections) may be represented by right lines, that are supposed to be always in the same proportion to each other as these quantities. They are represented by right lines in this manner in the Elements, in the general doctrine of proportion, and by right lines and figures in the *Data* of EUCLID*. In this method likewise, quantities of the same kind may be represented by right lines, and the velocities of the motions by which they are supposed to be generated, by the velocities of points moving in right lines. All the velocities we have mentioned are measured, at any term of the time of the motion, by the spaces which would be described in a given time, by these points, lines or surfaces, with their motions continued uniformly from that term.

11. The velocity with which a quantity flows, at any term of the time while it is supposed to be generated, is called its *Fluxion* which is therefore always measured by the increment or decrement that would be generated in a given time by this motion, if it was continued uniformly from that term without any acceleration or retardation: or it may be measured by the quantity that is generated in a given time by an uniform motion which is equal to the generating motion at that term.

12. Time is represented by a right line that flows uniformly, or is described by an uniform motion; and a moment or termination of time is represented by
57|58 a point or termination of | that line. A given velocity is represented by a given line, the same which would be described by it in a given time. A velocity that is accelerated or retarded, is represented by a line that increases or decreases in the same proportion. The time of any motion being represented by the base of a figure, and any part of the time by the corresponding part of the base; if the ordinate at any point of the base be equal to the space that would be described, in a given time, by the velocity at the corresponding term of the time

* See the preface to the *Data* by Marinus, near the end [*Euclid 1896*].

continued uniformly, then any velocity will be represented by the corresponding ordinate. The fluxions of quantities are represented by the increments or decrements described in the last article which measure them; and, instead of the proportion of the fluxions themselves, we may always substitute the proportion of their measures.

13. When a motion is uniform, the spaces that are described by it in any equal times are always equal. When a motion is perpetually accelerated, the spaces described by it in any equal times that succeed after one another, perpetually increase. When a motion is perpetually retarded, the spaces that are described by it in any equal times that succeed after one another, perpetually decrease.

14. It is manifest, conversely, that if the spaces described in any equal times are always equal, then the motion is uniform. If the spaces described in any equal times that succeed after one another perpetually increase, the motion is perpetually accelerated: For it is plain, that if the motion was uniform for any time, the spaces described in any equal parts of this time would be equal; and if it was retarded for any time, the spaces described in equal parts of this time that succeed after one another would decrease: both of which are against the supposition. In like manner it is evident, that a motion is perpetually retarded, when the spaces that are described in any equal times that succeed after one another perpetually decrease. The following Axioms are as evident as that a greater or less space is described in a given time, according as the velocicy of the motion is greater or less. | 58|59

AXIOM I.

15. *The space described by an accelerated motion is greater than the space which would have been described in the same time, if the motion had not been accelerated, but had continued uniform from the beginning of the time.*

AXIOM II.

The space described by a motion while it is accelerated, is less than the space which is described in an equal time by the motion that is acquired by that acceleration continued uniformly.

AXIOM III.

The space described by a retarded motion is less than the space which would have been described in the same time, if the motion had not been retarded, but had continued uniform from the beginning of the time.

AXIOM IV.

The space described by a motion while it is retarded, is greater than the space which is described in an equal time by the motion that remains after that retardation, continued uniformly.

3
Jean Lérond D'Alembert (1717–1783)

D'Alembert contributed several articles on mathematics to the famous *Encyclopedia* of which he and Denis Diderot were the principal editors (*Diderot et alii. 1751–72*); he also wrote the philosophical introduction, the *Discours préliminaire de l'Encyclopédie* (*D'Alembert 1751*), which contains a lengthy and influential treatment of the philosophical views of Descartes, Bacon, Newton, and Locke, and which became a manifesto for the *philosophes* of the French enlightenment.

D'Alembert was the illegitimate son of a French cavalry officer and a well-known salon hostess. He was abandoned at birth and raised in humble surroundings by an artisan and his wife; he had little formal education in mathematics, but studied on his own the works of Newton, L'Hospital, and the Bernoullis. In the 1740s D'Alembert published extensively and influentially on topics in mathematical physics. He gave a systematic treatment of the science of mechanics in his *Traité de dynamique* (1743); pioneered the use of partial differential equations in physics; and published important memoirs on celestial mechanics, the three-body problem, and vibrating strings.

The following excerpts from D'Alembert's articles in the *Encyclopedia* illustrate the state of reasoning about the metaphysics of the calculus in the middle of the eighteenth century. The article *differential* appeared in the *Encyclopedia* in 1754 (Vol. 4); the first half of the translation of that article is by William Ewald; the second half is translated by Dirk Struik and is reprinted from *Struik 1969*. The articles *infinite* and *limit* both appeared in the *Encyclopedia* in 1765 (Volumes 8 and 9 respectively); the translations are by William Ewald. D'Alembert's cross-references to other articles in the *Encyclopedia* have been left in the original French so that the articles can be more easily found.

A. DIFFERENTIAL (*D'ALEMBERT 1754*)

DIFFERENTIAL, *adj*. In higher *geometry* one calls a quantity that is infinitely small, or smaller than any assignable quantity, a *differential* quantity, or simply a *differential*.

One calls it a *differential* or a *differential* quantity because one usually considers it as the infinitely small difference of two finite quantities, where one surpasses the other by an infinitely small amount. Newton and the English call it a fluxion because they consider it as the momentary increase of a quantity. *See* FLUXION etc. ‖in this *Encyclopédie*‖. Leibniz and others also call it an *infinitely small quantity*.

DIFFERENTIAL CALCULUS is the method of differentiating quantities, that is to say, of finding the infinitely small difference of a finite variable quantity.

This method is one of the most beautiful and fertile in all of mathematics; Leibniz, who was the first to publish it, calls it the *differential calculus*, considering infinitely small quantities as the differences of finite quantities; this is why he expresses them by the letter d that he prefixes to the differentiated quantity; so the *differential* of x is expressed dx, that of y by dy, etc.

Newton calls the *differential* calculus the method of fluxions because, as we said, he takes the infinitely small quantities to be fluxions or momentary increases. For example, he considers a line as engendered by the fluxion of a point, a surface by the fluxion of a line, a solid by the fluxion of a surface, and instead of the letter d he designates fluxions by a point above the differentiated quantity. For example, for the fluxion of x he writes $\dot{x}$, for that of y, $\dot{y}$ etc.; and this is the solitary difference between the *differential* calculus and the method of fluxions. See FLUXION.

One can reduce all the rules of the *differential* calculus to the following:

1. The difference of the sum of two quantities is equal to the sum of their differences. So $d(x + y + z) = dx + dy + dz$.

2. The difference of xy is $ydx + xdy$.

3. The difference of x^m where m is a positive integer is $mx^{m-1}dx$.

There is no quantity one cannot differentiate by these three rules. Consider, for example, $x/y = xy^{-1}$. *See* EXPOSANT. The *differential* (*rule 2*) is $y^{-1}dx - xd(y^{-1}) =$ (*rule 3*)

$$\frac{dx}{y} - \frac{xdy}{y^2} = \frac{ydx - xdy}{y^2}.$$

The *differential* of $z^{(1/q)}$ is $(1/q)z^{(1/q-1)}\,dz$. For if $z^{(1/q)} = x$ one has $z = x^q$ and $dz = qx^{(q-1)}dx$ and

$$dx = \frac{dz}{q}x^{-q+1} = \frac{dz}{q}z^{-1+1/q}.$$

Similarly,

$$\sqrt{xx + yy} = \overline{xx + yy}^{\,1/2},$$

of which the *differential* is

$$\tfrac{1}{2}(2xdx + 2ydy)(xx + yy)^{-1/2} = \frac{xdx + ydy}{\sqrt{xx + yy}}$$

and similarly for the rest.

The above three rules are proven in an exceedingly clear manner in many works, above all in the first section of de l'Hospital's analysis of the *infinitely small*, to which we shall return. This section lacks the *differential* calculus of logarithms and exponents, which one can find in the first volume of the *Works* of Jean Bernoulli and in the first part of the younger de Baugainville's *Traité du calcul intégral*. One should consult these works, which are in the hands of all the world. *See* EXPONENTIEL. What concerns us most here is the metaphysics of the *differential* calculus.

This metaphysics, of which so much has been written, is even more important and perhaps more difficult to explain than the rules of this calculus themselves: various mathematicians, among them Rolle,[a] who were unable to accept the assumption concerning infinitely small quantities, have rejected it entirely, and have held that the principle was false and capable of leading to error. Yet in view of the fact that all results obtained by means of ordinary Geometry can be established similarly and much more easily by means of the *differential* calculus, one cannot help concluding that, since this calculus yields reliable, simple, and exact methods, the principles on which it depends must also be simple and certain.

Leibniz was embarrassed by the objections he felt to exist against infinitely small quantities, as they appear in the *differential* calculus; thus he preferred to reduce infinitely small to merely incomparable quantities. This, however, would ruin the geometric exactness of the calculations; is it possible, said Fontenelle,[b] that the authority of the inventor would outweight the invention itself? Others, like Nieuwentijt,[c] admitted only *differentials* of the first order and rejected all others of higher order. This is impossible; indeed, considering an infinitely small chord of first order in a circle, the corresponding abcissa or versed sine is infinitely small[d] of second order; and if the chord is of the second order, the abscissa mentioned will be of the fourth order, etc. This is proved easily by elementary geometry, since the diameter of a circle (taken as a finite quantity) is always to the chord as the chord to the corresponding abscissa.[e] Thus, if one admits the infinitely small of the first order, one must admit all the others, though in the end one can rather easily dispense with all this metaphysics of the infinite in the *differential* calculus, as we shall see below.

Newton started out from another principle; and one can say that the metaphysics of this great mathematician on the calculus of fluxions is very exact and illuminating, even though he allowed us only an imperfect glimpse of his thoughts.

---

[a] ‖Michael Rolle (1652–1719), member of the French Academy, is best known for the theorem in the theory of equations called after him. In 1700 he took part in a debate in the French Academy on the principles of the calculus; see (*Boyer 1939*, 241).‖

[b] ‖Bernard le Bovier de Fontenelle (1657–1757) was a predecessor of D'Alembert as *secrétaire perpétuel* of the Academy. (*Boyer 1939*, 241–2.)‖

[c] ‖Bernard Nieuwentijt (1654–1718), a physician burgomaster of Purmerend, near Amsterdam, expounded and opposed Leibniz's concept of the calculus; see above, p. 17.‖

[d] ‖Versed sin $a - 1 = \cos a = a^2/2! - a^4/4! + \ldots$ (D'Alembert still takes the dimension to be that of a chord, hence his vers a is really our R vers a).‖

[e] ‖$2R : 2R \sin a/2 = 2R \sin a/2 : R(1 - \cos a)$.‖

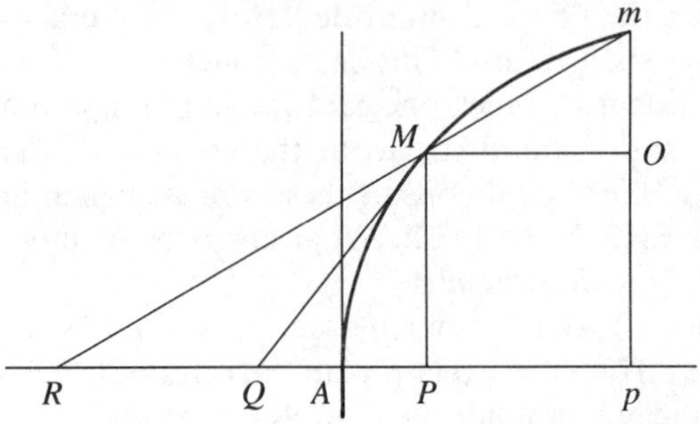

He never considered the *differential* calculus as the study of infinitely small quantities, but as the method of first and ultimate ratios, that is to say, the method of finding the limits of ratios. Thus this famous author has never differentiated quantities but only equations; in fact, every equation involves a relation between two variables and the differentiation of equations consists merely in finding the limit of the ratio of the finite differences of the two quantities contained in the equation. Let us illustrate this by an example which will yield the clearest idea as well as the most exact description of the method of the *differential* calculus.

Let AM be an ordinary parabola, the equation of which is $yy = ax$; here we assume that $AP = x$ and $PM = y$, and a is a parameter. Let us draw the tangent MQ to this parabola at the point M. Let us suppose that the problem is solved and let us take an ordinate pm at any finite distance from PM; furthermore, let us draw the line mMR through the points M, m. It is evident, *first*, that the ratio MP/PQ of the ordinate to the subtangent is greater than the ratio MP/PR or mO/MO which is equal to it because of the similarity of the triangles MOm, MPR; *second*, that the closer the point m is to the point M, the closer will be the point R to the point Q, consequently the closer will be the ratio MP/PR or mO/MO to the ratio MP/PQ; finally, that the first of these ratios approaches the second one as closely as we please, since PR may differ as little as we please from PQ. Therefore, the ratio MP/PQ is the limit of the ratio of mO to OM. Thus, if we are able to represent the ratio mO/OM in algebraic form, then we shall have the algebraic expression of the ratio of MP to PQ and consequently the algebraic representation of the ratio of the ordinate to the subtangent, which will enable us to find this subtangent. Let now $MO = u$, $Om = z$; we shall have $ax = yy$, and $ax + au = yy + 2yz + zz$. Then in view of $ax = yy$ it follows that $au = 2yx + zz$ and $z/u = a/(2y + z)$.

This value $a/(2y + z)$ is, therefore, in general the ratio of mO to Om, wherever one may choose the point m. This ratio is always smaller than $a/2y$: but the smaller z is, the greater the ratio will be and, since one may choose z as small as one pleases, the ratio $a/(2y + z)$ can be brought as close to the ratio $a/2y$ as we like. Consequently $a/2y$ is the limit of the ratio $a/(2y + z)$, that is to say, of the ratio mO/OM. Hence $a/2y$ is equal to the ratio MP/PQ, which we have found to be also the limit of the ratio of mO to Om, since two quantities that are the limits of the same quantity are necessarily equal to each other. To prove this, let X and Z be the limits of the same quantity Y. Then I say that $X = Z$; indeed, if they were to have the difference V, let $X = Z \pm V$: by hypothesis the quantity Y may approach X as closely as one may wish; that is to say, the difference between Y and X may be as small as one may wish. But, since Z differs from X by the quantity V, it follows that Y cannot approach Z closer than the quantity V and consequently Z would not be the limit of Y, which is contrary to the hypothesis. *See* LIMITE, EXHAUSTION.

From this it follows that MP/PQ is equal to $a/2y$. Hence $PQ = 2yy/a = 2x$. Now, according to the method of the *differential* calculus, the ratio of MP to PQ is equal to that of dy to dx; and the equation $ax = yy$ yields $a\,dx = 2y\,dy$ and $dy/dx = a/2y$. So dy/dx is the limit of the ratio of z to u, and this limit is found by making $z = 0$ in the fraction $a/(2y + z)$.

But, one may say, is it not necesssary also to make $z = 0$ and $u = 0$ in the fraction $z/u = a/(2y + z)$, which would yield $\frac{0}{0} = a/2y$? What does this mean? My answer is as follows. First, there is no absurdity involved; indeed $\frac{0}{0}$ may be equal to any quantity one may wish: thus it may be $= a/2y$. Secondly, although the limit of the ratio of z to u has been found when $z = 0$ and $u = 0$, this limit is in fact not the ratio of $z = 0$ to $u = 0$, because the latter one is not clearly defined; one does not know what is the ratio of two quantities that are both zero. This limit is the quantity to which the ratio z/u approaches more and more closely if we suppose z and u to be real and decreasing. Nothing is clearer than this; one may apply this idea to an infinity of other cases. *See* LIMITE, SÉRIE, PROGRESSION.

Following the method of differentiation (which opens the treatise on the quadrature of curves by the great mathematician Newton), instead of the equation $ax + au = yy + 2yz + zz$ we might write $ax + a0 = yy + 2y0 + 00$, thus, so to speak, considering z and u equal to zero; this would have yielded $\frac{0}{0} = a/2y$. What we have said above indicates both the advantage and the inconveniences of this notation: the advantage is that z, being equal to 0, disappears without any other assumption from the ratio $a/(2y + 0)$; the inconvenience is that the two terms of the ratio are supposed to be equal to zero, which at first glance does not present a very clear idea.

From all that has been said we see that the method of the *differential* calculus offers us exactly the same ratio that has been given by the preceding calculation. It will be the same with other more complicated examples. This should be sufficient to give beginners an understanding of the true metaphysics of the *differential* calculus. Once this is well understood, one will feel that the assumption made concerning infinitely small quantities serves only to abbreviate and simplify the reasoning; but that the *differential* calculus does not necessarily suppose the existence of those quantities; and that moreover this calculus merely consists in *algebraically determining the limit of a ratio, for which we already have the expression in terms of lines, and in equating those two expressions. This will provide us with one of the lines we are looking for*. This is perhaps the most precise and neatest possible definition of the *differential* calculus; but it can be understood only when one is well acquainted with this calculus, because often the true nature of a science can be understood only by those who have studied this science. *See* the Disc. Prélimin. page xxxvii.

In the preceding example the known geometric limit of the ratio of z to u is the ratio of the ordinate to the subtangent; in the *differential* calculus we look for the algebraic limit of the ratio z to u and we find $a/2y$. Then, calling s the subtangent, one has $y/s = a/2y$; hence $s = 2yy/a = 2x$. This example is sufficient to understand the others. It will, therefore, be sufficient to make oneself familiar with the previous example concerning the tangents of the parabola, and, since the whole *differential* calculus can be reduced to the problem of the tangents, it follows that one could always apply the preceding principles to various problems of this calculus, for instance to find *maxima and minima*, points of inflection, cusps, etc. *See* those words.

What does it mean, in fact, to find a maximum or a minimum? It consists, it is said, in setting the difference dy equal to zero or to infinity; but it is more precise to say that it means to look for the quantity dy/dx which expresses the limit of the ratio of finite dy to finite dx, and to make this quantity zero or infinite. In this way all the mystery is explained; it is not dy that one makes = to infinity: that would be absurd, since dy is taken as infinitely small and hence cannot be infinite; it is dy/dx: that is to say, one looks for the value of x that renders the limit of the ratio of finite dy to finite dx infinite.

We have seen above that in the *differential* calculus there are really no infinitely small quantities of the first order; that actually those quantities called u are supposed to be divided by other supposedly infinitely small quantities; in this state they do not denote either infinitely small quantities or quotients of infinitely small quantities; they are the limits of the ratio of two finite quantities. The same holds for the second-order differences and for those of higher order. There is actually no quantity in Geometry such as $d\,dy$; whenever $d\,dy$ occurs in an equation it is supposed to be divided by a quantity dx^2, or another of the same order. What now is $d\,dy/dx^2$? It is the limit of the ratio $d\,dy/dx$ divided by dx; or, what is still clearer, it is the limit of dz/dx, where $dy/dx = z$ is a finite quantity. ⟦D'Alembert concludes the article with a lengthy five-column exposition of the differential calculus.⟧

B. INFINITE
(*D'ALEMBERT 1765a*)

INFINITE, *adj.* (Metaphys.) ⟦D'Alembert's discussion of the infinite in metaphysics occupies two full columns of Vol. 8 of the *Encyclopédie*, pp. 702–3. He then turns to the infinite in mathematics:⟧ INFINITE, (*geomet.*) *Geometry of the infinite* is properly speaking the new geometry of the infinitely small, containing the rules of the differential and integral calculus. In 1727 de Fontenelle published a work entitled *Elémens de la géometrie de l'infini*. The author proposes to give the metaphysics of that geometry, and to deduce from that metaphysics, using scarcely any calculus, the majority of the properties of curves. Several geometers have written against the principles of that work; *see* the second volume of MacLaurin's *Traité des fluxions*. In a note, that author attacks the fundamental principle of de Fontenelle's work; *see also* the *Préface* to de Buffon's translation of Newton's *Méthode des fluxions*.

De Fontenelle seems to have believed that the differential calculus necessarily presupposes infinitely large actual quantities, and infinitely small quantities. Persuaded of this principle, he believed it was necessary to establish, at the start of his book, that one could always suppose the quantity to be really augmented or diminished *ad infinitum*, and this proposition is the foundation of the entire work; it is the proposition that MacLaurin believed he ought to attack in the treatise of which we have already spoken. Here is the reasoning of de Fontenelle, and the objections that in our view can be raised against it.

'The quantity being capable of augmentation without end, it follows,' he says, 'that one can suppose it to be really augmented without end; for it is impossible that the quantity capable of augmentation without end should be in the same case as if it were not capable of it without end. Now, if it were not capable of it without end, it would always remain finite; however, the essential property that distinguishes the quantity capable of augmentation without end from the quantity which is not capable without end, is that the latter necessarily remains always finite, and can never be supposed to be anything other than finite; but the first of these two species of quantity can be supposed to be actually *infinite*.'

The response to this argument is that a quantity that is not capable of augmentation without end not only remains always finite, but never exceeds a certain finite quantity, whereas the quantity capable of augmentation without end remains always finite, but can be augmented until it surpasses any finite quantity one chooses. So it is not the possiblity of becoming *infinite*, but the possibility of surpassing any given finite quantity (while nevertheless remaining always finite) that distinguishes the quantity capable of augmentation without end from the quantity that is not capable. If one reduces de Fontenelle's reasoning to a syllogism, one sees that the expression *is not in the same case* (which is there the middle term) is a vague expression that has many different senses; and thus that the syllogism violates the rule that the middle term be *one*. *See the article* DIFFERENTIEL, where one shows that the differential calculus, or the new geometry, does not strictly speaking presuppose quantities that are actually *infinite* or infinitely small.

The *infinite* quantity is properly one that is larger than any assignable quantity; and since no such quantity exists in nature, it follows that the *infinite* quantity properly speaking does not exist except in our mind ‖esprit‖, and does not exist in our mind except by a species of abstraction, in which we remove the idea of limits. So the idea we have of the *infinite* is entirely negative and comes from the idea of the finite, as the negative word *infinite* itself shows. *See* FINI. There is the following difference between *infinite* and *indefinite*, that in the idea of the *infinite* one abstracts from all limits, while in that of the *indefinite* one abstracts from this or that limit in particular. An *infinite* line is one that is imagined to have no limit; an *indefinite* line is one that is imagined to terminate wherever one wishes, without its length or consequently its limits being fixed.

One admits in geometry, at any rate as a manner of speaking, *infinite* quantities of the second, third, fourth order; for example, one says that, in the equation of a parabola $y = x^2/a$, if one takes x *infinite*, y will be *infinite* of the second order, that is, as *infinite* with respect to the *infinite* x as x is itself with respect to a. This manner of speaking is not terribly clear; for if x is *infinite*, how are we to conceive of y as infinitely greater?—Here is the response. The equation $y = x^2/a$ represents $y/x = x/a$, which shows that the relation of y to x always increases to the extent that x increases, so that by taking x large enough, the relation of y to x can be made greater than any given quantity: this is all that one means when one says that, if x is *infinite* of the first order, y is *infinite* of the second. This simple example will suffice to make the others intelligible. *See* INFINIMENT PETIT.

C. LIMIT
(*D'ALEMBERT 1765b*)

LIMIT, n. (*math.*) One says that one quantity is the *limit* of another quantity when the second can approach the first more closely than by a given quantity, as small as one can imagine, but such that the approaching quantity can never exceed the quantity that it approaches; so that the difference between such a quantity and its *limit* is absolutely unassignable.

For example, consider two polygons, one inscribed in a circle, and the other circumscribing it. It is evident that one can increase the number of sides as much as one wishes, and in this case each polygon approaches ever more closely to the circumference of the circle: the contour of the inscribed polygon increases, and that of the circumscribed polygon decreases; but the perimeter or contour of the former never surpasses the length of the circumference, and that of the latter will never be smaller than that same circumference; therefore, the circumference of the circle is the *limit* of the augmentation of the first polygon, and of the diminution of the second.[a]

1. If two quantities are the *limit* of the same quantity, the two quantities will be equal to each other.

[a] [The preceding two paragraphs were written by the Abbé de la Chapelle. The continuation of the entry is by D'Alembert.]

2. Let $A \times B$ be the product of the two quantities A, B. Suppose C is the *limit* of the quantity A, and D is the *limit* of the quantity B; I say that $C \times D$, the product of the *limits*, will necessarily be the *limit* of $A \times B$, the product of the two quantities A, B.

These two propositions, which one will find exactly proved in the *Institutions de Géometrie*, serve as principles for rigorously proving that one obtains the area of a circle when one multiplies the half-circumference by the radius. *See* the work cited, *vol. 2, pp. 331 ff.*

The theory of *limits* is the basis of the true metaphysics of the differential calculus. *See* DIFFERENTIEL, FLUXION, EXHAUSTION, INFINI. Strictly speaking, the *limit* never coincides with, or never becomes equal to, the quantity of which it is the *limit*; but it always approaches it more and more closely, and can differ from it by as little as one pleases. The circle, for example, is the *limit* of the inscribed and circumscribed polygons; for it is never strictly identical with them, although they can approach it *ad infinitum*. This notion can serve to clarify several mathematical propositions. For example, one says that the form of a decreasing geometric progression in which the first term is a and the second b is $aa/a - b$; but this value is properly speaking not the sum of the progression. It is the *limit* of that sum—that is to say, the quantity it can approach as closely as one wishes, without ever arriving there exactly. For if e is the last term of the progression, the exact value of the sum is $aa - be/a - b$, which is always less than $aa/a - b$ since in a geometric progression, even if it is decreasing, the last term e is never equal to 0: but since this term continually approaches zero, without ever arriving there, it is clear that zero is the *limit* and that consequently the *limit* of $aa - be/a - b$ is $aa/a - b$, by supposing $e = 0$—that is to say, by replacing e with its limit. *See* SUITE *ou* SÉRIE, PROGRESSION, etc.

4
Immanuel Kant (1724–1804)

Kant was the most influential philosopher of mathematics in modern times, and his *Critique of pure reason* to a considerable extent set the agenda for the philosophy of mathematics of the next century; the impact of his thought and of the reactions to it can be detected in most of the selections that follow. The importance, the influence, and the difficulty of his ideas can perhaps be gauged by noting that C.S. Peirce, a notoriously rapid learner, studied the *Critique of pure reason* 'for two hours a day . . . for more than three years, until I almost knew the whole book by heart, and had critically examined every section of it.' (*Peirce 1931–58*; Vol. i, p. ix.)

Kant's philosophy of mathematics can hardly be separated from the rest of his difficult thought, and his comments on mathematics are but a part of his investigation into epistemology and the foundations of natural science. In consequence, they are inseparably interwoven with his views on such topics as the nature of space and time; the theory of *a priori* knowledge; the distinction between analytic and synthetic judgments; the role of intuitions and concepts in perception, judgement, and reasoning; the nature of logic; and the status and possibility of metaphysical knowledge. The force of the Kantian doctrines comes from these interconnections, from the 'critical philosophy' as a whole: not from the individual fragments. So the selections that follow are merely an introduction to some of the central themes in Kant's philosophy of mathematics.

Although he was not a working scientist, Kant was deeply occupied throughout his career with problems of mathematics and the natural sciences. He regularly lectured on mathematics and physics at Königsberg, and he was well informed about the progress of current research. In two of his early publications he made original, though little-noticed, speculative contributions to the cosmological theory of the day. In his *Natural history and theory of the heavens* (*1755*), Kant proposed—four decades before Laplace—the now universally accepted 'nebular hypothesis' about the constitution of the universe; that is, the hypothesis that the solar system is part of a single galaxy of stars similar to the sun, and that the 'nebular stars' are in fact other galaxies of such suns. Kant attempted to demonstrate, on principles of Newtonian mechanics alone, how the universe could have arisen from an original, undifferentiated chaos of dispersed matter: 'Give me matter, and I shall show you how a world can arise therefrom' (*Preface*, p. Axxxv).

And in the following selection from Kant's first published work, *Thoughts on the true estimation of active forces* (*1747*; written and submitted to the faculty of philosophy at Königsberg in 1746), Kant argues that physical space is not necessarily three-dimensional, and that the physical laws of gravitation determine the geometric structure of space; he further contends that a science of the possible kinds of space 'would certainly be the highest geometry that a finite understanding could undertake'. These remarkably prescient ideas were not to resurface until the middle of the next century, in the work of Riemann and Grassmann on *n*-dimensional spaces. This passage is rarely mentioned in discussions of Kant's philosophy of mathematics; but it confutes the common assertion that Kant had no inkling of the possibility of other geometries.

The literature on Kant is scarcely surveyable; for a detailed recent study of Kant and the natural sciences, see *Friedman 1992*, which contains numerous further references.

The translation is by William Ewald; summary notes that originally appeared in the margins at the start of each paragraph have been converted into footnotes. References to *Kant 1747* should be to the section numbers, which appeared in the original edition.

A. *FROM* THOUGHTS ON THE TRUE ESTIMATION OF ACTIVE FORCES (*KANT 1747*)

§9.

[a] It is easy to show that neither space nor extension would exist if substances had no force ‖Kraft‖ to act outside themselves ‖außer sich zu wirken‖. For without this force there is no connection ‖Verbindung‖, without connection no order, and without order finally no space. But it is harder to see how the multiplicity of the dimensions of space follows from the law according to which this force of substances acts outside itself.

[b] Because it seems to me that Leibniz reasons in a circle when he bases his proof (somewhere in the *Theodicée*) on the number of lines, meeting in a point, that can be drawn orthogonally to one another, I thought of proving the three-dimensionality of extension from what one sees in the powers of numbers. The first three powers are completely simple and cannot be reduced to any other, but the fourth, as the square of the square, is nothing but the repetition of the second power. But although this property seemed to me a good way of explaining the three-dimensionality of space, it nevertheless did not work in practice. For in everything that we can imagine of space, the fourth power is a nonsense

[a] ‖Marginal note:‖ If the substances had no force to act outside of themselves, then there would be no extension, and also no space.

[b] ‖Marginal note:‖ The reason ‖Grund‖ for the three-dimensionality of space is still unknown.

‖ein Unding‖. In geometry you cannot multiply a square by itself, or a cube by its root; so the necessity of the three-dimensionality does not rest so much on the fact that, when you take several dimensions you do nothing other than to repeat the earlier ones (as is the case with the powers of numbers), but rather on some other necessity which I am not yet in a position to explain.

§10.

[c] Since everything that occurs among the properties of a thing must be derivable[d] from that which contains within itself the complete reason ‖Grund‖ of the thing, it follows that the properties of extension (and consequently also its three-dimensionality) will be based on the property of force which substances possess in respect to the things with which they are linked ‖verbunden‖. The force with which a substance acts in unison with others cannot be thought without a certain law which makes itself manifest in the manner of its action. Because the type of the law according to which the substances act on one another must also determine the type of union and combination of a multiplicity of them, it follows that the law by which a whole collection of substances (that is, a space) is measured or the dimension of the extension will stem from the laws by which the substances seek to unite in virtue of their essential forces.

[e] In consequence, I hold: that the substances in the existing world, of which we are a part, have essential forces of such a kind as to act on one another according to the inverse square of the distance; secondly, that the whole that arises in this way has, in virtue of this law, the property of being three-dimensional; third, that this law is arbitrary, and that God could have chosen a different one—for instance, the inverse cube law; finally, fourth, that from another law there would have resulted an extension with other properties and dimensions. A science of all these possible kinds of space would certainly be the highest geometry that a finite understanding could undertake. The impossibility that we find in ourselves to imagine a space of more than three dimensions seems to me to rest on the fact that our soul also receives impressions from outside in accordance with the inverse square of the distance, and because its nature is made, not merely to be so affected, but to act outside itself in this way.

---

[c] ‖Marginal note:‖ It is likely that the three-dimensionality of space originates from the law of the forces by which substances act on one another.

[d] ‖muß hergeleitet werden können.‖

[e] ‖Marginal note:‖ The three-dimensionality thus seems to originate from the fact that the substances in the existing world act on each other in such a way that the strength of the action ‖die Stärke der Wirkung‖ varies as the inverse square of the distance.

B. *FROM THE* TRANSCENDENTAL AESTHETIC (*KANT 1787*)

Kant's mature views on the nature of geometry are contained in his influential discussion of space and time in the *Transcendental aesthetic* passage of the *Critique of pure reason*, where Kant appears to have abandoned the empiricist view of geometry expressed in his *1747*.

Kant has surprisingly little to say in his philosophical writings about Euclid's Axiom of Parallels or about its relevance to his theory of geometry. He was surely aware that mathematicians had unsuccessfully attempted to prove the Axiom,[a] and that the absence of a proof was regarded as a notorious unsolved problem; but in the *Critique of pure reason* he does not discuss the Axiom or the possibility of alternative geometries.

The translation is by Norman Kemp Smith, and is reprinted from *Kant 1933*. References should be to pages B40–41 of the second German edition of the *Critique*, *Kant 1787*.

---

§3

THE TRANSCENDENTAL EXPOSITION OF THE CONCEPT OF SPACE

I understand by a transcendental exposition the explanation of a concept, as a principle from which the possibility of other *a priori* synthetic knowledge can be understood. For this purpose it is required (1) that such knowledge does really flow from the given concept; (2) that this knowledge is possible only on the assumption of a given mode of explaining the concept.

Geometry is a science which determines the properties of space synthetically, and yet *a priori*. What, then, must be our representation of space, in order that such knowledge of it may be possible? It must in its origin be intuition; for B 41

---

[a] Kant would have known of the difficulties surrounding the Axiom of Parallels from at least three sources. The contemporary *locus classicus* for the discussion of the Axiom was the widely read *Vorrede* to Kästner's *Anfangsgründe der Arithmetik* (*1758*); an annotated copy of this book was in Kant's library (*Warda 1922*). In addition, Kant corresponded with, and admired the works of, J.H. Lambert, who made a notable attempt (*Lambert 1786*, translated in part below) to prove the Axiom. (The Axiom is not, however, discussed in their surviving correspondence.) Moreover, Kant was a close associate of Johann Schultz, Professor of Mathematics in Königsberg, whom Kant called 'the best philosophical mind I know in these parts' (*Letter to Herz*, 21 Feb. 1772), and with whom he used to discuss mathematical and scientific problems (*Vorländer 1977*, Vol. II, p. 33). After the publication of the *Critique of pure reason*, Schultz became a quasi-official interpreter of the critical philosophy, about which he wrote two books (*1785* and *1789–92*); shortly before, in 1784, he had published a treatise on parallel lines, the *Entdeckte theorie der Parallelen nebst einer Untersuchung über den Ursprung ihrer bisherigen Schwierigkeit* (*1784*). An annotated copy of this work was also in Kant's library, apparently a gift from the author.

from a mere concept no propositions can be obtained which go beyond the concept—as happens in geometry (Introduction, V). Further, this intuition must be *a priori*, that is, it must be found in us prior to any perception of an object, and must therefore be pure, not empirical, intuition. For geometrical propositions are one and all apodeictic, that is, are bound up with the consciousness of their necessity; for instance, that space has only three dimensions. Such propositions cannot be empirical or, in other words, judgements of experience, nor can they be derived from any such judgements (Introduction, II).

How, then, can there exist in the mind an outer intuition which precedes the objects themselves, and in which the concept of these objects can be determined *a priori*? Manifestly, not otherwise than in so far as the intuition has its seat in the subject only, as the formal character of the subject, in virtue of which, in being affected by objects, it obtains *immediate representation*, that is, *intuition*, of them; and only in so far, therefore, as it is merely the form of outer *sense* in general.

Our explanation is thus the only explanation that makes intelligible the *possibility* of geometry, as a body of *a priori* synthetic knowledge. Any mode of explanation which fails to do this, although it may otherwise seem to be somewhat similar, can by this criterion be distinguished from it with the greatest certainty.

C. *FROM THE* DISCIPLINE OF PURE REASON (*KANT 1781*)

The following passage is the entirety of the section entitled *The discipline of pure reason in its dogmatic employment*; it is Kant's most extensive discussion of the relationship of mathematical knowledge to philosophical knowledge.

The translation is by Norman Kemp Smith, and is reprinted from *Kant 1933*. References should be to the pages of the original *A* and *B* German editions, *Kant 1781* and *1787*, as given in the margin.

THE DISCIPLINE OF PURE REASON IN ITS DOGMATIC EMPLOYMENT

Mathematics presents the most splendid example of the successful extension of pure reason, without the help of experience. Examples are contagious, especially as they quite naturally flatter a faculty which has been successful in one field,
A 713 ‖leading it‖ to expect the same good fortune in other fields. Thus pure reason
B 741 hopes to be able to extend its domain as successfully and securely in its transcendental as in its mathematical employment, especially when it resorts to the same

method as has been of such obvious utility in mathematics. It is therefore highly important for us to know whether the method of attaining apodeictic certainty which is called *mathematical* is identical with the method by which we endeavour to obtain the same certainty in philosophy, and which in that field would have to be called *dogmatic*.

Philosophical knowledge is the *knowledge gained by reason from concepts*; mathematical knowledge is the knowledge gained by reason from the *construction* of concepts. To *construct* a concept means to exhibit *a priori* the intuition which corresponds to the concept. For the construction of a concept we therefore need a *non-empirical* intuition. The latter must, as intuition, be a *single* object, and yet none the less, as the construction of a concept (a universal representation), it must in its representation express universal validity for all possible intuitions which fall under the same concept. Thus I construct a triangle by representing the object which corresponds to this concept either by imagination alone, in pure intuition, or in accordance therewith also on paper, in empirical intuition—in both cases completely *a priori*, without having borrowed the pat- A 714
tern from any experience. The single figure which we draw is empirical, and B 742
yet it serves to express the concept, without impairing its universality. For in this empirical intuition we consider only the act whereby we construct the concept, and abstract from the many determinations (for instance, the magnitude of the sides and of the angles), which are quite indifferent, as not altering the concept 'triangle'.

Thus philosophical knowledge considers the particular only in the universal, mathematical knowledge the universal in the particular, or even in the single instance, though still always *a priori* and by means of reason. Accordingly, just as this single object is determined by certain universal conditions of construction, so the object of the concept, to which the single object corresponds merely as its schema, must likewise be thought as universally determined.

The essential difference between these two kinds of knowledge through reason consists therefore in this formal difference, and does not depend on difference of their material or objects. Those who propose to distinguish philosophy from mathematics by saying that the former has as its object *quality* only and the latter *quantity* only, have mistaken the effect for the cause. The form of mathematical knowledge is the cause why it is limited exclusively to quantities. For it is the concept of quantities only that allows of being constructed, A 715
that is, exhibited *a priori* in intuition; whereas qualities cannot be presented in B 743
any intuition that is not empirical. Consequently reason can obtain a knowledge of qualities only through concepts. No one can obtain an intuition corresponding to the concepts of reality otherwise than from experience; we can never come into possession of it *a priori* out of our own resources, and prior to the empirical consciousness of reality. The shape of a cone we can form for ourselves in intuition, unassisted by any experience, according to its concept alone, but the colour of this cone must be previously given in some experience or other. I cannot represent in intuition the concept of a cause in general except in an example supplied by experience; and similarly with other concepts. Philosophy, as well

as mathematics, does indeed treat of quantities, for instance, of totality, infinity, etc. Mathematics also concerns itself with qualities, for instance, the difference between lines and surfaces, as spaces of different quality, and with the continuity of extension as one of its qualities. But although in such cases they have a common object, the mode in which reason handles that object is wholly different in philosophy and in mathematics. Philosophy confines itself to universal concepts; mathematics can achieve nothing by concepts alone but hastens at once to intuition, in which it considers the concept *in concreto*, though not
A 716 empirically, but only in an intuition which it presents *a priori*, that is, which
B 744 it has constructed, and in which whatever follows from the universal conditions
of the construction must be universally valid of the object of the concept thus constructed.

Suppose a philosopher be given the concept of a triangle and he be left to find out, in his own way, what relation the sum of its angles bears to a right angle. He has nothing but the concept of a figure enclosed by three straight lines, and possessing three angles. However long he meditates on this concept, he will never produce anything new. He can analyse and clarify the concept of a straight line or of an angle or of the number three, but he can never arrive at any properties not already contained in these concepts. Now let the geometrician take up these questions. He at once begins by constructing a triangle. Since he knows that the sum of two right angles is exactly equal to the sum of all the adjacent angles which can be constructed from a single point on a straight line, he prolongs one side of his triangle and obtains two adjacent angles, which together are equal to two right angles. He then divides the external angle by drawing a line parallel to the opposite side of the triangle, and observes that he has thus obtained an external adjacent angle which is equal to an internal
A 717 angle—and so on. In this fashion, through a chain of inferences guided through-
B 745 out by intuition, he arrives at a fully evident and universally valid solution of
the problem.

But mathematics does not only construct magnitudes (*quanta*) as in geometry; it also constructs magnitude as such (*quantitas*), as in algebra. In this it abstracts completely from the properties of the object that is to be thought in terms of such a concept of magnitude. It then chooses a certain notation for all constructions of magnitude as such (numbers),[a] that is, for addition, subtraction, extraction of roots, etc. Once it has adopted a notation for the general concept of magnitudes so far as their different relations are concerned, it exhibits in intuition, in accordance with certain universal rules, all the various operations through which the magnitudes are produced and modified. When, for instance, one magnitude is to be divided by another, their symbols are placed together, in accordance with the sign for division, and similarly in the other processes; and thus in algebra by means of a symbolic constructtion, just as in geometry by means of an ostensive construction (the geometrical construction of the

---

[a] [Reading, with Hartenstein and Erdmann (*Zahlen*), *als . . . Wurzeln usw*. for (*Zahlen, als . . . Subtraktion usw*).]

objects themselves), we succeed in arriving at results which discursive knowledge could never have reached by means of mere concepts.

Now what can be the reason of this radical difference in the fortunes of the philosopher and the mathematician, both of whom practise the art of reason, the one making his way by means of concepts, the other by means of intuitions A 718
which he exhibits *a priori* in accordance with concepts? The cause is evident B 746
from what has been said above, in our exposition of the fundamental transcendental doctrines. We are not here concerned with analytic propositions, which can be produced by mere analysis of concepts (in this the philosopher would certainly have the advantage over his rival), but with synthetic propositions, and indeed with just those synthetic propositions that can be known *a priori*. For I must not restrict my attention to what I am actually thinking in my concept of a triangle (this is nothing more than the mere definition); I must pass beyond it to properties which are not contained in this concept, but yet belong to it. Now this is impossible unless I determine my object in accordance with the conditions either of empirical or of pure intuition. The former would only give us an empirical proposition (based on the measurement of the angles), which would not have universality, still less necessity; and so would not at all serve our purpose. The second method of procedure is the mathematical one, and in this case is the method of geometrical construction, by means of which I combine in a pure intuition (just as I do in empirical intuition) the manifold which belongs to the schema of a triangle in general, and therefore to its concept. It is by this method that universal synthetic propositions must be constructed.

It would therefore be quite futile for me to philosophise upon the triangle, that is, to think about it discursively. I should not be able to advance a single A 719
step beyond the mere definition, which was what I had to begin with. There B 747
is indeed a transcendental synthesis ‖framed‖ from concepts alone, a synthesis with which the philosopher is alone competent to deal; but it relates only to a thing in general, as defining the conditions under which the perception of it can belong to possible experience. But in mathematical problems there is no question of this, nor indeed of existence at all, but only of the properties of the objects in themselves, ‖that is to say‖, solely in so far as these properties are connected with the concept of the objects.

In the above example we have endeavoured only to make clear the great difference which exists between the discursive employment of reason in accordance with concepts and its intuitive employment by means of the construction of concepts. This naturally leads on to the question, what can be the cause which necessitates such a twofold employment of reason, and how we are to recognise whether it is the first or the second method that is being employed.

All our knowledge relates, finally, to possible intuitions, for it is through them alone that an object is given. Now an *a priori* concept, that is, a concept which is not empirical, either already includes in itself a pure intuition (and if so, it can be constructed), or it includes nothing but the synthesis of possible A 720
intuitions which are not given *a priori*. In this latter case we can indeed make B 748

use of it in forming synthetic *a priori* judgements, but only discursively in accordance with concepts, never intuitively through the construction of the concept.

The only intuition that is given *a priori* is that of the mere form of appearances, space and time. A concept of space and time, as quanta, can be exhibited *a priori* in intuition, that is, constructed, either in respect of the quality (figure) of the quanta, or through number in their quantity only (the mere synthesis of the homogeneous manifold). But the matter of appearances, by which *things* are given us in space and time, can only be represented in perception, and therefore *a posteriori*. The only concept which represents *a priori* this empirical content of appearances is the concept of a *thing* in general, and the *a priori* synthetic knowledge of this thing in general can give us nothing more than the mere rule of the synthesis of that which perception may give *a posteriori*. It can never yield an *a priori* intuition of the real object, since this must necessarily be empirical.

Synthetic propositions in regard to *things* in general, the intuition of which does not admit of being given *a priori*, are transcendental. Transcendental propositions can never be given through construction of concepts, but only in accordance with concepts that are *a priori*. They contain nothing but the rule according to which we are to seek empirically for a certain synthetic unity of that which is incapable of intuitive representation *a priori* (that is, of percep-
A 721 tions). But these synthetic principles cannot exhibit *a priori* any one of their
B 749 concepts in a specific instance; they can only do this *a posteriori*, by means of experience, which itself is possible only in conformity with these principles.

If we are to judge synthetically in regard to a concept, we must go beyond this concept and appeal to the intuition in which it is given. For should we confine ourselves to what is contained in the concept, the judgement would be merely analytic, serving only as an explanation of the thought, in terms of what is actually contained in it. But I can pass from the concept to the corresponding pure or empirical intuition, in order to consider it in that intuition *in concreto*, and so to know, either *a priori* or *a posteriori*, what are the properties of the object of the concept. The *a priori* method gives us our rational and mathematical knowledge through the construction of the concept, the *a posteriori* method our merely empirical (mechanical) knowledge, which is incapable of yielding necessary and apodeictic propositions. Thus I might analyse my empirical concept of gold without gaining anything more than merely an enumeration of everything that I actually think in using the word, thus improving the logical character of my knowledge but not in any way adding to it. But I take the material body, familiarly known by this name, and obtain perceptions by means
A 722 of it; and these perceptions yield various propositions which are synthetic but
B 750 empirical. When the concept is mathematical, as in the concept of a triangle, I am in a position to construct the concept, that is, to give it *a priori* in intuition, and in this way to obtain knowledge which is at once synthetic and rational. But if what is given me is the *transcendental* concept of a reality, substance, force, etc., it indicates neither an empirical nor a pure intuition, but only the synthesis of empirical intuitions, which, as being empirical, cannot be given *a*

priori. And since the synthesis is thus unable to advance *a priori*, beyond the concept, to the corresponding intuition, the concept cannot yield any determining synthetic proposition, but only a principle of the synthesis* of possible empirical intuitions. A transcendental proposition is therefore synthetic knowledge through reason, in accordance with mere concepts; and it is discursive, in that while it is what alone makes possible any synthetic unity of empirical knowledge, it yet gives us no intuition *a priori*.

There is thus a twofold employment of reason; and while the two modes of A 723
employment resemble each other in the universality and *a priori* origin of their B 751
knowledge, in outcome they are very different. The reason is that in the ‖field of‖ appearance, in terms of which[b] all objects are given us, there are two elements, the form of intuition (space and time), which can be known and determined completely *a priori*, and the matter (the physical element) or content—the latter signifying something which is met with in space and time and which therefore contains an existent[c] corresponding to sensation. In respect to this material element, which can never be given in any determinate fashion otherwise than empirically, we can have nothing *a priori* except indeterminate concepts of the synthesis of possible sensations, in so far as they belong, in a possible experience, to the unity of apperception. As regards the formal element, we can determine our concepts in *a priori* intuition, inasmuch as we create for ourselves, in space and time, through a homogeneous synthesis, the objects themselves—these objects being viewed simply as *quanta*. The former method is called the employment of reason in accordance with concepts; in so employing it[d] we can do nothing more than bring appearances under concepts, according to their actual content. The concepts cannot be made determinate in this manner,[e] save only empirically, that is, *a posteriori* (although always in accordance with these concepts as rules of an empirical synthesis). The other method is the employment of reason through the construction of concepts; and since the concepts A 724
here relate to an *a priori* intuition, they are for this very reason themselves *a* B 752
priori and can be given in a quite determinate fashion in pure intuition, without the help of any empirical data. The consideration of everything which exists in space or time, in regard to the questions, whether and how far it is a quantum or not, whether we are to ascribe to it positive being or the absence of such, how far this something occupying space or time is a primary substratum or a mere determination ‖of substance‖, whether there be a relation of its existence to some other existence, as cause or effect, and finally in respect of its existence

* With the concept of cause I do really go beyond the empirical concept of an event (something happening), yet I do not pass to the intuition which exhibits the concept of cause *in concreto*, but to the time-conditions in general, which in experienee may be found to be in accord with this concept. I therefore proceed merely in accordance with concepts; I cannot proceed by means of the construction of concepts, since the concept is a rule of the synthesis of perceptions, and the latter are not pure intuitions, and so do not permit of being *given a priori*.

b ‖als wodurch.‖

c ‖Dasein.‖

d ‖Reading, with Erdmann, *in dem* for *indem*.‖

e ‖Reading, with Erdmann, *dadurch* for *darauf*.‖

whether it is isolated or is in reciprocal relation to and dependence upon others—these questions, as also the question of the possibility of this existence, its actuality and necessity, or the opposites of these, one and all belong altogether to knowledge obtained by reason from concepts, such knowledge being termed *philosophical*. But the determination of an intuition *a priori* in space (figure), the division of time (duration), or even just the knowledge of the universal element in the synthesis of one and the same thing in time and space, and the magnitude of an intuition that is thereby generated (number),—all this is the work of reason through construction of concepts, and is called *mathematical*.

The great success which attends reason in its mathematical employment quite naturally gives rise to the expectation that it, or at any rate its method, will have the same success in other fields as in that of quantity. For this method has the
A 725 advantage of being able to realise all its concepts in intuitions, which it can pro-
B 753 vide *a priori*, and by which it becomes, so to speak, master of nature; whereas pure philosophy is all at sea when it seeks through *a priori* discursive concepts to obtain insight in regard to the natural world, being unable to intuit *a priori* (and thereby to confirm) their reality. Nor does there seem to be, on the part of the experts in mathematics, any lack of self-confidence as to this procedure—or on the part of the vulgar of great expectations from their skill—should they apply themselves to carry out their project. For, since they have hardly ever attempted to philosophise in regard to their mathematics (a hard task!), the specific difference between the two employments of reason has never so much as occurred to them. Current, empirical rules, which they borrow from ordinary consciousness, they treat as being axiomatic. In the question as to the source of the concepts of space and time they are not in the least interested, although it is precisely with these concepts (as the only original quanta) that they are themselves occupied. Similarly, they think it unnecessary to investigate the origin of the pure concepts of understanding and in so doing to determine the extent of their validity; they care only to make use of them. In all this they are entirely in the right, provided only they do not overstep the proper limits, that is, the limits of the natural world. But, unconsciously, they pass from the field of sensibility to the precarious ground of pure and even transcendental con-
A 726 cepts, a ground (*instabilis tellus, innabilis unda*) that permits them neither to
B 754 stand nor to swim, and where their hasty tracks are soon obliterated. In mathematics, on the other hand, their passage gives rise to a broad highway, which the latest posterity may still tread with confidence.

We have made it our duty to determine, with exactitude and certainty, the limits of pure reason in its transcendental employment. But the pursuit of such transcendental knowledge has this peculiarity, that in spite of the plainest and most urgent warnings men still allow themselves to be deluded by false hopes, and therefore to postpone the total abandonment of all proposed attempts to advance beyond the bounds of experience into the enticing regions of the intellectual world. It therefore becomes necessary to cut away the last anchor of these fantastic hopes, that is, to show that the pursuit of the mathematical method cannot be of the least advantage in this kind of knowledge (unless it be in exhibiting more plainly the limitations of the method); and that mathematics[f] and philosophy, although in natural science they do, indeed, go hand in hand, are none the less so completely different, that the procedure of the one can never be imitated by the other.

The exactness of mathematics rests upon definitions, axioms and demonstrations. I shall content myself with showing that none of these, in the sense in which they are understood by the mathematician, can be achieved or imitated {A 727 B 755} by the philosopher. I shall show that in philosophy the geometrician can by his method build only so many houses of cards, just as in mathematics the employment of a philosophical method results only in mere talk. Indeed it is precisely in knowing its limits that philosophy consists; and even the mathematician, unless his talent is of such a specialised character that it naturally confines itself to its proper field, cannot afford to ignore the warnings of philosophy, or to behave as if he were superior to them.

1. *Definitions.*—To *define*, as the word itself indicates, really only means to present the complete, original concept of a thing within the limits of its concept.* If this be our standard, an *empirical* concept cannot be defined at all, but only *made explicit*. For since we find in it only a few characteristics of a certain species of sensible object, it is never certain that we are not using the word, in denoting one and the same object, sometimes so as to stand for more, {A 728 B 756} and sometimes so as to stand for fewer characteristics. Thus in the concept of *gold* one man may think, in addition to its weight, colour, malleability, also its property of resisting rust, while another will perhaps know nothing of this quality. We make use of certain characteristics only so long as they are adequate for the purpose of making distinctions; new observations remove some properties and add others; and thus the limits of the concept are never assured. And indeed what useful purpose could be served by defining an empirical concept, such, for instance, as that of water? When we speak of water and its properties, we do not stop short at what is thought in the word, water, but proceed to experiments. The word, with the few characteristics which we attach to it, is more properly to be regarded as merely a designation than as a concept of the thing; the so-called definition is nothing more than a determining of the word. In the second place, it is also true that no concept given *a priori*, such as substance, cause, right, equity, etc., can, strictly speaking, be defined. For I can never be certain that the clear representation of a given concept, which as given may still be confused, has been completely effected, unless I know that it is adequate to its object. But since the concept of it may, as given, include many obscure representations, which we overlook in our analysis, although we are

[f] [Messkunst.]

* *Completeness* means clearness and sufficiency of characterteristics; by *limits* is meant the precision shown in there not being more of these characteristics than belong to the complete concept; by *original* is meant that this determination of these limits is not derived from anything else, and therefore does not require any proof; for if it did, that would disqualify the supposed explanation from standing at the head of all the judgements regarding its object.

constantly making use of them in our application of the concept, the com-
A 729 pleteness of the analysis of my concept is always in doubt, and a multiplicity
B 757 of suitable examples suffices only to make the completeness *probable*, never to
make it *apodeictically* certain. Instead of the term, definition, I prefer to use the term, *exposition*, as being a more guarded term, which the critic can accept as being up to a certain point valid, though still entertaining doubts as to the completeness of the analysis. Since, then, neither empirical concepts nor concepts given *a priori* allow of definition, the only remaining kind of concepts, upon which this mental operation[g] can be tried, are arbitrarily invented concepts. A concept which I have invented I can always define; for since it is not given to me either by the nature of understanding or by experience, but is such as I have myself deliberately made it to be, I must know what I have intended to think in using it. I cannot, however, say that I have thereby defined a true object.[h] For if the concept depends on empirical conditions, as *e.g.* the concept of a ship's clock, this arbitrary concept of mine does not assure me of the existence or of the possibility of its object. I do not even know from it whether it has an object at all, and my explanation may better be described as a declaration of my project than as a definition of an object. There remain, therefore, no concepts which allow of definition, except only those which contain an arbitrary synthesis that admits of *a priori* construction. Consequently, mathematics is the only science that has definitions. For the object which it thinks it exhibits *a priori* in intuition, and this object certainly cannot contain either
A 730 more or less than the concept, since it is through the definition[i] that the con-
B 758 cept of the object is given—and given originally, that is, without its being neces-
sary to derive the definition[i] from any other source. The German language has for the ‖Latin‖ terms *exposition*, *explication*, *declaration*, and *definition* only one word, *Erklärung*,[j] and we need not, therefore, be so stringent in our requirements as altogether to refuse to philosophical explanations[k] the honourable title, definition. We shall confine ourselves simply to remarking that while philosophical definitions are never more than expositions of given concepts, mathematical definitions are constructions of concepts, originally framed by the mind itself; and that while the former can be obtained only by analysis (the completeness of which is never apodeictically certain), the latter are produced synthetically. Whereas, therefore, mathematical definitions *make* their concepts, in philosophical dcfinitions concepts are only *explained*. From this it follows:

(*a*) That in philosophy we must not imitate mathematics by beginning with definitions, unless it be by way simply of experiment. For since the definitions are analyses of given concepts, they presuppose the prior presence of the concepts, although in a confused state; and the incomplete exposition must precede

[g] ‖dieses Kunststück.‖

[h] ‖einen wahren Gegenstand.‖

[i] ‖Erklärung.‖

[j] ‖This term Kant usually employs in the sense of explanation; but, as above indicated, it is used in the preceding sentence in the sense of definition.‖

[k] ‖Erklärungen.‖

the complete. Consequently, we can infer a good deal from a few characteris-
tics, derived from an incomplete analysis, without having yet reached the com- A 731
plete exposition, that is, the definition. In short, the definition in all its precision B 759
and clarity ought, in philosophy, to come rather at the end than at the beginning of our enquiries.* In mathematics, on the other hand, we have no concept whatsoever prior to the definition, through which the concept itself is first given. For this reason mathematical science must always begin, and it can always begin, with the definition.

(*b*) That mathematical definitions can never be in error. For since the concept is first given through the definition it includes nothing except precisely what the definition intends should be understood by it. But although nothing incorrect can be introduced into its content, there may sometimes, though rarely, be a defect in the form in which it is clothed, namely as regards precision. Thus the
common explanation of the circle that it is a *curved* line every point in which
is equidistant from one and the same point (the centre), has the defect that the A 732
determination, curved, is introduced unnecessarily. For there must be a par- B 760
ticular theorem, deduced from the definition and easily capable of proof, namely, that if all points in a line are equidistant from one and the same point, the line is curved (no part of it straight). Analytic definitions, on the other hand, may err in many ways, either through introducing characteristics which do not really belong to the concept, or by lacking that completeness which is the essential feature of a definition. The latter defect is due to the fact that we can never be quite certain of the completeness of the analysis. For these reasons the mathematical method of definition does not admit of imitation in philosophy.

2. *Axioms.*—These, in so far as they are immediately certain, are synthetic *a priori* principles. Now one concept cannot be combined with another synthetically and also at the same time immediately, since, to be able to pass beyond either concept, a third something is required to mediate our knowledge. Accordingly, since philosophy is simply what reason knows by means of concepts, no principle deserving the name of an axiom is to be found in it. Mathematics, on the other hand, can have axioms, since by means of the construction of con-
cepts in the intuition of the object it can combine the predicates of the object
both *a priori* and immediately, as, for instance, in the proposition that three A 733
points always lie in a plane. But a synthetic principle derived from concepts B 761
alone can never be immediately certain, for instance, the proposition that everything which happens has a cause. Here I must look round for a third something, namely, the condition of time-determination in an experience; I cannot obtain

* Philosophy is full of faulty definitions, especially of definitions which, while indeed containing some of the elements required, are yet not complete. If we could make no use of a concept till we had defined it, all philosophy would be in a pitiable plight. But since a good and safe use can still be made of the elements obtained by analysis so far as they go, defective definitions, that is, propositions which are properly not definitions, but are yet true, and are therefore approximations to definitions, can be employed with great advantage. In mathematics definition belongs *ad esse*, in philosophy *ad melius esse*. It is desirable to attain an adequate definition, but often very difficult. The jurists are still without a definition of their concept of right.

knowledge of such a principle directly and immediately from the concepts alone. Discursive principles are therefore quite different from intuitive principles, that is, from axioms; and always require a deduction. Axioms, on the other hand, require no such deduction, and for the same reason are evident—a claim which the philosophical principles can never advance, however great their certainty. Consequently, the synthetic propositions of pure, transcendental reason are, one and all, infinitely removed from being as evident—which is yet so often arrogantly claimed on their behalf—as the proposition that *twice two make four*. In the Analytic I have indeed introduced some axioms of intuition into the table of the principles of pure understanding; but the principle[l] there applied is not itself an axiom, but serves only to specify the principle[m] of the possibility of axioms in general, and is itself no more than a principle[l] derived from concepts. For the possibility of mathematics must itself be demonstrated in transcendental philosophy. Philosophy has therefore no axioms, and may
A 734 never prescribe its *a priori* principles in any such absolute manner, but must
B 702 resign itself to establishing its authority in their regard by a thorough deduction.

3. *Demonstrations.*—An apodeictic proof can be called a demonstration, only in so far as it is intuitive. Experience teaches us what is, but does not teach us that it could not be other than what it is. Consequently, no empirical grounds of proof can ever amount to apodeictic proof. Even from *a priori* concepts, as employed in discursive knowledge, there can never arise intuitive certainty, that is, ‖demonstrative‖ evidence, however apodeictically certain the judgement may otherwise be. Mathematics alone, therefore, contains demonstrations, since it derives its knowledge not from concepts but from the construction of them, that is, from intuition, which can be given *a priori* in accordance with the concepts. Even the method of algebra with its equations, from which the correct answer, together with its proof, is deduced by reduction, is not indeed geometrical in nature, but is still constructive in a way characteristic of the science.[n] The concepts attached to the symbols, especially concerning the relations of magnitudes, are presented in intuition; and this method, in addition to its heuristic advantages, secures all inferences against error by setting each one before our eyes. While philosophical knowledge must do without this advantage,
A 735 inasmuch as it has always to consider the universal *in abstracto* (by means of
B 763 concepts), mathematics can consider the universal *in concreto* (in the single intuition) and yet at the same time through pure *a priori* representation, whereby all errors are at once made evident. I should therefore prefer to call the first kind *acroamatic* (discursive) *proofs*, since they may be conducted by the agency of words alone (the object in thought), rather than *demonstrations* which, as the term itself indicates, proceed in and through the intuition of the object.

From all this it follows that it is not in keeping with the nature of philosophy,

---

[l] ‖Grundsatz.‖

[m] ‖Prinzipium.‖

[n] ‖charakteristische Konstruction. The meaning in which Kant uses this phrase is doubtful. It might also be translated 'construction by means of symbols'.‖

especially in the field of pure reason, to take pride in a dogmatic procedure, and to deck itself out with the title and insignia of mathematics, to whose ranks it does not belong, though it has every ground to hope for a sisterly union with it. Such pretensions are idle claims which can never be satisfied, and indeed must divert philosophy from its true purpose, namely, to expose the illusions of a reason that forgets its limits, and by sufficiently clarifying our concepts to recall it from its presumptuous speculative pursuits to modest but thorough self-knowledge. Reason must not, therefore, in its transcendental endeavours, hasten forward with sanguine expectations, as though the path which it has traversed led directly to the goal, and as though the accepted premisses could be so securely relied upon that there can be no need of constantly returning to them and of considering whether we may not perhaps, in the course of the inferences, discover defects which have been overlooked in the princples, and which render it necessary either to determine these principles more fully or to change them entirely. A 736

I divide all apodeictic propositions, whether demonstrable or immediately certain, into *dogmata* and *mathemata*. A synthetic proposition directly derived from concepts is a *dogma*; a synthetic proposition, when directly obtained through the construction of concepts, is a *mathema*. Analytic judgements really teach us nothing more about the object than what the concept which we have of it already contains; they do not extend our knowledge beyond the concept of the object, but only clarify the concept. They cannot therefore rightly be called dogmas (a word which might perhaps be translated *doctrines*°). Of the two kinds of synthetic *a priori* propositions only those belonging to philosophical knowledge can, according to the ordinary usage of words, be entitled dogmas; the propositions of arithmetic or geometry would hardly be so named. The customary use of words thus confirms our interpretation of the term, namely, that only judgements derived from concepts can be called dogmatic, not those based on the construction of concepts. B 764

Now in the whole domain of pure reason, in its merely speculative employment, there is not to be found a single synthetic judgement directly derived from concepts. For, as we have shown, ideas cannot form the basis of any objectively valid synthetic judgement. Through concepts of understanding pure reason does, indeed, establish secure principles, not however directly from concepts alone, but always only indirectly through relation of these concepts to something altogether contingent, namely, *possible experience*. When such experience (that is, something as object of possible experiences) is presupposed, these principles are indeed apodeictically certain; but in themselves, directly, they can never be known *a priori*. Thus no one can acquire insight into the proposition that everything which happens has its cause, merely from the concepts involved. It is not, therefore, a dogma, although from another point of view, namely, from that of the sole field of its possible employment, that is, experience, it can be proved with complete apodeictic certainty. But though it needs proof, it should be entitled a *principle*, not a *theorem*, because it has the peculiar character that it makes possible

° [*Lehrsprüche.*]

the very experience which is its own ground of proof, and that in this experience it must always itself be presupposed.

Now if in the speculative employment of pure reason there are no dogmas, to serve as its special subject-matter,[p] all *dogmatic* methods, whether borrowed from the mathematician or specially invented, are as such inappropriate. For they only serve to conceal defects and errors, and to mislead philosophy, whose true purpose is to present every step of reason in the clearest light. Nevertheless
A 738 B 766 its method can always be *systematic*. For our reason is itself, subjectively, a system, though in its pure employment, by means of mere concepts, it is no more than a system whereby our investigations can be conducted in accordance with principles of unity, the material being provided by *experience* alone. We cannot here discuss the method peculiar to transcendental philosophy; we are at present concerned only with a critical estimate of what may be expected from our faculties—whether we are in a position to build at all; and to what height, with the material at our disposal (the pure *a priori* concepts), we may hope to carry the edifice.

D. FREGE ON KANT (*FREGE 1884)*

Frege's famous criticism of Kant in the *Grundlagen* (1884) is reproduced below, together with two of the passages criticized.

(i) From *Frege's* Grundlagen

The text is §5 of *Frege 1884*; the translation and editorial insertions are by John Langshaw Austin.

§5. We must distinguish numerical formulae, such as $2 + 3 = 5$, which deal with particular numbers, from general laws, which hold good for all whole numbers.

The former are held by some philosophers[1] to be unprovable and immediately

---

[p] |auch dem Inhalte nach.|

[1] Hobbes, Locke, Newton. Cf. Baumann, *Die Lehren von Zeit, Raum und Mathematik*, |Berlin 1868, Vol. I| pp. 241–42, 365 ff., 475–76. |Hobbes, *Examinatio et Emendatio Mathematicae Hodiernae*, Amsterdam 1668, Dial. I–III, esp. I, p. 19 and III, pp. 62–63; Locke, *Essay*, Bk. IV, esp. Cap. iv, §6 and cap. vii, §§6 and 10; Newton, *Arithmetica Universalis*, Vol. I, cap. i–iii, esp. iii, n. 24.|

self-evident like axioms. KANT[2] declares them to be unprovable and synthetic, but hesitates to call them axioms because they are not general and because the number of them is infinite. HANKEL[3] justifiably calls this conception of infinitely numerous unprovable primitive truths incongruous and paradoxical. The fact is that it conflicts with one of the requirements of reason, which must be able to embrace all first principles in a survey. Besides, is it really self-evident that

$$135\,664 + 37\,863 = 173\,527?$$

It is not; and KANT actually urges this as an argument for holding these propositions to be synthetic. Yet it tells rather against their being unprovable; for how, if not by means of a proof, are they to be seen to be true, seeing that they are not immediately self-evident? KANT thinks he can call on our intuition of fingers or points for support, thus running the risk of making these propositions appear to be empirical, contrary to his own expressed opinion; for whatever our intuition of 37 863 fingers may be, it is at least certainly not pure. Moreover, the term "intuition" seems hardly appropriate, since even 10 fingers can, in different arrangements, give rise to very different intuitions. And have we, in fact, an intuition of 135 664 fingers or points at all? If we had, and if we had another of 37 863 fingers and a third of 173 527 fingers, then the correctness of our formula, if it were unprovable, would have to be evident right away, at least as applying to fingers; but it is not.

KANT, obviously, was thinking only of small numbers. So that for large numbers the formulae would be provable, though for small numbers they are immediately self-evident through intuition. Yet it is awkward to make a fundamental distinction between small and large numbers, especially as it would scarcely be possible to draw any sharp boundary between them. If the numerical formulae were provable from, say, 10 on, we should ask with justice "Why not from 5 on? or from 2 on? or from 1 on?"

(ii) From *The critique of pure reason*

The first passage is from the *Introduction* to the second (or B) edition of the *Critique* (*Kant 1787*); it is followed by a passage from the chapter entitled *The schematism of the pure concepts of the understanding*, which appeared in both the A and B editions (*Kant 1781* and *1787*). The translation is by Norman Kemp Smith, and is reprinted from *Kant 1933*. References should be to the pages of the original German editions, *Kant 1781* and *1787*, as given in the margin.

---

[2] *Critique of Pure Reason*; Collected Works, ed. Hartenstein, Vol. III, p. 157 ⟦Original edns. A 164/B205⟧.

[3] *Vorlesungen über die complexen Zahlen und ihren Functionen*, p. 53.

We might, indeed, at first suppose that the proposition $7+5=12$ is a merely analytic proposition, and follows by the principle of contradiction from the concept of a sum of 7 and 5. But if we look more closely we find that the concept of the sum of 7 and 5 contains nothing save the union of the two numbers into one, and in this no thought is being taken as to what that single number may be which combines both. The concept of 12 is by no means already thought in merely thinking this union of 7 and 5; and I may analyse my concept of such a possible sum as long as I please, still I shall never find the 12 in it. We have to go outside these concepts, and call in the aid of the intuition which corresponds to one of them, our five fingers, for instance, or, as Segner[a] does in his *Arithmetic*, five points, adding to the concept of 7, unit by unit, the five given in intuition. For starting with the number 7, and for the concept of 5 calling in the aid of the fingers of my hand as intuition, I now add one by one
B 16 to the number 7 the units which I previously took together to form the number 5, and with the aid of that figure[b] ⟦the hand⟧ see the number 12 come into being. That 5 should be added to 7,[c] I have indeed already thought in the concept of a sum $= 7 + 5$, but not that this sum is equivalent to the number 12. Arithmetical propositions are therefore always synthetic. This is still more evident if we take larger numbers. For it is then obvious that, however we might turn and twist our concepts, we could never, by the mere analysis of them, and without the aid of intuition, discover what ⟦the number is that⟧ is the sum.

After what has been proved in the deduction of the categories, no one, I trust, will remain undecided in regard to the question whether these pure concepts of understanding are of merely empirical or also of transcendental employment; that is, whether as conditions of a possible experience they relate *a priori* solely to appearances, or whether, as conditions of the possibility of things in general, they can be extended to objects in themselves, without any restriction to our sensibility. For we have seen that concepts are altogether impossible,[a] and can have no meaning, if no object is given for them, or at least for the elements of which they are composed. They cannot, therefore, be viewed as applicable to things in themselves, independent of all question as to whether and how these may be given to us. We have also proved that the only manner in which objects
B 179 can be given to us is by modification of our sensibility; and finally, that pure
A 140 *a priori* concepts, in addition to the function of understanding expressed in the category, must contain *a priori* certain formal conditions of sensibility, namely, those of inner sense. These conditions of sensibility constitute the universal

[a] ⟦*Anfangsgründe der Arithmetik*, translated from the Latin, second edition, Halle, 1773, pp. 27, 79.⟧

[b] ⟦an jenem meinem Bilde.⟧

[c] ⟦Reading, with Erdmann, 5 zu 7.⟧

[a] ⟦Altered by Kant (*Nachträge* lviii) to: "are for us without meaning."⟧

condition under which alone the category can be applied to any object. This formal and pure condition of sensibility to which the employment of the concept of understanding is restricted, we shall entitle the *schema* of the concept. The procedure of understanding in these schemata we shall entitle the *schematism* of pure understanding.

The schema is in itself always a product of imagination. Since, however, the synthesis of imagination aims at no special intuition, but only at unity in the determination of sensibility, the schema has to be distinguished from the image. If five points be set alongside one another, thus, ..., I have an image of the number five. But if, on the other hand, I think only a number in general, whether it be five or a hundred, this thought is rather the representation of a method whereby a multiplicity, for instance a thousand, may be represented in an image in conformity with a certain concept, than the image itself. For with such a number as a thousand the image can hardly be surveyed and compared with the concept. This representation of a universal procedure of imagination B 180
in providing an image for a concept, I entitle the schema of this concept.

Indeed it is schemata, not images of objects, which underlie our pure sensible A 141
concepts. No image could ever be adequate to the concept of a triangle in general. It would never attain that universality of the concept which renders it valid of all triangles, whether right-angled, obtuse-angled, or acute-angled; it would always be limited to a part only of this sphere. The schema of the triangle can exist nowhere but in thought. It is a rule of synthesis of the imagination, in respect to pure figures in space. Still less is an object of experience or its image ever adequate to the empirical concept; for this latter always stands in immediate relation to the schema of imagination, as a rule for the determination of our intuition, in accordance with some specific universal concept.

5
Johann Heinrich Lambert (1728–1777)

Lambert's principal contribution to the philosophy of mathematics is his discussion of the status of the Axiom of Parallels—a discussion which raises issues that will be encountered in the selections from Gauss, Riemann, von Helmholtz, Klein, Poincaré, and Hilbert. But before we consider Lambert's work on this problem, some words of historical background on the Axiom of Parallels are in order.

Even before Euclid, Aristotle had spoken of the sophistries that occurred in the theory of parallel lines—in particular, the logical fallacy of *petitio principii*:

To beg and assume the point at issue is a species of failure to demonstrate the problem proposed; but this happens in many ways. A man may not deduce at all, or he may argue from premisses which are more unknown or equally unknown, or he may establish what is prior by means of what is posterior; for demonstration proceeds from what is more convincing and prior. Now begging the point at issue is none of these; but since some things are naturally known through themselves, and other things by means of something else (the first principles through themselves, what is subordinate to them through something else), whenever a man tries to prove by means of itself what is not known by means of itself, then he begs the point at issue. This may be done by claiming what is at issue at once; it is also possible to make a transition to other things which would naturally be proved through the point at issue, and demonstrate it through them, for example if *A* should be proved through *B*, and *B* through *C*, though it was natural that *C* should be proved through *A*; for it turns out that those who reason thus are proving *A* by means of itself. This is what those persons do who suppose that they are constructing parallel lines; for they fail to see that they are assuming facts which it is impossible to demonstrate unless the parallels exist. So it turns out that those who reason thus merely say a particular thing is, if it is: in this way everything will be known by means of itself. But that is impossible.[a]

Euclid's *Elements* removed the *petitio* by explicitly stating the Parallel Postulate as an axiom. Euclid's Parallel Postulate says:

If two straight lines are cut by a third straight line so that the interior angles on the same side are less than two right angles, then the two straight lines, if produced indefinitely, meet on that side on which the angles are less than the two right angles.

---

[a] *Prior analytics*, II 16, 65a 4 (translated by A.J. Jenkinson in *Aristotle 1984*). See also *Prior analytics*, II 17, 66a 11–15, and *Posterior analytics*, I 5, 74a 13–16. For a commentary on this passage see the discussion by Sir Thomas Heath in *Euclid 1925* (Vol. i, pp. 308–9) or in *Heath 1921* (Vol. i, pp. 335–48).

Already in antiquity this Postulate was criticized for not being self-evident; in particular, it was objected that the existence of asymptotic approximations shows that two curves can converge indefinitely without intersecting. (This objection is criticized by Lambert in the following selection.) Proclus, for example, in his *Commentary* on the Postulate, said:

This ought even to be struck out of the Postulates altogether; for it is a theorem involving many difficulties, which Ptolemy, in a certain book, set himself to solve, and it requires for the demonstration of it a number of definitions as well as theorems. And the converse of it is actually proved by Euclid himself as a theorem. It may be that some would be deceived and would think it proper to place even the assumption in question among the postulates as affording, in the lessening of the two right angles, ground for an instantaneous belief that the straight lines converge and meet. To such as these Geminus correctly replied that we have learned from the very pioneers of this science not to have any regard to mere plausible imaginings when it is a question of the reasonings to be included in our geometrical doctrine. For Aristotle says that it is as justifiable to ask scientific proofs of a rhetorician as to accept mere plausibilities from a geometer; and Simmias is made by Plato to say that he recognizes as quacks those who fashion for themselves proofs from probabilities. So in this case the fact that, when the right angles are lessened, the straight lines converge is true and necessary; but the statement that, since they converge more and more as they are produced, they will sometime meet is plausible but not necessary, in the absence of some argument showing that this is true in the case of straight lines. For the fact that some lines exist which approach indefinitely, but yet remain non-secant ‖*ἀσύμπτωτοι*‖, although it seems improbable and paradoxical, is nevertheless true and fully ascertained with regard to other species of lines. May not then the same thing be possible in the case of straight lines which happens in the case of the lines referred to? Indeed, until the statement in the Postulate is clinched by proof, the facts shown in the case of other lines may direct our imagination the opposite way. And, though the controversial arguments against the meeting of the straight lines should contain much that is surprising, is there not all the more reason why we should expel from our body of doctrine this merely plausible and unreasoned ‖hypothesis‖?

It is then clear from this that we must seek a proof of the present theorem, and that it is alien to the special character of postulates. But how it should be proved, and by what sort of arguments the objections taken to it should be removed, we must explain at the point where the writer of the Elements is actually about to recall it and use it as obvious. It will be necessary at that stage to show that its obvious character does not appear independently of proof, but is turned by proof into matter of knowledge. (Quoted in *Euclid 1925*, Vol. i, pp. 202–3. The passage is also translated in *Proclus 1970*, pp. 150–1.)

Moreover, in his commentary on Euclid's Proposition I: 29, Proclus considered the possibility (to be realized in Lobatchevsky's geometry) that the straight lines of Euclid's postulate might converge asymptotically for all interior angles *slightly* less than two right angles; that is to say, only after the interior angles had been sufficiently decreased would the lines actually intersect.

For more than two thousand years, mathematicians attempted to prove the Parallel Postulate; the literature is enormous, and one bibliography (*Riccardi 1887–93*) lists twenty pages of titles of works on the Parallel Postulate between

1607 and 1887. For a compendious account of the principal efforts to prove the Postulate, the reader is referred to T.L. Heath's discussion in *Euclid 1925* (Vol. i, pp. 202–20). Although these attempts to prove the Parallel Postulate did not succeed, they did clarify the logical implications of the Postulate within Euclidean geometry, and they led to such fruitful innovations as Desargues's 1639 explanation of parallel lines as lines which intersect at the same infinitely distant point. Most of the proofs involved assuming the Parallel Postulate in an equivalent but disguised form; for instance, John Wallis, in the *De postulato quinto*, proved the Parallel Postulate from the postulate of the existence of similar figures of arbitrary size (*Wallis 1695–9*, Vol. ii, pp. 665–78); similarly, Legendre, in the appendices to the twelve editions of his *Elements of geometry* (*1794*), gave many proofs of the Postulate from such starting-points as the assumption that the sum of the angles of a triangle equals two right angles (*Legendre 1833*). Elegant accounts of these proofs are given by Heath, *loc. cit.*

Another line of approach to the Axiom of Parallels commences with the work of Gerolamo Saccheri (1667–1733); Lambert's researches stand in this tradition. In his *Euclid freed from every flaw* (*1733*) Saccheri considered the quadrilateral ABCD in which AC = BD and in which CAB and DBA are both right angles.

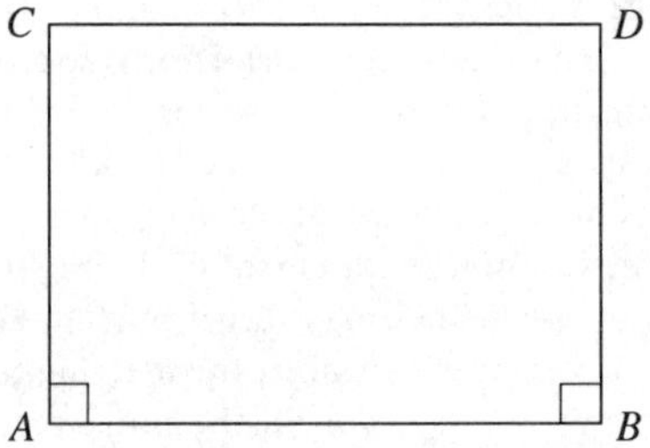

He easily showed that the angles at C and D are equal; if the parallel postulate were true, they would be right angles. Saccheri explored the consequences of assuming that the angles at C and D are acute and of assuming that they are obtuse—the 'acute hypothesis' and the 'obtuse hypothesis', as he called them. He hoped to derive a contradiction from each of these hypotheses, and thereby to prove the Parallel Postulate. Instead, he unwittingly derived many of the basic principles of non-Euclidean geometry.

The most important research on non-Euclidean geometry was eventually to be carried out in Germany during the nineteenth century, by mathematicians such as Gauss, Riemann, von Helmholtz, and Klein. But the Germans were latecomers to this field. Although Christoph Schlüssel's discussion of the Axiom of Parallels (in his 1574 commentary on Euclid) was mentioned by Wallis in the *De postulato quinto*, German mathematicians ignored the problem of parallel lines until 1758, when Abraham Gotthelf Kästner (1719–1800) called attention to its importance in the preface to his influential *Anfangsgründe der Arithmetik und Geometrie* (*Kästner 1758*). Kästner (who was later to be the teacher of Gauss) wrote:

The difficulty which arises in the theory of parallel lines has occupied me for many years. I used to believe that it had been entirely removed by Hausen's *Elementa matheseos* ‖1734‖. The former preacher to the French congregation in Leipzig, Mr. Coste, shook my complacency when, during one of the walks that he often granted me, he mentioned that, in the above-mentioned work by Hausen, an inference is made that does not follow. I soon discovered this mistake myself, and from that moment exerted myself either to remove the difficulty or to find an author who had removed it; but both efforts were in vain, although I soon assembled virtually a small library of individual writings or works on the first principles of geometry where this topic was considered especially closely. After the present work caused me to reflect on the subject anew, I have been able to find no expedient that comes closer to satisfying me than the one I have adopted in the corollary to theorem eleven and in theorem twelve.

(The 'expedient' mentioned by Kästner was essentially that of Wallis—i.e. assuming the existence of similar figures of arbitrary size.)

Kästner's 'small library' of writings on the Axiom of Parallels was used by his student Georg Simon Klügel to produce a Göttingen dissertation (*Klügel 1763*) containing the first extensive discussion of the history of the Parallel Axiom. Klügel examined some thirty attempts to prove the Axiom, and showed that each was a failure. He concluded 'To be sure, it might be possible that non-intersecting lines diverge from each other. We know that such a thing is absurd, not in virtue of rigorous inferences or clear concepts of straight and crooked lines, but rather through experience and the judgement of our eyes' (*Klügel 1763*, p. 16). Kästner himself is alleged to have despaired of ever finding a proof of the Axiom, and to have recommended that it simply be accepted as a given (see *Schweickart 1807*, p. 6). These inconclusive remarks appear to be the first tentative expressions of the view that the Axiom of Parallels might not be provable, and that its plausibility might rest on experience.

Klügel's dissertation—and in particular his discussion of Saccheri's attempted proof of the Axiom—seems to have inspired the work of Johann Heinrich Lambert, who cites it in §3 of the selection that follows. Lambert was a self-taught Swiss mathematician, astronomer, physicist, economist, logician, cartographer, meteorologist, and philosopher. At the suggestion of his fellow-countryman Leonhard Euler he emigrated to Berlin in 1764, and was elected to the Academy of Sciences in 1765. Lambert wrote voluminously on scientific topics, and published some 190 essays on philosophy, astronomy, mathematics, and the physics of light and heat. He also published 21 books, and his *Nachlass* contains a further 837 manuscripts. He developed a highly personal system of dress (multicoloured) and of etiquette (which required him to stand at right angles to his interlocutors). Frederick the Great, on meeting this strange creature for the first time, is said to have exclaimed that the biggest blockhead in the kingdom had been proposed to him for membership in the Prussian Academy. (Frederick later became Lambert's vigorous supporter.) In mathematics, Lambert is now best remembered for being the first to prove the irrationality of π and e; the historian of mathematics Moritz Cantor ranked him after Euler, Lagrange, and Laplace as one of the great mathematicians of his generation. Further details of his life and scientific publications can be found in *Scriba 1973*.

Lambert's scientific work was closely tied to his writings on philosophy. He published two important works of metaphysics, the *New organon, or Thoughts on the investigation and designation of truth and of the distinction between error and appearance* (*1764*) and the *Foundations of architectonic, or Theory of the simple and primary elements in philosophical and mathematical knowledge* (*1771*), in which he attempted to reform the systems of Locke, Leibniz, and Wolff. He began by investigating the origins and scope and interrelationships of the basic *a priori* concepts of metaphysics—to present them as the foundation for metaphysics and the empirical sciences. He then attempted to construct the propositions of metaphysics by deducing them from the basic concepts in analogy with the mathematical method. Once metaphysics had been constructed in this mathematically exact way, the deduced propositions were to be applied to the empirical subject-matter of the individual sciences, for which they would form the foundation. Lambert thus was led to discuss epistemological questions at the foundations of science; the nature of axioms; the possibility of a logical calculus; the relationship between mathematical knowledge and metaphysical knowledge; the nature of logic and its relationship to language and the theory of signs.

Lambert, like Leibniz, tried to develop an *ars characteristica combinatoria*, a logical calculus in which syllogistic reasoning would be represented by algebraic operations. Lambert represented the combination of two concepts a and b into a common concept by $a + b$; and the common part of the two concepts by ab. He then explored the laws governing these operations of 'addition' and 'multiplication,' and introduced inverse operations of 'subtraction' and 'division'. However, he had no effective way of expressing logical negation, and his system did not embrace the whole even of syllogistic logic; that task was not to be accomplished until Boole's *Mathematical analysis of logic* in 1847.

Lambert was greatly admired by Kant, especially for his celebrated *Cosmological letters* (*1761*) and for his *New organon*. Indeed, many of the insights of the *Critique of pure reason* can be found in embryo in the writings of Lambert. The two thinkers carried out a regular correspondence, and Kant, in a letter to Johann Bernoulli of 16 November 1781, acknowledged Lambert's influence on the discussion of space and time in the *Critique of pure reason*. (Indeed, at one point Kant intended to dedicate the *Critique* to Lambert; Lambert, however, died before the work was completed. This Johann Bernoulli, incidentally, was the nephew of the mathematician Johann Bernoulli (1667–1748).)

Lambert's most extensive treatment of the foundations of Euclidean geometry is his *Theory of parallel lines* (*1786*). This work was not published by Lambert himself, presumably because he was dissatisfied with its failure to solve the problem of the Axiom of Parallels. The manuscript was first published in 1786, after Lambert's death, by Johann Bernoulli, who gives the date of composition as 1766.

In this work, Lambert followed Saccheri's strategy for proving the Axiom of Parallels. (He does not, however, appear to have known Saccheri's work at first hand, but only the brief account of it given in Klügel's dissertation.) Lambert went much further than Saccheri in deducing the geometrical consequences of the obtuse and acute hypotheses and their implications for the sums of the angles of a triangle and for the measurement of area; in effect, his work amounted to an unintentional deduction of the basic properties of Riemannian and Lobatchevskian geometry. In particular, he observed (§82) that the obtuse hypothesis holds for triangles on the surface of a sphere, and he conjectured that the acute hypothesis holds on a sphere with imaginary radius. This was a remarkably prescient conjecture, and was only to be fully confirmed in 1829 with the first publication of Lobatchevsky's work on non-Euclidean geometry. But Lambert's attempt to prove the Axiom of Parallels in effect presupposed Bolyai's axiom (that it is always possible to draw a circle through any three points of the plane), an axiom which is equivalent to the Axiom of Parallels. The inadequacy of Lambert's proposed proof at the end of his treatise (§88) was quickly noted when the work was published, and was very probably obvious to Lambert as well—a fact which may explain his failure to publish the manuscript himself.

The passage translated below is the methodological introduction to the *Theory of parallel lines*, where Lambert discusses the logical and philosophical status of the Axiom of Parallels and of various attempts to prove it. After an analysis of Euclid's methodology and of the controversy surrounding the Axiom,[b] Lambert (§§4–8) criticizes Wolff's attempt to *define* parallel lines in such a way as to remove the difficulty. For (§5) the Wolffian definition is not obtained by abstraction from things of which we have experience—both because actual 'straight lines' are never precisely straight, and because we have no experience of lines that are infinitely extended. So Wolff's definition must be an 'arbitrarily conjoined concept'. But then Wolff must supply a justification for introducing such a concept; and this he conspicuously fails to do. Lambert concludes that Wolff has merely moved the difficulty from the axioms to the definitions.

Lambert's own strategy is set out in §10 and 11. He first argues that the *truth* of the Axiom of Parallels is not what is at issue: for its plausibility is established inductively by the plausibility of the consequences one is able to derive from it. Rather, the question is *whether the Parallel Postulate can be derived by rigorous, logical inferences from the other Euclidean axioms*. And here (§11) Lambert makes it clear that by 'derivability' he means a purely syntactic proof which abstracts altogether from the intended reference of terms like 'parallel line'. Lambert gives a remarkably clear early statement of what was later to be the kernel of Hilbert's approach to the axioms of geometry: 'It must be possible', Hilbert said, 'to replace in all geometic statements the words, *point*, *line*, *plane* by *table*, *chair*, *mug*.'[c]

---

[b] In a roughly contemporaneous letter to Baron Georg Johann von Holland (1742–84), Lambert discusses these matters further, and makes clear that the difficulty with parallel lines is that one must imagine them as being extended 'into the *infinite*' [ins *Unendliche*] in both directions. The letter, which is dated 11 April 1765, is reproduced in *Stäckel and Engel 1895* (pp. 141–2).

[c] (*Weyl 1944b*; reprinted in *Weyl 1968*, Vol. iv, p. 153.) Lambert's discussion of axioms and definitions should also be compared with Gergonne's important *Essay on the theory of definitions* (*Gergonne 1818*), where the notion of implicit definitions occurs for the first time.

Lambert thus held in his hands most of the tools necessary for justifying and developing non-Euclidean geometry: the idea that Saccheri's obtuse and acute hypotheses could be modelled on the surfaces of spheres (possibly of imaginary radius); a rigorous deduction of many of the basic properties of Riemannian and Lobatchevskian geometry; a clear understanding of the methodological issues; and the concept of an uninterpreted, formal system of axioms. The issues Lambert touches on in this short piece—the status of axioms and definitions, the problem of the infinite in geometry, the logical character of deductions within a formal axiom system, the nature of mathematical proof, the truth of Euclidean geometry, the reference of algebraic symbols—were to play a large role in the mathematics of the nineteenth century, and will recur in most of the selections that follow. But Lambert—unlike Gauss—never made the leap to conceiving of the possibility of alternative geometries, and his subtle methodological distinctions were intended instead to pave the way for a proof of the Axiom of Parallels, a proof whose possibility he appears never to have doubted. And in contrast to Hilbert, Lambert did not actually *use* the idea of a purely symbolic calculus to obtain mathematical theorems; and so his idea, despite the clarity of his formulation, lay neglected until Pasch and Hilbert independently rediscovered it at the end of the nineteenth century. (Even Stäckel and Engel, who edited and reprinted Lambert's treatise in their *1895*, attached no particular importance to his §11.)

The translation is by William Ewald from the version in *Stäckel and Engel 1895*; references to *Lambert 1786* should be to the section numbers, which appeared in the original edition.

A. *FROM THE* THEORY OF PARALLEL LINES (*LAMBERT 1786*)

1. Preliminary considerations

§1.

The present treatise deals with a difficulty that arises in the very beginnings of geometry, and that since the days of Euclid has been a stumbling-block to those who do not wish to believe the doctrines of this science on the testimony of others, but wish rather to be convinced by reasons, and never to forgo the rigour that they find in most proofs.

This difficulty leaps to the eye of anybody who reads Euclid's *Elements*—and does so at the very outset, for it does not arise in the theorems but in the axioms with which Euclid prefaces the First Book. The eleventh[a] axiom assumes

[a] [As the text makes clear, Lambert is here referring to the Axiom of Parallels—the Fifth Axiom in Heath's edition of Euclid (*Euclid 1925*).]

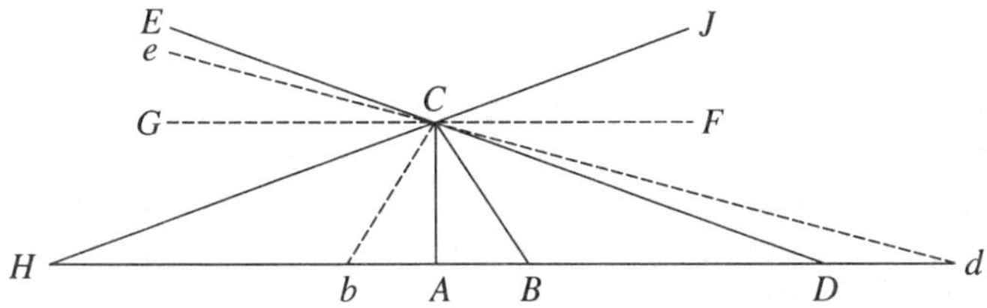

as something clear and in no need of proof, *that, when two lines CD, BD* (Fig. 1) *are intersected by a third, and the two inner angles DCB, DBC taken together, are less than two right angles, then the two lines CD, BD meet on the side of D, or the side where these angles are found.*

§2.

This axiom is incontestably neither as clear nor as evident as the others; and one's natural reaction is not merely to desire a proof of it—instead, one somehow feels that it is capable of proof, that there must exist a proof of it.

As I see the matter, this is the *first* reaction. But if one reads further in Euclid, one must not only admire the care and acuteness of his proofs, and a certain noble simplicity in his manner of proceeding; but one becomes even more perplexed about his eleventh axiom when one sees that he proves theorems that one would far more readily have conceded without proof.

It is sometimes said that Euclid did this in order to make his doctrines secure against even the most hair-splitting objections of the Sophists. But if this is so, then I confess that I am utterly unable to comprehend these Sophists, if Euclid could imagine that they would not contest his eleventh axiom, just because without it the majority of the theorems of geometry would collapse. One should rather think that Euclid and the Sophists (if these latter did not raise any objections in Euclid's day) must have had standards for judging the axioms and the manner of executing a geometrical proof which were quite different from the standards of those who thought about this matter in later times, or who made difficulties about the proofs that had been attempted by others.

Of these difficulties or objections, one that occurs to me is the following: that in order to prove the Euclidean axiom rigorously, or to establish ‖festzusetzen‖ geometry at all, one may neither *visualize* nor make *a representation of the thing itself*.[b] It is clear that with such a requirement one can also contest the twelfth Euclidean axiom, *that two straight lines do not enclose a space*.[c]

---

[b] ‖'weder *sehen* noch sich *von der Sache selbst* eine *Vorstellung* machen dürfe'. *Sache* in Lambert's usage sometimes means the subject-matter of Euclidean geometry, and sometimes a particular geometrical object.‖

[c] ‖This axiom was an interpolation into Euclid, and was criticized by Proclus as superfluous; in some editions of the *Elements* it appears as 'Common Notion 9'. See the discussion by Heath in *Euclid 1925*, p. 232.‖

§3.

But it seems to me equally clear that the Sophists of Euclid's day were less strict, and that they must have conceded the *representation* ‖*Vorstellung*‖ *of the thing*. But with this presupposition Euclid's manner of proceeding can be quite adequately justified (at least in the absence of an alternative, and because it is subject to fewer difficulties).

Specifically, one can put off the eleventh axiom until one comes to *Prop.* XXIX of the first book. Meanwhile, one learns *to know the thing itself* (of which the axiom speaks) and also to add in thought that which seems to be missing in the axiom and in its representation, even if one cannot express this in words. In the two immediately preceding Propositions XXVII and XXVIII one learns that, if the angles

$$FCB + CBD = 180°$$

or if the angles $FCB = CBA$, then the lines AB, CF do not intersect in the direction either of F or of G. One thereby learns that the 34th definition[d] is not an

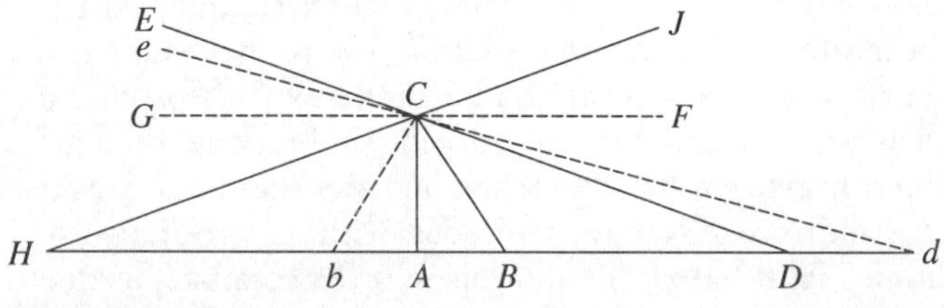

absurdity ‖Unding‖ or a figment of the imagination; but that non-intersecting straight lines actually occur in the realm of reality. For until now this definition remained open; and until now one could also allow the axiom to remain open, because it stands in close connection with the parallel lines, and as it were marks the boundary between parallel lines and lines that meet.

What one now further *imagines* ‖sich *vorstellt*‖ in order to convince oneself of the correctness and *thinkability* of the axiom comes in my opinion to this: one imagines *CF, AB* in Propositions XXVII or XXVIII as not converging, and imagines an arbitrary *straight* line *CD* drawn through the angle *BCF*; and then one knows that, no matter how small the angle *DCF* may be, necessarily

$$DBC + BCD < 180°$$

and consequently satisfies the condition of the axiom. If one should now in the same way imagine that *CD, BD* converge, then one will need to *imagine* the lines *CF, CD, AD* as *straight* lines. And this thought-experiment shows that as *CD* is extended it not only increases its distance from *CF* but also approaches *AD* in such a way that it must intersect it at some distance *BD*.

[d] ‖In *Euclid 1925* it is the 23rd.‖

Whoever at this point objects that *CD* could perhaps approach *AD* asymptotically (like, for instance, the hyperbola and other asymptotic bent lines) in my opinion changes what the logicians call the *statum quaestionis*, or he deviates from Euclid, where the talk is not about *proofs* but about *representation* and the *thinkability*[e] *of the thing*—because one can certainly assume of Euclid that he would not otherwise have counted or placed his proposition among the *axioms*. But if it is a matter of the *representation*[f] *of the thing*: then I do not see how in the representation of *straight* lines objections about hyperbolas can be made. One could equally well doubt that two straight lines cannot be placed together so as to enclose a space—as happens when two circular arcs of equal size are placed together.

I mention this only to show that, if you start by presupposing the actual representation of the thing, and if you do not simply demand only words, then Euclid's procedure can be justified; all the more so because his way of proceeding, so far as I am aware, even today encounters fewer difficulties than all the attempts that have been made since his time to proceed differently. Here one can read a short and concisely written dissertation by Herr Klügel, who illustrates, with ingenuity and moderation, the defects that occur in such attempts—often with hidden logical circles, gaps, leaps, paralogisms, and incorrectly used and gratuitously assumed definitions and axioms.

§4.

Although (as this dissertation relates) many such daring attempts have appeared in print in this century, there is no doubt that there would not have been so many, particularly in Germany, if Wolff (who, for a period of forty or more years, was *dux gregis* with respect to these geometrical writings—and who, to be sure, deserved to be, for many good reasons)—if Wolff, I say, had had a better sense of the above-mentioned difficulty, or, more importantly, if he had been more critical in his first principles. The latter course would for obvious reasons have elicited a multitude of writings on the subject. And the former, as I see the matter, would even have had a marked influence on Wolff's world-view.

The problem is not that Wolff did not know full well *that arbitrarily conjoined concepts must be established* ‖*erwiesen*‖. He emphasized the point in both of his treatises on reason, and even in his preliminary reports on mathematical method, and he illustrated it with examples from geometry. But I conclude from this that Wolff must not have regarded his definition of parallel lines as an *arbitrarily conjoined concept*, because I am confident that otherwise he would have thought about giving a proof of their possibility, or at least remembered that something remained to be done; or else he would have adhered to Euclid's method, and then the difficulties would have become apparent, as they did in Euclid.

---

e ‖*Gedenkbarkeit.*‖
f ‖*Vorstellung.*‖

But if we ask why Wolff, not thinking of anything arbitrary, contented himself with calling the parallel lines *equidistant*, we must then assume that he found this concept by applying his other method of finding concepts, namely, by *abstraction from individual examples*. He says of such concepts and definitions that they need no further proof. This I concede. But in the execution of the argument one must nevertheless play fair with the reader and show him how one has abstracted the concept. Otherwise the reader could legitimately suppose that a *vitium subreptionis* has slipped in. For concepts that one abstracts from examples are also to that extent always *a posteriori*; and one can regard them *a priori* only if, after one has found them, they are thinkable for themselves[g]—that is to say, are *simple*. Otherwise, if one wishes to fend off the suspicion of a *vitium subreptionis*, one must produce the examples for the reader and give an account of all the precautions one has taken in the abstraction.

Bülfinger[h] was well aware of the necessity of this procedure, and precisely for that reason he was in a better position than Wolff himself to lessen the difficulties that had arisen for the Wolffian world-view. But it would have been desirable if Wolff himself, in the central chapters of his two treatises on reason (where he deals in part with definition, and in part with the written presentation of dogmatic propositions), had shown, in detail and with full emphasis, both the manner whereby one can fend off the suspicion of a *vitium subreptionis* in working with definitions that have been found by abstraction, and the necessity for doing so.

§5.

Now, for the definition of parallel lines this would have been quite out of the question. For however many of these one draws, there remain two marked deficiencies. First, the drawing lacks geometrical precision. Second, it is absolutely impossible to continue both lines into the infinite. And so one does not get anywhere *a posteriori* and with abstraction; and the definition (or, to put it better, the possibility of the thing) must be established from other and simpler grounds that are thinkable for themselves.

Wolff incontestably did not make these reflections. And one finds in him indications from which one can clearly infer that he conceded too much to the definitions; and because he wanted to arrange them suitably for the subject-matter ‖Sache‖, he imported the difficulties that are in the subject-matter into the definitions.[i] That in the case of parallel lines they were often more hidden there than in the subject-matter itself, one could at any rate infer from the fact that in those times, when a widespread mania for demonstrations was the

[g] ‖für sich gedenkbar.‖

[h] ‖Georg Bernhard Bilfinger or Bülfinger (1693–1740), author of *Dilucidationes philosophicae*; for a discussion of his work, see *Zeller 1875* p. 231.‖

[i] ‖In Lambert's letter of 3 February 1766 to Kant, he remarks: 'Wolff assumed nominal definitions as it were *gratis*, and, without noticing it, he shoved or hid all the difficulties there.'‖

prevailing fashion, more of a fuss would have been made had Wolff retained the Euclidean procedure in his *Anfangsgründe der Messkunst*.

§6.

I just said that Wolff conceded too much to the definitions. Now, this happened in actual fact rather than expressly in words; and for many it became the fashion, without noticing it, *to believe that they had no concept whatsoever of a thing* ‖*Sache*‖ *if its name had not been defined*. Even all axioms had to be preceded by definitions, without which they supposedly could not be understood. So it is hardly surprising that the proposition: *every definition, until it has been proved, is an empty hypothesis*—that this proposition, which Euclid knew so well, and which he continually observed, was forgotten, even if it was not entirely lost.

I note this here all the more, because it had very detrimental consequences for the procedure of the philosophical sciences; also because it is precisely the point where Wolff lagged behind when he abstracted his method from Euclid; and finally because parallel lines give the most obvious example, that *a definition given in advance, so long as it has not itself been established, proves nothing*.

§7.

It is false that, before he had established the possibility of the thing, Euclid used any of his definitions other than as a *mere hypothesis*, or that he regarded it as a *categorical principium demonstrandi*. For him, the expression *per definitionem* means no more than *per hypothesin*. And if one looks more closely: he does not obtain the *categorical* features in his theorems from the *definitions* but actually and primarily from the *postulates*. It is to these that Cicero's words actually apply: *si dederis, danda sunt omnia*.[j]

Among the axioms, I find that only the eleventh contains a positive category which immediately concerns the figures. But this is precisely the axiom that one does not wish to count as valid. What is *categorical* in it is supposed to be extracted from the postulates by inferences. The others concern for the most part only the concept of *equality* and *inequality*, and for that reason, because they concern *relation-concepts*, they do not belong to the *matter*, but to the *form of the inferences* that Euclid makes in his proofs, where they always appear only as auxiliary propositions. The twelfth axiom, *that two straight lines enclose no space*, is negative, and, like the ninth, *that the whole is greater than its part*, is used by Euclid where the proof is *apagogic*, or where the truth of the proposition is established from the impossibility of the opposite.

---

[j] ‖The quotation occurs in Cicero's *De finibus bonorum et malorum*, Book 5, XXVIII, 83. Cicero is praising the cogency of an argument in moral philosophy, and remarks, *Ut in geometria, prima si dederis, danda sunt omnia*'—'As in geometry, if you have conceded the first ‖principles‖, all must be conceded'.‖

§8.

This is a brief sketch of the spirit of the Euclidean method; I find little or nothing of it in Wolff's theories of reason, and often the opposite in his procedure and arguments.

For example, Wolff, with several others, believes that one could remove the difficulty caused by Euclid's eleventh axiom if one were to alter his definition of parallel line. But this would neither remove it, nor avoid it, nor yet get around it in a clever way and, as it were, remove it indirectly. Rather (if all goes well) the difficulty is only *taken away from the axiom* and *brought into the definition*; so far as I can see, it is not in the process made any easier to remove. Indeed, Euclid's definition can be proved without regard for his eleventh axiom. Wolff's definition, on the other hand, either cannot be proved without this axiom, or, if it can, then the axiom is as good as established at the same time.

But actually it is not a question of the definition at all. One can leave it wholly out of consideration in Euclid; and in Props. XXVII and XXVIII one will replace the expression *parallelae lineae* with *lineae sibi non coincidentes*.[k] And, when one notices that this is a peculiar property, one will oneself hit on a short and convenient *term* for it, or give a *name* to such lines which never intersect no matter how far one extends them on either side. And one will be even more encouraged to do this when one subsequently sees that precisely these lines also remain the same distance from each other.

This is the genuine synthetic way of proceeding, in which one thinks of the *term* only after the thing ‖Sache‖ has been brought forward and only if it is significant enough to deserve a special name. There are countless examples in mathematics, and there should be others in any science where one can (or thinks one can) proceed *a priori*.

§9.

Proclus, who also found Euclid's eleventh axiom problematical, demanded a proof of it on the grounds that *its converse can be established*.

In fact, the converse proposition is proved in Bk. I, Prop. XVII. It seems to me quite right that for an axiom it must be clear for itself ‖für sich‖ what sort of a reason there is for the axiom or its converse. For precisely speaking an axiom should consist purely of simple concepts that are thinkable for themselves; and it must be evident immediately from the representation ‖Vorstellung‖ of the concepts whether and to what extent they can be conjoined.

So, for example, the eighth Euclidean axiom, *that extended quantities which are congruent with one another are equal (Quae sibi mutuo congruunt, sunt aequalia)*—this proposition is thinkable for itself. But the following is also in

---

[k] ‖'lines which do not intersect themselves'.‖

the same way thinkable for itself: that the converse holds only for straight lines and angles, while for figures one needs yet another condition, namely that of similarity, if the converse is to apply.

§10.

In order after these general reflections to come closer to the theory of parallel lines (where I intend both to make the difficulties apparent and to remove them) I shall first state the actual *statum quaestionis*.

First, the *question itself* concerns *neither the truth nor the thinkability of the Euclidean axiom*: things would have looked bad hitherto for the greater part of geometry if this were the question. As for *thinkability*, I have already shown (§3) the order in which it arises in the reading of Euclid. That the axiom is also thought of as *true* is entirely clear. But its truth is also established from all the consequences that one draws from it in every respect—established in such an illuminating and necessary way that one can regard these consequences, taken together, as an *induction* that is in many ways complete.

Then one also finds, in the many attempts that one can make to prove this axiom, that it almost always presupposes itself in the proof, and is a consequence of itself in many different ways, but that there is no way of refuting it.

This may also be a reason why Euclid, in the absence of a proof, included it among the axioms; especially since he chose the definition that could be established without recourse to this axiom, and which could be most immediately linked to it. For one sees quite clearly that his Prop. XXIX, where this axiom is used, serves principally only to prove that that there are no parallel lines other than the ones established in the two Props. XXVII and XXVIII. And in this respect a very small gap is thereby filled, because one can imagine without difficulty that only those among the non-intersecting lines remain to be excluded that make a *smaller* angle with *CF* (Fig. 1) than do all those lines *CD, Cd* whose intersection *D, d* can be given—that is, whose intersection has a finite distance from *A*. For, if one rotates *CF* around the point *C* down towards *D*, then Prof. Kästner remarks correctly that the first intersection point cannot be given because, wherever one locates it on *AD*, one can always find another one that is more distant. But in my opinion this has the consequence that, where the angles *DCF. dCF* are very small, the distances *AD, Ad* must increase in

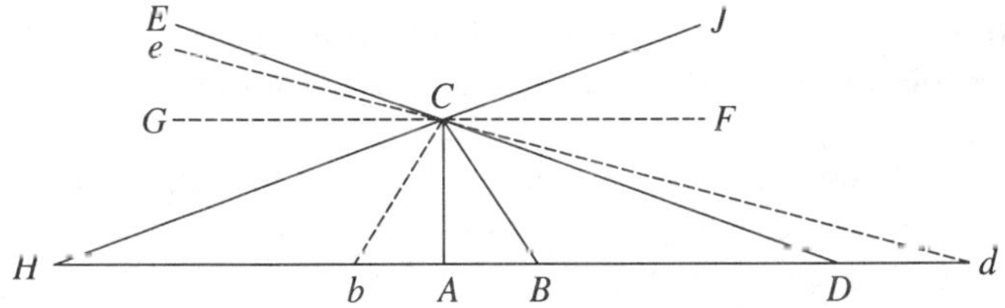

inverse proportion to the angles *DCF, dCF* (or a function of them that is not very different). For they cannot increase in direct proportion to the angles *ACD, ACd* (or a function of them) because otherwise, if

$$DAC + ACF = 180°$$

or even if it is greater, then *CF* would have to intersect the line *AD* at a finite distance from *A*; and this would contradict Euclid's Bk. I, Prop. XXVII.

However, I do not believe that the topic ‖Sache‖ can be discussed in this manner; although it can easily be shown, if the problem ‖Sache‖ has been correctly stated, that in order to bisect the angle *DCF* one need only set *Dd* = *DC*. But there are also other ways of conceiving the topic.[1]

For example, whoever regards the two non-intersecting lines *CF, AD* as making an angle that equals zero will easily be able to prove that every line *Cd* makes an angle with *Ad* that is greater than zero, and that these two lines accordingly intersect somewhere. The proof is precisely the one where one shows that *CDA* > *CdA* (Euclid, Bk. I, Prop. XVI). For if one rotates *CD* upwards around point *C*, then the angle *CDA* becomes ever smaller, and finally completely negative, as soon as *CD* comes over *CF*. It must therefore equal zero somewhere, and that this occurs in the position *CF* follows, in my opinion, from the *idea* ‖*Vorstellung*‖ that *AD, CF* are *straight* lines—an idea which cannot coexist with the idea of an *asymptotic approach*.

But whether this consideration of negative angles, and of such as are equal to zero, is proper in the first book of Euclid—that is a completely different question, which one can easily deny, maintaining that such a procedure is more algebraic than geometric.

§11.

I furthermore remark that the difficulties concerning Euclid's eleventh axiom essentially come down to the question: *whether it can be derived* ‖*hergeleitet*‖ *in proper order from the Euclidean postulates together with his other axioms.* Or, *if these are not sufficient, whether there cannot then be produced other postulates or axioms (or both) which have the same obviousness* ‖*Evidenz*‖ *as the Euclidean do, and from which his eleventh axiom could be proved?*

In the first part of this question, one can abstract from everything that I earlier called *representation of the thing*. And since Euclid's *postulata* and other axioms have been expressed in words, it can and should be demanded that the proof never appeal to the thing itself, but that the proof should be carried out purely symbolically—when this is possible. In this respect, Euclid's *postulata* are as it were like so many algebraic equations which one already has in front of oneself and from which one is to compute *x, y, z*, etc. without looking back to the thing itself. But since they are not exactly such formulas, one can concede the drawing of a figure as a guideline for the execution of the proof.

[1] ‖sich die Sache vorzustellen.‖

On the other hand, it would be absurd if in the other part of the question one were to forbid the contemplation and representation of the thing, and were to demand that the new postulates and axioms should be found without any thought about the thing—out of thin air, as it were. But I also do not see how one is fairer to Euclid if one rejects his axioms without asking the question I have asked at the beginning of this section. For since Euclid counts his proposition among the axioms, he thereby incontestably presupposes the representation of the thing; and onc can presume that, in the absence of a proof (which is yet to be found), he chose his own method of proceeding consciously.

Moreover, I have no doubt that Euclid did not himself think of a way of bringing his eleventh axiom among the theorems. At any rate, in the first book of his *Elements* there are several dim hints of this. How easily, for example, his Proposition XVII follows from Proposition XXXII once this latter has been proved! But Euclid specially proves the former proposition—probably to show how much one can determine about the angles of a triangle without invoking the eleventh axiom.

6
Bernard Bolzano (1781–1848)

Bolzano's writings mark a turning-point in research on the foundations of mathematics—a transition from the mathematical style of the eighteenth century to that of the nineteenth. The selections from Berkeley, Newton, and MacLaurin have shown how mathematicians, following what MacLaurin called the 'geometrical manner of the ancients', sought to ground the calculus in concepts derived from geometry or from the theory of motion. Even D'Alembert, in the articles translated above from the *Encyclopédie*, does not distinguish sharply between geometric considerations and arithmetical, and his conception of mathematical deduction remains loose and intuitive.

Bolzano was the first mathematician explicitly to reject the traditional geometric and spatial approach to foundations, calling instead, on explicitly *logical* grounds, for a 'purely analytic' grounding of the calculus—that is, a grounding in *arithmetic*. He thus stands at the head of two intertwined movements in nineteenth-century mathematics: the *arithmetization of mathematics*, a project that was to be carried forward by Cauchy, Gauss, Abel, Riemann, Dirichlet, Weierstrass, Heine, Cantor, Dedekind, and others; and the search for *logical foundations* that was pursued by Frege, Peirce, Peano, Russell, Brouwer, Hilbert, and Weyl.

Bolzano's methodology

Bolzano's call for arithmetical foundations had its origin in a new and fruitful conception of the nature of *axioms*, and in a demand for logical rigour that went beyond what had been required by previous mathematicians. Logical issues for Bolzano were never far removed from his work in arithmetic or the methodology of science, and much of the fascination of his writings lies in the fruitful way he combines philosophy, mathematics, and logic, using each to illuminate the others. Indeed, his *Beiträge* (*Bolzano 1810*) provides perhaps the clearest example of the way in which explicit philosophical reflection can prepare the ground for a major mathematical advance—in this case, for his celebrated paper on the intermediate value theorem, *Bolzano 1817a*.

The seeds of Bolzano's distinctive approach to mathematics seem to have been planted early. As a student at the University of Prague he studied Abraham Gotthelf Kästner's *Anfangsgründe der Arithmetik* (*Kästner 1758*); and in a revealing passage in his autobiography Bolzano praised Kästner because he proved what is generally passed over because everyone already knows it, i.e. he sought to make the reader clearly aware of the basis ‖Grund‖ on which his judgements rest. That was what I liked most of all. My special pleasure in mathematics rested therefore particularly on its purely speculative parts, in other words, I prized only that part of mathematics which was at the same time philosophy (*Bolzano 1836*, p. 64.).

'Proving what everyone already knows' and 'prizing that part of mathematics which was at the same time philosophy' are precisely the traits that were to be characteristic of Bolzano's own mathematical work; but he was to pursue both far more deeply than anything in Kästner.

Bolzano's new and more rigorous approach to axiomatics can already be detected in his earliest mathematical writings—for instance, in §3 of the *Preface* to his *1804*, where he declares that he is obliged to search for proofs even of *obvious* propositions, until he has 'unfolded all truths of mathematics down to their ultimate grounds'. Bolzano's predecessors, in contrast, had taken a more relaxed approach. In their view, the essential requirement for an axiom was that it be *certain*—an immediate and obvious truth on which the calculus of fluxions or Euclidean geometry could be founded.[a] *Bolzano 1804* gives three reasons for rejecting this conception of axioms and for pursuing even obvious truths 'down to their ultimate grounds': such a procedure will be conducive to thoroughness, to the ease of learning the subject, and to the discovery of new theorems.

In later writings Bolzano was to elaborate on the philosophical underpinnings of this conception of the axiomatic method. The crucial argument first occurs in §2 of Part II of the *Beiträge* (*Bolzano 1810*), where he distinguishes between the '*objective connection*' that holds between true judgements when one is a consequence of the other and our *subjective recognition* of such a connection. *Obviousness*, for Bolzano, belongs to the realm of the subjective; but in his eyes the purpose of a scientific exposition is rather to disclose the objective connections between propositions, to bring their true logical structure to light. Axioms therefore become for Bolzano the *logically* ultimate truths on which the truths of mathematics rest; they are not to be confused with the *psychologically* ultimate truths that strike us as simple or indubitable.[b] Bolzano was eventually to expand this non-psychologistic conception of mathematics to the whole of science, which, in the *Wissenschaftslehre* (*Bolzano 1837*), he treats as a realm of objectively true propositions—the *Sätze-an-sich*—which are independent of any knowing subject, and which are organized into a complex structure by their

[a] See, for example, *MacLaurin 1742*, p. 52; *Lambert 1786*, §9 and the first paragraph of §11; but compare *Lambert 1786*, §10.

[b] As is well known, Gottlob Frege and Bertrand Russell, many decades later, and apparently in ignorance of Bolzano's writings on the objectivity of scientific knowledge, were to make similar points about the non-psychological nature of logic. Students interested in pursuing the topic of anti-psychologism should compare Bolzano's arguments with Frege's numerous remarks on the subject (for example, in the Introduction to the *Grundlagen*, *Frege 1884*) or with Russell's distinction (contained, for instance, in the opening paragraphs of *Russell 1907*) between 'empirical premisses' and 'logical premisses'.

objective logical interrelationships. The *Sätze-an-sich* were to play an increasingly central role in Bolzano's later writings; indeed, he even put them to mathematical use in §§13 and 14 of the *Paradoxien des Unendlichen* (*Bolzano 1851*) to establish the existence of infinite sets. (Dedekind was to employ an analogous argument in Theorem 66 of §5 of *Was sind und was sollen die Zahlen?* (*Dedekind 1888*)—the notorious 'proof' that the realm of thoughts is infinite.)

Application of the methodology to mathematics

Bolzano's conception of logical methodology led him to deepen his studies in the foundations of mathematics, and yielded him a rich harvest of theorems. The process begins in §§4–6 of the *Preface* to his *1804*, where Bolzano applies his general methodological principles to the particular case of the foundations of geometry. He criticizes earlier mathematicians for importing conceptions from the theory of motion into their discussions of geometry, pointing out that the *theory of space* is logically antecedent to the *theory of the movement of objects in space*, and must therefore be developed without recourse to the latter theory. Bolzano explicitly criticizes Kant, Mercator, and Kästner on this point; but his remarks can equally well be read as a response to the selections above from MacLaurin and Newton.

Bolzano's argument about logical antecedence was to receive its fullest exposition in Part One of the *Beiträge* (*Bolzano 1810*)—the crucial application to mathematics being made in §11, where he argues that arithmetic (or, as he calls it, *general mathesis*) is logically antecedent to geometry. It was this step—in an essentially philosophical treatise—that led Bolzano to his 'purely analytic' investigations of the real continuum, to his rigorous definition of *limit*, and to his proofs of the intermediate value theorem and the Bolzano–Weierstrass theorem. (The philosophical sources of these deep mathematical discoveries are explicitly mentioned in the *Preface* to *Bolzano 1817a*.)

Bolzano's harvest of theorems did not end here. He also in later writings carried out pioneering studies of the real numbers, gave the first purely analytic definition of *continuity* and *differentiability*, and was the first to discover a continuous, nowhere-differentiable function.[c] His demand for rigour led him to seek proofs of previously unnoticed theorems—for instance, he was the first to state and attempt to prove that a simple closed curve divides the plane into two parts. (This theorem, the Jordan Curve Theorem, was not to receive a fully rigorous proof until Oswald Veblen's proof of 1905.)

Bolzano's philosophical methodology thus led him to introduce powerful new *concepts* and *techniques* and *conjectures* into mathematics. This is an important aspect of his work, and sets him apart from a thinker like Lambert, whose methodological observations on the Axiom of Parallels (*Lambert 1786*, §11)

---

[c] See *Bolzano 1930*, §75. *Bolzano 1930* was actually written in 1834, but not published until 1930. Weierstrass independently discovered a continuous, nowhere-differentiable function in the 1870s, forty years after Bolzano.

were as shrewd as anything in Bolzano, but who was unable to put them to any actual use in the proving of new theorems. Indeed, the history of mathematics is strewn with similar examples of unexploited anticipations of great advances—recall, for example, Kant's observations on the possibility of alternative geometries, or D'Alembert's discussion of the concept of a *limit*, or Leibniz's dream of a mathematical logic, or Lambert's remarks on formal axiom systems. Such insights, unless they can be shown to perform some actual mathematical work, tend to be sterile, and are only noticed years later when somebody else has demonstrated their significance. Bolzano managed both to have a crucial insight, and to show how to develop it into new branches of mathematics; unfortunately this accomplishment was no guarantee against being ignored, and the circumstances of his life kept his work from becoming widely known.

Life

Bolzano was born in Prague, the youngest son of an Italian father (an art dealer) and a German mother. He entered the University of Prague in 1796, where he was educated in philosophy, mathematics, and physics. In philosophy, he read the *Metaphysica* of the Wolffian philosopher, Alexander Gottlieb Baumgarten (*Baumgarten 1739*); in mathematics, he was particularly influenced by his close study of Eudoxus, Euler, and Lagrange, as well as of Kästner's *Anfangsgründe der Arithmetik* (*Kästner 1758*).

In 1800 Bolzano took up the study of theology; he was called to the new chair of religion at the University of Prague in 1805. The chair had been established by the Emperor Franz I of Austria to shore up the position of the conservative Catholic hierarchy against the tide of freethinking and republicanism that had been rising in Central Europe since the French Revolution. From the point of view of the political and religious authorities, the appointment of Bolzano was not a happy choice. Although his appointment was confirmed in 1807, his own social, ethical, and religious sympathies inclined to the cause of Enlightenment, and he found himself in perpetual trouble with the authorities. (Among the doctrines that caused him difficulty was his publicly-expressed conviction that one day men would live without kings.) Bolzano was a popular lecturer, and in 1818 was elected head of the philosophy faculty; nevertheless, in 1819 he was dismissed from his professorship, forbidden to publish, and placed under police supervision. For the remaining decades of his life he lived in the countryside, writing on ethics, religion, politics, logic, and the foundations of science.

Despite the clarity of his arguments, the power of his theorems, and the fruitfulness of his techniques, and although his *Paradoxien des Unendlichen* was known and admired by Peirce, Cantor, and Dedekind, Bolzano's work in real analysis—the work of an obscure theologian, most of it published by equally obscure Bohemian publishers—seems to have remained entirely unnoticed until Otto Stolz called attention to it in *Stolz 1881*. But by this time Bolzano's most important results had been independently discovered by Weierstrass and his school. His voluminous *Nachlass*, much of it written in shorthand, is at present divided between the Handschriftensammlung of the Oesterreichische Nationalbibliothek in Vienna and the Literary Archive of the Museum of Czech Literature in Prague, and is only now in the process of being published in the Bolzano *Gesamtausgabe* (*Bolzano 1969–*). The quantity and quality of Bolzano's writings are impressive, and his thought has by no means been fully plumbed by subsequent scholars; certainly it should be emphasized that the selections that follow represent only a fragment of the thought of this exceptionally rich and original thinker.

The translation of the following selection from *Bolzano 1804* is by Stephen Russ, and is printed here for the first time. References should be to the paragraph numbers, which have been added in this translation.

A. *PREFACE TO* CONSIDERATIONS ON SOME OBJECTS OF ELEMENTARY GEOMETRY (*BOLZANO 1804*)

⟦1⟧ It is quite well known that in addition to the widespread use which its *application* to practical life yields, mathematics also has a second use which, while not so obvious, is no less useful. This is the exercise and sharpening of the mind: the beneficial development of a *thorough way of thinking*. This the use which the State chiefly has in view when it requires every academic to study this science. As I could no longer suppress the bold desire to contribute something to the further progress of this splendid subject, I have—following my personal preferences—considered for the most part only the improvement of theoretical mathematics, i.e., mathematics in so far as it may fulfil the second use mentioned above.

⟦2⟧ It is necessary to mention here a pair of rules which in my opinion apply to this work as well as to others.

⟦3⟧ *Firstly*, I stipulate the rule that the *obviousness* ⟦*Evidenz*⟧ *of a proposition* does not absolve me from the obligation still to look for a proof of it, at least until I clearly realize why absolutely no proof could ever be required. If it is true that ideas are easier to grasp when they are everywhere clear, correct, and connected in the most perfect order, than when they are scattered, confused, and incorrect, then one must regard the endeavour of unfolding all truths of mathematics down to their ultimate grounds, and thereby providing all concepts of this science with the greatest possible clarity, correctness, and order, as an endeavour which will not only promote the *thoroughness* of education but also make it *easier*. Furthermore, if it is true that if the first ideas are clearly and correctly grasped then much more can be deduced from them than if they remain confused, then this endeavour can be credited with a *third* possible use—the *extending* of the science. The whole of mathematics offers the clearest examples of this. Something could at one time have appeared superfluous, as when Thales (or whoever was the discoverer of the first geometric proofs) took much trouble to prove that the angles at the base of an isosceles triangle are equal, for this is obvious to common sense; but Thales did not doubt *that* it was so, he only wanted to know *why* the mind makes this necessary statement. Note that because he drew out the elements of an implicit deduction and became clearly aware of it, he thereby obtained the key to new truths which were not clear to common sense. The application is easy.

⟦4⟧ *Secondly*, I must point out that I believed I could not be satisfied with a completely strict proof *if it were not even derived from concepts* which the thesis to be proved contained, but rather made use of some fortuitous alien, *intermediate concept* ⟦*Mittelbegriff*⟧, which is always an erroneous *μετάβασις εἰς ἄλλο γένος*.[a] Here I counted it an error in geometry if all propositions about angles and the ratios of straight lines to one another (in triangles) are proved by means of *considerations of the plane*, for which there is no suggestion in the theses. I also include here the concept of *motion*, which some mathematicians have used to prove theorems in pure geometry. Even *Kästner* is one of these mathematicians (for example *Geometrie*, Pt. II, Grundsatz von der Ebne). *Nikolaus Mercator*, who sought to derive a particularly systematic geometry, accepted the concept of motion in it as essential.[b] Finally *Kant* also claimed that motion as the *describing* of a space belonged to geometry. His distinction (*Kritik der reinen Verrunft* p. 155)[c] in no way removes my doubt about the necessity, or even just the admissibility of this concept in pure geometry, for the following reasons.

⟦5⟧ *Firstly*, I at least cannot see how the idea of motion is to be possible without the idea of a *movable object* in space (although only imagined) which is to be distinguished from space itself. This is because, in order to obtain the idea of motion one must imagine not only infinitely many *equal* spaces next to one another, but one must assume *one and the same thing* to be successively in different positions. Now even if *Kant* regards the concept of an object as an *empirical* concept, or even if it is admitted that the concept of a thing *distinguished* from space is alien to a subject which *merely* deals with *space* itself, then the concept of motion may not be permitted as valid in geometry.

⟦6⟧ *On the other hand*, I think the theory of motion already presupposes that of space, i.e. if one had to prove the possibility of a certain motion which had been assumed with reference to a geometrical theorem, then one would have to have recourse to just this geometrical proposition. An example is the above-mentioned axiom of the plane (of Kastner). Now because the assumption of any motion presupposes, for the proof of its possibility (which one has a duty to

---

[a] ⟦'crossing to another kind'. A similar phrase occurs Aristotle's *Posterior Analytics* 75a 38: 'One cannot, therefore, prove by crossing from another kind—for example something geometrical by arithmetic'.⟧

[b] ⟦The reference is to *Mercator c.1678*.⟧

[c] ⟦Second edition.⟧

give), some theorems of space, there must be a science of the latter which precedes all concepts of the former. This is now called pure geometry.

[7] In favour of my opinion I have *Schultz*, who, in his valued *Anfangsgründe der reinen Mathesis*, Königsberg 1790, assumed no idea of motion.

[8] In the present pages I supply no *œuvre achevé* but only a small sample of my investigations so far, which only concerns *the very first propositions of pure geometry*.

[9] If this paper is not received completely unfavourably then a second on the first principles of mechanics might follow it shortly. I would especially like to have the judgement of those familiar with contemporary geometrical ideas. That is the reason that I have put something into print on this difficult material rather than on another subject (which would certainly have been possible). Now something more about this paper.

[10] It is obvious that for a correct theory of the straight line—I mean for the proofs of the propositions: the possibility of a straight line, its determination by two points, the possibility of its being infinitely extended, and some others—no considerations of *triangles or planes* can be employed; rather, conversely, the theory of the latter must be based on the former. So I have set out in the *first section* an attempt to prove the *first propositions of the theory of triangles and parallels* merely from the presupposition of *the theory of the straight line*. As far as I am aware this has not been done before, because in all other places various *axioms of the plane* have been presupposed—axioms which, if one had to prove them, would require just that theory of triangles. Therefore in my view the first theorems of geometry have been proved only *per petitionem principii*;[d] and even if this were not so, the *probatio per aliena et remota*[e] (as already mentioned) must not be allowed.

[11] I regard the theory of the straight line itself, although *provable*—independently of the theory of triangles and planes—yet still so little *proved*, that in my view it is at present the most difficult part of geometry. In the *second section* I shall put forward those of my own thoughts on the matter which seem to come nearest to the first principles, although they still do not reach the [true] basis. This I do only to find out whether I should continue on this path.

B. CONTRIBUTIONS TO A BETTER-GROUNDED PRESENTATION OF MATHEMATICS (*BOLZANO 1810*)

The following selection, one of the most penetrating essays ever written on the methodology of mathematics, sets forth in detail Bolzano's programme for

[d] [‘by begging the question’.]

[e] [‘proof by foreign and remote (ideas)’.]

research in the foundations of mathematics, and lays the groundwork for his important mathematical discoveries of 1816 and 1817.

Bolzano begins in the *Preface* by noting the confusion that reigns in mathematics—in the differential calculus, in the Axiom of Parallels, in the theory of negative and positive numbers; and he proposes to correct these deficiencies by a reform of the methodology of mathematics.

The *Beiträge* itself is divided into two Parts and an Appendix on Kant. Part One is concerned with determining the nature and structure of mathematics; Part Two with methodology. Bolzano begins Part One by seeking a precise *definition* of mathematics. He rejects (§§1–3) as either too narrow or too wide the standard definition of mathematics as 'the science of quantity'. He also (§§5–6) rejects the Kantian definition of mathematics as the 'science of the construction of concepts', but he postpones his detailed criticism of Kant to the Appendix. In §§8 and 9 he proposes his own definition of mathematics as 'the science of the general laws (forms) to which things must conform in their existence'. The strength of this definition was that it allowed Bolzano to think of mathematics as being arranged in a logical, hierarchical structure ordered by the kinds of 'things' studied: so he did not have to level all of mathematics to the homogeneous 'science of quantity'.[a] §§11–20 of Part One are accordingly devoted to a classification of the various branches of mathematics. The details are apt to strike a modern eye as peculiar, with 'things' being classified according to whether they are free or not-free, in space, in time, or open to the senses; and Bolzano counts among the chief branches of mathematics disciplines such as 'temporal aetiology', which contains such theorems as 'Every effect is simultaneous with its cause', and which applies 'not only [to] spatial, material things, but also [to] spiritual forces'. But despite the manifest imperfections of this scheme, it led Bolzano, in the Note to §11, to classify *arithmetic* as logically prior to *geometry*, a classification that was the taproot of his work on the arithmetization of real analysis.

Part Two of the *Beiträge* is devoted to the methodology of mathematics, and discusses in turn the nature of definitions, axioms, proofs, and theorems. Bolzano (for example in §§2, 11, 12 of Part Two) draws a sharp distinction between the (subjective) obviousness of a proposition and the (objective) grounds for its truth, and he argues (§2) that the purpose of a scientific exposition is 'to represent this objective connection of judgements, i.e. to choose a set of judgements and arrange them one after another so that a consequence is represented as such and conversely'. In contrast to previous mathematicians, for whom an axiom was an obvious and immediately evident proposition, axioms for Bolzano become the objective and unprovable foundation of the—possibly more obvious—theorems of the rest of mathematics. And by the same token a proof of a given theorem is not merely a persuasive argument, but (§12) 'the representation of its objective dependence on other truths'.

---

[a] Bolzano continued to hold this definition for most of his career, but towards the end of his life, in his *Über die Einteilung der schönen Künste* (*Bolzano 1849*), he retracted his earlier definition and returned to the traditional definition of mathematics as the 'science of quantity'.

Bolzano's views mark a sharp break from traditional accounts of mathematical method. His objective conception of mathematics, his insistence on seeking proofs even of obvious truths, and his conviction that arithmetic was logically prior to geometry were to be the starting-point for his subsequent analytic investigations of the foundations of the calculus; the importance of these broadly philosophical reflections to his mathematical research is explicitly acknowledged in *Bolzano 1817a*.

The translation of *Bolzano 1810* is by Steven Russ, and is printed here for the first time; references should be to the Part and Section numbers, which appeared in the original edition.

Multum adhuc restat operis, multumque restabit: nec ulli nato post mille saecula praecludetur occasio aliquid adhuc adjiciendi.—Multum egerunt, qui ante nos fuerunt, sed non peregerunt: suspiciendi tamen sunt, et ritu Deorum colendi.[a]

Senec. Epist. 64.

PREFACE

Every impartial critic must admit that of all the sciences mathematics comes nearest to the ideal of perfection. In the most ordinary textbook of mathematics there is more precision and clarity in the concepts, and more certainty and conviction in the judgements, than will be found at present in the most complete textbook of metaphysics. However, undeniable as that is, the mathematician should never forget that what is quoted above about all human knowledge is also true of his science, 'much work still remains'. The greatest experts in this science have always maintained not only that the structure of their science is still unfinished and incomplete in itself, but also that even the first foundations of this structure (which in other ways is magnificent) are still not completely strong and in order; or, to speak without metaphor, that some omissions and imperfections are still to be found in even the most elementary theories of all mathematical disciplines.

We shall now give some reasons for this opinion. Have not the greatest mathematicians of modern times recognized that in *arithmetic the theory of opposite quantities*, together with all that depends on it, is still not clear? Is there not a different presentation of this theory in almost every arithmetic textbook?—

[a] ['Much work still remains, and much will remain; he who shall be born a thousand ages hence will not be barred from the opportunity of adding something further. . . . Our predecessors have done much, but have not done everything. Yet they should be revered, and should be worshipped with a divine ritual.'—Seneca, Epistle 64, 'On the philosopher's task'.]

The chapter on irrational and imaginary quantities is even more unstable, and in places it is full of contradictions. I do not want to mention anything here about the defects in *higher algebra and the differential and integral calculus.* It is well known that up till now there has not been any agreement on the concept of a differential. Only at the end of last year the *Royal Jablonovsky Society of Sciences* at Leipzig gave as their prize-question *the analysis of different theories of the infinitesimal calculus and the decision as to which of these is preferable.*

None the less, it seems to me that arithmetic is still by far the most complete of the mathematical disciplines; geometry has much more important defects which are more difficult to remove. At present a precise definition is still lacking for the important concepts of *line, surface, and solid.* It has not even been possible to agree on the definition of a straight line (which could perhaps be given before the concept of *a line in general*). Some years ago, *Grashof* (*Theses sphaereologicae, quae ex sphaerae notione veram rectae lineae sistunt definitionem, omnisque geometriae firmum jaciunt firmamentum.* Berlin 1806) presented us with a completely new definition, which, however, was hardly satisfactory. But the most striking defect concerns the *theory of parallels.* As far as we know, people have been concerned with the improvement of this theory since the time of *Proclus* and very probably even long before *Euclid.* So many attempts have been made in the past, and yet no one has beem successful enough to enjoy general acclaim.

In *mechanics* the concepts of *speed* and *force* are almost as much of a stumbling-block as the concept of the straight line is in geometry. It has also been agreed for a long time that the two most important theorems of this science, namely that of the *parallelogram of forces* and that of the *lever* have still not been rigorously proved. For this reason the *Royal Society of Sciences at Copenhagen* made a better-grounded *theory of the parallelogram of forces* a prize-problem in 1807. Since I have still not seen the paper of Prof. *de Mello* which gained the prize, I cannot be sure whether the attempt which I intend to give in *these* pages will be something new. As for the theory of the *lever*, it is of course thought that *Kästner*'s proof removes all the difficulties, but I believe that in the present work I shall show the contrary.[b]

Finally, in *all* parts of mathematics, but especially in geometry, the *imperfection in the order of arrangement* has been criticized since the time of *Ramus.* In fact, what kind of dissimilar objects are not dealt with in the individual theorems in *Euclid*? Firstly *triangles*, that are already accompanied by *circles* which intersect in certain points, then *angles*, adjacent and vertically opposite

---

[b] [In spite of the phrases, 'the present work', and, in the previous sentence, 'these pages', the problems referred to are not in fact treated here at all. Bolzano refers to the entire projected work of which only this first issue was published. However, at least some of the material intended for the second issue has now appeared in the *Bolzano Gesamtausgabe* (*1969–*) Vol. 2A5 (1977).]

angles, then the *equality* of triangles, and only much later their *similarity*, which however, is derived by an atrocious detour, from the consideration of parallel lines, and even of the *area* of triangles, etc.! But if one considers, *ταυθ' ὅπως* ⟦*μεν*⟧ *γέγραπται τοῖς καιροῖς καὶ ταῖς ἀκριβείαις*,[c] and if one reflects how every successive proposition, with the proof with which *Euclid* understands it, necessarily requires that which precedes it, then one must surely come to the conclusion that the reason for that disorder must be fundamental: the entire method of proof which *Euclid* uses must be incorrect.

Now the purpose of the present pages is not only to make some contributions to the improvement of what has just been criticized, but also to remove some other defects of mathematics whose presence can only be proved subsequently. I may reasonably be asked *how competent I am for this*. I wish to state here quite frankly what I know there is to say for or against myself in this respect.

For about fifteen years—I have not known mathematics for longer—this science has always been one of my favourite studies, particularly its theoretical part, as a branch of philosophy and as a means of training in correct thinking. As soon as theoretical mathematics became known to me (which was through *Kästner*'s splendid textbook), I was struck by one and then another defect, and I occupied myself in my leisure time with their removal, certainly not from vanity, but from an intrinsic interest that I found in such speculations. With further consideration the number of defects which I believed I had discovered increased. Indeed, I gradually succeeded in removing them one after another; but I did not trust the solution straight away from fear of deceiving myself, because I loved the truth more than the pleasure of a supposed discovery. Only if I examined my opinion from all sides and always found it confirmed would I put more confidence in it. Meanwhile, as far as my other studies (and for five years my teaching commitments and other circumstances) allowed, I also examined *those books* which have been written with a view to perfecting the scientific system of mathematics. In these I found some things had already been carried out to which I had been led by my own thinking. On the other hand there are some things I have still not found anywhere. However, since I could not obtain a complete knowledge of the mathematical literature, it is possible that some of the things which I regard as new have already been said somewhere; but this will certainly not be the case with everything.

However, I am *perfectly aware* that it is an altogether *risky* undertaking to wish to change and improve anything in the first foundations of mathematics. Kästner says somewhere, with historical truth, '*hitherto all those who wanted to surpass Euclid have come to grief themselves*'. Does not a similar fate also await me, especially as prejudice and obstinacy will resist me even when I should

---

[c] ⟦'how well this discourse has been composed with respect to appropriateness and finish of style'. From the concluding sentence of Isocrates' letter to Philip of Macedon.⟧

have truth on *my* side? But it certainly does not follow from the failure of *several* attempts that all the *rest* must fail. Also, the method which I adopt here is very different from the methods attempted previously. I therefore regard it as my duty to submit it for the judgement of the expert.

In the year 1804 I published a small sample of my alterations under the title, *Betrachtungen über einige Gegenstände der Elementargeometrie*. But the small extent of the pamphlet, its uninformative title, the far too laconic style, the anonymity of the author, and many other circumstances, were certainly not favourable for securing attention to it. Therefore nothing further followed than that it was announced in some learned journals (e.g. in the Leipziger Jahrg. 1805 Jul. Pt. 95; in the Jen. 1806 Feb. No. 29) without an *obvious* mistake being indicated in the theory of parallels which is expounded in it. But I have naturally advanced in my ideas since that time, and therefore I believe many things are now presented better and more correctly than they were then. Therefore in these *Contributions* I intend to deal with the individual *a priori* disciplines of mathematics one at a time, according to the order set out in the present work in Sec.I §20. The contributions are to appear in small issues like the present one at indefinite intervals, and I cannot decide in advance how many there will be. Most of the alterations, and the most important ones, will concern geometry. I shall therefore state these as soon as possible, so that through the criticism of experts either my views will be confirmed or my errors explained and I shall not lose any more time on a wrong course.[d]

—ἐί γὰρ τίς μοι ἀνὴρ ἅμ ἕποιτο καὶ ἄλλος,
Μᾶλλον *θαλπωρὴ, καὶ θαρσαλεώτερον ἔσται.*
—μοῦνος δ' εἴ πέρ τι νοήσῃ,
Ἀλλά τέ οἱ βράσσων τε νόος, λεπτὴ δέ τε μῆτις

—Iliad X, 222.[e]

I. On the concept of mathematics and its classification

§1.

It is well known that the oldest mathematics textbook, Euclid's *Elements* (which in some ways is still unsurpassed), does *not even contain a definition* of the

---

[d] [See Note b above: the *Gesamtausgabe* Vol. 2A5 contains (p. 135), *Eine neue Theorie der Parallelen* (1813).]

[e] ['—I wish another man could go with me. I should feel more comfortable, and also more inclined to take a risk. [Two men together seize advantages that one would miss;] whereas a man on his own is liable to hesitate, if he does see a chance, and make stupid mistakes' (trans. E. V. Rieu). Diomedes is speaking to Nestor about a proposed night raid on the camp of some Thracian allies of Troy. The bracketed words are omitted in Bolzano's quotation.]

science with which it is concerned. Whether its immortal author did this from a kind of wilfulness or because he did not know any valid definition, I shall not venture to decide. By contrast, in all *modern* textbooks of mathematics the definition is put forward: *mathematics is the science of quantities. Kant* has already found fault with this definition in his *Kritik der reinen Vernunft* (see the 2nd edition, p. 742), because as he says, '*no essential characteristic* of mathematics is given, and also *the effect* is mistaken for *the cause*'.

§2.

Naturally everything here depends on what is understood by the word '*quantity*'. The anonymous author of the book, *Versuch, das Studium der Mathematik durch Erläuterung einiger Grundbegriffe und durch zweckmässigere Methoden zu erleichtern*, Bamberg & Wurzburg 1805 (p. 4), puts forward the following definition of quantity. '*A quantity is something that exists and can be perceived through some sense.*' This definition is either too wide or too narrow according to whether the author takes the words, '*exists and can be perceived*' in their widest sense, when they mean a purely *ideal existence* and a *capacity of being thought*, or in their narrower and proper sense, in which they hold only of an *actually existing object open to the senses*. In the first case quantity would be *every conceivable thing without exception*, and if we then defined mathematics as the science of quantities we would basically bring all sciences into the domain of this one. On the other hand, in the second case, if quantities were only *sensible objects*,[f] the domain of mathematics would obviously then be excessively restricted—for objects not open to the senses, for example spirits and spiritual forces, can also become objects of mathematics and in particular of numerical calculation.

§3.

However, this explanation of quantity (§2), interpreted as above or otherwise, is in fact quite contrary to the ordinary use of language. I have only mentioned it here to show subsequently that this author may also have had in his mind (although only dimly) a certain idea which seems to me true. If we do not wish to move too far away from the ordinary use of language (something which of course we should never do even in the sciences without necessity), then we must understand by quantity *a whole in so far as it consists of several equal parts*, or, still more generally, *something which can be determined by numbers*. Presupposing this meaning of the word 'quantity', the usual definition of mathe-

[f] ‖Sometimes 'sinnlicher Gegenstand' has been translated 'object open to the senses' and sometimes, as here, by 'sensible object'.‖

matics as a science of quantities is of course defective and indeed *too narrow*. For quantity is considered *by itself and in abstracto* only in pure *general mathesis*, i.e. in *logistics*[g] or *arithmetic*, but it does not exhaust the content of even *this* science. The concept of quantity or of number does not even appear in many problems of the *theory of combinations* (this very important part of general mathesis). For example, if one puts the question: *which permutations* (*not how many*) *of given things a,b,c . . . are admissible*? In the *particular* parts of mathematics, *chronometry*, *geometry* etc., as the names already suggest, some object *other* than the concept of *quantity* (for example, time, space, etc.) appears everywhere, and the concept of quantity is merely *frequently applied* to it. So that in all these disciplines there are several axioms and theorems which do not even contain the concept of quantity. Thus, for example, in chronometry the proposition that *all moments are similar to each other*, and in geometry that *all points are similar to each other*, must be stated. Such propositions, which do not contain the concept of quantity at all, could not even be stated in mathematics if it were merely a *science of quantities*.

§4.

It has been easy for us to criticize and object to the usual definition ‖Erklärung‖; we shall find it harder to put a better one in its place. We have already observed that those *special objects* which appear in the individual parts of mathematics, alongside the concept of quantity, are of such a nature that the latter can easily be applied to them. This could perhaps lead to the idea of defining mathematics as *a science of those objects to which the concept of quantity is especially applicable*. And it really seems that those who adopted the definition quoted in §1 basically intended nothing but this to be understood. However, a more careful consideration shows that even this definition is objectionable. The concept of quantity is *applicable* to *all* objects, even to *objects of thought*. Therefore if one wanted to consider the mere *applicability of the concept of quantity* to an object as a sufficient reason for counting the theory of that object among the mathematical disciplines, one would in fact have to count all sciences as mathematics—for example, even the science in which the proposition is proved that there are only four (or as *Platner* correctly states only *two*) syllogistic figures; or the science which states that there are *no more and no less than four sets of three* pure simple concepts of the understanding (categories), etc. Therefore to preserve this definition one would have to take into account the *difference between more frequent and rarer applicability*, i.e. count only those objects for mathematics to which the concept can be applied *often*

[g] ‖At this time 'logistics' in German and English could mean general arithmetical computation or 'logistical arithmetic', which meant the arithmetic of sexagesimal fractions still used by astronomers.‖

and in many ways. But anyone can see that this would be an extremely vague, and not at all scientific, determination of the boundaries of the domain of mathematics. We must therefore look for a *better* definition.

§5.

The *critical philosophy* seems to promise us one. It claims to have discovered a definite and characteristic difference between the two main classes of all human *a priori* knowledge, philosophy and mathematics. Namely that *mathematical knowledge* is capable of adequately presenting, i.e. *constructing*, all its concepts in a *pure intuition*, and on account of this is also *able to demonstrate* its theorems; while, on the other hand, *philosophical knowledge*, devoid of all intuition, must be content with purely *discursive concepts*. Consequently the essence of mathematics may be expressed most appropriately by the definition *that it is a science of the construction of concepts* (cf. Kant's *Kritik der reinen Vernunft*, p. 712).[h] Several mathematicians who adhere to the critical philosophy have actually adopted this definition. Among others there is *Schultz*, who deserves much credit for the foundation of pure mathematics in his *Anfangsgründe der reinen Mathesis*, Königsberg 1791.

§6.

For my part I wish to admit openly that I have not yet been able to persuade myself of the truth of many doctrines of the critical philosophy, and especially of the correctness of the Kantian assertions about *pure intuitions* and about the *construction of concepts through them*. Furthermore, I believe that in the *concept of a pure* (i.e. *a priori*) *intuition* there already lies an intrinsic contradiction. Much less can I persuade myself that the concept of a *number* must necessarily be constructed in *time*, and consequently that the intuition of time belongs essentially to arithmetic. Since I say more about this in the appendix to this paper, I content myself here only to add that among the independent thinkers in Germany there are many who agree just as little as I do with these assertions of Kant. There are even some who at first had been inclined to the Kantian definition but subsequently found it necessary to abandon it. Such, for example, was *Michelsen* in his *Beiträge zur Beförderung des Studiums der Mathematik*, Berlin 1790; see Vol. 1, Part 5.

§7.

But more instructive to me than what *Michelsen* says in his paper was what I came across in the general *Leipz. Literatur-Zeitung* (1808 Jul., Pt. 81). The learned reviewer condemns the usual definition of mathematics as a *science of quantities*, and says about it, 'Quantity is only an object of mathematics because it is the most *general form, to be finite*, but in its nature *mathematics* is a *general theory of forms*. Thus, for example, is *arithmetic*, in so far as it considers *quantity* as

---

[h] ‖This reference should be p. 741 in the 2nd edition, *Kant 1787* (the edition specified earlier), or p. 713 in the 1st edition, *Kant 1781*. The passage is translated above, p. 136.‖

the general *form of finite things*; *geometry*, in so far as it considers *space* as the general *form of Nature*; the *theory of time*, in so far as it considers the general *form of forces*; the *theory of motion*, in so far as it considers the general form of *forces acting in space*.'—I do not know whether I understand these definitions quite in the sense of their author, but I must admit that they helped me to develop and further improve the following definition and classification of pure mathematics, which I had already sketched in their main outline.

§8.

I therefore think that mathematics could best be defined as a *science which deals with the general laws (forms) to which things must conform* ‖*sich richten nach*‖ *in their existence*. By the word '*things*' I understand here not merely those which possess an *objective existence* independent of our awareness, but also those which simply exist among our *ideas*, either as *individuals* (i.e. *intuitions* ‖*Anschauungen*‖, or simply as *general concepts*, in other words, *everything at all which can be an object of our perception*. Furthermore, if I say that mathematics deals with *the laws to which these things conform in their existence*, this indicates that our science is concerned not with the proof of the *existence* of these things, but only with the *conditions of their possibility*. When I call these laws *general*, I mean it to be understood that mathematics never deals with a single thing as an individual, but always with whole *genera*. These genera can of course be higher or lower, and on this will be based the classification of mathematics into *individual disciplines*.

§9.

The definition given here will certainly not be *found too narrow*, for it clearly covers everything that has previously been counted in the domain of mathematics. But I am more afraid that it might be *found rather too wide*, and the objection might be made that it leaves too little for *philosophy* (metaphysics), as the latter will be limited by my definition to the single concern of proving, from *a priori* concepts, the *real existence* of certain objects. *Mathematics and metaphysics*, the two main parts of our *a priori* knowledge, would, by this definition, be contrasted with each other so that the *first* deals with the general conditions under which the existence of things is *possible*; the latter, on the other hand, seeks to prove *a priori* the *reality* of certain objects (such as the freedom of God and the immortality of the soul). Or, in other words, the *former* concerns itself with the question, *how must things be made in order that they should be possible*? The *latter* raises the question, *which things are real*—and indeed (because it is to be answered *a priori*)—*necessarily real*? Or still more briefly, *mathematics* would deal with *hypothetical** *necessity*, *metaphysics* with *absolute necessity*.

---

* However, not all its propositions have this hypothetical *form*, because the condition, especially in chronometry and geometry, where it is the same for all propositions, is tacitly assumed.

§10.

If I come upon some ideas which are new to me I am accustomed to always putting the question to myself, whether anyone before me has had the same opinion. If I find this is the case, then naturally I gain *conviction*. Now as far as the above definition is concerned, I need hardly say how closely what the astute reviewer has said (§7) coincides with my formulation, that is, if it does not amount to exactly the same thing. This idea also seems to have been in the mind of the author of the book mentioned in §2, although only dimly. In this book he defines quantity, or the object of mathematics, *as that which is*; he seems to have felt that mathematics is concerned with *all* the forms of things, not merely with their *capacity to be compounded out of equal parts* (*their countability*).—*Kant* defines *pure natural science* (which has always been regarded under the name of *mechanics*, as a part of mathematics) as *a science of the laws which govern the existence of things* (of phenomena). By this definition one can very easily be led to our definition given above. *Time* and *space* are also *two conditions* which govern the *existence of* appearances; therefore *chronometry* and *geometry* (which consider the properties of these two forms *in abstracto*) deal likewise, though only *indirectly*, with the laws which govern the existence of things (i.e. things open to the senses). Finally *arithmetic*, which deals with the laws of *countability*, thereby develops the most *general* laws according to which things must be regulated in their existence, even in their ideal existence.

§11.

Now let us try to derive from this definition of mathematics a *logical classification* of this science into several single disciplines. If we succeed fairly naturally with this classification then this may even be a new confirmation of the validity of that definition. According to it *mathematics* is to be a *science of the laws to which things must conform* ‖*sich richten nach*‖ *in their existence*. Now these laws are so general that either they apply to *all things without exception*, or not. The *former* laws, gathered together and ordered scientifically, will accordingly constitute the main part of mathematics. It can be called *general mathesis*; everything else is then *particular mathesis*.

Note: To this *general* mathesis belongs, as we shall see below, *arithmetic*, *theory of combinations*, and several other ‖branches‖. These parts of mathematics must therefore not be considered as *co-ordinate with the rest* (chronometry, geometry, etc.); rather, the latter are *subordinate* to the general mathesis as a whole, as species to the genus. Because the concept of *number* is one of those of *general* mathesis it will also appear frequently in all these particular parts, but it will not exhaust their content.

§12.

Now in order to obtain the *particular* or *special* parts of mathematics we must gather into certain *classes* the *things themselves* with whose general forms mathematics is concerned. Before we do this let us draw attention to a certain concept of our understanding which (as far as I can see) is not *completely applicable to all things* and therefore, in all strictness, should not be included in *general mathesis*. On the other hand, it is applicable to things of *such different kinds* that it would hardly be suitable for a classification of mathematics into single disciplines. This is the *concept of opposition* ‖*Entgegensetzung*‖. I do not believe that to *every* thing there is an *opposite*; but the *past* and *future* in *time*, *being on this side* and *that side* in *space*, *forces* which act in opposite directions in *mechanics*, *credit* and *debit* in the calculation of *accounts*, *pleasure* and *displeasure* in *feelings*, *good* and *evil* in the *free decisions of the will*, and so forth, are clear examples of *opposition* which sufficiently prove how widely this remarkable concept applies. At the same time we see from these examples how little the concept is suited as a basis for a main division of the mathematical disciplines into those to which it would be applicable and those to which it would not be applicable. On the contrary, the classifications which actually exist (which we may not completely destroy) are made according to a quite different basis of classification, which we can only vaguely imagine. Included in every individual discipline is whatever specifically arises by applying the concept of opposition to the object of the discipline. The general part, however, which can deal with all things without exception which are capable of opposition, certainly deserves a separate consideration. It can be expounded (as to a certain extent has already been done) in a special *appendix to the general mathesis*.

§13.

Everything which we may ever think of as *existing* we must think of as being one or the other: either *necessary* or *free* (i.e. not necessary) in its existence.* That which we think of as something *free* is subject to no conditions and laws in its *becoming or existence*,** and is therefore not an object of mathematics.*** *That* which we think of as *necessary* in its existence is so either *simply* (i.e. *in itself*) or only *conditionally* (i.e. on the presupposition of something else). The *necessary in itself* is called *God*, and is considered in *metaphysics* not as a merely *possible* object but as an *actual* object. There remains therefore only the hypothetically necessary object which we consider as *produced through some ground*.[i] Now there are certain *general conditions* according to which everything which is produced through a ground (*in* or *out* of time) must be regulated in its

* An example of the first kind is the *speed* of a moving body; an example of the second kind, every *decision of the human will*.

** In case it does not become, but only *is*, as for example the *free working of the Deity* as far as we are not considering it in *time*.

*** But it is perhaps the object of *morality*, which investigates the question: *how that* which occurs (or exists) freely should occur (or exist).

[i] ‖Bolzano himself distinguishes 'ground' (Grund) and 'cause' (Ursache); see §15, foot-note. I have generally translated 'Grund' as 'ground', since this conveys the substantial and metaphysical idea that Bolzano seems to have had in mind better than 'reason'.‖

becoming or *existence*. These conditions taken together and ordered scientifically will therefore constitute the first main part of mathesis, which I call, for want of a better name, *theory of grounds* ‖*Grundlehre*‖ or *aetiology*.

Note: This part of mathematics contains the theorems of ground and consequence ‖Grund und Folge‖ some of which are usually also presented in ontology, for example that *similar grounds* have *similar consequences*. It is similar with the theories known under the name of the *calculus of probability*. (This latter I only mention here so that it will not be thought that we have completely ignored this important part of mathematics.) Moreover, in a scientific exposition aetiology must precede chronometry and geometry, because the latter appeal to certain theorems of the former, as we shall see more clearly in due course.

§14.

Everything which we not only *think of* as real but is *perceived* as real must be perceived in *time* and—if in addition we perceive it as a thing *outside ourselves*—also in *space*. In other words, *time* and *space* are the two *conditions* which must govern all things *open to the senses* ‖*sinnlichen Dinge*‖, i.e. all things which *appear* to us as real. Therefore if we develop the properties of *time and space in abstracto* and order them scientifically these sciences must also be counted as mathematics in that they also deal, albeit indirectly, with the conditions *to which things must conform in their existence*. We therefore have the *second* and *third* components of particular mathesis, the *theory of time* (chronometry), and the *theory of space* (geometry).

Note: It really does not matter which of these two we put before the other in the system, as the properties of time and space are completely independent of each other. However, because the concept of time is applicable to *more* objects than that of space it seems expedient to allow chronometry to precede geometry.

§15.

If, finally, time and space are not to be considered merely *in abstracto* but as filled with *actual things*, and indeed with such as are not free in their existence but are subject to the laws of causality, then two *new sciences* appear, which are, as it were, compounded equally from the one mentioned in §14 with that in §13. Namely:

(a) The general laws to which *unfree things* which are in time must conform in their existence (and in their changes), constitute the content of a proper science which I call, for lack of a more suitable name, *theory of causes* ‖*Ursachenlehre*‖ or *temporal aetiology*.*

* I thus distinguish the words *ground* ‖*Grund*‖ and *cause* ‖*Ursache*‖. The latter means for me a *ground* which acts in *time*.

(b) The general laws which govern *unfree things* which are *in both time and space* constitute the content of that mathematical discipline which is called *pure natural science*, otherwise also *theory of motion* or *mechanics*.

Note: To the domain of *temporal aetiology* there belong, for example, the theorems: every effect is simultaneous with its cause, the size of the effect which originates from a constant cause varies as the product of the degree of the cause and the time of its acting, and so forth. These theorems are indeed so general that they hold not only for spatial material things but also for spiritual forces, our ideas, and generally for all things which appear in time and are subject to the laws of causality.—*Pure natural science* is already known.

§16.

A science has often been spoken of in which there should appear the general laws *of the possibility of motion without regard to a moving force*—therefore the concepts of *time*, *space*, and *matter* without that of a *cause*. *Hermann*, *Lambert*, and *Kant* have called this science *Phoronomy*, and Kant considers it as a part of pure natural science, though it would not belong there according to our definition above. Herr *E.G. Fischer* in his *Untersuchung über den eigentlichen Sinn der Höheren Analysis nebst einer idealen Übersicht der Mathematik und Naturkunde nach ihrem ganzen Umfange*, Berlin, 1808, also puts forward this science under the name *Phorometry*, and lets it follow *geometry* as the *second* main part of *spatial mathematics*. However, provided the views which I intend to give subsequently are not entirely incorrect, such a science cannot even exist. For all propositions which have been put forward in it so far are in fact only provable with the aid of the concept of cause.

§17.

One is accustomed to dividing the individual disciplines of mathematics into *elementary* and *higher*. However, I do not know any strictly *scientific basis for this classification*. How far such a basis applies simply to a single discipline (e.g. geometry) will be discussed more appropriately in the treatment of this discipline. Here therefore we deal only with those bases of classification which are to apply throughout the whole domain of mathematics. This is the case for all those which have been suggested for the *general mathesis*. Classifications which are made in this way must therefore pervade all particular parts of mathematics. In *Michelsen*'s *Gedanken über den gegenwärtigen Zustand der Mathematik, usw.*, Berlin 1789, the elementary and higher mathesis are so divided that the one has for its objects *constant quantities*, and the other *variable quantities* (or in general, *things*). I believe this classification cannot be accepted because the *assumption* on which it is tacitly based (which many have adopted even in the definition of mathematics) is, that it is the sole business of mathematics to find *quantities which are not given from others which are given*. If this proposition were correct, then all propositions of mathematics should take the form of *problems. Axioms and theorems*, etc., could not really appear in mathematics at all. However, before it can be asked what follows from given things, one must *first* have shown, or have accepted as a *postulate*, that these things *can* be given, i.e. *are possible.*—A completely different classification was suggested by *Michelsen* in his above mentioned *Beiträge* in Vol. 1, pt. 2 (*Über den Begriff der Mathematik und ihre Teile*). Here he assumes three main parts of general mathesis:

1. the *lower*, which allows quantities consisting of *equal* components;

2. the *higher*, in which quantities are considered as compounded partly from *equal* components and partly from *unequal* (theory of differences and sums); and,

3. the *transcendental*, in which the components of quantities are proper *elements* or *units of magnitude* ‖*Grösseneinheiten*‖ in the strictest sense (differential and integral calculus). I do not know if I understand this classification correctly. For it seems to me that in the sense in which calculations of sums and differences can be said to consider quantities compounded partly from equal and partly from unequal quantities, elementary arithmetic already does this if it views, for example, $2 + \frac{1}{2}$ as a whole. Still less do I see how differentials can be considered as *units of magnitude* in the strictest sense of the word, when one also attributes a magnitude to them, at least in the sense of an intensive quantity.

The best procedure might well be to count as *higher* mathematics only that in which the concept of an *infinity* (whether infinitely great or small), or of a *differential*, appears. However, this concept has not been sufficiently explained at the present time. If in the future it should be decided that the *infinite* or the *differential* is nothing but a *symbolic expression*, just like $\sqrt{-1}$ and suchlike, and if also it turns out that the method of proving truths by merely symbolic inventions is a method of proof which (although *quite special*) is always correct and logically admissible, then I believe it would be most expedient in the domain of higher mathematics to continue to refer to the concept of *infinity* and any other equally symbolic concept. *Elementary* mathesis would then be that which accepts only *real concepts or expressions* in its exposition—*higher* mathesis that which also accepts merely *symbolic* ones.

§18.

We also have something to say on the classification of mathematics into *pure* and *applied*. If, as commonly happens, one understands by *applied mathematics* the same as *empirical mathematics*, then we cannot, without contradicting ourselves, even admit the *existence* of such a thing, because in our above definition we counted the *whole* of mathematics among the pure *a priori* sciences. However, it need not be feared that in this way we shall lose a substantial part of the mathematical disciplines. The history of mathematics shows increasingly that whatever has been accepted at first merely from experience is subsequently derived from concepts, and therefore treated as part of a pure *a priori* mathesis. And this may be ground enough for making no scientific distinction between pure and empirical mathematics. For example, because *we* do not know* how to derive *a priori* the existence of an attractive force, and the law that it acts in inverse proportion to the square of the distance, does it follow from this that it will also never be known by our descendants and that it is absolutely underivable *a priori*?—Furthermore, what one accepts from *experience* in the so-called applied parts of mathematics does not make these disciplines fundamentally *empirical*. For mathematics deals in general not with what actually *occurs* but with the *conditions* or *forms* which something must have *if it is to occur*. Therefore it is only necessary to present those propositions which experience offers, purely hypothetically, and then to derive by *a priori* deductions, on the one hand, the *possibility* of these hypotheses, and on the other hand, the *consequences* which follow from them. Hence no empirical judgement appears in the whole exposition, and the science is therefore *a priori*. So, for example, for the exposition of optics one in no way needs to borrow from *experience* the law that light bends in going from air to glass in proportion 2 : 3 and so forth; it is sufficient to make intelligible *a priori* just the *possibility* of anything like light and its bending in passing through different media. On this basis one sets up in hypothetical form the statement, '*if one supposes something . . . then this and that consequence must follow from it*.' But that *possibility* can never be difficult to prove, in that everything which is perceptible as real in experience must already be recognized as possible.

§19.

Therefore in so far as one wishes to understand by *applied* mathematics something which is *essentially based on some propositions borrowed from experience*, I do not believe its existence can be justified. But one can also understand by the name *applied* mathematics something quite different, something which I prefer to call *practical*—or, with a term borrowed from the critical philosophy and perhaps more specific, *technical mathematics. This is an exposition of the mathematical disciplines specially set up for useful application to ordinary life*. Such an exposition is clearly distinguished from the *purely scientific* by the difference in *purpose*; that of the *latter* is the greatest possible *perfection of scientific form* and thereby also the *best possible exercise in correct thinking*, while that of the *former*, in contrast, *is direct usefulness for the needs of life*. Therefore in the practical exposition all the excessively *general views* which are not absolutely essential to the application can be dropped, while many *examples* and *special references to actual cases* can well be inserted. There is not the compulsion to note these actual cases as mere *possibilities* (as must happen in a

* *Kant* has nevertheless attempted it.

purely scientific exposition), but they are put forward straight away as *realities* proved by experience. Moreover, I need hardly say that in most current textbooks of mathematics there is a basically *mixed* approach which aims at combining those two purposes, the purely scientific and the practical. However, it is not my opinion that this is generally a *fault* in those textbooks. A completely expedient textbook composed according to this *mixed* approach would in fact be a far more useful work than a purely scientific one. Only I believe the first cannot be achieved until the *purely scientific* system has been completed. Whoever works for the perfection of the latter can be allowed, for the time being, to put the *second* purpose completely out of his mind so as to fasten his attention solely on the *first*, the scientific perfection.

§20.

From all this it is now clear, I believe, that in a scientifically ordered mathematics there are only the *main parts* mentioned §11–15. The following table sets these out in a convenient summary. The bracketed words indicate the object of each discipline.

A.

General Mathesis
(things in general)

B.

Particular mathematical disciplines

(particular things)

I.

Aetiology

(things which are not free)

II.

(things which are not free but are open to the senses)

a.

(Form of these things in *abstracto*)

α.	β.
Theory of time (time)	Theory of space (space)

b.

(things (*in concreto*) which are open to the senses)

α.	β.
Temporal aetiology (things in time which are open to the senses)	Pure natural science (things in space and time which are open to the senses)

II. On mathematical method

§1.

The method which mathematicians employ for the exposition of their science has always been praised on account of its high degree of perfection, and it has also been believed* up to the time of *Kant* that its *essential* features can be applied to *every* scientific subject. I personally still firmly adhere to this opinion. The so-called *methodus mathematica* is, in its *essence*, not in the least different *from any scientific exposition*. On this assumption a work on mathematical method would be basically nothing but logic, and not even belong to mathematics at all. However, I shall be permitted to put forward some remarks here, in the briefest possible way, on individual parts of this method—especially since everything which we shall say here refers only to mathematics, and chiefly to the removal of some of its imperfections.

§2.

I do not know whether it will cause a certain class of my readers to lose all faith in me, but out of love for the truth I must admit at the outset that I am not completely clear myself on the true nature of a scientific exposition. The reason for this will be understood better subsequently. But this much seems to me certain: in the realm of truth, i.e. in the sum total of all true judgements, a certain *objective connection* prevails which is independent of our accidental and *subjective recognition* of it. As a consequence of this some of these judgements are the grounds of others and the latter are the consequences of the former. To represent this objective connection of judgements, i.e. to choose a set of judgements and arrange them one after another so that a consequence is represented as such and conversely, seems to me to be the proper *purpose*

* Especially in the *Leibniz–Wolff* school.

to pursue in a scientific exposition. Instead of this, the purpose of a scientific exposition is *usually* imagined to be the greatest possible *certainty* and *strength of conviction*. It therefore happens that one discounts the obligation to prove propositions which in themselves already have complete certainty. This is a procedure which, where we are concerned with the practical purpose of certainty, is quite correct and praiseworthy; but it cannot possibly be valid in a scientific exposition, because it contradicts its essential purpose. However, I believe that Euclid and his predecessors were in agreement with me; and they did not regard the mere *increase in certainty* as part of their method. This can be seen clearly enough from the trouble which these men took to provide many a proposition (which in itself had complete certainty) with a proper proof, although it did not thereby become any more certain. For example, whoever was made more *certain* since he read *Elem. Bk. I. Prop.* 5 that in an isosceles triangle the angles at the base are equal? No, the most immediate and direct purpose which all strictly philosophical people had in their scientific investigations was nothing but the desire to see the ultimate grounds of their judgements. And this desire then had the *further purpose, on the one hand*, of putting them in the position of deriving *some* of our judgements, perhaps also some *new* judgements and truths, from these clearly recognized grounds; and *on the other hand*, providing an *exercise* in correct and orderly thinking which should then *indirectly* contribute to greater *certainty* and strength in *all* our convictions.—This has been a preliminary remark on the purpose of the mathematical method. Now to its individual parts.

A. *On descriptions, definitions and classifications*

§3.

It is usually said that *the mathematician must always begin with definitions*. Let us see if this does not require rather too *much*, for it is easy to see that *exaggerated* demands often do just as much harm in the sciences as a general *looseness*. The logician understands by a *definition*, in the truest sense of this word, the *statement of the most immediate components* (two or more) *out of which a given concept is compounded*.

Note: The *general form* which comprehends all definitions is (if the letters a, α, A designate concepts) as follows: *a, which is* α, is A; or (*a cum* α)$=A$. If a concept is determined by a negative characteristic *non*-α, then one can instead write (*a cum non*-α), even more briefly (*a sine* α). But in each case neither a, nor α, *alone* can be$=A$.

§4.

From this it is now clear that there are true definitions only for *compound*, and therefore also *divisible*, concepts; but for such as these they do always exist. *Simple concepts*, i.e. such as cannot be divided into two or more components different from each other and from the original concept itself, provided such concepts exist, *cannot be defined*. However, in my opinion, the existence of such simple concepts cannot be derived from anything but our own consciousness. Since a concept is called *simple* only if *we ourselves* can distinguish nothing of *multiplicity* in it, then it would follow from the opposite assertion (that there are *no* simple concepts *at all*), that each of our concepts could be split up *ad infinitum*; and yet we are certainly not aware of this.

§5.

Accordingly, to decide whether a given concept is compound or simple ultimately depends on our consciousness, on our ability or inability to analyse it. We mention the following as a *few rules* which could facilitate this decision in future cases:

(a) *If we think of an object as compound, then precisely for this reason the concept of it is not a simple one*. For the *concept* of an object is nothing but what we think when we *think* of the object.*

If this remark is correct it follows immediately that the concepts of the *straight line*, the *plane* and several others, which have so often been taken as simple concepts, are not such at all, and that therefore one cannot entirely discount the obligation to explain them. For obviously the straight line and the plane are *objects of a compound kind* in which we have in mind, for example, innumerably many points, as well as particular relationships which these points must have to some given ones.

(b) *Not every concept which is subordinate to a more general one ceases on this account to be simple*. That is to say, a concept only ceases to be simple if it is decomposable. But for a *decomposition* there must be the statement of at least *two* components which are each *conceivable in themselves*. Now we actually consider the *more general* concept as a *component* of the narrower one which is subordinate to it. But it could possibly be that to this *first* component (*genus proximum*) no second one (*differentia specifica*) can be discovered, i.e. that that which must *be added to* the general concept, to produce from it the narrower one, is not in itself *conceivable*. So, for example, the concept of a point is indeed narrower than that of a *spatial object*, the latter is narrower than that of an *object in general*, and the *most general* of all human concepts is, of course, that of an *idea in general*. But it certainly does not follow from this that all previous concepts of a spatial object as *genus proximum* $= a$ reduce, in the form of a definition, to the concept of a point $= A$. It will be seen that the

* That which one thinks when one *thinks* of an object (i.e. as already *contained* in it) is something quite different from that which one *can* (if desired) *add to it mentally*, or *think of connected with it*. The clearer this difference seems to me the more surprised I am that the great Lambert has made an error about it. In his *Deutscher gelehrter Briefwechsel* Vol. I. p. 348 he writes to Kant: "*The simple concepts are individual concepts. For genera and species contain the fundamenta divisionum et subdivisionum in themselves, and are therefore the more compound the more abstract and general they are. The concept* ens *is the most compound of all concepts*." Lambert often returns in his writings to this assertion, which is so completely opposed to our definition above, and he derives from it the strange consequence that in metaphysics one must not *begin* with the concept *ens* but rather *end* with it. The whole thing seems to me an error in which the *compound nature* of a concept has been confused with its *capacity to be compounded*.

characteristic that must be added to *a* so as to obtain *A* is none other than the *concept of a point itself* = *A*, which one wanted to define. Therefore in *general* if one wishes to ascertain whether a certain concept is *simple* or *divisible* then one assumes a *genus proximum* for it and tries to think of some *differentia specifica* to add to it which is not itself already identical with the concept to be defined. If this cannot be done in any way, the concept concerned is a simple one.

Note: The question arises here of *whether one and the same concept may admit of several definitions*. We believe this must be denied in the same way as we deny below (§30) the similar question of *whether there are several proofs for one truth*. One and the same concept consists only of the same simple parts. If it has *more than two* simple component concepts and one does not go back in the definition to the *ultimate components*, then it is certainly possible to divide it in different ways into two parts and to this extent there are several definitions of the same concept. But the difference in these definitions lies only in the *words*, and so it is merely subjective, not objective and not *scientific*. The usual distinction between *nominal*, *real*, *genetic* and other definitions seems to me therefore unacceptable. What was falsely regarded as a *definition* was very often a proper *theorem*. For example, the proper *definition of beauty* might not really be difficult to find, but what is sought is not the *concept* but a *theorem* which shows us how that which produces the feeling of beauty in us must be constituted.

§6.

Propositions in which it is stated that one intends in future to allot this or that particular *symbol* to a certain concept are called *arbitrary propositions* ‖willkürliche Sätze‖. Therefore also *definitions*, in so far as they are expressed in *words* and attribute a particular *word* to the compound concept, are a *kind of arbitrary proposition*.—But there are also arbitrary propositions which are nothing but definitions, for example the proposition: *the sign for addition is* +.

Note: One should not be misled by the *name* of arbitrary propositions and think that it is *absolutely arbitrary* what kind of symbol is chosen to denote this or that concept. Semiotics[j] prescribes definite rules here. The symbol must be

---

[j] ‖The term *semiotics* was used by John Locke (*An essay concerning human understanding*, IV.xxi) to refer to '*the doctrine of signs*; the most usual whereof being words, it is aptly enough termed also . . . *logic*: the business whereof is to consider the nature of signs, the mind makes use of for the understanding of things, or conveying its knowledge to others.'

C.S. Peirce was the first philosopher to pursue semiotics as an independent discipline, more general than logic; he wrote at length on the topic, and saw it as important both for his work in logic and for pragmatism. Peirce's terminology, taken out of its original context, filtered into twentieth-century philosophical usage *via* the writings of Charles W. Morris; for a recent account, see Umberto Eco, *A theory of semiotics* (1976). Today the term is often used interchangeably with *semantics*.

In eighteenth-century Germany *Semiotik* (as in Locke) designated the theory of signs and symbols; in *Lambert 1764* it refers to the construction of a *characteristica universalis*—a language of symbols that would not be prey to the ambiguities of ordinary discourse. This appears to be the sense in which Bolzano is using the term here. The word also appears in the novels of Aldous Huxley (*Genius and goddess*, p. 42): 'He kissed her—kissed her with an intensity of passion . . . for which the semiotics and the absent-mindedness had left her entirely unprepared.'‖

easily recognized, possess the greatest possible similarity with the concept denoted, be convenient to represent, and most important, it must not be in contradiction to any symbols already used or cause any ambiguity. Several mathematical symbols could be chosen more expediently in this respect. For example the symbol for the sėparation of decimal fractions, a stroke to the *right* after the units, is clearly defective. Why to the right? It would be just as good to stand to the *left*. It should rather stand *above* or *below* the place of the *units*, so that numerals which are equally far from the *sign*, to left or right, would refer to an equal positive or negative power of ten. The difficulty which this mistake in notation causes will be clearly observed when instructing a beginner. So before a sign is put forward one should prove in a previous proposition that it has the properties mentioned above and is therefore suitable. In the theory of exponential quantities we shall make particular use of these rules. Now if arbitrary propositions are not completely *arbitrary* it is possible to judge in what sense the proposition, *definitions are arbitrary*, has to be understood. Really nothing is arbitrary in definitions but the *word* which is chosen for the denotation of the new compound concept, though it is obvious here that one should not do unnecessary violence to the use of language. On the other hand, it is not arbitrary which *concepts* are combined into a single one. For *firstly* these combinations must occur according to the law of *possibility*, and *secondly* one must select from the *possible* combinations only those whose considerations can be of *use*.

§7.

From what has just been said the *place* where definitions should be presented in a scientific exposition can now be determined. They obviously *should not appear at the very beginning*. One must first have seen that a certain combination of two or more *words* (and the concept designated by it) produces a new and genuine concept; only then is it worth while to give this combination a particular *name*. This process should also enable us to realize the purpose for which this combination is made and considered—and if not to realize it perfectly clearly, at least to anticipate it partially. Therefore it is an error against good method when *Euclid* gathers all his definitions at the beginning; *Ramus* has already justifiably criticized him about this.

§8.

After all this, and certainly after the example quoted in §5(b), there is no more question whether the mathematician can be required to let a definition precede all the concepts which he puts forward. He *cannot* do so, for among the concepts with which he is concerned there are several which are absolutely simple. *But how does he begin to inform his readers about such simple concepts and the word that he chooses for their denotation*? This is not a great difficulty. For either his readers already use certain words or ways of speech to denote this concept, and then he need only indicate these, for example 'I call *possible* that of which you say that it *could be*'; or they have no particular sign for the concept he is introducing, in which case he assists them by stating several propositions in which the concept to be introduced appears in *different combinations* and is denoted by its own particular word. From the comparison of these propositions the reader himself then abstracts which particular concept the unknown word denotes. So, for example, from the propositions: *the point* is the *simple object* in space; it is the boundary of a line and itself no *part* of the line; it has neither extension in length, breadth, nor depth, etc., anyone can derive which concept is denoted by the word 'point'. This is well known as the means by which we come to know the first meanings of words in our mother tongue. Moreover, concepts which are *completely simple* are needed only rarely in social life, and so either have *no* designation or a *very variable* one.* In a scientific exposition which begins with simple concepts it is generally superfluous that one first explains the precise notation for these concepts by one of those two methods. Such explanations one could call—owing to the difference from a proper definition—*denotations* or *descriptions*. They also belong to the class of the *arbitrary propositions*, in so far as they only serve to provide a particular symbol for a certain concept. They would thus be the *first* thing with which any scientific exposition must begin, in so far as it has simple concepts.

§9.

The *classifications* ‖*Einteilungen*‖ also belong to a scientific exposition, to produce order in it and make it easy to *survey*. Concerning these I believe that every genuine classification can only be a *dichotomy*.† A genuine scientific classification only arises if for a certain concept *A* (to be *classified*), there is a certain second concept *B* (the *basis of classification*) which must be consistent with *A* and which can be adjoined to *A* or excluded from it. Accordingly the general form of all classifications would be: *All things which are contained under concept A are either contained under the concept (A* cum *B) or under the concept (A* sine *B)*. From this it may be seen immediately that the concepts obtained through classification [(*A cum B*), (*A sine B*)] are always *compound*, and therefore *definable*, concepts. Also it may be seen that definitions with *negative* characteristics cannot be absolutely rejected, in that there certainly are, and must be, concepts with negative characteristics [(*A sine B*)] (§3 note).

Note: An objection made to mathematics, often not without justification, is that it makes hardly any use of *classifications*, and from this arises the striking *disorder* which is found in mathematical disciplines. It is in fact extremely difficult to remove this disorder and to introduce a true and natural order (rather than a merely *apparent* order). For this, of course, one must be clear about

* An example, among many others, is the concept just mentioned of the mathematical point. It surprises me that the astute *Locke* could maintain just the opposite on this.

† *Kant* mentions it as noteworthy that *trichotomy* appears in his table of categories. However, in my view there is no true classification there because otherwise the categories could not be *simple, fundamental concepts*.

all the simple concepts and axioms of these disciplines and already know exactly which premisses each axiom needs or does not need for its logically correct proof. Until this has been done all efforts to remove the confusion depend only on good luck, and it is not surprising if they fail. For example, if the concepts which we intend to put forward below in geometry are correct, then all the attempted classifications in this science made so far by *Schultz* and others are unusable.

B. *On axioms and postulates*

§10.

In the usual mathematics textbooks, and even in many logic books, it is said about axioms that 'they are to be propositions which on account of their intuitiveness (obviousness) require no proof, or whose truth is recognized as soon as their meaning is understood'. Hence the characteristic of an axiom would lie in its intuitive nature. However, with some thought it will easily be seen that this property is hardly suitable to provide a *firm basis for the classification* of all truths into two classes, that is, into axioms and theorems. For *firstly*, being intuitive is one of those properties which allow of innumerable differences in *degree*. One will never therefore be able to determine precisely what degree would be sufficient for an axiom. *Furthermore*, the intuitive nature of a truth depends on all sorts of very fortuitous circumstances, for example whether, by education or experience, we have been brought to recognize it often or only rarely. *Finally*, for this reason the degree of intuitiveness is also *very different* for different people; what for one person is perfectly obvious, often appears obscure to another. The greatest mathematicians, as we have already noted (§2), seem to have always dimly felt this, in that they themselves classed the obvious truths (provided they knew how to find a proof for them) as theorems. *Euclid* and his predecessors proved what they *could* prove, and the notorious parallel postulate, together with some other propositions, were certainly only put among the so-called *κοιναὶ ἔννοιαι*[k] because they still did not know how to prove them.*

§11.

The lack of knowledge of *how to prove a truth* certainly makes it into a *κοινὴ ἔννοια*, i.e. *common and naïve* ‖*ungelehrten*‖ *knowledge* (for thus knowledge

---

[k] ‖'common notions or axioms'.‖

* *Michelsen* puts forward the conjecture in his *Gedanken über den gegenwärtigen Zustand der Mathematik*, etc. that *Euclid's postulates* and *axioms* were originally perhaps nothing but certain aids to the memory for the discovery of solutions and proofs for a beginner. It is indeed *possible* that they had this *origin*, because they are certainly *very useful* for that purpose. Yet in *Euclid*'s time this alleged purpose was surely not in mind, or many other propositions would have been added to them, e.g. that of the square of the hypotenuse, which appears in proofs just as frequently as the *11th axiom*. However that may be, this purpose of axioms would be nothing less than scientific and to be imitated by us.

is called which is not based on clearly recognized grounds). It would also be good if such truths were set out, separated from all the rest at the beginning of the textbook, so as to separate the proved from the unproved and draw the attention of the scholar especially to the latter. But this would then be only a purely subjective classification of propositions, made not for the *science in itself* but only for the benefit of *its authors*. What one person puts forward today as a *κοινὴ ἔννοια* another would find a proof for tomorrow, and it would therefore be struck off that *list of liabilities*. If therefore the word *axiom* is to be taken in an *objective* sense we must understand by it a *truth* which *we not only do not know how to prove but which is in itself unprovable*.

Note: Also the *common knowledge* can be used for the proofs of other propositions in a systematic exposition, provided one is convinced that the unknown proof of the first propositions will proceed without the assumption of just those propositions for whose proof one wishes to use them. For example in the *theory of triangles* all theorems *of the straight line* can be presupposed as *κοιναὶ ἔννοιαι*.

§12.

Now the further question arises, what should properly be understood by the *proof* of a truth? One often calls every sequence of judgements and inferences by which the truth of a certain proposition is made generally *recognizable and clear*, a *proof of the proposition*. In this *widest* sense, *all* true propositions, of whatever kind they may be, can be proved. We must therefore take the word in a *narrower* sense, and by the *scientific proof* of a truth we understand the representation of the *objective dependence* of it on other *truths*, i.e. the derivation of it from such truths as are to be considered as *the ground for it*—not fortuitously, but *in themselves and necessarily*; while the truth itself must be considered as their *consequence*. Axioms are therefore *propositions* which in an *objective* respect can only be considered as *ground* and never as *consequence*. Here, of course, it should now be discussed at length *how many simple, and essentially different kinds of inference* ‖*Schlussarten*‖ *there are*, i.e. how many ways there are that a truth can be dependent on other truths. It is not without hesitation that I proceed to put forward my opinion, which is so very different from the usual one. Firstly, concerning the *syllogism*, I believe there is only a single, simple form of this, namely *Barbara* or *Γραμματα* in the *first* figure. But I would like to modify it, in that I would put the *minor* before the *major*, so that in this way the three concepts, *S, M, P* proceed with regard to content from the particular to the general. I find it more natural to argue: *Gaius is a man, All men are mortal, therefore Gaius is mortal*, than the more usual order: *All men are mortal, Gaius is a man, therefore Gaius is mortal*. However, these are small matters.—Every other figure and form of the syllogism seems to me to be either not essentially different from *Barbara* or not completely simple. But on the other hand, I believe that there are some *simple kinds of inference* apart from the syllogism. I shall indicate briefly those which have occurred to me so far.

(a) If one has the two propositions:

A is (or *contains*) *B*, and
A is (or *contains*) *C*;

then by a proper kind of inference there follows from these the third proposition:

A is (or *contains*) [*B* et *C*].

This proposition is obviously different from the first two, considered each for itself, for it contains a different predicate. It is also not the same as their *sum*, for the latter is not a single proposition but a collection of *two*. Finally, is is also obvious that according to the necessary laws of our thinking the first two propositions can be considered as *ground* for the third, and not conversely.

(b) In the same way it can also be shown that from the two propositions,

A is (or *contains*) *M*, and
B is (or *contains*) *M*,

the third proposition:

[*A* et *B*] *is* (or *contains*) *M*

follows by a simple inference.

(c) Again, it is another simple inference which from the two propositions

A is (or *contains) M*, and
(A cum *B) is possible* or *A can contain B*,

derives the third:

(A cum *B) is* (or *contains*) *M*.

This inference has much similarity with the syllogism, but is nevertheless to be distinguished from it. The syllogistic form would really arrange the second of the two premisses thus: *(A* cum *B) is (*or *contains) A*. But *this* proposition first needs, for its verification, the proposition, *(A* cum *B) is possible*. But if this is assumed one can dispense with the first as being merely analytic, or take it in *such* a sense that in fact they both mean the same thing and only differ verbally.—All these kinds of inference, including the syllogism, have the common property that from *two* premisses they derive only *one* consequence. On the other hand, the following could seem like an example of another kind of inference, whereby *two* consequences come from *one* premiss:

A is (or *contains*) [*B* cum *C*]

therefore,

A is (or *contains*) *B*, and
A is (or *contains*) *C*.

But I do not believe that this is an *inference* in that sense of the word which we established at the beginning of this paragraph. I can perhaps *recognize* subjectively from the truth of the *first* of these three propositions the truth of the two others, but I cannot view the first *objectively* as the *ground* of the others.— I cannot allow myself here to enter into a detailed discussion of all these claims.

§13.

Now the two questions arise: 'whether there are *generally* truths which are in themselves unprovable, and further whether there are definite *characteristics* of this unprovability?' Both must be answered affirmatively if there are to be axioms in the sense of the word given above (§11). Since it will always be doubted by some people whether there are any judgements in the realm of truth which absolutely cannot be proved, it seems to me worth the trouble to attempt a short proof of this claim here. Every provable judgement is to be viewed, according to the definition given in §12, as a *consequence*, and its combined premisses as its *ground*. Therefore to claim that all judgements are provable means to accept a *series of consequences* in which there is no *first*, i.e. there is no *ground* which itself is not in turn a consequence. But this is absurd. On the contrary, therefore, one must necessarily accept some judgements—at least two (§12)—which are themselves not consequences but basic judgements ⟦Grundurteile⟧ in the strictest sense of the word, i.e. *axioms*.

Note: As the contradictory nature of a *series of consequences without a first ground* is quite clear with a *finite number* of terms one could try to make it less noticeable by extending the series into *infinity*. However, it can easily be shown that this does not remove the contradiction at all. For this does not rest on the number of terms but only *on the fact that* the denial of a *first ground* (according to our definition of this term given above) is the postulation of a *consequence* which has no *ground*. If one assumes that the sequence stretches back to infinity then it only follows from this that whoever starts to count from a given term backwards never reaches that contradiction. However, although he does not find it straight away in *this* way, he must none the less think of it as *existing*. It is a rather similar matter to refute that notorious objection to the possibility of motion called the *argumentum Achilleum*. In the way which the author of this objection deliberately makes his attack, one can of course never arrive at the moment in which Achilles reaches the tortoise. But it does not follow from this that this moment does not occur: it can actually be found in another way very easily. One may compare on this *Cochius*' well-known paper: *Ob jede Folge einen Anfang habe* in *Hissmanns Magazin, Vol. 4.*

§14.

But how does one recognize that a proposition is unprovable? To answer this question properly we shall have to go back somewhat further, namely to the concept of a judgement and its different kinds. The usual definition of *judgement*, as a *combination of two concepts*, is obviously too wide because every *compound concept* is also a *combination of two* (or more) *concepts*. Of course, it is *one* combination by which we join two concepts into a new compound *concept* and *another one* through which we join two concepts into a *judgement*, but *both* combinations are in my opinion *simple*, *indefinable* actions of our mind. *Kant* certainly claims to have given us a *precise, fixed definition of judgement* (*Metaph. Anfangsgründe der Naturwissenschaft, 3rd ed., Leipz. p. XVIII Preface; likewise in his Logik)*, as *an action through which given ideas first become cognitions of an object*. But provided I understand this definition correctly (by the *given ideas* I understand the *predicate*, and by the *object* the so called *subject* of the judgement), then the whole concept of judgement lies here in the word '*cognition*'. And the phrases 'to consider an idea (namely the predicate) *as the cognition of an object* (otherwise called the subject)' or, 'to view an idea as a *criterion or characteristic of another one*' etc., are only different *descriptions*, but not proper *definitions*, i.e. not an *analysis* of the concept of judgement. This assertion is important to me because in the opposite case, if the concept of judgement were a compound one, then the concepts of the different *kinds* of judgement would also have to be compound concepts, and we could therefore not simply *enumerate* them, as we shall immediately be doing, but we would have to *classify* them according to a logical basis of classification.

§15.

As far as I know logicians have hitherto assumed that all judgements could be traced back to the form *A* is *B*, where *A* and *B* are to express the two *concepts* to be combined and the little word 'is' (called the '*copula*') expresses the *way* in which the understanding *combines A* and *B* in the judgement. Now it seems to me that this *way of combining* the two concepts is not the same for all judgements, and therefore it should not be designated by the same *word*. The most *substantial distinction between judgements* seems to me to lie in the variation in this way of combining ‖concepts‖, and accordingly the following kinds of judgements have occurred to me so far.

1. Judgements which can be traced back to the form: *S is a kind of P*, or what amounts to the same, *S contains the concept P*, or, *the concept P belongs to the thing S*. The combining concept in these judgements is the concept *of the belonging* of a certain *property*, or what is just the same, of the *inclusion* of a certain thing, as *individual or kind*, in a certain *genus*. This concept, although it is expressed here with *several words*, seems to me nevertheless to be a *simple one*, and if not identical with the concept of *necessity* it is *preferably* included under it. Therefore, so as to give them a proper name, I would like to call *this* class of judgements, *judgements of necessity*. An example of such a judgement of necessity is the proposition: *two lines which cut the arms of an angle in disproportional parts meet when sufficiently produced*. It is properly expressed thus: *the concept of two lines which cut the arms of an angle in two*

disproportional parts (=S)—is a kind of—the concept of two lines which have a point in common (=P) ⟦*(=S)—ist eine Art—von dem Begriffe zweyer Linien*⟧. Moreover, those judgements can be affirmative or negative—which is also true of the subsequent classes.

2. Judgements which *state a possibility*, and are included under the form: *A can be a kind of B*. Their combining concept is the concept of *possibility*, so I call them *judgements of possibility*. An example is the proposition: *There are equilateral triangles*. It is properly expressed thus: *The concept of a triangle (=A)—can be a kind of—the concept of an equal-sided figure (=B)*.

3. Judgements which express a *duty*, and are included under the form: *You, or generally N, should do X*. The combining concept here is that of *obligation* or *duty*, and the *subject N* is essentially a *free rational being*. These are called *practical judgements*.

4. Judgements which express some mere *existence*, without necessity, and can be included under the form: *I perceive X*. Here *perception* is taken in its widest sense, in which one can perceive not only ideas through the senses but generally *all* one's ideas. Its essential subject is '*I*'. We call them *empirical judgements of perception* or *judgements of reality*.

5. Finally (it seems to me) judgements of probability also form a proper class of judgements, whose combining concept is that of *probability*. Yet I am still not clear about their proper nature.

§16.

The most *substantial matter* on which I differ from others in this enumeration (§15) consists in my taking certain concepts into the *copula* of the judgement which otherwise are put into the predicate or subject. I must therefore indicate briefly what has caused me to make this change. It was chiefly the *judgements of possibility* and *duty*. I believe I have found that all judgements whose subject or predicate are compound concepts must be *provable* judgements (see below, §20). Now if *judgements of possibility* are expressed (in the usual method) so that the concept of possibility seems to form the *predicate*, then the subject would essentially be a compound concept, for it is well known that it is superfluous to assert the possibility of a *simple* concept. *(A* cum *B) is possible*, would then be the general form of all judgements of this kind, where *(A* cum *B)* would represent the subject, and *possible* the predicate. From the remark just made, therefore, all these judgements must be *provable*; nevertheless, it is easily seen that there must be some absolutely unprovable judgements of this kind, because every judgement of possibility, if it is to be proved, presupposes a premiss in which the concept of possibility is already present, i.e. *another* judgement of possibility. But if we take the concept of possibility into the copula there can be judgements of possibility whose subject and predicate are both completely simple concepts and which we can, therefore, without hesitation, allow to be regarded as unprovable judgements.

That is, if *A* and *B* are simple concepts then it is not unreasonable to assume that the *judgement 'A can be B'* is unprovable, because its subject and predicate are simple. It is the same with the *practical judgements*: if one transfers the concept of *obligation* or *duty* to the predicate or subject, they must always be compound judgements. And nevertheless there must be a first practical judgement (namely the highest moral law) which is absolutely unprovable.

Note: It can at least be seen that it is often not so easy to determine what properly belongs to the subject and what to the predicate in a judgement. The appearance is deceptive. Thus, for example, in the proposition, *In every isosceles triangle are the base angles equal*, it would clearly be wrong to make the words before 'are' the subject, and those following, the predicate. For here the subject would be repeated in the predicate because the concept *of the base-angles* tacitly includes that *of the isosceles triangle*, because only those angles which are *opposite equal sides* and of which equality can be asserted are called the *base angles*. The proposition must therefore be expressed: *The concept of the relation of the two base angles in an isosceles triangle (=S)—is a kind of—concept of the equality of two angles (=P)* ‖*(=S)—ist eine Art—von dem Begriffe der Gleichheit*‖.

§17.

One of the classifications of judgements which is quite different from those considered so far, and which has become especially important since *Kant*, is the classification into *analytic* and *synthetic* judgements. In our so-called *necessity judgements*, §15 no. 1., the subject appears as a *species* whose *genus* is the predicate. This relation of species to genus can be of two kinds: either there is a characteristic which can be thought of and stated in itself, which is added in thought as a *differentia specifica* to the *genus* (predicate *P*) to produce the *species* (subject *S*), or not. In the first case the judgement is called *analytic*; in every other case, which may be any of the classes mentioned in §15, it is called *synthetic*. In other words an *analytic* judgement is one of which the predicate is contained directly or indirectly in the definition of the subject, and every other judgement is *synthetic*.

§18.

From this definition it follows immediately that analytic judgements can never be considered as *axioms*. Indeed in my opinion they do not even deserve the name of *judgements*, but only that of *propositions*, because they teach us *something new* only as *propositions*, i.e. in so far as they are expressed in words, but not as *judgements*. In other words that which *one can learn from them* never concerns *concepts and things in themselves*, but at most only their *designations*. Therefore they do not even deserve a place in a scientific system, and if they are used it is only to recall the concept which is designated with a certain word, just as with the arbitrary propositions. In any case, it is determined even by the usual views that no analytic judgements are *axioms*, for their truth is not recognized from them themselves, but from the definition of the subject.

§19.

Therefore, if all our judgements were analytic there could be no unprovable judgements, i.e. axioms, at all. And because this opinion still actually has some support we wish to try to demonstrate, in a way independent of §15, that there really are *synthetic judgements*.

1. *All judgements whose subject is a simple concept are thereby already synthetic*. This is clear without further proof from the definition in §17. But no one will doubt that there really are such judgements with simple subjects unless they deny the existence of simple concepts (§4). For judgements must be able to be formed about every concept, of whatever kind it may be, because for each concept every other one must either belong or not belong to it as predicate.
2. *All negative judgements, if their subject is a positive concept, are also synthetic*. For if the subject is a *positive* concept, i.e. either completely simple or composed of several simple ones by pure *affirmation* in which no negation appears, then one can never prove from its mere definition that this predicate could not belong to it as subject. Now there really are such negative judgements, for example '*A point has no magnitude*'.

Note: There is an *essential* difference between positive and negative concepts which does not merely rest on the arbitrary *choice of words*. We have already defined the *positive* concepts; the *negative* concepts are those which contain some *denial* (i.e. an *exclusion*, not merely a *non-being*). But of course it is not always immediately apparent *from the word* which designates a concept whether it is affirmative or negative.

However, if one attempts the definition of the concept and proceeds to the simple characteristics, it will always be discovered whether the concept contains a denial or not. For example, the concept of a *right angle* is a positive one, for it is the concept of *an angle which is equal to its adjacent angle*. On the other hand, the concept of an *oblique angle*, i.e. *not a right angle*, is obviously negative; but the concepts of an *acute* and an *obtuse* angle are again positive, etc. Among the negative concepts the *simplest* are those which arise from a mere denial or exclusion of a certain positive concept A without thereby putting anything definite forward. They are of the form: '*Everything which is not A*'. I denote them (Π sine A) and call them *indeterminate or infinite concepts* (*terminos indefinitos*). They appear with the *converse of affirmative propositions*. For if one has the proposition, '*M is A*', for example, then it follows that: *Everything which is not A* = (Π sine A) *is also not M*. From this example one sees straight away what is *really* intended by the expression, '*everything which* *is not A.*' Namely, the word '*everything*' is not taken collectively, so that it would designate the *totality of the things which are not A*, but *distributively*, i.e. so that one understands by it *this or that object which is undetermined except that it may not be A*. In as much as every concept must be a concept of *something*, the infinite concepts are *of* something quite indeterminate, except that it may not be A; hence their name, *indeterminate concepts*. The judgements in which they appear as *predicates* are called *limiting* or *infinite judgements*. They are not different from *negative ones*, in so far as the two are *equipollent*; but the former appear essentially as the *minor* in *negative* syllogisms. Such concepts can only occur as *subject* in *negative* judgements. For nothing affirmative and general can be predicated of *something which is not A* except that it is *something that is not A*, which is a merely *identical* proposition. But one can certainly form negative propositions in which (Π sine *A*) is the subject. If one knows, for example, that *a is always A*, then one can say (Π sine *A*) *is never a*. On the other hand, the other negative concepts, which have some positive determination *(M* sine *A)*, can certainly lead to affirmative judgements. For example, *if two numbers are not equal to each other then one of them is greater than the other.*

§20.

After these preliminary matters we can now answer the question we put in §14 in the following way:

(a) *Judgements whose subject is a compound concept, if they are knowable* a priori *as true, are always provable propositions*. It is of course clear that a compound and its properties must be dependent on those simple things of which it is composed and on their properties. Therefore if any subject is a compound concept then its properties, i.e. the *predicates* which can be attributed to it, must be dependent on those individual concepts of which it is composed and on their properties, i.e. on those *judgements* which can be formed about these simple concepts. Thus every proposition whose subject is a compound concept is a proposition *dependent* on several other propositions and thus (inasmuch as it is knowable *a priori*) it is actually also a *derivable, i.e. a provable, proposition*, and therefore can in no way be regarded as an *axiom*.

(b) *Judgements whose predicate is a compound concept, if they are knowable* a priori *as true, are always provable propositions*. That a certain predicate belongs to a certain subject depends as much on the subject as on the predicate and its properties. Now if the latter is a compound concept, then its properties depend on those individual concepts of which it is composed and on their properties, i.e. on those judgements which can be formed about these concepts. Therefore the truth of a judgement whose predicate is a compound concept depends on several other judgements, and so, as before, it is clear that it cannot be an *axiom*.

(c) Hence it now follows that the strictly unprovable propositions, or axioms, are only to be sought in the class of those judgements in which both subject and predicate are completely simple concepts. And because, in general, there *are* axioms (§13), they *must* be found within this class. The following, third, theorem can provide us with a more precise determination. *For every simple concept there is at least one unprovable judgement in which it appears as subject*. There are, in general, *judgements* about it. Therefore let '*A is B*' be such a judgement. If we now suppose it is *provable*, then *on account of its simple subject and predicate it can only be proved by a syllogism whose premisses are of the form 'A is X' and 'X is B'*. If the premiss '*A is X*' should again be provable, then in the same way this presupposes another of the form '*X is Y*', etc. Therefore if one did not wish to concede that there is at least one axiom in the form '*A is M*', one would have to accept an infinite series of consequences without a first ground.

Note: In the proof of this last proposition the assertion which is printed in *italics* seems to need a further elucidation. The question could of course be raised why I limited this assertion only to propositions with a simple subject and predicate. If it is *generally* valid *too much* follows from my proof, namely, that to *every compound* concept there is an unprovable judgement in which it appears as predicate, which contradicts the *first theorem*. Therefore for the complete verification of the former assertion *three things* will really be needed: (1) that propositions with simple concepts are provable only through syllogisms; (2) that these syllogisms always presuppose a premiss whose subject is one and the same simple concept *A*; and, (3) that at least one of these two things holds exclusively of propositions with *simple* concepts—and not with *compound concepts*. Now I have already mentioned in §12 that I do not regard the syllogism as the only simple kind of inference. Moreover, from the example quoted there it may be seen that for propositions with compound concepts, but only for these, there are at least *three other* simple forms of inference. On the other hand, how propositions with *simple concepts* could be proved other than through a syllogism, I really do not know. I therefore assume that (1) and (3) are correct, and this only leaves (2). It is quite clear to me that (2) holds at least for propositions with simple concepts, but whether it is a different matter for propositions with compound concepts I shall not at present venture to decide. Of course the question arises here (one which is also interesting in another respect) whether each judgement can be expressed in only *one way*, i.e. whether it retains the same subject and predicate. This is obvious for judgements with simple concepts; but for other kinds it might be thought that one and the same judgement could be changed into a *new one* containing a quite different subject and predicate, without changing its sense, merely by taking some characteristics of the subject over to the predicate (or conversely). Now if this should be the case,* then a compound judgement could be proved through a series of syllogisms without each of these syllogisms containing a premiss whose subject was the same unaltered

* But I readily admit that the *contrary* seems to me much more probable.

subject of the judgement. But this would then be a *second* reason why my above assertion could also extend, not against my will, to *compound* judgements.

§21.

If by the foregoing we have now proved that every judgement which is to be an axiom must consist of only simple concepts, it is nevertheless not proved, conversely, that every judgement which consists of simple concepts is an axiom. Therefore in order to prove that a given proposition '*A* is *B*' is an axiom it is not enough to show that the concepts *A* and *B* are both completely simple, but one must show furthermore that there are no two propositions of the form '*A* is *X*' and '*X* is *B*' from which '*A* is *B*' could be inferred. This will in most cases require a *special consideration* ‖*eine eigene Betrachtung*‖ to which, to distinguish it from a proper *proof* (or a *demonstration*), I give the definite name of a *derivation* (or *deduction*). Axioms will therefore not be *proved*, but they will be *deduced, and these deductions are an essential part of a scientific exposition*, because without them one could never be certain whether those propositions which are used as axioms really are axioms.

Note: If it is felt that what we said about simple *concepts* in §8 also holds for the *axioms*, then it will not be expected that all axioms should appear in our minds with perfect vividness. On the contrary, our most vivid and clear judgements are obviously *inferred*. The proposition *that a curved line between two points is longer than the straight line*, is far clearer and more intuitive than some of those from which it must be laboriously derived. The proposition (to give, for once, an example from another science), '*You should not lie*', is far clearer and more obvious than that principle from which it follows, '*You should further the common good*'.—Indeed, it could even be that an axiom may appear *dubious* and *doubtful*, especially from a misunderstanding of its words, or because we do not see immediately that what can be derived from it we recognize at once as true. (Thus some people find the merely identical proposition, '*Follow reason*', dubious because they understand it as though it nullified the obligation to obey divine commands or those of a legitimate authority.) In such cases the deduction of the axiom must first inspire us with confidence in its truth, and this will happen if it proceeds from some generally accepted and obviously clear propositions which are basically nothing but *inferences*, and even judgements *inferred* from that axiom which we wish to deduce. By making this connection apparent we will become convinced of the truth of the axiom itself.

§22.

If the foregoing is correct the question can now be answered, '*whether mathematics also has axioms*?' Of course, if all mathematical concepts were *definable concepts* then there could be no axioms in the mathematical disciplines. However, since there are *simple* concepts which belong specifically to mathematics (§8), one certainly has to acknowledge real *axioms* in mathematics. The domain of the axioms stretches as far as that of the pure simple concepts: where the latter ends and the *definitions* begin, there also the axioms cease and the *theorems* begin.*

§23.

Mathematicians are in the habit of introducing, as a special kind of axiom, *postulates*, by which they understand those axioms which assert the *possibility* of a certain object. According to §16 there are, and there must be, postulates (unprovable judgements of possibility). However, they occur, at most, only with concepts which are composed of *two* simple ones. The possibility of a concept composed of *three* or more simple components is a *provable* proposition. On the other hand, the possibility of *completely simple* concepts is really not even a judgement, for it lacks the predicate which the word 'possible' in the verbal expression of the proposition only apparently represents. Possibility, just like impossibility, only occurs with compound concepts.

C. *On theorems, corollaries, consequences, and their proofs*

§24.

It is well known that provable propositions are sometimes called *theorems*, and sometimes either *corollaries* or *consequences* ‖*Folgerungen*‖. It seems that mathematicians have regarded it as worth their while to distinguish the former from the latter. What one of them sets up as a *theorem* another is to be seen treating as a *corollary* or as a *consequence*. However, I should think it would not be a disadvantage to science if it were to be made clear what difference one really intended to indicate by the difference in these names. Most mathematicians hitherto have, through some vague feeling, accorded the character of theorem to the *more noteworthy* propositions. But it is clear that this property is not very suitable to provide a reliable means of distinguishing propositions, since it is purely relative and extremely vague. I would therefore propose the following basis for the distinction. All theorems and corollaries have this in common, that they are judgements inferred from previous propositions. But in order to be able to derive this consequence from the previous proposition one may either require the help *of a proper axiom in this science* (or what amounts to the same thing, a proposition already inferred from an axiom), or not. In the *first* case one may call the inferred proposition a *theorem*, in the *second*, i.e. if it arises from the previous proposition merely by means of an axiom of *another* science or merely by reference to an existing definition or something similar, then it may be called a *corollary* or *consequence*.

---

* From this one sees how wrong it is to say, as the usual text-books of mathematics do, '*the axioms follow the definitions*'.

Note: According to this rule, for example, if it has first been proved *that the arithmetical square on the hypotenuse is equal to the sum of the squares on the other two sides*, then the proposition *that the side of a square is in the proportion 1: $\sqrt{2}$ to its diagonal* would be a mere corollary. The same is true for the proposition *that all irrational ratios of the form $\sqrt{n}$: $\sqrt{m}$ can be represented by the ratio of two straight lines*, and similar propositions. For all these propositions require, for their derivation from that first one, only particular applications of arithmetic, but no new geometrical axiom. On the other hand, the proposition *that the area of the square on the hypotenuse equals the areas of the squares on the other two sides* would be a new *theorem*, because it can only be proved by the aid of a new geometrical axiom, or what amounts to the same thing, a new theorem proceeding from such, namely, the proposition *that the areas of similar figures are in proportion to the arithmetical squares of their corresponding sides*.

§25.

If one also wanted to allow a difference between *corollaries* and *consequences* then one could perhaps retain the distinction already made in the ordinary use of language. This is the custom that one calls *corollaries* (equally, *additional propositions*) those judgements which are preceded by some true *proposition*, i.e. a *theorem*. In contrast other propositions, e.g. those which follow directly on a *definition*, cannot on this account properly be called *corollaries*: they are therefore called mere *consequences*. We can thus combine this definition with the previous one: *consequences are those provable judgements which follow from a mere definition, but corollaries those which follow from a preceding theorem without the help of any new axiom of this science*.

§26.

From the above discussions several consequences can be derived about the *organization and succession of propositions* in a scientific system and about the nature of their proofs. We wish to mention some of these, (though they will contain nothing that is basically new or that has not been said before) *because* they are still not being *followed*, and it seems they are not sufficiently recognized as being *indispensable*.

1. *If several propositions appearing in a scientific system have the same predicate, then the proposition with the narrower subject must follow that with the wider subject, and not conversely*. For if the two judgements, *S is P*, and *Σ is P*, are valid, where *Σ is narrower than S*, then either Σ = *(S* cum *s)* or the proposition, *Σ contains S*, is valid. In the first case one knows that the proposition, *Σ is P*, is to be considered as a consequence from the two: *S is P*, and *(S* cum *s) is possible*, according to the third method of inference in §12. In the second case the proposition, *Σ is P*, is to be viewed as a consequence (through

the syllogism) of the two propositions, *Σ contains S* and *S is P*. Moreover, the truth of this assertion has always been known and expressed in the following phrase, '*in a scientific account one must always proceed from the general to the particular*'. For this means nothing else but that the proposition with a *narrower* subject must always follow the proposition with a *wider* subject.

Note: However, a proposition can sometimes be so expressed that from its words it seems to have a greater generality than a certain other one, without this in fact being the case. One must therefore not make a mistake on this account. The following proposition provides an example: *In every triangle bac,* $ab^2 = ac^2 + bc^2 \pm 2ac.cd$*, according as the perpendicular ad falls outside or inside cb.* This proposition seems to have a *wider subject* than the *well-known theorem of Pythagoras*, from which nevertheless it is only derived. Yet in fact it is otherwise. First of all notice that there are basically two propositions here, which are put together to look like a single one. There are really *two* subjects and *two* predicates, according as the perpendicular mentioned falls outside or inside the side *cb*. Euclid was therefore quite correct in forming two separate theorems from it. Furthermore, it must not be thought that the *theorem of Pythagoras* can be considered as contained under one of these two propositions. Because the *case* when the perpendicular meets the vertex of the included angle is a completely *special* case in which nothing can be said of the rectangle *ac.cd*. Thus the subjects of these three propositions are not subordinate to one another, but are co-ordinate concepts.

§27.

2. *If several propositions appearing in a scientific system have the same subject, then the proposition with the more compound predicate must follow that with the simpler predicate and not conversely*. For in the proposition, *S contains* (*P* cum Π), the proposition, *S contains P*, is presupposed in such a way that one definitely has to think of the latter before the former and not conversely (§12). This truth has forced itself particularly clearly upon those who have thought about the nature of a scientific exposition. '*One must*', they have said, '*always teach more, not less, in the later propositions than in the preceding ones.*' Moreover, it is obvious here that we cannot extend our assertion *further*, and instead of the expression, '*the proposition with the more compound predicate*', put the more general one, '*the proposition with the narrower predicate*'. For whenever we make an inference by a syllogism one of the premisses (namely the so-called *major*), with just the same subject as the conclusion, has a predicate (namely the *terminus medius*) which is *narrower* than that of the conclusion: *S contains M*, *M contains P*, therefore *S contains P*, where the concept *M* must obviously be narrower than *P* because otherwise the proposition, *M contains P*, could not be true, and yet the judgement, *S contains M*, must be considered as one preceding the judgement, *S contains P*.

§28.

With regard to the proofs which must be supplied for all provable propositions in a scientific system, we shall content ourselves with mentioning just *two properties* which are required for their correctness as *conditio sine qua non*.

1. *If the subject (or the hypothesis) of a proposition is as wide as it can be for the predicate (or the thesis) to be applicable to it, then in any correct proof of this proposition all characteristics of the subject must be used, i.e. they must be applied in the derivation of the predicate; and if this does not happen the proof is incorrect*. For in such a proposition the *whole subject* (not only one of its component concepts) is seen as the sufficient basis ‖Grund‖ for the presence of the predicate. But if one were not to *use* some characteristic of the subject at all in the proof, i.e. derive no consequences from it, then the predicate would appear independent of this one characteristic. Thus no longer the whole subject but only a part of it would be considered as the basis for the predicate. Therefore if some characteristic of the subject is not used in the proof this is a sure sign that either the theorem itself must be expressed too narrowly and contains superfluous restrictions or, if this is not the case, that the proof itself contains some hidden error. As the well-known maxim of the logician says: *quod nimium probat, nihil probat*.[1]

Note: Though simple and clear in itself, this observation has often been overlooked in practice. Sometimes futile effort is made to find a proof for a certain theorem without seeing how all the conditions present in the *hypothesis* will be used in the proof. At other times, proofs are put forward which, because they do not use all conditions of the *hypothesis*, must obviously be defective. An example of the *first* is the notorious *parallel postulate*, in which the intersection of the two lines clearly only holds under the condition that both lines lie in the same plane. However, very few people concerned with discovering a proof of this proposition have considered how this condition could be used in the proof. This would actually have caused them to seek it in quite a different way. An example of the *second* case is offered by *Kästner's theory of the lever*, which is usually considered the best. After *Kästner* has very correctly demonstrated the *theorem* in his *Mathematischen Anfangsgründe, Part II, 1st Sec.*, in 4th ed. (16): '*that equal lever arms and forces produce equilibrium*', he concludes in the corollary (18): '*Loads which are not to sink must be borne up. Here there is nothing to do the bearing except the support of A. Therefore this support takes the complete load 2P = 2Q*, i.e. (according to a definition of this phrase which he gives in 29) *if one wanted to hold this weighted lever on a thread AZ then one must pull along AZ with a force F = 2P*'. In this proof there is no mention of the condition that the two forces must act perpendicular to the lever, or at least *in parallel directions*, a condition which is nevertheless

[1] ‖'What proves too much, proves nothing'.‖

necessary for the above conclusion. Therefore this proof is clearly false, for it proves too much.

§29.

2. As well as the characteristics of the subject several other *intermediate concepts* could also appear in the proof. However, if the proof is to contain nothing superfluous, then *for an affirmative proposition there should only appear intermediate concepts which are not narrower than the subject and not wider than the predicate. For a negative proposition they should only be wider than the subject or wider than the predicate.* From §17 it may be seen that in all *a priori* judgements, at least in all judgements of necessity and possibility (such as all mathematical judgements are), if they are affirmative the predicate is a concept which if not *wider is at least as wide** as the subject.

Now if such an affirmative judgement, for example *A contains B*, is to be proved then it will require in addition two other similarly affirmative judgements as premisses.

In whichever of the four *simple forms of inference* mentioned above (§12) it can be proved, it is apparent that the intermediate concepts used are always narrower than *A* and wider than *B*. From the following table, in which the signs > and < mean *wider* and *narrower*, this will be seen without any trouble if one makes use of the relationship which has been asserted between the extents of the two concepts of an affirmative judgement.

* This seems to be contradicted by what *Selle* (*De la réalité et de l'idéalité des objets de nos connaissances*, in the Mémoires de l'Académie de Berlin 1787, p. 601) claims to have discovered, that the real difference between analytic and synthetic judgements does not lie in what we said above in agreement with *Kant*, but only in the fact that with analytic judgements the predicate is contained in the subject, while in synthetic judgements the latter is contained in the former. Now the phrase, 'a concept *A* is contained in another *B*' is *ambiguous* in French just as it is in German, and can mean that either *A* or *B* is the *narrower* of the two. In each case this means that there are judgements in which the predicate is *narrower* than the subject, which to me at least seems absurd with the two classes of judgement mentioned. Firstly, for *negative* judgements, the predicate is obviously neither narrower nor wider than the subject, but both completely exclude each other. However, for *positive* judgements, if they are *particular* and *disjunctive* it could be possible for it to seem at first as if their subject were wider than their predicate. Thus one could say that in the *particular* judgement: '*some quadrilaterals are squares*' the subject '*some quadrilaterals*' is obviously wider than the predicate '*squares*'. However, with more careful consideration one may see that the form of particular judgements is not even pure *a priori*, but is *empirical*. For that some quadrilaterals *are really* squares is, when expressed in that way, an empirical assertion. The only assertion which can be called pure *a priori* is: *the concept quadrilateral can contain the concept of a figure with equal sides and angles*. This judgement belongs to the class of pure *a priori* judgements of possibility (§15), in which the predicate is obviously a wider concept than the subject. The *disjunctive* judgements are really all of the form: *A is either B or not B*. This judgement really amounts to: *the concept of the sum of (A* cum *B) and (A* sine *B) contains the concept of all A*. Now here the subject, '*the sum of (A* cum *B) and (A* sine *B)*' is again obviously a *narrower* concept, only a *kind* of concept of the predicate, 'the *totality of things which are A*'.

First Form

A is a kind of M
A is a kind of N

A is a kind of (M *et* N) = B
| $M > A$ and $< B$
| $N > A$ and $< B$

Second Form

M is a kind of B
N is a kind of B

(M *et* N) = A is a kind of B
| $M > A$ and $< B$
| $N > A$ and $< B$

Third Form

M is a kind of B
M can be a kind of N

(M *cum* N) = A is a kind of B
| $M > A$ and $< B$
| $N > A$ and $< B$

Fourth Form

A is a kind of M
M is a kind of B

A is a kind of B
| $M > A$ and $< B$

In a *negative* judgement subject and predicate are mutually exclusive, so one cannot say of either of the two that it is *wider* or *narrower* than the other. But it is clear from the following that the *intermediate concepts* which are required in the proof of such negative judgements *A is not a kind of B*, must always be *wider than the subject A or wider than the predicate B*:

First Form

A is not a kind of M
A is not a kind of N

A is not a kind of (M *et* N) = B
| $M > B$
| $N > B$

Second Form

M is not a kind of B
N is not a kind of B

(M *et* N) = A is not a kind of B
| $M > A$
| $N > A$

Third Form

Whatever is a kind of M is not a kind of B
M can be a kind of N

(M *cum* N) = A is not a kind of B
| M > A
| N > A

Fourth Form

Whatever is a kind of M is not a kind of B
A is a kind of M

A is not a kind of B
| M > A

Note: Therefore if there appear in a proof *intermediate concepts* which are, for example, *narrower* than the subject, then the proof is obviously defective; it is what is usually otherwise called a *μετάβασις εἰς ἄλλο γένος*.[m] Sometimes it is of course clear at first glance that a certain concept does not belong to a proof of a proposition and is of a completely alien kind. For example, in *Théorie des fonctions analytiques, No. 14*, the important claim that the function $f(x + i)$, with i continuously variable, can in general be expressed

$$f(x + i) = (x) + ip + i^2 q + i^3 r + \ldots$$

is derived from a *geometrical* consideration: namely from the *fact* that a continuously curved line which cuts the x-axis has no smallest ordinate. Here one is in a real *circulus vitiosus*, because only on the assumption of the purely arithmetical assertion about to be proved can it be shown that every equation of the form $y = fx$ gives a continuously curved line. But in other cases it can only be decided whether some superfluous concept is involved or not by a detailed analysis of the whole proof into its simple inferences, and by dividing all its concepts into their simple components.

§30.

We must give our opinion here about the question of *whether there can possibly be several proofs for one proposition*. It depends on what one counts as the *essence* of a proof. If one counts the *order of the propositions (which make use of certain premisses which may, or may not, be expressed in words)* as part of the essence of a proof, then a proof will be called different from another one if the propositions in it merely follow each other in another order, and some intermediate propositions introduced explicitly in one are omitted in the other. Then there is no doubt there could be *several* proofs of the one proposition.

---

[m] ['Crossing to another kind'.]

On the other hand, if one regards the essential matter of a proof as lying in those *judgements* on which the conclusion to be proved is based, like a consequence on its grounds (irrespective of whether these judgements are all explicitly stated or whether some of them are merely tacitly assumed, or indeed whether they follow one another in this or that order), then for every true judgement there is only a *single* proof. For although in general not every consequence determines its ground and equal consequences can sometimes proceed from unequal grounds, it is nevertheless a different matter with the *grounds of our knowledge*. It is clear from the foregoing that the one or two intermediate concepts which are required for every single proof are always *determined*, and cannot arbitrarily be taken in different ways. We shall show this, for lack of space, for only one of the kinds of inference, for example the syllogistic kind, in such a way that it will be seen how the same also holds for the others. Let M, N, $O, \ldots$ be *intermediate concepts* between A and B, i.e. they are $> A$ and $< B$; then one can use each of them for a syllogism from which the judgement, *A is a kind of B*, is to be inferred. That is, one can set up the syllogisms, *A is a kind of M, M is a kind of B, therefore A is a kind of B*. In the same way, *A is a kind of N, N is a kind of B, therefore A is a kind of B*, etc. So far it may seem as though there really were several proofs of the judgement, *A is a kind of B*. However, if between A and M there is some intermediate concept L, so that the judgement, *A is a kind of M*, is itself provable, then in a complete proof in which nothing is to be omitted, the syllogism, *A is a kind of L, L is a kind of M, therefore A is a kind of M*, must be put first. So in other words there must be as many syllogisms set up as there are intermediate concepts between A and B. And since the number of these intermediate concepts is obviously *determined*, so also the number and form of these syllogisms, and therefore the whole proof, is *determined*.

§31.

From these considerations it follows how one has to decide on the classification of proofs into *analytic* and *synthetic*. The whole difference between these two kinds of proof is based simply on the *order and sequence* of the propositions in the exposition. This is exactly like the remark made by the admirable *Platner* (in his *Philosoph. Aphorismen*, Vol. 1, §554, 2nd ed.) about the distinction between the fourth and first syllogistic figures, in which the two premisses are simply interchanged. Now such distinctions certainly cannot establish an objectively valid or scientific classification, but nevertheless they are not therefore to be dismissed completely. Unless I am mistaken the real difference between theorems and problems rests on this distinction, and we shall say more about it presently.

§32.

Finally we should give our opinion briefly about the *apagogic kind of proof*, since it is used so frequently in mathematics. First of all, though, we must put in a remark on the so-called *converse of a proposition*. If the affirmative judgement, *A is B* (or, more definitely, *A is a kind of B*) holds then it is well known that there also holds, as *propositio inversa*, the judgement, '*what is not B is also not A*' (or more definitely, *what is not a kind of B is also not a kind of A*). Usually the latter is considered as a kind of *consequence* of the former. But should the converse proposition be derived from its affirmative counterpart by a *proper inference*, i.e. should it be so derived that the *former* would have to be considered as *ground*, and the *latter* as *consequence*? Cannot the former be derived from the latter just as well as the latter from the former, so that we are therefore equally justified in viewing the former as consequence and the latter as ground? I therefore do not think that the formation of the *converse of a proposition* should be counted in the class of *kinds of inference* of which we spoke above (§12).

§33.

Having said this, two particular kinds of apagogic proofs have been distinguished. The first kind is as follows. The proposition, *A is B*, is proved by assuming the opposite, *A is not B*, and deriving from it something impossible, i.e. some contradiction with a proposition, *A is C*, which has already been proved. Now in this kind of proof the *essential point* (I believe) does not lie in the false hypothesis, *A is not B*, but solely in the fact that premisses appear of the form, '*what is not M is also not N*', so that in my opinion *only negative propositions essentially require an indirect proof*—affirmative propositions can always be proved *directly*. The usual form of the apagogic proof goes like this: *A* is *B*, for, *suppose A were not B, then also A would not be C, which is absurd*. This reasoning could also have been arranged in the following form: *whatever is C is always B; A is C; and so therefore A is also B*. Here no negative proposition appears at all. On the other hand, if the proposition to be proved were negative, for example *A is not B*, then one would proceed by the usual method thus: *suppose A were B, then it would follow that A is also C, which is absurd*. This can however also be presented: *what is not C is also not B; A is not C; A is therefore also not B*.

§34.

The *second* kind of indirect proof derives from the *false* hypothesis, *A is not B*, the true proposition, *A is B*. Now this was justifiably found objectionable until *Wolff*, *Lambert*, and others showed that it is not really the false premiss, *A is not B*, on which one builds, but that at the beginning of the proof it is *just left undetermined* whether *A* is a kind of *B* or not. Therefore, unless any other converse proposition appears as an essential premiss in it, this kind of proof is not really apagogic.

D. *On problems, solutions, notes, etc.*

§35.

A particular kind of mathematical proposition are the *problems*, with their *solutions*. Concerning their true nature the definitions of the mathematicians are still not in complete agreement. Most *recent authors* define the problems as those '*provable propositions which state the possibility of a concept*'. It is in their favour that *theorems* and *problems* are thus distinguished in just the same way as *axioms* and *postulates*. However, if we look at the *use* which the mathematicians always *actually* make of their problems, we find it does not correspond to this concept at all. For it is not merely *propositions of possibility* that are presented under the title of problem, but also many *propositions of necessity*. For example, '*To find the third side of a triangle from two sides and the included angle.*' In this there is no mention of any *possibility* at all. Similar propositions which do not state any possibility also appear in *Euclid* under the title of *problems*, for example *Elem. Book II, Prop. 14*; *Book III, Prop. 1*; *Book VII, Prop. 2*, and several others. Conversely, on the other hand, several *propositions of possibility* are set up under the title of *theorems*. Such is *Elem. Book I Prop. 7 Theor. 4: It is impossible, on the same straight line, in the same plane and towards one side, to draw two equal straight lines to more than one point*—a negative proposition of possibility. I am not against making the distinction, already established among *unprovable propositions*, of axioms and postulates, and also, in conformity with this, distinguishing the *provable propositions* into those which express a *necessity* and those which express a *possibility*, and devising suitable terminology for these. Indeed, this is very much to be *recommended*, because the propositions of possibility are also to be proved from quite different grounds from the propositions of necessity: namely, the former from postulates, but the latter from axioms. However, I believe the distinction between *problems* and *theorems* to be yet *another* distinction, and it deserves to be retained, although it is not *objectively scientific* and only concerns the mere *manner of presentation*. Of course, provable propositions can be presented in a scientific system in *two ways*: either one first just *states* them (i.e. first one only makes known their *sense*) and allows the assurance of their truth and the representation of their objective connection ⟦Zusammenhang⟧ with other propositions to *follow*, or one does not do this. The *first* way gives the form of the *theorem*, the *second* way that of the *problem*. *The theorem therefore has two components*: the *proposition* (*thesis*), in which the new judgement is merely *expressed*, and the *proof*, in which its objective connection with other truths is shown. However, the *problem also has two components*: the *question* (otherwise also called the *problem* in the *narrower* sense), in which one only determines the object about which one now intends to state something new, and the *solution*, in which, proceeding from unprovable, or already proved truths, one reaches the new truth that was sought. This therefore shows that it is a misuse to provide the solution with a proper *proof*, for thereby the problem reverts again into the form of a *theorem*. Instead, solution and proof should, through the use of the synthetic method, be amalgamated into one.

§36.

Assuming these concepts we can now determine more precisely which truths are more suited to the form of a theorem and which to that of a problem. For those propositions whose statement *does not cause surprise*, and whose truth one can grasp immediately, even though dimly, from what went before, the form of the *theorem* is suitable. *On the other hand, for those which would never have entered one's mind*, the form of a *problem* is more appropriate. So, for example, the proposition *that factors taken in a different order give one and the same product* is suited, by its nature, to being a *theorem*. On the other hand, the proposition about *how the highest common factor of two numbers is to be found* is always more appropriately presented as a problem.

Note: With regard to the place that is given to the problems it is to be noted that care must be taken to see the *possibility of solving them* by the preceding propositions. For example, before one puts forward the problem: '*to calculate the area of a triangle from its three sides*', one must have proved that this area is *determined* by those three sides.

§37.

This seems to be the right place to denounce a certain arbitrary *restriction* which mathematicians have introduced especially in geometry. Namely, never to accept an object as *real* before having shown the method by which it can be constructed with certain instruments. It is well known that in *Euclidean* geometry no spatial object is accepted as real unless its construction has first been demonstrated by means of *plane*, *circle*, and *straight line*. This restriction betrays its *empirical* origin clearly enough. *Board*, *compass*, and *ruler* are of course the simplest instruments which were needed initially for the drawing. However, considered in themselves the *straight line*, *the circle*, and also *the plane* are such compound objects that their possibility cannot be accepted in any way as a *postulate*, but on the contrary it must first be proved from the possibility of just *those* things which *Euclid* teaches us how to construct through those three. For example, the proposition that between every two points lies a *mid-point* is far simpler than the proposition that between every two points a *straight line* can be drawn. Nevertheless, *Euclid* proves the former from the latter and several others. It is sufficient for the *theoretical* exposition of mathematics (Sec. I, §18) that one proves the *possibility* of every conceptual connection which is put forward. How, and in what way, an *object* analogous to the concept can be produced in *reality* belongs to *practical mathematics*. For example, it is enough to prove *that to every three straight lines there must be a fourth proportional line*; we do not need to show the way in which it can be found.

§38.

Finally, there is one more term which is useful in the *a priori* disciplines of mathematics, namely the *note*. This refers to remarks which do not belong to the science in an objective respect, but have only a subjective purpose: for example historical remarks, explanations, other proofs, examples, applications, warnings of misunderstandings, and similar things. But why have *reminders*, *introductions*, *transitions*, and similar terms not also been taken into mathematical terminology, since they are already useful in other scientific expositions? Would not mathematical exposition thereby lose some of its rigidity and be able to be more flexible, illuminating, and popular?—If everything that has been said so far is now gathered together, then the *complete mathematical apparatus* (to give a brief summary of it) consists of the following parts:

1. *Notation* (for simple concepts)
2. *Definitions* (analyses of compound concepts into their simple components)
3. *Arbitrary propositions* (notations for simple as well as compound concepts)
4. *Classifications* (which always give compound concepts)
5. *Postulates* (unprovable propositions which state a *possibility*)
6. *Axioms* (unprovable propositions which state a necessity)
7. *Theorems* (provable propositions inferred from axioms of the same science which are either (a) *propositions of possibility* or (b) *propositions of necessity*)
8. *Consequences* (provable propositions which are inferred from definitions with the help of alien axioms)
9. *Corollaries* (provable propositions which are inferred from theorems with the help of alien axioms)
10. *Problems with their solutions* (provable propositions which are presented according to the synthetic method)
11. *Introductions, transitions, notes*
12. *Common knowledge* (provable propositions whose truth one is convinced of but whose systematic proof is not yet known)

APPENDIX

ON THE KANTIAN THEORY OF THE CONSTRUCTION OF CONCEPTS THROUGH INTUITIONS

§1.

To have first correctly drawn attention to the important difference that exists between the analytic and synthetic parts of our knowledge remains a *service* which *Kant* has rendered to us, even if we cannot justify, and agree with, everything that this philosopher has claimed about the intrinsic nature of our synthetic judgements. It is certain that the truth of analytic judgements rests on a quite *different* basis from that of the synthetic. If, in fact, the *former* deserve the name of *true judgements* (which I do not admit without some hesitation*), then they are all based on that *one* general proposition which is expressed in the formula: *(A* cum *B) is a kind of A*. If this is called the *law of identity* or of *contradiction*,then one can always say that the *law of contradiction is the general source of all analytic judgements*. However, it is entirely different with the *synthetic judgements*: these obviously cannot be derived from that axiom. *Kant* therefore posed the question, 'what is the basis which determines our understanding to attribute to a certain subject a predicate which is certainly not contained in the *concept* (in the *definition*) of that subject?' And he believed he had found that this basis could be nothing but an *intuition* which we connect with the concept of the subject and which at the same time contains the predicate. Accordingly, to all concepts about which we can form synthetic judgements there must correspond *intuitions*. But if these intuitions are always merely *empirical*, then also the judgements arrived at by means of them are always *empirical*. Now since there are also *a priori* synthetic judgements (such as mathematics and pure natural science undeniably contain) there must also be, strange as it may sound, *a priori intuitions*. Once it has been decided that there can be such things one is easily convinced that with respect to mathematics and pure natural science they are *time* and *space*.

§2.

We may reasonably ask here what *Kant* understands by an *intuition*. From his *Logik* (ed. *Jasche*), for example, and from many other places in his writings (for example *Kritik der reinen Vernunft*, p. 47, etc.) we receive the answer: All ideas ‖Vorstellungen‖ are either *intuitions* ‖Auschauungen‖, i.e. ideas of an individual, or *concepts*, i.e. ideas of something general. If we then ask what a *pure a priori intuition* is meant to be, then it seems to me at least, that no other answer is possible than: *an intuition which is combined with the awareness of the necessity that it must be so and not otherwise*. For only if this awareness of the necessity is contained in the intuition can it also lie in the connection, made by means of the intuition, between the subject and the predicate, i.e. in the *judgement*.

§3.

Several people have of course already taken exception to these *a priori intuitions* of the *critical* philosophy. For my part, I readily admit that there has to be a certain *basis*, quite different from the law of contradiction, by which the understanding connects the predicate of a synthetic judgement with the concept of

* See Sec. II, §18.

the subject. But how this basis can be, and be called, *intuition* (and even, with *a priori* judgements, *pure* intuition) I do not find clear. Indeed, if I am to be really honest, all this seems to me to rest on a distinction which is not thought out clearly enough, between that which is called *empirical*, and that which is called *a priori* in our cognitions. The *Kritik der reinen Vernunft* begins with this distinction, but it gives no proper *definition* of these things, and I already found this unsatisfactory on my first acquaintance with this book. Now how can this defect be made good? Since the two concepts *empirical* and *a priori* are in mutual contradiction, it would be enough to determine only one of them properly, for example that of the *empirical*; then the determination of the other would simply be given by the opposite proposition. What therefore do we properly call *empirical*? One will not want to give the answer, '*the empirical is what we obtain through the five senses—or through an external object.*' As philosophers we may certainly not presuppose what the *five senses* are, nor that there are *external objects*.

§4.

In my view, the distinction between the empirical and *a priori* in our cognition applies originally only to our *judgements*, and it is only through these that it can also be indirectly extended to our *concepts* or *ideas*. That is, I am conscious of possessing judgements of the form, '*I perceive X*'; I call these judgements *empirical judgements of perception* or *judgements of reality*, and the *X* in them I call an *intuition* or, if one prefers, an *empirical idea*. The essential *copula* of this judgement is the concept of *perceiving*, which I hold to be a *simple*, and therefore *indefinable, concept*. But in order to *describe* it and to guard against misunderstanding one could say it is the concept of an *existence* ‖*Sein*‖: (a) of a mere, *pure* existence without necessity; (b) of an existence not of an *external object* as such, but only of a mere *idea in me* (namely the *intuition*).* Now the *rest* of my judgements, namely those which express (a) a *necessity*, (b) a *possibility*, or (c) an *obligation* (cf. §15 of Sec. II), I call *a priori*, and the concepts which appear in them as subject or predicate, I call *a priori concepts*.

§5.

By the *principle of sufficient reason* ‖*Satz vom Grunde*‖ I am of course bound to look for a certain ground for all my judgements; but for the empirical judgements this is something quite different than for the *a priori* ones. The former, the so-called *judgements of reality*, have the peculiarity that I seek their ground in *that which is* (in something real, in *things*), and indeed, according to circumstances, partly in that which I call, '*the particular nature of my perceptive faculty*', and partly in certain '*things different from me, i.e. external*

* For it must first be *proved* that an external object may correspond, as basis, to the idea.

things', which (as the phrase goes) '*affect my perceptual faculty*'.—It is not so with my *a priori judgements*, for which I can assume that the ground on account of which I attribute the predicate to the subject cannot possibly lie anywhere other than in the *subject itself* (and in the specific nature of the predicate).* We have already done this above in Sec. II, §20.—Here *intuitions* are, and, in my opinion, can be, of no use: this will perhaps be made clearer from the following paragraphs.

§6.

There is actually one kind of judgement, the so-called *judgements of experience* or *probability* (see Sec. II, §15) with which the combining of the predicate with the subject is in fact brought about *through intuitions*. For if I have only the *judgements of perception*, '*I perceive the intuitions X and Y and indeed never X without Y*', then I derive from these by means of the principle of sufficient reason the judgement of probability, '*the thing which is the ground of the intuition X is probably connected with the thing which is the ground of Y, like a cause with an effect*'. In my view all our so-called *judgements of experience* are of this form. If, for example, we say, *the sun warms the stone*, then this basically means nothing but: the object (sun) which is the *cause of the intuition X* (namely the illuminating disc of the sun), *is also the ground of the intuition Y* (namely that of a warm stone). But all these judgements have, in their nature, only a *probability*.

§7.

But how could judgements which are *absolutely certain*, such as all *a priori* judgements, result from the connection with intuitions?† *Kant* seems to want to say:

> If I combine the general concept, for example of a *point* or of a *direction* or *distance*, with an *intuition*, i.e. imagine a *single* point, a *single* direction or distance, then I find that this or that predicate belongs to these *single* objects and feel at the same time that this *is also the case* with *all other* objects which belong under this concept.

If this is the opinion of *Kant* and his followers, then I now ask: how do we come, from the intuition of that *single object*, to the *feeling that what we observe in it also belongs to every other one*? Through that which is *single* and individual in this object, or through that which is *general*? Obviously only through the latter, i.e. through the *concept*, not through the *intuition* (§2).

---

* For if one says this basis lies in the *absolute necessity of the thing* or in the *special nature of* our understanding, then these are, I believe, *empty phrases* which in the end say no more than *it is so because it is so*.

† All *a priori* judgements are absolutely certain; but this is also true of empirical judgements. It is not only judgements of *necessity* which have absolute certainty, as one usually imagines, but also *judgements of possibility*, *of actuality*, and of *duty*. Briefly, all our judgements apart from those which we have spoken of in the previous paragraph. These accordingly deserve the characteristic name of *judgements of probability*.

§8.

It is particularly clear how dubious the Kantian theory of intuition is if it is extended to other propositions not belonging to *geometry*. The principle of sufficient reason and the majority of propositions of arithmetic are, according to Kant's correct observation, *synthetic propositions*. However, who does not feel how artificial it is, that Kant, in order to carry through his theory of intuitions generally, has to assert that even *these* propositions are based on intuition, in fact the *intuition of time* (for what else should it be)? Nevertheless, the *principle of sufficient reason* holds also where there is no time, and it was only as a result of *this* proposition (according to a remark that has often been made) that *Kant* himself accepted the existence of *noumena* which are not in time. The propositions of *arithmetic* do not need the intuition of time in any way. We will only analyse a single example. Kant quoted the proposition, $7 + 5 = 12$, instead of which, for an easier review, we shall take the shorter $7 + 2 = 9$. The proof of this proposition offers no difficulty as soon as one assumes the general proposition, $a + (b + c) = (a + b) + c$, i.e. that with an *arithmetic* sum one only looks at the *set* of terms, not their *order* (a definitely wider concept than *sequence in time*). This proposition excludes the concept of *time* rather than presupposing it. But having accepted it, the proof of the above proposition can be carried out in the following way: the statements $1 + 1 = 2$, $7 + 1 = 8$, $8 + 1 = 9$ are mere *definitions* and *arbitrary propositions*. Therefore, $7 + 2 = 7 + (1 + 1)$, (*per def.*) $= (7 + 1) + 1$, (*per propos. praeced.*) $= 8 + 1$ (*per def.*) $= 9$, (*per def.*).

§9.

'However', it will be said, 'it is true at least in *geometry* that there are certain underlying intuitions. For in fact, however much one may *think of* only the concept *point* there is also the *intuition* of a point before our eyes.' But of course this *picture* accompanying our pure *concept* of the point is not connected with it *essentially*, but only through the association of ideas, because we have often thought both of them together. Therefore also the nature of this picture is different with different people, and determined by thousands of fortuitous circumstances. For example, whoever had always seen only rough and thickly-drawn *lines* or whoever had always represented a straight line by chains or sticks, would have in mind with the idea of a line, the image of a chain or a stick. With the word '*triangle*' one person always has in mind an *equilateral* triangle, another a *right-angled* triangle, a third perhaps an *obtuse-angled* triangle. I therefore do not understand at all how Kant has been able to find such a great difference between the intuition which some triangle actually *sketched* in front of us produces, and that produced by a triangle *constructed only in the imagination*, that he declares the first as altogether superfluous and insufficient for the proof of an *a priori*, synthetic proposition, but the latter as necessary and sufficient. According to my ideas it is of course *unavoidable* that with the thought of some often-seen spatial object our imagination paints a *picture* of it for us. It is also *useful and good* for the easier assessment of the object that this picture is in our mind, but I do not regard it as being absolutely *necessary* for this assessment. There are in fact theorems in geometry for which we have no intuitions at all. The proposition that every straight line can be extended to infinity has no intuition behind it: the lines which our imagination can picture are not infinitely long. In *stereometry* we are often concerned with such complex spatial objects that even the most lively imagination is not able to imagine them clearly any more; but we none the less continue to operate with our *concepts* and find truth.

§10.

But if it is not the *intuitions* which make for the essential distinction between mathematics and the other sciences, from where does the former derive its great *certainty* and *obviousness*?—I would answer that it is because one can very easily *test* the results of mathematics by *intuition* and *experience*. For example, that the straight line really is the shortest one between two points is proved by everybody by innumerable experiments a long time before he can prove it by deductions. Also the well-known obviousness of mathematics gradually disappears where the experience is lacking. In the same way, propositions which are inferred often have a far higher degree of intuitiveness than genuine axioms (cf. Sec. II, §21, note).

§11.

'Would this therefore allow no distinction between those intuitions which Kant called *a priori* and the empirical ones? All objects must have a *form*, but they need not possess *colour*, *smell* and such like.'—I would answer that not *all* objects which can make an *appearance* to us must possess a *form*, but only those which we conceive as *external* to us, i.e. in *space*. But even these must then also have something which *occupies* ‖erfüllt‖ this form; and this, due to the particular nature of our perceptual faculties, can only be one of the following five things, either a *colour* or a *smell*, etc. Therefore, colour, smell, etc. are also *a priori* forms in the same sense of the word as space and time are; only, the sphere to which the former apply is narrower than that of the latter, just as the form of *space* has a narrower *sphere* than that of *time*. Among *concepts* there are none (this is our final decision) to justify a distinction by which they could be divided into *empirical* and *a priori*: they are all *a priori*.

> Tu, si quae nosti rectius istis,
> Candidus imperti! si non: his utere mecum.[n]

---

[n] ‖The quotation comes from the end of Horace's Epistle to Numicius:

> Vive, vale! si quid novisti rectius istis,
> Candidus imperti; si non, his utere mecum.

'Live long, farewell! If you have ever known something better than these ‖Stoic precepts‖, share it with me frankly; if not, join me in following these.'—Horace, *Epistles*, I, vi, 67–8.‖

C. PURELY ANALYTIC PROOF OF THE THEOREM THAT BETWEEN ANY TWO VALUES WHICH GIVE RESULTS OF OPPOSITE SIGN THERE LIES AT LEAST ONE REAL ROOT OF THE EQUATION (*BOLZANO 1817a*)

Bolzano's most significant contribution to mathematics is his epoch-making paper on the foundations of real analysis, the *Rein analytischer Beweis* of 1817. In the years following the appearance of Berkeley's *Analyst*, mathematicians had made various attempts to put the calculus on a firmer foundation. The most common approaches (some of which are recorded above) were to base the calculus on one of the following ideas: on *motion* (Newton, MacLaurin); on *limits* (D'Alembert, L'Huilier); on *ratios of zeros* (Euler); on *infinitesimals* (Leibniz and—with reservations—Carnot). In perhaps the most radical proposal, James (= Jacob/Jacques) Bernoulli proposed amending the laws of logic by abandoning, for infinitesimals, the Euclidean 'common notion' that, if equals are subtracted from equals, the remainders are equal (*Bernoulli, Jacob 1744*, Vol. ii, p. 7). Perhaps the most influential approach was that of Lagrange, who, in his *Théorie des fonctions analytiques* (*1797*), assumed the existence of a Taylor-series expansion for every function; the full title of his work—*Théorie des fonctions analytiques contenant les principes du calcul différentiel, dégagés de toute considération d'infiniment petits, d'évanouissans, de limites et de fluxions, et réduits à l'analyse algébrique des quantités finies*—shows the scepticism with which he, and many other mathematicians, regarded the notion of limits, and his desire to reduce the calculus to 'l'analyse algébrique'. Bolzano's paper was not the first to attempt to find an analytic foundation for the calculus, nor was it the first to employ the notion of limits (as is shown above by the selections from D'Alembert). But, in contrast to his predecessors, Bolzano employed a limit-concept that was not based upon motion, and that was analytically defined;[a] and, more importantly, he was the first actually to *use* this definition to prove significant mathematical theorems.

Bolzano's announced aim is to prove the intermediate value theorem—in his formulation, that if f and ϕ are continuous functions such that $f(\alpha) < \phi(\alpha)$

[a] In a footnote in the *Preface* to his *1817b*, Bolzano gives the following definition of an analytic procedure:

> A *purely analytic* (also purely arithmetic or algebraic) *procedure* is one by which a certain function is derived from one or more other functions just through certain changes and combinations which are expressed *by a rule completely independent* of the nature of the designated quantities. Thus, for example, the way the function $(1 + x)^n$ is derived from $(1 + x)$ is said to be a purely analytic procedure, for $(1 + x)^n$ is obtained from $(1 + x)$ by making certain changes and combinations in the latter which are given by a rule which is completely independent of the nature of the quantity denoted by $1 + x$.

while $f(\beta) > \varphi(\beta)$, then for some x, $\alpha < x < \beta$, $f(x) = \phi(x)$. This theorem he eventually proves in §15. But he begins with an important critique of previous proofs, and in Part II of the *Preface* he gives the first precise definition of a continuous function. His definition is essentially the same as that given by Cauchy in his *Cours d'analyse* in 1821; whether Cauchy knew of Bolzano's work is uncertain. (Bolzano improved on his definition in his unpublished *Functionenlehre*, written in 1834; there he gives a definition of pointwise continuity and distinguishes between left and right continuity.)

In §7, Bolzano states and attempts to prove the sufficiency of the 'Cauchy condition' for the convergence of an infinite series. A rigorous proof requires a precise definition of the real numbers, which Bolzano did not possess; he himself (in the *Functionenlehre*) later admitted that the §7 proof was incomplete. Similarly, although Bolzano had a precise definition of continuity, he did not have the modern notion and definition of *function*. Lagrange, in the *Théorie*, had indeed defined a function of one or several quantities to be 'any mathematical expression in which those quantities appear in any manner, linked or not with some other quantities that are regarded as having given and constant values, whereas the quantities of the function may take all possible values'; but in practice he and his successors treated functions as equations. The modern conception did not enter mathematics until Dirichlet's paper, *Über die Darstellung ganz willkürlicher Functionen*, in 1837.

Having defined continuity and stated the Cauchy condition, Bolzano proceeds (§12) to prove a lemma that was eventually to become the cornerstone of the theory of real numbers. This lemma (the greatest lower bound principle) is the first published version of the Bolzano–Weierstrass theorem, which, in modern terminology, says that every bounded infinite point-set has an accumulation point.

Although Bolzano's proofs are incomplete, and although they are somewhat clumsily presented, this paper is a milestone in the history of real analysis. It was the first successful attempt to free the calculus from infinitesimals, and it is the starting point for the modern theory of the continuum; the precision of Bolzano's definitions and the rigour of his deductions mark a break with the mathematics of the past. The project of putting the theory of the real line on a solid, arithmetical foundation was to be carried forward, largely in ignorance of Bolzano's work, throughout the nineteenth century—most notably by Cauchy, Abel, Dirichlet, Weierstrass, Cantor, and Dedekind.

It is important to appreciate the role Bolzano's objective conception of axioms, described above in the *Beiträge* (*Bolzano 1810*), played in the *Rein analytischer Beweis*. Bolzano was not driven by scepticism or by fear of paradox. He makes it clear that he did not doubt the *truth* of the intermediate value theorem; and he was not attempting to place it on firmer or more obvious foundations—for the greatest lower bound principle is, if anything, less evident than the theorem he is trying to prove. Similarly, his criticism of the proofs based upon motion is not that intuitions of motion are unreliable, but the logical objection that the proofs beg the question. Bolzano's ambition was not so much to attain some superior brand of mathematical certainty as to reveal the objective reasons for the truth of the intermediate value theorem—to uncover its true logical foundations. And *a fortiori* Bolzano's quest for rigour in his *1817a* was not prompted by the 'challenge to geometric intuition' presented by the discovery of continuous nowhere-differentiable functions; on the contrary, it was Bolzano's rigour that made his subsequent discovery of such counter-intuitive phenomena possible. (The discovery is in *Bolzano 1930*, §75; the passage was written in 1834.)

The translation of *Bolzano 1817a* is by Stephen Russ; references should be to the section numbers, which appeared in the original edition.

PREFACE

There are two propositions in the theory of equations of which it could still be said, until recently, that a completely correct proof was unknown. One is the proposition: *that between every two values of the unknown quantity, which give results of opposite sign, there must always lie at least one real root of the equation*. The other is: *that every algebraic rational integral function of one variable quantity can be decomposed into real factors of first or second degree*. After several unsuccessful attempts at proving the latter proposition by *D'Alembert*, *Euler*, *de Foncenex*, *Lagrange*, *Laplace*, *Klügel*, and others, finally *Gauss* supplied, last year, two proofs which hardly leave anything to be desired. Indeed, this outstanding scholar had already presented us with a proof of this proposition in 1799;* but it had, as he admitted, the defect that it proved a purely analytic truth on the basis of a *geometrical consideration* ‖*Betrachtung*‖. But his two most recent proofs** are quite free of this defect; the *trigonometric functions* which occur in them can, and must, be understood in a purely analytical sense.

The other proposition which we mentioned above is not one which has concerned scholars so far to any great extent. Nevertheless, we do find mathematicians of great repute concerned with the proposition and already *different* kinds of proof have been attempted. To be convinced of this one need only compare the various treatments of the proposition which have been given by, for exam

* *Demonstratio nova Theorematis, omnem functionem algebraicam rationalem integram unius variabilis in factores reales primi vel secundi gradus resolvi posse. Helmstadii 4° 1799.*

** *Demonstratio nova altera etc.*, and *Demonstratio nova tertia*: both 1816.

ple, *Kästner,** *Clairaut,*** *Lacroix,*****Metternich,*+ *Klügel,*++ *Lagrange,*+++ *Rösling,*++++ and several others.

However, a more careful examination very soon shows that none of these proofs can be viewed as adequate.

I. The most common kind of proof depends on a truth borrowed from *geometry*, namely, *that every continuous line of simple curvature of which the ordinates are first positive and then negative* (or conversely) *must necessarily intersect the x-axis somewhere at a point that lies in between those ordinates*. There is certainly no objection against the *correctness*, nor indeed against the *obviousness* of this geometrical proposition. But it is equally clear that it is an intolerable offence against *correct method* to derive truths of *pure* (or general) mathematics (i.e. arithmetic, algebra, analysis) from considerations which belong to a merely *applied* (or special) part, namely *geometry*. Indeed, have we not felt and recognized for a long time the incongruity of such *μετάβασις εἰς ἄλλο γένος*?[a] Have we not already avoided this whenever possible in hundreds of other cases, and regarded this avoidance as a merit?* So if we wish to be consistent must we not try and do the same here?—For in fact, if one considers that the proofs of the science should not merely be *confirmations* ‖*Gewissmachungen*‖, but rather *justifications* ‖*Begründungen*‖, i.e. presentations of the objective reason for the truth concerned, then it is self-evident that the strictly scientific proof, or the objective reason, of a truth which holds equally for *all* quantities, whether in space or not, cannot possibly lie in a truth which holds merely for quantities which are in *space*. On this view it may, on the contrary, be seen that such a *geometrical* proof is, in this as in most cases, really circular. For while the geometrical truth to which we refer is (as we have already said), extremely *evident*, and therefore needs no *proof* in the sense of *confirmation*, it none the less needs *justification*. For its component concepts are obviously so combined that one cannot hesitate for a moment to say that it is not one of those simple truths which are called *basic propositions* ‖*Grundsätze*‖, or *basic truths*, because they are only the *ground* for other truths and never themselves consequences. On the contrary, it is a *theorem* or *consequential truth* ‖*Folgewahrheit*‖, i.e. such a truth as has its basis in certain other truths; and

* *Anfangsgründe der Analysis endlicher Grössen*. 3rd ed. §316 ‖*Kästner 1794*‖.
** *Elémens d'Algèbre*. 5th ed. Supplémens. Chap. I no. 16 ‖*Clairault 1797*‖.
*** *Elémens d'Algèbre*. 7th ed. ‖cf. *Lacroix c.1805*‖.
+ In his *translation* of the above work of Lacroix. Mainz. 1811. §211 ‖*Lacroix 1811*‖.
++ In his *Mathematisches Wörterbuch* Vol. 2, p. 447 ff. ‖Klügel 1803–31‖.
+++ *Traité de la résolution des équations numériques de tous les degrés*. Paris. 1808 ‖Lagrange 1808‖.
++++ *Grundlehren von den Formen, Differenzen, Differentialien und Integralien der Functionen*. Part 1, §49 ‖Rösling 1805‖.
[a] ‖'crossing to another kind' A similar phrase occurs in Aristotle's *Posterior Analytics* 75a 38: 'One cannot, therefore, prove by crossing from another kind—for example, something geometrical by arithmetic.'‖
*The papers of Prof. Gauss quoted before provide an example.

therefore, in the science, it must be proved by a derivation from these other truths.* Consider now the objective reason why a line in the above-mentioned circumstances intersects the x-axis. Everyone will, no doubt, see very soon that this reason lies in nothing other than that general truth, as a result of which every continuous function of x which is positive for one value of x, and negative for another, must be zero for some intermediate value of x. And this is just the truth which is to be proved. It is therefore quite wrong to have allowed the latter to be derived from the former (as happens in the kind of proof we are examining). Rather, conversely, the former must be derived from the latter if we wish to represent the truths in the science in the same way as they are linked to each other in their objective coherence.

II. No less objectionable is the proof which some have constructed from the concept of the *continuity* of a function with the inclusion of the concepts of *time* and *motion*. 'If two functions fx and ϕx,' they say, 'vary according to the law of continuity and if for $x = \alpha$, $f\alpha < \phi\alpha$, but for $x = \beta$, $f\beta > \phi\beta$, then there must be some value u, lying between α and β for which $fu = \phi u$. For if one imagines that the variable quantity x in both these functions gradually takes all values between α and β, and the same value is always taken by them both at the same moments, then at the *beginning* of this continuous change in x, $fx < \phi x$, and at the *end*, $fx > \phi x$. But since both functions, by virtue of their continuity, must first go through all intermediate values before they can reach a higher value, there must be some *intermediate moment* at which they are both equal to one another.'—This is further illustrated by the example of the *motion* of two bodies, of which one is initially *behind* the other and later *ahead* of the other. It necessarily follows that at one time it must have been going *beside* the other.

No one will deny that the concepts of *time* and *motion* are just as foreign to general mathematics as that of *space*. Nevertheless, if these two concepts were only introduced here for the sake of *clarification* we would have nothing against them. For we are in no way party to such an exaggerated *purism* as demands, in order to keep the science free from everything alien, that in its exposition one can never use an *expression* borrowed from another field, even if only in a metaphorical sense and with the purpose of describing a fact more briefly and clearly than could be done by a strictly literal description: not even if it is used just to avoid the jarring of the constant repetition of the same word, or so as to remember, by the mere name given to a thing, an example which would serve to confirm the assertion. Thus it may be noted that we do not regard *examples* and *applications* as detracting in the least from the perfection of a scientific exposition. On the other hand, we strictly require only this: that examples never be put forward instead of *proofs*, and that the essence

* Compare on all this my *Beyträge zu einer begründeteren Darstellung der Mathematik*, *1st edition*, Prague 1810. Section II §§2, 10, 20, 21, where the logical concepts, which I have assumed here as known, are developed further.

⟦Wesenheit⟧ of a deduction never be based on the merely metaphorical use of phrases or on their accompanying ideas, so that the deduction itself would become void as soon as these were changed.

In accord with these views, the inclusion of the concept of *time* in the above proof may still perhaps be excused, because no conclusion is based on phrases which contain it, which would not also hold without it. But in no way can the last *illustration* about the *motion* of a body be viewed as anything more than a mere *example*, which does not prove the proposition but rather is only to be proved by it.

(a) So let us drop this example and examine the rest of the reasoning. Let us *first* notice that this is based on an incorrect concept of *continuity*. According to a *correct definition*, the expression *that a function fx varies according to the law of continuity for all values of x inside or outside certain limits** means just that, *if x is some such value, the difference* $f(x + \omega) - fx$ *can be made smaller than any given quantity provided* ω *can be taken as small as we please.*[b] With the notation I introduced in §14 of *Der Binomische Lehrsatz*, etc. (Prague 1816) this is, $f(x + \omega) = fx + \Omega$. But, as assumed in this proof, the continuous function is one which never reaches a higher value without first going through all lower values, i.e. $f(x + n\Delta x)$ can take every value between fx and $f(x + \Delta x)$ as n takes arbitrary values between 0 and 1. That is certainly a very *true* assertion, but it cannot be viewed as a *definition* of the concept of continuity: it is rather a *theorem* about continuity. Indeed, it is a theorem which can only be proved on the assumption of the proposition which here it was desired to prove by the theorem. For if M is some quantity between fx and $f(x + \Delta x)$, then the assertion that there is a value of n between 0 and 1 for which $f(x + n\Delta x) = M$ is only a special case of the general truth that if $fx < \phi x$, and $f(x + \Delta x) > \phi(x + \Delta x)$, then there must be some intermediate value $x + n\Delta x$ for which $f(x + n\Delta x) = \phi(x + n\Delta x)$. The first assertion comes from this general truth in the case when the function ϕx has constant value M.

(b) But even supposing one could prove this proposition in another way, the proof which we are examining would have yet another error. From the fact that $f\alpha > \phi\alpha$ and $f\beta < \phi\beta$ it would only follow that, if u was some value lying between α and β for which ϕu is $> \phi\alpha$ but $< \phi\beta$, then fx becomes equal to ϕu in going from $f\alpha$ to $f\beta$, i.e. for *some* x lying between α and β, $fx = \phi u$. But

* There are functions which vary continuously for *all* values of their root, e.g. $\alpha x + \beta x$. ⟦Probably a typesetter's misprint for $\alpha + \beta x$.⟧ But there are others which are continuous only for values of their root inside or outside certain limits. Thus $x + \sqrt{(1 - x)(2 - x)}$ is continuous only for values of $x < +1$ *or* $> +2$ but not for values between $+1$ and $+2$.

[b] ⟦There was no standard symbol at this time for absolute value or modulus. Bolzano always regards inequalities as applying to the magnitudes of each side regardless of sign. Thus $x > -1$ in Bolzano's usage would be expressed today by $x < -1$.⟧

whether this happens for *just the same* value of x which $= u$, i.e. (because u can be any arbitrary value lying between α and β which makes $\phi u > \phi\alpha$ and $< \phi\beta$) whether there is some value of x lying between α and β for which both functions fx and ϕx are equal to one another, this would still not follow.

(c) The deceptiveness of the whole proof rests mainly on the inclusion of the concept of *time*. For if this were omitted it would be seen straight away that the proof was nothing but a repetition in different words of the proposition to be proved. For to say that the function fx, before it passes from the state of being smaller than ϕx to that of being greater, must first go through the state of being equal to ϕx, is to say, without the concept of time, that among the values that fx takes if one puts in every value of x between α and β there is one that makes $fx = \phi x$. This is exactly the proposition to be proved.

III. Others prove our proposition on the basis of the following (either quite without proof or just supported by some examples borrowed from geometry): '*Every variable quantity can pass from a positive state to a negative one only through the state of being zero or infinity.*' Now since the value of an equation cannot be *infinitely large* for any finite value of the root, they conclude that any transition here must occur through *zero*.—

(a) If we wish to detach from the above proposition the metaphorical idea of a *transition*, which contains the concept of a change in *time* and *space*, thereby also omitting the senseless expression of *a state of non-existence*, then we eventually arrive at the following proposition: '*If a variable quantity which depends on another quantity x is positive for $x = \alpha$, and negative for $x = \beta$, there is always a value of x lying between α and β for which the quantity is zero or one for which it is infinite.*' Now everyone surely perceives that such a compound assertion is not a basic truth, but would have to be proved, and that its proof could hardly be easier than the very proposition which we wish to establish.

(b) Indeed, a closer examination shows that the assertion is fundamentally *identical* with the proposition. For it must not be forgotten that this assertion is actually true only if it refers to *quantities which vary continuously*. Thus for example, the function $x + \sqrt{(x-2)(x+1)}$ is *positive* for $x = +2$ and negative for $x = -1$, yet because it does not vary within these limits according to the law of continuity there is no value of x between 2 and -1 for which the function is zero or infinite. However, if we confine the assertion simply to quantities which vary continuously we must also exclude those functions which become *infinite* for a certain value of their root. For a function such as $\dfrac{a}{b-x}$ does not actually vary continuously for *all* values of x, but only for all values which are $>$ or $< b$. For the function does not have any *determinate* value when $x = b$, but it becomes what is called *infinitely great*. So it cannot be said that the values which it takes for $x = b + \omega$, all of which are *determinate*, can come as close as desired to the value it has for $x = b$. And this is a part of the concept of continuity (II.a). Now let us add the concept of continuity to the above assertion while omitting the case of the function's *becoming infinite*. It then becomes, word for word, the original proposition we had to prove: namely that every continuously variable function of x which is positive for $x = \alpha$, and negative for $x = \beta$, must be zero for some value between α and β.

IV. In some works the following conclusion may be found: '*Because fx is positive for $x = \alpha$ and negative for $x = \beta$, there must be between α and β, two quantities a and b at which the transition from the positive values to the negative values of fx takes place, so that between a and b no more values of x occur for which fx would still be positive or negative.*' Etc.—This error scarcely needs refuting, and would not be introduced here if it did not serve to prove how unclear are the concepts of even some reputable mathematicians on this subject. It is well enough known that between any two nearby values of an *independent variable*, such as the root x of a function, there are always infinitely many *intermediate* values. And in the same way that there is no *last x* for a continuous function which makes it positive, and no *first x* which makes it negative, so there are no numbers *a* and *b* as here described!

V. The failure of these attempts to prove *directly* the proposition with which we are concerned, leads to the idea of deriving it from the *second* proposition which we mentioned at the beginning, namely from *the divisibility of every function into certain factors*. There is also no doubt, that if the latter is conceded, the former can be concluded from it. But the fact is that such a derivation could not be called a strictly scientific *justification*, in that the second proposition clearly expresses a much *more complex* truth than the first. The second can therefore certainly be based on the first, but not, conversely, the first on the second. No one has really succeeded yet in proving the second without presupposing the first. With regard to the proofs, the unacceptability of which *Gauss* has already shown in his paper of 1799, it is because they have already been shown to be unacceptable that it is unnecessary to investigate whether or not they are based on our present proposition. The proof of *Laplace,** likewise, has its faults, which we need not point out here, because it is explicitly based on our present proposition. And in the same way, we need pay no regard to the *first* proof of *Gauss*, because it relies on *geometrical* considerations. Moreover, it would be easy to show that even in that proof our proposition is implicitly accepted, in that the geometrical considerations which are employed in it are quite similar to those which we mentioned in I.—So it all depends on the *Demonstratio nova altera* and *tertia* of *Gauss*. The former refers explicitly to our proposition, when it presupposes on p. 30: *aequationem*

---

* In the Journal de l'école normale, or also in *Lacroix, Traité du Calcul différéntiel et intégral.* T.I. no. 162, 163.

ordinis imparis certo solubilem esse.[c] An assertion which is well known to be nothing but an easy consequence of our proposition. It is not so obvious that the *Demonstratio nova tertia* depends on our proposition. It is based among other things on the following theorem: *If a function remains positive for all values of its variable quantity x which lie between α and β, then its integral taken from $x = \alpha$ to $x = \beta$ has a positive value.* Now in the proof given by *Lagrange*[†] for this theorem no explicit reference to our proposition is to be found. However, this proof of Lagrange still has a gap. It is required that the quantity i be taken sufficiently small that,

$$\frac{f(x+i)-fx}{i}-f'x<\frac{f'x+f'(x+i)+f'(x+2i)+\ldots+f'(x+(n-1)i)}{n}$$

where the product $i.n$ is to remain equal to a *given* quantity and the well-known notation $f'x$ represents the first derived function of fx. Now the question arises here whether it is possible to satisfy this requirement? However small one takes i, so as to diminish the difference $\dfrac{f(x+i)-fx}{i}-f'x$, the divisor of the right-hand side, n, must be taken that much greater if $i.n$ is to remain constant. Now also the set of terms in the numerator increases, but whether this increase in the numerator grows in proportion with the denominator, or whether the value of the whole fraction is decreased by the decrease in i, by as much as, or by more than, the expression $\dfrac{f(x+i)-fx}{i}-f'x$—this is yet to be shown. Now if this gap is to be filled, it can surely only be done with reference to our present proposition, since we had to refer to the latter for the proof of the theorem,* which although much simpler, is related to this one of *Lagrange*.

Thus all the proofs so far of the proposition which forms the title of this paper are defective. Now the one which I put forward here for the judgement of scholars contains, I flatter myself, not a mere *confirmation*, but the objective *justification* of the truth to be proved, i.e. it is strictly scientific.**

The following is a short summary of the method adopted.

The truth to be proved, that between any two values α and β which give results of opposite sign there always lies at least one real root, clearly rests on

---

[c] ‖‘an equation of odd degree is certainly soluble.’‖

[†] *Leçons sur le Calcul des fonctions. Nouvelle Edition, Paris 1806* Leç. 9, p. 89 ‖*Lagrange 1806*‖.

* Namely the proposition §29 in the paper: *Der Binomische Lahrsatz*, etc.

** Yet it is not to be expected that I comply here with *all* the rules which I myself set out in the *Beyträge zu einer begründeteren*, etc. (Pt. II) for the construction of a *strictly scientific* exposition. For though I am completely convinced of the correctness of these rules, to follow them precisely is only possible when one begins the exposition of a science from its *first* propositions and concepts, but not where one is only dealing with some theories taken out of the context of the whole. This observation obviously also applies to the paper *on the binomial theorem*.

the more general truth, that, if two continuous functions of x, fx and ϕx, have the property that for $x=\alpha$, $f\alpha<\phi\alpha$, and for $x=\beta$, $f\beta>\phi\beta$, there must always be some value of x lying between α and β for which $fx = \phi x$. However, if $f\alpha < \phi\alpha$, then by the law of continuity it is possible that $f(\alpha + i) < \phi(\alpha + i)$, if i is taken small enough. *The property of being smaller* therefore belongs to the function of i represented by the expression $f(\alpha + i)$, for all values smaller than a certain value. Nevertheless this property does not hold for *all* values of i without restriction, namely not for an i which is $= \beta - \alpha$, for $f\beta$ is already $> \phi\beta$. Now the *theorem* holds that whenever a certain property M belongs to all values of a variable quantity i which are smaller than a given value, and yet not for *all values in general*, then there is always some greatest value u, for which it can be asserted that all i which are $<u$ possess the property M. For this value of i itself $f(\alpha + u)$ cannot be $<\phi(\alpha + u)$ because then by the law of continuity $f(\alpha + u + \omega) < \phi(\alpha + u + \omega)$ if ω were taken small enough. And consequently it would not be true that u is the greatest of the values for which the assertion holds that all lower values of i make $f(\alpha + i) < \phi(\alpha + i)$; for $u + \omega$ would be a still greater value for which this holds. But still less can it be that $f(\alpha + u) > \phi(\alpha + u)$, for then also $f(\alpha + u - \omega) > \phi(\alpha + u - \omega)$ would be true if ω is taken sufficiently small, and consequently it would not be true that for all values of i which are $<u, f(\alpha + i) < \phi(\alpha + i)$. So therefore it must be that $f(\alpha + u) = \phi(\alpha + u)$, i.e. there is a value of x lying between α and β, namely $\alpha + u$, for which the functions fx and ϕx are equal to one another. It is now only a question of the proof of the *theorem* mentioned. This we prove by showing that those values of i for which it can be asserted that all smaller values possess property M and those of which this cannot be asserted can be brought as near one another as desired. Whence it follows for everyone who has a correct concept of *quantity*, that the idea of an i, which is the greatest one of which it can be said that all below it possess property M, is the idea of a real, i.e. *actual*, quantity.

Before I finish this preface I may be allowed to make a confession and a request which apply not only to the *present*, but to all my writings, also, God willing, to my *future* ones.

A careful reader could already have gathered from the few writings which have appeared *so far*, but particularly from that outline of a *new logic* which is supplied in the first issue of *Beyträge zu einer begründeteren Darstelling der Mathematik* in its *second* section, headed, '*Über die mathematische Methode*', that I hold certain views which, if they are not found to be completely incorrect, must lead to a *complete reorganization in all pure* a priori *sciences*. The greatest and most important part of these views I have already examined for such a long time and with so much impartiality, that it is not premature for me to venture to speak more openly about them. Views which embrace the whole domain of one or more sciences can be made known in two ways. They can either be stated once in a connected form or partially in single papers. The *first* way is, up till now, by far the most usual, and it is without doubt the way any one must adopt if he just wants to achieve in the shortest time a reputation among the learned section of his contemporaries. But for the perfection of the sciences, I regard the *second* procedure to be much more advantageous for the following reasons:

Firstly, because in this way the discoverer of new views runs much less danger of being rushed. For the partial presentation of his opinions allows him to postpone to a later time his explanation of points on which he was initially in doubt himself. He can learn from the criticisms which work already published receives, and still correct some things which had been incorrect.

Secondly, with such a serial development of his views, he can also expect a far stricter examination on the part of the reader. For whoever presents an already completed system, offers for our attention all at once, a larger number of new assertions than we could hope anyone to examine with as much care as if they had been presented singly. Whoever supplies a complete theory, shows, or at least, should show, how those truths which common sense knows with undeniable certainty can be derived from his *unusual* premisses. But this reconciles us to those premisses straight away, and we concede to them much more rapidly than if he had presented them singly and had allowed us to be in doubt as to whether, and to what extent, they agree with the rest of what we hold to be true. Finally, it is surely not to be denied that the mere sight of a *large, thick book* which promises a complete system of this or that science, instils into us a kind of reverence before we have even read it. Now if we discover, through reading it, a certain coherence in the assertions, if the structure of human knowledge which is sketched out here has a pleasing form, if everything is laid out according to size and number and symmetry, then our judgement is affected, and we even begin to wish that here, at last, might be *that single correct system* which we have sought for so long. And the least that occurs is that because of the coherence observed, we imagine that we are only free either to accept or reject the entire system, when in fact neither the one nor the other should happen!

These were the main reasons for which I decided in 1804 not to begin in any science with the publication of a *complete textbook*, but to first make known these unusual concepts of mine in single papers. And if, after much correction, these have found favour with a part of the public, only then should the preparation of an entire system be considered. That is, unless death does not oblige us to leave this latter task to others.

I began my written output with a paper concerning *mathematics* under the title, *Betrachtungen über einige Gegenstände der Elementargeometrie* (Prague, C. Barth, 1804). In this I put forward, as well as several other views, *a new theory of parallels*.* Some years later I resolved to publish all my views in the area

* This theory might deserve attention for at least two reasons: *firstly*, because it is the only one in which no obvious error has been detected; *then* because the greatest living French geometer, *Legendre*, hit upon just the same view of things quite independently of me in the *tenth* edition of his *Elémens de Géométrie*, Paris, 1813 ‖cf. *Legendre 1794*‖.

of *mathematics* in serial form under the title, *Beyträge zu einer begründeteren Darstellung der Mathematik*. But the *first* of these issues, (Prague, C. Widtmann, 1810) had the misfortune, with all the importance of its contents, of not even being announced and reviewed in some learned journals, and in others only very superficially. This forced me to postpone the continuation of these contributions to a later time, and meanwhile just to attempt to make myself better known to the learned world by publishing some papers which, by their titles, would be more suited to arouse attention. For this purpose there appeared in 1816 the paper already mentioned, *Der binomische Lehrsatz*, etc. (Prague, Enders) ⟦*Bolzano 1816*⟧. My wish is that the present paper should also serve this purpose; besides, its publication was necessary because in that earlier paper I already referred to the proposition proved here. Some *other* papers are also written out ready for printing, for example one with the following title, *Die drey Probleme der Rectification, der Complanation und der Cubirung, ohne Betrachtung des unendlich Kleinen, ohne die annahmen des Archimedes und ohne irgend eine nicht streng erweisliche Voraussetzung gelöst*. These still await their publisher.[d]

If I should continue this way, which seems to me the most advantageous, then the only *favour* of the public for which I must *ask* is that this single paper be not overlooked on account of its small extent, but rather examined with all possible strictness, and the results of this examination made known publicly. So that what is perhaps unclear may be clearly explained and what is quite incorrect will be retracted. The sooner truth and correctness gain general assent, the better.

---

§1.

Convention. Suppose that for a *series of quantities* the special case does not occur that all the terms after a certain term are *zero*, as happens for example in the *binomial series* after the $(n + 1)$th term for every positive integer exponent n. Then it is obvious that the *value of this series*, that is the quantity resulting from summing its terms, cannot always remain the same if the set of terms is arbitrarily increased. On the contrary, this value must certainly change every time the number of terms is increased by a *single one* which is not zero. Hence the value of a series depends not only on the *rule* which determines the construction of the individual terms but also on their *number*. So the value represents a *variable quantity*, even though the *form* and *magnitude* of the individual terms remain unchanged. With this in mind, we denote a *function* of x, which consists of an arbitrarily long series of terms, and whose value therefore depends, apart from x, on the *number of terms*, r, by $\overset{(r)}{F}(x)$ or $\overset{r}{F}x$. So, for example, $A + Bx + Cx^2 + \ldots + Rx^r = \overset{r}{F}x$, and on the other hand $A + Bx + Cx^2 + \ldots + Rx^r + \ldots + Sx^{r+s} = \overset{(r+s)}{F}x$.

[d] ⟦The paper quoted was published later in 1817 in Leipzig. One of the others referred to must have been *Versuch einer Begründung der Lehre von den drey Dimensionen des Raumes*, published in 1845, but according to its Preface already written in 1815.⟧

§2.

1st Lemma. The *change in value*, i.e. the *increase* or *decrease* in value, which occurs in a series by increasing its number of terms by a *definite set* (for example by one) can be, according to circumstances, a *constant* quantity (if, namely, the terms of the series are all equal), but also may be variable. In the latter case, the change may be a quantity which increases for a while and decreases for a while, or one which increases or decreases steadily. Thus the change in the series

$$1 + 1 + 1 + 1 + 1 + 1 + \ldots$$

if it is increased by *one* term is a *constant* quantity. The change in the series

$$a + ae + ae^2 + ae^3 + \ldots$$

by the increase of *one* term is a variable quantity, provided e is not $= 1$. It becomes ever larger if $e > \pm 1$, and ever smaller if $e < \pm 1$.

§3.

2nd Lemma. If the change (*increase* or *decrease*) in a series due to the increase in its set of terms by a definite number (for example one) always remains the *same* or even always increases—and if in both cases it retains the *same sign*—then it is clear that the *value* of this series will become *greater than any given quantity* if it is continued far enough. For suppose the growth in the series by the increase of every n terms is $=$ or $>d$ and it is required to make the series larger than a given value D. Then one need only take a whole number r which is $=$ or $>D/d$ and extend the series by $r.n$ terms, thereby obtaining an increase which is

$$= or > \; (r.d = or > \; (D/d).d = D).$$

§4.

3rd Lemma. But there are also series whose value, however far they may be continued, *never exceeds a certain quantity*. Of this kind is the series,

$$a - a + a - a + \ldots$$

whose value, however far it is continued, is always either 0 or a, and therefore never exceeds a.

§5.

4th Lemma. Particularly interesting among such series is the class of those series having the property that the *change* in value (*increase* or *decrease*) *due to any further continuation* of terms, always remains *smaller* than a certain quantity, which itself can be taken *as small as desired* provided the series has been continued far enough beforehand. That there *are* such series is proved not only by those whose terms after a certain point are all *zero* (and which therefore really have no continuation *after* this term, and are no more capable of changing value than the *binomial series* of §1); but also by the fact that also of this kind are all series in which the terms decrease at the same rate as, or faster than, the terms of a *geometric progression* with ratio a proper fraction. For the value of the geometric series,

$$a + ae + ae^2 + \ldots + ae^r$$

is well known to be $= a.\dfrac{1 - e^{r+1}}{1 - e}$. And if this series is extended by s terms, then the increase is

$$ae^{r+1} + ae^{r+2} + ae^{r+3} + \ldots + ae^{r+s} = ae^{r+1}.\frac{1 - e^s}{1 - e}.$$

Now if $e < \pm 1$ and r is taken sufficiently large, then this increase remains smaller than any given quantity, however large s becomes. For because e^s always remains $< \pm 1$ then $ae^{r+1}.\dfrac{1 - e^s}{1 - e}$ is obviously always *smaller* than $ae^{r+1}.\dfrac{2}{1 - e}$. But this latter can be made smaller than any given quantity by increasing r, because the value it takes for the next larger value of r is just the previous result multiplied by e, a constant proper fraction (see *Der binomische Lehrsatz*, §22). Therefore every geometric progression whose ratio is a proper fraction can be continued so far that the increase caused by every further continuation must remain smaller than some given quantity. So much more must this hold for series whose terms decrease more rapidly than those of a decreasing geometric progression.

§6.

5th Lemma. If the values of the sums of the first $n, n + 1, n + 2, \ldots, n + r$ terms of a series like those of §5 are denoted (§1) respectively by $\overset{n}{F}x$, $\overset{n+1}{F}x$, $\overset{n+2}{F}x, \ldots, \overset{n+r}{F}x$, then we regard the quantities

$$\overset{1}{F}x, \overset{2}{F}x, \overset{3}{F}x, \ldots, \overset{n}{F}x, \ldots, \overset{n+r}{F}x, \ldots$$

as a *new* series (called the *series of sums* of the previous one). The assumption has been made in this of the special property that the difference between its nth term $\overset{n}{F}x$ and every later term $\overset{n+r}{F}x$ (no matter how far from that nth term) stays smaller than any given quantity if n has been taken large enough in the first place. This difference is the increase to the *original* series by the continuation beyond the nth term, and this increase can remain, by assumption, as small as desired if n has been taken large enough in the first place.

§7.

Theorem. If a series of quantities

$$\overset{1}{F}x,\ \overset{2}{F}x,\ \overset{3}{F}x, \ldots, \overset{n}{F}x, \ldots, \overset{n+r}{F}x, \ldots$$

has the property that the difference between its nth term $\overset{n}{F}x$ and every later one $\overset{n+r}{F}x$, however far from the former this is, remains smaller than any given quantity if n has been taken large enough, then there is always a certain *constant quantity*, and indeed only *one*, which the terms of this series approach, and to which they can come as near as desired if the series is continued far enough.

Proof. That such a series as the theorem describes is possible is clear from §6. The hypothesis that there is the quantity X which the terms of this series approach as close as desired through ever further continuation certainly contains nothing impossible provided it is not assumed that this quantity be *unique* and *invariable*. For if it is a quantity which can vary then it can, of course, always be taken so that it is suitably near the term $\overset{n}{F}x$ with which it is being compared—even exactly the same as it. But the assumption of an *invariable* quantity with this property of proximity to the terms of our series contains no impossibility, because with this assumption it is possible to determine the quantity as accurately as desired. For suppose it is required to determine X so accurately that the difference from the true value of X does not exceed a small given quantity d. Then one simply looks in the given series for a term $\overset{n}{F}x$ with the property that every term $\overset{n+r}{F}x$ following it has a difference less than $\pm d$. There must be such an $\overset{n}{F}x$ by assumption. I now say the value of $\overset{n}{F}x$ differs from the true value of X by at most $\pm d$. For if r is increased arbitrarily, with the same n, the difference $X - \overset{n+r}{F}x = \pm\,\omega$ can be made as small as desired, but the difference $\overset{n}{F}x - \overset{n+r}{F}x$ always remains $< \pm\, d$, however large r is taken. Therefore the difference,

$$X - \overset{n}{F}x = (X - \overset{n+r}{F}x) - (\overset{n}{F}x - \overset{n+r}{F}x)$$

must always remain $< \pm\,(d + \omega)$. But since for the same n this is a *constant* quantity, and ω can be made as small as required by increasing r, then $X - \overset{n}{F}x$ must be = or $< \pm d$. For if it were *greater*, for example $= \pm\,(d + e)$, it would be impossible for the relation $d + e < d + \omega$, i.e. $e < \omega$, to hold if ω is further reduced. The true value of X therefore differs from the value which the term $\overset{n}{F}x$ has, by at most d, and can thus be determined as accurately as required, since d can be made arbitrarily small. There *is* therefore a *real quantity* which the terms of the series approach as closely as desired if it is continued far enough. But there is only *one* such quantity. For suppose that besides X there is another *constant* quantity Y which the terms of the series approach as close as desired if it is continued far enough, then the differences $X - \overset{n+r}{F}x = \omega$ and $Y - \overset{n+r}{F}x = \overset{\text{I}}{\omega}$, can be made as small as desired if r is taken large enough. Therefore this also holds for their own difference, i.e. for $X - Y = \omega - \overset{\text{I}}{\omega}$, which, if X and Y are *constant* quantities, is impossible unless one assumes $X = Y$.

§8.

Remark. If one tries to determine the value of the quantity X in the way described in the previous paragraph, namely through one of the terms from which the given series is composed, then one will never determine X completely *accurately* unless the terms of this series from some point on are all equal. But one must beware of concluding from this that the quantity X is always *irrational*. For if we consider the series

$$0.1;\ 0.11;\ 0.111;\ 0.1111;\ \ldots$$

(which is the series of sums of the geometrical progression

$$1/10;\ 1/100;\ 1/1000;\ 1/10\,000;\ \ldots)$$

then the quantity which the terms approach as close as desired is not irrational, but the fraction 1/9. So from the fact that a quantity cannot be determined accurately in *a certain way*, it does not follow that it cannot be completely determined in any *other* way, and is therefore *irrational.*

§9.

Lemma. If therefore some given series has the property that every single term is *finite*, but the change which it undergoes on every further continuation is smaller than any given quantity, provided only that the number of terms taken in the first place is large enough, then there is always one and only *one constant quantity* which comes as close to the value of this series as desired if it is continued far enough. For such a series is of the kind described in §5, and hence the values which are the sums of the n, $n + 1$, $n + 2$, . . . terms form a series like those of §6 and §7: therefore such a series has also the property proved in §7.

§10.

Remark. It is not to be thought that in the above proposition of §9 the condition, '*that the change*, (increase or decrease) *which the series undergoes on every continuation must remain smaller that any given quantity if it has been continued far enough beforehand*', is superfluous, and that the proposition could perhaps be expressed with greater generality: '*If it is possible to make the terms of a series when continued ever smaller and as small as desired, then there is always a constant quantity which the value of the series, when continued, approaches as near as desired.*' This assertion would be contradicted immediately by the following example. The terms of the series

$$1/2 + 1/3 + 1/4 + 1/5 + \ldots$$

can be made as small as desired, and yet it is a well-known truth, from the properties of a *rectangular hyperbola* (but also derivable from purely arithmetic considerations), that the value of this series can become greater than any given quantity if it is continued far enough.

§11.

Preliminary note. In investigations of applied mathematics the case often arises that one learns that a definite property M belongs to *all* values of a *variable quantity* x which are *smaller* than a certain u, without at the same time knowing whether this property does or does not belong to all values which are *greater* than u. In such a case there may perhaps be some $\overset{1}{u}$ that is $> u$ for which, in the same way as it holds for u, all values of x lower ⟦than $\overset{1}{u}$⟧ possess property M. This property M may even belong to *all values of x without exception*. But if *this* is known, *that M does not belong to all x in general*, then by combining these two conditions it will be correct to conclude: *there is a certain quantity U which is the greatest of those for which it is true that all smaller values of x possess property M*. This is proved in the following theorem.

§12.

Theorem. If a property M does not belong to *all* values of a variable x but does belong to *all* values which are *smaller* than a certain u, then there is always a quantity U which is the greatest of those of which it can be asserted that all smaller x have property M.

Proof. 1. Because the property M holds for all x *smaller* than u but not for *all x in general* there is some quantity $V = u + D$ (where D represents something positive) for which it can be said that M does not belong to all x which are $< V = u + D$. If I then ask the question, whether *M belongs to all x which are* $< u + D/2^m$ where the exponent m is first 0, then 1, then 2, then 3, etc.: I am sure the *first* of my questions will *have to be denied*. For the question whether M belongs to all x which are $< u + D/2^0$ is the same as whether M belongs to all x which are $< u + D$, which is denied by assumption. So it is a matter of whether all the *succeeding* questions as m is put larger and larger will be denied. Should this be the case, it is clear that u *itself* is the greatest value for which the assertion holds that all smaller x have property M. For if there were a greater, for example $u + d$, the assertion would hold that all x which are $< u + d$ have the property M; but then it is clear that if I take m large enough,

$u + D/2^m$ will at one time be = or $< u + d$, and consequently if M belongs to all x which are $< u + d$, it also belongs to all x which are $< u + D/2^m$; therefore we would not have denied this question, but must have affirmed it. Thus it is proved that in this case (when all the above questions are denied) there is a certain quantity U (namely u itself) which is the greatest for which the assertion holds that all x below it possess the property M.

2. However, if one of the above questions is *affirmed* and m is the definite value of the exponent for which it is *first* affirmed (m can be 1 but, as we have seen, not 0), then I now know that the property M belongs to all x which are $< u + D/2^m$ but not to all x which are $< u + D/2^{m-1}$. But the difference between $u + D/2^m$ and $u + D/2^{m-1}$ is $= D/2^m$. If I therefore do the same with this as before with the difference D, i.e. I ask the question whether M belongs to all x which are $< u + D/2^m + D/2^{m+n}$, and here the exponent n is first 0, then 1, then 2, etc.: then I am again sure that at least the *first* question will have to be denied. For to ask whether M belongs to all x which are $< u + D/2^m + D/2^{m+0}$ is just the same as to ask whether M belongs to all x which are $< u + D/2^{m-1}$, which had already been denied. But if all my *succeeding* questions are denied as I make n larger and larger, then as before $u + D/2^m$ is that greatest value, or the U, for which the assertion holds that all x below it possess the property M.

3. However, if one of these questions is affirmed and this happens first for the definite value n, then I now know M belongs to all x which are $< u + D/2^m + D/2^{m+n}$, but not to all x which are $< u + D/2^m + D/2^{m+n-1}$. The difference between these two quantities is $D/2^{m+n}$, and I repeat with this as before with $D/2^m$, etc.

4. If I continue this way as long as desired it may be seen that the result that I finally obtain must be one of two things.

(a) I find a value of the form $u + D/2^m + D/2^{m+n} + \ldots + D/2^{m+n+\ldots+r}$ which is the greatest for which the assertion holds that all x below it possess a property M. This happens in the case when the questions whether M belongs to all x which are

$$< u + D/2^m + D/2^{m+n} + \ldots + D/2^{m+n+\ldots+r+s}$$

are denied for every value of s.

(b) I find that M belongs to all x which are

$$< u + D/2^m + D/2^{m+n} + \ldots + D/2^{m+n+\ldots+r}$$

but not to all x which are

$$< u + D/2^m + D/2^{m+n} + \ldots + D/2^{m+n\ldots+r-1}$$

Here I am always free to make the number of terms in these two quantities even greater through new questions.

5. Now if the *first* case occurs the truth of the theorem is already proved. In the *second* case it may be noticed that the quantity

$$u + D/2^m + D/2^{m+n} + \ldots + D/2^{m+n \ldots +r}$$

represents a series whose number of terms I can increase arbitrarily, and which belongs to the class described in §5. Because, according as $m, n, \ldots, r$ are either all $= 1$, or some are greater than 1, it decreases at the same rate, or more rapidly than, a geometric progression whose ratio is the proper fraction $\frac{1}{2}$. From this it follows that it has the property of §9, i.e. there is a certain *constant quantity* which it can approach as near as desired if the set of its terms is sufficiently increased. Let this quantity be U; then I claim the property M holds for all x which are $< U$. For if it did not hold for some x that is $< U$, for example for $U - \delta$, then the quantity

$$u + D/2^m + D/2^{m+n} + \ldots + D/2^{m+n+\ldots+r},$$

must always keep the distance δ from U, because for all x that are smaller than it, the property M should apply. For every x that is

$$= u + D/2^m + D/2^{m+n} + \ldots + D/2^{m+n+\ldots+r} - \omega,$$

however small ω is, possesses the property M. On the other hand, M does not belong to the $x = U - \delta$, therefore,

$$U - \delta > u + D/2^m + D/2^{m+n} + \ldots + D/2^{m+n+\ldots+r} - \omega$$

or

$$U - [u + D/2^m + D/2^{m+n} + \ldots + D/2^{m+n+\ldots+r}] > \delta - \omega.$$

Hence the difference between U and the series cannot be made as small as desired, since $\delta - \omega$ cannot be made smaller than any given quantity.—But just as little can M hold for all x which are $< U + \varepsilon$. For the value of the series

$$u + D/2^m + D/2^{m+n} + \ldots + D/2^{m+n+\ldots+r-1}$$

can be brought as close to the value of the series

$$u + D/2^m + D/2^{m+n} + \ldots + D/2^{m+n+\ldots+r}$$

as desired because the difference between the two is only $D/2^{m+n+\ldots+r}$. Further, because the value of the latter series can be brought as close as desired to U, so also can the value of the first series come as close to U as desired. Therefore

$$u + D/2^m + D/2^{m+n} + D/2^{m+n+\ldots+r-1}$$

can certainly become $< U + \varepsilon$. But now by assumption M does not hold for all x which are $< u + D/2^m + D/2^{m+n} + \ldots + 2^{m+n+\ldots r-1}$; so much less therefore does M hold for all x which are $< U + \varepsilon$. Therefore U is the greatest value for which the assertion holds that all x below it possess the property M.

§13.

1st Remark. The last theorem is of the greatest importance, and is used in all areas of mathematics, as much in analysis as in the applied parts, in geometry, chronometry, and mechanics. Not infrequently in the past this false proposition has served instead: '*If a property M holds not for all x but for all smaller than a certain value, then there is always some greatest x to which the property M belongs*.' This I say is *false* in consequence of the theorem just proved. For if there is some quantity U which is the greatest of those of which it can be said that all x below them have property M then there is *no greatest* x to which this property belongs *provided x is either a freely or a continuously variable quantity*. For it is well known that for a quantity that varies *freely or according to the law of continuity* there is never a *greatest* value that is smaller than a certain limit U, because however close it may be to this limit it can always be brought closer.—In order to illustrate this by an example, consider a *rectangular hyperbola*, and take one of its asymptotes as the x-axis and take the origin, not at the centre c, but at another point a on this asymptote which is at a distance D from c. Now let us define the direction ac as the *positive x-axis* and the direction ab which is the *perpendicular* ordinate at a as the positive y-axis. Then *every* x-coordinate which is *smaller* than a certain one, say smaller than $D/2$, has the property that *its corresponding ordinate is positive*. However, this property (M) will not hold for *all* positive x-coordinates, namely not for those which are greater than D. Now is there here a *greatest* x-coordinate, or a greatest value of x, to which the property M belongs? In no way, but there is certainly a U, i.e. an x-coordinate which is the greatest among those of which it can be said that all smaller than it have positive ordinates, i.e. possess the property M. This x-coordinate is $+D$.

§14.

2nd Remark. Perhaps someone might have the idea that the proof of the theorem in §12 could have been accomplished quite briefly in the following way: 'If there were no greatest U of which the assertion holds that all x below it possess the property M, one would always be able to take *u greater and greater* and so as great as one desired and consequently M must hold for *all x without exception*.'—However, this would be a very mistaken conclusion, because it is based on the concealed assumption: '*that a quantity that can always be taken greater than it already is can become as great as desired*'.—How false this is may be shown, for example, by the well known series $\frac{1}{2} + \frac{1}{4} + \frac{1}{8} + \ldots$, whose value can always be made greater than it already is and nevertheless always remains < 1!—We would not even mention such an easily discernible error if it did not happen from time to time that mathematicians are guilty of it—as one was only recently in his '*complete theory of parallels*'.

§15.

Theorem. If two functions of x, fx and ϕx, vary *according to the law of continuity* either for all values of x or just for all which lie between α and β, and if $f\alpha < \phi\alpha$ and $f\beta > \phi\beta$ then there is always a certain value of x between α and β for which $fx = \phi x$.

Proof. We must remember that in this theorem the values of the functions fx and ϕx are to be compared with one another simply in their *absolute* values, i.e. without regard to signs or as though they were quantities incapable of being of opposite signs. But the signs which α and β have are important.

I. 1. Firstly assume that α and β are both positive and that (without loss of generality) β is the *greater* of the two, so $\beta = \alpha + i$, where i denotes a positive quantity. Now because $f\alpha < \phi\alpha$, if ω denotes a positive quantity which can be taken as small as desired, then also $f(\alpha + \omega) < \phi(\alpha + \omega)$. For because fx, ϕx vary continuously for all x between α and β, and $\alpha + \omega$ lies between α and β whenever $\omega < i$, then $f(\alpha + \omega) - f\alpha$ and $\phi(\alpha + \omega) - \phi\alpha$ must be able to be made as small as desired if ω is taken small enough. Hence if Ω and Ω' denote quantities which can be made as small as desired, $f(\alpha + \omega) - f\alpha = \Omega$, and $\phi(\alpha + \omega) - \phi\alpha = \Omega'$. Hence,

$$\phi(\alpha + \omega) - f(\alpha + \omega) = \phi\alpha - f\alpha + \Omega' - \Omega.$$

But $\phi\alpha - f\alpha$ equals, by assumption, some positive quantity of constant value A. Therefore,

$$\phi(\alpha + \omega) - f(\alpha + \omega) = A + \Omega' - \Omega,$$

which remains positive if Ω and Ω' are taken small enough, i.e. if ω is given a very small value, and this also holds for all smaller values. Therefore for *all* values of ω which are *less* than a certain one it can be asserted that the two functions $f(\alpha + \omega)$ and $\phi(\alpha + \omega)$ stand in the relation of smaller quantity to larger. Let us designate this property of the variable ω by M. Then we can say that *all* ω that are smaller than a certain value possess the property M. But nevertheless it is clear that this property M does not belong to all values of ω, namely not for the value $\omega = i$, because $f(\alpha + i) = f\beta$ which by assumption is not $<$ but $>$ $\phi(\alpha + i) = \phi\beta$. Consequently the theorem of §12 gives a certain U which is the greatest of those for which it can be asserted that all ω which are $< U$ have the property M.

2. This U must lie *between* 0 and i. For *firstly* it cannot be $= i$, because this would mean $f(\alpha + \omega) < \phi(\alpha + \omega)$, whenever $\omega < i$, and however near it comes to the value i. But in exactly the same way that we have just proved that from the assumption $f\alpha < \phi\alpha$ the consequence $f(\alpha + \omega) < \phi(\alpha + \omega)$ may be drawn if ω is taken small enough, so one can also prove that from the assumption $f(\alpha + i) > \phi(\alpha + i)$, the consequence $f(\alpha + i - \omega) > \phi(\alpha + i - \omega)$ follows if ω is taken small enough. Therefore it is not true that the two functions fx and ϕx stand in the relation of smaller to greater for all values of x which are $<\alpha + i$.—*Secondly*, still less can it be the case that $U > i$, because then i would be one of the values of ω which are $< U$, and hence also $f(\alpha + i) < \phi(\alpha + i)$, which directly contradicts the assumption of the theorem. Therefore, since it is *positive*, U surely lies between 0 and i, and consequently $\alpha + U$ lies between α and β.

3. It may now be asked what relation holds between fx and ϕx for the value $x = \alpha + U$? In the *first* place, it cannot be that $f(\alpha + U) < \phi(\alpha + U)$, for this would also give $f(\alpha + U + \omega) < \phi(\alpha + U + \omega)$ if ω is taken small enough, and so $\alpha + U$ would not be the greatest value for which it can be asserted that all x below it have the property M.—Just as little, *secondly*, can it be that $f(\alpha + U) > \phi(\alpha + U)$, because this would also give $f(\alpha + U - \omega) > \phi(\alpha + U - \omega)$ if ω is taken small enough, and would therefore be against the assumption that the property M is preserved for all x which are below $\alpha + U$. There therefore remains nothing else than that $f(\alpha + U) = \phi(\alpha + U)$, and so it is proved that there is a value of x lying between α and β, namely $\alpha + U$, for which $fx = \phi x$.

II. The same proof is also applicable to the case when α and β are both *negative* if one takes ω, i, and U as negative quantities, because then in the same way $\alpha + \omega$, $\alpha + i$, $\alpha + U$, $\alpha + U - \omega$ represent quantities between α and β.

III. If $\alpha = 0$ and β is *positive* then just take $i(=\beta)$, ω, U positive; and if β is *negative* take these others negative and the proof (I). can be used word for word.

IV. Finally, if α and β are of opposite sign and (without loss of generality) for example α is negative and β positive, then the assumption of the theorem in respect of the continuity of the functions fx and ϕx states that this continuity refers to all values of x which, if negative, are $<\alpha$, and, if *positive* are $<\beta$. Then among these is the value $x=0$. One therefore investigates what relationship holds between $f(0)$ and $\phi(0)$. If $f(0) = \phi(0)$ then the theorem is already proved. But if $f(0) > \phi(0)$, then since $f\alpha < \phi\alpha$ we have by III. a value between 0 and α, and if $f(0) < \phi(0)$ a value between 0 and β, for which $fx = \phi x$. Therefore in every case there is a value of x between α and β which makes $fx = \phi x$.

§16.

Remark. It is in no way claimed here that there is *only a single value* of x which makes $fx = \phi x$. Namely, if $f\alpha < \phi\alpha$ and $f(\alpha + U) = \phi(\alpha + U)$ we must indeed have $f(\alpha + U + \omega) > \phi(\alpha + U + \omega)$ if ω is taken small enough, i.e. the function fx which before was *smaller* than ϕx must, after they are *equal* to one another, soon become greater than ϕx. However, with ever *greater* increase in ω it is certainly possible that before $\alpha + U + \omega$ is made $= \beta$, there is a value for which *again* $fx < \phi x$. In such a case, it follows directly from our theorem, that there must be two other values of x apart from U, between α and β which make $fx = \phi x$. For if $f(\alpha + U + \chi) < \phi(\alpha + U + \chi)$, then because $f(\alpha + U + \omega)$ was already $> \phi(\alpha + U + \omega)$, there must be a value of x between $\alpha + U + \omega$ and $\alpha + U + \chi$, i.e. between α and β, for which $fx = \phi x$. And in the same way, because $f(\alpha + i)$, or $f\beta$, is again $> \phi\beta$, there is also a value of x between $\alpha + U + \chi$ and β which makes $fx = \phi x$. In this way it becomes clear that there must in general always be *an odd number* of values of x which make $fx = \phi x$.

§17.

Theorem. Every function of the form $a + bx^m + cx^n + \ldots + px^r$, in which $m, n, \ldots, r$ designate whole positive exponents, is a quantity which varies *according to the law of continuity* for all values of x.

Proof. If x changes to $x + \omega$ the change in the function is clearly

$$=b[(x+\omega)^m - x^m] + c[(x+\omega)^n - x^n] + \ldots + p[(x+\omega)^r - x^r]$$

—a quantity for which it can easily be shown that it can be made as small as desired if ω is taken small enough. For by the *binomial theorem*, whose validity for whole positive powers (§8 of *Der binom. Lehrs.*) is independent of the investigations with which the present paper is concerned, this quantity is

$$= \omega \left\{ \begin{array}{l} mbx^{m-1} + m.\dfrac{m-1}{2}bx^{m-2}\omega + \ldots + b\omega^{m-1} \\ + ncx^{n-1} + n.\dfrac{n-1}{2}cx^{n-2}\omega + \ldots + c\omega^{n-1} \\ + \ldots\ldots\ldots\ldots\ldots\ldots\ldots\ldots\ldots\ldots\ldots\ldots \\ + rpx^{r-1} + r.\dfrac{r-1}{2}px^{r-2}\omega + \ldots + p\omega^{r-1} \end{array} \right\}$$

The set of terms, of which the factor contained in the brackets consists, is known to be always *finite* and *independent* of the values of the quantities x and ω. Since these occur only with *positive powers* the value of every single term, and consequently the whole expression, for every value of x (also for $x = 0$) is always *finite*. But if for the same x, the value of ω is diminished, then the terms in which ω appears decrease, while the others remain unchanged. Let us therefore designate by S the quantity arrived at by putting a definite value, say $\overset{1}{\omega}$, for ω in all the single terms of the expression, then summing as if they all had the same sign. Then the actual value this expression has for the same $\overset{1}{\omega}$ is certainly not $> S$, but that which it takes for every *smaller* ω is surely $< S$. Hence if it is required to make the change which the function $a + bx^m + cx^n + \ldots + px^r$ undergoes $< D$, then take an ω that is $< \overset{1}{\omega}$ and also $< D/S$. Then $\omega.S$, and even more so the product of ω with a quantity which is $< S$, must be $< D$.

§18.

Theorem. If a function of the form

$$x^n + ax^{n-1} + bx^{n-2} + \ldots + px + q,$$

in which n denotes a whole positive number, is *positive* for $x = \alpha$ and *negative* for $x = \beta$ then the equation

$$x^n + ax^{n-1} + bx^{n-2} + \ldots + px + q = 0$$

has at least one *real root* lying between α and β.

Proof. 1. If α and β both have the same sign (either both positive or both negative) then it is clear that just the same terms of the function which are positive or negative for $x = \alpha$ keep this sign also for $x = \beta$ and for all values of x which lie in between α and β. Suppose now the value of the function is positive for $x = \alpha$ but negative for $x = \beta$. This change can only arise because the sum of the positive terms in it turn out *greater* than that of the negative ones for $x = \alpha$ but *smaller* than that of the negative ones for $x = \beta$. But the sum of the former, as well as of the latter, is of the form

$$a + bx^m + cx^n + \ldots + px^r$$

of §17, i.e. a continuous function. Let us therefore designate the one by ϕx and the other by fx. Then because $f\alpha < \phi\alpha$ and $f\beta > \phi\beta$, *by* §15 there must be some value of x lying between α and β for which $fx = \phi x$. But for this value $fx - \phi x$, i.e. the given function, is zero; therefore this value is a root of the equation

$$x^n + ax^{n-1} + bx^{n-2} + \ldots + px + q = 0$$

2. But if α and β are of opposite sign, consider the value the given function takes for $x = 0$. If this is zero, then this already shows that the given equation has a real root lying between α and β, namely $x = 0$. But if this value (the quantity q) is *positive* then it is now known that the given function is positive for $x = 0$ but negative for $x = \beta$, and because the same terms which are positive or negative for $x = \beta$ retain these signs also for all values lying between 0 and β one can prove, through the same arguments as in no. 1, that there must be a value of x lying between 0 and β which makes the function zero. Finally, if q is *negative*, then what we have just said holds if α is put instead of β. Now since a value lying between 0 and β or between 0 and α also lies between α and β, if they are of opposite sign, then the truth of our theorem is proved for every case.

D. *FROM* PARADOXES OF THE INFINITE (*BOLZANO 1851*)

The following selection comprises §§1–37 of Bolzano's *Paradoxien des Unendlichen*. These paragraphs are a milestone in the foundations of mathematics, and raise many issues that were to loom large in the research of the 1880s. In contrast to Bolzano's other writings, which went largely unnoticed during the nineteenth century, the *Paradoxien* were read and admired by Cantor, Peirce, Dedekind, and Husserl.

In §§1–10, Bolzano treats the notions of *number* and of *infinity* in terms of the more general and abstract notion of *set* (*Menge*). He briefly (§2) notes the connection between sets and logical particles, and in §§6–8 makes one of the earliest attempts to explicate the natural numbers in terms of *sets* and *order properties*. These ideas, as will emerge below, were to be further (and more rigorously) explored by Frege (starting with Part III of *Frege 1879*), by Peirce (*Peirce 1881*), and especially by Dedekind in his *Was sind und was sollen die Zahlen*? (*Dedekind 1888*).

Bolzano's set-theoretical conception led him to break with the traditional mathematical view of the infinite. As we have seen, writers like D'Alembert thought of the infinite in terms of 'quantities capable of augmentation without end', from which it was but a short step to thinking of the infinite as a mere *façon de parler*. But Bolzano—just as he had refused to allow the concept of motion to figure in the 'purely analytic' foundations of the calculus—purified the notion of infinity from notions of growth and change, and treated it instead as a property of *sets*: for example he points out (§11) that a line infinitely extended in both directions is *infinite* but not *variable*.[a] He accordingly rejects the view of the infinite as a *façon de parler*, treating it instead as an objective property of sets; and in §§11 and 12 he defends his objective conception against the rival views of Hegel, Cauchy, and Spinoza. In §§13–17 he espouses an anti-psychologistic conception of objective propositions (the *Sätze-an-sich*) which he uses to distinguish between the objective existence of (finite or infinite) sets and the subjective mental representation of such a set, arguing that only the objective existence is of importance in mathematics.[b]

The set-theoretical conception of the infinite and the defence of the actual infinite were also to be central philosophical themes in the writings of Georg Cantor; and indeed Bolzano's *Paradoxien* is the most significant philosophical contribution to the theory of the infinite before Cantor's *Grundlagen* (*Cantor 1883d*). Both thinkers were inspired by a mixture of mathematical and theological considerations, and both attempted to answer the traditional metaphysical

[a] Bolzano had earlier tentatively expressed support for such a view in §17 of Part I of the *Beiträge* (*Bolzano 1810*).

[b] Bolzano in §13 advances an argument to establish that the class of *Sätze-an-sich* is infinite; it should be compared with Dedekind's argument showing that the realm of thoughts is infinite—Theorem 66 in §5 of *Was sind und was sollen die Zahlen*? (*Dedekind 1888*).

objections to the completed infinite: so the two selections can usefully be read together.

Mathematically, too, Bolzano partially anticipated some of the accomplishments of Cantor and Dedekind. In §20 of the *Paradoxien* Bolzano observes that an infinite set can be put into a strict one-to-one correspondence with a proper subset; this observation was later to be adopted by Peirce, and independently by Dedekind, as the definition of an infinite set.[c] And in §§29–33 Bolzano defends and explores the view that there exist infinite quantities of different magnitudes. These were remarkably perceptive insights, and were to be the cornerstones of the theories of the infinite of Dedekind and Cantor. However, in contrast to his earlier accomplishments in real analysis, Bolzano put these observations to no actual mathematical use, and used them to prove no new mathematical theorem. As Dedekind pointed out in the *Preface* to the second edition of *Dedekind 1888*, Bolzano's observation in §20 was merely recorded as a curiosity about infinite collections: he did not erect a theory on it. And as Cantor pointed out in §7 of the *Grundlagen*, Bolzano lacked both a precise definition of the *cardinality* of a set, and a precise definition of a (finite or infinite) *ordinal number*—the two fundamental tools for Cantor's mathematical theory of the infinite.

The translation of *Bolzano 1851* is by Donald Steel, and is reprinted from *Bolzano 1950*. References should be to the section numbers, which appeared in the original edition.

---

§1

Certainly most of the paradoxical statements encountered in the mathematical domain—though not, as *Kästner* would have it, all of them—are propositions which either immediately contain the idea of the *infinite*, or at least in some way or other depend upon that idea for their attempted proof. Still less is it open to dispute that this category of mathematical paradoxes includes precisely those which merit our closest scrutiny, inasmuch as a satisfactory refutation of their apparent contradictions is requisite for the solution of very important problems in such other sciences as physics and metaphysics.

This, then, is the reason why I address myself in the present treatise exclusively to the consideration of the paradoxes of the infinite. As is readily understood, however, it would be impossible to recognize the appearance of contradiction in these paradoxes for what it is, namely a mere appearance, unless we first

[c] See *Peirce 1881*, and, somewhat more clearly, §IV of *Peirce 1885*. Dedekind's equivalent definition is in §5 of *Dedekind 1888*.

become quite clear what precise concept we attach to the term *infinite*. Consequently, this point shall be taken first.

§2

The very word *infinite* already shows that we contrast the *infinite* with everything merely *finite*. Again, the derivation of the former *name* from the latter betrays the additional fact that we consider the *concept* of the infinite to arise from that of the finite by, and only by, the adjunction of a new element; for such in fact is the bare idea of *negation*. And finally: it is already one reason against denying the application of both ideas to *sets* (more precisely, to *multitudes*, that is, sets of unities) and hence to *quantities* as well, that *mathematics*, the doctrine of quantity, is the very place where we most often speak of the infinite; for we there select as matter for consideration and even for computation both finite and infinite multitudes, as also both *finite* and *infinite* magnitudes, the latter including not only the *infinitely great* but also the *infinitely small*. Quite independently of the hypothesis that these two ideas, of the finite and of the infinite, are applicable only to objects which in some respect or other exhibit *quantity* and *multiplicity*, we are already entitled to hope that a rigorous investigation of the circumstances in which we pronounce a set to be finite or to be infinite will also afford us information about the nature of the *infinite as such*.

§3

For this purpose, however, we must go back to one of the simplest concepts in our minds and seek agreement on the meaning of the word by which we propose to designate it. It is the concept underlying the conjunction *And*, whose most appropriate expression, enabling it to come to the forefront as clearly as the aims both of philosophy and of mathematics demand in numberless cases, is found to my best belief in the words: 'An aggregate of well-defined objects' ⟦ein Inbegriff gewisser Dinge⟧ or: 'A whole composed of well-defined members'. We intend these words to be interpreted so widely that, whenever the conjunction *And* is customarily employed, for example in the sentences:

'*The sun, the earth, and the moon act upon one another*,'
'*The rose, and the conception of a rose, are two very different things*,'
'*The names* 'Socrates' *and* 'son of Sophroniscus' *designate one and the same person*,'

we can say that the subject of these sentences is a *well-defined aggregate of objects*, or *a whole composed of well-defined members*, and say in the first example, that it is the whole composed of the sun, the earth, and the moon which we pronounce to be a whole whose members act upon one another; in the second example, that it is the aggregate composed of the object 'the rose' and the object 'the concept of a rose' as members which we pronounce to be an aggregate whose members are entirely distinct entities; and so forth. These few remarks will presumably be quite sufficient for an understanding of the conception here dealt with, at least if we add this statement: that every arbitrary object A whatever can be united with any other arbitrary objects B, C, D . . . whatever to form an aggregate, or (to speak still more rigorously) that these objects already form an aggregate without our intervention ‖an sich selbst schon‖ and one about which numerous truths of different degrees of importance can be stated—provided only that A, B, C, D . . . really represent, each and every one of them, a *different* object; that is, provided only that none of the propositions 'A is identical with B,' 'A is identical with C,' 'B is identical with C' and so forth be true. For if A were the same thing as B, it would of course be absurd to speak of an aggregate composed of the things A and B.

§4

There exist aggregates which, although they contain the same members A, B, C, D, . . ., nevertheless present themselves as *different* when seen under different aspects ‖Gesichtspunkte‖ or under different concepts ‖Begriffe‖; and this kind of difference we call 'essential.' For example: an unbroken tumbler and a tumbler broken in pieces, considered as a drinking vessel. We call the ground of distinction between two such aggregates the *mode of combination* ‖*Art der Verbindung*‖ or the *arrangement* ‖*Anordnung*‖ of their members. An aggregate whose basic conception renders the arrangement of its members a matter of indifference, and whose permutation therefore produces no essential change from the current point of view, I shall call a *set* ‖*Menge*‖, and a set whose members are considered as *individuals* ‖*Einheiten*‖ of a stated species A (that is, as objects subsumable under the concept A) is called a *multitude* ‖*Vielheit*‖ of A.

§5

Some aggregates, it is well known, have members which are themselves composite, themselves aggregates in turn. Among these again, some are regarded from a viewpoint which renders it indifferent whether or not we take the members of the subaggregates as members of the main aggregate ‖die Teile der Teile als Teile des Ganzen selbst auffassen‖. I borrow a term from mathematics, and call such aggregates *sums*. For it is the essential nature of a sum that $A + (B + C) = A + B + C$.

§6

If we regard an object as belonging to a category of entities such that each two of them, say M and N, can stand in no other mutual relation than that either they are *equal*, or else one of them can be represented as a sum whose augend is equal to the other of them (in other words, either $M = N$, or else $M = N + \nu$

or $N = M + \mu$, where ν and μ must again be either equal or such that one can be considered as a part contained in the other), then we regard that object as a *quantity*.

§7

The name *series* ‖*Reihe*‖ shall be given to a proposed aggregate of objects . . . $A, B, C, D, E, F, \ldots, L, M, N, \ldots$ when it is possible to assign to any one member M exactly one other member N by a *law applying uniformly* to all members and *determining*, either N through its relation to M, or M through its relation to N. The members shall be called the *terms* ‖*Glieder*‖. The law whereby M determines N or conversely shall be called the *principle of construction* ‖*Bildungsgesetz*‖ of the series. One of the two terms (whichever you like, but without wishing to imply any temporal or spatial sequence) shall be called the *antecedent* ‖*vorderes Glied*‖ and the other the *consequent* ‖*hinteres Glied*‖. Every term M which has both an antecedent and a consequent (and which therefore both takes its rise by the principle of construction from another member, and in turn gives rise to a third) shall be called an *inner term* ‖*inneres Glied*‖ of the series. Hence, finally, it will be plain which terms (if any of the kind exist) shall be called *outer*, which *first*, and which *last*.*

§8

Let us imagine a *series* whose *first* term is an *individual* of the species A, and whose every subsequent term is derived from its predecessor by adjoining a fresh individual of the species A to form a sum with that predecessor. Then clearly all the terms of this series, with the exception of the first, which was a *mere individual* of the species A, will be *multitudes of the species A*. Such multitudes I call *finite* ‖*endlich*‖ or *countable* ‖*zählbar*‖ *multitudes*, or quite boldly: *numbers*; and more specifically: *whole* numbers—under which the first term shall also be comprised.

§9

When we vary the nature of the concept here denoted by A, we shall find that the objects subsumed under it, the individuals of the species A, form now a more numerous and now a less numerous set; and the series under consideration will contain now a more numerous and now a less numerous set of terms. In particular, the series may contain so many terms that it cannot, compatibly with taking in and exhausting *all* the individuals of that species, be conceded to have

* More detailed explanations of these ideas, and of some others which have occurred in previous paragraphs, will be found in the *Wissenschaftslehre*.

a *last term*; a point which we shall handle in greater detail in the sequel. Assuming it for the present, I propose the name *infinite multitude* for one so constituted that every single finite multitude represents only a part of it.

§10

I hope it will be conceded that this definition of a *finite* and an *infinite* multitude will distinguish between them exactly in the sense intended by those who have used these terms rigorously. It will also be granted that no vicious circle lurks in the definitions. It only remains to ask, therefore, whether a definition merely of what is to be called an infinite *multitude* can put us in a position to determine what the *infinite* is in itself. That would be the case if it came to light that multitudes were the only things to which the idea of the infinite could be applied in its strict sense—in other words, if it came to light that infinitude were, strictly speaking, a property only of multitudes; or again in other words, if everything we judge *infinite* is so judged solely because, and solely in so far as, we find it possessing a property which can be considered as infinite multitude. Now to my thinking, this really is the case. Mathematicians use the word plainly in no other meaning: for what they are occupied in determining is hardly ever anything else than quantity, and they do so by first choosing one object of the species as *unit*, and then employing the idea of number. If they find a quantity greater than any finite number of the assumed units, they call it *infinitely great*; if they find one so small that its every finite multiple is smaller than the unit, they call it *infinitely small*; nor do they recognize any other kind of infinitude than these two, together with the quantities derived from them as being infinite to a higher order of greatness or smallness, and thus based after all on the same idea.

§11

But especially in our day some philosophers, such as *Hegel* and his followers, are not satisfied with this mathematically familiar infinitude. They contemptuously dub it the 'bad infinity,' and claim knowledge of a true, a vastly superior, a *qualitative infinity*, to be found in *God* particularly, and speaking generally only in the *absolute*. Now I agree for my part with *Hegel*, *Erdmann*, and others so long as they are thinking of the mathematically infinite only as a *variable* quantity knowing no limit to its growth (a definition adopted, as we shall soon see, even by many mathematicians) and so long as they are finding fault with this notion of a quantity which is always *growing* into the infinite but never *reaching* it. In fact, a *truly infinite* quantity (for example, the length of a straight line unbounded in either direction, meaning: the magnitude of the spatial entity containing all the points determined solely by their abstractly conceivable relation to two fixed points) does not by any means need to be variable; and in the adduced example it is in fact not variable. Conversely, it is quite possible for a quantity merely capable of being taken greater than we have already taken it, and of becoming larger than any one pre-assigned (finite) quantity, nevertheless to remain at all times merely finite: which holds in particular of every numerical quantity 1, 2, 3, 4, . . .; what I refuse to admit is only this: that the philosopher knows any object to which he is entitled to attach the predicate of infinitude without having first established that in some respect or other that object exhibits infinite quantity, or at least infinite multitude. Now if I can once show that in God himself, the being whom we regard as the most perfectly one, aspects can be found under which we see even in him an infinite multitude, and if I can show that we attribute infinitude to him under those aspects alone, then it will scarcely be necessary to go on and show that similar considerations lie at the bottom of all the other cases where the idea of the infinite holds good. I say then: we call God infinite because we are compelled to admit in him more than one kind of force possessing infinite magnitude. Thus, we must attribute to him a power of knowledge which is true omniscience, and which therefore comprises an infinite set of truths, to wit, all truths—and so forth. And what idea of the truly infinite would people impose upon us other than the one here set up? They say it is that All which comprises every Something whatever, the absolute All outside of which there exists nothing more. Yet even on this formulation it would still be an infinite which according to our definition included an infinite multitude. It would be an aggregate not alone of all actual entities, but also of all things having no actuality, of all absolute propositions and truths ⟦Sätzen und Wahrheiten an sich⟧. Consequently, there appear to be no grounds for departing from our conception of the infinite and embracing theirs—to say nothing of all the other errors they have woven into this doctrine of the All.

§12

On the other hand, I am obliged to reject as erroneous many other definitions of the infinite, even such as have been set up by mathematicians in the belief that they were doing nothing but exhibit the elements of this one same concept.

1. As I mentioned only a short while ago, some mathematicians including even *Cauchy* (in his '*Cours d'analyse*' and several other works), and the author of the article 'Infinite' in *Klügel's* Dictionary imagined they were defining the infinite when they described it as a variable quantity whose value increases *without limit* and in keeping therewith is capable of surpassing *any one preselected quantity* ⟦*gegebene Grösse*⟧ *no matter how large* a one has been chosen ⟦*noch so gross*⟧. The *limit* of this unlimited growth is alleged to be the *infinitely great quantity*. In this manner the tangent of a right angle, regarded as a continuous quantity, is supposed to be unlimited, unterminated, *infinite in the strict sense*. How mistaken this opinion is can be seen from the simple fact that what mathematicians call a *variable quantity* is not properly a quantity at all, but only the idea and notion of a quantity, and a notion, moreover, which comprises under itself not a fixed individual quantity at all, but rather an infinite set of quantities, distinct from one another, and differing from one another in

their values—which is precisely the same thing as saying that they differ from one another in *quantity*. The thing they call infinite is not, however, any one of the *distinct* values taken on, for example, by *tan* (ϕ) for distinct values of ϕ, but rather the single value which they imagine (wrongly as it here happens) *tan* (ϕ) to take on for $\phi = \frac{1}{2}\pi$. Then again, it is surely another contradiction to speak of the limit of an unlimited growth, and when defining the infinitely small to speak of the limit of an unlimited dimunition; and if the former is declared to be the infinitely great, then analogy demands that the latter, namely zero pure and simple (just Nothing) be declared to be the infinitely small: which is surely incorrect, and which neither *Cauchy* nor *Grunert* ventures to assert.

2. The definition just considered was too wide. Another definition is too narrow, namely the one adopted by Spinoza and many other philosophers, as well as mathematicians, according to which *those things alone are infinite which are incapable of further increase*, or to which nothing can be adjoined or added. Mathematicians hold themselves entitled, in fact, to add other quantities to any quantity whatever, even to one infinitely great, and not only to add finite ones, but also to add other quantities which are already themselves infinite, and even to multiply one infinity by another infinity, and so forth. And granting that the legitimacy of this process is disputed by some, what mathematician is there who, if he allows infinity of any kind, is not forced to concede that the length of a straight line bounded on one side but stretching to infinity on the other is infinitely great and nevertheless capable of being increased by additions to the first side?

3. No more satisfactory is the definition given by those who lean on the derivation of the word and say: the infinite is that *which has no end*. If they are thinking in this definition only of an end in time, only of a cessation, then no other things could be infinite but those subject to temporal flux, whereas we also ask of things not so subject, like lines or abstract quantities, whether they be finite or infinite. But if they take the word *end* in a wider sense, say as equivalent with *limit* as such, then I call attention to two points:

Firstly, there are many objects which, while we cannot well prove them to possess a limit without foisting on that word a highly vacillating and thoroughly confusing significance, are nevertheless classified by nobody whatever as infinite. For example, a simple point of time or space has no such thing as a limit, but rather is itself the limit of an interval of time or of a line, most people defining it precisely as such, just as if this were its very essence; but for all that, it never occurred to anybody (with the possible exception of Hegel) to espy infinitude in a mere point. Or for another example: mathematicians know of no bounding point in the circular periphery, of no bounding point or line in the numerous other closed curves and surfaces, and consider them for all that to be finite objects—unless indeed they come to speak of the infinite set of points contained in them, from which point of view, however, the like infinitude must be recognized in every bounded rectilinear segment.

Secondly, I observe that plenty of objects are undeniably bounded and yet are still regarded as belonging to the class of infinite quantities. Such is the case not only with the unilaterally unbounded straight line mentioned above, but also with the space between a pair of infinitely long parallels, or between the arms of a plane angle, and with much besides. In rational psychology too, we shall call a power of knowledge infinite (though short of omniscience) as soon as it is able to envisage an infinite set of truths, say those enunciating in turn the infinitely many digits in the decimal representation of the single quantity $\sqrt{2}$.

4. The commonest formulation of the infinitely great is: 'that which is greater than any *assignable* quantity'. The first necessity here is to determine more exactly what is meant by '*assignable*'. Is it to signify no more than that something is *possible* (that is, *capable* of actuality) or to signify no more than that it is *non-self-contradictory*? In the former alternative, we confine the idea of the *finite* exclusively to the category of things *actual*—either actual at all times, or actual at certain past times, or actual at certain future times, or at least *capable* of attaining actuality at some time or other. This seems in fact to be the meaning attached to the infinite by *Fries* (*Naturphilosophie*, §47) when he calls it the *uncompletable*. Yet linguistic custom applies the two notions, of the finite and of the infinite, not only to objects enjoying actuality (as above all to God) but to others in whose case we cannot so much as speak of existence. Of this kind are the absolute propositions and truths together with their components, the absolute presentations ⟦Vorstellungen an sich⟧: for we assume both finite and infinite sets thereof. In the latter alternative, however, when we understand by the assignable everything barely *non-self-contradictory*, then we insinuate the non-existence of the infinite into its very definition, since a quantity supposed greater than any non-self-contradictory quantity would have to be greater than itself, which is of course absurd. Now in reality there is a third meaning in which the word '*assignable*' could be taken, to wit: as applying to all and only those things which can *in some way or other be given to us*, i.e. become the objects of *our experience*. Here I must really ask people whether they do not in every case use the words *finite* and *infinite* in such a sense (and whether their useful employment in science do not make it necessary to use them in such a sense) that they refer to a definite *internal* property of the thing so described, and by no means exclusively to a relation they bear to our *power of knowledge*, and still less to a relation they bear to our *sensitive faculties*—and this, quite irrespectively of whether we are able or unable to collect *experience* of them. As a consequence, the question whether a given object be finite or infinite certainly cannot depend on whether its quantity does or does not fall within our perception, or on whether we are able or unable to command a view of it.

§13

Now that we are agreed on the idea to be attached to the word *infinite*, and now that we have made ourselves clearly aware of the elements composing that idea, the next question to be asked is that concerning its *objective existence*—that is, whether there exist objects to which it can be applied, whether there exist sets which we may judge to be infinite in the sense here declared. This I venture to answer with a decided *affirmative*. Even in the *realm of things which do not claim actuality, and do not even claim possibility*, there exist beyond dispute sets which are infinite. *The set of all absolute propositions and truths* ⟦*Die Menge der Sätze und Wahrheiten an sich*⟧ is easily seen to be infinite. For if we fix our attention upon any truth taken at random, say the proposition that there exist such things as truths, or any other proposition, and label it *A*, we find that the proposition conveyed by the words '*A is true*' is distinct from the proposition *A* itself, since it has the complete proposition *A* for its own subject. Now by the same law which enabled us to derive from the proposition *A* another and different one, which we shall call *B*, we are further enabled to derive a third proposition *C* from *B*, and so forth without end. The aggregate of all these propositions, every one of which is related to its predecessor by having the latter for its own subject, and the latter's truth for its own assertion, comprises a set of members (each member a proposition) which is greater than any particular finite set. The reader does not need to be reminded of the similarity borne by this series together with its principle of construction to the *series of numbers* considered in §8. The similarity consists in the fact that to each member of the former there corresponds a member of the latter; in the fact that howsoever large an integer be chosen, there exists a set of just so many among the above propositions; and finally in the fact that we can always continue the construction of such propositions—or rather, that such further propositions exist whether we construct them or not. Whence it follows that the aggregate of all the above propositions enjoys a multiplicity surpassing every individual integer, and is therefore infinite.

§14

For all the simplicity and clearness of the demonstration just proffered, a large number of acute and learned men hold the thesis itself to be not merely paradoxical, but even false in fact. They deny the *existence of anything infinite*. To nothing whatever—neither among the things possessing actuality, nor among the rest, neither to an individual nor to any aggregate—is it possible, according to their assertion, to attribute in any respect whatever an infinite set of members. The arguments which they bring against anything infinite in the realm of the actual will be considered later, for only later shall we bring our own arguments for the existence of such an infinite. Let us examine at the present moment, therefore, only the grounds upon which it is proposed to show that no infinite exists even among the things which lay no claim to actuality.

Firstly: 'Nowhere can an infinite set exist,' they say, 'for the simple reason that an infinite set can *never be united to form a whole, never be collected together in thought*.' I must stigmatize this assertion as a mistake, and as a mistake engendered by the false opinion that a whole consisting of certain objects *a*, *b*, *c*, *d*, . . . cannot be constructed in thought unless one first forms separate mental representations ⟦Vorstellungen⟧ of its separate component objects. This is by no means true. I can think of the set, of the aggregate, or if you prefer it, the *totality* of the inhabitants of Prague or of Peking without forming a separate representation of each separate inhabitant. So indeed do I act every time that I speak of this set and put forward my estimate, in the case of Prague, that the population lies between 100 000 and 120 000. As soon, in fact, as we possess a representation A which represents the objects a, b, c, d, ... and no others, it is extremely easy to arrive at a representation which represents the *aggregate* of all these objects taken together. Nothing more is needed than to combine the idea denoted by the word *aggregate*, and the notion A, in the manner expressed by the words '*the aggregate of all A*.' This single remark, whose correctness I trust will be evident to all, removes all the difficulties made against the idea of a set comprising infinitely many members: provided only that a specific concept is present, under which every member, and no non-member, can be subsumed. This is fulfilled for *the set of all absolute propositions and truths*, the requisite specific concept being no other than the one already before us, to wit: 'an absolute proposition or truth.'

Nor ought I to let a *second* error in this first objection pass unreprehended. It is the opinion that 'sets do not exist unless there be somebody engaged in *thinking* of them.' In order to be as consistent as is possible while maintaining an error, those who make this assertion should not only deny that there are *infinitely many* absolute propositions and truths, but also deny that there are any absolute propositions and truths *at all*. For if we have worked our way to a clear idea of absolute propositions and truths and entertain no doubt of their objective existence, we shall not easily stumble into assertions like the one just quoted; or if we do stumble into them, we shall not easily persist in them. To the end that everybody may see this clearly, I take leave to raise the question whether there be no fluid or solid bodies at the poles of the earth, no air or water or stones and the like, whether these bodies do not act upon one another according to fixed laws, in such wise for example that the velocities communicated at their collision are inversely proportional to their masses, and whether all this happens even when no human or other thinking being is present to observe it? If this is granted—and who is not obliged to grant it?—then there exist absolute propositions and truths which record these events completely, without there being anyone to think about them or to know them. In these propositions, moreover, there is frequent mention of wholes and sets: for every physical body is a whole, and produces very many of its effects only through the multitude of parts of which it is composed. Consequently, there exist wholes and sets even in the absence of a being to think of them. Were this not so, were these sets not objectively present, how could the judgements we pass upon them be true? Or rather, what content could these judgements have, if the presence of an observer of these processes were a necessary condition of the truth of these judgements? If I say: 'This boulder came loose from that rock before my very eyes, and hurtled down, cleaving the air as it fell,' then what I am supposed to mean must be something like this: 'As a result of my combining in my imagination certain simple substances at such a spot and such a height, there was formed a composite entity called by me a boulder; and this composite entity moved away from certain other simple substances which, as a result of my combining them too in my imagination, coalesced into a whole called by me a rock, and so on and so forth.'

Secondly: Someone may urge the following: 'Whether or not we choose to unite certain simple objects into an aggregate, it is only an operation *on our part*, and in most cases a very arbitrary operation; and relations arise between them only after we have carried out the operation. The midmost atom in this coat-button of mine, and the midmost atom in yonder steeple-knob, have absolutely nothing to do with one another, and no connection; only by my passing act of thinking them in combination is some sort of connection established between them.'—This too I am obliged to contradict. Even before a thinking being conjoined his mental representations, the two atoms were already exerting an influence upon one another, for example, by their force of attraction; and except in the event of that thinking being's proceeding in consequence of his thoughts to actions which do bring about a change in the relationships of the two atoms, it is altogether untrue that his thinking of them in combination is the sole cause of relations arising between them which otherwise would not obtain. If I am to judge correctly that this atom is the lower and that atom the higher, and judge correctly that as a result the former is attracted a little upwards by the latter, then all these things must take place even did I not think of them. And so forth.

Thirdly: others again would say: 'For the subsistence of an aggregate it is not requisite that it be *actually in the mind* of some thinking being, but it is necessary that it be *potentially* there. And since it is impossible for a being to exist, which is capable of representing to itself each separate one of an infinite set of objects, and then of combining these representations together, it must be impossible to have an aggregate embracing as members an infinite set of objects.'

Our answer to the first objection has already evinced the falseness of the hypothesis, here several times repeated, that the thought of an aggregate calls for a separate thought of each separate member. Nor are we under the necessity of appealing to an omniscient being as one for whom the distributive apprehension of an infinite set of objects causes no difficulty. For quite apart from these points, we cannot permit ourselves to concede even the first hypothesis, namely that the capacity of *becoming an object of thought* could be a necessary condition for the subsistence of an aggregate. For nothing can derive its capacity for existing from its *capacity for being thought of*. On the very contrary, its capacity for existing is the prior ground upon which a rational being (in so far as not involved at the time in error) deems a thing *possible*, or as we somewhat inappropriately say, *thinkable*. The reader will become still more firmly persuaded of the correctness of this remark, and of the complete untenability of the very widespread opinion I am here combating, if he tries to make clear to himself the elements which go to compose this extremely important concept of 'possibility.' To say that those things are *possible* which *can exist* is no genuine analysis of that concept; for that concept is contained entire in the meaning of the word 'can'. Still more faulty would it be to set up as a definition that just those things are possible *which can be thought of.* In the strict sense of the word, which includes mere *mental representation*, we can even 'think of' the impossible, and do really think of it as often as we make statements about it, declaring this or that to be impossible, for example, when we declare that there exists and could exist no quantity represented by 0 or $\sqrt{-1}$. Indeed, even if we exclude mere mental representation, and let 'thinking' only mean 'holding to be true', it is still false that everything is possible which we can hold to be true. Error sometimes causes us to hold the impossible to be true: for example, to believe that we have squared the circle. The true formulation, which I anticipated above, must thus be as follows: 'Just those things are possible which a thinking being, if it judges in accordance with the truth, judges capable of existing,' in other words, judges to be 'possible'; and this definition obviously contains a vicious circle! We are hence constrained, when defining the possible, to abandon altogether the reference to a thinking being, and to search for some other characteristic property. Sometimes we are told that the possible is *that which does not contradict itself.* Now it is true that everything which contains a contradiction within itself is impossible, for example, that a sphere be not a sphere. On the other hand, however, not everything impossible is impossible in such a way that the contradiction lies between the components out of which its notion is put together. It is impossible, for example, to have a polyhedron with seven congruent faces—but the contradictoriness of this is not visible in the bare collocation of these words. We are thus forced to extend our definition. But were we to say that the impossible is that which stands in contradiction to any truth at all, we should be classifying as impossible everything which is non-existent—if only because the corresponding existential proposition would stand in contradiction to the truth of its non-existence. We should in this event make no sort of distinction between the actual and the possible, nor indeed between these and the necessary, whereas we all do make this distinction. From this we learn that the impossible does not contradict all truths indiscriminately, but only a definite category of truths; and it can now scarcely escape us, what category it is. It is the category of purely conceptual truths ‖reine Begriffswahrheiten‖. Whatever contradicts a purely conceptual truth must be called an *impossibility*; and conversely, those things are possible which conflict with no purely conceptual truth. To anyone who has seen that this is the correct notion of possibility, it can hardly occur to assert that a thing only becomes possible by being actually thought of, that is, when some thinking being, without erring in his judgement, is looking upon it as possible. To say this would be to say as follows: 'For a proposition to contradict no purely conceptual truth, it is requisite that it should contradict no purely conceptual truth to have a thinking being which judges non-erroneously that the proposition contradicts no purely conceptual truth.' Who does not see how entirely irrelevant is here the intervention of a thinking being? And once it is decided that the *act of thought* ‖das *Denken*‖ is not the source of possibility, what justification remains for denying the existence of infinite sets of objects on the alleged ground that they cannot be *assembled in thought*?

§15

The existence of infinite sets, at least with non-actual members, is something which I now regard as sufficiently proved and defended; as also, that the set of *all absolute truths* is an infinite set. Arguments like those in §13 will win assent for the statement that the set of *all numbers* is infinite—that is, the set of all the so-called natural or whole numbers, as defined in §8. Yet this statement also sounds *paradoxical*, and we may regard it as the *first paradox* to appear in the realm of mathematics—for the one just considered belongs, properly speaking, to a science more general than that of quantity.

'If each number,' it might be protested, 'is by definition a merely finite set, how can the set of *all* numbers be infinite? If we contemplate the series of the natural numbers

$$1, 2, 3, 4, 5, 6, \ldots,$$

we become aware that the numbers of the series lying between the first (unity) and any particular one of them form a set which is enumerated by that particular one. For example, those from one to six form a set of six numbers. Consequently, the set of *all* numbers must be enumerated by the *last* number, and being therefore itself a number, not be infinite.'

The fallacy of this argument disappears at once if we remember that in the natural series of natural numbers no term occupies the *last place*, and that the notion of a last or highest number is an empty notion, being a self-contradictory one. In fact, the *principle of construction* of this series, as explained in §8, assigns to each of its terms a *following* term. This single remark must therefore be regarded as solving the present paradox.

§16

If the set of all *numbers* (namely the so-called *integers*) is infinite, then with all the greater certainty is the set of all quantities (as explained in §6 above, and in §87 of my *Wissenschaftslehre*) an *infinite* set. For according to that explanation, not only are all numbers also quantities, but there are far more quantities in existence than there are numbers: because the fractions $\frac{1}{2}$, $\frac{1}{3}$, $\frac{2}{3}$, $\frac{1}{4}$, ..., and likewise the so-called *irrational* expressions $\sqrt{2}$, $\sqrt[3]{2}$, ..., π, e, ... denote quantities. Indeed, it is consistent with that explanation to speak of quantities *infinitely great* and others *infinitely small*, on condition that by an *infinitely great* quantity we only mean one which, after a unit has been chosen, presents itself as a whole of which every finite set of those units is only a part; and by an *infinitely small* quantity only one such that, the same unit being taken, this unit now presents itself as a whole of which every finite multiple of the proposed quantity is only a part. The set of all numbers manifests itself immediately as an indisputable example of an infinitely great *quantity*. I say advisedly, as an example of a quantity; and certainly not as an example of an infinitely great *number*; for this infinitely great multitude cannot, as we remarked in the previous paragraph, be given the name of number. On the other hand, if we take a quantity which appears as *infinitely great* relative to a preselected unit, and then make that quantity into a differently chosen new unit, then a renewed comparison shows that the old unit realizes the definition of the *infinitely small.*

§17

Space and *time*—which again do not belong to the domain of the actual, though they can be *determinations* of the actual—form a very important category of infinitely great quantities. Neither space nor time is an actual object, for neither is a *substance* and neither is a *quality* ‖*Beschaffenheit*‖ of a substance; but both appear only as determinations ‖Bestimmungen‖ of all incomplete *substances*—that is, of limited, finite, or (what comes to the same thing) dependent, created substances. Each of these, in fact, must always possess a definite position in space and in time, in such a manner that every simple substance must occupy a definite point in space at each definite *point in time.* Now the set of spatial points of which space is composed and the set of temporal points of which time is composed are both *infinite.* Indeed: not only is the set of all points in every region of space or of time an infinite set, but so too is the set of all the instants between two fixed instants howsoever proximate, or the set of all the spatial points between two fixed points howsoever proximate again. I have little need to involve myself in a defence of these propositions, because hardly any mathematician who admits the infinite at all fails to admit this as well. To escape conceding so manifest an infinite as this one, the opponents of infinity *in any form* take refuge in the pretext that 'whereas we can admittedly always *think of more* instants or points than we have thought of so far, yet the set of those *actually* present remains constantly finite.' My answer to this is, that neither time nor space, and therefore neither instants nor points, are anything actual at all; so in consequence it is absurd to speak of a finite set of them as *actually* existing; and even more absurd to imagine that these points receive their actuality as a result of our *thought.* For it would be deducible from this that the qualities of time and space depend on our acts of thought and judgement: and as a particular consequence, that the ratio of the diameter of a circle to its circumference was rational as long as we erroneously believed it to be rational. Furthermore, all the properties of space which we are destined to discover in the future will only accrue to space when that future arrives!

Even if the adversaries are given a chance to correct their wording and say rather that only such thinking as corresponds with the truth goes to determine the true properties of time and space, the only outcome is a sheer tautology—namely that those things are true, which are true. From this, however, it is impossible to deduce anything at all against our assertion of the infinitude of time and space. At all events, it is absurd ‖abgeschmackt‖ to say that time and space contain only so many points as we happen to be thinking of.

§18

Although the capacity of being regarded as a whole consisting of an infinite set of parts in some respect or other is a necessary condition for a quantity, or any other object, to be regarded by us in that same respect as infinite, yet it is not by itself sufficient; for not every quantity which we consider as the sum of an infinite set of other quantities each separately finite need itself be infinite. It is universally recognized, for example, that irrational quantities such as $\sqrt{2}$ are finite relative to their standard unit, despite the fact that they can be looked upon as composed from an infinite set of fractions of the form

$$\frac{14}{10}+\frac{1}{100}+\frac{4}{1000}+\frac{2}{10\,000}+\ldots,$$

with integer numerators and denominators; as also that the sum of an infinite series of addends having the form

$$a + ae + ae^2 + \ldots \textit{in inf.} = \frac{a}{1-e}$$

amounts to the finite quantity $\frac{a}{1-e}$ whenever $e < 1$ holds.* In consequence

---

* Since the customary proof for the summation of this series appears to be not quite rigorous, perhaps I may be allowed the opportunity to make the following suggestions. Let us take $a = 1$ and e positive, since the modification for other cases is obvious, and let us write the symbolical equation

$$S = 1 + e + e^2 + \ldots \textit{in inf.} \tag{1}$$

Then this much is certain, that S denotes a positive quantity, be it finite or infinite. But for every integer value of n whatever, we have

$$S = 1 + e + e^2 + \ldots + e^{n-1} + e^n + e^{n+1} + \ldots \textit{in inf.}$$

or again,

$$S = \frac{1-e^n}{1-e} + e^n + e^{n+1} + \ldots \textit{in inf.} \tag{2}$$

for which we can also write

$$S = \frac{1-e^n}{1-e} + P' \tag{3}$$

if we denote by P' the value of the infinite series

$$e^n + e^{n+1} + \ldots \textit{in inf.},$$

whereby again this much is certain, that P' denotes a quantity which, whether measurable or not, depends on e and n and is at any rate positive. But the same infinite series can be expressed as

$$e^n + e^{n+1} + \ldots \textit{in inf.} = e^n\left(1 + e + \ldots \textit{in inf.}\right),$$

and the sum bracketed on the right side, consisting of infinitely many addends, now wears exactly the same appearance as the series put equal to S in the symbolic equation (1). Nevertheless, it cannot be regarded as identical therewith, since the set of its addends and the set of those in (1), while of this, the assertion that a sum of infinitely many separately finite quantities can nevertheless present itself as a merely finite quantity contains no contradiction; for if it did, we should not have found it possible to demonstrate its truth. The semblance of paradox which some may see in it originates in forgetfulness of the fact that the addends become smaller and smaller. Nobody can be surprised if a

---

agreeing in being infinite, fail to be identical, in that the former has indisputably fewer terms than the latter. With full confidence, therefore, we can do no more than put

$$\left(1 + e + e^2 + \ldots \textit{in inf.}\right) = S - P'',$$

whereby we may at any rate suppose that P'' is always a positive quantity dependent on n. We thus have

$$S = \frac{1 - e^n}{1 - e} + e^n\left(S - P''\right) \tag{4}$$

or alternatively

$$S\left(1 - e^n\right) = \frac{1 - e^n}{1 - e} - e^n P'',$$

and finally

$$S = \frac{1}{1 - e} - \frac{e^n}{1 - e^n} P'' \tag{5}$$

A comparison of equations (3) and (5) yields

$$\frac{-e^n}{1 - e} + P' = \frac{-e^n}{1 - e^n} P'',$$

or alternatively

$$P' + \frac{e^n}{1 - e^n} P'' = + \frac{e^n}{1 - e},$$

from which it can be inferred that if we choose n arbitrarily great and thus depress the value of

$$\frac{e^n}{1 - e^n}$$

below any given quantity

$$\frac{1}{N}$$

however small this be, each of the two quantities

$$P' \text{ and } \frac{e^n}{1 - e^n} P''$$

must separately sink below any given value. This being so, each of the equations (3) and (5) indicates, in view of the fact that the value of S depends on e but not on n, that this

$$S = \frac{1}{1 - e}.$$

sum of addends each of which halves its predecessor can never surpass the double of the initial addend, since at however late a term in the series we halt, exactly so much is wanting to make up that double, as the term in question has value.

§19

Even in the examples of the infinite so far considered, it could not escape our notice that not all infinite sets can be deemed equal *with respect to the multiplicity of their members*. On the contrary, many of them are *greater* (or *smaller*) than some other in the sense that the one includes the other as a part of itself (or stands to the other in the relation of part to whole). Many consider this as yet another *paradox*, and indeed, in the eyes of all who define the infinite as that which is incapable of increase, the idea of one infinite being greater than another must seem not merely paradoxical, but even downright *contradictory*. We on the other hand have now seen that this opinion rests on a notion of the infinite which is altogether discordant with linguistic usage. Our own definition—which agrees not only with linguistic usage, but also with the purposes of scientific knowledge—does not tempt anyone to think it contradictory, or even astonishing, that one infinite set be greater than another. Who can fail to see that the length of the straight line taken from *a* unboundedly towards *R* and beyond is infinite? But who can fail to see in addition, both that the unilaterally unbounded line from *b* rightwards exceeds that from *a* rightwards by the piece *ba*, and also that the bilaterally unbounded line taken both from *a* to *R* and beyond and from *a* to *S* and beyond, is greater still by an excess which is itself infinite? And so forth.

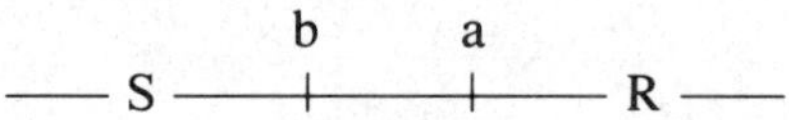

§20

We now pass on to consider a very remarkable peculiarity which can occur in the relations between two sets *when both are infinite*. Properly speaking, it always does occur—but, to the disadvantage of our insight into many a truth of metaphysics and physics and mathematics, it has hitherto been overlooked. Even now, when I come to state it, it will sound so paradoxical that we shall do well to linger somewhat over its investigation. My assertion runs as follows: When two sets are both infinite, they can stand in such a relation to one another that:

(1) it is possible to couple each member of the first set with some member of the second in such a way that, on the one hand, no member of either set fails to occur in one of the couples; and on the other hand, not one of them occurs in two or more of the couples; while at the same time,

(2) one of the two sets can comprise the other as a mere part of itself, in such a way that the multiplicities to which they are reduced, when we regard all

their members as interchangeable individuals, can stand in the most varied relationships to one another.

I shall conduct the proof of my assertion by means of two examples in which the above situation undeniably occurs.

EXAMPLE I: let us choose any two abstract quantities, say 5 and 12. Then the set of all quantities between zero and 5 (or less than 5) is clearly infinite, as is also the set of all quantities less than 12. With no less certainty the latter set is greater than the former, seeing that the former constitutes a mere part of the latter. We could even alter the quantities 5 and 12 into any others, and still be forced to admit that the corresponding two sets need by no means always stand in the same relation, but rather are able to enter into a great variety of relations. But no less true than all this is the following: if x denotes any arbitrary quantity between zero and 5, and if we fix the ratio between x and y by the equation

$$5y = 12x,$$

then y is a quantity lying between zero and 12; and conversely, whenever y lies between zero and 12, x lies between zero and 5. The above equation also entails that to every value of x there belongs only a single value of y, and conversely. From these two premisses it follows that to every quantity x in the set between zero and 5 there corresponds a quantity y in the set between zero and 12 such that, on the one hand, no constituent of either set remains uncoupled, and on the other hand, none appears in two or more of the couples.

EXAMPLE II: we intend to take a spatial object for our second example. Those, to be sure, who already know that the qualities of space rest on those of time, and those of time on those of abstract numbers and quantities, need no example to show them that infinite sets like the above one in the realm of quantity must also exist in time and space. But for the sake of rightly conducting the application of our theorem in the sequel it is necessary all the same to study in detail at least one case in which sets of that kind exist. Let then a, b, c be three arbitrary points in a straight line, the ratio $ab : ac$ being arbitrary save for the condition that ac shall be the greater of the two distances.

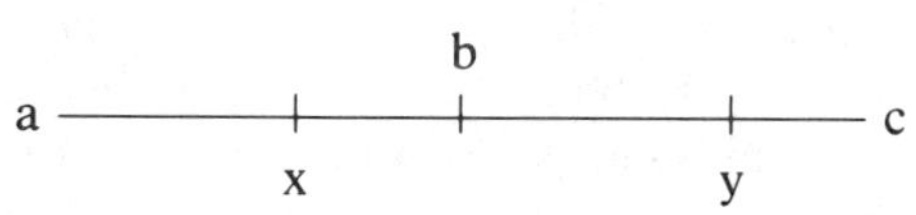

This being so, the set of points in ab and the set in ac will both be infinite, and, this notwithstanding, the set of points in ac will surpass the set in ab, because the former contains all the latter together with all the points found in bc but not in ab. Nor can we escape the admission that arbitrary changes in the ratio $ab : ac$ would produce great changes in the relations between the two sets. Yet with respect to the couples that can be formed with one term from each, the same can be said of this new pair of sets as was said of the previous pair of sets, in which the one consisted of all the quantities between zero and 5, and the other of all between zero and 12. For let x be any point in ab; and let y be produced in the direction ax by means of the proportion

$$ab : ac = ax : ay.$$

Then y will be a point in ac. And conversely, if y is a point in ac, and if x is specified from it by the same proportion, then x will be a point in ab. Moreover, different x will yield different y, and different y diffferent x. From these two premisses it follows similarly that to every point in ab there corresponds one in ac, and to every point in ac one in ab, again in such a manner that we can say of the couples formed out of points thus corresponding, both that no point in either set fails to occur in some one couple, and that no point in either set makes two or more appearances.

§21

The mere fact, therefore, that two sets A and B are so related that every member a of A corresponds by a fixed rule to some member b of B in such a way that the set of these couples $(a + b)$ contains every member of A or B once and only once, never justifies us, we now see, in inferring the *equality of the two sets, in the event of their being infinite*, with respect to the multiplicity of their members—that is, when we abstract from all individual differences. On the contrary, and in spite of their entering symmetrically into the above relation with one another, the two sets can still stand in a relation of inequality, in the sense that the one is found to be a whole and the other a part of that whole. This identity of multiplicities ⟦Gleichheit der Vielheiten⟧ cannot be inferred until some further grounds are supplied, for example, an identical mode of specification ⟦Bestimmungsgrund⟧ or of generation ⟦Entstehungsweise⟧ for the two sets.

§22

As I am far from denying, an air of paradox clings to these assertions; but its sole origin is to be sought in the circumstance that the above and oft-mentioned relation between two sets, as specified in terms of couples, really does suffice, in the case of *finite* sets, to establish their perfect equi*multiplicity* in members. Whenever, in fact, two finite sets are constituted so that every object a in the one corresponds to another object b in the other which can be paired off with it, no object in either set being without a partner in the other, and no object occurring in more than one pair: then indeed the two finite sets are always equal in respect of multiplicity. The illusion is therefore created that this ought to hold when the sets are no longer finite, but infinite instead.

The illusion, I say—for a closer study reveals the fact that no such neccesity exists, because the grounds upon which this holds for finite sets are bound up precisely with the finitude, and become inoperative for infinite sets. Suppose namely that both sets, A and B, are finite—or suppose something even less, yet still sufficient, namely that we have ascertained the finitude only of the one set A. Moreover, let us abstract from all individual differences, so as to compare the sets exclusively with respect to their multiplicities.

To begin, let us mark with 1 any arbitrary object in the set A, with 2 any arbitrary other object, and generally, every following object with the number of objects so far considered, the new one included. Sooner or later we must come to a member of A, after the marking of which we find in A no more objects to mark. This much follows immediately from the idea of a finite or countable multitude. If the above-mentioned last member of A was marked n, the enumeral ⟦Anzahl⟧ of the members of A will be n.

Turning now to B, let us give to each object in B the same mark as to its partner in A, according to our hypothesis as to the couples. It must then turn out that the number of objects from B which we have used up in this manner is likewise n: since each of them received a mark showing how many had been used up so far. This makes it clear that the number of objects in B is certainly not less than n; for one of them, namely the last to be used up, actually does carry this number as its mark. On the other hand, that number is also not greater than n; for if there existed even a single one over and above those now used up, this single one would lack a partner in A, contrary to the hypothesis. Consequently, the number of objects in B is neither less than nor greater than n, therefore equal to n. Both sets therefore have one and the same multiplicity, or as we may also express it, an *equal* multiplicity. This argument plainly fails so soon as the set of objects in A is an *infinite* one; for in this case not only do we who *do the counting* never arrive at a last term in A, but such a *last* term is prevented from existing at all by the very force of the definition of an infinite set. In other words, however many objects have hitherto been marked, there are still others yet unmarked. For this reason, the mere fact that B never runs short of objects to couple with those of A offers no justification for concluding that the multiplicities of the two sets are identical.

§23

What we have just said shows indeed that our oft-mentioned relation between two sets entails their equality when they are finite, but *not when they are infinite*; but it omits to explain how and why two sets that are infinite can still fail to be equal. For an explanation of this we must go back to the adduced examples. They show us that the members a and b which make up a couple *do not play exactly the same part in their respective sets*. For if a' and b' form a second couple, and if we compare the relationships between a and a' as members of A with those of b and b' as members of B, we discover quite soon that they are different. Take the first example, and choose any two members of the set of quantities between 0 and 5, say 3 and 4. Their partners in B are then clearly

$$\frac{12}{5} \times 3 \text{ and } \frac{12}{5} \times 4, \text{ that is, } 7\tfrac{1}{5} \text{ and } 9\tfrac{3}{5}.$$

If we mean by the *relationship* of two things, as we ought to mean, the aggregate of *all* the qualities manifest in their conjunction, then among all the relationships between 3 and 4 as members of the one set, and between $7\frac{1}{5}$ and $9\frac{3}{5}$ as members of the other, we do wrong to confine our attention exclusively to what is called *geometrical ratio*. We should pay heed to everything that belongs to them, in particular to the *arithmetical differences*. That between 3 and 4, which is 1, differs entirely from that between $7\frac{1}{5}$ and $9\frac{3}{5}$, which is $2\frac{2}{5}$. Thus, although every quantity in *A* or *B* allows of coupling with one and only one in *B* or *A*, yet the set of quantities in *B* is other and greater than in *A*, since the *distance* between two quantities in *B* is other and greater than the *distance* between the corresponding quantities in *A*. One natural consequence is, that two quantities in *B* include between them a set of quantities other and greater than the set of quantities included between their partners in *A*. No wonder, therefore, that the *whole* set of quantities in *B* is other and greater than the set of quantities in *A*.

In the second example the situation is altogether similar. We intend, therefore, to say no more about it than this: that the points in *ab* which are partnered with points in *ac* lie uniformly *closer* to one another than do their respective partners in *ac*, because the distances between the former are to the distances between the latter in the ratio *ab* : *ac*.

§24

Now that the theorem in §20 can be considered as sufficiently proved and explained, one of the next deductions from it is this: *When two quantities are sums of infinite sets of addends equal in corresponding pairs, we are not entitled for that sole reason to equate them*; but only if we can first convince ourselves that the infinite multitude of those addends is identical in the two sums. It is not disputed, to be sure, that addends determine their sum, and that equal addends yield equal sums. This holds not only for finite but also for infinite sets of summands. In the case of the latter, however, it is necessary to make sure that the infinite set of summands in the one sum really is identical with the infinite set of summands in the other sum; seeing, namely, that there are different kinds of infinite set. And to make sure of that point, we see from our theorem how altogether insufficient it is to be able to pair off the terms in the one sum with those in the other. The conclusion will be unsafe unless *the two sets have identical terms of specification* ⟦*gleiche Bestimmungsgründe*⟧. The sequel will bring many examples of the absurdities in which a calculation with the infinite involves us if we fail to pay attention to this point.

§25

I now proceed to the assertion that there exists an infinite *even in the realm of the actual*, and not merely among the things which make no claim to actuality. Anyone who had arrived at the momentous conviction (whether by a chain of reasoning from purely conceptual truths or otherwise) *that there exists a God*, a Being whose existence is grounded in that of no other being, and precisely for this reason is a *universally perfect Being*, uniting in himself all powers and perfections which are compatible with one another at all, and each of them in the highest degree of which it is capable—such a person, I say, agrees by this very fact upon the existence of a Being possessed of infinitude in more than one respect; with respect to his *knowledge*, in that he *knows infinitely much*, to wit, the sum of all truths; to his *volition*, in that he *wills infinitely much*, to wit, the sum of every single possible good; and to his might, or *action ad extra*, in that he *confers actuality*, in virtue of his power of action *ad extra*, to *everything that he wills*. From this last attribute of God follows the existence of beings other than God, *creatures*, which we contrast with him and call merely *finite beings*, but in which for all that many a trace of infinitude can be found. For the *set* of such beings must already be an infinite one, as also the set of all the *conditions* experienced by any single one of them during no matter how short an interval of time—because every such interval contains infinitely many instants. We therefore encounter infinites even in the realm of the actual.

§26

Even among well-instructed people who would not think of rejecting an infinite among things possessed of no actuality (such as absolute propositions and truths) there are several who refuse assent to my last thesis. In their eyes, an infinite in the realm of the actual is disallowed by the immemorial *principle of the universal determinateness of the actual*. As against this I believe that I have already shown in my *Wissenschaftslehre* (Volume I, §45) that this principle is true of the non-actual in just the same sense in which it is true of the actual. It holds in all cases solely in the sense that, given any one object whatever, and any two *contradictory* qualities, one of the qualities must be affirmable of that object, and the other quality deniable of it. If it really conflicted with this principle to assume an infinite in the realm of the actual, then it would not be lawful to speak of an infinite among the unactual objects of our reflection, and not lawful to admit an infinite set of absolute truths or of abstract numbers. But by pronouncing an object to be infinite, we are still very far from violating the principle in question. We only say, after all, that the object exhibits in some respect or other a multiplicity of parts or members which is greater than any particular number whatever—admittedly, therefore, a multiplicity which *does not allow of determination by a mere number* ‖*durch eine blosse Zahl*‖. It does not follow from this at all that the multiplicity *cannot be determined in any manner whatever*. Nor does it at all follow that a single pair of opposite qualities *b* and non-*b* exist, both of which would have to be denied of it. A thing insusceptible of colour obviously cannot be specified by its colour, nor a thing incapable of emitting sound be specified by its note, and so forth; yet such things are by no means indeterminable, and constitute no exception to the rule that of the two predicates *b* and non-*b* (blue and non-blue, euphonious and non-euphonious) one must be attributable to each chosen object ‖and the other unattributable‖—provided only that we interpret them in the manner neccessary for them to remain contradictory. Just as 'non-blue' and 'non-fragrant' are predicable, though to be sure very remotely, of the Theorem of Pythagoras, exactly so is the bare statement that 'the set of points between *m* and *n* is infinite' one of the things predicable of this set. Frequently enough, a very few statements are sufficient to determine such an infinite set of objects *completely*, i.e. in such a manner that *all* its attributes are deducible from the few actually named. Thus, we have completely determined the last-mentioned set of points between *m* and *n* the moment we determine the two points *m* and *n* themselves, say by an intuition ‖Anschauung‖ referring to them. For these few words suffice to decide with certainty whether any other proposed point does or does not belong to this set.

§27

Hitherto, I have had reason to defend the acceptance of an infinite against unjustified opposition; but now I must acknowledge with equal candour that many of the learned, especially in the ranks of the mathematicians, have gone too far in the other direction, and accepted sometimes an infinitely great, sometimes an infinitely small, in cases where it is my firmest persuasion that none exists.

1. If an *infinite interval of time* were understood as one which had no beginning, or had no termination, or had neither beginning nor termination, and thus meant the whole of time or the aggregate of all instants, then I should have no objection to raise. At the same time, I hold it necessary to think of the *quantitative ratio* between two intervals of time, each of which lies between terminal instants, as a merely finite quantitative ratio, and determinable completely in terms of pure concepts. I would never allow the hypothesis that one duration, enclosed between terminal instants, be infinitely greater or less than another such duration. Yet too many mathematicians, as is well known, sometimes do this by speaking of infinitely great intervals of time which are nevertheless terminated at both ends, but more often by speaking of *infinitely small intervals of time* in comparison with which every finite interval of time, such as a second, is supposed to have to be avowed infinitely great.

2. The same thing happens with the *distances between pairs of points in space*, which in my opinion always bear a merely finite ratio to one another, a ratio determinable completely in terms of pure concepts; whereas nothing is more common among our mathematicians than to speak of *infinitely great* and *infinitely small distances*.

3. And finally, the same thing happens again with the *forces* which both metaphysics and physics assume to act in the universe. We ought to suppose of them too, that none is ever infinitely greater or smaller than another, and that all of them stand in ratios determinable completely in terms of pure concepts; though, to be sure, we quite often permit ourselves to act otherwise. I shall not be able, of course, to explain the grounds of all these assertions on the present occasion to anyone who is still wholly unaware of what meaning

I give to the words *Intuition* ‖*Anschauung*‖ and *Concept* ‖*Begriff*‖, the *Derivability* of one proposition from another, the *Objective derivation* ‖*objektive Abfolge*‖ of one truth from others, and of what *definition* I lay down for Time and Space. However, the following demonstration will not be altogether incomprehensible to those who have read at least the two treatises *An Essay towards an Objective Foundation of the Doctrine of the Composition of Forces** and *An Essay towards an Objective Foundation of the Doctrine of the Three Dimensions of Space.*†

The definitions of space and time lead immediately to the conclusion that all *dependent* (that is, created) substances continually act upon one another; as also to the conclusion that, given any two instants α and β, however near to one another or however far apart, it is permissible to regard the condition of the universe at α as a *cause*, α being the earlier instant; and the condition of the universe at the later instant β as an *effect*, if only as a mediate effect—provided always that any immediate interventions of God taking place in the interval $\alpha\beta$ are reckoned with the causes. A further deduction is, that given the two instants α and β, given all the *forces* with which created substances are endowed at the moment α, given the *position* each occupies in space, and given finally what divine interventions take place during $\alpha\beta$, we can infer the *force* with which these substances are endowed at the moment β, together with the *positions* they then occupy, in just the same way as an *effect*, whether mediate or immediate, must be deducible from its complete cause. This necessitates in turn that all the qualities of the effect should permit of derivation from those of the cause by means of a major premiss whose terms are pure concepts, and which takes the form: 'Every cause with the qualities u, u', u'', ... has an effect with the qualities w, w', w'', ...'; and a simple corollary of this, which we need for our purpose, is that 'Every particularity ‖Umstand‖ in the cause, under whose variation the effect is not invariant, must *allow of complete determination* by means of pure concepts taken exclusively from among those required for the determination of the effect.'

After these preliminary remarks, our previous assertions are easily established:

1. Did but two instants α and β exist, whose distance were infinitely greater or smaller than the distance of two others γ and δ, we should be faced with the absurd result that the condition of the universe at β would be utterly impossible to determine from its condition at α together with a knowledge of the divine interventions and the duration $\alpha\beta$. The laying down of a standard unit of time is necessary, in fact, for the determination of the condition in which created beings are placed at a given moment, and even for the sole determination of the *magnitude of their forces*; for since these forces are nothing but *forces for the effecting of change* ‖*Veränderungskräfte*‖ their

* Prague 1842, by commission with Kronberger and Rziwnas ‖*Bolzano 1842*‖.
† Prague 1843, by commission with Kronberger and Rziwnas ‖*Bolzano 1843*‖.

magnitude can only be judged by observing the changes they bring about in a given stretch of time. Now take, as we must be allowed to take, the time interval $\gamma\delta$ as the unit. Then even in the most favourable case of all—that in which all the forces with which created substances are endowed at the instant α can be determined in terms of this unit, and in which everything else belonging to the complete cause of the condition of the universe at the instant β can be exactly determined as well—yet the distance between this instant itself and the instant α could not be expressed in terms of the chosen unit, inasmuch as it would turn out to be simply infinitely great or infinitely small. *Conversely*, if it is to be possible to consider every particular condition of the universe as the cause of every particular later condition, under the conditions now repeatedly mentioned, then we cannot have two instants α and β whose distance would turn out to be infinitely great or small in comparison with the distance of two other instants γ and δ.

2. Did but two points *a* and *b* exist in space, whose distance were infinitely great or small in comparison with that of two other points *c* and *d*, then the determination of the condition of the universe at a specified instant α would require among other things the magnitude of the force of attraction or repulsion to be known, which the substance placed at *a* then exercises on the substance placed at *b*. But take, as is certainly permissible, the distance *cd* as the unit of length. Then even in the most favourable case of all—that in which we were succcssful with all other forces—it would still be impossible to succced with this particular force. For even if, indeed precisely because, the force of attraction exercised over the chosen unit of length *cd* by the substance *A* on the substance *B* (or on one otherwise completely similar to it) is a completely determinate quantity, the magnitude of that force would become indeterminate if the ratio *ab* : *cd*, on which it certainly also depends, were infinite and consequently itself indeterminate.

3. Finally: did but a single force *k* present itself as infinitely great or small in comparison with another force *l*, and this at an instant we shall label α, then even in the most favourable case—that in which all other forces proved finite when measured by their respective units of time and space, and in which, therefore, *l* too would be finite—the quantity *k* would still turn out to be infinitely great or small, that is, indeterminate. On this ground, however, the total situation of the universe at the instant α would come out indeterminate, and we should be faced with the impossibility of tracing any subsequent situation of the universe as an effect produced by the situation first mentioned.

§28

I am persuaded that the foregoing paragraphs lay down the fundamental rules for judging all the strange-sounding doctrines which have to be advanced in the sequel, the rules by which it must be decided whether they are errors due for abandonment, or theorems due for retention because they are true in spite of their appearance of preposterousness. The sequence in which we set forth these paradoxes shall be settled by what sphere of knowledge they belong to, and by their greater or lesser importance.

The first and most comprehensive science in which we encounter paradoxes of the infinite is the *general theory of quantity*, as several examples have already taught us, and such paradoxes are not wanting even in the *doctrine of number* itself. We intend in consequence to begin with these.

I confess that the mere *idea* of a *calculation with the infinite* has the appearance of contradicting itself. For to try to *calculate* anything means, after all, to attempt a *determination of it* in terms of number. But how can we possibly hope to determine the infinite by means of number—that very infinite which by our own definition must admit of being conceived as a set with infinitely many members, a set greater than any particular number and thus incapable of determination by a merely numerical datum? But this scruple vanishes when we reflect that a correctly conducted calculation with the infinite is not a numerical determination of what is therein not numerically determinable (namely, not a numerical determination of the infinite multitude as such) but only aims at determining the *ratio* ⟦des *Verhältnisses*⟧ between one infinite and another; a thing which can, in fact, be done in certain cases, as we hope to show by several examples.

§29

All who admit the existence of multitudes that are infinite, and hence also of infinite quantities, must go on to admit that infinite quantities exist which exhibit manifold differences with respect to their actual magnitude or bigness. If the sequence of the natural numbers is set down in the form

$$1, 2, 3, 4, \ldots, n, n+1, \ldots \textit{ in inf.},$$

then the symbol ⟦Zeichnung⟧

$$1 + 2 + 3 + 4 + \ldots + n + (n+1) + \ldots \textit{ in inf.}$$

will represent ⟦darbieten⟧ the *sum* of these natural numbers; and the following symbol

$$1^0 + 2^0 + 3^0 + 4^0 + \ldots + n^0 + (n+1)^0 + \ldots \textit{ in inf.},$$

whose addends are all simple unities, will represent the bare *set* of all the natural numbers. If we denote the latter by N_0, construct the merely symbolic equation

$$1^0 + 2^0 + 3^0 + 4^0 + \ldots + n^0 + (n+1)^0 + \ldots \textit{ in inf.} = N_0, \qquad (1)$$

and denote the set of natural numbers from $(n+1)$ onwards similarly by N_n, constructing the similar equation

$$(n+1)^0 + (n+2)^0 + (n+3)^0 + \ldots \textit{ in inf.} = N_n, \qquad (2)$$

subtraction yields us the quite irreproachable equation

$$1^0 + 2^0 + 3^0 + 4^0 + \ldots + n^0 = n = N_0 - N_n, \tag{3}$$

whence we learn how two infinite quantities N_0 and N_n can sometimes have an altogether definite finite difference.

If on the contrary we denote the quantity representing the *sum* of all the natural numbers by S_0, or write down the merely symbolic equation

$$1 + 2 + 3 + 4 + \ldots + n + (n+1) + \ldots \textit{ in inf.} = S_0, \tag{4}$$

we shall no doubt apprehend at once that S_0 must be far greater than N_0; but it will be less easy for us exactly to ascertain the difference between these two infinite quantities, or even their (geometrical) *ratio* to one another. For if we tried to do what many indeed have tried to do, and set down the equation

$$S_0 = \frac{N_0(N_0+1)}{2},$$

we could justify it on scarcely any other grounds than that the equation

$$1 + 2 + 3 + 4 + \ldots + n = \frac{n(n+1)}{2}$$

holds for every finite set of terms, whence it appears to follow that, when we pass to the whole infinite set of numbers, n simply becomes N_0. But that is not the case, for it is absurd to speak of an infinite series as having a last term with the value N_0.

The merely symbolic equation (4) ‖*sic*‖ having meanwhile been laid down, it will indeed be permissible to derive the following two equations from it by letting N_0 multiply each side:

$$1^0 N_0 + 2^0 N_0 + 3^0 N_0 + \ldots \textit{ in inf.} = N_0^2,$$

$$1^0 N_0^2 + 2^0 N_0^2 + 3^0 N_0^2 + \ldots \textit{ in inf.} = N_0^3,$$

and so forth, which convinces us that there exist infinite quantities of so-called *higher orders* as well, such that one exceeds the other infinitely many times. The existence, moreover, of infinite quantities bearing to one another any prescribed ratio, rational or irrational, say $\alpha : \beta$, already follows from the fact that if N_0 only denotes any constant infinite quantity, then αN_0 and βN_0 are a pair of quantities which are also infinite and bear to one another the prescribed ratio $\alpha : \beta$.

It will presumably be not less evident that the entire *set* (or multiplicity) of quantities lying between two given ones, say 7 and 8, depends only on the distance 8–7 between those terminals and is hence equal ‖gleich‖ and necessarily equal to any other for which that distance is equal—and this, despite the fact that it is an *infinite* set, and thus incapable of determination by any number however great. On this supposition, and denoting the set of all quantities lying between a and b by

$$\text{mult}\,(b - a),$$

there must be numberless equations of the form

$$\text{mult}(8\text{–}7) = \text{mult}(13\text{–}12),$$

as also of the form

$$\text{mult}(b-a) : \text{mult}(d-c) = (b-a) : (d-c),$$

against whose validity there is no sound objection.

§30

Now that the possibility of *calculating with the infinitely great* has been vindicated by these few examples, we assert the like for the *infinitely small.* For if N_0 is infinitely great,

$$\frac{1}{N_0}$$

necessarily represents an infinitely small quantity, and we shall have no reason for denying objective reference to such an idea, at any rate in the *general* theory of quantity. To give a single example: if the probability be asked that anyone should shoot off a bullet at random, and yet have its midpoint pass exactly through the midpoint of yonder apple on yonder tree, then all must agree that the set of all the possible cases, with their various degrees of probability, is infinite: whence it follows that the degree of probability in question has a value equal to or less than $1/\infty$. This alone suffices to show that we have infinitely many infinitely small quantities, with one standing to another in any prescribed ratio, and in particular, such that one is infinitely greater than another; as also that among the infinitely small quantities we have, just as among the infinitely great ones, an infinite number of different orders. Furthermore, if certain rules are kept, it will indeed be possible to find very many correct equations between quantities of this sort.

Suppose we have ascertained, for example, that the value of a variable quantity y depends upon another x in such a way that the equation

$$y = x^4 + ax^3 + bx^2 + cx + d$$

constantly holds between them, and that it is consistent with the nature of the particular class of quantities to which x and y belong that they should become infinitely small and thus be susceptible of infinitely small increments. Then, if we let x increase by the infinitely small amount denoted by dx, and denote by dy the change which y then undergoes, the following equation necessarily holds:

$$y + dy = (x + dx)^4 + a(x + dx)^3 + b(x + dx)^2 + c(x + dx) + d,$$

which incontestably entails the following:

$$\frac{dy}{dx} = (4x^3 + 3ax^2 + 2bx + c) + (6x^2 + 3ax + b)dx + (4x + a)dx^2 + dx^3;$$

and this in turn exhibits the two infinitely small quantities as depending not only on a, b, c, and x, but in addition on the value of the variable dx.

§31

But the majority of the mathematicians who ventured on a calculation with the infinite went much farther than is permissible on the principles here laid down. To assume an infinitely great and an infinitely small among quantities whose nature does not allow of them—examples of which will be given later—was not the only liberty which they took without hesitation. They also made bold to declare equal to one another, or greater or less than one another, quantities arising from the summation of an infinite series, on the sole ground that the terms of the series were equal in pairs or unequal in pairs, and despite their obvious inequality when considered as sets. They dared to make the statement not only that every infinitely small quantity *vanishes like an ordinary zero* when added to a finite one, and not only that every one of a *higher* order does so in conjunction with one of *lower* order, but even that every infinitely great quantity of lower order does so in conjunction with another of *higher* order. And in order to find some justification for this method of calculation they had recourse to the statement that it is permissible to divide by zero, and that the quotient

$$1/0$$

is in reality nothing else than an *infinitely great quantity*, and the quotient

$$0/0$$

a quantity *completely indeterminate*. We must show how false and misleading these ideas are, because they are more or less in vogue even today.

§32

As recently as 1830, a writer signing himself M.R.S. tried to prove in *Gergonne's Annales de Mathématiques* (Volume 20, Number 12) that the well-known infinite series

$$a - a + a - a + a - a + \dots \textit{ in inf.}$$

has the value $a/2$. Setting its value equal to x, he believed himself entitled to deduce that

$$x = a - a + a - a + \dots \textit{ in inf.} = a - (a - a + a - a + \dots \textit{ in inf.}),$$

and that the bracketed series, being identical with that whose value is being sought, could again be set equal to x, yielding

$$x = a - x,$$

and consequently

$$x = a/2.$$

The fallacy does not lie deep. The bracketed series no longer has the same identical set of terms as the one originally put equal to x, because it lacks the initial a. If it had a value at all, that value would have to be denoted by $x - a$, and this would give us the mere identity

$$x = a + x - a.$$

Some one may perhaps urge that 'there is something paradoxical in the thought that this series, which is assuredly not infinitely great, should have no exactly determinable and measurable value at all—and this all the more, in that it is generated by the infinitely continued division of a by 2 in the form $1 + 1$: a fact which speaks for the correctness of the assumption that its true value is after all none other than $a/2$.'

In answer to this I recall first that there is nothing intrinsically impossible in our having *quantitative expressions* ‖*Grössenausdrücke*‖ which denote *no actual quantity* ‖*keine wirkliche Grösse* bezeichnen‖, and of these, zero is and must be a recognized example.

In particular: once we declare that we intend to consider a *series* only as a quantity, viz. only as the *sum* of its terms, then in virtue of the *definition* of a sum (which classes it with sets, and therefore with those aggregates from the order of whose members we abstract) it must be such as to undergo no variation in its value however we vary the sequence of its terms. Among quantities it is necessary to have

$$(A + B) + C = A + (B + C) = (A + C) + B.$$

This characteristic, however, is a clear proof to us that the symbol in question, to wit

$$a - a + a - a + a - a + \ldots \textit{ in inf.}$$

is not the expression of an actual quantity. For if a quantity were really represented by it at all, that quantity would suffer no alteration when we changed the symbol to:

$$(a - a) + (a - a) + (a - a) + \ldots \textit{ in inf.}, \tag{1}$$

because we should be doing nothing more than combine each consecutive pair of terms into a partial sum, a thing which must be possible, because the given series really has no *last* term. By this process, however, we obtain

$$0 + 0 + 0 + \ldots \textit{ in inf.},$$

which can obviously be equal only to 0.

Again, if a quantity were really represented by our expression, it would suffer equally little alteration on being transformed as follows:

$$a + (-a + a) + (-a + a) + (-a + a) + \ldots \textit{ in inf.}, \tag{2}$$

where, setting aside the first term, we combine into a partial sum each consecutive pair of following terms; or again on being transformed as follows:

$$-a + (a - a) + (a - a) + (a - a) + \dots \textit{ in inf.}, \qquad (3)$$

which is what becomes of (1) when we transpose its terms in successive pairs and then effect the alteration which led from (1) to (2). Now if the expression under investigation were not *devoid of objective reference*, the symbols (1), (2),and (3) would have to denote one and the same quantity; because it is evident that the representation ⟦Vorstellung⟧ of the sum of one and the same set of quantities cannot represent several quantities *different from one another*, as does happen for example with the representations

$$\sqrt{(+1)},\ \text{arc sin}\ (\tfrac{1}{2}),$$

and many others. Even the quantitative representation

$$1 - 1 + 1 - 1 + 1 - 1 + \dots \textit{ in inf.},$$

unless wholly devoid of objective reference, would have an equal right to be put equal to zero (which is usually called a quantity, though in an improper sense) and to $+a$ and to $-a$, which is thoroughly absurd, and entitles us to conclude that we actually do have before us a representation devoid of objective reference.

It is true, again, that the series under review does appear as the quotient in a continued division of a by 2 in the form $1 + 1$; but none of the series generated in this fashion can yield the true value of the quotient (in our case $a/2$) unless the remainders arising on further division become smaller than any chosen small quantity: the easily understood reason being that the division continually leaves a remainder (in our case, alternately $-a$ and $+a$). The condition just mentioned was fulfilled by the series studied in §18, which is generated on division of a by $1 - e$, provided $e < 1$. But when $e = 1$, as in the case now before us, and still more when $e > 1$, which causes the remainders to increase at every step in the division process, nothing is easier to understand than that the value of the series cannot be equated to the quotient

$$\frac{a}{1-e}.$$

How, indeed, could we possibly put the series,

$$1 - 10 + 100 - 1\,000 + 10\,000 - 100\,000 + \dots \textit{ in inf.},$$

whose terms alternate in sign, and which is generated on division of 1 by $1 + 10$, equal to $1/11$? And who, to crown it all, would care to estimate at $-1/9$ the value of the series

$$1 + 10 + 100 + 1\,000 + 10\,000 + \dots \textit{ in inf.},$$

all of whose terms are positive, on the sole ground that the development of the fraction

$$\frac{1}{1-10}$$

leads to that series? Nevertheless, the above-mentioned M.R.S. undertakes to defend such summations, and regards the equation

$$1 - 2 + 4 - 8 + 16 - 32 + 64 - 128 + \ldots \textit{ in inf.} = \tfrac{1}{3}$$

as justified on the sole ground that, as he alleges,

$$\begin{aligned} x &= 1 - 2 + 4 - 8 + 16 - 32 + 64 - \ldots \\ &= 1 - 2(1 - 2 + 4 - 8 + 16 - 32 + \ldots) = 1 - 2x; \end{aligned}$$

whereby he fails to notice that the bracketed series is by no means identical with the one first taken, because it no longer contains the identical set of terms. The lack of an objective reference for this quantitative expression is manifested in the same way as for the one previously considered, namely by its leading to contradictory results. On the one hand, in fact, we must have

$$\begin{aligned} &1 - 2 + 4 - 8 + 16 - 32 + 64 - \ldots \\ &= 1 + (-2 + 4) + (-8 + 16) + (-32 + 64) + \ldots \\ &= 1 + 2 + 8 + 32 + 64 + \ldots, \end{aligned}$$

and on the other, with equal certainty,

$$\begin{aligned} &= (1 - 2) + (4 - 8) + (16 - 32) + (64 - 128) + \ldots \\ &= -1 - 4 - 16 - 64 - \ldots, \end{aligned}$$

so that by two legitimate processes two values are found for the same expression, one of them infinitely great and positive, the other infinitely great and negative.

§33

In order not to go astray in our calculations with the infinite, we must therefore never allow ourselves, upon two infinite quantities' arising from the summation of two infinite series, to put the first quantity equal to (or greater than, or less than) the other quantity, solely because each term in the one series is equal to (or greater than, or less than) some one corresponding term in the other series. Nor ought we even to declare the first sum greater, solely because it contains all the addends of the second sum together with others beside—no, not even if these excess terms are (all positive and) infinite in number. For despite all this it still remains possible that the first sum is smaller, or indeed infinitely smaller, than the second sum. One example is to hand in the well-known sum of the *squares* of all the natural numbers, compared with the sum of their *first powers*. No one, assuredly, can contest that every term in the series of *squares*

$$\left.\begin{array}{l} 1^2 + 2^2 + 3^2 + 4^2 + 5^2 + 6^2 + 7^2 + 8^2 + 9^2 + 10^2 + \ldots \textit{ in inf.} = \\ 1 + 4 + 9 + 16 + 25 + 36 + 49 + 64 + 81 + 100 + \ldots \textit{ in inf.} = \end{array}\right\} S_2,$$

being itself one of the natural numbers, also occurs in the series of first powers of the natural numbers

$$1 + 2 + 3 + 4 + 5 + 6 + 7 + 8 + 9 + 10 + 11 + 12 + 13 + 14 + 15 + 16 + \ldots \textit{ in inf.} = S_1;$$

or that the latter series S_1 contains all the terms of S_2 and many, indeed infinitely many, more which are wanting in S_2 because they are not square numbers. Nevertheless, the sum S_2 of the square numbers is not equal to S_1, the sum of the first powers, but incontestably greater. For in the first place the *set of terms* in the two series (not yet considered as sums at all, and hence not partitionable into arbitrary subsets of members) is the same, in spite of all appearances to the contrary. The mere raising to the second power of each separate term in S_1 certainly alters the quality and amount of these terms, but it does not alter their multitude. But if the set of terms in S_1 and S_2 is the same, then it is evident that S_2 must be much greater than S_1 in that, apart from the *first* term in each series, each subsequent term in S_2 is decidedly greater than the homologous term in S_1; in such a manner that, considered as a quantity, S_2 contains the whole of S_1 as a mere part of itself; indeed, over and beyond this mere part, it contains as a second part yet another infinite series, equinumerous in terms with S_1, to wit:

$$0, 2, 6, 12, 20, 30, 42, 56, \ldots, n(n-1), \ldots \textit{ in inf.},$$

in which, apart from the first *two* terms, all the following terms exceed their homologues in S_1, so that the sum of this whole new series is again incontestably greater than S_1. If then we subtract the series S_1 for the second time from the above remainder, we obtain as our *second* remainder a series equinumerous in terms:

$$-1, 0, 3, 8, 15, 24, 35, 48, \ldots, n(n-2), \ldots \textit{ in inf.},$$

in which, apart from the first *three* terms, once again all the following terms exceed their homologues in S_1; so that this third ⟦*sic*⟧ remainder can without fear of contradiction be deemed greater than S_1. Now since this argument can be continued without end, it is clear that the sum S_2 is infinitely many times greater than the sum S_1, for in the general case:

$$S_2 - mS_1 = (1^2 - m) + (2^2 - 2m) + (3^2 - 3m) + (4^2 - 4m) + \ldots + (m^2 - m^2) + \ldots + n(n-m) + \ldots \textit{in inf.},$$

a series in which only a finite set of terms (the first $m - 1$) are negative, the mth is zero, and all the following ones are positive and increase to infinity.

§34

We cannot cast an appropriate light on the incorrectness of the other assertions mentioned in §31 until we define the idea of *zero* somewhat more precisely than is customary.*

It is beyond dispute that all mathematicians desire to attach to the symbol 0 an idea of such a character that, when A is any quantitative expression whatever (irrespective of whether it corresponds to an actual quantity or lacks objective reference) the two equations

$$\text{I.} \quad A - A = 0, \qquad \text{II.} \quad A \pm 0 = A$$

can always be legitimately written down. Now everybody will admit at this point that this can only be done if we look upon the symbol 0 itself not as the representation of an actual quantity, but as the absence, the bare absence of any quantity; and upon the symbol $A \pm 0$ as a command, neither to add to nor to subtract from the quantity denoted by A. On the other hand, it would be an error to believe that the simple declaration 'zero is a quantitative representation devoid of objective reference' sufficiently determines the idea which mathematicians attach to this symbol. For there exist other quantitative notations of common usage in mathematics, such as the now important symbol $\sqrt{-1}$ in analysis, which are equally devoid of objective reference, and which for all that are by no means to be regarded and treated as equivalent to 0. But if we define the meaning of the symbol 0 more precisely by requiring that the two equations I and II shall universally hold, then we set up a concept which is, on the one hand, quite as wide as past usage and the interests of science demand, and on the other hand, restricted enough to prevent its abuse.

Upon closer scrutiny, the postulate that equations I and II be universally valid does more than just demarcate the idea of zero: it also confers upon the idea of *adding* and *subtracting* (which appear under the signs + and −) a peculiar kind of extension which redounds greatly to the advantage of science.

That same advantage of science requires in addition that we conceive of *multiplication* too in a sense so wide that, whatever A is (whether a finite quantity, an infinitely great quantity, an infinitely small quantity, or like $\sqrt{-1}$ merely a quantitative representation lacking objective reference, or finally zero itself) we may in all cases write down the equation

$$\text{III.} \quad 0 \times A = A \times 0 = 0.$$

Lastly, and again in the interests of science, we must demand so wide a conception of *division* as we possibly can without coming into conflict with one of the previous three equations, and hence allow the symbol B in the equation

* I pay willing tribute to the merits of M. Ohm for being the first to draw the attention of the mathematical public to the difficulties inherent in the idea of zero, in his valuable work: *Versuch eines vollkommen consequenten Systems der Mathematik* (2nd edition, Berlin, 1828).

$$\text{IV.}\quad B \times \left(\frac{A}{B}\right) = \left(\frac{A}{B}\right) \times B = A$$

as wide a range as is ever compatible with the universal validity of those three equations. Now though they permit B to denote any arbitrary finite or infinitely great or infinitely small quantity, and even the imaginary $\sqrt{-1}$, they simply do not permit B to be put equal to 0, in other words, they do not permit us ever to use zero, or any expression equivalent to zero, as a *divisor*. For equation III requires $0 \times A = 0$ for every A, and if we put $B = 0$ in equation IV, we should have to have

$$B\left(\frac{A}{B}\right) = 0,$$

and this would agree with the requirement in IV that

$$B\left(\frac{A}{B}\right) = A$$

only in the solitary case of $A = 0$. Lest we fall into contradiction, therefore, we must lay down the rule *that zero, or an expression equivalent to zero, may never be used as a divisor in an equation which purports to be more than a mere identity*, such as

$$\frac{A}{0} = \frac{A}{0}.$$

The need for observing this rule is shown not only by what has just been said, but also by a large number of very absurd consequences which ensue from entirely correct premisses as soon as we permit ourselves to divide by zero.

Let a be any real quantity whatever. Then the well-known and certainly quite correct method of division yields, so soon as we are allowed to divide by the expression $1 - 1$, which is equivalent to zero, the following equation:

$$\frac{a}{1-1} = a + a + \ldots + a + \frac{a}{1-1},$$

where the number of summands a may be any we desire. If we now subtract the same quantitative expression $\frac{a}{1-1}$ from both sides, then we obtain the highly absurd equation

$$a + a + \ldots + a = 0.$$

Again, if a and b are a pair of different quantities, the pair of identities

$$a - b = a - b,$$

$$b - a = b - a$$

will hold, and by their addition we obtain in turn

$$a - a = b - b,$$

$$a(1 - 1) = b(1 - 1).$$

Were it permissible, however, to divide both sides of an equation by a factor equivalent to zero, we should obtain the absurd result that $a = b$, irrespective of the values of a and b. And yet it is a matter of common knowledge how easy it is in longer calculations to hit upon a false result, if we remove a common factor from both sides of an equation without first ascertaining that it does not vanish.

§35

It is now easy to prove the incorrectness of the assertion made by so many people that, in the course of addition or subtraction, both an infinitely small quantity of higher order when conjoined with another of lower order or with a finite quantity, and an infinitely large quantity of any order when conjoined with another of higher order, and a finite quantity when conjoined with an infinitely large one, *vanish after the manner of a mere zero*. If this is to be understood in the sense that, when M is infinitely greater than m, we can simply leave m out of the compound expression $M + m$, even when M itself vanishes in the course of the calculation (say by subtraction of another equal to it) then I have no need to begin demonstrating the erroneousness of this rule—and in the common expositions, moreover, which are in some degree even more careless than the phraseology I have used above, no effort is made to prevent such a misunderstanding.

The rejoinder will be made that such is not the true meaning of what was said. It will be urged that when the quantities M and $M \pm m$ are declared equal it is not suggested that they give the same result when entering into further additions and subtractions, but only that they give the same result on going through a process of measurement in terms of a quantity N of equal rank with them, that is, standing in a finite and completely determinate ratio to one of them, say to M. So much at the very least are we justified in expecting to find in the definition of the phrase that two quantities are *equal in magnitude*. But do M and $M \pm m$ fulfil even this requirement? ‖Let us see.‖ If one of them, say M, stands in an irrational ratio to the measure N, it can certainly turn out to be the case that (as in the commonest type of measurement) given any ‖whole‖ number q, however great, another ‖whole‖ number p can be found to render

$$\frac{p}{q} < \frac{M}{N} < \frac{p+1}{q},$$

and it can so happen that $\frac{M \pm m}{N}$ constantly remains within the same bounds, in other words, that we also have

$$\frac{p}{q} < \frac{M \pm m}{N} < \frac{p+1}{q}.$$

But in the alternative case where the ratio $M:N$ is rational, there exists a number q for which, alongside

$$\frac{M}{N} = \frac{p}{q},$$

we have either

$$\frac{M \pm m}{N} < \frac{p}{q} \text{ or else } \frac{M \pm m}{N} > \frac{p}{q};$$

so that in this alternative case the difference between the two quantities betrays itself even through the medium of abstract *numbers* (finite quantities). What right, then, have we to call them equal?

§36

Following in the footsteps of *Euler*, a number of mathematicians sought to avoid such contradictions by taking refuge in the declaration that infinitely small quantities are in reality *mere zeros*, and infinitely great quantities are the quotients arising from finite quantities upon division by a mere zero. This declaration, it is true, more than justified the vanishing or the rejection of an infinitely small quantity as the addend to a finite augend; but it only increased the difficulty of making intelligible the existence of infinitely great quantities, the emergence of a finite quantity from the division of two infinitely small or of two infinitely great quantities, and the occurrence of infinitely great and small quantities of higher orders. For the infinitely great quantities make their appearance, on this view, as a result of division by zero or by a quantitative expression equivalent to zero (something devoid, strictly speaking, of objective reference) and hence in a manner forbidden by the laws of calculation; while all the finite or infinite quantities which were supposed to result from the division of one infinite quantity by another bore the stain of multiply illegitimate birth.

The most plausible plea for this calculation with zeros seems to be the method of computing a quantity y which depends on a variable x by means of the equation

$$y = \frac{F(x)}{\Phi(x)}$$

in cases where a particular value $x = a$ reduces either the denominator alone, or both denominator and numerator, to zero. In the former alternative, when $\Phi(x) = 0$ but $F(x)$ continues to be a finite quantity, people infer that y has become *infinitely great*; and in the latter alternative, when both $\Phi(x) = 0$ and $F(x) = 0$, that the two expressions $\Phi(x)$ and $F(x)$ both contain the factor $(x - a)$ once or oftener, and hence have the form

$$\Phi(x) = (x-a)^m\phi(x), \quad F(x) = (x-a)^n f(x),$$

where $\phi(x)$ or $f(x)$ may possibly denote a constant. If thereupon $m > n$, people infer that, even after cancellation of the factor common to numerator and denominator—which leaves the value of the fraction $F(x)/\Phi(x)$ unaltered—the denominator still becomes zero, and continue to argue that $x = a$ yields an infinitely great y. But if $m = n$, they regard the finite quantity expressed by $f(a)/\phi(a)$ as the true value of y, on the grounds that $F(x)/\Phi(x) = f(x)/\phi(x)$ must hold. Finally, if $m < n$, they argue from the fact that in this event

$$\frac{F(x)}{\Phi(x)} = \frac{(x-a)^{n-m}f(x)}{\phi(x)}$$

becomes zero for $x = a$, that the choice $x = a$ makes y vanish.

My verdict on this process is as follows. When the value of y for $x = a$ is declared in various foregoing cases to be infinitely great, this can only be true if the quantity y belongs to a kind *capable* of becoming infinitely great; and even then, only by accident; for the given expression, here requiring division by zero, is not the source from which the result springs. This is the inescapable fact. From the mere circumstance that the value of y is said always to be the one yielded by the expression $F(x)/\phi(x)$, we can deduce the properties of the quantity y only for all those values of x which represent an actual quantity, but not for those which render the expression *devoid of objective reference*, as happens if its numerator, or even just its denominator (still more if its numerator and its denominator together) should become zero. We can indeed say that in the first-mentioned case, where $\phi(x)$ alone vanishes, the quantity y becomes *greater* than any already given quantity; in the second case, where $F(x)$ alone vanishes, that y becomes *smaller* than any already given quantity; and lastly in the third case, where the fraction $F(x)/\phi(x)$ has equally many factors $(x - a)$ in numerator and denominator, that y comes arbitrarily near to $f(a)/\phi(a)$ as we push x arbitrarily near to the value a; but nothing follows from all this concerning the nature of this value at places where the expression $F(x)/\phi(x)$ *loses objective reference* (in other words, represents no value at all) because it then assumes either the form 0 itself, or the form $c/0$, or else even the form $0/0$. For the theorem that the value of a fraction is not altered by cancellation, though valid in all other cases, is invalid when the cancelled factor *vanishes*. Otherwise, once we had a right to say that

$$\frac{2\times 0}{3\times 0} = \frac{2}{3},$$

we should have an equal right to say of any arbitrary quantity, say of the number 1000, that

$$1000 = \frac{2}{3}.$$

For $3000 \times 0 = 0$ is doubtlessly just as true as $2 \times 0 = 0$. If we may write

$$\frac{2 \times 0}{3 \times 0} = \frac{2}{3},$$

then we may also write

$$\frac{2 \times (3000 \times 0)}{3 \times (2 \times 0)} = \frac{(2 \times 3000) \times 0}{(3 \times 2) \times 0} = \frac{2 \times 3000}{3 \times 2} = 1000.$$

The fallacy which here is so evident is less so in the former case, because we then divided by the zero-equivalent factor $(x - a)$ in a form which concealed the fact of its vanishing. And our confidence in being allowed to do so is increased by the circumstance that the process, allowable as it is in every other case, here yields precisely the value which we consider ourselves entitled to expect, namely: if it is *finite* at all, then just what the law of continuity demands; zero, if the neighbouring values decrease towards zero; and infinitely great if the neighbouring values increase to infinity. We are forgetting, however, that the law of continuity is far from being observed by all variable quantities. We are forgetting that a quantity which becomes arbitrarily small when we bring x arbitrarily close to a need not for that sole reason become zero for $x = a$, and that one which increases to infinity while x is approaching a has as little need to become really infinite for $x = a$. Numerous geometrical quantities know of no law of continuity, for example: the magnitudes of the lines and angles which go to determine the perimeters and areas of polygons and polyhedra, and many others.

§37

Many as the faults are, with which I believe we have not unjustifiably reproached past expositions of the *doctrine of the infinite*, it is nevertheless a matter of common knowledge that we *usually obtain quite correct results* if we follow the generally accepted rules of infinitary calculation with suitable precautions. Such results could never have appeared had there not existed some unimpeachable mode of conceiving and conducting these calculations; and I am quite ready to believe that this unimpeachable mode was hovering in the minds of the sagacious discoverers of the method, and at bottom only this mode, even if it be true that they were not yet in a position to set forth their thoughts with all the desirable clearness: an aim which in difficult cases usually requires for its attainment a number of preliminary experiments.

May I be allowed, therefore, to sketch here in rough outline how I think it necessary to conceive this method of calculation, in order to justify it completely. It will suffice to speak of the process followed in the so-called *differential* and *integral calculus*, for the methods of calculation with the infinitely great follow at once by mere antithesis, particularly after all that *Cauchy* has achieved in the matter.

I have no need, then, of so restrictive a hypothesis in this matter as the one so often considered necessary, to wit: that the quantities to be calculated with can become *infinitely small*. By such a restriction we should be excluding at the outset all bounded temporal quantities from the scope of this method, all bounded spatial quantities, all the forces of bounded substances, in short, all the quantities whose determination matters most to us. I ask one thing only: that these quantities, when they are *variable*, and not independently variable but *dependently* upon one or more other quantities, should possess a *derivative*, or what Lagrange calls 'une fonction dérivée'—if not for every value of their *determining* variable, then at least for all the values to which the process is to be validly applied. In other words: when x designates one of the independent variables, and $y = f(x)$ designates a variable dependent upon it, then, if our calculation is to give a correct result for all values of x between $x = a$ and $x = b$, the mode of dependence of y upon x must be such that for all values of x between a and b the quotient

$$\frac{\Delta y}{\Delta x} = \frac{f(x + \Delta x) - f(x)}{\Delta x}$$

(which arises from the division of the increase in y by the increase in x) can be brought as close as we wish to some constant, or to some quantity $f'(x)$ depending solely upon x, by taking Δx sufficiently small; and subsequently, on our making Δx smaller still, either remains as close thereto or comes closer still.*

Once an equation between x and y is given it is usually a very easy and well-known matter to find this derivative of y. If for example

$$y^3 = ax^2 + a^3, \tag{1}$$

then we should have for every Δx other than zero

$$(y + \Delta y)^3 = a(x + \Delta x)^2 + a^3, \tag{2}$$

whence by the known rules

$$\frac{\Delta y}{\Delta x} = \frac{2ax + a\Delta x}{3y^2 + 3y\Delta y + \Delta y^2} = \frac{2ax}{3y^2} + \frac{3ay^2\Delta x - 6axy\Delta y - 2ax\Delta y^2}{9y^4 + 9y^3\Delta y + 3y^2\Delta y^2},$$

and the *derived function* of y, or in Lagrange's notation y', would be discovered to be

$$\frac{2ax}{3y^2},$$

a function obtained from the expression for

---

* It can be shown that all *dependently variable quantities* which can be *determined* at all are subject to this law, with the proviso that exceptional values, though they may occur to an infinite number, can only occur at *isolated values* of the *independent variable*.

$$\frac{\Delta y}{\Delta x}$$

by first suitably developing it, namely into a fraction whose numerator and denominator separate the terms multiplying Δx and Δy from the terms which do not, and then putting Δx and Δy equal to zero in the expression

$$\frac{2ax + a\Delta x}{3y^2 + 3y\Delta y + \Delta y^2}$$

thus arrived at.

I have no need to speak of the manifold advantages of finding this *derivative*, or of how by its means we can calculate the finite increment of y corresponding to a finite increment of x, or yet of how, when only the derivative $f'(x)$ is given, we can identify the primitive function $f(x)$ save for an ⟦additive⟧ constant.

Having now seen that the derived function of a dependent quantity y with respect to its variable x can be obtained by first developing

$$\frac{\Delta y}{\Delta x}$$

in such wise that neither Δx nor Δy appear as divisors, and then putting Δx and Δy both equal to zero, we shall not deem it so very inappropriate to symbolize the derivative by

$$\frac{dy}{dx},$$

provided we make two things clear: (i) that all the Δx and Δy which occur in the development of $\Delta x : \Delta y$ (or if you like, the dx and dy written in their stead) are to be regarded and treated as *mere zeros*; and (ii) that the symbol $dy : dx$ shall not be regarded as the *quotient* of dy by dx, but expressly and exclusively as a *symbol* for the derivative of y with respect to x.

It is evident that such a process cannot be reproached with assuming ratios between inexistent quantities (of zero to zero), for the above symbol is not put forward for any other interpretation whatever than as a *mere symbol*.

It will be equally irreproachable to denote by

$$\frac{d^2y}{dx^2}$$

the *second* derived function of y with respect to x, meaning that quantity depending on x (or possibly constant) to which the quotient

$$\frac{\Delta^2 y}{\Delta x^2}$$

comes as close as we like, provided we can take Δx as small as we like; and to interpret this as follows: (i) that the Δx and $\Delta^2 y$ which occur in the development of

$\Delta^2y{:}\Delta x^2$ are to be regarded and treated as *mere zeros*; and (ii) that we must see in the symbol $d^2y{:}dx^2$ not a division of zero by zero, but only the *symbol* of the function obtained from the developments of $\Delta^2y{:}\Delta x^2$ by making the alteration just demanded.

Once these meanings of the symbols dy/dx and d^2y/dx^2 . . . have been laid down in advance, we can rigorously prove that every variable quantity which depends in a definable manner on another independent variable x,

$$y = f(x),$$

is compelled to satisfy the equation

$$f(x + \Delta x) = f(x) + \Delta x \frac{df(x)}{dx} + \frac{\Delta x^2}{1.2}\frac{d^2f(x)}{dx} + \frac{\Delta x^3}{1.2.3}\frac{d^3f(x)}{dx} + \ldots + \frac{\Delta x^n}{1.2.3\ldots n}\frac{d^nf(x + \mu\Delta x)}{dx^n},$$

with $\mu < 1$, save at the very most for certain isolated values of x and Δx.*

Everybody knows how many important truths of the general theory of quantity, and above all of the so-called higher analysis, can be established by means of this one single equation. But this same equation also clears the way to the solution of the most difficult problems in the application of the theory of quantity—in the doctrine of space or geometry, in the doctrine of forces or statics and mechanics and so forth, in the rectification of curves, in the complanation of surfaces, in the cubature of solids—and all this without any supposition of the infinitesimal, which would in these cases be a contradiction, and without any other alleged principle such as the well-known one of Archimedes, and so forth.

Now if it is legitimate to set up such equations as that for the rectification of a curve in rectangular coordinates:

$$\frac{ds}{dx} = \sqrt{\left[1 + \left(\frac{dy}{dx}\right)^2 + \left(\frac{dz}{dx}\right)^2\right]}$$

in the sense described above, then we shall incur no danger of error in setting up equations of the following kind:

$$d(a + bx + cx^2 + dx^3 + \ldots) = bdx + 2cxdx + dx^2dx + \ldots, \text{ ‖}sic\text{‖}$$

$$ds^2 = dx^2 + dy^2 + dz^2;$$

or if r denotes the radius of curvature of a plane curve,

$$r = -\frac{ds^3}{d^2y.dx},$$

* A proof of this theorem for all modes of dependence of y upon x, whether capable or incapable of representation by symbols hitherto in use, has long since been written out by the author and will perhaps be published in the near future. ‖Note by the posthumous editor, Dr. Přihonský.‖

and so forth; whereby we regard the symbols dx, dy, dz, ds, d^2y and so forth never as the symbols of actual quantities, but always as equivalent to zero, and consider the entire equation to be nothing but a *compound symbol so constituted that* (i) *if we carry out only such changes as algebra allows with the symbols of actual quantities* (in this case, therefore, also divisions by dx and the like) *and* (ii) *if we finally succeed in getting rid of the symbols dx, dy and so forth on both sides of the equation, then no false result will ever be the outcome.*

This fact and its necessity are easily understood. For if the equation

$$\frac{ds}{dx} = \sqrt{\left[1 + \left(\frac{dy}{dx}\right)^2\right]}$$

is irreproachable, for example, why not also the equation

$$ds^2 = dx^2 + dy^2,$$

since the former is derived from the latter by just that process?

No effort is needed, finally, to notice that we are safe from error in dealing with equations which contain the symbols dx, dy, . . . if we abbreviate the process and omit at the very outset all the addends of which we know for certain that they will disappear in the end as being equivalent to zero. So soon, for example, as any sort of calculation has led us from (1) and (2) to the equation

$$3y^2\Delta y + 3y\Delta y^2 + \Delta y^3 = 2ax\Delta x + a\Delta x^2,$$

which the passage to zero-equivalent symbols transforms into

$$3y^2dy + 3ydy^2 + dy^3 = 2axdx + adx^2,$$

we perceive immediately that the addends containing higher powers dy^2, dy^3, dx^2 will in any event disappear at the finish, so that we may as well write

$$3y^2dy = 2axdx$$

at once, and obtain at once the desired derivative

$$\frac{dy}{dx} = \frac{2ax}{3y^2}$$

with respect to x.

Let us close by putting the whole process in a nutshell: it rests on principles altogether similar to those behind our calculations with the so-called *imaginary quantities* (which like our dx, dy, . . . are mere symbols) or to those behind the recently invented short methods of division and other abridgements of calculation. It is sufficient in all these cases, as in our own case above, to give notice that the symbols we introduce,

$$dx, \frac{dy}{dx}, \frac{d^2y}{dx^2}, \ldots, \sqrt{-1}, (\sqrt{-1})^3, \frac{\sqrt{-1}}{-\sqrt{-1}}, \ldots, \text{etc.},$$

shall be given only such a meaning, and shall be subject only to such changes, that when the non-objective symbols give place to others denoting quantities, the two sides of the equation always end by being really equal to one another.

7
Carl Friedrich Gauss (1777–1855)

Although Gauss is said to have studied Kant closely (see, for example, *Gauss 1863–1929*, Vol. xii, p. 63), he did not have a high opinion of philosophers, and wrote little about philosophical questions. In a letter to H.C. Schumacher (1 November 1844) he remarks:

That you believe a philosopher *ex professo* to be free of confusion in concepts and definitions is something I find almost astonishing. Nowhere else are they more common than in philosophers who are not mathematicians, and Wolff was no mathematician, though he put together many compendiums. Just look around at the modern philosophers, at Schelling, Hegel, Nees von Esenbeck and consorts—don't their definitions make your hair stand on end? Read in the history of ancient philosophy what the men of the day, Plato and others (I except Aristotle), gave as explanations. And even in Kant matters are often not much better; his distinction between analytic and synthetic propositions seems to me to be either a triviality or false (*Gauss 1863–1929*, Vol. xii, pp. 62–3).

Gauss's philosophical and methodological observations are scanty, and must be culled from his correspondence or from asides in his papers. The following selections concern the nature of mathematics, non-Euclidean geometry, and the geometric representation of the complex numbers. The first selection, *On the metaphysics of mathematics*, is an early draft, unpublished by Gauss, and dated by the editors of his collected works as 'early nineteenth century'; it states the traditional view, criticized above by Bolzano in Part I of the *Beiträge* (*Bolzano 1810*), that mathematics is the science of quantity. Compared with Gauss's mathematical writings or with the philosophical productions of a Kant or a Bolzano, it is a slight work; but it shows an anachronistically early concern with the foundations of elementary arithmetic.

The translation of *Gauss 1929* is by William Ewald; references should be to the paragraph numbers, which appeared in the original printing in *Gauss 1863–1929*.

A. ON THE METAPHYSICS OF MATHEMATICS (*GAUSS 1929*)

1. Mathematics has for its object all extensive quantities (those of which parts can be thought); intensive quantities (all non-extensive quantities) only to the

extent that they depend on the extensive. To the first sort of quantities belong: space or the geometric quantities (which include lines, surfaces, bodies, and angles), time, number; to the latter: speed, density, hardness, height and depth of tones, strength of tones and of light, probability, etc.

2. *One* quantity in itself cannot be the object of a mathematical investigation: mathematics considers quantities only in their relation to one another. The relation of quantities to one another that they have only in so far as they are quantities, one calls an arithmetical relation; for geometric quantities there is also a relation with respect to location, and one calls this a geometric relation. It is clear that geometric quantities can also have arithmetical relationships to one another.

3. Now, mathematics really teaches general truths concerning the relations of quantities, and the goal is to represent ‖darzustellen‖ quantities which have known relations *to known quantities* or *to which known quantities* have known relations—i.e. to make possible an idea ‖Vorstellung‖ of this. But we can have an idea of a quantity in two ways, either by immediate intuition ‖unmittelbare Anschauung‖ (an immediate idea), or by comparison with other quantities given by immediate intuition (mediate idea). The duty of the mathematician is accordingly either actually to represent the sought-for quantity (geometric representation or construction), or to indicate the way and manner in which, from the idea of an immediately given quantity, one can achieve the idea of the sought quantity (arithmetical representation). This happens by means of *numbers*, which show how many times one must imagine the immediately given quantity reiterated[1] if one is to obtain an idea of the sought quantity. One calls the former quantity the *unit*, and the procedure itself *measurement*.

4. These different relations of quantities and the different means of representing quantities are the foundation of the two primary mathematical disciplines. Arithmetic considers quantities in arithmetical relations, and represents them arithmetically; geometry considers quantities in geometric relations, and represents them geometrically. To represent geometrically quantities that have arithmetical relations—as was so common among the ancients—is no longer so common today; otherwise one would have to regard this as a part of geometry. On the contrary, one applies the arithmetical manner of representation extremely frequently to quantities in geometric relation, for example in trigonometry, and also in the theory of bent lines, which one considers a geometric discipline. That the moderns so strongly prefer the arithmetical manner of representation to the geometric is not without a reason, especially since our method of counting (by tens) is so much easier than that of the ancients.

5. Since there can be a great difference among the arithmetical relations of quantities to one another, the parts of mathematical science are of a very diverse nature. The most important circumstance is whether these relations presuppose the concept of infinity or not; in the former case, they belong to the realm of

[1] Occasionally too, how often one must imagine a part of it as reiterated, which then gives the concept of a fraction.

higher mathematics; in the latter, to common or lower mathematics. I pass over the more remote subdivisions that can be derived from the foregoing concepts.

6. In arithmetic one accordingly determines all quantities by indicating how many times one must repeat or put together a known quantity (the unit) or an aliquot part of the unit in order to obtain a quantity equal to it. That is, one expresses the quantity by a number, and thus the proper object of arithmetic is the *number*. But so that it becomes possible to abstract from the meaning of the unit, there must be a means of reducing quantities that are given by different units to a common unit: this problem will be solved in the sequel.

7. Since the proper object of mathematics is the relations of quantities, we have to make ourselves familiar with the most important of these relations, and especially with those that, on account of their simplicity, can be regarded as the elements of the others—although in fact even here the first (addition and subtraction) underlie the others (multiplication and division).[2]

8. The simplest relation between quantities is incontestably that between *wholes* and their *parts*, which is already an immediate consequence of the concept of extensive quantity. The chief theorem of this relation, which one can regard as an *axiom*, is, that *the parts, if they are united in any order, and if none is omitted, are equal to the whole*. The first mode (species) of calculation, addition, shows how to find the whole from the parts; the second m. c., subtraction, shows how, given the whole and a part, one finds the other. With respect to addition the parts are called the *quantities summed* and the whole the *sum* or the *aggregate*; with respect to subtraction the whole is called the *major* or *minuend*, the known part the *minor*, and the sought part the *difference* or the *remainder*. It is clear that minor and difference must be interchangeable with one another.

9. In addition to the relation between the whole and its parts, one has to notice the relation of the simple and the multiple ‖des Einfachen und Vielfachen‖, which also yields two modes of calculation. In this relation we have to consider three quantities, the simple, the multiple, and the number which indicates what sort of a multiple it is. Multiplication shows how to find the second from the first and the third; division, how to find the third from the first two: with respect to multiplication, the simple is called the *multiplicand*, the number that determines the sort of multiplicity, the *multiplier*, both the *factors*, and the multiple the *product*. With respect to division the simple is called the divisor, the number that determines the sort of multiplicity the *quotient*, and the multiple the *dividend*.

10. The principal truths of multiplication are the following:

(1) Multiplying the multiplier by the multiplicand yields the same product as multiplying the latter by the former, i.e. the factors can be exchanged: $a.b = b.a$.

---

[2] Although the following truths are valid for fractions as well as for integers, nevertheless they will here be proved only for integers; and the explanations as well in what follows will need only a small alteration to be applicable to fractions.

(2) If the multiplier is a product, then instead of *m*-ing the *m*-and by the *m*-er, one can multiply the *m*-and by one factor of the *m*-er, and then multiply the resulting product by the second factor: $(a.b).c = a.(b.c)$.

(3) A product of several factors remains unchanged, regardless of the order in which one takes these factors:

$$a.b.c.d = a.d.c.b = c.b.a.d, \text{ etc.}$$

(4) It does not matter whether one multiplies the *m*-and all at once by the *m*-er, or multiplies its parts individually by the *m*-er and adds the resulting products: $(a + b).c = ac + bc$.

(5) It does not matter whether one multiplies the *m*-and all at once by the *m*-er, or multiplies it by the parts of the *m*-er and unites the products:

$$a(b + c) = ab + ac.$$

11. Division shows how to find from the multiple and the simple the quantity which determines the sort of multiplicity. So here three quantities are in precisely the same relation to one another as in multiplication, and what was proved for them there must be valid here as well—except that one instead uses the names that are customary for this mode of calculation, instead of the ones that are usual for multiplication. When it is there shown that multiplier and *m*-and can be exchanged (i.e. that the simple can be regarded as a determining quantity of the multiple, and the determining quantity of the multiple as a simple) this amounts here to saying that quotient and divisor can be exchanged; consequently, if the quotient and dividend are given, one finds the divisor by precisely the same operation as if the divisor and dividend were given. So one sees that, although three combinations are possible, nevertheless only two modes of calculation arise.

B. GAUSS ON NON-EUCLIDEAN GEOMETRY

The historical background to Gauss's work on the foundations of geometry has already been discussed in the notes accompanying the selections from Kant and Lambert. During the late eighteenth and early nineteenth centuries the problem of proving the Axiom of Parallels was one of the chief open problems of mathematics, and numerous mathematicians—among them, Fourier, Lagrange, Laplace, and especially Legendre—attempted to provide a definitive solution. 'The explanation and the properties of straight and of parallel lines', wrote D'Alembert in an essay on the elements of geometry in 1759, 'are the reef and the scandal of elementary geometry.'[a] This sentiment was widespread, as was

[a] Similar views, and a discussion of the existing problems in the foundations of geometry, can be found in D'Alembert's article *Parallèle* in the *Encyclopédie*.

the conviction that the Axiom of Parallels must be capable of rigorous proof. As late as 1833, in the conclusion to his masterly essay on parallel lines, Legendre could claim to have provided such a proof, and to declare that, after two thousand years of fruitless effort, the theory of parallels had finally reached a satisfactory conclusion.

Gauss seems to have been the first mathematician explicitly to doubt the Axiom of Parallels, and to have conceived of the possibility of a non-Euclidean geometry. But he wrote little (and published nothing) on the subject, and the development of his ideas must be inferred from brief remarks in his correspondence. The facts, in brief chronological outline, are as follows.

1. Gauss worked on the theory of parallels from 1792 onwards—as we know from his letter to Bessel of 27 January 1829, and from his letters to Schumacher of 17 May 1831 and of 28 November 1846.[b] This fact is confirmed by a much earlier letter to Wolfgang Bolyai from the end of 1799. In that letter, the relevant portions of which are translated below, Gauss says that his own investigations tended less to prove the Axiom of Parallels than to cast doubt on the truth of geometry.[c]

2. Although Gauss did not publish his investigations (or even, apparently, write them down until 1831), he dropped some pale hints of his true opinion in two reviews published in the *Göttingische gelehrte Anzeigen* in 1816 and 1822, of which the relevant portions are translated below. The hints do little more than point out that nobody had yet provided a satisfactory proof of the Axiom of Parallels—a fact that was in any case common knowledge at the time.

3. However, in his correspondence with Bessel and Schumacher in the years between 1829 and 1846, Gauss plainly stated his position with respect to the Axiom of Parallels. The most important portions of these letters are translated below. In his letter to Schumacher of 17 May 1831, Gauss wrote, 'Some of my own meditations ⟦on parallel lines⟧ are already forty years old; but I have never written them down, and thus have been compelled to rethink some matters three or four times from the beginning. But a few weeks ago I began to write something down. *I did not wish it to disappear with me.*' Despite this assertion, no polished exposition of Gauss's research on parallel lines was found among his papers after his death.

4. Gauss's ideas were very probably communicated not only to Bessel and Schumacher, but also to his friends Wolfgang Bolyai (the father of János Bolyai, and a friend of Gauss since the two were fellow-students in Göttingen) and Johann Martin Bartels (the teacher of Nikolai Lobatchevsky in Kazan).

---

[b] It should be noted that Gauss's teacher of mathematics was Abraham Gotthelf Kästner, who in the preface to his *1758* had been the first to call the attention of German mathematicians to the difficulties in the theory of parallels. Kästner's contributions in this area are discussed above in the introductory note to Lambert. However, it is unclear whether Kästner prompted Gauss to work on the theory of parallels. Gauss does not say how his attention was drawn to this problem, and he could have been inspired by sources other than Kästner, of whose mathematical abilities he in any case had a low opinion.

[c] '. . . führt nicht so wohl zu dem Ziele, das man wünscht, als vielmehr dahin, die Wahrheit der Geometrie zweifelhaft zu machen.'

Both Lobatchevsky and the younger Bolyai published their research on non-Euclidean geometry in the early 1830s. Lobatchevsky's writings first appeared in Russian, and so did not come to the attention of Gauss until *Lobatchevsky 1840*, which was written in German. It is unclear to what extent Lobatchevsky and János Bolyai were aware of Gauss's discoveries; but the mere knowledge that Gauss believed other geometries than the Euclidean to be possible, coupled with an understanding of the existing results in the mathematical literature on the Axiom of Parallels, would have supplied both young mathematicians with the essential raw ingredients for their research. It seems particularly likely that János Bolyai would have seen Gauss's letters to his father. In any event, Gauss said that he had found nothing new—to him—in their writings (although he praised the style of the exposition in *Lobatchevsky 1840*). His somewhat tepid reaction to the discoveries of János Bolyai can be found in a letter of 1832 to Wolfgang Bolyai (reprinted in *Gauss 1863–1929*, Vol. viii, pp. 220–1); his reaction to Lobatchevsky can be found in his letter to Schumacher of 28 November 1846, the relevant portions of which are translated below.

5. Only after Gauss's death in 1855 did his views on the Axiom of Parallels become more widely known. In 1856 Sartorius von Waltershausen, professor of mineralogy at Göttingen and a friend of Gauss's, wrote in an obituary memoir, 'Gauss considered geometry to be only a consistent structure in which the theory of parallels is conceded as an axiom at the outset; however, he became convinced that this proposition could not be proved, but that one knows from experience, for example of the angles of the triangle Brocken, Hohenhagen, Inselberg, that it is approximately correct ‖näherungsweise richtig‖. On the other hand, if one does not wish to concede the mentioned axiom, then another, wholly independent, geometry arises, which he occasionally pursued, and designated by the name *anti-Euclidean geometry*.'[d]

6. The volume of the Gauss–Schumacher correspondence containing their letters of 1831 was published in 1860; the volume containing Gauss's letter to Schumacher of 1846 was published in 1863. Gauss's correspondence with Bessel from 1829 and 1830 was published (with errors) in Vol. 22 of the *Göttinger Abhandlungen* in 1877. In this way, Gauss's opinions on the Axiom of Parallels finally became publicly available in his own words. By this time, the work of Riemann and of von Helmholtz was becoming generally known; but the great authority of Gauss helped speed the acceptance of the new geometry.

Gauss's correspondence on non-Euclidean geometry has been published in various places; so references should be to the dates of the individual letters. The most convenient and complete collection of his writings on this topic can be found in his *Werke*, *Gauss 1863–1929* (Vol. viii, pp. 157–268), which also includes the letters received by him. A shorter collection with a useful introduction is to be found in *Stäckel and Engel 1895* (pp. 211–36).

[d] The passage from von Waltershausen is reproduced in *Gauss 1863–1929* (Vol. viii, pp. 267–68) ‖= *von Waltershausen 1856*‖.

The translations of the reviews from the *Göttingische gelehrte Anzeigen* and of Gauss's correspondence are by William Ewald.

(i) From the letter of Gauss to Wolfgang von Bolyai, end of 1799

I am very sorry that I did not use our earlier, greater proximity to learn *more* about your research on the first principles of geometry; I would then certainly have saved myself a great deal of vain effort and have become calmer than anybody like me can be when so much remains to be wished for in such a subject.

I myself have made a great deal of progress in my research in this area (although my other quite heterogeneous tasks leave me little time for it)—only *the* path I have taken does not so much lead to the goal one desires, but rather casts doubt on the truth of geometry. To be sure, I have come across matters that most people would accept as a proof, but in my eyes they prove as good as *nothing*.

For example, if one could prove that ‖for any given surface‖ a straight-lined triangle is possible whose area is greater than the given surface, then I am in a position to prove the whole of geometry with full rigour.

Most people would let this stand as an axiom; not me; for it might be that, no matter how far from each other in space one assumes the angles of the triangle to be, nevertheless the area always remains under a given bound.

I have several such theorems, but in none of them do I find anything satisfying.

(ii) From the Review of J.C. Schwab and Matthias Metternich (Gauss 1816)

There are few subjects in mathematics about which so much has been written as about the gaps in the beginnings of geometry concerning the foundation ‖Begründung‖ of the theory of parallel lines. Seldom does a year go by without the appearance of some new attempt to fill these gaps; and yet we cannot say, if we speak honestly and openly, that we have made any progress beyond Euclid two thousand years ago. Such a candid and honest confession seems to us more appropriate to the dignity of science than the vain struggle to conceal the gaps one cannot fill behind an untenable web of pseudo-proofs.

‖Gauss goes on to criticize Schwab's attempt to ground the theory of parallel lines on the concept of 'identity of place'; the specific criticisms are of little interest today. He continues:‖

A great part of the text turns on the contention against Kant that the certainty of geometry is not based on intuition but on definitions and on the *principium identitatis* and the *principium contradictionis*. Kant certainly did not wish to deny that use is constantly made in geometry of these logical aids to the presentation and linking of truths: but anybody who is acquainted with the essence of geometry knows that they are able to accomplish nothing by themselves, and that they put forth only sterile blossoms unless the fertilizing living intuition ⟦Anschauung⟧ of the object itself prevails everywhere. Schwab's contradiction seems moreover to rest in part only on a misunderstanding: at any rate, it seems to us that in the sixteenth paragraph of his work (which uses precisely the faculty of intuition from beginning to end, and which claims at the end to prove that 'postulata Euclidis in generaliora resolui posse, non sensu et *intuitione*, sed *intellectu* fundata'[a]) Schwab must have imagined something else than did the Königsberg philosopher in this terminology for two different branches of the faculty of knowledge.

⟦Gauss ends his review by pointing out a number of errors in the *Vollständige Theorie der Parallel-Linien* of Matthias Metternich.⟧

(iii) From the Review of Carl Reinhard Müller (Gauss 1822)

The reviewer already six years ago expressed his opinion in these pages that all previous attempts rigorously to prove the theory of parallel lines or to fill the gaps in Euclidean geometry have brought us no nearer to this goal, and he is constrained to extend this judgement as well to all attempts that have subsequently become known to him. However, many such attempts, even though they fail to achieve their goal, nevertheless exhibit such discernment that they remain worthy of the attention of the friends of geometry; and in this regard the reviewer believes himself obliged to single out this little essay, which came to his attention on the occasion of a school examination. To recount here in detail the entire, ingenious train of thought of the author would take too much space for these pages, and would also be superfluous, since the essay itself deserves to be read; but like other attempts it has its weak spots, and the aim of this review is to indicate where they lie.

⟦Gauss continues with a number of specific criticisms of Müller's reasoning; the details of his criticisms are of little interest today.⟧

[a] ⟦'The postulates of Euclid can be resolved into more general ones, based not on sensation and *intuition*, but on *understanding*.'⟧

(iv) From the letter of Gauss to Bessel, 27 January 1829

I have also occasionally in my spare time thought about another subject that has concerned me for nearly forty years, namely, the first principles of geometry; I do not know if I have ever spoken to you about my views on this matter. Here too I have consolidated many ideas yet further, and my conviction that we cannot establish geometry entirely *a priori* has, if possible, become even firmer. But I shall not for a long time work up for public consumption my very *extensive* investigations into these matters, and this will possibly not occur in my lifetime, since I fear the cry of the Boeotians if I were to express my opinion *completely*.—But it is strange that *besides* the known gaps in Euclid's geometry that people have tried in vain to fill, and will never fill, there is yet another defect which so far as I know nobody has previously criticized, and which is not at all easy to correct (though it is possible). This is the definition of the *plane* as a surface which *completely* contains the straight line connecting *any two* points. This definition contains *more* than is necessary to determine the surface, and tacitly involves a theorem which must be proved.

(v) From the letter of Bessel to Gauss, 10 February 1829

I should regret it very much if you were to refrain from explaining your views on geometry because of the 'cry of the Boeotians'. Because of what Lambert said, and what Schweikardt[b] said orally, it has become clear to me that our geometry is incomplete and needs a hypothetical correction that disappears when the sum of the angles of a plane triangle is 180°. That would be the *true* geometry, the Euclidean, the *practical*, at least for figures on the earth.

[b] [Ferdinand Karl Schweikardt (or Schweickardt, or Schweikart), 1780–1857, was a professor of jurisprudence who pursued mathematical research in his spare time. He was influenced by the work of Saccheri and Lambert to investigate the Axiom of Parallels, and in 1818 he sent Gauss a one-page memorandum in which he distinguished two geometries: the Euclidean (which he called 'a geometry in the strict sense'), and a geometry in which the sum of the angles of a triangle is not equal to two right angles. The latter-geometry he called *astral geometry*, because he conjectured that it might be true for lines drawn between fixed stars. (The memorandum is translated in *Bonola 1955*, p. 76.) Gauss replied in March 1819. He complimented Schweikardt on his work, expressed agreement with the memorandum, and observed that he had extended Schweikardt's results further. Schweikardt did not publish his memorandum, but encouraged his nephew, Franz Adolf Taurinus (1794–1874) to continue the study of astral geometry. In his *Geometriae prima elementa* (1826) Taurinus deepened Schweikardt's results, and showed that the formulae of astral geometry hold on a sphere of imaginary radius; however, he also argued that only Euclid's geometry could be true of physical space.]

(vi) From the letter of Gauss to Bessel, 9 April 1830

I was delighted by the ease with which you entered into my views on geometry, particularly because so few have an open mind on the subject. It is my deepest conviction that the theory of space has a completely different position in our *a priori* knowledge than does the pure theory of quantity. Our knowledge of the former utterly lacks the complete conviction of necessity (and also of absolute truth) that belongs to the latter; we must in humility grant that, if number is *merely* the product of our mind, space also possesses a reality outside our mind, and that we cannot entirely prescribe its laws *a priori*.

(vii) From the letter of Gauss to Schumacher, 12 July 1831

As for the parallel lines, I should gladly have sent you my opinion in the first letter; but I assumed that you would not derive much benefit without a complete development. But such a complete development, if it were to be truly convincing, would require a lengthy analysis in reply to something which you have only hinted at in a few lines; and at present I do not have the necessary mental cheerfulness for such a task. However, as a token of my goodwill I shall set down the following.

You apply the essential point immediately to every triangle; but you would basically be using the same reasoning if you were to apply it to the simplest case and state the theorem:

(1) In every triangle whose first side is finite and whose second (and therefore third) side is infinite, the sum of the two angles on that side is 180°.

Proof in your manner:

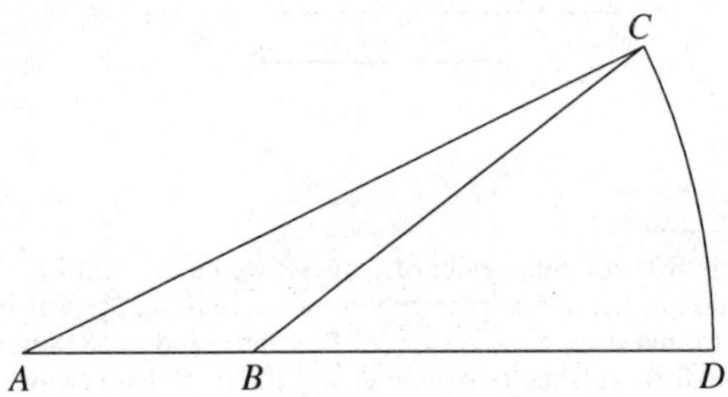

The circular arc *CD* is just as much the measure of the angle *CAD* as of *CBD*, for in a circle of infinite radius a finite displacement of the centre is to be considered as 0. So

$$CAD = CBD$$

$$CAD + CBA = CBD + CBA = 180°.$$

The remainder follows easily. Namely by this theorem:

$$\left.\begin{aligned} \alpha + \beta + \delta &= 180 \\ 180 &= \varepsilon + \delta \\ \gamma + \varepsilon &= 180 \end{aligned}\right\} \begin{array}{c} \text{so} \\ \alpha + \beta + \gamma = 180. \end{array}$$

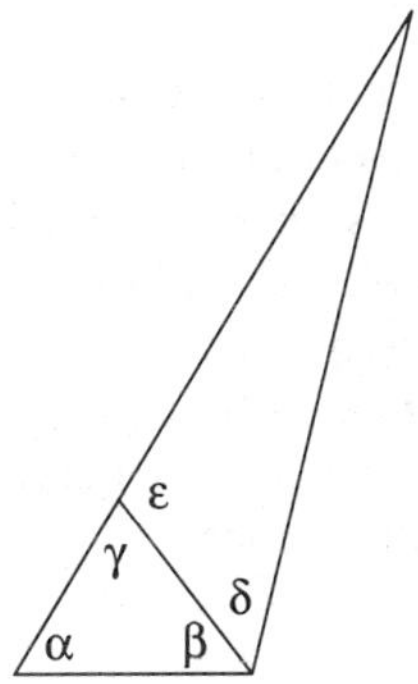

But as for your proof of (1), I protest in the first place against the use of an infinite quantity as something *completed* ‖*vollendet*‖, which is never allowed in mathematics. *The infinite is but a façon de parler* in that one actually speaks of limits to which certain relations ‖Verhältnisse‖ come as close as one desires, while others are allowed to increase without bound. In this sense the non-Euclidean geometry contains absolutely nothing contradictory, although at the start many of its results will be held to be paradoxical. But to hold them to be contradictory would be a self-deception, caused by the earlier habit of holding the Euclidean geometry to be *strictly* true.

In the non-Euclidean geometry there are absolutely no similar figures unless they are equal. For example, the angles of an equilateral triangle are not merely of 2/3 R, but also, depending on the length of the sides, different from one another; and if one lets the side grow beyond all limits, they can become as small as one wishes. It is therefore already a contradiction to want to *depict* ‖*zeichnen*‖ such a triangle by a smaller; one can basically only *designate* ‖*bezeichnen*‖ it.

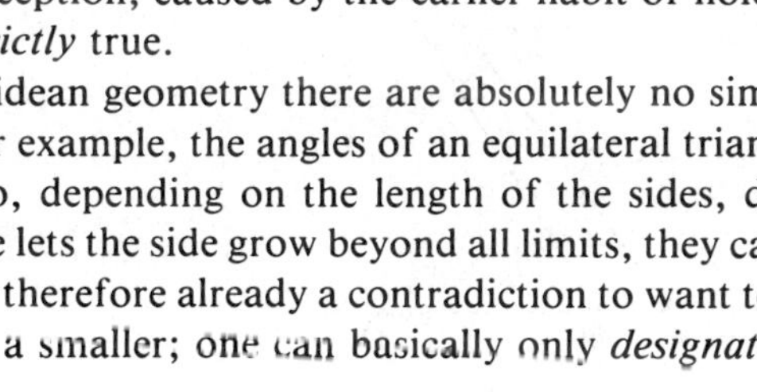

The designation of the infinite triangle in this sense would be, in the end,

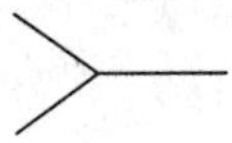

In the Euclidean geometry nothing has an absolute size, but not so in the non-Euclidean; this is precisely its essential character, and those who do not admit this are *eo ipso* already presupposing all of Euclidean geometry; but as I said, in my opinion this is mere self-deception. Now for the case in question there is absolutely nothing contradictory in what follows: when the points A, B and the direction AC have been given, where C can grow without bound, then although DBC always comes closer to DAC, nevertheless the difference can never be brought under a certain finite difference.

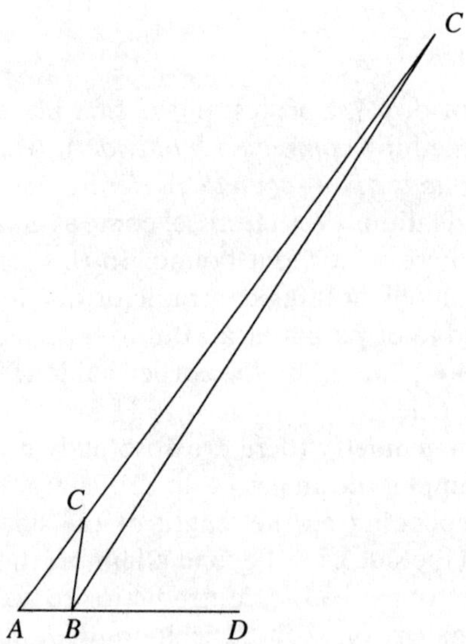

Your invocation of the arc CD makes the conclusion much more captious, but if one clearly develops what you have only hinted at, it would go like this:

We have $CAB : CBD = \dfrac{CD}{ECD} : \dfrac{CD'}{E'CD'}$, and since AC grows into the infinite, CD and CD' (on the one hand) and ECD, $E'CD'$ (on the other) come ever closer to equality.

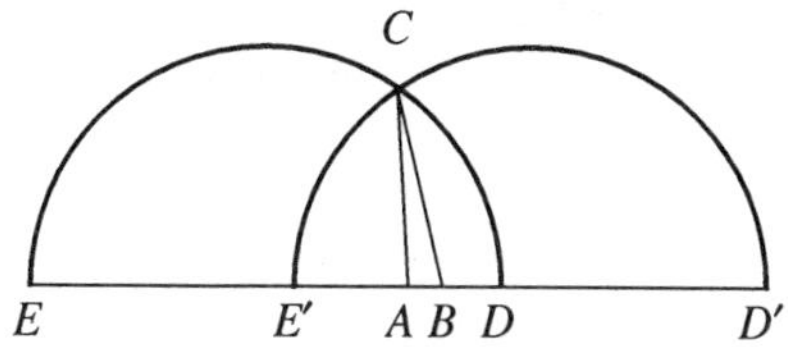

Neither is true in the non-Euclidean geometry, if one understands it to mean that their geometrical relationships come as close to equality as one wishes. Indeed, in non-Euclidean geometry the half-circumference of a circle with radius r equals

$$\frac{1}{2}\pi k(e^{r/k} - e^{-r/k})$$

where k is a constant of which we know by experience that it must be extraordinarily large in comparison with everything we can measure. In Euclid's geometry it is infinite.

In the metaphorical language of the infinite one would therefore have to say that the peripheries of two infinite circles whose radii differ by a finite quantity are themselves different by a quantity that has a finite relationship to them.

But there is nothing contradictory in all this if finite man does not presume to wish to contemplate something infinite as something that is given and that is to be encompassed by him with his ordinary intuition ⟦Anschauung⟧.

You see that in fact the point at issue touches directly on metaphysics.

But enough for now.

(viii) From the letter of Gauss to Schumacher, 28 November 1846

I recently had occasion to look through the booklet by Lobatchevsky again (*Geometrische Untersuchungen zur Theorie der Parallellinien*. Berlin 1840, G. Funcke. Four signatures long). It contains the main features of the geometry that would have to occur, and that perfectly consistently ⟦streng consequent⟧ could occur, if the Euclidean geometry is not true. A certain Schweikardt[1] called such a geometry *astral geometry*; Lobatchevsky, *imaginary geometry*. You know that already for 54 years (since 1792) I have had the same conviction (with a certain later extension which I shall not mention here). So I did not find material in Lobatchevsky's work that was new to me, but the development follows a different path than the one I have taken, and indeed is executed by

[1] Formerly in Marburg, now professor of jurisp. in Königsberg.

Lobatchevsky in a masterly fashion and in a true geometric spirit. I thought I should call your attention to the book, which will certainly furnish you with a quite exquisite pleasure.

C. NOTICE ON THE THEORY OF BIQUADRATIC RESIDUES (*GAUSS 1831*)

The following selection contains a summary by Gauss of his work on the theory of biquadratic residues. His *Disquisitiones arithmeticae*, written in Latin in 1798 (when Gauss was twenty), had inaugurated the modern theory of numbers—systematizing the notation, classifying the problems, introducing new techniques, and exploiting the ideas of congruences, algebraic numbers, and forms to prove powerful new theorems. This work, published in 1801, was the first of a series of major writings by nineteenth-century Göttingen mathematicians on the theory of numbers—a series that includes Dirichlet's *Lectures on number-theory* (*Dirichlet 1863*), Dedekind's numerous supplements to *Dirichlet 1863*, and Hilbert's *Zahlbericht* (*Hilbert 1897*). In the *Disquisitiones*, Gauss gave the first rigorous proof of 'the gem of arithmetic'—the law of quadratic reciprocity. (He was ultimately to give eight distinct proofs of this celebrated theorem, whose statement can be found in any standard text on the theory of numbers.) In a series of papers published between 1808 and 1817 Gauss worked on reciprocity laws for congruences of higher degree, and in two papers published in 1828 and 1832 stated (but did not prove) the law of biquadratic reciprocity.[a] The two papers themselves were written in Latin and communicated to the Academy of Sciences in Göttingen in 1825 and 1831 respectively. The following *Notice*, written in German and published in the *Göttingische gelehrte Anzeigen* for 23 April 1831, contains a description of the principal results in the paper that was to appear in 1832.

In studying the congruence

$$x^4 \equiv k \pmod{p}$$

Gauss concentrated on the central case where the modulus p is a prime of the form $4n + 1$, and found himself obliged to examine the complex factors into which prime numbers of such a form can be decomposed. He was thereby led to widen his investigations from the ordinary integers to the complex integers, i.e. to numbers of the form $a + bi$, where a and b are real integers. Gauss

[a] The two papers by Gauss are reprinted in *Gauss 1863–1929* (Vol. ii, pp. 65–92 and 93–148). The first proofs of the biquadratic reciprocity theorem were provided by Jacobi and Eisenstein. The first two of Eisenstein's five published proofs are in the *Journal für die reine und angewandte Mathematik*, **28**, 53–67 and 223–45 (1844). Morris Kline (*1972*, p. 817) reports that Jacobi had earlier given a proof during his lectures in Königsberg in 1836–7.

showed that many of the properties of real integers are shared by the complex integers—for example that each complex integer has a unique prime factorization (modulo the four units, ± 1 and $\pm i$); that the Euclidean algorithm for finding the greatest common divisor of two integers carries over to the complex case; that Fermat's theorem has a complex analogue; and the like. These results are described in §§1–15 of the following selection, as are his chief theorems on biquadratic residues.

The principal methodological innovation in these investigations was the widening of the number concept to embrace the complex integers. Mathematicians in the seventeenth and eighteenth centuries had been almost as distrustful of imaginary numbers as of infinitesimals, and even Leibniz could write:

> The nature of things, the mother of eternal manifolds, or rather the divine spirit, is more jealous of its splendid multiplicity than to allow everything to be herded together under a common genus. Therefore it found a sublime and wonderful refuge in that miracle of analysis, the *monstrum* of the ideal world, almost an *amphibium* between being and non-being, which we call the imaginary roots (*Leibniz 1849–63*, Vol. v, p. 357).

Gauss was not the first to conceive of representing complex numbers by points in the plane: according to Oskar Becker (*1964*, p. 213) the idea had already occurred to John Wallis (in the *Treatise of algebra*, *Wallis 1685*), to Caspar Wessel in 1792, and to Jean Robert Argand in 1806; and William Rowan Hamilton was later to popularize the conception of a complex number as an ordered couple of real numbers. But Gauss was the first to use complex integers in a systematic way in his work on biquadratic residues. His remarks on the complex integers touch upon a central theme in nineteenth-century mathematics: the widening of the number concept, and the growth of abstract algebra. (This theme, which has already been broached in Berkeley's *Analyst* (§§8, 36, 37), in §11 of *Lambert 1786*, and in *Bolzano 1810*, will reappear in many of the selections that follow—notably in the young Dedekind's 'On the introduction of new functions into mathematics,' which was delivered as an *Habilitation* lecture in the presence of Gauss in 1854.)

The translation of *Gauss 1831* is by William Ewald; references should be to the paragraph numbers, which have been added in this translation.

⟦1⟧ A lecture submitted by Privy Counsellor Gauss of the Royal Society: *Theoria residuorum biquadraticorum. commentatio secunda*, is the continuation of the treatise published in the sixth volume of the *Commentationes novae*, of which a notice has already appeared in our pages. Although this continuation is more than twice as long as the first treatise, it does not yet exhaust the extraordinarily rich subject, and the work will be completed only in a third treatise.

⟦2⟧ Although we could presuppose that anybody who has studied higher arithmetic is acquainted with the basic concepts in these theories and with the contents of the first treatise, we shall nevertheless briefly recall them for the benefit of those friends of this part of mathematics who do not have the first treatise directly at hand. An integer k is called a biquadratic residue with respect to an arbitrary integer p if there are numbers of the form $x^4 - k$ that are divisible by p; otherwise, k is a biquadratic non-residue of p. It suffices if we confine our attention to the case where p is a prime number of the form $4n + 1$ and k is not divisible by p; all other cases are either clear or reduce to this one.

⟦3⟧ For such a given value of p, all numbers not divisible by p fall into *four* classes: the first contains the biquadratic residues, the second, the biquadratic non-residues that are quadratic residues of p; into the remaining two classes are distributed the biquadratic non-residues that are also quadratic non-residues. The principle of this distribution is, that always either $k^n - 1$ or $k^n + 1$ or $k^n - f$ or $k^n + f$ is divisible by p, where f is an integer that makes $ff + 1$ divisible by p. Everybody who knows the elementary terminology can see how to use it to dress up this verbal explanation.

⟦4⟧ The theory of this classification is completely worked out in the first treatise, not only for the superficial case $k = -1$, but also for the subtler cases $k = \pm 2$. The beginning of the present treatise proceeds to larger values of k: but here one need only consider numbers that are themselves prime; and trial and error shows that the results appear most readily if one takes the values to be positive or negative according as they are of the form $4m + 1$ or $4m + 3$. Induction yields with the greatest ease a rich harvest of new theorems, of which a few shall be mentioned here. The numbering of the classes by 1, 2, 3, 4 is determined by whether k^n is congruent to the numbers $1, f, -1, -f$; here the number f always takes that value which makes $a + bf$ divisible by p if $aa + bb$ represents the decomposition of p into an even and an odd square. Thus one finds by induction that the number -3 always belongs to class 1, 2, 3, 4 according as $b, a + b, a, a - b$ is divisible by 3; that the number $+5$ belongs to the appropriate class according as $b, a - b, a, a + b$ is divisible by 5; that the number -7 falls into class 1 if a or b, into class 2 if $a - 2b$ or $a - 3b$, into class 3 if $a - b$ or $a + b$, and into class 4 if $a + 2b$ or $a + 3b$ is divisible by 7. Similar theorems hold for the numbers -11, $+13$, $+17$, -19, -23, etc. But easy though it is to discover such special theorems by induction, it is extremely difficult to discover in this way a general law for these forms—even though many similarities leap to the eye; and it is even harder to find proofs of these theorems. The methods used in the first treatise for the numbers $+2$ and -2 cannot be applied here, and although other methods might serve to settle the question about the first and third classes, these methods do not suffice for *complete* proofs.

⟦5⟧ Accordingly, one quickly sees that one can make progress in this rich domain of higher arithmetic only by employing methods that are entirely new. The author already gave a hint in the first treatise that for this end it is necessary to widen the entire field of higher arithmetic; but he did not then say what this widening would consist in. The present treatise illuminates this subject.

⟦6⟧ This is nothing less than the following. For the true grounding of the theory of biquadratic residues, one must extend the field of higher arithmetic, which has hitherto been confined to the real integers, into the imaginary integers, and must concede to the latter the same legitimacy as the former. As soon as one has seen this, that theory appears in an entirely new light, and its results acquire a startling simplicity.

⟦7⟧ But before the theory of biquadratic residues can be developed in this widened domain of numbers, the doctrines of higher arithmetic, which hitherto have been explored only for real numbers, must be extended as well. Here we can mention only a few points about these preliminary investigations. The author calls every quantity $a + bi$, where a and b denote real quantities and where i is an abbreviation for $\sqrt{-1}$, a complex integer, provided that a and b are integers. So the complex quantities are not opposed to the real, but contain them as a special case where $b = 0$. For the sake of simplicity it was necessary to give names to several concepts that are based on the complex quantities, but in this notice we shall avoid using them.

⟦8⟧ Just as in the arithmetic of real numbers one speaks only of two units, the positive and the negative, so in the arithmetic of complex numbers we have four units: $+1$, -1, $+i$, $-i$. A complex integer is said to be *composite* if it is the product of two whole factors different from unity; but a complex number that does *not* admit such a decomposition into factors is said to be a complex prime number. For example, the real number 3, considered as a complex number, is prime, while 5 is composite $= (1 + 2i)(1 - 2i)$. Just as in the higher arithmetic of real numbers, the prime numbers play a leading role in the widened field of this science.

⟦9⟧ When a complex integer $a + bi$ is taken as a modulus, then it is possible to find $aa + bb$ (and no more) complex numbers which are pairwise incongruent, and such that any given complex integer is congruent to one of them; one can call them a complete system of incongruent residues. The so-called smallest and absolutely smallest residues in the arithmetic of real numbers have their precise analogues here. For example, for the modulus $1 + 2i$ the complete system of the absolutely smallest residues consists of the numbers 0, 1, i, -1, and $-i$. Almost all of the investigations of the first four sections of the *Disquisitiones arithmeticae* find a place in the extended arithmetic (sometimes with a few modifications). For instance, the famous theorem of Fermat assumes the following form: If $a + bi$ is a complex prime number and k is a complex number not divisible by it, then $k^{aa+bb-1} = 1$ for the modulus $a + bi$. But it is particularly noteworthy that the fundamental theorem for quadratic residues in arithmetic has its perfect, but simpler, counterpart here; namely, if $a + bi$ and $A + Bi$ are complex prime numbers such that a and A are odd, b and B even, then the first is a quadratic residue of the second, if the second is a quadratic residue of the first, whereas the first is a quadratic non-residue of the second if the second is a quadratic non-residue of the first.

⟦10⟧ Because the treatise proceeds from these preliminary investigations to the theory of biquadratic residues, one first distributes the numbers that are not divisible by the modulus into four classes, instead of merely distinguishing between biquadratic residues and non-residues. Namely, if the modulus is a complex prime number $a + bi$, where it is always presupposed that a is odd and b even, and where $aa + bb$ is abbreviated by p, and where k is a complex number not divisible by $a + bi$, then $k^{\frac{1}{4}(p-1)}$ is always congruent with one of the numbers $+1$, $+i$, -1, $-i$, and thereby justifies a distribution of all numbers not divisible by $a + bi$ into four classes, which are assigned, in sequence, the biquadratic characters 0, 1, 2, 3. Clearly character 0 pertains to the biquadratic residues, the others to the biquadratic non-residues, and in such a way that quadratic residues correspond to character 2, while quadratic non-residues correspond to characters 1 and 3.

⟦11⟧ One easily sees that it is primarily a matter of being able to determine this character for values of k that are themselves complex prime numbers, and here induction leads at once to extremely simple results.

⟦12⟧ If we set $k = 1 + i$ then it turns out that the character of this number is always $\frac{1}{8}(-aa + 2ab - 3bb + 1)$ (mod 4), and similar expressions appear for the cases $k = 1 - i$, $k = -1 + i$, $k = -1 - i$.

⟦13⟧ On the other hand, if $k = \alpha + \beta i$ is a prime number where α is odd and β even, then one easily discovers by induction a law of reciprocity that is entirely analogous to the fundamental theorem of quadratic residues. This law can be most easily expressed as follows:

⟦14⟧ If both $\alpha + \beta - 1$ and $a + b - 1$ are divisible by 4 (to which case all the others can easily be reduced) and if the character of the number $\alpha + \beta i$ with respect to the modulus $a + bi$ is designated by λ, while the character of $a + bi$ with respect to the modulus $\alpha + \beta i$ is designated by l, then $\lambda = l$ provided that one or both of the numbers β, b is divisible by 4, otherwise, $\lambda = l + 2$.

⟦15⟧ These theorems contain all that is essential in the theory of biquadratic residues: but easy though it was to discover them, it is difficult to give rigorous proofs—especially for the second, the fundamental theorem of biquadratic residues. Because of the great length of the present treatise, the author was compelled to leave the proof of the latter theorem (which he has possessed for twenty years) to a future third treatise. But the present treatise contains the complete proof of the first theorem concerning the number $1 + i$ (on which the theorems for $1 - i$, $-1 + i$, $-1 - i$ depend), and this proof gives some idea of the complexity of the whole.

⟦16⟧ We now have to add some general remarks. To transfer the theory of biquadratic residues into the domain of the complex numbers might seem offensive and unnatural to many a person who is unfamiliar with the nature of the imaginary quantities and who is prejudiced by false ideas about them. And it might cause the opinion that the investigation is thereby, as it were, built upon air—that it takes on an uncertain posture, and is removed entirely from clarity and concreteness. Nothing would be more unfounded than such an opinion. On the contrary, the arithmetic of the complex numbers is capable of the most intuitive sensible representation ⟦der anschaulichsten Versinnlichung fähig⟧, and although the author in the present account has pursued a purely arithmetical treatment, he nevertheless also gave the necessary hints for this sensible representation—a sensible representation which makes the insight come to life, and is therefore to be recommended; these hints will suffice for readers who think for themselves. Just as the absolute integers are represented by a sequence of points, ordered, at equal intervals, in a straight line, in which the starting-point represents the number 0, the next the number 1, and so on; and just as the representation of negative numbers requires only an unbounded extension of this sequence on the opposite side of the starting-point: so for the representation of the complex integers we need only add that the sequence is to be considered as located in a determinate unbounded plane, and that parallel with it on both sides there is an unlimited number of similar sequences at equal distances from each other, so that instead of a sequence of points we have before us a system of points which can be ordered in two ways as a sequence of sequences, and which serve to divide the entire plane into precisely equal squares. The point next to 0 in the first adjacent sequence on the one side of the sequence which represents the real numbers is then correlated to the number i, just as the point next to 0 in the first adjacent sequence on the other side is correlated to $-i$, and so on. By this device the effect of the arithmetical operations on the complex quantities becomes capable of a sensible representation that leaves nothing to be desired.

‖17‖ Moreover, in this way the true metaphysics of the imaginary quantities is placed in a bright new light.

‖18‖ Our general arithmetic, whose scope so greatly outstrips the geometry of the ancients, is entirely the creation of modern times. It started from the concept of absolute integers, and has gradually extended its territory; the fractions have been added to the integers, the irrationals to the rationals, the negatives to the positives, and the imaginaries to the reals. But these advances always began with timid and hesitant steps. The first algebraists still called the negative roots of equations false roots—which is just what they are, if the task they are related to is presented in such a manner that the sought-after quantities admit of no opposite. But just as one has so few scruples about taking the fractions into *general* arithmetic (even though there are so many countable things where a fraction is without sense), so the negative numbers may not be denied equal rights with the positive just because innumerable things admit no opposite: the reality of the negative numbers is sufficiently justified because they have an adequate substrate in innumerable other cases. To be sure, one has been in the clear about these matters for a long time: only the imaginary quantities—which are contrasted with the real ones, and which were formerly, and are still occasionally (although improperly) called *impossible* are still merely tolerated rather than fully accepted, and therefore appear more like a game with symbols, in itself empty of content, to which one unconditionally denies a thinkable substrate—without, however, wishing to scorn the rich rewards which this game with symbols achieved for our understanding of the relationships of the real quantities.

‖19‖ The author has for many years considered this most important part of mathematics from a different point of view, whereby an object can just as well be attributed ‖unterlegt‖ to the imaginary quantities as to the negative. But hitherto there has been no occasion to express this view publicly, although attentive readers will readily find traces of it in the 1799 article on equations and in the prize essay on the transformation of surfaces. In the present article I state the principles of such a view; they consist in the following.

‖20‖ Positive and negative numbers can find an application only where that which is counted has an opposite, so that the thought of them as united is to be equated with annihilation. Precisely regarded, this condition occurs only where what is counted is not substances (objects thinkable in themselves) but relations between any two objects. It is then supposed that these objects are ordered into a sequence in a determinate way, for example $A, B, C, D, \ldots$, and that the relation of A to B can be regarded as equal to the relation of B to C, etc. Now here the concept of opposition is just that of the *exchange* of the members of the relation, so that if the relation (or the transition) of A to B counts as $+1$, then the relation of B to A must be represented by -1. So to the extent that such a sequence is unbounded on both sides, every real integer represents the relation of an arbitrary member chosen as the origin and a determinate member of the sequence.

‖21‖ But suppose the objects are of such a sort that they cannot be ordered in a sequence (even if it is unbounded) but only in sequences of sequences; or, what comes to the same thing, suppose that they form a manifold of two dimensions. Then if the relations of one sequence to another or the transitions from one into the other behave in a similar manner to the transitions just mentioned from one term of a sequence to another term of the same sequence, then clearly for the measurement of the transition from one term of the system to another we need besides the previous units $+1$ and -1 two others, also inverse to each other, $+i$ and $-i$. But clearly one must also postulate that the unit i always designates the transition from a given member of a sequence to a *determinate* member of the immediately adjoining sequence. In this way the system can be ordered in two ways into sequences of sequences.

‖22‖ The mathematician abstracts totally from the nature of the objects and the content of their relations; he is concerned solely with the counting and comparison of the relations among themselves: just as he regards the relations designated by $+1$ and -1, in themselves, as similar, so is he entitled to regard all four elements $+1$, -1, $+i$, $-i$ as similar.

‖23‖ These relationships can be brought to intuition ‖Anschauung‖ only by a representation in space, and the simplest case is when there is no reason to arrange the symbols of the objects in any other way than quadratically. That is, one divides an unbounded plane into squares by means of two systems of parallel lines which cross each other at right angles, and one assigns the points of intersection to the symbols. Every such point A has four neighbours, and if one designates the relation of A to a neighbouring point by $+1$, then the point to be designated by -1 is thereby determined, and one can choose either of the two others for $+i$ or can take the point related to $+i$ to be *right* or *left* as one wishes. This difference between right and left is *in itself* fully determinate, as soon as we have (by choice) established forward and backward in the plane, and up and down with respect to the two sides of the plane; although we can communicate our intuition of this difference to others *only* by appeal to really existent material things.[1] But if one has also decided about the latter, one sees that it is up to our discretion which of the two sequences that cross at a given point we shall choose as the main sequence, and which direction we shall regard as correlated with the positive numbers; one also sees that if one takes the relation previously designated by $+i$ to be $+1$, then one must take the relation previously designated by -1 to be $-i$. But that means, in the language of the mathematicians, that $+i$ is a geometric mean between $+1$ and -1, or corresponds to the sign $\sqrt{-1}$: we intentionally do not say *the* geometric mean, for clearly $-i$ has the same claim. Here, therefore, an intuitive meaning of $\sqrt{-1}$ is completely established, and one needs nothing further to admit this quantity into the domain of the objects of arithmetic.

⟦24⟧ We have believed that we were doing the friends of mathematics a favour by this account of the principal parts of a new theory of the so-called imaginary quantities. If one formerly contemplated this subject from a false point of view and therefore found a mysterious darkness, this is in large part attributable to clumsy terminology. Had one not called $+1$, -1, $\sqrt{-1}$ positive, negative, or imaginary (or even impossible) units, but instead, say, direct, inverse, or lateral units, then there could scarcely have been talk of such darkness. The author has reserved the right to treat this subject, which in the present treatise is only occasionally touched upon, at greater length later. Then too the question will be answered, why the relations between things that form a manifold of more than two dimensions cannot supply yet another type of number that is admissible in higher arithmetic.

---

[1] Kant already made both observations, but one cannot understand how this acute philosopher could have believed that the first gives a proof of his opinion that space is *only* the form of our outer intuition, since the second so clearly proves the opposite, namely, that space must have a real significance independent of our mode of intuition.

8
Duncan Gregory (1813–1844)

We have already seen in earlier texts that in the eighteenth and early nineteenth centuries many important mathematical problems—the justification of the calculus, the status of the axioms of geometry, the legitimacy of equations containing $\sqrt{-1}$—were intermingled with philosophical concerns about language and about the reference of algebraic expressions. For example, Berkeley, in *The analyst*, criticizing the foundations of the differential calculus, says:

> Nothing is easier than to devise expressions or notations, for fluxions and infinitesimals of the first, second, third, fourth, and subsequent orders, proceeding in the same regular form without end or limit $\dot{x}$. $\ddot{x}$. $\dddot{x}$. $\ddddot{x}$. *&c.* or *dx. ddx. dddx. ddddx &c.* These expressions indeed are clear and distinct, and the mind finds no difficulty in conceiving them to be continued beyond any assignable bounds. But if we remove the veil and look underneath, if, laying aside the expressions, we set ourselves attentively to consider the things themselves which are supposed to be expressed or marked thereby, we shall discover much emptiness, darkness, and confusion; nay, if I mistake not, direct impossibilities and contradictions. (*Berkeley 1734*, §8.)

The issue of the reference of mathematical symbols has arisen also in Kant's account of mathematics as the 'science of the construction of concepts' (*Kant 1781*, pp. 713–38); in Lambert's observation that the Axiom of Parallels should be proved from the other axioms 'purely symbolically', without regard to the intended meaning of the geometrical terms (*Lambert 1786*, §11); in Bolzano's definition of mathematics as the 'science of the general laws to which things must conform in their existence' (*Bolzano 1810*, Part I, §8); and in Gauss's remarks on the geometric interpretation of the complex integers (*Gauss 1831*, §§16–24).

In the early years of the nineteenth century, as mathematicians deepened their use of the calculus and of negative and complex numbers, the justification of these powerful tools became more pressing, and controversy about proper methodology became more intense. Some mathematicians argued that the only legitimate algebraic symbols were those that referred, directly or indirectly, to the positive real numbers: on this view, negative and imaginary numbers were meaningless, and deserved to be banished from mathematics. Others, like Argand (in 1806) and Gauss (in 1831), sought to justify the negative and complex numbers by supplying them with a geometric interpretation. Yet others, like François-Joseph Servois (1767–1847) justified the complex numbers, not by giving them a geometric reference, but by pointing out their usefulness in making calculations. For Servois, algebra was a rule-governed calculus of uninterpreted symbols, and he criticized Argand's geometric interpretation of the complex numbers as being 'a geometric mask applied to algebraic forms, the direct use of which seems to me simple and more expeditious' (*Servois 1814a*, p. 230).

At stake in these debates was not only the legitimacy of extending the number-concept beyond the positive numbers, but also (as the quotation from Berkeley makes clear) the justification of the methodology of the differential calculus. In the writings of the British algebraists of the nineteenth century the philosophical issues about the reference of mathematical expressions become entangled with a further cluster of mathematical problems—most notably, with the algebraic analysis of logic and with the beginnings of abstract algebra; and progress on the mathematical front was in part made possible by a new-found clarity on the philosophical issues.

The first systematic attempt in England to grapple with these problems was made by George Peacock (1791–1858), a Fellow of Trinity College, Cambridge, and (after 1837) Lowndean Professor of Geometry and Astronomy at Cambridge. Peacock was also an ordained clergyman who served as dean of Ely Cathedral. In his student years at Cambridge, Peacock, with his fellow undergraduates Charles Babbage and John Frederick William Herschel, founded the Analytical Society, which advocated the adoption of a Continental approach to the calculus.[a] A few years later, the three friends published a translation of Sylvestre François Lacroix's *An elementary treatise on the differential and integral calculus* (*Peacock 1816*); Peacock subsequently published his own supplementary text, *A collection of examples of the application of the differential and integral calculus* (*Peacock 1820*). These works, though now forgotten, had a considerable historical importance, for they introduced into British mathematics the powerful new techniques developed by French and German analysts in the eighteenth century. In undertaking this project, Peacock had to struggle against the entrenched British mistrust of algebra, and to champion the notation of Leibniz against that of Newton (which to that time had been held almost sacred). It is important to be aware of this fact and of the obstacles Peacock faced—for only then is it possible to understand the motives behind his pursuit of a new theory of algebra, and to appreciate the difficulty and the mathematical significance of his accomplishment.

Ever since MacLaurin's *Treatise of fluxions* (*MacLaurin 1742*) British mathematicians had clung to Newton's geometric style of proof and to his method

[a] The 'irascible genius' Charles Babbage (1792–1871), Lucasian Professor of Mathematics at Cambridge, was a pioneer in computing logic and computing technology. He worked on the mathematics of difference equations, and devoted much of his life to an unsuccessful effort to build an 'analytical engine' that would mechanically carry out complex mathematical computations. J.W.F. Herschel (1792–1871), the son of the great astronomer William Herschel, was an astronomer, physicist, and chemist; he was the most celebrated British scientist of his day. Herschel, like Peacock and Babbage, vigorously championed Leibniz's calculus notation against Newton's—a fact which did not preclude him from being buried next to Newton in Westminster Abbey.

of fluxions, while resisting the more algebraic methodology employed on the Continent. This resistance is often ascribed to the dispute between Newton, Leibniz, and their followers over priority in the discovery of the calculus; but patriotic loyalties were only a small part of the story, and in fact (as we shall see) the critics of Continental methodology showed no reluctance to criticize Newton and MacLaurin for straying from the path of strict mathematical rigour. The problem was rather that the 'geometric method of the ancients' defended and employed by MacLaurin in the *Treatise of fluxions* seemed to British mathematicians a more reliable methodology than that of the Continentals. The Archimedean style of proof had been the paradigm of mathematical rigour for over two thousand years; its foundations were well understood; and, as Peacock himself conceded, its premisses were more precise and its conclusions supported by better evidence than those obtained by the use of algebra (*Peacock 1833*, p. 188).

In contrast, the symbolic manipulations of the Continentals—with their application of operations like addition and multiplication to strange new objects like d/dx or $\sqrt{-1}$—seemed to hang in the air. Even Cauchy, in the *Cours d'analyse* (*Cauchy 1821*) had had difficulty in justifying such fundamental matters as the application of the familiar arithmetical operations to the negative integers. Cauchy's work was the high-water mark of French analytical rigour; but Peacock cites several examples of 'the extraordinary vagueness of the reasoning which is employed to establish these theorems.' For instance, Cauchy had written

If a and b be whole numbers, it may be proved that ab is identical with ba; *therefore*, ab is identical with ba, whatever a and b may denote, and whatever may be the interpretation of the operation which connects them.

Or again:

One represents those magnitudes which are to serve as increases by numbers preceded by the sign +, and those magnitudes which are to serve as diminutions by numbers preceded by the sign −. This being done, one can understand the signs + and − placed in front of numbers (as has already been said) as adjectives placed next to their substantives (*Peacock 1833*, p. 193).

With arguments like these being used to justify the foundations of Continental analysis, it is not surprising that British mathematicians were mistrustful of the new techniques—a mistrust that was reinforced by the prevailing British hostility towards even relatively unproblematic portions of the algebra of the negative integers. One of the most influential eighteenth-century English writers on elementary mathematics was Francis Maseres (1731–1824), a lawyer and civil servant who wrote *A dissertation on the use of the negation sign in algebra* (*Maseres 1758*). Although Newton, Taylor, and MacLaurin—as well as Euler, Lambert, Lagrange, and Laplace—had employed negative and 'impossible' numbers to secure some of their most important results, Baron Maseres declared their methodology to be illegitimate:

A single quantity can never be marked with either of these signs, or considered as either affirmative or negative; for if any single quantity, as *b*, is marked either with the sign + or with the sign − without affecting some other quantity, as *a*, to which it is to be added, or from which it is to be subtracted, the mark will have no meaning or signification: thus if it be said that the square of −5, or the product of −5 into −5, is equal to +25, such an assertion must either signify no more than 5 times 5 is equal to 25 without any regard to signs, or it must be mere nonsense or unintelligible jargon.

Even though Maseres was himself devoid of mathematical depth (*Maseres 1758* contains an appendix on 'Mr. Machin's quadrature of the circle') his doctrines influenced the teaching of mathematics in Britain for several decades, as can be seen from the textbooks *Manning 1796–8*, *Vilant 1798*, and the five editions of *Ludlam 1809*.

The doctrines of Maseres were taken up by a Cambridge clergyman, William Frend (1757–1841), in his two-volume treatise on *The principles of algebra* (*Frend 1796–9*). Like Maseres, Frend did not hesitate to scold MacLaurin and even Newton when he considered them to be confused. A quotation from the *Preface* to *Frend 1796–9* will illustrate the substance of his views:

The first error in teaching the first principles of algebra is obvious on perusing a few pages only of the first part of Maclaurin's Algebra. Numbers are there divided into two sorts, positive and negative: and an attempt is made to explain the nature of negative numbers, by allusions to book debts and other arts. Now when a person cannot explain the principles of a science, without reference to a metaphor, the probability is, that he has never thought accurately on the subject. A number may be greater or less than another number: it may be added to, taken from, multiplied into, or divided by, another number; but in other respects it is very intractable; though the whole world should be destroyed, one will be one, and three will be three, and no art whatever can change their nature. You may put a mark before one, which it will obey; it submits to be taken away from another number greater than itself, but to attempt to take it away from a number less than itself is ridiculous. Yet this is attempted by algebraists, who talk of a number less than nothing, of multiplying a negative number into a negative number, and thus producing a positive number, of a number being imaginary. Hence they talk of two roots to every equation of the second order, and the learner is to try which will succeed in a given equation: they talk of solving an equation which requires two impossible roots to make it soluble: they can find out some impossible numbers, which being multiplied together produce unity. This is all jargon, at which common sense recoils; but from its having been once adopted, like many other figments it finds the most strenuous supporters among those who love to take things upon trust and hate the colour of a serious thought.

From the age of Vieta, the father, to this of Maseres, the restorer of algebra, many men of the greatest abilities have employed themselves in the pursuit of an idle hypothesis, and have laid down rules not founded in truth, nor of any sort of use in a science admitting in every step of the plainest principles of reasoning. If the name of Sir Isaac Newton appears on this list, the number of advocates for error must be considerable. It is, however, to be recollected, that for a much longer period, men scarcely inferiour to Newton in genius, and his equals, probably, in industry, maintained a variety of positions in philosophy, which were overthrown by a more accurate investigation of nature; and if the name of Ptolemy can no longer support his epicycles, nor that of Des Cartes his vortices, Newton's dereliction of the principles of reasoning cannot establish the fallacious notion, that every equation has as many roots as it has dimensions.

This notion of Newton and others is founded upon precipitation. Instead of a patient examination of the subject, an hypothesis which accounts for many appearances is formed; where it fails, unintelligible terms are used; in those terms indolence acquiesces: much time is wasted on a jargon which has the appearance of science, and real knowledge is retarded. Thus volumes upon volumes have been written on the stupid dreams of Athanasius, and on the impossible roots of an equation of *n* dimensions.

As might be expected, the attempt by Frend, Maseres, and others to pursue algebra without the negative and complex numbers did not get very far; Peacock's dry comment on Baron Maseres was that 'He seems generally to have forgotten that any change had taken place in the science of algebra between the age of Ferrari, Cardan, Des Cartes, and Harriot, and the end of the 18th century' (*Peacock 1833*, p. 191). (It is perhaps worth mentioning that Frend's daughter, Sophia Elizabeth Frend, married Augustus De Morgan in 1837. De Morgan's dissent from his father-in-law's views on algebra can be found in the *Budget of paradoxes* (*De Morgan 1872*, pp. 117–25); although, as will be seen below in the selections from De Morgan, he, too, in his early years had difficulty in providing a satisfactory justification of the negative and the complex numbers.)

Peacock's own theory of 'symbolical algebra' was intended to provide Continental analysis with a rigorous justification—to put it on an equal footing with the geometric method Newton and MacLaurin had inherited from Archimedes. His theory rested on a distinction between what he called *arithmetical algebra* and *symbolical algebra*.

In *arithmetical algebra* (which Peacock also calls *universal arithmetic*) the 'general symbols' (i.e. the variables) represent positive real numbers. The signs '+' and '−' cannot be used independently in expressions like '+5' or '−3,' but only when preceded by some other numerical expression; subtraction is impossible when the subtrahend is greater than the minuend; and multiple positive roots of equations are prohibited, as are negative and 'impossible' roots (*Peacock 1833*, p. 189.).

Although Peacock was aware of the existence of irrational numbers, he does not treat them as especially problematic; in his understanding of the reals he tarries far behind *Bolzano 1817a*. His principal concern, both in *Peacock 1833* and in his later textbook, *Arithmetical algebra* (*1842–5*), was with techniques of calculation—with describing various methods of finding solutions to simple algebraic equations. His methodology is loose and intuitive. He does not explicitly state the principles that would justify his calculations; still less does he attempt to prove theorems about the real numbers from a rigorous set of axioms. In consequence, the problematic aspects of the irrationals appear never to have forced themselves on his attention.

Arithmetical algebra, says Peacock, as developed by Maseres and others, is unquestionably a logical and complete system; but it is unable to justify a great multitude of algebraic propositions that use negative and impossible quantities—

propositions that have proved to be both consistent and mathematically useful. Arithmetical algebra does, however, suggest a new and distinct and more general science, which Peacock calls *symbolical algebra*. In symbolical algebra, one abandons the assumption that the 'general signs' represent positive real numbers, and one abandons the restrictions on the symbols for operations like subtraction, allowing them to occur between *any* numerical expressions.

Peacock makes the crucial observation that symbolical algebra, though it has been *suggested* by arithmetical algebra, is strictly speaking independent of it; and he says that, by following out the suggestions of arithmetical algebra,

> we do necessarily arrive at a new science much more general than arithmetic, whose principles, however derived, may be considered as the immediate, though not the ultimate foundation of that system of combinations of symbols which constitutes the science of algebra. It is more natural and philosophical, therefore, to assume such principles as independent and *ultimate*, as far as the science itself is concerned, in whatever manner they may have been suggested, so that it may thus become essentially a science of symbols and their combinations, constructed upon its own rules, which may be applied to arithmetic and to all other sciences by interpretation: by this means, interpretation will *follow*, and not *precede*, the operations of algebra and their results (*Peacock 1833*, pp. 194–5).

Peacock further observes that, in symbolical algebra, it is not necessary that there exist *any* interpretation of the operation-symbols. But at this point Peacock begins to pull back from the implications of his argument. He argues that, although it would be *possible* to develop a system of symbolical algebra in which neither the general signs nor the operation-symbols have an interpretation, such a purely formal calculus would be of no mathematical interest. He therefore takes arithmetical algebra to be what he calls a *subordinate science of suggestion* to guide the development of symbolical algebra:

> It is evident that a system of symbolical algebra might be formed, in which the symbols and the *conventional* operations to which they were required to be subjected would be perfectly general both in value and application. If, however, in the construction of such a system, we looked to the assumption of such rules of operation or of combination only, as would be sufficient, and not more than sufficient, for deducing equivalent forms, without any reference to any subordinate science, we should be without any means of interpreting either our operations or their results, and the science thus formed would be one of symbols only, admitting of no application whatever. It is for this reason that we adopt a subordinate science as a science of suggestion, and we frame our assumptions so that our results shall be the same as those of that science, when the symbols and the operations upon them become identical likewise: and in as much as arithmetic is the science of calculation, comprehending all sciences which are reducible to measure and to number; and in as much as arithmetical algebra is the immediate form which arithmetic takes when its digits are replaced by symbols and when the fundamental operations of arithmetic are applied to them, *those symbols being general in form, though specific in value*, it is most convenient to assume it as the subordinate science, which our system of symbolical algebra *must be required to comprehend in all its parts* (*Peacock 1833*, p. 200).

For Peacock, the relationship between arithmetical algebra and symbolical algebra is as follows. The rules of symbolical algebra are not *deducible* from arithmetical algebra, and therefore not *founded* upon arithmetical algebra; strictly speaking, symbolical algebra is an independent science, a purely formal calculus. However, the rules of arithmetical algebra *suggest* the rules of symbolical algebra; and, conversely, the rules of symbolical algebra may be *applied* to arithmetical algebra *by interpretation*. (Peacock does not spell out the precise details of how the suggestion takes place; but in practice his methodology is, first, to disinterpret the variables in arithmetical algebra; and, second, to take closures of appropriate classes of symbols under given operation-symbols—in his terminology, he assumes 'that the symbols in symbolical algebra are perfectly general and unlimited both in value and representation, and that the operations to which they are subject are equally general likewise' (*Peacock 1833*, p. 195). For Peacock it is of the essence that symbolical algebra and arithmetical algebra should be in complete agreement wherever they overlap—i.e., whenever the expressions of symbolical algebra have an arithmetical interpretation. This fundamental principle Peacock calls the *principle of the permanence of equivalent forms*. He states it as follows:

Direct proposition:

Whatever form is algebraically equivalent to another when expressed in general symbols, must continue to be equivalent, whatever those symbols denote.

Converse proposition:

Whatever equivalent form is discoverable in arithmetical algebra considered as the science of suggestion, when the symbols are general in their form, though specific in their value, will continue to be an equivalent form when the symbols are general in their nature as well as in their form (*Peacock 1833*, pp. 198–9).

(Roughly speaking, in modern terminology an 'equivalent form' is an equation; symbols are 'general in their form' when they are variables rather than numerals; general symbols are 'specific in their value' when they are subjected to certain restrictions on how they can be interpreted (specifically, when they are restricted to represent positive real numbers); general symbols are 'general in value' when those restrictions are removed.)

Despite the murkiness of Peacock's theory of algebra, his writings introduced two ideas of cardinal importance into British mathematics. First, the idea of algebra as a purely symbolic calculus—an autonomous discipline whose rules might be suggested by some other science, but which is ultimately governed by its own syntactic principles; and second, the idea that the techniques of the Continental analysts could be justified by regarding them as precisely such a symbolical algebra. Peacock won his battle to introduce Continental methodology into Britain, thus paving the way for the researches of Stokes, Maxwell, Heaviside, and others; and the algebraic accomplishments of Gregory, De Morgan, and Boole would scarcely have been possible without the new syntactic theory of algebra. Peacock, it is true, was not a great creative mathematician. His writings ramble, and are for the most part expositions of the work of others; his 'symbolical algebra' is clumsily presented, and was more an *ad hoc* defence of the work of the Continental analysts than an attempt to explore new waters. Peacock himself is unlikely to have understood the implications of his own conception. But it is easy in retrospect to underrate the originality and the difficulty of his idea of a purely symbolic calculus. More than a century after *Peacock 1833*—long after the world of mathematics had absorbed non-Euclidean geometries, transfinite numbers, and continuous, nowhere-differentiable functions—Bertrand Russell, in the second edition of *The principles of mathematics*, could still say of Hilbert:

> The formalists have forgotten that numbers are needed, not only for doing sums, but for counting. Such propositions as "There were 12 apostles" or "London has 6,000,000 inhabitants" cannot be interpreted in their system. For the symbol "0" may be taken to mean any finite integer, without thereby making any of Hilbert's axioms false; and thus every number-symbol becomes infinitely ambiguous. The formalists are like a watchmaker who is so absorbed in making his watches look pretty that he has forgotten their purpose of telling the time, and has therefore omitted to insert any works (*Russell 1937*, p. vi).

Unfortunately, however, Peacock's principle of the permanence of equivalent forms was a strait-jacket on his symbolical algebra, and led him to develop its rules in too close analogy with the laws of arithmetic. He did not conceive clearly of algebras in which addition would be non-commutative, or in which the operation-symbols would represent geometric or logical or combinatorial operations. His overriding purpose was to justify the extensions of the concept of *finite real number* that had been exploited by the Continental mathematicians; and for this purpose he had no need to consider any algebraic system beyond the complex numbers.

From this same fact flow other shortcomings of his system. He makes no attempt to develop a strictly presented formalistic calculus, and indeed he is not fully consistent in distinguishing between symbols and operations. Nor does he attempt to lay down a body of axioms from which the other propositions of symbolical algebra could be rigorously derived. His methodology remains loose and intuitive; and his principle of the permanence of equivalent forms is too obscure to provide mathematicians with effective guidance. Had he not known in advance which algebraic formulae he intended to justify, the principle of the permanence of equivalent forms would hardly have led him to their discovery.

The next important step in the development of algebra in Britain was taken by Duncan Farquharson Gregory. Gregory (a great-great-grandson of the mathematician James Gregory) was born in Edinburgh to a professor of medicine at the University of Edinburgh; his grandfather and great-grandfather had held the same position. He studied in Geneva and the University of Edinburgh before, in 1833, entering Trinity College, Cambridge, where Peacock was at the time Fellow and Tutor in Mathematics. Gregory graduated in 1837, and became himself a Fellow of Trinity College in 1840. In 1838 he founded and edited the *Cambridge Mathematical Journal*; among the authors whom he befriended and whose work he published was the young and then unknown George Boole. Before Gregory's death at the age of 31 he had published two mathematical textbooks on the calculus and on the application of analysis to solid geometry (*Gregory 1841* and *1845*); he also published a number of research articles, most of which are collected in *Gregory 1865*.

His most important original contribution was to the theory of algebra. In his early papers on the differential calculus and on the calculus of finite differences he had treated symbols for *operations*—symbols like d/dx or Δ—as though they were symbols of quantity, freely adding them together or multiplying them by constants. This technique had been used by other mathematicians before Gregory, but its legitimacy was never sufficiently justified, and even those who used it regarded it with suspicion—as a mere analogy between operations and numbers. The technique—the *calculus of operations*, as it was known—was also important to Boole's early mathematical research on the calculus of finite differences; and Boole's knowledge of Gregory's defence of the technique underlies his algebraic analysis of logic in *Boole 1847*.

In attempting to understand and to justify the calculus of operations Gregory was led to the conception of algebra which he presents in the selection that follows: where Peacock's algebraic variables had stood for *numbers*, Gregory's would stand for *operations*. Gregory builds principally on the work of Peacock and Servois (whose *1814a*, as Gregory notes, was little known in England at the time, and appears to have been unknown to Peacock when he wrote *Peacock 1833*). Anybody who has examined the long-winded volumes of Maseres and Frend, or struggled to understand Peacock's explanation of the 'principle of the permanence of equivalent forms', will appreciate the economy and clarity of Gregory's exposition:

> The light, then, in which I would consider symbolical algebra, is, that it is the science which treats of the combination of operations defined not by their nature, that is, by what they are or what they do, but by the laws of combination to which they are subject. And as many different kinds of operations may be included in a class defined in the manner I have mentioned, whatever can be proved of the class generally, is necessarily true of all the operations included under it. This, it may be remarked, does not arise from any analogy existing in the nature of the operations, which may be totally dissimilar, but merely from the fact that they are all subject to the same laws of combination.

Algebra has here been cut loose from the 'principle of the permanence of equivalent forms' and from a too-close dependence on the arithmetic of the positive real numbers; and Gregory is now free to interpret his algebraic laws as applying to *geometric* operations as well as to the operations of elementary arithmetic. Gregory adopts Servois's terms *distributive* and *commutative* to characterize the 'laws of combination' which certain important operations obey; and he gives several examples of each. (He sometimes slips, using 'permutative' for 'commutative'.) He does not, however, discuss non-commutative operations, or even note the possibility of their existence. That insight was first to occur to others—to Hermann Grassmann in his *Ausdehnungslehre* (*1844*), and to William Rowan Hamilton, who conceived of quaternions while walking beside the Royal Canal in Dublin in October 1843. Nor does Gregory work within even an informal axiomatic framework, laying down explicit rules and attempting to deduce their consequences. But his singling out of the distributive and commutative properties, and his view of algebra as the study of such 'laws of combination', was an important milestone for British mathematics, and opened the door to the algebraic researches of the mid-century.

Gregory 1840 was presented to the Royal Society of Edinburgh on 7 May 1838, and printed in that Society's *Transactions* in 1840. References should be to the section numbers, which were marked with Roman numerals in the original edition.

A. ON THE REAL NATURE OF SYMBOLICAL ALGEBRA (*GREGORY 1840*)

THE following attempt to investigate the real nature of Symbolical Algebra, as distinguished from the various branches of analysis which come under its dominion, took its rise from certain general considerations, to which I was led in following out the principle of the separation of symbols of operation from those of quantity. I cannot take it on me to say that these views are entirely new, but at least I am not aware that any one has yet exhibited them in the same form. At the same time, they appear to me to be important, as clearing up in a considerable degree the obscurity which still rests on several parts of the elements of symbolical algebra. Mr PEACOCK is, I believe, the only writer in this country who has attempted to write a system of algebra founded on a consideration of general principles, for the subject is not one which has much attraction for the generality of mathematicians. Much of what follows will be found to agree with what he has laid down, as well as with what has been written by the Abbé BUEE and Mr WARREN; but as I think that the view I have taken of the subject is more general than that which they have done, I hope that the following pages will be interesting to those who pay attention to such speculations.

The light, then, in which I would consider symbolical algebra, is, that it is the science which treats of the combination of operations defined not by their nature, that is, by what they are or what they do, but by the laws of combination to which they are subject. And as many different kinds of operations may be included in a class defined in the manner I have mentioned, whatever can be proved of the class generally, is necessarily true of all the operations included under it. This, it may be remarked, does not arise from any analogy existing in the nature of the operations, which may be totally dissimilar, but merely from the fact that they are all subject to the same laws of combination. It is true that these laws have been in many cases suggested (as Mr PEACOCK has aptly termed it) by the laws of the known operations of number; but the step which is taken from arithmetical to symbolical algebra is, that, leaving out of view the nature of the operations which the symbols we use represent, we suppose the existence of classes of unknown operations subject to the same laws. We are thus able to prove certain relations between the different classes of operations, which, when expressed between the symbols, are called algebraical theorems. And if we can show that any operations in any science are subject to the same laws of combination as these classes, the theorems are true of these as included in the general case: Provided always, that the resulting combinations are all possible in the particular operation under consideration. For it may very well, and does actually happen, that, though each of two operations in a certain branch of science may be possible, the complex operation resulting from their combination is not equally possible. In such a case, the result is inapplicable to that branch of science. Hence we find, that one family of a class of operations may have a more general application than another family of the same class. To make my meaning more precise, I shall proceed to apply the principle I have been endeavouring to explain, by shewing what are the laws appropriate to the different classes of operations we are in the habit of using.

Let us take as usual F and f to represent any operations whatever, the natures of which are unknown, and let us prefix these symbols to any other symbols, on which we wish to indicate that the operation represented by F or f is to be performed.

I. We assume, then, the existence of two classes of operations F and f, connected together by the following laws.

$$(1.)\ \mathrm{F}\,\mathrm{F}\,(a) = \mathrm{F}\,(a). \qquad (2.)\ ff(a) = \mathrm{F}(a).$$
$$(3.)\ \mathrm{F}\,f(a) = f(a). \qquad (4.)\ f\mathrm{F}(a) = f(a).$$

Now, on looking into the operations employed in arithmetic, we find that there are two which are subject to the laws we have just laid down. These are the operations of addition and subtraction; and as to them the peculiar symbols of + and − have been affixed, it is convenient to retain these as the symbols of the general class of operations we have defined, and we shall therefore use them instead of F and f. As it is useful to have peculiar names attached to each class, I would propose to call this the class of *circulating* or *reproductive* operations, as their nature suggests.

Again, on looking into geometry, we find two operations which are subject to the same laws. The one corresponding to + is the turning of a line, or rather transferring of a point, through a circumference; the other corresponding to − is the transference of a point through a semicircumference. Consequently, whatever we are able to prove of the general symbols + and − from the laws to which they are subject, without considering the nature of the operations they indicate, is equally true of the arithmetical operations of addition and subtraction, and of the geometrical operations I have described. We see clearly from this, that there is no real analogy between the nature of the operations + and − in arithmetic and geometry, as is generally supposed to be the case, for the two operations cannot even be said to be opposed to each other in the latter science, as they are generally said to be. The relation which does exist is due not to any identity of their nature, but to the fact of their being combined by the same laws. Other operations might be found which could be classed under the general head we are considering. Mr PEACOCK and the Abbé BUEE consider the transference of property to be one of these; but as there is not much interest attached to it in a mathematical point of view, I shall proceed to the consideration of other operations.

II. Let us suppose the existence of operations subject to the following laws:

$$(1.)\ f_m(a).f_n(a) = f_{m+n}(a). \quad (2.)\ f_m f_n(a) = f_{mn}(a).$$

Where f_m, f_n are different species of the same genus of operations, which may be conveniently named index-operations, as, if we define the form of f by making $f_1(a) = a$, and suppose m and n to be integer numbers, we have those operations which are represented in arithmetical algebra by a numerical index. For if m and n be integers, and the operation a^m be used to denote that the operation a has been repeated m times, then, as we know

$$a^m.a^n = a^{m+n}. \qquad (a^m)^n = a^{mn}.$$

We have now to consider whether we can find any other actual operations besides that of repetition which shall be subject to the laws we have laid down. If we suppose that m and n are fractional instead of integer, we easily deduce from our definition that the notation $a^{\frac{p}{q}}$ is equivalent to the arithmetical operation of extracting the qth root of the pth power of a, or generally the finding of an operation, which being repeated q times, will give as a result the operation a^p. Thus we find, as might have been expected, a close analogy existing between the meanings of a^m when m is integer, and when it is fractional. Again, we might ask the meaning of the operation a^{-m}; and we find without difficulty, from the law of combination, that a^{-m} indicates the inverse operation of a^m, whatever the operation a may be. When, instead of supposing m to be a number integer or fractional, we suppose it to indicate any operation whatever, I do not know of any interpretation which can be given to the rotation, excepting in the case when it indicates the operation of differentiation, represented by the symbol d. For we know by TAYLOR'S theorem, that

$$\varepsilon^{h\frac{d}{dx}} f(x) = f(x+h)$$

Or,

$$a^{\frac{d}{dx}} f(x) = f(x + \log a).$$

In the case of negative indices, we have combined two different classes of operations in one manner, but we may likewise do it in another. What meaning, we may ask, is to be attached to such complex operations as $(+)^m$ or $(-)^m$? When m is an integer number, we see at once that the operation $(+)^m$ is the same as $+$, but $(-)^m$ becomes alternately the same as $+$ and as $-$, according as m is odd or even, whether they be the symbols of arithmetical or geometrical operations. So far there is no difficulty. But if it be fractional, what does

$(+)^m$ or $(-)^m$ signify? In arithmetic, the first may be sometimes interpreted, as because $(+)^m = +$ when m is integer, $(+)^{\frac{1}{m}}$ also $= +$, and as $(-)^{2m} = +$, also $(+)^{\frac{1}{2m}} = -$: But the other symbol $(-)^m$ has, when m is a fraction with an even denominator, absolutely no meaning in arithmetic, or at least we do not know at present of any arithmetical operation which is subject to the same laws of combination as it is. On the other hand, geometry readily furnishes us with operations which may be represented by $(+)^{\frac{1}{m}}$ and $(-)^{\frac{1}{m}}$, and which are analogous to the operations represented by $+$ and $-$. The one is the turning of a line through an angle equal to $\frac{1}{m}$th of four right angles, the other is the turning of a line through an angle equal to $\frac{1}{m}$th of two right angles. Here we see that the geometrical family of operations admits of a more extended application than the arithmetical, exemplifying a general remark we had previously occasion to make. Whether when the index is any other operation, we can attach any meaning to the expression, has not yet been determined. For instance, we cannot tell what is the interpretation of such expressions as $(+)^{\frac{d}{dx}}$ or $(-)^{\frac{d}{dx}}$, or $(+)^{\log}$.

III. I now proceed to a very general class of operations, subject to the following laws:

$$(1.)\ f(a)+f(b)=f(a+b).$$
$$(2.)\ f_{\prime}f(a) \quad =ff_{\prime}(a).$$

This class includes several of the most important operations which are considered in mathematics; such as the numerical operation usually represented by a, b, &c., indicating that any other operation to which these symbols are prefixed is taken a times, b times, &c.; or as the operation of differentiation indicated by the letter d, and the operation of taking the difference indicated by Δ. We therefore see what an important part this class of functions plays in analysis, since it can be at once divided into three families which are of such extensive use. This renders it advisable to comprehend these functions under a common name. Accordingly, SERVOIS, in a paper which does not seem to have received the attention it deserves, has called them, in respect of the first law of combination, *distributive* functions, and in respect of the second law, *commutative* functions. As these names express sufficiently the nature of the functions we are considering, I shall use them when I wish to speak of the general class of operations I have defined.

It is not necessary to enter at large here, into the demonstration that the symbols of differentiation and difference are subject to the same laws of combination as those of number. But it may not be amiss to say a few words on the effect of considering them in this light. Many theorems in the differential calculus, and that of finite differences, it was found might be conveniently expressed by separating the symbols of operation from those of quantity, and treating the former like ordinary algebraic symbols. Such is LAGRANGE's elegant theorem, the first expressed in this manner, that

$$\Delta^n u_x = (\varepsilon^{h\frac{d}{dx}} - 1)^n u_x;$$

or the theorem of LEIBNITZ, with many others. For a long time these were treated as mere analogies, and few seemed willing to trust themselves to a method, the principles of which did not appear to be very sound. Sir JOHN HERSCHEL was the person in this country who made the freest use of the method, chiefly, however, in finite differences. In France, SERVOIS was, I believe, the only mathematician who attempted to explain its principles, though BRISSON and CAUCHY sometimes employed and extended its application: and it was in pursuing this investigation that he was led to separate functions into distributive and commutative, which he perceived to be the properties which were the foundation of the method of the separation of the symbols, as it is called. This view, which, so far as it goes, coincides with that which it is the object of this paper to develope, at once fixes the principles of the method on a firm and secure basis. For, as these various operations are all subject to common laws of combination, whatever is proved to be true by means only of these laws, is necessarily equally true of all the operations. To this I may add, that when two distributive and commutative operations are such that the one does not act on the other, their combinations will be subject to the same laws as when they are taken separately; but when they are not independent, and one acts on another, this will no longer be true. Hence arises the increased difficulty of solving linear differential equations with variable coefficients; but for more detailed remarks on this, as well as for examples of a more extended use of the method of the separation of symbols than has hitherto been made, I refer to the *Cambridge Mathematical Journal*, Nos. 1, 2, and 3.

As we found geometrical operations which were subject to the laws of circulating operations, so there is a geometrical operation which is subject to the laws of distributive and permutative operations, and therefore may be represented by the same symbols. This is transference to a distance measured in a straight line. Thus if x represent a point, line, or any geometrical figure, $a(x)$ will represent the transference of this point or line; and it will be seen at once that

$$a(x) + a(y) = a(x + y);$$

or the operation a is distributive. What, then, will the compound operation $b(a(x))$ represent? If x represent a point, $a(x)$, which is the transference of a point to a rectilinear distance, or the tracing out of a straight line, will stand for the result of the operation; and then $b(a(x))$ will be the transferring of a line to a given distance from its original position. In order to effect this, the line must be moved parallel to itself, the effect of which will be the tracing out of a parallelogram. The effect will be the same if we suppose a to act on $b\ (x)$, since in this, as in the other case, the same parallelogram will be traced out: that is to say,

$$a(b(x)) = b(a(x))$$

or a and b are commutative operations.

The binomial theorem, the most important in symbolical algebra, is a theorem expressing a relation between distributive and commutative operations, index operations, and circulating operations. It takes cognizance of nothing in these operations except the six laws of combination we have laid down, and, as we shall presently shew, it holds only of functions subject to these laws. It is consequently true of all operations which can be shewn to be commutative and distributive, though apparently, from its proof, only true of the operations of number. The difficulties attending the general proof of this theorem are well known, and much thought has been bestowed on the best mode of avoiding them. The principles I have been endeavouring to exhibit appear to me to shew in a very clear light the correctness of EULER's very beautiful demonstration. Starting with the theorem as proved for integer indices, which he uses as a suggestive form, he assumes the existence of a series of the same form when the index is fractional or negative, which may be represented by $f_m(x)$. He then considers what will be the form of the product $f_m(x) \times f_n(x)$. This form must depend only on the laws of combination to which the different operations in the expression are subject. When x is a distributive and commutative function, and m and n integer numbers, we know that $f_m(x) \times f_n(x) = f_{m+n}(x)$. Now integer numbers are one of the families of the general class of distributive and permutative functions; and if we actually multiplied the expressions $f_m(x)$ and $f_n(x)$ together, we should, even in the case of integers, make use only of the distributive and permutative properties. But these properties hold true also of fractional and negative quantities. Therefore, in their case, the form of the product must be the same as when the indices are integer numbers. Hence $f_m(x) \times f_n(x) = f_{m+n}(x)$ whether m and n be integer or fractional, positive or negative, or generally if m and n be distributive and permutative functions.

The remainder of the proof follows very readily after this step, which is the key-stone of the whole, so that I need not dwell on it longer. I will only say, that this mode of considering the subject shews clearly, that not only must the quantities under the vinculum be distributive and commutative functions, but also the index must be of the same class,—a limitation which I do not remember to have seen any where introduced. Therefore the binomial theorem does not apply to such expressions as $(1 + a)^{\log}$ or $(1 + a)^{\sin}$; and, though it does apply to $(1 + a)^{\frac{d}{dx}}$, since both a and $\frac{d}{dx}$ are distributive and commutative operations, it does not apply to $(1 + f(x))^{\frac{d}{dx}}$, as $f(x)$ and $\frac{d}{dx}$ are not relatively commutative.

Closely connected with the binomial theorem is the exponential theorem, and the same remarks will apply equally to both. So that, in order that the relation

$$\varepsilon^x = 1 + x + \frac{x^2}{1.2} + \frac{x^3}{1.2.3} + \&c.$$

may subsist, it is necessary, and it suffices, that x should be a distributive and commutative function. On this depends the propriety of the abbreviated notation for TAYLOR's theorem

$$f(x+h) = \varepsilon^{h\frac{d}{dx}} f(x).$$

Properly speaking, however, the symbol ε ought not to be used, as it implies an arithemtical relation, and instead, we ought to employ the more general symbol of $\log^{-1}$. But this depends on the existence of a class of operations on which I may say a few words.

IV. If we define a class of operations by the law

$$f(x) + f(y) = f(xy),$$

we see that, when x and y are numbers, the operation is identical with the arithmetical logarithm. But when x and y are any thing else, the function will have a different meaning. But so long as they are distributive and commutative functions, the general theorems such as

$$\log(1+x) = x - \frac{x^2}{2} + \frac{x^3}{3} - \&\text{c}.$$

being proved solely from laws we have laid down, are true of all symbols subject to those laws. It happens that we are not generally able to assign any known operation to which the series is equivalent when x is any thing but a number, and we therefore say that $\log(1+x)$ is an abbreviated expression for the series $x - \frac{x^2}{2} + \frac{x^3}{3} - \&\text{c}.$ But there may be distinct meanings for such expressions as $\log\left(1 + \frac{d}{dx}\right)$, or $\log\left(\frac{d}{dx}\right)$, as there are for $\varepsilon^{h\frac{d}{dx}}$, that is $\log^{-1}\left(\frac{d}{dx}\right)$. In the case of another operation, Δ, we know that $\log(1+\Delta) = \frac{d}{dx}$.

V. The last class of operations I shall consider is that involving two operations connected by the conditions

$$(1) \quad a\mathrm{F}(x+y) = \mathrm{F}(x)f(y) + f(x)\mathrm{F}(y)$$

and

$$(2) \quad af(x+y) = f(x)f(y) - c\mathrm{F}(x)\mathrm{F}(y).$$

These are laws suggested by the known relation between certain functions of elliptic sectors; and when a and c both become unity, they are the laws of the combinations of ordinary sines and cosines, which may be considered in geometry as certain functions of angles or circular sectors, but in algebra we only know of them as abbreviated expressions for certain complicated relations between the first three classes of operations we have considered. These relations are,

$$\mathrm{Sin}\, x = x - \frac{x^3}{1.2.3} + \frac{x^5}{1.2.3.4.5} \&\text{c}.$$

$$\mathrm{Cos}\, x = 1 - \frac{x^2}{1.2} + \frac{x^4}{1.2.3.4} \&\text{c}.$$

The most important theorem proved of this class of functions is that of DEMOIVRE, that

$$(\cos x + (-)^{\frac{1}{2}} \sin x)^n = \cos nx + (-)^{\frac{1}{2}} \sin nx.$$

It is easy to see that, in arithmetical algebra, the expression $\cos x + (-)^{\frac{1}{2}} \sin x$ can receive no interpretation, as it involves the operation $(-)^{\frac{1}{2}}$. In geometry, on the contrary, it has a very distinct meaning. For if a represent a line, and $a \cos x$ represent a line bearing a certain relation in magnitude to a, and $a \sin x$ a line bearing another relation in magnitude to a, then $a (\cos x + (-)^{\frac{1}{2}} \sin x)$ will imply, that we have to measure a line $a \cos x$, and from the extremity of it we are to measure another line $a \sin x$; but in consequence of the sign of operation $(-)^{\frac{1}{2}}$, this new line is to be measured, not in the same direction as $a \cos x$, but turned through a right angle. As, for instance, if AB $= a \cos x$, and BC′ $= a \sin x$, we must not measure it in the prolongation of AB, but turn it round to the

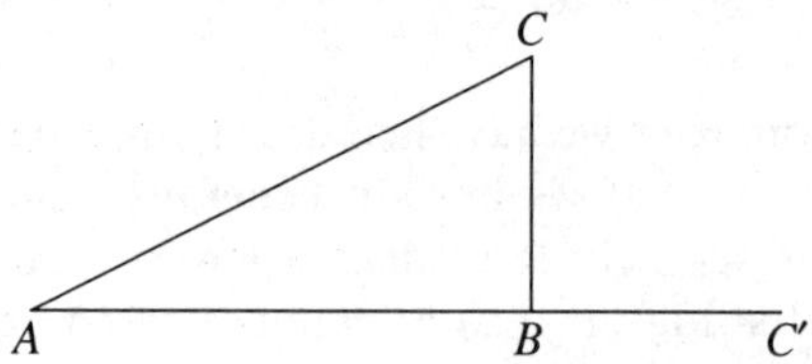

position BC; and thus, geometrically, we arrive at the point C. Also, from the relation between sin x and cos x, we know that the line AC will be equal to a, and thus the expression $a (\cos x + (-)^{\frac{1}{2}} \sin x)$ is an operation expressing that the line whose length is a, is turned through an angle x. Hence, the operation indicated by $\cos \frac{2\pi}{n} + (-)^{\frac{1}{2}} \sin \frac{2\pi}{n}$ is the same as that indicated by $(+)^{\frac{1}{n}}$, the difference being, that, in the former, we refer to rectangular, in the latter to polar co-ordinates. Mr PEACOCK has made use of the expression cos $x + (-)^{\frac{1}{2}} \sin x$ to represent direction, while Mr WARREN has employed one which, though disguised under an inconvenient and arbitrary notation, is the same as $(+)^{\frac{1}{n}}$. The connection between these expressions is so intimate, that, being subject to the same laws, they may be used indifferently the one for the other. This has been the case most particularly in the theory of equations. The most general form of the root is usually expressed by $a (\cos \theta + (-)^{\frac{1}{2}} \sin \theta)$, while the more correct symbolical form would be $(+)^{\frac{p}{q}} a$, since the expression

$$x^n + P_1 x^{n-1} + P_2 x^{n-2} + \&c. + P_n = 0$$

does not involve any sine or cosine, but may be considered as much a function of + as of x, so that the former symbol may be easily supposed to be involved in the root. Hence, instead of the theorem that every equation must have a root, I would say every equation must have a root of the form $(+)^{\frac{p}{q}} a$, p and q being numbers, and a a distributive and commutative function.

9

Augustus De Morgan (1806–1871)

We have seen above that in the eighteenth and early nineteenth centuries many central mathematical problems—the justification of the calculus, the status of the axioms of geometry, the legitimacy of equations containing $\sqrt{-1}$—were intertwined in the writings of mathematicians with broadly philosophical problems about language and the reference of mathematical expressions; recall, for example, Berkeley's *Principles* (*1710*, §§6–25), his *Analyst* (*1734*, §§8, 36, 37), Lambert's remarks on the nature of an axiom system (*1786*, §11), Kant's theory of mathematics as the 'science of the construction of concepts' (*1781*), Bolzano's paper on the proper presentation of mathematics (*1810*), or the concluding paragraphs of Gauss's memoir on the status of the complex numbers (*1831*). In the writings of the British algebraists of the nineteenth century these philosophical issues become entangled with a further range of mathematical problems, most notably, with the algebraic analysis of logic and with the beginnings of abstract algebra; and progress on the mathematical front was in part made possible by a new-found clarity on the philosophical issues. De Morgan, one of the most prolific algebraists and logicians of the period, exemplifies the new trend.

The preceding selection and its accompanying Note has described the first stirrings of a new conception of algebra in British mathematics. The pioneers were George Peacock and Duncan Gregory: Peacock introducing the idea of a purely syntactic *symbolic algebra*, and Gregory introducing the idea that the algebraic symbols need not be taken to stand for *numbers*, but could equally well represent *operations*.

Augustus De Morgan took the next important step. He was the first mathematician to appreciate the importance of the new algebra for the analysis of logic, and the first to provide a reasonably complete and explicit description of a formal calculus. He viewed algebra with the eyes of a logician and logic with the eyes of an algebraist, seeking both to understand the principles of reasoning that were implicit in the new algebra as a form of mathematical inference, and to use the techniques of algebra to improve on the traditional theory of logic.[a] Within logic proper, De Morgan made important strides beyond the

[a] De Morgan, like Boole, frequently insisted on the close relationship between algebra and logic; for instance, in the two concluding paragraphs of his important paper, 'On the syllogism, No. IV, and on the logic of relations' (*De Morgan 1864b*): 'It is to *algebra* that we must look for the most habitual use of logical forms. Not that the onymatic relations are found in frequent occurrence:

Aristotelian syllogism by introducing the ideas of 'quantifying the predicate' and by studying the logic of relations, thus freeing logic from its analysis of all propositions into the subject–predicate form.

Although De Morgan was not a powerfully original mathematician like Hamilton or Boole or Cayley, he was an important systematizer and expositor, and in his day exerted a considerable influence on British mathematics. De Morgan was educated at Trinity College, Cambridge, where he studied under George Peacock and William Whewell; he took his degree in 1827. In 1828 he was appointed to the chair of mathematics at University College, London, where he remained until 1866. (He resigned his chair twice over issues of principle; the second resignation was permanent.) Among his students at University College was James Joseph Sylvester. De Morgan was a founding member and the first president of the London Mathematical Society, giving its inaugural lecture in 1865; he was also active in the Society for the Diffusion of Useful Knowledge, which published many of his books. He wrote voluminously and influentially, producing elegant textbooks on arithmetic (1830), spherical trigonometry (1834), algebra (1835), number and magnitude (*De Morgan 1836*), trigonometry (1837), probability and insurance (1838), logic and geometry (*De Morgan 1839*), and the differential and integral calculus (1842); he also wrote hundreds of expository articles for a general audience—over 850 in the *Penny Cyclopaedia* (later called the *English Cyclopaedia*) alone. He coined the term 'mathematical induction' in 1838 for a type of reasoning that had been used at least since Euclid's proof that there are infinitely many prime numbers; introduced into logic the idea of a 'universe of discourse'; was the first mathematician to state the four-colour conjecture as a problem in need of solution; gave in his calculus textbook a precise, analytical definition of a *limit*; and

---

but so soon as the syllogism is considered under the aspect of combination of relations, it becomes clear that there is more of syllogism, and more of its variety, in algebra than in any other subject whatever, though the matter of the relations—pure quantity—is itself of small variety. And here the general idea of relations emerges, and for the first time in the history of knowledge, the notions of relation and *relation of relation* are symbolized. And here again is seen the scale of gradations of form, the manner in which what is difference of *form* at one step of the ascent, is difference of *matter* at the next. But the relation of algebra to the higher developments of logic is a subject of far too great extent to be treated here. It will hereafter be acknowledged that, though the geometer did not think it necessary to throw his ever-recurring *principium et exemplum* into an imitation of '*Omnis homo est animal, Socrates est homo, &c.*', yet the algebraist was living in the higher atmosphere of syllogism, the unceasing composition of relation, before it was admitted that such an atmosphere existed.

I expect agreement in what I have said neither from the logicians nor from the algebraists; but, for reasons given in my last paper, I do not submit myself to either class. Not that I by any means take it for granted that all those who have cultivated both sciences will agree with me. When two countries are first brought by the navigators into communication with each other, it is found that there are two kinds of perfect agreement, and one case of nothing but discordance. All the inhabitants of each of the countries are quite at one in believing a huge heap of mythical notions about the other. At first, the only persons who though similarly circumstanced nevertheless tell different stories are the very mariners who have passed from one land to the other. This will go on for a time, and for a time only: *multi pertransibunt, et augebitur scientia* ‖many will make the crossing, and knowledge will be increased‖.'

invented a new convergence test for infinite series. He wrote biographies of Newton and of Halley, published an index to the scientific correspondence of the seventeenth century, and compiled one of the first bibliographies of mathematics. He also served as a nineteenth-century Martin Gardner, writing the exuberant *Budget of paradoxes* (*1872*), which deals with such topics as spiritualism, the history of π, astrology, circle-squarers, and selected issues in the history of science and of mathematics. C.S. Peirce's obituary notice of De Morgan in *The Nation* (13 April 1871) declares that

> As a writer and a teacher, he was one of the clearest minds that ever gave instruction, while his genial and hearty manners in private and in the school-room strongly attached to him all who came in contact with him. He was a man of full habit, much given to snuff-taking; and those who have seen him at the blackboard, mingling snuff and chalk in equal proportions, will not soon forget the singular appearance he often presented.

De Morgan's chief scientific contributions were to logic; many of his results are presented in the *Formal logic: or, The calculus of inference, necessary and probable* (*1847*) and in the articles collected in *On the syllogism and other logical writings* (*De Morgan 1966*). Although the following selections are concerned with the foundations of algebra, De Morgan's main logical accomplishments deserve a brief mention.

One of his most important discoveries, dating from the mid-1840s, is the notion of 'quantifying the predicate'.[b] De Morgan observed that, although in the Aristotelian syllogistic no inference can be drawn from 'Some *Y*s are *X*s' and 'Some *Y*s are *Z*s', nevertheless the following is a valid inference:

Most *Y*s are *X*s.
Most *Y*s are *Z*s.
―――――――――――
Therefore, some *X*s are *Z*s.

De Morgan stated the inference quantitatively: if there are y *Y*s, and if z of the *Y*s are in Z and if x of the *Y*s are in X, then at least $(x + z - y)$ *X*s are *Z*s. This method of 'quantifying the middle term' (or 'quantifying the predicate') allowed him to generalize the traditional syllogistic reasoning; it also embroiled him in an acrimonious public dispute in 1847 with the Scottish philosopher Sir William Hamilton (not to be confused with the Irish mathematician, Sir William Rowan Hamilton), who had published similar but cruder ideas a few years earlier, and who accused De Morgan of plagiarism. The resulting argument sprawls over a series of increasingly intemperate letters and pamphlets. The documents relevant to the argument are listed (with a twenty-five page

[b] De Morgan's 'quantification of the predicate' is a step on the road to the modern theory of quantification; and indeed, De Morgan's terminology led C.S. Peirce to coin the word 'quantifier'. From the writings of Peirce and Schröder the term passed into general use. See *Peirce 1885* (reproduced below), footnote to §3.

commentary by De Morgan) in an appendix to *De Morgan 1847*.[c] One unintended and welcome consequence of this dispute was to attract the attention of George Boole to logic; Boole acknowledged the stimulus in the preface to his *Mathematical analysis of logic (1847)*, reproduced below.

In addition to investigating the 'quantification of the predicate', De Morgan was also the first logician to make a systematic study of the logic of relations. He pointed out that Aristotelian logic, because it analysed every proposition into the subject-predicate form, could not demonstrate the validity of the inference:

All men are animals.

Therefore, the head of a man is the head of an animal.

The topic of relations had been touched upon sporadically since the time of Aristotle, but the first sustained investigation is contained in De Morgan's 'On the syllogism, No. IV, and on the logic of relations' (*De Morgan 1864b*).

In this paper he declared that the syllogism, on which his own work had previously been focused, was merely a special case of the composition of relations, and he introduced a now obsolete symbolism to handle the logic of simple relational expressions. In his notation, '$X..LY$' stood for 'X is an L of Y'. (The curious dots were derived from an earlier system of notation designed by De Morgan for the analysis of the syllogism.) '$X.LY$' was the denial of the original proposition, 'X is not an L of Y.' The *contrary* relation 'not-L' was represented by lower-case l, so that '$X.LY$' and '$X..lY$' are equivalent. The *converse* relation of L he represented by L^{-1}, so that '$X..LY$' and '$Y..L^{-1}X$' are equivalent. In his system, relative terms could be combined; thus, '$X..LMY$' stood for 'X is an L of one of the Ms of Y.' De Morgan proceeded to ponder various ways in which pairs of relative terms could be combined, and showed that any combination could be expressed in terms of one of the three patterns: (i) 'Aristotle is a friend of a disciple of Socrates'; (ii) 'Aristotle is a friend of every disciple of Socrates'; and (iii) 'Aristotle is a friend of none but a disciple of Socrates.'

De Morgan's study of relations broke important new ground, but his system had a number of serious shortcomings. He did not consider more elaborate combinations of larger numbers of relational terms; his system was burdened

---

[c] By the final stages of the dispute, Sir William was denouncing De Morgan's notation as 'horrent with mysterious spiculae'—in contrast to his own, which he described as 'easy, simple, compendious, all-sufficient, consistent, manifest, precise, complete'. He compares De Morgan to 'an owl in daylight', and warns that 'mathematics and dram-drinking tell, especially in the long run. For a season, I admit, Toby Philpot may be the Champion of England, and Warburton testifies,—"It is a thing notorious, that the *oldest mathematician* in England is the *worst reasoner* in it." So much for Mathematical Logic; so much for Cambridge Philosophy' (*W. Hamilton 1852*, pp. 705–6). C.S. Peirce accurately described the controversy as follows: 'The reckless Hamilton flew like a dor-bug into the brilliant light of De Morgan's mind in a way which compelled the greatest formal logician that ever lived to examine and report upon ‖Hamilton's‖ system. There was a considerable controversy; for Hamilton and several of his pupils were as able in controversy as they were impotent in inquiry—but De Morgan's final and unanswerable paper will be found in ‖*De Morgan 1864c*‖' (*Peirce 1931–58*, Vol. ii, §534).

with two kinds of negation; he had no convenient way of representing n-placed relations for $n \geqslant 3$; his notation for the basic logical operations was cumbrous and inelegant. De Morgan's writings were but the start of the work done on the logic of relations in the nineteenth century, and in retrospect his entire system of logic cries out for the apparatus of quantification theory. The task of exploring the subject in depth—of improving the notation, expanding the kinds of relations studied, and introducing the quantifiers—was to devolve upon others, most notably upon C.S. Peirce and Ernst Schröder.[d]

Ironically, De Morgan is today best remembered for what are called 'De Morgan's laws'—in modern notation

$$(A \cap B)' = A' \cup B'$$
$$(A \cup B)' = A' \cap B'.$$

De Morgan stated these rather obvious laws in his *1864a*, but he did not in fact discover them. (Indeed, they were well known to the Schoolmen. See, for instance, Occam's *Summa logicae*, ii, §32.)

De Morgan was acquainted with most of the leading algebraists of the day, and was a prolific letter-writer; those wishing to delve further into the history of algebra and logic in Britain in the nineteenth century should consult his voluminous scientific correspondence. His correspondence with Rowan Hamilton (dealing with quaternions, logic, real analysis, complex numbers, political reform, astronomy, religion, convergent series, poetry, anagrams, and the history of algebra) is partially reproduced in the third volume of the biography of Hamilton, *Graves 1882–9*; excerpts are printed below. De Morgan's mathematical correspondence with Boole has been published as *Boole and De Morgan 1982*. No complete bibliography of De Morgan's scientific writings exists; the fullest and most reliable is in *Boole and De Morgan 1982*.

---

[d] Both Peirce and Schröder had studied De Morgan's writings, and gave him credit for opening the new subject. Peirce describes his encounter with De Morgan's work on relations as follows: 'It must have been in 1866 that Professor De Morgan honored the unknown beginner in philosophy that I then was (for I had not earnestly studied it for more than ten years, which is a short apprenticeship in this most difficult of subjects), by sending me a copy of his memoir "On the Logic of Relations, etc." [*De Morgan 1864b*]. I at once fell to upon it; and before many weeks had come to see in it, as De Morgan had already seen, a brilliant and astonishing illumination of every corner and every vista of logic. Let me pause to say that no decent semblance of justice has ever been done to De Morgan, owing to his not having brought anything to its final shape. Even his personal students, reverent as they perforce were, never sufficiently understood that his was the work of an exploring expedition, which every day comes upon new forms for the study of which leisure is, at the moment, lacking, because additional novelties are coming in and requiring note. He stood indeed like Aladdin (or whoever it was) gazing upon the overwhelming riches of Ali Baba's cave, scarce capable of making a rough inventory of them' (*Peirce 1931–58*, Vol. i, §562). Incidentally, Peirce and De Morgan met in London in 1870, shortly before De Morgan's death (*Peirce 1982–*, Vol. ii, p. xxxii).

A. ON THE FOUNDATION OF ALGEBRA (*DE MORGAN 1842a*)

De Morgan's early writings on algebra display traces of the distrust of negative and imaginary numbers that British mathematicians had inherited from writers like Francis Maseres and William Frend, whose daughter De Morgan was to marry in 1837. (The algebraic theories of Maseres and Frend are discussed above in the Introductory Note to Duncan Gregory.) The following quotation from De Morgan was published in 1831—the same year, incidentally, as Gauss's paper on biquadratic residues and the complex numbers. De Morgan was twenty-six, and as perplexed by the imaginary numbers as his future father-in-law:

> We have shown the symbol $\sqrt{-a}$ to be void of meaning, or rather self-contradictory and absurd. Nevertheless, by means of such symbols, a part of algebra is established which is of great utility. It depends upon the fact, which must be verified by experience, that the common rules of algebra may be applied to these expressions without leading to any false results. . . .
>
> The imaginary expression $\sqrt{-a}$ and the negative expression $-b$ have this resemblance, that either of them occurring as the solution of a problem indicates some inconsistency or absurdity. As far as the real meaning is concerned, both are equally imaginary, since $0 - a$ is as inconceivable as $\sqrt{-a}$ (*De Morgan 1831*, Ch. 10).

The next few years saw a flurry of activity among British mathematicians working on the foundations of algebra. George Peacock's *Report on certain branches of analysis* appeared in 1833; Rowan Hamilton's essay on algebra as the science of pure time appeared in 1837; Duncan Gregory's 'On the real nature of symbolical algebra' was read to the Royal Society of Edinburgh in 1838 (although it was not printed until 1840). The following selection, read to the Cambridge Philosophical Society on 9 December 1839, shows De Morgan grappling with the new ideas; it is the first of four papers 'On the foundations of algebra' read to that Society between 1839 and 1844 (*De Morgan 1842a*, *1842b*, and *1849a*). He explicitly mentions Peacock and Rowan Hamilton, and his reference in §1 to the 'separation . . . of the symbols of operation and quantity' suggests he knew of Gregory's work as well. Following in the footsteps of Peacock and Gregory, De Morgan (§3) distinguishes between algebra as a purely symbolic calculus and algebra as the method of interpreting the symbols: in his terminology, between *technical algebra* and *logical algebra*. De Morgan does not distinguish the two as carefully as he was to do later; nor does he attempt to describe a formal calculus in precise, fully syntactic terms. He seems to have been aware (§§3–6) that his distinction between technical and logical algebra raises difficult questions about the reference of linguistic expressions generally (an observation which his reading of the Schoolmen would have prepared him to make); but he does not follow up on the observation in the present essay. After remarking that 'A symbol is not the representation of an external object absolutely, but of a state of the mind in relation to that object', he notes that different people can have different mental representations of the same mathematical concept; but, unlike Bolzano or Frege, he does not discuss the tension between these two observations, nor discuss their implications for a general theory of meaning, nor grapple with the general issue of psychologism in logic and mathematics. Despite these shortcomings, *De Morgan 1842a* contains the first inkling that the new algebra being developed in Britain had deep connections to logic and to the philosophies of language and mind.[a]

In more narrowly mathematical terms, the following selection shows De Morgan trying to use the new algebraic ideas to give an account of the fundamental operations of arithmetic. In §8 he makes a passing reference to the possibility of other technical algebras, but shows no explicit awareness of the possibility of (say) a non-commutative operation of addition. *That* idea was first to occur to Hermann Grassmann (*1844*) and independently to William Rowan Hamilton in 1843; although, as Hamilton acknowledged, his discovery of quaternions was in part inspired by De Morgan's work on 'double algebra' (essentially the algebra of the complex numbers), and by the problem mentioned here by De Morgan in §11 (*Hamilton 1853*, Preface, §43).

References to *De Morgan 1842a* should be to the paragraph numbers, which have been added in this reprinting.

---

[1] The extent to which explanation of the meaning of the symbolical results of Algebra has been carried within the last half century; the complete interpretation of all which formerly appeared incongruous; the separation, as it was called, of the symbols of operation and quantity, which amounts to the use of an algebra in which the symbols represent something more than simple magnitude;—will for some time to come suggest inquiry into the *logic* of this many-handled instrument of reasoning, which seems to be capable of presenting, under fixed laws of operation, all the results which arise from very distinct primary conceptions as to the things operated upon.

[2] When several different hypotheses lead to results which admit of a common mode of expression, we are naturally led to look for something which the hypotheses have in common, and upon which the sameness of the method of

[a] This topic was to take on ever greater importance in De Morgan's theorizing about algebra, and by 1860, in his entry on 'Logic' in the *English Cyclopaedia*, De Morgan could write: 'Mr Boole's generalization of the forms of logic is by far the boldest and most original of those of which we have to treat. It cannot be separated from Mathematics, since it not only demands algebra, but such taste for thought about the notation of algebra as is rarely acquired without much and deep practice. When the ideas thrown out by Mr Boole shall have borne their full fruit, algebra, though only founded on ideas of number in the first instance, will appear like a sectional model of the whole form of thought. Its forms, considered apart from their matter, will be seen to contain all the forms of thought in general. The anti-mathematical logician says that it makes thought a branch of algebra, instead of algebra a branch of thought. It *makes* nothing; it *finds*; and it finds the laws of thought symbolized in the forms of algebra' (*De Morgan 1860b*).

expression depends. A comparison of the properties of the ellipse and hyperbola would bewilder the imagination, under any of the distinct definitions which might be given of the two curves; nor would the mind rest satisfied until it had discovered the reason of the similarity which exists between these properties.

⟦3⟧ Algebra now consists of two parts, the technical, and the logical. Technical algebra is the art of using symbols under regulations which, when this part of the subject is considered independently of the other, are prescribed as the definitions of the symbols. Logical algebra is the science which investigates the method of giving meaning to the primary symbols, and of interpreting all subsequent symbolic results. It is desirable that the word *definition* should not enter in two distinct senses, and I should propose to retain it as used in the *art* of algebra, applying the terms *explanation* and *interpretation* to denote the preparatory and terminal processes of the *science*. Thus a symbol is *defined* when such rules are laid down for its use as will enable us to accept or reject any proposed transformation of it, or by means of it. A simple symbol is *explained* when such a meaning is given to it as will enable us to accept or reject the application of its definition, as a consequence of that meaning: and a compound symbol is interpreted, when, having occurred as a result of explained elements, used under prescribed definitions, a necessary meaning can be given to it; the necessity arising from the tacit supposition that the compound symbol, considered as a new simple one, must still be subject to the prescribed definitions, when it subsequently comes in contact with other symbols. The last words may need the remark, that though we sometimes appear to interpret a symbol merely for the purpose of explaining a result, yet we know that such interpretation would be subsequently rejected, if the use of the symbol, under the prescribed definitions, were not found to be logically admissible.

⟦4⟧ A symbol is not the representation of an external object absolutely, but of a state of the mind in regard to that object; of a conception formed, for the formation of which the mind knows that it is or was indebted to the presence, bodily or ideal, of the object. Those who do not remember this, the real use of a symbol, are apt to dogmatize,* declaring one or another explanation of a symbol, that is, the signification by it of one or another impression produced on their own minds, to be real, true, natural, or necessary: it being neither one nor the other, except with reference to the particular mind in question. To take a very simple case, and one which bears upon our subject, let us imagine that we form successively a conception of the absence of all definite magnitude, followed by one of the existence of a certain magnitude, say a line of given length. The mind of one person may pass from the one to the other by imagining the given length to be instantaneously generated, no one portion of it coming into the thoughts before or after another; that of a second may make the transition by imagining a point to move from one extremity to the

* Of course, I use this word in its primitive sense, without any censure implied; the very sentence in which the word occurs is, and is meant to be, dogmatical.

other: while that of a third may dwell rather on the relative position of the two extremities, and may think more of *B* attained by motion from *A*, than of the quantity of length in *AB*. All three would use, perhaps, the same modes of expression: and I suspect* that there could be detected, among persons who think about first principles, a very considerable degree of variety in the points of view under which fundamental words suggest their objects; while as much exists, but could not as easily be found, among those who have studied the exact sciences, without paying particular attention to their foundations.

⟦5⟧ A symbol may thus denote either magnitude, operation, by which magnitude is attained, or the conception of one extreme arrived at, the other having been the previous object of contemplation. The earlier† algebraists most certainly dwelt on the first notion; $a + b$ is with them the result of an operation, in which the method of obtaining it is so completely forgotten, that the *result* $a + b$ is actually obtained by a distinct operation.

⟦6⟧ It seems to me that Sir William ⟦Rowan⟧ Hamilton, in his very original and methodical memoir on algebra as the science of pure time, has adopted a view of the third kind. I cannot see why the whole paper might not be as easily applied to succession of points in a line, as to succession of epochs in time. Succession, that is to say *continuous* succession, might be made the fundamental conception in both cases; and if such were the author's intention in the use of the word *time*, I should be very glad to maintain after him that *one* of the explanations which suffice to convert technical into logical algebra, has been fully established in his memoir. But, if any thing more *physical*‡ be intended by the distinguished author, and if some of his phrases are to be interpreted as of his asserting algebra to be *the* science of pure time, I should then cite him as an instance of the *dogmatism* already alluded to: and the more readily, that by the association of the word with his labours, I may claim to have purified it, for the purposes of this paper, from the dyslogistic associations usually connected with it.

⟦7⟧ The modern algebraists usually dwell on the second notion, namely that of operation; and this I shall adopt in the present paper, not only as the most common mode of conception, but also as being equally capable of connexion

* In a short biographical account (which I have before me, in a private communication) of the late Mlle Sophie Germain, whose papers on the theory of elastic surfaces are well known, it is asserted that she could never form the conception of space, except by the means of time: this was her own mode of expressing, to the writer of the notice, a state of mind by which he accounts for another fact, namely, that she had very little aptitude for pure geometry, and a great attachment to the theory of numbers.

† See my Calculus of Functions, sect. 245.

‡ This word is here improperly used; but I refer to the notion of those who would have made geometry a part of mixed mathematics: that is, if the algebra of Sir W. Hamilton would, in the opinion of those just alluded to, also have been a part of their mixed mathematics, and if Sir W. Hamilton should admit that they have as much reason, his terms being understood in his own sense, for their location of his algebra as for that of geometry, I should then say that the word used in the text is allowable.

with either of the other two. Imagine the process, whatever it may be, by which we pass from the contemplation of 0 to that of a; then if a represent a line, we can consider, as a result of our process, either the position of one extremity with respect to the other, or the quantity of length intercepted between the two.

⟦8⟧ I separate the following maxims from the rest as being equally applicable to the symbolical algebra which we have, and to any other which we might have. For it must never be forgotten that, though our present inquiry includes only the possible explanations of one given technical algebra, the subject may and probably must end in the investigation of others, or at least in the extension of the present one.

1. A simple symbol is the representative of one process, and of one only.

2. All processes, how many soever, may be looked at in their united effect as one process, and may be represented by one symbol.

3. Every process by which we can pass from one object of contemplation to another, involves a second by which we can reinstate the first object in its position: or every direct process has another which is its inverse. To complete the separation of these maxims from all others, I propose some considerations connected with the possible extensions of technical algebra.

⟦9⟧ The system of explanations which proceeds on the supposition that length affected by direction is the primary object of contemplation in algebra, is well known as to its history by Professor Peacock's Report to the British Association, and as to its present state by the Treatise on Algebra of the same author.* But in this branch of logical algebra the lines must be all in one plane, or at least affected by only one modification of direction: the branch which shall apply to a line drawn in any direction from a point, or modified by two distinct directions, is yet to be found.

⟦10⟧ It is obvious that our power of making the preceding application of algebra is co-ordinate with that of assigning a symbol Ω, such that

$$a + b\Omega = a_1 + b_1\Omega \text{ gives } a = a_1 \text{ and } b = b_1.$$

⟦11⟧ An extension to geometry of three dimensions is not practicable until we can assign two symbols, Ω and ω, such that

$$a + b\Omega + c\omega = a_1 + b_1\Omega + c_1\omega \text{ gives } a = a_1,\ b = b_1 \text{ and } c = c_1:$$

and no *definite* symbol of ordinary algebra will fulfil this condition. Again, in passing from x to $-x$ by two operations, we make use in ordinary algebra of one particular solution of

* Professor Peacock is the first, I believe, who distinctly set forth the difference between what I have called the technical and the logical branches of algebra. The second term, I am aware, is a very bad one, and I should be glad to see a better one proposed; but I prefer *technical* to *symbolical*, because the latter word does not distinguish the use of symbols from the explanation of symbols.

$$\phi^2 x = -x, \text{ namely } \phi x = \sqrt{-1}\,.\,x.$$

An extension to three dimensions would require a solution of the equation $\phi^3 x = -x$, containing an arbitrary constant, and leading to a function of triple value, totally unknown at present.

⟦12⟧ A general solution of $\phi^2 x = ax$ can be expressed when any particular solution $\bar{\omega} x$ is known. For if $f\bar{\omega}f^{-1}x$ be the general solution, we have

$$\phi^2 x = f\bar{\omega}^2 f^{-1}x = faf^{-1}x = ax, \text{ or } fax = afx:$$

so that it is only necessary that f and a should be convertible. Since then $(-1)^{\frac{1}{2}}x$ is a particular solution of $\phi^2 x = -x$, a general solution is $f\,\{\overline{-1}|^{\frac{1}{2}}f^{-1}x\}$ where $f(-x) = -fx$. But with our very limited knowledge of the laws of inversion, no solution which we can now express in finite terms will afford any help. Our means of expression must be augmented before we can hope to overcome this difficulty: or, as in most other cases of the kind, our difficulties recur in a circle; the means which we have used to propound a possible method require the problem itself to be solved before they can be successfully used.

⟦13⟧ Let the object of contemplation be simple magnitude of any one kind, as in the arithmetic of concrete quantity. The process which must precede all others is what we call selecting one magnitude for consideration. Previously to this step, we have no object under our perceptions, and may write 0 as the representative of this preceding state, and as the recognition of its existence. This first magnitude we may call 1, and the operation of transition from one state to the other we may denote by 0 + 1. The contemplation of simple existence, and of the possibility of expressing it by a spoken symbol, suggested the earliest definition of unity—ΜΟΝΑΣ *ἐστι, καθ ἥν ὅ ἕκαστον τῶν ὄντων ἕν λέγεται.*[a] If we represent our present state by (0 + 1), we may consider that with respect to any other possible magnitude our position is what it was when we denoted it by 0. If we now denote it by 0′, we may, as before, make the transition from 0′ to 0′ + 1, which implies that we have further taken into consideration a new magnitude of the same amount.

⟦14⟧ This result, (0 + 1) + 1, we may, if we please to consider it as attained by one operation, signify by 0 + 2: and so on. Using the symbol − to denote the process by which we retrace our steps, we have all that is necessary to express addition and subtraction. The principle which I wish here to enforce is, that *addition is connected with the symbol* 0 *in a manner which requires us to imagine that we start from one magnitude, as it were from a new* 0, *and renew*[*] *the process by which we passed from the first* 0 *to that magnitude.*

⟦15⟧ Let us now suppose that modified magnitude is under contemplation, and let the simple symbol a denote a line measured in a given direction from

---

[a] ⟦'Unity is that by virtue of which each of the things which exist is said to be one' (Euclid, *Elements*, Book 7, Def. 1).⟧

[*] Any one who doubts the justness of this fundamental position should add six to four on his fingers, having previously refreshed his notions of six and four by the same process.

the zero point 0. In this zero of space, which admits of an infinite number of positions, we seize more clearly than before that notion which, as to simple magnitude, is not easily admitted as necessary, and may seem rather fanciful: namely, that every magnitude attained may, as to future addition, be considered as a new zero. We are now to assume that,

1. Parallelism and sameness of direction are meant to be identical terms; that is to say, the two directions conceivable on any one of two parallels are severally the same as the two directions on the other.

2. Every simple symbol represents a line given in length and direction: thus $a = b$ means that the lines a and b, equal in length, have also the same direction. And the process implied in $0 + a$ is the transference of a point from the position 0 to a given length in a given direction.

⟦16⟧ We can now find the necessary meaning of $(0 + a) + b$; *necessary*, on the supposition that the technical algebra is to become logical on the explanation of the symbols before us. Let $0A$ and $0B$ represent the lines symbolized by a and b: if then we take A, at which we arrive by the process $0 + a$, as a new zero, and proceed with it in the same manner as in performing $0 + b$ on the old zero, we draw AC parallel and equal to $0B$, whence $0C$ being symbolized by c, we have with reference to the first zero,

$$0 + c = (0 + a) + b = (0 + b) + a.$$

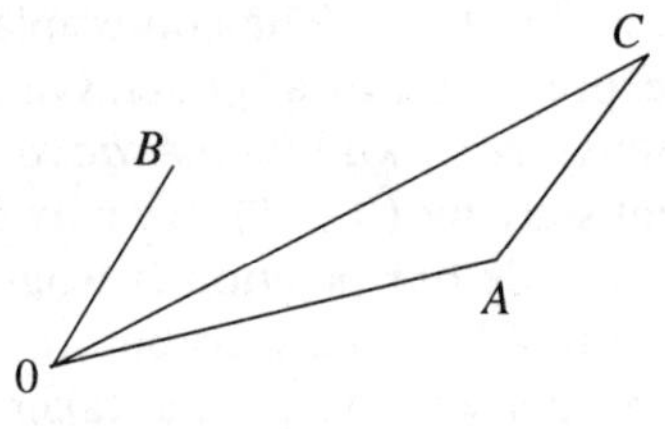

⟦17⟧ I need not further dwell on the connection of addition and subtraction in arithmetic with the processes called by the same names in this explanation. I shall only here suggest that perhaps the words *direct zero process* and *inverse zero process* might occasionally be found useful.*

⟦18⟧ The usual method of defining the process of addition by reference to the diagonal of a parallelogram is convenient, but destructive of all true analogy. The fundamental theorem of statics suffers from the same method of statement.

⟦19⟧ I now proceed to the process of multiplication, which will readily be seen to be connected with *unity* in precisely the same manner as is addition with zero. If b be formed from unity by the train of processes $0 + 1 + 1 + 1$, we

* In my *Calculus of Functions* (sect. 12, 13, 17) will be found some analogies connecting simple addition with zero, and multiplication with unity.

consider a as a new unit, and let the symbol ba represent the same operation on this new unit, or $0 + a + a + a$. Similarly, if by the line 1 we mean a line having the length and direction 1, and $0A$ and $0B$ by a and b, and if we take $0A$ as a new unit, and perform on it the operations by which we pass from 01 to $0B$, that is, take an angle $A0C$ equal to $10B$, and let $0C$ be in length the result of the arithmetical operation on $0A$ and $0B$,—then $0C$ must be represented by ab. The processes of multiplication and division might be called the direct and inverse *unit processes*.

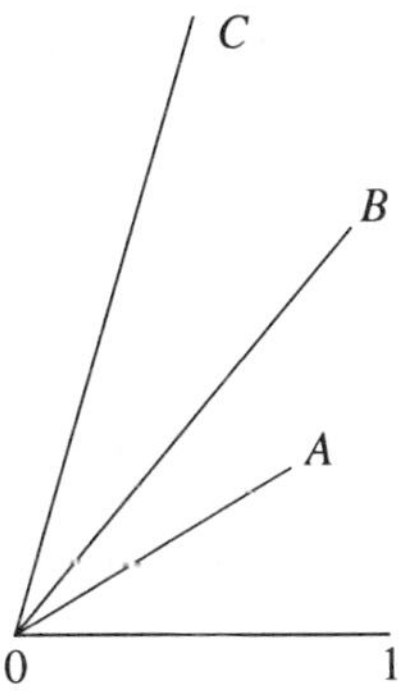

⟦20⟧ There is now nothing particular to be said about the four operations, or the simple powers, with positive or negative, whole or fractional, real exponents, or any combinations of them. The interpretation of $a + b\sqrt{-1}$ follows in the usual manner.

⟦21⟧ In illustration of the propriety of considering symbols as functions of zero or unity for purposes of addition or multiplication, it may be advanced that unless we do so, we change the meaning of the terms direct and inverse as we proceed from the lower to the higher parts of the science. Unquestionably, if ever we have a right to assume a clear conception of this distinction, it is in the comparison of addition with subtraction, and of multiplication with division; but for all that, $a + x$ and $a - x$ are not inverse functions, considered with respect to x, though they are so with respect to a. And similarly of ax and $a \div x$. When we come to the symbol x^n, then, and then only, do we begin to describe inversion correctly: for we usually consider this as a function of x and not of n, when we assert $x^{\frac{1}{n}}$ to be the inverse. But if we considered this as a function of n, the inverse would be $\log n : \log x$.

⟦22⟧ The separation, as it is called, of the symbols of operation and quantity, is a method of explaining technical algebra as simple in its character as the preceding. Let the fundamental object of conception be $\phi(x - nh)$, n being infinite, which stands in the place hitherto occupied by 0. Let* $\Sigma\phi(x + ah)$

* In the common method of treating this subject, the inverse symbol is made to precede the direct one. Several adaptations of notation are necessary before we can exactly represent the common methods.

represent the train of operations by which we pass from $\phi(x-\infty h)$ to $\phi(x+\overline{a-1}h)$, or

$$\phi(x-\infty h)+\ldots\ldots+\phi(x-h)+\phi x+\phi(x+h)+\ldots\ldots+\phi(x+\overline{a-1}h).$$

⟦23⟧ The inverse operation, or rather the, operation by which $\phi(x+ah)$ is obtained from $\Sigma\phi(x+ah)$, is either $\Sigma\{\phi(x+\overline{a+1}h)-\phi(x+ah)\}$, or $\Sigma\phi(x+\overline{a+1}h)-\Sigma\phi(x+ah)$, and may be symbolized either by $\Delta\Sigma\phi(x+ah)$ or $\Sigma\Delta\phi(x+ah)$.

⟦24⟧ The proper way, however, of considering this class of extensions may not be as a simple explanation of technical algebra, (though it might be regarded in that point of view,) but as an extension of technical algebra itself, in which new explanations of the direct and inverse unit process are used co-ordinately with the one already established. If we agree to signify ∇^0, ∇^1, ∇^2, &c. a new progression of operations, in which the zero and its processes remain subject to the usual definitions, nothing prevents us from supposing that the prescribed definitions of the unit process may remain true if ∇^0 be made the unit, ∇^2 being derived from ∇^1 by the same train of operations as ∇^1 from ∇^0, and so on. Neither is it impossible that the same laws of convertibility and distribution may exist between compound operations, in which different units are employed, as are laid down in the prescribed definitions relatively to the different unit processes suggested by simple magnitudes.

⟦25⟧ Let $\nabla^0=\phi x$, and

$$\nabla^1=a_0\phi x+a_1\phi(x+h)+a_2\phi(x+2h)+\ldots\ldots$$

where a_0, a_1, &c. may be functions of h, but not of x. Technical algebra may be carried to its full length under these explanations, and many developements may be and have been simplified by their means. It is not my intention here to write a treatise on this subject: my object is, to point out *that the logic of each and all of these explanations is the same; no mode of arriving at any one explanation differing from that of any other in the fundamental, and what we may call the arithmetical, part of the subject*. It is certain that the discovery of inverse operations is not yet complete: this must be reserved until such time as the branches, which adopt length modified by direction as the explanation of simple symbols, are properly connected with that technical algebra, in which various unit processes are used co-ordinately with the same zero process.

⟦26⟧ It may perhaps be worthy of note that the series

$$a_0+a_1x+a_2x^2+\ldots\ldots$$

may be considered as $\nabla\varepsilon^{\upsilon}$ when $\upsilon=0$ in the equation

$$\nabla\varepsilon^{\upsilon}=a_0\varepsilon^{\upsilon}\times a_1\varepsilon^{\upsilon+\log x}+a_2\varepsilon^{\upsilon+2\log x}+\ldots\ldots$$

⟦27⟧ I now return to the purely algebraical question. It is in our power to avoid all ambiguity in results, by simply prescribing that every symbol shall express not merely the length and direction of a line, but also, the quantity of revolution by which a line, setting out from the unit line, is supposed to attain that direction. When this is done, I shall use a double sign of equality to denote it. Thus, if we denote by (a, θ) a line of a length a, which has made the revolution θ, it is allowable to write

$$(a, \theta) = (a, \theta + 2\pi) = (a, \theta + 4\pi), \ldots\ldots$$

but not

$$(a, \theta) = = (a, \theta + 2\pi) = = (a, \theta + 4\pi) \ldots\ldots.$$

‖28‖ As long as we neglect this additional prescription, great care will be requisite to prevent our falling into error. While exponents transform lengths into lengths, and directions into directions, no great caution is requisite: but when, as we shall presently see, an exponential process causes the exponent of a length to affect that of direction, or *vice-versâ*, the following fallacy of a continental analyst, mentioned by Professor Peacock in his Report, is frequently likely to occur. Stripped of unnecessary details, it as follows:

$$\varepsilon^{2\pi n\sqrt{-1}} = 1,\ (\varepsilon^{2\pi n\sqrt{-1}})^{2\pi n\sqrt{-1}} = 1^{2\pi n\sqrt{-1}} = 1,$$

$$\text{or } \varepsilon^{-4\pi^2 n^2} = 1, \text{ an absurd result.}$$

‖29‖ The answer is very simple: if no extension of explanations be contemplated, $1^{2\pi n\sqrt{-1}}$ is not necessarily $= 1$, since it may have an infinite number of values. If the extensions be made, and if $=$ merely denote sameness of direction, the same thing is true; for $1^{2\pi n\sqrt{-1}}$ or $(\varepsilon^{2\pi n\sqrt{-1}})^{2\pi n\sqrt{-1}}$ has an infinite number of values, of which one only $(\varepsilon^0)^{2\pi n\sqrt{-1}}$ is $= 1$: and the same fallacy might be thus propounded;

$$\sqrt{x^2} = +x,\ \sqrt{x^2} = -x, \text{ therefore } x = -x.$$

‖30‖ But if $=$ imply sameness of revolution, it is not true that $\varepsilon^{2\pi n\sqrt{-1}} = 1$ except in length.

‖31‖ The interpretation of $A^{\sqrt{-1}}$ might be easily attained from prescribed definitions, and from their necessary result

$$\varepsilon^{\theta\sqrt{-1}} = \cos\theta + \sin\theta\sqrt{-1};$$

nor would this step be logically objectionable. It would, however, be more satisfactory if something like an *à priori* interpretation, or simple explanation, could be given. I do not consider the following as complete, but it is, as far as it goes, of a new character.

‖32‖ Conformably to definitions, we must have

$$\{(\log a, \theta)^{\sqrt{-1}}\}^{\sqrt{-1}} = \{\log a, \theta\}^{-1} = (-\log a, -\theta),$$

where by $(\log a, \theta)$ is meant a line of the length a, and amount of revolution θ. Now we cannot suppose that the first operation changes the sign of log a only, and the second that of θ only: for this would be to make the operation $(\ )^{\sqrt{-1}}$ mean different things in different places. We must propose some operation of permanent form, which being twice performed will make the alteration required.

[33] From the definitions, it follows that

$$(\log a,\ 0) \times (0,\ \theta) = (\log a,\ \theta),$$

whence $(\log a, \theta)$ must be the product of two functions, one of a and the other of θ, the first of which is known, being $\varepsilon^{\log a}$ or a, and the second of which must be of the form E^{θ}, since by definition

$$(0,\ \theta) \times (0,\ \theta') = (0,\ \theta + \theta').$$

[34] Hence aE^{θ}, or $a(0,\ 1)^{\theta}$, is the representative of a line a, inclined at an angle θ. If then we make $\cos\theta$ and $\sin\theta$ mean nothing more than the projecting factors of a length inclined at the angle θ upon the axis of the unit line and its perpendicular, we have

$$(\cos 1 + \sqrt{-1}\sin 1)^{\theta} = \cos\theta + \sqrt{-1}\sin\theta.$$

[35] The definition does not differ from that of $\cos\theta$ and $\sin\theta$ in geometry, and this equation is an *à priori* property of these functions, deducible immediately from the definition, in any system which gives meaning to $\sqrt{-1}$ from its commencement.

[36] The hardest and most delicate part of this investigation is the connexion of $\varepsilon^{\theta\sqrt{-1}}$ with a unit inclined at an angle θ; or generally to show that the operation $(\ )^{\sqrt{-1}}$ changes the exponent of length into one of direction, and *vice versâ*, without the necessity of inferring this from interpretation. If we assume beforehand that $\varepsilon^{\sqrt{-1}}$ is *real*, under the extended definitions, it would be difficult to imagine what other office $(\ )^{\sqrt{-1}}$ could perform; but such an assumption would not be a proper one, since all the associations of preceding algebra would lead us to suppose that each extension removes only one class of inexplicables, and leaves, or perhaps introduces, others. I cannot complete this part of the subject satisfactorily, but the following considerations will show that the most simple mode of attaining, upon an explanation, the technical end of the operation $(\ )^{\sqrt{-1}}$ is precisely that which answers to the above.

[37] Required an operation which repeated n times upon a function of n quantities shall end by changing the sign of all. Take four quantities, a, b, c, and d. Sucessive changes of sign made upon one after the other will be really different successive operations; but if we change the sign of a given one, say the first, and at the same time make a set of periodic interchanges, writing b for a, c for b, d for c, and a for d, we shall have an operation which repeated four times will produce the desired effect. Thus we have successively,

$$\phi(b,\ c,\ d,\ -a),\ \phi(c,\ d,\ -a,\ -b),\ \phi(d,\ -a-b,\ -c),$$

$$\phi(-a,\ -b,\ -c,\ -d).$$

Thus we see in the succession $(\log a,\ \theta)$, $(-\theta,\ \log a)$, $(-\log a,\ -\theta)$ a method of passing from A to A^{-1} at two similar steps, which does not involve the use of $\sqrt{-1}$. We see the same in $(\log a,\ \theta)$, $(\theta,\ -\log a)$, and $(-\log a,\ -\theta)$. If then we assume, as a suggestion,

$$(\log a,\ \theta)^{\sqrt{-1}} = (-\theta,\ \log a),\ (\log a,\ \theta)^{-\sqrt{-1}} = (\theta,\ -\log a),$$

we find, making $A = (\log a,\ \theta)$, the following equations;

$$(A^{\sqrt{-1}})^{\sqrt{-1}} = A^{-1},\ (A^{-\sqrt{-1}})^{\sqrt{-1}} = A^{-1},\ (A^{\sqrt{-1}})^{-\sqrt{-1}} = A,$$

$$(A^{\frac{1}{\sqrt{-1}}})^{\frac{1}{\sqrt{-1}}} = A^{-1},\ (A^{-\frac{1}{\sqrt{-1}}})^{-\frac{1}{\sqrt{-1}}} = A^{-1},\ (A^{\frac{1}{\sqrt{-1}}})^{-\frac{1}{\sqrt{-1}}} = A,$$

in perfect fulfilment of all the fundamental conditions which prescribed definitions impose. The assumption gives

$$(aE^{\theta})^{\sqrt{-1}} = E^{\theta\sqrt{-1}}\,.\,\varepsilon^{\log a.\sqrt{-1}},$$

where $E^{\theta\sqrt{-1}}$ must be a symbol of length, and $\varepsilon^{\log a\sqrt{-1}}$ of a unit inclined at the angle $\log a$. Consequently $\varepsilon^{\theta\sqrt{-1}}$ must signify a unit inclined at an angle θ.

⟦38⟧ It might be asked whether there is anything in the preceding process which restricts us to the use of the base ε rather than any other, I answer, nothing whatever: but at the same time there is nothing which binds us to the use of any particular method of measuring angles. It may be deduced from the preceding that the base ε must be used co-ordinately with that mode of measurement which I call *theoretical*.* This connexion depends entirely upon the purely numerical process by which the equation $\varepsilon^{2\pi\sqrt{-1}} = 1$ is proved to be satisfied when ε and π have their usual meanings. If for any reason we prefer the base a, the measure of two right angles must be $\pi \times \{\log \varepsilon$ to the base $a\}$.

⟦39⟧ I think it cannot be disputed that interpretation should be avoided where explanation can be given. If where the latter cannot be obtained suggestion upon such analogies as present themselves were to take its place, the former would be also replaced by verification. In the present instance, the attainment of

$$\varepsilon^{\theta\sqrt{-1}} = \cos\theta + \sqrt{-1}\sin\theta \text{ from } E^{\theta} = \cos\theta + \sqrt{-1}\sin\theta$$

is the verification.

⟦40⟧ I now come to the theory of logarithms. It is a circumstance which I hold to be not a little remarkable, that the ancient form of algebra was only saved from being convicted of incapacity to produce its own legitimate results, but very little time before such an escape would have been rendered impossible by its receiving the necessary accession from the more extended form. Mr GRAVES has admitted that his view of the new logarithms should rather have been that of an extension imperatively required than of a correction to already existing formulæ: and in this view I perfectly agree. If we define $\log x$, or rather λx, (reserving $\log x$ for the numerical logarithm of the length) to be any legitimate solution of $\varepsilon^{\lambda x} = x$, it is plain that the logarithm of n inclined at an angle ν, (or of N) to the base b inclined at an angle β, (or B) is to be derived

* In those works on Trigonometry which use the arc and angle indiscriminately, this mode of measurement is said to be *in parts of the radius*. A term is much wanted which shall not imply this confusion between arcs and angles; and I propose that the angle which subtends an arc equal to the radius shall be called the *theoretical unit*.

(avoiding ambiguity) from

$$(b\varepsilon^{\beta\sqrt{-1}})^x = = n\varepsilon^{\nu\sqrt{-1}},$$

$$\text{or } \lambda_B N = = \frac{\log n + \nu\sqrt{-1}}{\log b + \beta\sqrt{-1}}$$

This result is real when $\frac{\log n}{\log b} = \frac{\nu}{\beta}$; nor is it more surprising that an impossible quantity (hitherto so called) should have a possible logarithm, than that exponential operations not containing $\sqrt{-1}$, or not interchanging exponents of length and direction, should in certain cases enable us to pass from one line to another. I need not enter into details of the properties of the preceding equation. If we admit all symbols to be algebraical (in the old sense) which denote lines drawn in the unit line or its continuation, whatever may be the number of complete revolutions after which they rest there, we must then admit that the logarithm of a unit which is in its position for the $(m + 1)^{\text{th}}$ time, with respect to ε which is in its position for the $(n + 1)^{\text{th}}$ time is

$$\lambda_{(0,\ 2n\pi)}(0,\ 2m\pi) = \frac{2m\pi\sqrt{-1}}{1 + 2n\pi\sqrt{-1}}$$

the form proposed by Mr Graves.

⟦41⟧ In a work of M. Cauchy, and perhaps in other writings which I am not acquainted with, mention is made of a singular point in curves which he calls *point d'arrêt*, at which the branch suddenly stops. Such a point has long been admitted in the spiral of Archimedes and other curves, owing to the neglect of making those extensions with regard to the sign of the radius vector which were necessary to complete the connexion* of polar and rectangular co-ordinates; and from the assumption of the impossibility of which (I speak from memory) D'Alembert drew those instances in which he contended that the negative quantity is not *always* the contrary of the positive quantity. Disregarding such *points d'arrêt*, there is another sort which frequently occurs (but only in exponential or logarithmic curves), in which the abruptness of the termination is better marked. Thus in $y = (1 - x) \log (1 - x)$, there is, in our present system, an absolute cessation of the curve when $x = 1$ and $y = 0$. Here, when the requisite extensions of the logarithmic theory are made, it will be seen that there is not an absolute abrupt termination, but the commencement of what French writers have called a *branche pointillée*, a part of a curve, which I do not remember to have seen mentioned in any English work, except Professor Peacock's Report.

A. DE MORGAN.

UNIVERSITY COLLEGE, LONDON,
October 16, 1839.

---

* On this subject I may be allowed to refer to page 341 of my Treatise on the Differential Calculus.

B. TRIGONOMETRY AND DOUBLE ALGEBRA (*DE MORGAN 1849b*)

Reprinted here are the first two chapters of Book II of De Morgan's textbook, *Trigonometry and double algebra*. Book I ('Trigonometry') is a straightforward exposition of the principles of elementary trigonometry; De Morgan uses $\sqrt{-1}$ in an 'experimental' way—trying to accustom the student to working with the complex numbers, but providing no definitive justification. Book II takes up the task of justification, and is devoted to 'Double algebra'—'double algebra' being De Morgan's name for what is in essence the algebra of the complex numbers.

As William Rowan Hamilton acknowledged (*Hamilton 1853*, Preface, §43) the discovery of quaternions in part grew out of an unsuccessful attempt to generalize De Morgan's double algebra into a 'triple algebra' for the three dimensions of physical space. (Hamilton's discovery of quaternions occurred in October, 1843, and in turn prompted De Morgan to devote the last of his four papers 'On the foundations of algebra' (*De Morgan 1849a*) to the topic of triple algebra; this paper was read to the Cambridge Philosophical Society in October, 1844.)

In Chapter I of the present selection De Morgan is concerned with a central problem for British algebraists in the early nineteenth century: the justification of the negative and the complex numbers. By 1849 the lessons of Peacock and Gregory had been absorbed, and De Morgan crisply distinguishes between *symbols*, their *rules of operation*, and their *meanings*. A *symbolic calculus* ignores meanings and is concerned solely with symbols and their rules of operation; indeed (as De Morgan notes in §9 of Chapter I and in §23 of Chapter II) the formal, syntactic manipulations could perfectly well be carried out by a machine. As for the negative and complex numbers, the symbolic calculi of 'single algebra' and 'double algebra' are (he points out) susceptible of various interpretations; but his concern here is solely with exploring the symbolic calculi themselves.

The basic idea of invoking a purely symbolic calculus to justify the negative and the complex numbers harks back to François-Joseph Servois, George Peacock, and Duncan Gregory; even earlier, Johann Heinrich Lambert had had the idea of regarding the axioms of geometry as a symbolic calculus (*Lambert 1786*, §11). But De Morgan's exposition is clearer and more exact than those of his predecessors. (Indeed, his analysis of the relationships between symbols, syntactic rules, and meanings even leads him in §13 to a brief, partial anticipation of Quine's problem of the indeterminacy of translation.) Moreover, whereas De Morgan's predecessors had spoken in generalities about the possibility of a formal system, De Morgan, in Chapter II, became the first mathematician to try to give a precise, purely syntactic description of a symbolic algebra (in essence, a field)—explicitly gathering together and describing, in a single place, its symbols and its rules of syntax. By twentieth-century standards the exposition contains many gaps and imprecisions. For instance, De Morgan does not mention the associativity of the basic operations; he does not clearly distinguish between axioms, definitions, and axiom schemata; he does not give a precise definition of the well-formed formulas; he does not state formal rules of inference. Nevertheless, the spirit of his exposition is remarkably modern, and in point of rigour De Morgan's presentation of a formal calculus was not to be surpassed until the appearance of Frege's *Begriffschrift* in 1879.

References to *De Morgan 1849b* should be to the paragraph numbers, which have been added in this reprinting.

CHAPTER I.
DESCRIPTION OF A SYMBOLIC CALCULUS.

⟦1⟧ THE object of this book is the construction of Algebra upon a basis which will enable us to give a meaning to every symbol *and combination of symbols* before it is used, and consequently to dispense, first, with all unintelligible combination, secondly, with all search after interpretation of combinations subsequently to their first appearance.

⟦2⟧ In arithmetic and in ordinary algebra we use *symbols* of previously assigned *meaning*, from which meaning, by self-evident notions of number, &c., are derived *rules of operation*. The student must understand by *symbols*, the *peculiar* symbols of arithmetic and algebra: strictly speaking, the written or spoken words by which meaning is conveyed are themselves symbols. And symbols must be explained by other symbols, except when they denote external objects or actions, in which case the symbol may be explained by pointing to the object present or the action taking place. Language itself is a science of symbols (namely, words) having meanings (which are described in the *dictionary* by words of the same or another language) and rules of combination (laid down in its *grammar*).

⟦3⟧ No science of symbols can be fully presented to the mind, in such a state as to demand assent or dissent, until its peculiar symbols, their meanings, and the rules of operation, are *all* stated. In this case we have but to ascertain—first, whether the peculiar symbols be distinguishable from each other; secondly, whether the meanings are capable of being distinctly apprehended, each symbol having either one only, or an attainable and intelligible choice; thirdly, whether the given rules of operation be necessary consequences of the given meanings as applied to the given symbols. If these inquiries produce as many affirmative answers, the basis of the science is *so far unobjectionable;* and all intelligible conclusions which are drawn from a correct and intelligible use of the rules of operation, are true. But yet it may be *imperfect*.

⟦4⟧ First, it may be *incomplete in its peculiar symbols*. There may be a want of symbols which those already in use suggest, but which are not made to appear. This is not the incompleteness to which algebra is most liable: it suffers more from its symbolic combinations growing much faster than the ordinary language in which they are, if possible, to be occasionally expressed.

⟦5⟧ Secondly, it may be *incomplete in its meanings*. For example, it may be capable of applying, with the same symbols, to more subjects than its actual meanings take in. This is one possible incompleteness, of a very obvious character. Another, of a much less obvious character, and which probably nothing but actual experience of it would have suggested, is this: symbols, defined in a manner which makes them separately intelligible, may be unintelligible in combination; their separate definitions may involve what, in the attempt* to combine them, produces contradiction. The second case may be a consequence of the first, or it may not: contradictory combinations may arise from limitation of meaning, and may cease to be contradictory under extended meanings; or it may happen, either that no such abolition of contradiction is possible in the case thought of, or else that every extension of meaning which destroys contradiction in one combination creates it in another.

⟦6⟧ Thirdly, it may be *incomplete in its rules of operation*. This incompleteness may amount either to an absolute privation of results, or only to the imposition of more trouble than, with completeness, would be requisite. Every rule the want of which would be a privation of results, may be called *primary*: all which might be dispensed with, except for the trouble that the want of them would give, may be treated merely as consequences of the primary rules, and called *secondary*.

⟦7⟧ Each of the three great objects of consideration, peculiar symbols, assigned meanings, and rules of operation, may then be defective, independently of the rest. Can we carry the defect to far as to imagine one or more of them to be entirely wanting? The cases of absolute deficiency, which it may be worth while to notice here, principally to accustom the student to the idea of the separation, are as follows:—

⟦8⟧ 1. *Meanings and rules without peculiar symbols.* Unquestionably algebra *might* be deprived of its peculiar symbols, ordinary words taking their places. There is no more truth, no more meaning, and no more possibility of drawing consequences in

$$(a^2 - b^2) \div (a - b) = a + b,$$

than in 'the difference of the products of two numbers, each multiplied by itself,

* The student may be surprised at my saying that we should never have imagined such a result in algebra without actual experience of it: for it may strike him immediately that in ordinary language we may have not merely unmeaning, but contradictory, combinations. But the answer is that we are so accustomed to contradictory combinations, used in some emphatic sense, that they are recognised idioms: it even happens that they express more and better meaning while they are fresh, and before use makes the contradiction wear off, than afterwards. When General Wolfe first used the expression 'choice of difficulties', which was contradiction, choice then meaning voluntary election, he made those to whom he wrote see his position with much more effect than could have been produced a second time by the same words. Ordinary language has methods of instantaneously assigning meaning to contradictory phrases: and thus it has stronger analogies with an algebra (if there were such a thing) in which there are preorganized rules for explaining new contradictory symbols as they arise, than with one in which a single instance of them demands an immediate revision of the whole dictionary.

divided by the difference of those numbers, gives the sum for a quotient'. Before the time of Vieta, algebra had always been much retarded by the want of a sufficient use of peculiar symbols.

‖9‖ 2. *Peculiar symbols, and meanings, without rules of operation.* In this case the only process must be one of unassisted reason, thinking on the objects which the symbols represent; as in geometry, which has its peculiar symbols (as *AB*, signifying a line joining two points named *A* and *B*). But no science of *calculation** can proceed without rules; and these geometry does† not possess.

‖10‖ 3. *Peculiar symbols, and rules of operation, without assigned meanings.* Nothing can be clearer than the possibility of dictating the symbols with which to proceed, and the mode of using them, without any information whatever on the meaning of the former, or the purpose of the latter. A corresponding process takes place in every manual art in which an assistant obeys directions, without understanding them. The use of such a process, as an exercise of mind, must depend much (but not altogether) upon the value of the meanings which we suppose are to be ultimately assigned. A person who should learn how to put together a map of Europe dissected before the paper is pasted on, would have symbols, various shaped pieces of wood, and rules of operation, directions to put them together so as to make the edges fit, and the whole form an oblong figure. Let him go on until he can do this with any degree of expertness, and he has no consciousness of having learnt anything: but paste on the engraved paper, and he is soon made sensible that he has become master of the forms and relative situations of the European countries and seas.

‖11‖ As soon as the idea of acquiring symbols and laws of combination, without given meaning, has become familiar, the student has the notion of what I will call a *symbolic calculus*; which, with certain symbols and certain laws of combination, is *symbolic algebra*: an art, not a science; and an apparently useless art, except as it may afterwards furnish the grammar of a science. The proficient in a symbolic calculus would naturally demand a supply of meaning. Suppose him left without the power of obtaining it from without: his teacher is dead, and he must *invent meanings* for himself. His problem is, Given symbols and laws of combination, required meanings for the symbols of which the right to make those combinations shall be a logical consequence. He tries, and succeeds; he invents a set of meanings which satisfy the conditions. Has he then supplied what his teacher would have given, if he had lived? In one particular, certainly: he has turned his *symbolic* calculus into a *significant* one. But it does

---

* A *calculus*, or *science of calculation*, in the modern sense, is one which has organized processes by which passage is made, or may be made, mechanically, from one result to another. A *calculus* always contains something which it would be *possible* to do by machinery.

† Those who introduce *algebraical* symbols into elementary geometry, destroy the peculiar character of the latter to every student who has any mechanical associations connected with those symbols; that is, to every student who has previously used them in ordinary algebra. Geometrical reasoning, and arithmetical process, have each its own office: to mix the two in elementary instruction, is injurious to the proper acquisition of both.

not follow that he has done it in the way which his teacher would have taught him, had he lived. It is possible that many* different sets of meanings may, when attached to the symbols, make the rules necessary consequences. We may try this in a small way with three symbols, and one rule of connection. Given symbols M, N, $+$, and one sole relation of combination, namely that $M + N$ is the same result (be it of what kind soever) as $N + M$. Here is a symbolic calculus: how can it be made a significant one? In the following ways, among others. 1. M and N may be *magnitudes*, $+$ the sign of addition of the second to the first. 2. M and N may be *numbers*, and $+$ the sign of multiplying the first by the second. 3. M and N may be *lines*, and $+$ a direction to make a rectangle with the antecedent for a base, and the consequent for an altitude. 4. M and N may be *men*, and $+$ the assertion that the antecedent is the brother of the consequent. 5. M and N may be nations, and $+$ the sign of the consequent having fought a battle with the antecedent: and so on.

‖12‖ We may also illustrate the manner in which too limited or too extensive a meaning interferes with the formation of the most complete significant calculus. In (1), limitation to *magnitude* is not necessary, unless *ratio* and *number* be signified under the term. In (2), if M (only) were allowed to signify number, $N + M$ would be intelligible, but $M + N$ would be unintelligible; an *impossible* symbol of this calculus. In (3), $(M + N)$ signifying the rectangle, $(M + N) + P$ would be unintelligible at first: further examination would show that the explanation is not complete; and that the proper *extension* is that $M + N + P$ should signify the formation of the *right solid* (rectangular parallelepiped) with the sides M, N, P. But $M + N + P + Q$ will be always unintelligible, as space has not four dimensions. In (4), the extension of M and N to signify *human beings*, would spoil the applicability of the rule, unless the meaning of $+$ were at the same time extended to signify the assertion that the antecedent was *brother or sister* (as the case might be) of the consequent.

‖13‖ But when the symbols are many, and laws of combination various, is it to be thought possible such a number of coincidences should occur, as that the same symbolic combinations (unlimited in number) which express truths under one set of meanings, express other truths under another? Could two different languages be contrived, having the same words and grammar, but in which the words have different meanings, in such manner that any sentence which has a true meaning in the first, should also have a true, but a different, meaning in the second? This last question may almost certainly be answered in the negative: the thousands of arbitrary terms which a language presents, and the hundreds of grammatical junctions, present a possible variety of combinations of which it would be hopeless to expect an equal number of coincidences of the kind required. But Algebra has few symbols and few combinations,

* Most inverse questions lead to multiplicity of answers. But the student does not fully expect this when he asks an inverse question, unless he be familiar with the logical character of the predicate of a proposition. A always gives B: what gives B? answer, A always, and, for aught that appears, many other things.

compared with a language: more explanations than one are practicable, and many more than have yet been discovered may exist. And the student, if he should hereafter inquire into the assertions of different writers, who contend for what each of them considers as *the* explanation of $\sqrt{-1}$, will do well to substitute the indefinite article.

⟦14⟧ We can now form some idea of the object in view; and we must ask, first, what are the steps through which we have gone, to arrive at algebra as it stands in the mind of the student who commences this book. They are, very briefly, as follows:

⟦15⟧ Beginning with *specific* or *particular arithmetic*, in which every symbol of number has one meaning, we have invented signs, and investigated rules of operation. An easy ascent is made to *general* or *universal arithmetic*, in which general symbols of number are invented, the letters of the alphabet being applied to stand for numbers, each letter having a numerical meaning, known or unknown, on each occasion of its use. And thus, omitting many circumstances which have no particular reference to our present subject, we arrive at a calculus in which the actual perfomance of computations is deferred until we come to the time when the values of the letters are found or assigned. Accordingly, whereas in particular arithmetic every computation is completed as it arises, or declared impossible, in universal arithmetic we have a calculus of forms of computation, in which each numerical computation is only signified, and not performed; the proviso, *if possible*, being annexed by a reasoner to every step of every process in which a chance of impossibility occurs.

⟦16⟧ Out of a few cases of difficulty, there is selected one, which appears at first sight destined always to make the proviso above mentioned an essential part of most processes of universal arithmetic. It is the *impossible subtraction*; the constant appearance in problems of a demand to take the greater from the less, to say how many units there are in 6 – 20, for instance. An examination of the circumstances under which such phenomena occur shows, inductively, that their producing cause is always this, that either in the statement of the problem, or in its treatment, some one quantity is supposed to be of a kind diametrically opposite to that which it ought to be.

⟦17⟧ Simple number, the subject of abstract arithmetic, be it particular or universal, fails to show any acknowledgement of a distinction which strikes us in almost every notion of concrete magnitude. Measure 10 feet from a given point on a given line: the command is ambiguous until we are told which of two directions to take. A sum of money in the concerns of A and Co. is incapable of being entered in their books until we know whether it be gain or loss. A weight is generally of one kind, but not always: the weight of a balloon is a tendency in the direction opposite to that of most weights; or rather, the word weight being by usage not allowed a double signification, we say a balloon has no weight, but something which is the direct opposite of weight. A time, one extreme epoch of which is mentioned, is not sufficiently described until we know whether it is all before, or all after, the epoch. And so on. In every one of these cases, the numerical quantity of a concrete magnitude, described by means of a standard unit, is not a sufficient description; it is necessary to specify to which of two opposite kinds it belongs. This specification must be made by something not numerical: number is wholly inadequate.

‖18‖ The first suggestion would be, it might be thought, to invent signs of distinction: but universal arithmetic makes a suggestion which forces attention, before the necessity for distinction is more than barely perceived. Should we ever suppose that the result of a problem is gain, or distance in one direction, or time after an epoch, &c., when it is in reality, say 4 of loss, or of distance in the opposite direction, or of time before the epoch, &c., the answer always presents itself as $0 - 4$, or $m - (m + 4)$, or as some version of the attempt to take away 4 more than there are to be taken away. It is then judged convenient (that the convenience amounts to a necessity is hardly seen at that period) to make -4 the symbol of 4 units of a kind directly opposite to those imagined in 4, or $0 + 4$. And this is the first of the steps by which universal arithmetic becomes common, or *single* algebra. See Algebra, pp. 12–19 and 44–66, for more detail. ‖*De Morgan 1837*‖.

‖19‖ This word *single*, as applied to algebra, is derived from space of *one* dimension, or length, in which it is always possible to represent the effect of every intelligible operation of single algebra, and the interpretation of every result which admits of any interpretation at all. When we reckon time, gain and loss, &c., it is always possible to translate our reckoning into terms of length, as follows:

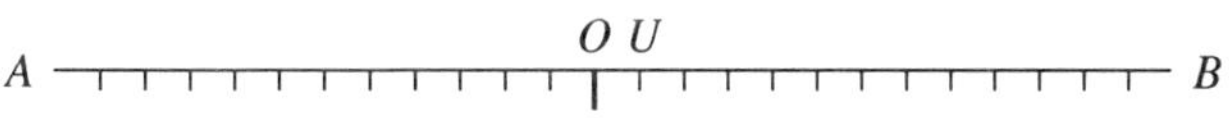

‖20‖ Take any point O, in a straight line, which call the *zero-point*, from which all measurement is to begin. Let OU represent the unit of any particular magnitude, and let magnitudes of one kind, say gains, be measured towards A, and losses towards B. Successive gains and losses may be taken off, and the final balance exhibited, by the compasses. As long as the result is always of one kind, so that an assumption to that effect would never render the processes of pure arithmetic unintelligible, the successive results always appear on one side of O: but the moment a result of the contrary kind appears, (which, unless the arithmetical computer were aware of it, and had provided accordingly, would leave him with an attempt at impossible subtraction on his hands,) it is indicated on the opposite side of O.

‖21‖ The convention as to the meaning of $+1$ and -1, namely, that they shall represent units of diametrically opposite kinds, is a very bold one: not merely because it takes up signs which are originally intended for nothing but addition and subtraction, and fixes another signification on them; but because it still employs them to connect quantities, *and by a new kind of connexion*. The signs in fact are used in two senses, the *directive*, and the *conjunctive*. $+(-3)$ tells us, by virtue of $-$, what we are talking of, and by virtue of $+(\ )$

how we are to join what we talk of to the rest. As conjunctive signs, + means *junction*, or putting on what we speak of; and − means *removal*. Thus, if + and − in the directive sense indicate gain and loss, the question, What is

$$(-3) + (+8) - (-7) + (-4) - (+3)?$$

is the following:—A man loses 3, and gets a gain of 8, with the removal of a loss of 7, the accession of a loss of 4, and the removal of a gain of 3: what is the united effect of all these actions on his previous property? The answer is, the accession of a gain of 5, $+ (+5)$.

⟦22⟧ The mere beginner is allowed to slide into single algebra from universal arithmetic in a manner which leads him to underrate the magnitude of the change. I do not see that it can be otherwise: but, at this period, my reader may be made to observe that the process by which we shall pass from single to double algebra, is the surest and most demonstrative (perhaps the only demonstrative) mode of passing from universal arithmetic to single algebra. It is not until he can drop all meanings, collect the laws of combination of the symbols, and so form a purely symbolic calculus, and then proceed to furnish that calculus with extended meanings, that he becomes fully master of the change. But the close resemblances, which make the slide above referred to so easy, might make it doubtful whether he would be fit to take proper note of this case of *reduction and restoration** until he has seen a more striking form of the same process, namely, that which is exhibited in the transition from single to double algebra.

⟦23⟧ When the earlier algebraists first began to occupy themselves with questions expressed in general terms, the difficulties of subtraction soon became obvious, inasmuch as the greater would sometimes demand to be subtracted from the less. The science has been brought to its present state through three distinct steps. The first was tacitly to contend for the principle that human faculties, at the outset of any science, are judges both of the extent to which its results can be carried, and of the form in which they are to be expressed. *Ignorance*, the necessary predecessor of knowledge, was called *nature*; and all conceptions which were declared unintelligible by the former, were supposed to have been made impossible by the latter. The first who used algebraical symbols in a general sense, Vieta, concluded that subtraction was a defect, and that expressions containing it should be in every possible manner avoided. *Vitium negationis*, was his phrase. Nothing could make a more easy pillow for the mind, than the rejection of all which could give any trouble; but if Euclid had altogether dispensed with the *vitium parallelorum*, his geometry would have been confined to twenty-six propositions of the first book.

* Algebra, *al jebr e al mokābala*, restoration and reduction, got its Arabic name, I have no doubt, from the *restoration* of the term which completes the square, and *reduction* of the equation by extracting the square root. The solution of a quadratic equation was the most prominent part of the Arabic algebra. Alter the order of the words, and the phrase may well represent the final mode of establishing algebra: *reduction* of universal arithmetic to a symbolic calculus, followed by *restoration* to significance under extended meanings.

⟦24⟧ The next and second step, though not without considerable fault, yet avoided the error of supposing that the learner was a competent critic. It consisted in treating the results of algebra as necessarily true, and as representing some relation or other, however inconsistent they might be with the suppositions from which they were deduced. So soon as it was shewn that a particular result had no existence as a quantity, it was permitted, by definition, to have an existence of another kind, into which no particular inquiry was made, because the rules under which it was found that the new symbols would give true results, did not differ from those previously applied to the old ones. A symbol, the result of operations upon symbols, either meant quantity, or nothing at all; but in the latter case it was conceived to be a certain new kind of quantity, and admitted as a subject of operation, though not one of distinct conception. Thus, $1 - 2$, and $a - (a + b)$, appeared under the name of negative quantities, or quantities less than nothing. These phrases, incongruous as they always were, maintained their ground, because they always produced a true result, whenever they produced any result at all which was intelligible: that is, the quantity less than nothing, in defiance of the common notion that all conceivable quantities are greater than nothing, and the square root of the negative quantity, an absurdity constructed upon an absurdity, always led to truths when they led back to arithmetic at all, or when the inconsistent suppositions destroyed each other. This ought to have been the most startling part of the whole process. That contradictions might occur, was no wonder; but that contradictions should uniformly, and without exception, lead to truth in algebra, and in no other species of mental occupation whatsoever, was a circumstance worthy the name of a mystery.

⟦25⟧ Nothing could prevail against the practical result that theorems so produced were true; and at last, when the interpretation of the abstract negative quantity shewed that a part at least of the difficulty admitted of rational solution, the remaining part, namely that of the square root of a negative quantity, was received, and its results admitted, with increased confidence.

⟦26⟧ The single algebra, when complete, leads to an unintelligible combination of symbols, $\sqrt{-1}$: not more unintelligible than was -1 when it first presented itself; for there are no degrees of absurdity in absolute contradiction of terms. The use of $\sqrt{-1}$, which leads to a variety of truths (page 41), points out that it "must have a logic" (page 41, note). I now proceed (page 92) to collect the symbols and laws of combination of algebra, or to describe *Symbolic Algebra*. ⟦Page 92 = §§10–11 of the present chapter.⟧

CHAPTER II.
ON SYMBOLIC ALGEBRA.

⟦1⟧ In abandoning the meanings of symbols, we also abandon those of the words which describe them. Thus *addition* is to be, for the present, a sound void of sense. It is a mode of combination represented by $+$; when $+$ receives its meaning, so also will the word *addition*. It is most important that the student should bear in mind that, *with one exception*, no word nor sign of arithmetic or algebra has one atom of meaning throughout this chapter, the object of which is *symbols, and their laws of combination*, giving a *symbolic algebra* (page 92) which may hereafter become the grammar of a hundred distinct *significant algebras*. If any one were to assert that + and − might mean reward and punishment, and A, B, C, &c. might stand for virtues and vices, the reader might believe him, or contradict him, as he pleases—but not out of *this* chapter.

⟦2⟧ The one exception above noted, which has some share of meaning, is the sign = placed between two symbols, as in $A = B$. It indicates that the two symbols have the same resulting meaning, by whatever different steps attained. That A and B, if quantities, are the same amount of quantity; that if operations, they are of the same effect, &c.

⟦3⟧ The following laws are not all unconnected: but the unsymmetrical character of the exponential operation, and the want of the connecting process of + and x, pointed out in the last chapter, renders it necessary to state them separately.

I. The *fundamental* symbols of algebra are

$$0,\ 1,\ +,\ -,\ \times,\ \div,\ (\ )^{(\ )},\ \text{and letters.}$$

⟦4⟧ In $(\ )^{()}$ there is the best mode of expressing the peculiar case in which the symbol consists in position; as in A^{B}, in which the distinctive symbolical force of the form lies in writing B over A.

⟦5⟧ II. It is usual to call + and − *signs*, and them only: but in laying down the laws of symbolic algebra, the close connexion existing between + and − on the one hand, and × and ÷ on the other, requires that the latter should also be called *signs*. Let the former be called *term-signs*, the latter *factor-signs*. It is to insist on this connexion that I do not (for a while) introduce the more common synonymes for $A \times B$ and $A \div B$, namely AB and $\dfrac{A}{B}$.

⟦6⟧ III. A symbol preceded by + or − is a *term*; by × or ÷ a *factor*. In A^{B}, A is the *base*, B the *exponent*. When an expression consists of terms, let them be called *co-terms*; when of factors, *co-factors*.

⟦7⟧ IV. Let 0 and 1 be a *co-term* and *co-factor* of every symbol, + and × being the connecting signs of the symbol, but either + or −, either × or ÷, those of 0 and 1. As seen in

$$\begin{aligned} A &= 0 + A = 1 \times A, \\ &= A + 0 = A - 0 = A \times 1 = A \div 1, \\ &= 0 + 1 \times A. \end{aligned}$$

Thus 0 and 1 are a kind of initial or starting symbols, the first of terms, the second of factors.

⟦8⟧ It is seen that + and ×, placed before a symbol, do not alter it: $\times A$

is A, having reference to 1 understood, as in $1 \times A$; and $+A$ is A, having reference to 0 understood, as in $0 + A$.

⟦9⟧ V. Co-terms and co-factors which differ only in sign, are equivalent to term 0 and factor 1.

$$+A - A = 0, \quad \times A \div A = 1.$$

The more usual form of the last is $1 \times A \div A = 1$. The starting symbol is frequently used in factors, but rarely in terms. The student is well accustomed to $+A$ and $-A$, in abbreviation of $0 + A$ and $0 - A$: but not to $\times A$ and $\div A$ for $1 \times A$ and $1 \div A$. But he must use the latter a little, if he would see the complete analogy of the term and factor signs.

⟦10⟧ VI. A symbol is said to be *distributive* over terms or factors when it is the same thing whether we combine that symbol with each of the terms or factors, or whether we make it apply to the compound term or factor. Thus, looking at

$$\overbrace{A\ B\ C\ D}^{*} \text{ and } \overset{*}{A}\ \overset{*}{B}\ \overset{*}{C}\ \overset{*}{D},$$

we see the * of the first distributed in the second.

⟦11⟧ VII. Term-signs are distributive over terms, and factor-signs over factors: as in

$$+(+A - B) = +(+A) + (-B), \quad \div(\times A \div B) = \div(\times A) \div (\div B),$$

at full length $0 + (0 + A - B) = 0 + (0 + A) + (0 - B)$,

$$1 \div (1 \times A \div B) = 1 \div (1 \times A) \div (1 \div B).$$

⟦12⟧ VIII. The term-signs of factors may belong, each one of them, to any factor of the compound, or to the compound.

$$-A \times -B = -(-A) \times B = -(-)(A \times B).$$

⟦13⟧ IX. Like term-signs in combination produce $+$; unlike, $-$. Like factor-signs in combination produce $\times$; unlike, $\div$. As in

$$+(-A) = -A, \ -(-A) = +A, \ \times(\div A) = \div A, \ \div(\div A) = \times A.$$

⟦14⟧ X. Terms and factors are convertible in order, terms with terms, factors with factors. As in

$$+A - B = -B + A, \quad \times A \div B = \div B \times A.$$

⟦15⟧ XI. Factors are distributive over the terms of any cofactor with the sign $\times$. (The corresponding law for $\div$ factors can be deduced, and is not to be set down as fundamental). As in

$$(+A) \times (+B - C) = (+A) \times (+B) + (+A) \times (-C)$$
$$= +A \times B - A \times C,$$

and

$$\times (B - C) \div A = B \div A - C \div A.$$

⟦16⟧ XII. The relations of the starting symbols 0 and 1, as exponents, are $A^0 = 1$, $A^1 = A$.

⟦17⟧ XIII. The exponent is distributive over factors with $\times$ (the case of $\div$ is deducible). As in

$$(\times A \times B)^C = \times A^C \times B^C.$$

⟦18⟧ XIV. The operations of $\times$ and the exponential operation $(\ )^{(\ )}$, successively repeated with same base, are reducible to the lower operations + *and* $\times$ performed with the exponents. As in

$$A^B \times A^C = A^{B+C},\ (A^B)^C = A^{B \times C}.$$

⟦19⟧ Any system of symbols which obeys these rules and no others,—except they be formed by combinations of these rules—and which uses the preceding symbols and no others—except they be new symbols invented in abbreviation of combinations of these symbols—is *symbolic algebra*. Ordinary algebra contains all these symbols and all these rules, but its assigned meanings do not make *all* results significant. I now proceed to combined symbols, and to a sufficient amount of proof by instance, that one who admits these rules admits, as consequences, all the combinations of ordinary algebra.

⟦20⟧ Let 1 + 1 be abbreviated into 2; 2 + 1 into 3; 3 + 1 into 4, and so on. Now introduce the abbreviations of $A \times B$ and $A \div B$, namely, AB and $\frac{A}{B}$.

⟦21⟧ We have then $A + A = 2A$; for (IV), $A + A$ is $1 \times A + 1 \times A$ or (XI) $(1 + 1)A$ or $2A$. Similarly, $A + A + A = 3A$. Again, $4A \div 7$ is $\frac{4}{7}A$: for (X), $1 \times 4 \times A \div 7$ is $1 \times 4 \div 7 \times A$, or $A \times 4 \div 7$, or (VII), (VIII), $A \times (\times 4 \div 7)$, or $\frac{4}{7}A$;

$$(A - B)(C - D) \text{ is (XI) } (A - B)C - (A - B)D, \text{ or, (XI) again,}$$
$$AC - BC - (AD - BD), \text{ or (VII)}, AC - BC - (+AD) - (-BD),$$
$$\text{or (IX)}, AC - BC - AD + BD;$$

$$\frac{A}{B} = \frac{AC}{BC}, \text{ for } \times A \times C \div (B \times C) \text{ is (VII)}, \times A \times C \div B \div C, \text{ or}$$
$$\text{(X)}, \times A \div B \times C \div C, \text{ or (V)}, \times A \div B, \text{ or } \frac{A}{B};$$

$A \times 0 = 0$; for (V) $A \times 0$ is $A(+B - B)$, or (XI) $+AB - AB$, which (V) is 0.

⟦22⟧ From what precedes $\frac{A}{B + C}$ is $\frac{1}{\frac{B}{A} + \frac{C}{A}}$. This is an instance of the deducible part of (XI); it is $\times A \div (B + C) = \div (B \div A + C \div A)$. The complete rule XI, in all its parts, fundamental and deducible, is this:—A factor may be distributed over the terms of its cofactor, with its factor-sign or the contrary, according as the receiving cofactor is $\times$ or $\div$. Thus

$$\div A \div (B + C) \text{ is } \div (A \times B + A \times C);$$

$\frac{A}{B} \pm \frac{C}{D}$ has been shewn to be $\frac{AD}{BD} \pm \frac{CB}{BD}$, or (XI) $\frac{AD \pm BC}{BD}$.

$$A^{\mathrm{B}} \times A^{-\mathrm{B}} \text{ is (XIV) } A^{\mathrm{B}+(-\mathrm{B})}, \text{ or (IX) } A^{\mathrm{B}-\mathrm{B}}, \text{ or (V) } A^0, \text{ or (XII) } 1.$$

$$\text{So that } A^{-\mathrm{B}} = \frac{1}{A^{\mathrm{B}}}; \text{ and } \frac{A^{\mathrm{B}}}{A^{\mathrm{C}}} = A^{\mathrm{B}}A^{-\mathrm{C}} = A^{\mathrm{B}-\mathrm{C}};$$

$$\left(\frac{A}{B}\right)^{\mathrm{C}} = (A^1B^{-1})^C = A^{\mathrm{C}}B^{-\mathrm{C}}(\text{XIV}) = \frac{A^{\mathrm{C}}}{B^{\mathrm{C}}};$$

$$A^2 \text{ is } A^{1+1}, \text{ or } A^1A^1, \text{ or } AA; A^3 \text{ is } AAA, \text{\&c.}$$

$$A^{\frac{1}{3}} \text{ gives } (A^{\frac{1}{3}})^3 = A^1 = A \text{ (XII), or } A^{\frac{1}{3}}A^{\frac{1}{3}}A^{\frac{1}{3}} = A;$$

$$-A \times -B \text{ is (VIII)} - (-)\, A \times B, \text{ or } +AB, \text{ or } AB;$$

$$A \times (BC) \text{ is } A \times (\times B \times C), \text{ or (VII)} A \times (\times B) \times (\times C),$$

$$\text{or (IX) } A \times B \times C.$$

[23] In this way the student must examine narrowly a large number of fundamental operations, satisfying himself that he could produce them from the *rules alone*, independently of every notion of meaning. The question is this,— Might a machine, which could, when guided, make introductions and alterations by the preceding rules and no others, be made to turn one of the alleged equivalent combinations into the other.

[24] It will be exceedingly convenient to reserve the small letters a, b, c, &c. most strictly to signify pure combinations of the unit-symbol 1, with any term or factor-signs, as $+2$, $-\frac{3}{8}$, &c.: and to use the capitals A, B, C, &c. for other cases. With the exception of ε I shall use Greek letters only for angles.

10
William Rowan Hamilton (1805–1865)

Despite Gauss's defence of the legitimacy of the complex numbers (*Gauss 1831*, §§ 16–24), both *i* and even the negative integers remained a source of mystery to many mathematicians. The sense of mystery was compounded by two independent and almost simultaneous events: the publication in Germany of Hermann Grassmann's *Ausdehnungslehre* or 'Theory of extension' (*H. Grassmann 1844*), which described *n*-dimensional geometries and hypercomplex number systems; and the discovery by the Irish mathematician Sir William Rowan Hamilton[a] of *quaternions*—new 'numbers' which did not obey the familiar commutative law of ordinary arithmetic. The vector algebras described by Grassmann were even more general than those of Hamilton. But his presentation was obscure, and the importance of his ideas was not immediately appreciated; in the short run Hamilton's quaternions had the greater influence on mathematical research.

Quaternions today are little more than a historical curiosity. But they were the first of a series of strange objects that were to confront British mathematicians in the middle decades of the nineteenth century: octonians, Clifford algebras, linear associative algebras, matrices, vector spaces. James Joseph Sylvester, calling himself the 'mathematical Adam', was to delight in inventing names for the newly-discovered flora and fauna of the mathematical world—*combinants*, *reciprocants*, *concomitants*, *discriminants*, *zetaic multipliers*, *skew pantographs*, and dozens more of the like; Hamilton himself, in his *Lectures on quaternions*, used new terms like *vector*, *vehend*, *vection*, *vectum*, *revector*, *revehend*, *revection*, *revectum*, *provector*, *transvector*, and so on. By 1860, Benjamin Peirce (the father of the logician, Charles Sanders Peirce, and one of the earliest and most vigorous champions of quaternions) had classified 162 separate multiplication tables for linear associative algebras: thirty years after Gauss's defence of the complex integers, the number-concept had been widened far beyond the bounds known to previous mathematicians.

The discovery of these new, non-standard algebraic structures was almost as startling to traditional conceptions as the discovery of non-Euclidean geometry, and the impact on mathematics was equally great. Mathematicians were forced

[a] The Irish mathematician Sir William Rowan Hamilton should not be confused with the Scottish philosopher and logician Sir William Hamilton—a figure now largely forgotten, with whom Augustus De Morgan had a celebrated controversy concerning priority in discovering the 'quantification of the predicate', and whose metaphysical writings John Stuart Mill scrutinized in *An examination of Sir William Hamilton's philosophy*.

to confront the issue of the relationship between mathematical *symbols* and mathematical *objects*, and to distinguish more clearly than they had hitherto done between the manipulation of *algebraic equations* and the manipulation of *numbers*. The new, more flexible conception of algebra that emerged in the middle decades of the nineteenth century made possible both the extension by Boole, Peirce, and Schröder of algebraic techniques to symbolic logic, and the development of abstract algebra at the hands of Cayley, Sylvester, Clifford, Gibbs, Dedekind, and their colleagues.

Rowan Hamilton was born in Dublin in 1805; his father was a solicitor. A precocious child, he exhibited a strong early interest in languages and mathematics—in particular, in mathematical astronomy. In his early youth he studied Newton's *Principia* and Laplace's *Mécanique céleste*; he began making astronomical observations with his own telescope and writing papers on optics and on the properties of curves and surfaces. He entered Trinity College, Dublin in 1823, where he received a solid grounding in the methods of the Continental mathematicians. (In this respect, Ireland was in advance of the leading English universities, which still favoured Newton's method of fluxions.) From 1827 until his death he was Astronomer Royal at Dunsink Observatory (near Dublin) and Andrews Professor of Astronomy at Trinity College, Dublin; he was still an undergraduate, aged twenty-two, at the time of his initial appointment.

This was the start of an age of discovery in the foundations of algebra. We have seen that in the 1830s a series of English algebraists associated with the University of Cambridge turned their efforts towards justifying the negative and imaginary numbers, and towards answering the criticisms that had been voiced by such writers as Francis Maseres and William Frend. In the process, they developed a new, syntactic conception of algebra. *Peacock 1833* put forward the idea of a *symbolical algebra*; *Gregory 1840* added the observation that the symbols of symbolical algebra need not be taken to stand for numbers, but could equally well stand for operations; and *De Morgan 1842a* and *1849b* extended the new algebra to the logic of relations and improved on the description of a formal calculus.

During approximately the same time, Hamilton was carrying out his own investigations into the foundations of algebra. His research was largely independent of the Cambridge school, and in important respects based on a different set of presuppositions; indeed, his deepest results were to some extent obtained by swimming against the historical tide.

The sources of Hamilton's ideas can best be appreciated by considering their development chronologically. In 1827, at the time of his appointment to the chair of astronomy, Hamilton began to work on mathematical physics—work that was to continue throughout the period of his research on algebra.[b] In 1827 he

[b] Indeed, Hamilton is today best remembered for his work as a mathematical physicist, and in particular for his contributions to geometrical optics and to dynamics. Hamilton himself believed that quaternions were his most important contribution to physics, and that they would one day prove as useful in the description of the physical world as the calculus itself. This hope proved

presented to the Royal Irish Academy his first major paper on optics, the 'Theory of a system of rays.' This paper introduced his central idea in mathematical physics—the function V that he called the *characteristic function* of a system of rays. V was a single function that completely described the behaviour of the optical system; in Hamilton's words, the characteristic function contained 'the whole of mathematical optics' (*Hamilton 1931–67*, Vol. i, p. 168). The idea of compressing as much information as possible into a single equation was powerfully attractive to Hamilton. He was to use the same strategy when he turned to dynamics, there using the characteristic function V to describe the action of a dynamical system in moving from its initial to its final point in configuration space. In practice, Hamilton's system of dynamics was no easier to work with than Lagrange's system of numerous differential equations; but as Hamilton remarked,

> The difficulty is at least transferred from the integration of many equations of one class to the integration of two of another: and even if it should be thought that no practical facility is gained, yet an intellectual pleasure may result from the reduction of the most complex and, probably, of all researches respecting the forces and motions of body, to the study of one characteristic function, the unfolding of one central relation (*Hamilton 1931–67*, Vol. ii, p. 105).

During this period Hamilton was preoccupied with his discoveries in optics; but his friend, the mathematician John T. Graves, spurred him to think about the foundations of algebra. Graves in 1826 and 1827 was trying to construct logarithms for the negative and complex numbers, and regularly corresponded with Hamilton about the problem. Graves submitted a memoir on imaginary logarithms to the Royal Society in 1828; it was coolly received by Peacock and Herschel, but published after Hamilton interceded on Graves's behalf. This episode was the beginning of Hamilton's interest in the complex numbers, and in a letter to Graves of 20 October 1828 he lamented the confusion that existed in the foundations of algebra:

---

mistaken; however, Hamilton's work in other areas turned out to have unanticipated physical applications. His exploration of the analogy between optics and dynamics influenced Schrödinger and De Broglie in their formulation of wave mechanics; his work on linear algebra supplied Heisenberg with some of the fundamental mathematical tools for matrix mechanics; his ideas about quaternions inspired others—Sylvester, Cayley, Clifford, Benjamin Peirce, Gibbs, Heaviside—to study matrices, vectors, and their physical applications. Schrödinger was explicit in his praise:

> I daresay not a day passes—and seldom an hour—without somebody, somewhere on this globe, pronouncing or reading or writing Hamilton's name. That is due to his fundamental discoveries in general dynamics. The Hamiltonian Principle has become the cornerstone of modern physics, the thing with which a physicist expects every physical phenomenon to be in conformity. . . .
>
> The modern development of physics is constantly enhancing Hamilton's name. His famous analogy between optics and mechanics virtually anticipated wave mechanics, which did not have much to add to his ideas and had only to take them more seriously (*Schrödinger 1945*, p. 82).

The details of Hamilton's contributions to mathematical physics lie beyond the compass of the present work; for a detailed account of his contributions to mathematical optics, see *Synge 1937*; for an account of his contributions to dynamics, see *Cayley 1857*; for a history of the early development of linear algebra, see *Crowe 1967*. *Hankins 1980* is a fine full-length intellectual biography of Hamilton.

Your remarks on developments in general are interesting, and the whole subject is one very well worth pursuing. For my own part I have always been greatly dissatisfied with the phrases, if not the reasonings, of even very eminent analysts, on a variety of subjects, of which the Theory of Developments is one. I have often persuaded myself that the whole analysis of infinite series, and indeed the whole logic of analysis (I mean of algebraic analysis) would be worthy of ‖ra‖dical revision. But it would be ‖right‖ for a person who should attempt this to go to the root of the matter, and either to discard negative and imaginary quantities, or at least (if this should be impossible or unadvisable, as indeed I think it would be) to explain by strict definition, and illustrate by abundant example, the true sense and spirit of the reasonings in which they are used. An algebraist who should thus clear away the metaphysical stumbling-blocks that beset the entrance of analysis, without sacrificing those concise and powerful methods which constitute its essence and its value, would perform a useful work and deserve well of Science (*Graves 1882–9*, Vol. i, pp. 303–4).

(Hamilton's objections to the traditional use of complex numbers are stated in more detail in §4 of the first selection below, *Hamilton 1837*.)

Graves suggested that Hamilton read John Warren's recently published *Treatise on the geometrical representation of the square roots of negative quantities* (*Warren 1828*), a work which described the Argand representation of a complex number as a point on the real plane. The Argand representation immediately caught Hamilton's eye. For if one regards the complex numbers as vectors in the plane, then the elementary arithmetic operations have a natural geometric interpretation—with complex addition corresponding to vector addition, and complex multiplication to a vector rotation *cum* scalar multiplication. It was natural for Hamilton to wonder whether generalized 'complex numbers' can be defined that will represent rotations and displacements of vectors in three-dimensional space. And, if so, it was natural to hope that such numbers would be a powerful tool for formulating the basic laws of physics and for describing the motion of rigid bodies in space. Just as the characteristic function contained in a single, compact representation 'the whole of mathematical optics', so the new numbers would contain all relevant information about spatial vectors, replacing the artificial device of three-dimensional rectangular coordinates. Hamilton worked on this problem for fifteen years before suddenly seeing the solution in October, 1843, as he was walking with his wife along the Royal Canal in Dublin.

In the years immediately after he read *Warren 1828*, Hamilton tried to find some other, more fundamental algebraic representation of the complex numbers than that provided by the Argand diagram. To modern ways of thinking, Argand's geometric interpretation of the complex numbers is an adequate justification of their legitimacy. But like almost all mathematicians of his day, Hamilton thought of algebra as the science of *number*, of *quantity*, just as geometry was the science of physical space. Argand's diagram did not fit well with the traditional paradigm of algebra: it did not explain 'imaginary quantities' as quantities, or make clear their relationship to the idea of numerical magnitude. Hamilton in the 1820s had discovered for himself (independently of *Cauchy 1821*) the Cauchy–Riemann equations and the representation of a function of a complex variable by a pair of functions of a real variable; he called the Cauchy–Riemann equations *equations of conjugation*, and the two real functions *conjugate functions*. This discovery suggested to him that complex numbers could be viewed as ordered pairs or 'couples' of real numbers, and he used these new number-couples to investigate Graves's problem of imaginary logarithms. But Hamilton now had to explain how the number-couples were related to ordinary real numbers—and this problem led him to search for some deeper concept that underlay both kinds of number. He persuaded himself already in the late 1820s—before he studied Coleridge and Kant—that the foundations of algebra were to be sought in the *ordinal* properties of the whole numbers, and that these ordinal properties were deeply connected to the experience of time.

Hamilton was to bring his ideas to full development in the early 1830s. The years 1830–3 were a time of ferment in the development of his thought, and show him working simultaneously on physics, algebra, and philosophy. He continued his research on optics and the characteristic function, publishing three important supplements to 'A theory of a system of rays' between 1830 and 1832. He also began to study metaphysics, paying particular attention to discussions of time and of the foundations of physical science. In 1830 he read closely the works of George Berkeley and of the Croatian philosopher of science Rudjer Bošković.[c] The following year he read Samuel Taylor Coleridge's *Aids to reflection*. Hamilton, like Boole and Sylvester, was deeply interested in poetry, and, had he not been dissuaded from such a course by William Wordsworth, would have abandoned his scientific research for a career in literature. Hamilton continued as a scientist (and as a writer of undistinguished verse); but he steeped himself in the literary, aesthetic, and religious theories of the romantic movement. He viewed his scientific work as a form of poetry, in later life describing the quaternions as 'a curious offspring of a quaternion of parents, say of geometry, algebra, metaphysics, and poetry' (*Hankins 1980*, p. 247). Coleridge's philosophical writings were largely an exposition of the ideas of the German idealists, and led Hamilton, at Coleridge's urging, in October, 1831, to turn to the original source and make a close study of Kant's *Critique of pure reason*—a work which had not yet been translated into English, and which was still largely unknown to British scientists. These philosophical studies impressed him deeply, and throughout his subsequent career he was to portray Kant and Bošković as major influences on his scientific thought.

---

[c] Bošković (1711–87) was a Jesuit priest who lived successively in Rome, Paris, London, Istanbul, Warsaw, Vienna, Venice, Pavia, Paris, and Milan. His writings were as varied as his travels. He wrote on theology, archaeology, astronomy, geodesy, architecture, meteorology, optics; he worked as an engineer, draining swamps and designing harbours; he was in his later years professor of mathematics at Pavia and Milan. He published extensively on mathematics and celestial mechanics. His most influential contribution was to the foundations of physics, where he developed a theory of spatial point-atoms, and showed how such an atomistic theory could be combined with Newton's physical system. This work on atoms was well-known in England in Hamilton's day (the *Encyclopedia Britannica* of 1801 offered a 14-page description), and was studied, not only by Hamilton, but also by Faraday, Maxwell, and Kelvin.

While he was studying the philosophers, Hamilton turned his attention to the study of dynamics, and showed how to apply his characteristic function to the motions of the planets and the comets; in the introduction to his celebrated 'On a general method in dynamics' (*Hamilton 1834*) he declared his adherence to Bošković's atomic theory of matter. He also deepened his investigations into the foundations of algebra. On 4 November 1833 he presented to the Royal Irish Academy a paper on 'The theory of conjugate functions, or algebraic couples'—that is, on ordered pairs of real numbers. He defined operations of addition and multiplication for couples, and demonstrated that the square of the couple (0, 1) is the couple (1, 0), so that (0, 1) may be regarded as $\sqrt{-1}$; he then worked out the basic arithmetic properties of couples, and, in effect, described the properties of a field.[d] Two years later, on 1 June 1835, Hamilton presented a second paper to the Royal Irish Academy, entitled 'A preliminary and elementary essay on algebra as the science of pure time' in which he tried to ground both real numbers and number-couples in the intuition of time; the two papers were combined and published with some 'General introductory remarks' (written in 1835) as *Hamilton 1837*. The introductory remarks are reprinted below.

Although Hamilton was the first mathematician to treat complex numbers as ordered pairs, it is important to observe that the spirit of his analysis is different both from that of modern textbooks and from the approach of his contemporaries in Cambridge. The modern emphasis is on ontological parsimony, on eliminating superfluous entities: as one says, the complex numbers are *reduced* to ordered pairs of reals. Similarly, the thrust of the Cambridge algebraists was to eliminate the reference of algebraic expressions altogether—to regard 'symbolical algebra' as merely a system of rules for manipulating meaningless symbols. But such nominalistic approaches are alien to Hamilton. He comes to algebra with the interests and attitudes of a physicist, with the geometric interpretation of the complex numbers firmly before his eyes. He wants algebra to be a *science*—a body of *truths* about some external reality; and in the following selection (§§5, 6) he rejects the merely 'philological' or syntactic conception of algebra as inadequate. His ambition was to establish an objective reference for the equations of complex analysis—to *find* the numbers that were being referred to, rather than to *eliminate* them. In this spirit, he argues that, just as geometry is the science of space, so algebra is the science of time (§12); in later writings (for example in §3 of *Hamilton 1853*) he was to invoke Kant in support of his view. (The depth of Hamilton's understanding of Kant on this point may well be doubted; for sceptical remarks, see *Peirce 1898*, reproduced below.)

Thus Hamilton and the Cambridge algebraists were in important respects marching in divergent directions, with the Cambridge school taking the route of nominalism, and Hamilton that of scholastic realism. In this sense, Hamilton

[d] Gauss in an 1837 letter to Wolfgang Bolyai remarked that already in 1831, the year of his paper on biquadratic residues, he had had the idea of representing complex numbers as ordered pairs of real numbers (*Kline 1972*, p. 776). Hamilton, however, was the first to publish the idea.

was swimming against the historical tide, and his researches stand apart from the main line of the development of British algebra.

His approach had undeniable advantages—and equally undeniable disadvantages. Had he thought of the complex numbers purely syntactically and algebraically he would have been less likely to discover the quaternions. For if one is solely interested in solving algebraic equations, then there is no need to look beyond the complex numbers: every algebraic equation with real or complex coefficients has its solutions within the complex field, and there is no obvious reason to widen the number concept yet further. But if one possesses a strong sense of the geometric and physical interpretation of the complex plane, it is natural to wonder about three-dimensional analogues; and in this way Hamilton was led to discover the first non-commutative algebra.[c] On the other hand, his conception of algebra as *the* 'science of pure time'—quite apart from its questionable philosophical underpinnings—would have been an inadequate foundation for the development of algebra in the nineteenth century. Hamilton appears to have believed (he is not explicit on the point) that the nature of time required all algebraic operations to be associative, and multiplicative inverses always to exist. (For example, he regarded it as a telling defect that the eight-dimensional octonians discovered by Cayley and independently by J.T. Graves in 1845 do not obey the associative law of multiplication (*Hamilton 1853*, §61).) But these requirements would have imposed as much of a strait-jacket on the development of British algebra as Peacock's 'principle of the permanence of equivalent forms' since (as is now known) the only real associative division algebras are the real numbers themselves, the complex numbers, and the quaternions. It is clear in retrospect that the exploration of a wide range of algebraic structures needed the more flexible conception of algebra developed by Gregory, De Morgan, and Boole.

Hamilton's most important published and unpublished mathematical papers are collected in the three volumes of *Hamilton 1931–67*. Further unpublished material can be found in *Graves 1882–9*, with a full bibliography of the published papers at the end of volume three. The Hamilton papers are for the most part stored in Dublin, at the Library of Trinity College and at the National Library of Ireland.

The first selection below is the *General introductory remarks* from *Hamilton 1837*; these remarks should be read in conjunction with Gauss's roughly contemporary observations on the negative and complex numbers in *Gauss 1831*, as well as with the selections above from Duncan Gregory and Augustus De Morgan. Although the *Introductory remarks* were first printed in 1837, they were written in 1835; the memoir itself was read to the Royal Irish Academy on 4

[c] De Morgan, to be sure, in §11 of *De Morgan 1842a*, raised the question of a three-dimensional analogue to the complex numbers; this fact was explicitly acknowledged by Hamilton (*Hamilton 1853*, Preface, §43). But the problem never became a central preoccupation to De Morgan in the way it did to Hamilton.

November 1833 and 1 June 1835. References to *Hamilton 1837* should be to the paragraph numbers, which have been added in this reprinting.

A. *FROM THE* THEORY OF CONJUGATE FUNCTIONS, OR ALGEBRAIC COUPLES; WITH A PRELIMINARY AND ELEMENTARY ESSAY ON ALGEBRA AS THE SCIENCE OF PURE TIME (*HAMILTON 1837*)

General introductory remarks

‖1‖ The Study of Algebra may be pursued in three very different schools, the Practical, the Philological, or the Theoretical, according as Algebra itself is accounted an Instrument, or a Language, or a Contemplation; according as ease of operation, or symmetry of expression, or clearness of thought, (the *agere*, the *fari*, or the *sapere*,) is eminently prized and sought for. The Practical person seeks a Rule which he may apply, the Philological person seeks a Formula which he may write, the Theoretical person seeks a Theorem on which he may meditate. The felt imperfections of Algebra are of three answering kinds. The Practical Algebraist complains of imperfection when he finds his Instrument limited in power; when a rule, which he could happily apply to many cases, can be hardly or not at all applied by him to some new case; when it fails to enable him to do or to discover something else, in some other Art, or in some other Science, to which Algebra with him was but subordinate, and for the sake of which and not for its own sake, he studied Algebra. The Philological Algebraist complains of imperfection, when his Language presents him with an Anomaly; when he finds an Exception disturb the simplicity of his Notation, or the symmetrical structure of his Syntax; when a Formula must be written with precaution, and a Symbolism is not universal. The Theoretical Algebraist complains of imperfection, when the clearness of his Contemplation is obscured; when the Reasonings of his Science seem anywhere to oppose each other, or become in any part too complex or too little valid for his belief to rest firmly upon them; or when, though trial may have taught him that a rule is useful, or that a formula gives true results, he cannot prove that rule, nor understand that formula: when he cannot rise to intuition from induction, or cannot look beyond the signs to the things signified.

‖2‖ It is not here asserted that every or any Algebraist belongs *exclusively* to any *one* of these three schools, so as to be *only* Practical, or *only* Philological, or *only* Theoretical. Language and Thought react, and Theory and Practice help each other. No man can be so merely practical as to use frequently the rules of Algebra, and never to admire the beauty of the language which expresses those rules, nor care to know the reasoning which deduces them. No man can be so merely philological an Algebraist but that things or thoughts will at some times intrude upon signs; and occupied as he may habitually be with the logical building up of his expressions, he will feel sometimes a desire to know what they mean, or to apply them. And no man can be so merely theoretical or so exclusively devoted to thoughts, and to the contemplation of theorems in Algebra, as not to feel an interest in its notation and language, its symmetrical system of signs, and the logical forms of their combinations; or not to prize those practical aids, and especially those methods of research, which the discoveries and contemplations of Algebra have given to other sciences. But, distinguishing without dividing, it is perhaps correct to say that every Algebraical Student and every Algebraical Composition may be referred upon the whole to one or other of these three schools, according as one or other of these three views habitually actuates the man, and eminently marks the work.

⟦3⟧ These remarks have been premised, that the reader may more easily and distinctly perceive what the design of the following communication is, and what the Author hopes or at least desires to accomplish. That design is *Theoretical*, in the sense already explained, as distinguished from what is Practical on the one hand, and from what is Philological upon the other. The thing aimed at, is to improve the *Science*, not the Art nor the Language of Algebra. The imperfections sought to be removed, are confusions of thought, and obscurities or errors of reasoning; not difficulties of application of an instrument, nor failures of symmetry in expression. And that confusions of thought, and errors of reasoning, still darken the beginnings of Algebra, is the earnest and just complaint of sober and thoughtful men, who in a spirit of love and honour have studied Algebraic Science, admiring, extending, and applying what has been already brought to light, and feeling all the beauty and consistence of many a remote deduction, from principles which yet remain obscure, and doubtful.

⟦4⟧ For it has not fared with the principles of Algebra as with the principles of Geometry. No candid and intelligent person can doubt the truth of the chief properties of *Parallel Lines*, as set forth by Euclid in his Elements, two thousand years ago; though he may well desire to see them treated in a clearer and better method. The doctrine involves no obscurity nor confusion of thought, and leaves in the mind no reasonable ground for doubt, although ingenuity may usefully be exercised in improving the plan of the argument. But it requires no peculiar scepticism to doubt, or even to disbelieve, the doctrine of Negatives and Imaginaries, when set forth (as it has commonly been) with principles like these: that a *greater magnitude may be subtracted from a less*, and that the remainder is *less than nothing*; that *two negative numbers*, or numbers denoting magnitudes each less than nothing, may be *multiplied* the one by the other, and that the product will be a *positive* number, or a number denoting a magnitude greater than nothing; and that although the *square* of a number, or the product obtained by multiplying that number by itself, is therefore *always positive*, whether the number be positive or negative, yet that numbers, called *imaginary*, can be found or conceived or determined, and operated on by all the rules of positive and negative numbers, as if they were subject to those rules, *although they have negative squares*, and must therefore be supposed to be themselves neither positive nor negative, nor yet null-numbers, so that the magnitudes which they are supposed to denote can neither be greater than nothing, nor less than nothing, nor even equal to nothing. It must be hard to found a SCIENCE on such grounds as these, though the forms of logic may build up from them a symmetrical system of expressions, and a practical art may be learned of rightly applying useful rules which seem to depend upon them.

‖5‖ So useful are those rules, so symmetrical those expressions, and yet so unsatisfactory those principles from which they are supposed to be derived, that a growing tendency may be perceived to the rejection of that view which regarded Algebra as a SCIENCE, *in some sense analogous to Geometry*, and to the adoption of one or other of those two different views, which regard Algebra as an *Art*, or as a *Language*: as a System of Rules, or else as a System of Expressions, but not as a System of *Truths*, or Results having any other validity than what they may derive from their practical usefulness, or their logical or philological coherence. Opinions thus are tending to substitute for the Theoretical question,—'Is a Theorem of Algebra *true*?' the Practical question,—'Can it be *applied as an Instrument*, to do or to discover something else, in some research which is not Algebraical?' or else the Philological question,—'Does its *expression harmonise*, according to the Laws of Language, with other Algebraical expressions?'

‖6‖ Yet a natural regret might be felt, if such were the destiny of Algebra; if a study, which is continually engaging mathematicians more and more, and has almost superseded the Study of Geometrical Science, were found at last to be not, in any strict and proper sense, the Study of a Science at all: and if, in thus exchanging the ancient for the modern Mathesis, there were a gain only of Skill or Elegance, at the expense of Contemplation and Intuition. Indulgence, therefore, may be hoped for, by any one who would inquire, whether existing Algebra, in the state to which it has been already unfolded by the masters of its rules and of its language, offers indeed no rudiment which may encourage a hope of developing a SCIENCE of Algebra: a Science properly so called; strict, pure, and independent; deduced by valid reasonings from its own intuitive principles; and thus not less an object of priori contemplation than Geometry, nor less distinct, in its own essence, from the Rules which it may teach or use, and from the Signs by which it may express its meaning.

‖7‖ The Author of this paper has been led to the belief, that the Intuition of TIME is such a rudiment.

‖8‖ This belief involves the three following as components: First, that the notion of Time is connected with existing Algebra; Second, that this notion or intuition of Time may be unfolded into an independent Pure Science; and Third, that the Science of Pure Time, thus unfolded, is co-extensive and identical with Algebra, so far as Algebra itself is a Science. The first component judgement is the result of an induction; the second of a deduction; the third is a joint result of the deductive and inductive processes.

‖9‖ 1. The argument for the conclusion that *the notion of Time is connected with existing Algebra*, is an induction of the following kind. The History of Algebraic Science shows that the most remarkable discoveries in it have been made, either expressly through the medium of that notion of *Time*, or through the closely connected (and in some sort coincident) notion of *Continuous Progression*. It is the genius of Algebra to consider what it reasons on as *flowing*, as it was the genius of Geometry, to consider what it reasoned on as *fixed*. Euclid* defined a tangent to a circle, Apollonius† conceived a tangent to an ellipse, as an indefinite straight line which had only one point in common with the curve; they looked upon the line and curve not as nascent or growing, but as already constructed and existing in space; they studied them as *formed* and *fixed*, they compared the one with the other, and the proved exclusion of any second common point was to them the essential property, the constitutive character of the tangent. The Newtonian Method of Tangents rests on another principle; it regards the curve and line not as *already* formed and fixed, but rather as *nascent*, or in process of generation: and employs, as its primary conception, the thought of a *flowing point*. And, generally, the revolution which Newton‡ made in the higher parts of both pure and applied Algebra, was founded mainly on the notion of *fluxion*, which involves the notion of *Time*.

‖10‖ Before the age of Newton, another great revolution, in Algebra as well as in Arithmetic, had been made by the invention of *Logarithms*; and the 'Canon Mirificus' attests that Napier* deduced that invention, not (as it is commonly

---

* *Εὐθεῖα κύκλου ἐφάπτεσθαι λέγεται, ἥτις ἁπτομένη τοῦ κύκλου καὶ ἐκβαλλομένη οὐ τέμνει τὸν κύκλον.*—Euclid, Book III. Def. 2. Oxford Edition, 1703. ‖'A straight line is said to touch a circle if in meeting the circle and being extended it does not cut the circle.'‖

† *Εὰν ἐν κώνου τομῇ ἀπὸ τῆς κορυφῆς τῆς τομῆς ἀχθῇ εὐθεῖα παρὰ τεταγμένως κατηγμένην ἐκτὸς πεσεῖται τῆς τομῆς.—ἐκτὸς ἄρα πεσεῖται, διόπερ ἐφάπτεται τῆς τομῆς.*—Apollonius, Book I. Prop. 17. Oxford Edition, 1710. ‖'If a straight line be drawn through the extremity of the diameter of any conic section parallel to the ordinates of that diameter, the straight line will fall outside the conic section. It will do this because it touches the section.'‖

‡ Considerando igitur quod quantitates aequalibus temporibus crescentes et crescendo genitae, pro velocitate majori vel minori qua crescunt ac generantur evadunt majores vel minores; methodum quaerebam determinandi quantitates ex velocitatibus motuum vel incrementorum quibus generantur; et has motuum vel incrementorum velocitates nominando *Fluxiones*, et quantitates genitas nominando *Fluentes*, incidi paulatim annis 1665 et 1666 in Methodum Fluxionum qua hic usus sum in Quadratura Curvarum—*Tractatus de Quad. Curv.*, Introd., published at the end of Sir I. Newton's Opticks, London 1704. ‖'Therefore, considering that quantities which increase over equal periods of time and which are created by increasing become greater or less in proportion to the greater or lesser velocities by which they increase and are generated, I sought a method of determining the quantities from the velocities of the motions or increments by which they are generated. By calling these velocities of motions and increments *fluxions*, and the quantities created *flowing points*, in 1665 and 1666 I gradually came upon the 'Method of Fluxions' which I have used here in *The quadrature of curves*'.‖

* Logarithmus ergo cujusque sinus, est numerus quam proxime definiens lineam, quae aequaliter crevit interea dum sinus totius linea proportionaliter in sinum illum decrevit, existente utroque motu synchrono, atque initio aequiveloce. Baron Napier's *Mirifici Logarithmorum Canonis Descriptio*, Def. 6, Edinburgh 1614.—Also in the explanation of Def. 1, the words *fluxu* ‖by flowing‖ and *fluat* ‖it flows‖ occur. ‖'The logarithm therefore of any sine is a number very neerly expressing the line, which increased equally in the meane time, whiles the line of the whole sine decreased proportionally into that sine, both motions being equal-timed, and the beginning equally swift'—*A description of the admirable table of logarithms*, Waterson, London, 1618; trans. Edward Wright.‖

said) from the arithmetical properties of powers of numbers, but from the contemplation of a *Continuous Progression*; in describing which, he speaks expressly of *Fluxions*, *Velocities* and *Times*.

‖11‖ In a more modern age, Lagrange, in the Philological spirit, sought to reduce the Theory of Fluxions to a system of operations upon symbols, analogous to the earliest symbolic operations of Algebra, and professed to reject the notion of time as foreign to such a system; yet admitted[†] that fluxions might be considered only as the velocities with which magnitudes vary, and that in so considering them, abstraction might be made of every mechanical idea. And in one of his own most important researches in pure Algebra, (the investigation of limits between which the sum of any number of terms in Taylor's Series is comprised,) Lagrange[‡] employs the conception of *continuous progression* to show that a certain variable quantity may be made as small as can be desired. And not to dwell on the beautiful discoveries made by the same great mathematician, in the theory of singular primitives of equations, and in the algebraical dynamics of the heavens, through an extension of the conception of *variability*, (that is, in fact, of *flowingness*,) to quantities which had before been viewed as *fixed* or constant, it may suffice for the present to observe that Lagrange considered Algebra to be the *Science of Functions*,[§] and that it is not easy to conceive a clearer or juster idea of a *Function* in this Science, than by regarding its essence as consisting in a *Law connecting Change with Change*. But where *Change* and *Progression* are, there is TIME. The notion of Time is, therefore, inductively found to be connected with existing Algebra.[‖]

‖12‖ 2. The argument for the conclusion that *the notion of time may be unfolded into an independent Pure Science*, or that *a Science of Pure Time is possible*, rests chiefly on the existence of certain a priori intuitions, connected with that notion of time, and fitted to become the sources of a pure Science; and on the actual deduction of such a Science from those principles, which the author conceives that he has begun. Whether he has at all succeeded in *actually effecting* this deduction, will be judged after the Essay has been read; but that such a deduction is *possible*, may be concluded in an easier way, by an appeal to those intuitions to which allusion has been made. That a moment of time respecting which we inquire, as compared with a moment which we know, must

[†] Calcul des Fonctions, Leçon Premiere, page 2. Paris 1806.

[‡] Donc puisque *V* devient nul lorsque *i* devient nul, il est clair qu' en faisant croître *i* par degrés insensibles depuis zéro, la valeur de *V* croîtra aussi insensiblement depuis zéro, soit en plus ou en moins, jusqu' à un certain point, après quoi elle pourra diminuer,—Calcul des Fonctions, Leçon Neuvième, page 90. Paris 1806. An instance still more strong may be found in the First Note to Lagrange's *Equations Numeriques*, Paris, 1808.

[§] On doit regarder l'algèbre comme la science des fonctions.—Calc. des Fonct., Leçon Premiere.

[‖] The word 'Algebra' is used throughout this whole paper, in the sense which is commonly but improperly given by modern mathematical writers to the name 'Analysis,' and not with that narrow signification to which the unphilosophical use of the latter term (Analysis) has caused the former term (Algebra) to be too commonly confined. The author confesses that he has often deserved the censure which he has here so freely expressed.

either coincide with or precede or follow it, is an intuitive truth, as certain, as clear, and as unempirical as this, that no two straight lines can comprehend an area. The notion or intuition of ORDER IN TIME is not less but more deep-seated in the human mind, than the notion or intuition of ORDER IN SPACE; and a mathematical Science may be founded on the former, as pure and as demonstrative as the science founded on the latter. There is something mysterious and transcendent involved in the idea of Time; but there is also something definite and clear: and while Metaphysicians meditate on the one, Mathematicians may reason from the other.

⟦13⟧ 3. That the *Mathematical Science of Time*, when sufficiently unfolded, and distinguished on the one hand from all actual Outward Chronology (or collections of recorded events and phenomenal marks and measures), and on the other hand from all Dynamical Science (or reasonings and results from the notion of cause and effect), will ultimately be found to be co-extensive and identical with Algebra, so far as Algebra itself is a Science: is a conclusion to which the author has been led by all his attempts, whether to *analyse* what is *Scientific in Algebra*, or to *construct a Science of Pure Time*. It is a joint result of the inductive and deductive processes, and the grounds on which it rests could not be stated in a few general remarks. The author hopes to explain them more fully in a future paper; meanwhile he refers to the present one, as removing (in his opinion) the difficulties of the usual theory of Negative and Imaginary Quantities, or rather substituting a new Theory of *Contrapositives* and *Couples*, which he considers free from those old difficulties, and which is deduced from the Intuition or Original Mental Form of Time: the opposition of the (so-called) Negatives and Positives being referred by him, *not* to the opposition of the operations of increasing and diminishing a *magnitude*, but to the simpler and more extensive contrast between the relations of *Before* and *After*,* or between the directions of *Forward* and *Backward*; and *Pairs of Moments* being used to suggest a *Theory of Conjugate Functions*,† which gives reality and meaning to conceptions that were before Imaginary,‡ Impossible, or Contradictory, because Mathematicians had derived them from that

* It is, indeed, very common, in Elementary works upon Algebra, to allude to *past and future time*, as one among many *illustrations* of the doctrine of negative quantities; but this avails little for Science, so long as *magnitude* instead of PROGRESSION is attempted to be made the *basis* of the doctrine.

† The author was conducted to this Theory many years ago, in reflecting on the important symbolic results of Mr. Graves respecting Imaginary Logarithms, and in attempting to explain to himself the theoretical meaning of those remarkable symbolisms. The Preliminary and Elementary Essay on Algebra as the Science of Pure Time, is a much more recent development of an Idea against which the author struggled long, and which he still longer forbore to make public, on account of its departing so far from views now commonly received. The novelty, however, is in the view and method, not in the results and details: in which the reader is warned to expect little addition, if any, to what is already known.

‡ The author acknowledges with pleasure that he agrees with M. Cauchy, in considering every (so-called) Imaginary Equation as a symbolic representation of two separate Real Equations: but he differs from that excellent mathematician in his method generally, and especially in not introducing the sign $\sqrt{-1}$ until he has provided for it, by his Theory of Couples, a possible and real meaning, as a symbol of the couple (0, 1).

bounded notion of *Magnitude*, instead of the original and comprehensive thought of ORDER IN PROGRESSION.

B. PREFACE TO THE *LECTURES ON QUATERNIONS* (*HAMILTON 1853*)

As we have just seen, Hamilton already in 1833 possessed the interpretation of the complex numbers as ordered pairs of real numbers; this interpretation supplied him with a unified theory of the real and the complex numbers, in which both were subsumed under the theory of *order in progression* rather than of mere *magnitude* (*Hamilton 1837*, §13). Ever since 1829 (when he first read John Warren's treatise on the representation of complex numbers as points in the plane) Hamilton had sought a three-dimensional analogue to the complex numbers; but a solution to this problem still eluded him. In 1843 he suddenly saw the answer in a famous flash of insight:

> Tomorrow will be the fifteenth birthday of the Quaternions. They started into life, or light, full grown, in the 16th of October, 1843, as I was walking with Lady Hamilton to Dublin, and came up to Brougham Bridge. That is to say, I then and there felt the galvanic current of thought closed, and the sparks which fell from it were the fundamental equations between I, J, K; *exactly such* as I have used them ever since. I pulled out, on the spot, a pocketbook, which still exists, and made an entry, on which, *at the very moment*, I felt that it might be worth my while to expend the labour of at least ten (or it might be fifteen) years to come. But then it is fair to say that this was because I felt a *problem* to have been at that moment *solved*, an intellectual *want relieved*, which had *haunted* me for at least *fifteen years* before (*Kline 1972*, p. 779; citing *The North British Review*, Vol. xiv, p. 57 (1858)).

(In another description of this incident, Hamilton said he became so excited that he at once scratched the fundamental equations for quaternions into the stone of Brougham Bridge.) The following selection contains his account of the reasoning that led him to his discovery.

Two obstacles stood in the way of obtaining the desired generalization of the complex numbers. First, Hamilton originally believed that, since complex numbers could be represented by *pairs* of real numbers, three-dimensional numbers should be represented by *triples* of real numbers. But Hamilton was unable to construct any division algebra based on triples that would follow the laws of multiplication; and Frobenius eventually proved that no such division algebra exists. If one thinks of the problem geometrically, as a search for an operation of multiplication that will represent rotations around an axis in three-dimensional space (just as complex multiplication represents rotations in two-dimensional space) then it is clear that triples are inadequate. For to represent a three-dimensional rotation one needs to know not only (as in the two-dimensional case) the angle of rotation and the scalar displacement, but also the plane through which the rotation takes place; and this plane requires two real numbers for its specification. So one needs altogether four real numbers to describe three-dimensional rotations: hence 'quaternions'.

Second, Hamilton originally believed that the new numbers would satisfy the familiar laws of traditional algebra; but it turned out to be necessary to abandon the commutative law of multiplication for the new objects. (In fact the quaternions form what is today called a *skew field*: that is, they satisfy all the axioms of a field except that multiplication is not commutative.) This step was a large conceptual leap; but that it was necessary can be seen by considering the simpler case of rotations of a sphere, where rotation α followed by rotation β does not in general yield the same result as β followed by α.

Hamilton's enthusiasm for his new numbers was without bounds, and he believed that the quaternions would prove as important for physics as Newton's calculus. But the quaternions turned out to be less important than he had hoped. Essentially the problem was their four-dimensionality: quaternions were cumbersome to work with, and physicists preferred to adhere to the more readily intelligible three-dimensional Cartesian coordinates.

However, the idea of a non-commutative algebraic system and the idea of studying vector transformations in space were both to bear practical fruit, notably in Cayley's studies of the algebra of matrices, in Clifford's generalizations of the quaternion calculus to describe rotations of n-dimensional vectors, in the linear associative algebras investigated by Benjamin Peirce, and in the theory of vectors developed in the 1880s by Josiah W. Gibbs and Oliver Heaviside—tools that were to prove invaluable to the physicists of the twentieth century.

In the following selection, Hamilton makes some concessions to the 'philological' school (see in particular his long footnote to paragraph 19), and no longer rejects it quite as firmly as in *Hamilton 1837*. De Morgan's accusation of 'dogmatism' in §6 of his essay *On the foundations of algebra* (*De Morgan 1842a*, reproduced above) may have contributed to the mellowing; and Hamilton seems in the interval to have become better acquainted with the writings of Peacock, De Morgan, and Gregory. But he continues to hold to his theory of algebra as 'the science of pure time'.

References to *Hamilton 1853* should be to the paragraph numbers, which appeared (in single brackets) in the original edition.

Preface

‖1‖ The volume now offered to the public is designed as an assistance to those persons who may be disposed to study and to employ a certain new mathematical method, which has, for some years past, occupied much of my own attention, and for which I have ventured to propose the name of the Method or Calculus of Quaternions. Although a copious analytical index, under the form of a Table of Contents, will be found to have been prefixed to the work, yet it seems proper to offer here some general and preliminary* remarks: especially as regards that conception from which the whole has been gradually evolved, and the motives for giving to the resulting method an appellation not previously in use.

⟦2⟧ The difficulties which so many have felt in the doctrine of Negative and Imaginary Quantities in Algebra forced themselves long ago on my attention; and although I early formed some acquaintance with various views or suggestions that had been proposed by eminent writers, for the purpose of removing or eluding those difficulties (such as the theory of direct and inverse quantities, and of indirectly correlative figures, the method of constructing imaginaries by lines drawn from one point with various directions in one plane, and the view which refers all to the mere play of algebraical operations, and to the properties of symbolical language), yet the whole subject still appeared to me to deserve additional inquiry, and to be susceptible of a more complete elucidation. And while agreeing with those who had contended that negatives and imaginaries were not properly *quantities* at all, I still felt dissatisfied with any view which should not give to them, from the outset, a clear interpretation and *meaning*; and wished that this should be done, for the square roots of negatives, without introducing considerations *so expressly geometrical*, as those which involve the conception of an *angle*.

⟦3⟧ It early appeared to me that these ends might be attained by our consenting to regard ALGEBRA as being no mere Art, nor Language, nor *primarily* a Science of Quantity; but rather as the Science of Order in Progression. It was, however, a part of this conception, that the *progression* here spoken of was understood to be *continuous* and *unidimensional*: extending indefinitely *forward* and *backward*, but not in any *lateral* direction. And although the successive *states* of such a progression might (no doubt) be represented by *points upon a line*, yet I thought that their simple *successiveness* was better conceived by comparing them with *moments of time*, divested, however, of all reference to *cause* and *effect*; so that the "time" here considered might be said to be abstract, ideal, or *pure*, like that "space" which is the object of geometry. In this manner I was led, many years ago, to regard Algebra as the SCIENCE OF PURE TIME: and an Essay,† containing my views respecting it as such, was

* Some readers may find it convenient to pass over for the present these prefatory remarks, and to proceed at once to the Volume, of which a large part has been drawn up so as to suppose less of previous and technical preparation than some of the paragraphs of this Preface. Indeed, great pains have been taken to render the early Lectures as elementary as the subject would allow; and it is hoped that they will be found perfectly and even easily intelligible by persons of moderate scientific attainments. It is true that some of the subsequent portions of the Course (especially parts of the concluding Lecture) may possibly appear difficult, from the novel nature of the calculations employed: but perhaps on that very account those later portions may repay the attention of more advanced mathematical students.

† Theory of Conjugate Functions, or Algebraic Couples; with a Preliminary and Elementary Essay on Algebra as the Science of Pure Time. (Read November 4th, 1833, and June 1st, 1835).—Transactions of the Royal Irish Academy, Vol. XVII., Part II. (Dublin, 1835), pages 293 to 422.

published† in 1835. If I now reproduce a few of the opinions put forward in that early Essay, it will be simply because they may assist the reader to place himself in that *point of view*, as regards the first elements of *algebra*, from which a passage was gradually made by me to that comparatively *geometrical* conception which it is the aim of this volume to unfold. And with respect to anything unusual in the *interpretations* thus proposed, for some simple and elementary notations, it is my wish to be understood as not at all insisting on them as *necessary*,* but merely proposing them as consistent among themselves, and preparatory to the study of the quaternions, in at least one aspect of the latter.

[4] In the view thus recently referred to, if the letters A and B were employed as *dates*, to denote, any two *moments* of time, which might or might not be distinct, the case of the coincidence or *identity* of these two moments, or of *equivalence* of these two dates, was denoted by the equation,

$$\mathrm{B} = \mathrm{A};$$

which symbolic assertion was thus interpreted as not involving any *original* reference to *quantity*, nor as expressing the result of any comparison between two *durations* as *measured*. It corresponded to the conception of simultaneity or *synchronism*; or, in simpler words, it represented the thought of the *present* in time. Of all possible answers to the general question, "*When*," the *simplest* is the answer, "Now:" and it was the *attitude of mind*, assumed in the making of this answer, which (in the system here described) might be said to be originally symbolized by the *equation* above written. And, in like manner, the

---

† I was encouraged to entertain and publish this view, by remembering some passages in Kant's Criticism of the Pure Reason, which appeared to justify the expectation that it should be *possible* to construct. *à priori*, a Science of Time, as well as a Science of Space. For example, in his Transcendental Aesthetic, Kant observes:—"Zeit und Raum sind demnach zwey Erkenntniss-quellen, aus denen *à priori* verschiedene synthetische Erkenntnisse geschöpft werden können, wie vornehmlich die reine Mathematik in Ansehung der Erkenntnisse vom Raume und dessen Verhältnissen ein glänzendes Beyspiel gibt. Sie sind nämlich beide zusammengenommen reine Formen aller sinnlichen Anschauung, und machen dadurch synthetische Sätze *a priori* möglich." Which may be rudely rendered thus:—"Time and Space are therefore two knowledge-sources, from which different synthetic knowledges can be *à priori* derived, as eminently in reference to the knowledge of space and of its relation a brilliant example is given by the pure mathematics. For they are, both together [space and time], pure forms of all sensuous intuition, and make thereby synthetic positions *à priori* possible." (Critik der reinen Vernunft, p. 41. Seventh Edition. Leipzig: 1828).

* For example, the usual identity $(B - A) + A = B$, which in the older Essay was interpreted with reference to *time*, as in paragraph [8] of this Preface, the letters A and B denoting *moments*, is in the present work (Lecture I., article 25) interpreted, on an analogous plan indeed, but with a reference to *space*, the letters denoting *points*. Still it will be perceived that there exists a close connexion between the two views; a *step*, in each, being conceived to be applied to a *state* of a progression, so as to generate (or conduct to) another state. And generally I think that it may be found useful to compare the interpretations of which a sketch is given in the present Preface, with those proposed in the body of the work.

two formulæ of *non*-equivalence,

$$B>A, \quad B<A,$$

were interpreted, without any *primary* reference to quantity, as denoting the two contrasted relations of *subsequence* and of *precedence*, which answer to the thoughts of the *future* and the *past* in time; or as expressing, simply, the one that the moment B is conceived to be *later* than A, and the other that B is *earlier* than A: without *yet* introducing even the *conception of a measure*, to determine *how much later*, or how much earlier, one moment is than the other.

[5] Such having been proposed as the *first* meanings to be assigned to the three elementary marks = > <, it was next suggested that the *first* use of the mark −, in constructing a *science of pure time*, might be conceived to be the forming of a complex symbol B − A, to denote the *difference between two moments*, or the *ordinal relation* of the moment B to the moment A, whether the relation were one of identity or of diversity; and if the latter, then whether it were one of subsequence or of precedence, and in whatever degree. And *here*, no doubt, in attending to the *degree* of such diversity between two moments, the conception of *duration*, as *quantity* in time, was introduced: the *full* meaning of the symbol B − A, in any particular application, being (on this plan) not known, until we know *how long after*, or how long before, if at all, B is than A. But it is evident that the notion of a certain *quality* (or *kind*) of this diversity, or interval, enters into this conception of a *difference* between moments, at least as fully and as soon as the notion of *quantity*, amount, or duration. The contrast between the Future and the Past appears to be even earlier and more fundamental, in human thought, than that between the Great and the Little.

[6] After *comparing moments*, it was easy to proceed to *compare relations*; and in this view, by an *extension* of the recent signification [4] of the sign =, it was used to denote *analogy* in time; or, more precisely, to express the *equivalence of two marks of one common ordinal relation*, between *two pairs* of moments. Thus the formula,

$$D - C = B - A,$$

came to be interpreted as denoting an *equality between two intervals in time*; or to express that the moment D is *related* to the moment C, *exactly as* B *is to* A, with respect to identity or diversity: the *quantity and quality* of such diversity (when it exists) being here *both* taken into account. A formula of this sort was shewn to admit of *inversion* and *alternation* (C − D = A − B, D − B = C − A); and generally there could be performed a number of *transformations* and *combinations* of equations such as these, which all admitted of being *interpreted* and *justified* by this mode of viewing the subject, but which *agreed* in all respects with the received *rules* of algebra. On the same plan, the two contrasted formulæ of inequalities of differences,

$$D - C > B - A, \quad D - C < B - A,$$

were interpreted as signifying, the one that D was *later*, *relatively* to C, than B to A; and the other that D was *relatively earlier*.

⟦7⟧ Proceeding to the mark +, I used this sign *primarily* as a mark of combination between a symbol, such as the smaller Roman letter a, of a *step in time*, and the symbol, such as A, of the moment *from* which this *step* was conceived to be made, in order to form a complex symbol, a + A, *recording this conception of transition*, and denoting the moment (suppose B) *to* which the step was supposed to conduct. The step or transition here spoken of was regarded as a *mental act*, which might as easily be supposed to conduct *backwards* as *forwards* in the progression of time; or even to be a *null step*, denoted by 0, and producing *no effect* (0 + A = A). Thus, with these meanings of the signs, the notation

$$\mathrm{B} = \mathrm{a} + \mathrm{A},$$

denoted the conception that the moment B might be *attained*, or mentally *generated*, by making (in thought) the step a from the moment A. And it appeared to me that without ceasing to regard the symbol B − A as denoting, in one view ⟦5⟧, an *ordinal relation* between two moments, we might *also* use it in the *connected sense* of denoting this *step from one to another*: which would allow us (as in ordinary algebra) to write, with the recent suppositions,

$$\mathrm{B} - \mathrm{A} = \mathrm{a};$$

the two members of this new equation being here symbols for one common step.

⟦8⟧ The usual identity,

$$(\mathrm{B} - \mathrm{A}) + \mathrm{A} = \mathrm{B},$$

came thus to be interpreted as signifying *primarily* (in the Science of Pure Time) a certain conceived *connexion* between the operations, of *determining* the difference between two moments as a *relation*, and of *applying* that difference as a *step*. And the two other familiar and connected identities,

$$\mathrm{C} - \mathrm{A} = (\mathrm{C} - \mathrm{B}) + (\mathrm{B} - \mathrm{A}), \quad \mathrm{C} - \mathrm{B} = (\mathrm{C} - \mathrm{A}) - (\mathrm{B} - \mathrm{A}),$$

were treated, on the same plan, as originally signifying certain *compositions* and *decompositions* of ordinal relations or of steps in time. A special symbol for *opposition* between any two such relations or steps was proposed; but it was remarked that the more usual notations, +a and −a, for the step (a) itself, and for the opposite of that step, might, in full consistency with the same general view, be employed, if treated as abridgments for the more complex symbols 0 + a, 0 − a: the latter notation presenting *here* no difficulty of interpretation, nor requiring any attempt to conceive the *subtraction* of a *quantity* from *nothing*, but merely the *decomposition* of a *null step* into *two opposite steps*. But *operations on steps*, conducted on this plan, were shewn to agree in all respects with the usual *rules* of algebra, as regarded Addition and Subtraction.

⟦9⟧ One *time-step* (b) was next compared with another (a), in the way of algebraic *ratio*, so as to conduct to the conception of a certain complex *relation* (or *quotient*), determined partly by their *relative largeness*, but partly also by their *relative direction*, as similar or opposite; and to the closely connected conception of an algebraic *number* (or *multiplier*), which *operates* at once on the quantity and on the direction of the one step (a), so as to *produce* (or mentally *generate*) the quantity and direction of the other step (b). By a combination of these two conceptions, the usual identity,

$$\frac{\mathrm{b}}{\mathrm{a}} \times \mathrm{a} = \mathrm{b}, \text{ or } \mathrm{b} = a \times \mathrm{a}, \text{ if } \frac{\mathrm{b}}{\mathrm{a}} = a,$$

received an interpretation; the factor a being a *positive* or a *contra-positive* (more commonly called *negative*) *number*, according as it *preserved* or *reversed* the *direction* of the step on which it operated. The four primary operations, for combining any two such ratios or numbers or factors, a and b, among themselves, were *defined* by four equations which may be written thus, and which were indeed *selected* from the usual formulæ of algebra, but were employed with new *interpretations*:

$$(b + a) \times \mathrm{a} = (b \times \mathrm{a}) + (a \times \mathrm{a}); \quad (b - a) \times \mathrm{a} = (b \times \mathrm{a}) - (a \times \mathrm{a});$$

$$(b \times a) \times \mathrm{a} = b \times (a \times \mathrm{a}); \qquad b \div a = (b \times \mathrm{a}) \div (a \times \mathrm{a}).$$

⟦10⟧ *Operations* on algebraic *numbers* (positive or contrapositive) were thus made to depend (in thought) on operations of the same names on *steps*; which were again conceived to involve, in their ultimate analysis, a reference to comparison of *moments*. These conceptions were found to conduct to results agreeing with those usually received in algebra; at least when 0 was treated as a symbol of a *null number*, as well as of a null step ⟦7⟧, and when the symbols, $0 + a$, $0 - a$, were abridged to $+a$ and $-a$. In this view, there was no difficulty whatever, in interpreting the *product* of *two negative numbers*, as being equal to a *positive* number: the result expressing simply, in this view of it, that *two successive reversals restore* the direction of a step. And other difficulties respecting the *rule of the signs* appeared in like manner to fall away, more perfectly than had seemed to me to take place in any view of algebra, which made the thought of quantity (or of magnitude) the *primary* or *fundamental* conception.

⟦11⟧ This theory of algebraic numbers, as ratios of steps in time, was applied so as to include results respecting powers and roots and logarithms: but what it is at present chiefly important to observe is, that because, for the reason just assigned, the *square* of *every* number is *positive*, therefore *no number*, whether positive or negative, could be a *square root of a negative* number, in *this* any more than in *other* views of algebra. At least it was certain that no *single* number, of the kinds above considered, could possibly be such a root: but I thought that without going out of the same *general class* of interpretations, and especially without ceasing to refer all to the notion of *time*, explained and guarded as above, we might conceive and compare *couples of moments*; and so derive a conception of *couples of steps* (in time), on which might be founded a theory of *couples of numbers*, wherein no such difficulty should present itself.

‖12‖ In this extended view, the symbols A_1 and A_2 being employed to denote the two moments of one such pair or couple, and B_1, B_2 the two moments of another pair, I was led to write the formula,

$$(B_1, B_2) - (A_1, A_2) = (B_1 - A_1, B_2 - A_2);$$

and to explain it as expressing that the *complex ordinal relation* of one *moment-couple* (B_1, B_2) to another moment-couple (A_1, A_2) might be regarded as a *relation-couple*; that is to say, as a *system of two ordinal relations*, $B_1 - A_1$ and $B_2 - A_2$, between the *corresponding moments* of those two moment-couples: the *primary moment* B_1 of the one pair being compared with the primary moment A_1 of the other; and, in like manner, the *secondary moment* B_2 being compared with the secondary moment A_2. But, instead of this (analytical) *comparison* of moments with moments, and thereby of *pair with pair*, I thought that we might also conceive a (synthetical) *generation* ‖7‖ of one pair of moments from another, by the *application* of a *pair of steps* ‖11‖, or by what might be called the *addition* (see again ‖7‖), of a *step-couple* to a *moment-couple*; and that an *interpretation* might thus be given to the following *identity*, in the theory of couples here referred to:

$$(B_1, B_2) = \{(B_1, B_2) - (A_1, A_2)\} + (A_1, A_2).$$

And other results, respecting the compositions and decompositions of *single ordinal relations*, or of *single steps in time*, such as those referred to in paragraph ‖8‖ of this Preface, were easily extended, in like manner, to the corresponding treatment of *complex relations*, and of *complex steps*, of the kinds above described.

‖13‖ There was no difficulty in interpreting, on this plan, such formulæ of *multiplication* and *division*, as

$$a \times (\mathrm{a}_1, \mathrm{a}_2) = (a\mathrm{a}_1, a\mathrm{a}_2); \; (a\mathrm{a}_1, a\mathrm{a}_2) \div (\mathrm{a}_1, \mathrm{a}_2) = a;$$

where the symbols a_1, a_2 denote any two steps in time, and a any number, positive or negative. But the question became less easy, when it was required to interpret a symbol of the form

$$(\mathrm{b}_1, \mathrm{b}_2) \div (\mathrm{a}_1, \mathrm{a}_2),$$

where b_1, b_2 denoted two steps which could not be derived from the two steps a_1, a_2, through multiplication by *any single number*, such as a. To meet this case, which is indeed the general one in this theory, I was led to introduce the conception ‖11‖ of *number-couples*, or of *pairs of numbers*, such as (a_1, a_2); and to regard every *single* number (a) as being a *degenerate form* of such a number-couple, namely of $(a, 0)$; so that the recent formula, for the *multiplication of a step-couple by a number*, might be thus written:

$$(a_1, 0)\,(\mathrm{a}_1, \mathrm{a}_2) = (a_1\mathrm{a}_1, a_1\mathrm{a}_2).$$

It appeared proper to establish also the following formula, for the *multiplication of a primary step*, by an arbitrary number-couple:

$$(a_1,\ a_2)\,(\mathrm{a}_1,\ 0) = (a_1\mathrm{a}_1,\ a_2\mathrm{a}_1);$$

and to regard every such number-couple as being the *sum* of two others, namely, of a *pure primary* and a *pure secondary*, as follows:

$$(a_1,\ a_2) = (a_1,\ 0) + (0,\ a_2):$$

the analogous decomposition of a step-couple having been already established.

⟦14⟧ The difficulty of the *general* multiplication of a step--couple by a number-couple came thus to be reduced to that of assigning the product of one pure secondary by another: and the spirit of this whole theory of couples led me to conceive that, for such a product, we ought to have an expression of the form,

$$(0,\ a_2)\,(0,\ \mathrm{a}_2) = (\gamma_1 a_2\mathrm{a}_2,\ \gamma_2 a_2\mathrm{a}_2);$$

the coefficients γ_1 and γ_2 being some two constant numbers, independent of the *step* a_2, and of the *number* a_2: which two coefficients I proposed to call the *constants of multiplication*. These constants might be variously assumed: but reasons were given for adopting the following *selection** of values, as the basis of all subsequent operations:

$$\gamma_1 = -1;\ \gamma_2 = 0.$$

In this way, the required *law of operation*, of a general number-couple on a general step-couple, as multiplier on multiplicand, was found, with this choice of the *constants*, to be expressed by the formula:

$$(a_1,\ a_2)\,(\mathrm{a}_1,\ \mathrm{a}_2) = (a_1\mathrm{a}_1 - a_2\mathrm{a}_2,\ a_2\mathrm{a}_1 + a_1\mathrm{a}_2).$$

And in fact it was easy, with the assistance of this formula, to *interpret the quotient* ⟦13⟧ *of two step-pairs*, as being always equal to a *number-pair*, which could be definitely assigned, when the ratios of the four single steps were given.

⟦15⟧ With these conceptions and notations, it was allowed to write the two following equations:

$$(1,\ 0)\,(\mathrm{a},\ \mathrm{b}) = (\mathrm{a},\ \mathrm{b});\quad (0,\ 1)\,(\mathrm{a},\ \mathrm{b}) = (-\mathrm{b},\ \mathrm{a});$$

and I thought that these two factors, (1, 0), and (0, 1), thus used, might be called respectively the *primary unit*, and the *secondary unit*, of number. It was proposed to establish, by definition, for the chief *operations on number-pairs*, a few rules which seemed to be natural extensions of those already established for the corresponding operations ⟦9⟧ on *single numbers*: and it was seen that because

$$(0,\ 1)\,(-\mathrm{b},\ \mathrm{a}) = (-\mathrm{a},\ -\mathrm{b}) = (-1,\ 0)\,(\mathrm{a},\ \mathrm{b}),$$

* In some of my unprinted investigations, other selections of these constants were employed.

we were allowed, as a consequence of those rules, or of the conception which had suggested them, namely, (compare ‖33‖), by a certain *abstraction* of operators from operand, to establish the formula,

$$(0,\ 1)^2 = (-1,\ 0) = -1.$$

A new and (as I thought) clear *interpretation* was thus assigned, for that well-known expression in algebra, *the square root of negative unity*: for it was found that we might consistently write, on the foregoing plan,

$$(0,\ 1) = (-1,\ 0)^{\frac{1}{2}} = (-1)^{\frac{1}{2}} = \sqrt{-1};$$

without anything obscure, impossible, or *imaginary*, being in any way involved in the conception.

‖16‖ In words, if after *reversing* the direction of the *second* of any two steps, we then *transpose* them, as to order; thus making the old but reversed second step the *first* of the *new* arangement, or of the new step-couple; and making, at the same time, the old and unreversed first step the *second* of the same new couple; and if we then *repeat* this complex process of reversal and transposition, we shall, upon the whole, have *restored* the *order* of the two steps, but shall have *reversed* the *direction* of *each*. Now, it is the *conceived operator*, in this process of *passing from one pair of steps to another*, which, in the system here under consideration, was denoted by the celebrated symbol $\sqrt{-1}$, so often called IMAGINARY. And it is evident that the process, thus described, has no special reference whatever to the notion of *space*, although it has a reference to the conception of PROGRESSION. The symbol -1 denoted that NEGATIVE UNIT of number, of which the effect, as a *factor*, was to change a *single step* $(+a)$ to its own *opposite step* $(-a)$; and because *two* such reversals *restore*, therefore (see ‖10‖) the usual algebraic equation,

$$(-1)^2 = +1,$$

continued to subsist, in *this* as in other systems. But the symbol $\sqrt{-1}$ was regarded as *not at all less real* than those other symbols -1 or $+1$, although *operating on a different subject*, namely, *on a pair of steps* (a, b), and changing them to a *new pair*, namely, the pair (−b, +a). And the *form* of this well-known symbol, $\sqrt{-1}$, as an *expression* (in the system here described) for what I had previously written as (0, 1), and had called (see ‖15‖) the SECONDARY UNIT of number, was justified by shewing that the effect of its *operation*, when *twice* performed, *reversed each step* of the pair.

‖17‖ The more general expression of algebra, $a_1 + \sqrt{-1}\ a_2$, for any (so called) *imaginary root* of a quadratic or other equation was, on this plan, interpreted as being a symbol of the *number-couple* which I had otherwise denoted by (a_1, a_2); and of which the law of *operation on a step-couple* had already ‖14‖ been assigned: as also the analogous law, thence derived,* of its *multi-*

* The principles of such derivation were only hinted at in the Essay of 1835 (see page 403 of the Volume above cited): but it was perhaps sufficiently obvious that they depended on the "separation of symbols", or on the abstraction of a common operand. (Compare paragraphs ‖15‖, ‖33‖, of the present Preface.)

plication by another number-couple, namely, that which is expressed by the formula,

$$(b_1, b_2)(a_1, a_2) = (b_1 a_1 - b_2 a_2, b_2 a_1 + b_1 a_2).$$

In this view, instead of saying that the usual quadratic equation,

$$x^2 + ax + b = 0,$$

where a and b are supposed to denote two positive or negative numbers, has generally two roots, *real or imaginary*, it would be said that this *other form* of the same equation,

$$(x, y)^2 + (a, 0)(x, y) + (b, 0) = (0, 0),$$

is generally satisfied by *two* (real) *number-couples*; in which, according to the values of a and b, the secondary number (y) might or might not be zero. An equation of this sort was called a *couple-equation*, and was regarded as equivalent to a *system of two equations*† *between numbers*: for example, the recent *quadratic couple-equation* breaks itself up into the two following separate equations,

$$x^2 - y^2 + ax + b = 0,\ 2xy + ay = 0,$$

which always admit of real and numerical solutions, whether $\frac{1}{4}a^2 - b$ be a positive or a negative number; the difference being only that in the former case we are to take the factor $y = 0$, of the second equation of the pair, whereas in the latter case we are to take the *other factor* of that equation, and to suppose $2x + a = 0$. And similar remarks might be made on equations of higher orders: all notion of anything *imaginary*, *unreal*, or *impossible*, being quite excluded from the view.

⟦18⟧ The same view was extended, so as to include a theory of powers, roots, and logarithms of number-couples; and especially to confirm a remarkable conclusion which my friend John T. Graves, Esq., had communicated to me (and I believe to others) in 1826, and had published in the Philosophical Transactions for the year 1829: namely, that *the general symbolical expression for a logarithm is to be considered as involving two arbitrary and independent integers*;* the *general logarithm of unity*, to the Napierian base, being, for example, susceptible of the form,

† M. Cauchy, in his Cours d'Analyse (Paris, 1821, page 176), has the remark:—"Toute équation imaginaire n'est que la représentation symbolique de deux équations entre quantités réelles." That valuable work of M. Cauchy was early known to me: but it will have been perceived that I was induced to look at the whole subject of algebra from a somewhat different point of view, at least on the metaphysical side. As to the word "numbers", see a note to ⟦33⟧.

* It is proper to mention, that results substantially the same, respecting the entrance of two arbitrary whole numbers into the general form of a logarithm, are given by Ohm, in the second volume of his valuable work, entitled: "Versuch eines vollkommen consequenten Systems der Mathematik, vom Professor Dr. Martin Ohm" (Berlin, 1829, Second Edition, page 440. I have not seen the first Edition). For other particulars respecting the history of such investigations, on the subject of *general logarithms*, I must here be content to refer to Mr. Graves's subsequent Paper, printed in the Proceedings of the Sections of the British Association for the year 1834 (Fourth Report, pp. 523 to 531. London, 1835).

$$\log 1 = \frac{2\omega'\pi}{2\omega\pi - \sqrt{-1}},$$

where ω, ω' denote *any two whole numbers*, positive or negative or null. In fact, I arrived at an equivalent expression, in my own theory of number-couples, under the forms,

$$\overset{\omega'}{\underset{\omega(e,0)}{\log}} . (1, 0) = \frac{(0, 2\omega'\pi)}{(1, 2\omega\pi)};$$

and generally an expression for the *logarithm-couple*, with the *order* ω, and *rank* ω', of any proposed *number-couple* (y_1, y_2), to any proposed *base-couple* (b_1, b_2), was investigated in such a way as to confirm[†] the results of Mr. Graves.

⟦19⟧ After remarking that it was he who had proposed those names, of *orders and ranks of logarithms*, that early Essay of my own, of which a very abridged (although perhaps tedious) account has thus been given, continued and concluded as follows:—

But because Mr. GRAVES employed, in his reasoning, the usual principles respecting *Imaginary Quantities*, and was content to prove the symbolical necessity without shewing the interpretation, or inner meaning, of his formulæ, the present *Theory of Couples* is published to make manifest that hidden meaning: and to shew, by this remarkable instance, that expressions which seem, according to common views, to be merely symbolical, and quite incapable of being interpreted, may pass into the world of thoughts, and acquire reality and significance, if Algebra be viewed as not a mere Art or Language, but as the Science of Pure Time.* The author hopes to publish hereafter many other

[†] Another confirmation of the same results, derived from a peculiar theory of conjugate functions, had been communicated by me to the British Association at Edinburgh in 1834, and may be found reported among the Proceedings of the Sections for that year, at pp. 519 to 523 of the Volume lately cited. The partial differential "equations of conjugation", there given, had, as I afterwards learned, presented themselves to other writers: and the Essay on "Conjugate Functions, or Algebraic Couples", there mentioned, was considerably modified, in many respects, before its publication in 1835, in the Transactions of the Royal Irish Academy.

* Perhaps I ought to apologize for having thus ventured here to reproduce (although only historically, and as marking the progress of my own thoughts) a view so little supported by scientific authority. I am very willing to believe that (though not unused to calculation) I may have habitually attended too little to the *symbolical* character of Algebra, as a Language, or organized system of *signs*: and too much (in proportion) to what I have been accustomed to consider its *scientific* character, as a Doctrine analogous to Geometry, through the Kantian parallelism between the *intuitions* of Time and Space. This is not a proper opportunity for seeking to do justice to the views of others, or to my own, on a subject of so great subtlety: especially since, in the *present* work, I have thought it convenient to adopt throughout a *geometrical basis*, for the exposition of the theory and calculus of the Quaternions. Yet I wish to state, that I do not despair of being able hereafter to shew that my own old views respecting Algebra, perhaps modified in some respects by subsequent thought and reading, are not fundamentally and irreconcileably opposed to the teaching of writers whom I so much respect as Drs. Ohm and Peacock. The "Versuch," &c., of the former I have cited (the date of the first Volume of the Second Edition is Berlin, 1828): and it need scarcely be said (at least to readers in these countries) that my other reference is to the *Algebra* (Cambridge, 1830); the *Report on Certain Branches of Analysis*, printed in the Third Report of the British Association for the Advancement of Science (London, 1834); the *Arithmetical Algebra* (Cambridge, 1842); and the *Symbolical Algebra* (Cambridge, 1845): all by the Rev. George

applications of this view; especially to Equations and Integrals, and to a Theory of Triplets and Sets of Moments, Steps, and Numbers, which includes this Theory of "Couples".*

Peacock ‖*Peacock 1830, 1833, 1842-5*‖. I by no means dispute the possibility of constructing a consistent and useful system of algebraical calculations, by starting with the notion of *integer number*; unfolding that notion into its necessary consequences; expressing those consequences with the help of *symbols*, which are already general in *form*, although supposed at first to be limited in their signification, or *value*: and then, by *definition*, for the sake of *symbolic generality*, removing the *restrictions* which the original notion had imposed; and so resolving to *adopt*, as perfectly *general* in *calculation*, what had been only *proved* to be *true* for a certain subordinate and limited extent of *meaning*. Such seems to be, at least in part, the view taken by each of the two original and thoughtful writers who have been referred to in the present Note: although Ohm appears to dwell more on the study of the *relations* between the fundamental *operations*, and Peacock more on the *permanence* of equivalent *forms*. But I confess that I do not find myself able to frame a distinct *conception* of *number*, without *some* reference to the thought of *time*, although this reference may be of a somewhat abstract and transcendental kind. I cannot fancy myself as *counting* any set of things, without first *ordering* them, and treating them as *successive*: however *arbitrary* and *mental* (or *subjective*) this assumed succession may be. And by consenting to *begin* with the abstract notion (or pure intuition) of TIME, as the *basis* of the exposition of those axioms and inferences which are to be expressed by the symbols of algebra, (although I grant that the commencing with the more familiar conception of *whole number* may be more convenient for purposes of elementary instruction,) it still appears to me that an advantage would be gained: because the necessity for any merely *symbolical extension* of formulæ would be at least considerably *postponed* thereby. In fact (as has been partly shewn above), *negatives* would then present themselves as easily and naturally as positives, through the fundamental contrast between the thoughts of *past* and *future*, used *here* as no mere *illustration* of a result otherwise and symbolically deduced, without any clear comprehension of its meaning, but as the very *ground* of the reasoning. The ordinary *imaginaries* of algebra could be *explained* (as above) by *couples*; but might *then*, for convenience of calculation, be *denoted* by *single letters*, subject to all the ordinary *rules*, which rules would *follow* (on this plan) from the combination of *distinct conceptions* with *definitions*, and would offer no result which was not perfectly and easily *intelligible*, in strict consistency with that *original* thought (or intuition) of time, from which the whole theory should (on this supposition) be evolved. The doctrine of the n roots of an equation of the nth degree (for example) would thus suffer no attaint as to *form*, but would acquire (I think) new clearness as to *meaning*, without any assistance from geometry. The *quaternions*, as I have elsewhere shewn (in Vol. XXI., Part II., of the Transactions of the Royal Irish Academy), and even the *biquaternions* (as I hope to shew hereafter), might have their laws explained, and their symbolical results interpreted, by comparisons of *sets of moments*, and by operations on *sets of steps* in time. Thus, in the phraseology of Dr. Peacock, we should have a very wide "science of suggestion" (or rather, suggestive science) as our *basis*, on which to build up afterwards a new structure of purely *symbolical generalization*, if the *science of time* were adopted, instead of merely Arithmetic, or (primarily) the doctrine of *integer number*. Still I admit fully that the actual *calculations* suggested by this, or by any other view, must be performed according to some fixed *laws of combination of symbols*, such as Professor De Morgan has sought to reduce, for ordinary algebra, to the smallest possible compass, in his Second Paper on the Foundation of Algebra (Camb. Phil. Trans., Vol. VII., Part III), and in his work entitled "Trigonometry and Double Algebra" (London, 1849): and that in following out such *laws* to their symbolical consequences, uninterpretable (or at least uninterpreted) *results* may be expected to arise. In the present Volume (as has been already observed), I have thought it expedient to present the quaternions under a *geometrical aspect*, as one which it may be perhaps more easy and interesting to contemplate, and more immediately adapted to the subsequent applications, of geometrical and physical kinds. And in the passage which I have made (in the Seventh Lecture), from *quaternions* considered as *real* (or as geometrically *interpreted*), to *biquaternions* considered as *imaginary* (or as geometrically *uninterpreted*), but as symbolically *suggested* by the generalization of quaternion formulæ, it will be perceived, by those who shall do me the honour to read this work with attention, that I have employed a *method of transition*, from *theorems proved* for the *particular* to *expressions assumed* for the *general*, which bears a very close *analogy* to the methods of Ohm and Peacock: although I have *since* thought of a way of *geometrically interpreting the biquaternions* also.

* Trans. R.I.A., Vol. XVII., Part II., page 422.

⟦20⟧ The theory of *triplets* and *sets*, thus spoken of at the close of the Essay of 1835, had in fact formed the subject of various unpublished investigations, of which some have been preserved: and a brief notice of them here (especially as relates to triplets†) may perhaps be useful, by assisting to throw light on the nature of the passage, which I gradually came to make, from *couples* to *quaternions*.

Without departing from the same general view of algebra, as the science of pure time, it was obvious that no necessity existed for any limitation to *pairs*, of moments, steps, and numbers. Thus, instead of comparing, as in ⟦12⟧, *two moments*, B_1 and B_2, with two other moments, A_1 and A_2, it was possible to compare *three* moments, B_1, B_2, B_3, with three *other* moments, A_1, A_2, A_3; that is, more fully, to compare (or to conceive as compared) the *homologous* moments of these two *triads*, primary with primary, secondary with secondary, and tertiary with tertiary; and so to obtain a certain system or *triad of ordinal relations*, or a *triad of steps* in time, which might be denoted (compare ⟦5⟧, ⟦7⟧, ⟦12⟧) by either member of the following equation:

$$(B_1, B_2, B_3) - (A_1, A_2, A_3) = (B_1 - A_1, B_2 - A_2, B_3 - A_3).$$

And on the same plan (compare ⟦7⟧, ⟦8⟧, ⟦12⟧), if we denote the three *constituent steps* of such a triad as follows,

$$B_1 - A_1 = a_1, \; B_2 - A_2 = a_2, \; B_3 - A_3 = a_3,$$

it was allowed to write,

$$(B_1, B_2, B_3) = (a_1, a_2, a_3) + (A_1, A_2, A_3);$$

a triad of steps being thus (symbolically) *added* (or applied) to a triad of moments, so as to conduct (in thought) to another triad of moments. It appeared also convenient to establish the following formula, for the *addition of step-triads*,

$$(b_1, b_2, b_3) + (a_1, a_2, a_3) = (b_1 + a_1, b_2 + a_2, b_3 + a_3),$$

as denoting a certain *composition* of two such triads of steps, answering to that *successive application* of them to any given triad of moments (A_1, A_2, A_3), which conducts ultimately to a *third triad* of moments, namely, to the triad (C_1, C_2, C_3), if

$$C_1 - B_1 = b_1, \; C_2 - B_2 = b_2, \; C_3 - B_3 = b_3.$$

Subtraction of one step-triad from another was explained (see again ⟦8⟧) as answering to the analogous decomposition of a given step-triad into others; or to a system of *three distinct decompositions* of so many single steps, each into two others, of which one was given; and it was expressed by the formula,

† These remarks on *triplets* are now for the first time published.

$$(\mathrm{c}_1,\ \mathrm{c}_2,\ \mathrm{c}_3) - (\mathrm{a}_1,\ \mathrm{a}_2,\ \mathrm{a}_3) = (\mathrm{c}_1 - \mathrm{a}_1,\ \mathrm{c}_2 - \mathrm{a}_2,\ \mathrm{c}_3 - \mathrm{a}_3):$$

while the usual rules of algebra were found to hold good, respecting *such* additions and subtractions of triads.

⟦21⟧ *Multiplication* of a step-triad by a positive or negative *number* (a) was easy, consisting simply in the multiplication of *each constituent step* by that number; so that I had the equation,

$$a(\mathrm{a}_1,\ \mathrm{a}_2,\ \mathrm{a}_3) = (a\mathrm{a}_1,\ a\mathrm{a}_2,\ a\mathrm{a}_3):$$

and conversely it was natural (compare ⟦13⟧) to establish the following formula for a certain *case of division of step-triads*,

$$(a\mathrm{a}_1,\ a\mathrm{a}_2,\ a\mathrm{a}_3) \div (\mathrm{a}_1,\ \mathrm{a}_2,\ \mathrm{a}_3) = a.$$

But in the more general case (compare again ⟦13⟧), where the steps b_1, b_2, b_3 of one triad were *not proportional* to the steps a_1, a_2, a_3, it seemed to me that the *quotient* of these two step-triads was to be interpreted, on the same general plan, as being equal to a certain triad or *triplet of numbers*, a_1, a_2, a_3; so that there should be conceived to exist generally two equations of the forms,

$$(\mathrm{b}_1,\ \mathrm{b}_2,\ \mathrm{b}_3) \div (\mathrm{a}_1,\ \mathrm{a}_2,\ \mathrm{a}_3) = (a_1,\ a_2,\ a_3);$$

$$(\mathrm{b}_1,\ \mathrm{b}_2,\ \mathrm{b}_3) = (a_1,\ a_2,\ a_3)\,(\mathrm{a}_1,\ \mathrm{a}_2,\ \mathrm{a}_3):$$

the *three* (positive or negative) *constituents* of this *numerical triplet* $(a_1,\ a_2,\ a_3)$ depending, according to some definite laws, on the *ratios* of the *six steps*, $\mathrm{a}_1\mathrm{a}_2\mathrm{a}_3\mathrm{b}_1\mathrm{b}_2\mathrm{b}_3$.

⟦22⟧ In this way there came to conceived *three distinct and independent unit-steps*, a primary, a secondary, and a tertiary, which I denoted by the symbols,

$$1_1,\ 1_2,\ 1_3;$$

and also *three unit-numbers*, primary, secondary, and tertiary, each of which might *operate*, as a species of *factor*, or multiplier, on each of these three steps, or on their system, and which I denoted by these other symbols,

$$\times_1,\ \times_2,\ \times_3:$$

or sometimes more fully thus,

$$(1,\ 0,\ 0),\ (0,\ 1,\ 0),\ (0,\ 0,\ 1).$$

A *triad of steps* took thus the form,

$$r1_1 + s1_2 + t1_3,$$

where r, s, t were *three numerical coefficients* (positive or negative), although $1_1 1_2 1_3$ were still supposed to denote *three steps in time*; and any *triplet factor*, such as $(m,\ n,\ p)$, by which this *step-triplet* was to be multiplied, or *operated* upon, might be put under the analogous form,

$$m\times_1 + n\times_2 + p\times_3.$$

Continuing then to admit the *distributive* property of multiplication, it was only necessary to fix the significations of the *nine products*, or combinations, obtained by operating separately with *each* of the three units of number on *each* of the three units of step: every such product, or result, being conceived, in this theory, to be itself, in general, a *step-triad*, of which, however, some of the component steps might vanish. Hence, after writing

$$\times_1 1_1 = 1_{1,1};\ \times_1 1_2 = 1_{2,1};\ \ldots\ldots\ \times_3 1_2 = 1_{2,3};\ \times_3 1_3 = 1_{3,3},$$

I proceeded to develope these *nine step-triplets* into *nine trinomial expressions* of the forms,

$$1_{f,g} = 1_{f,g,1} 1_1 + 1_{f,g,2} 1_2 + 1_{f,g,3} 1_3,$$

where the *twenty-seven* symbols of the form $1_{f,g,h}$ represented certain *fixed numerical coefficients*, or *constants of multiplication*, analogous to those denoted by γ_1 and γ_2 in ⟦14⟧, and like them requiring to have their values *previously assigned*, before proceeding to multiplication, if it were demanded that the operation of a given triplet of numbers on a given triplet of steps should produce a perfectly a perfectly *definite step-triad* as its result.

⟦23⟧ Conversely, when once these numerical *constants* had been assigned, I saw that the equation of multiplication,

$$(m\times_1 + n\times_2 + px_3)\,(r1_1 + s1_2 + t1_3) = x1_1 + y1_2 + z1_3,$$

was to be regarded as breaking itself up, on account of the supposed *mutual independence* of the three unit-steps, into *three ordinary algebraical equations, between the nine numbers, m, n, p, r, s, t, x, y, z*; namely, between the *coefficients* of the multiplier, multiplicand, and product. These three equations were *linear*, relatively to *m*, *n*, *p* (as also with respect to *r*, *s*, *t*, and *x*, *y*, *z*); and therefore while they gave, *immediately*, expression for the coefficients *xyz* of the *product*, and so resolved *expressly* the problem of *multiplication*, they enabled me, through a simple system of three linear and ordinary equations, to resolve also the *converse* problem ⟦21⟧ of the *division* of one triad of steps by another: or to determine the coefficients *mnp* of the following *quotient* of two such triads,

$$m\times_1 + n\times_2 + p\times_3 = (x1_1 + y1_2 + z1_3) \div (r1_1 + s1_2 + t1_3).$$

⟦24⟧ Such were the most essential elements of that *general* theory of triplets, which occurred to me in 1834 and 1835: but it is clear that, in its *applications*, everything depended on the *choice of the twenty-seven constants of multiplication*, which might *all be arbitrarily assumed*, *before* proceeding to *operate*, but were *then* to be regarded as *fixed*. It was *natural*, indeed, to consider the *primary number-unit* $\times_1$ as producing *no change* in the step or triad on which it operates; and it was *desirable* to determine the constants so as to satisfy the condition,

$$\times_3\times_2 = \times_2\times_3,$$

for the sake of conforming to analogies of algebra. Accordingly, in one of several triplet-systems which I tried, the constants were so chosen as to satisfy these conditions, by the assumptions,

$$\begin{aligned}&\times_1 1_1 = 1_1,\ \times_1 1_2 = 1_2,\ \times_1 1_3 = 1_3,\\ &\times_2 1_1 = 1_2,\ \times_2 1_2 = 1_1 + (b - b^{-1})1_2,\ \times_2 1_3 = b1_3,\\ &\times_3 1_1 = 1_3,\ \times_3 1_2 = b1_3,\ \times_3 1_3 = 1_1 + b1_2 + c1_3;\end{aligned}$$

which still involved two arbitrary numerical constants, b and c, and gave, by a combination of *successive operations*, on any arbitrary *step-triad* (such as $r1_1 + s1_2 + t1_3$, whatever the *coefficients* r, s, t of this *operand triad* might be), the following *symbolic equations*,* expressing the *properties of the assumed operators*, $\times_2$, $\times_3$, and the laws of their mutual combinations:

$$\begin{aligned}&\times_2^2 = (b - b^{-1})\times_2 + 1;\\ &\times_2\times_3 = \times_3\times_2 = b\times_3;\\ &\times_3^2 = c\times_3 + b\times_2 + 1;\end{aligned}$$

while the factor $\times_1$ was suppressed, as being simply equivalent, in this system, to the factor 1, or to the ordinary unit of number. But although the symbol $\times_2$ appeared thus to be given by a *quadratic* equation, with the *two real roots* b and $-b^{-1}$, I saw that it would be improper to *confound* the *operation* of this *peculiar* symbol $\times_2$ with that of *either* of these two *numerical roots*, of that quadratic but *symbolical equation*, regarded as an *ordinary* multiplier. It was not *either*, *separately*, of the two operations $\times_2 - b$ and $\times_2 + b^{-1}$, which, when performed on a *general step-triad*, reduced that triad to another with every step a *null* one: but the *combination* of these two operations, successively (and in either order) performed.

⟦25⟧ In the same particular triplet system, the three general equations ⟦23⟧ between the nine numerical coefficients, of multiplier, multiplicand, and product, became the following:

$$\begin{aligned}&x = mr + ns + pt;\\ &y = ms + nr + (b - b^{-1})ns + bpt;\\ &z = mt + pr + b(nt + ps) + cpt;\end{aligned}$$

whence it was possible, *in general*, to determine the coefficients m, n, p, of the quotient of any two proposed step-triads. The same three equations were found to hold good also, when the *number-triplet* (x, y, z) was considered as the *symbolical product of the two number-tripets*, (m, n, p) and (r, s, t); *this* product being obtained by a certain *detachment* (or separation) of the symbols of the *operators* from that of a common *operand*, namely here an arbitrary *step-triad*. In other words, the *same algebraical equations* between the nine numerical coefficients, xyz, mnp, rst, expressed *also* the conditions involved in the formula of symbolical multiplication,

* These symbolic equations are copied from a manuscript of February, 1835.

$$(x, y, z) = (m, n, p)(r, s, t),$$

regarded as an *abridgment* of the following *fuller* formula:

$$(x, y, z)(\mathrm{a}_1, \mathrm{a}_2, \mathrm{a}_3) = (m, n, p)(r, s, t)(\mathrm{a}_1, \mathrm{a}_2, \mathrm{a}_3);$$

where a_1, a_2, a_3 might denote *any three steps* in time. Or they might be said to be the conditions for the correctness of this other *symbolical equation*,

$$x\times_1 + y\times_2 + z\times_3 = (m\times_1 + n\times_2 + p\times_3)(r\times_1 + s\times_2 + t\times_3),$$

interpreted on the same plan as the symbols $\times_2{}^2$, $\times_2\times_3$, $\times_3\times_2$, $\times_3{}^2$, in ‖24‖.

‖26‖ All the peculiar properties of the lately mentioned triplet system might be considered to be contained in the three ordinary and algebraical equations, ‖25‖, which connected the nine coefficients with each other (and in this case with two arbitrary constants). And I saw that these equations admitted of the three following combinations, by the ordinary processes of algebra:

$$x - b^{-1}y = (m - b^{-1}n)(r - b^{-1}s);$$
$$x + by + az = (m + bn + ap)(r + bs + at);$$
$$x + by + a'z = (m + bn + a'p)(r + bs + a't);$$

where a, a' were the two real and unequal roots of the ordinary quadratic equation,

$$a^2 = ca + b^2 + 1.$$

Here, then, was an *instance* of what occurred in *every other triplet system* that I tried, and seemed indeed to be a general and necessary consequence of the *cubic form* of a certain function, obtained by elimination between the three equations mentioned in ‖23‖, at least if we still (as is natural) suppose that $\times_1 = 1$: namely, that *the product of two triplets may vanish, without either factor vanishing*. For if (as *one* of the ways of exhibiting this result), we assume

$$n = bm, \; r = -bs, \; t = 0,$$

the recent relations will then give

$$x = 0, \; y = 0, \; z = 0;$$

so that, whatever values may be assigned to m, p, s, we have, in this system, the formula:

$$(m, bm, p)(-bs, s, 0) = (0, 0, 0).$$

For the same reason, there were *indeterminate cases*, in the operation of *division of triplets*: for example, if it were required to find the coefficients mnp of a quotient, from the equation

$$(m, n, p)(-bs, s, 0) = (x, y, z),$$

we should only be able to determine the function $m - b^{-1}n$, but not the numbers m and n themselves; while p would be entirely undetermined: at least if $x + by$ and z were each $=0$, for otherwise there might come *infinite* values into play.

⟦27⟧ The foregoing reasonings respecting triplet systems were quite independent of any sort of *geometrical interpretation*. Yet it was natural to interpret the results, and I did so, by conceiving the three sets of coefficients, (m, n, p), (r, s, t), (x, y, z), which belonged to the three triplets in the multiplication, to be the *co-ordinate projections*, on three rectangular axes, of *three right lines* drawn from a common origin; which *lines* might (I thought) be said to be, respectively, in this system of interpretation, the multiplier line, the multiplicand line, and the product line. And then, in the particular triplet system recently described, the formulæ of ⟦26⟧ gave easily a simple rule, for *constructing* (on this plan) the *product of two lines in space*. For I saw that if *three fixed and rectangular lines*, A, B, C, distinct from the original axes, were determined by the three following pairs of ordinary equations in co-ordinates:

$$
\begin{aligned}
&x + by = 0, \ z = 0, \text{ for line } A;\\
&y - bx = 0, \ z - ax = 0, \ \ldots \ B;\\
&y - bx = 0, \ z - a'x = 0, \ \ldots \ C;
\end{aligned}
$$

we might then enunicate this *theorem*:*

If a line L'' be the product of two other lines, L, L', then on whichever of the three rectangular lines A, B, C we project the two factors L, L', the product (in the ordinary meaning) of their two projections is equal to the product of the projections (on the same), of L'' and U, U being the primary unit-line (1, 0, 0).

⟦28⟧ I saw also that it followed from this theorem, or more immediately from the equations lately cited ⟦26⟧, from which the theorem itself had been obtained, that if we considered *three rectangular planes*, A', B', C', perpendicular respectively to the three lines A, B, C, or having for their equations,

$$y - bx = 0, \ (A'); \ x + by + az = 0, \ (B'); \ x + by + a'z = 0, \ (C');$$

then *every line* in *any one* of these three fixed planes gave a *null product line*, when it was multiplied by a line *perpendicular* to that fixed plane: the line A, for example, as a factor, giving a null line as the product, when combined with any factor line in the plane A'. For the same reason (compare ⟦26⟧), although the *division* of one line by another gave *generally* a determinate *quotient-line*, yet if the *divisor-line* were situated in any one of the three planes A', B', C', this quotient-line became then *infinite*, or *indeterminate*. And results of the same general character, although not all so simple as the foregoing, presented themselves in my examinations of various *other* triplet systems: there being, in

* This theorem is here copied, without any modification, from the manuscript investigation of February, 1835, which was mentioned in a former note.

all those which I tried, at least *one* system of line and plane, analogous to (A) and (A'), but not always *three* such (real) systems, not always at *right angles* to each other.

⟦29⟧ These speculations interested me at the time, and some of the results appeared to be not altogether inelegant. But I was dissatisfied with the departure from ordinary analogies of algebra, contained in the *evanescence* ⟦26⟧ ⟦28⟧ of a *product* of two triplets (or of two lines), in certain cases when neither *factor* was null; and in the connected *indeterminateness* (in the same cases) of a *quotient*, while the *divisor* was different from zero. There seemed also to be too much room for *arbitrary choice of constants*, and not any sufficiently decided reasons for finally preferring *one* triplet system to another. Indeed the assumption of the symbolic equation ⟦24⟧, $\times_1 = 1$, which it appeared to be convenient and *natural* to make, although *not essential* to the theory, determined immediately the values of *nine* out of the twenty-seven constants of multiplication; and *six* others were obtained from the assumptions, which also seemed to be *convenient* (although in *some* of my investigations the latter was not made),

$$\times_2 1_1 = 1_2, \ \times_3 1_1 = 1_3.$$

The supposed *convertibility* (see again ⟦24⟧), of the *order* of the two operations $\times_2$ and $\times_3$ gave then the three following conditions,

$$\times_3 \times_2 1_1 = \times_2 \times_3 1_1, \ \times_3 \times_2 1_2 = \times_2 \times_3 1_2, \ \times_3 \times_2 1_3 = \times_2 \times_3 1_3,$$

of which the first was seen at once to establish *three* relations between six of the twelve remaining coefficients of multiplication, namely (if the subscript commas be here for conciseness omitted),

$$1_{231} = 1_{321}, \ 1_{232} = 1_{322}, \ 1_{233} = 1_{323}.$$

The two other equations between step-triads, given by the recent conditions of convertibility, resolved themselves into six equations between coefficients, which were, however, perceived to be not all independent of each other, being in fact all satisfied by satisfying the *three* following:

$$1_{321} = 1_{223} 1_{332} - 1_{233} 1_{322};$$

$$1_{221} = 1_{233}(1_{233} - 1_{222}) + 1_{223}(1_{322} - 1_{333});$$

$$1_{331} = 1_{332}(1_{233} - 1_{222}) + 1_{322}(1_{322} - 1_{333});$$

of which the two former presented themselves to me under forms a little simpler, because, for the sake of preserving a *gradual ascent* from couples to triplets, or for preventing a *tertiary term* from appearing in the product, when no such term occurred in either factor, I assumed the value,

$$1_{223} = 0.$$

There still remained *five* arbitrary coefficients,

$$1_{222}, \ 1_{322}, \ 1_{323}, \ 1_{332}, \ 1_{333}$$

which it seemed to be permitted to choose at pleasure: but the decomposition of a certain *cubic function* [26] of r, s, t into *factors*, combined with *geometrical considerations*, led me, for the sake of securing the *reality* and *rectangularity* of a certain system of *lines* and *planes*, to assume the three following relations between those coefficients:

$$1_{222} = 1_{323} - 1_{323}^{-1},\ 1_{322} = 0,\ 1_{332} = 1_{323};$$

which gave also the values,

$$1_{221} = 1,\ 1_{321} = 0,\ 1_{331} = 1.$$

But the two constant coefficients 1_{323} and 1_{333} still seemed to remain wholly arbitrary,* and were those undetermined elements, denoted by b and c, which entered into the formulæ of triplet multiplication [25], already cited in this Preface.

[30] I saw, however, as has been already hinted [19] [20], that the same general *view* of algebra, as the science of pure time, admitted easily, at least in thought, of an *extension* of this whole theory, not only from couples to triplets, but also from triplets to *sets*, of moments, steps, and numbers. Instead of *two* or even *three* moments (as in [12] or [20]), there was no difficulty in conceiving a system or *set* of n such moments, $\mathrm{A}_1, \mathrm{A}_2, \ldots \mathrm{A}_n$, and in supposing it to be compared with another *equinumerous momental set*, $B_1, B_2, \ldots B_n$, in such a manner as to conduct to a new complex ordinal relation, or *step-set*, denoted by the formula,

$$(\mathrm{B}_1, \mathrm{B}_2, \ldots \mathrm{B}_n) - (\mathrm{A}_1, \mathrm{A}_2, \ldots \mathrm{A}_n) = (\mathrm{B}_1 - \mathrm{A}_1, \mathrm{B}_2 - \mathrm{A}_2, \ldots \mathrm{B}_n - \mathrm{A}_n).$$

Such step-sets could be *added* or *subtracted* (compare [20]), by adding or subtracting their *component steps*, each to or from its own corresponding step, as indicated by the double formula,

$$(\mathrm{b}_1, \mathrm{b}_2, \ldots \mathrm{b}_n) \pm (\mathrm{a}_1, \mathrm{a}_2, \ldots \mathrm{a}_n) = (\mathrm{b}_1 \pm \mathrm{a}_1, \mathrm{b}_2 \pm \mathrm{a}_2, \ldots \mathrm{b}_n \pm \mathrm{a}_n);$$

and a step-set could be *multiplied* by a *number* (a), or *divided* by *another step-set*, provided that the component steps of the one were *proportional* to those of the other (compare [13] [21]), by the formulæ:

$$a(\mathrm{a}_1, \mathrm{a}_2, \ldots \mathrm{a}_n) = (a\mathrm{a}_1, a\mathrm{a}_2, \ldots a\mathrm{a}_n);$$

$$(a\mathrm{a}_1, a\mathrm{a}_2, \ldots a\mathrm{a}_n) \div (\mathrm{a}_1, \mathrm{a}_2, \ldots \mathrm{a}_n) = a.$$

[31] But when it was required to divide one step-set by another, in the more general case (compare [13] [14] [21]), where the components or *constituent steps* $\mathrm{a}_1, \mathrm{a}_2, \ldots \mathrm{a}_n$ of the one set were *not* proportional to the corresponding

* The system of constants $b = 1$, $c = 1$, might have deserved attention, but I do not find that it occurred to me to consider it. In some of those old investigations respecting triplets, the symbol $\sqrt{-1}$ presented itself as a coefficient: but this at the time appeared to me unsatisfactory, nor did I see how to interpret it in such a connexion.

components $b_1, b_2, \ldots b_n$ of the other set, a difficulty again arose, which I proposed still to meet on the same general plan as before, by conceiving that a *numeral set*, or set or *system of numbers*, $(a_1, a_2, \ldots a_n)$, might *operate* on the *one* set of steps, $(a_1, a_2, \ldots a_n)$, in a way *analogous to multiplication*, so as to *produce* or generate the *other* given step-set, as a result which should be *analogous to a product*. Instead of *three* distinct and independent unit-steps, as in ‖22‖, I now conceived the existence of *n* such *unit-steps*, which might be denoted by the symbols,

$$1_1, 1_2, \ldots 1_n;$$

and instead of *three unit-numbers* (see again ‖22‖), I conceived *n* such *unit-operators*, which in those early investigations I denoted

$$\times_1, \times_2, \ldots \times_n,$$

and of which I conceived that *each* might operate on *each* unit-step, as a species of *multiplier*, or *factor*, so as to produce (generally) a *new step-set* as the result. There came thus to be conceived a number, $=n^2$, of such resultant step-sets, denoted, on the plan of ‖22‖, by symbols of the forms:

$$\times_g 1_f = 1_{f,g,1} 1_1 + 1_{f,g,2} 1_2 + \ldots + 1_{f,g,n} 1_n;$$

where the n^3 symbols of the form $1_{f,g,h}$ denoted so many *numerical coefficients*, or *constants of multiplication*, of the kind previously considered in the theories of couples ‖14‖, and of triplets ‖22‖, which *all* required to have their values *previously assumed*, or assigned, *before* proceeding to *multiply* a step-set by a number-set, in order that this operation might give generally a *definite step-set* as the result.

‖32‖ Conversely, on the plan of ‖23‖, when the n^3 numerical *values* of these coefficients or constants $1_{f,g,h}$ had been once fixed, I saw that we could then definitely interpret a *product* of the form,

$$(m\times_1 + \ldots + m_g \times_g + \ldots m_n \times_n)\ (r_1 1_1 + \ \ldots + r_f 1_f + \ldots + r_n 1_n),$$

where $m_1, \ldots m_g, \ldots m_n$ and $r_1, \ldots r_f, \ldots r_n$ were any $2n$ given numbers, as being equivalent to a certain new or *derived* step-set of the form,

$$x_1 1_1 + \ldots + x_h 1_h + \ldots + x_n 1_n;$$

where $x_1, \ldots x_h, \ldots x_n$ were *n* new or *derived numbers*, determined by *n* expressions such as the following:

$$x_h = \Sigma m_g r_f 1_{f,g,h};$$

the summation extending to all the n^2 combinations of values of the indices *f* and *g*. And because these expressions might in general be treated as a system of *n linear equations* between the *n* coefficients m_g of the multiplier set, I thought that the *division of one step-set by another* (compare ‖14‖ ‖23‖), might thus in general be accomplished, or at least conceived and interpreted, as being the process of *returning to that multiplier*, or of *determining the numeral set* which would *produce the dividend step-set*, by *operating on the divisor step-set*, and which might therefore be denoted as follows:

$$m_1 \times_1 + \ldots + m_g \times_g + \ldots m_n \times_n = (x_1 1_1 + \ldots + x_h 1_h + \ldots + x_n 1_n) \div (r_1 1_1 + \ldots r_f 1_f + \ldots + r_n 1_n);$$

or more concisely thus,

$$\Sigma m_g \times_g = \Sigma x_h 1_h \div \Sigma r_f 1_f:$$

while the numeral set thus found might be called the *quotient* of the two step-sets.

[33] It may be remembered that even at so early a stage as the interpretation of the symbol $b \times a$, for the algebraic product of two positive or negative *numbers*,* it had been proposed to conceive a reference to a *step* (a), which should be first *operated on* by those two numbers *successively*, and then *abstracted from*, as was expressed by the elementary formula [9],

$$(b \times a) \times \mathrm{a} = b \times (a \times \mathrm{a}).$$

Thus to interpret the product -2×-3 as $= +6$, I conceived that some time-step (a) was first tripled in length and reversed in direction; then that the new step (-3a) was doubled and reversed; and finally that the last resultant step ($+6$a) was *compared* with the original step (a), in the way of algebraic *ratio* [9], thereby conducting to a result which was *independent* of that original step. All this, so far, was no doubt extremely easy; nor was it difficult to extend the same mode of interpretation to the case [17] of the multiplication of two *number couples*, and to interpret the product of two such couples as satisfying the condition,

$$(b_1, b_2)\ (a_1, a_2) \times (\mathrm{a}_1, \mathrm{a}_2) = (b_1, b_2) \times (a_1, a_2)\ (\mathrm{a}_1, \mathrm{a}_2);$$

the arbitrary *step-couple* (a_1, a_2) being first operated on, and afterwards abstracted from. In like manner, in the theory of *triplets*, it was found possible [24] [25] to *abstract from an operand step-triad*, and thereby to obtain formulæ for the symbolic *multiplication* of the *secondary* and *tertiary number-units*, $\times_2$, $\times_3$, and more generally of any two *numerical* triplets among themselves. But when it was sought to extend the same view to the still more general *multiplication of numeral sets*, new difficulties were introduced by the essential complexity of the subject, on which I can only touch in the briefest manner here.†

* This word "number", whether with perfect propriety or not, is used throughout the present Preface and work, not as contrasted with *fractions* (except when accompanied by the word *whole* or *integer*), nor with incommensurables, but rather with those *steps* (in time, or on one axis), of some *two* of which it represents or denotes the *ratio*. In short, the *numbers* here spoken of, and elsewhere denominated "*scalars*" in this work, are simply those *positives* or *negatives*, on the *scale* of progression from $-\infty$ to $+\infty$, which are commonly called *reals* (or real quantities) in algebra.

† A fuller account of this theory of *sets*, with a somewhat different notation (the symbols $c_{r,s,t}$ and $n_{r,r',r''}$ being employed, for example, to denote the coefficients which would here be written

‖34‖ After operating on an arbitrary step-set $\Sigma r_f 1_f$ by a number-set $\Sigma m_g \times_g$, and so obtaining ‖32‖ another step-set, $\Sigma x_h 1_h$, we may conceive ourselves to operate on the same general plan, and with the same particular constants of multiplication, on this new step-set, by a *new number-set*, such as $\Sigma m'_{g'} \times_{g'}$, and so to obtain a *third step-set*, such as $\Sigma x'_{h'} 1_{h'}$: which may then be supposed to be *divided* (see again ‖32‖) *by the original step-set* $\Sigma r_f 1_f$, so as to conduct to a *quotient*, which shall be *another numeral set*, of the form $\Sigma m''_{g''} \times_{g''}$. Under these conditions, we may certainly write,

$$\Sigma m'_{g'} \times_{g'} (\Sigma m_g \times_g . \Sigma r_f 1_f) = \Sigma m''_{g''} \times_{g''} . \Sigma r_f 1_f;$$

but in order to justify the subsequent *abstraction of the operand step-set*, or the *abridgment* (compare ‖25‖) of this formula of *successive operation* to the following,

$$\Sigma m'_{g'} \times_{g'} . \Sigma m_g \times_g = \Sigma m''_{g''} \times_{g''},$$

which may be called a formula for the (symbolic) *multiplication of two number-sets*, certain *conditions of detachment* are to be satisfied, which may be investigated as follows.

‖35‖ Conceive that the required *separation of symbols* has been found possible, and that it has given, by a generalization of the process for triplets in ‖24‖, a system of n^2 symbolic equations of the form,

$$\times_{g'} \times_g = \Sigma 1'_{g,g',g''} \times_{g''};$$

where $1'_{g,g',g''}$ is one of a *new system of n^3 numerical coefficients*, and the sum involves n terms, answering to n different values of the index g''. Under the same conditions, the recent formula for the multiplication of numeral sets breaks itself up into n equations, of the form,

$$m''_{g''} = \Sigma m_g m'_{g'} 1'_{g,g',g''};$$

the summation here extending to n^2 terms arising from the combinations of the values of the indices g and g'. For all such combinations, and for each of the n values of f, we are to have (if the required detachment be possible) the following equation between step-sets:

$$\times_{g'} . \times_g 1_f = \times_{g'} \times_g . 1_f;$$

and conversely, if we can satisfy these n^3 equations between step-sets, we shall thereby satisfy the *conditions of detachment* ‖34‖, which we have at present in view. But *each* of these n^3 equations between *sets* resolves itself generally into

---

as $1_{t,r,s}$ and $1_{r,r',r''}$), and with a special application to the theory of *quaternions*, will be found in an Essay entitled: "Researches respecting Quaternions. First Series." Trans. R.I.A. Vol. XXI., Part II. Dublin: 1848. Pages 199 to 296. (Read November 13th, 1843.) This Essay was not fully printed till 1847, but several copies of it were distributed in that year, especially during the second Oxford Meeting of the British Association. The discussion of that portion of the subject which is here considered is contained chiefly in pages 225 to 231 of the volume above cited.

n equations between *numbers*: and thus there arise in general no fewer than n^4 *numerical equations*, as expressive of the conditions in question, which may all be represented by the formula,*

$$\Sigma 1_{f,g,h} 1_{h,g',h'} = \Sigma 1'_{g,g',h} 1_{f,h,h'};$$

all combinations of values of the indices f, g, g', h' (from 1 to n for each) being permitted, and the summation in each member being performed with respect to h. Now to satisfy these n^4 equations of condition, there were only $2n^3$ coefficients, or rather their ratios, disposable: and although the theories of couples and triplets already served to exemplify the *possibility* of effecting the desired *detachment*, at least in certain *cases*, yet it was by no means *obvious* that any *such extensive reductions*† were likely to present themselves, as were required for the accomplishment of the same object, in the more general theory of SETS. And I believe that the compass and difficulty, which I thus perceived to exist, in that very *general* theory, deterred me from pursuing it farther at the time above referred to.

⟦36⟧ There was, however, a motive which induced me then to attach a special importance to the consideration of *triplets*, as distinguished from those more general *sets*, of which some account has been given. This was the desire to connect, in some new and useful (or at least interesting) way, *calculation* with *geometry*, through some undiscovered *extension*, to *space of three dimensions*, of a method of *construction* or representation ⟦2⟧, which had been employed with success by Mr. Warren** (and indeed also by other authors,†† of whose

* A formula equivalent to this, but with a somewhat different notation, will be found at page 231 of the Essay and Volume referred to in a recent Note.

† On the subject of such general reductions, some remarks will be found at page 251 of the Essay and Volume lately cited.

** "Treatise on the Geometrical Representation of the Square Roots of Negative Quantities. By the Rev. John Warren, A.M., Fellow and Tutor of Jesus College, Cambridge." (Cambridge, 1828.) To suggestions from that Treatise I gladly acknowledge myself to have been indebted, although the interpretation of the symbol $\sqrt{-1}$, employed in it, is entirely distinct from that which I have since come to adopt in the geometrical applications of the quaternions.

†† Several important particulars respecting such authors have been collected in the already cited "Report on certain Branches of Analysis" (see especially pp. 228 to 235), by Dr. Peacock, whose remarks upon their writings, and whose own investigations on the subject, are well entitled to attention. As relates to the method described above (in paragraph ⟦36⟧ of this Preface), if *multiplication* (as well as *addition*) *of directed lines* in one plane be regarded (as I think it ought to be) as an *essential element* thereof, I venture here to state the impression on my own mind, that the true inventor, or at least the *first definite promulgator* of that method, will be found to have been Argand, in 1806: although his "Essai sur une Manière de représenter les Quantités Imaginaires," which was published at Paris in that year, is known to me only by Dr. Peacock's mention of it in his Report, and by the account of the same Essay given in the course of a subsequent correspondence, or series of communications (which also has been noticed in that Report, and was in consequence consulted a few years ago by me), carried on between Français, Servois, Gergonne, and Argand himself; which series of papers was published in Gergonne's Annales des Mathématiques, in or about the year 1813. My recollection of that correspondence is, that it was admitted to establish fully the priority of Argand to Français, as regarded the method ⟦36⟧ of (not merely *adding*, but) *multiplying* together directed lines in one plane, which is briefly described above: and

writings I had not then heard), for *operations on right lines in one plane*: which method had given a species of *geometrical interpretation* to the usual and well-known *imaginary symbol* of algebra. In the method thus referred to, *addition of lines* was performed according to the same rules as *composition of motions*, or of forces, by drawing the diagonal of a parallelogram; and the *multiplication* of two lines, in a given plane, corresponded to the construction of a species of *fourth proportional*, to an assumed line in the same plane, selected as the representative of *positive unity*, and to the two proposed *factor-lines*: such fourth proportional, or *product-line*, being *inclined* to *one* factor-line at the *same angle*, measured in the *same sense*, as that at which the *other* factor-line was inclined to the assumed *unit-line*; while its *length* was, in the old and usual signification of the words, a fourth proportional to the lengths of the unit-line and the two factor-lines. Subtraction, division, elevation to powers, and extraction of roots, were explained and constructed on the same general principles, and by processes of the same general character, which may easily be conceived from the slight sketch just given, and indeed are by this time known to a pretty

---

which was afterwards independently reproduced, by Warren in 1828, and in the same year by Mourey, in a work entitled: "La Vraie Théorie des Quantités Négatives, et des Quantités prétendues Imaginaires" (Paris, 1828). If the list of such independent re-inventors of this important and modern method of constructing by a *line* the *product of two directed lines in one fixed plane* (from which it is to be remarked, in passing, that my own mode of representing by a *quaternion* the product of two directed lines *in space* is altogether different) were to be continued, it would include, as I have lately learned, the illustrious name of Gauss, in connexion with his Theory of Biquadratic Residues (Göttingen, 1832) |*Gauss 1832*|. On the other hand, I cannot perceive that *any distinct anticipation* of this method of *multiplication of directed lines* is contained in Buée's vague but original and often cited Paper, entitled "Mémoire sur les Quantités Imaginaires," which appeared in the Philosophical Transactions (of London) for 1806, having been read in June, 1805. The ingenious author of that Paper had undoubtedly formed the notion of *representing the directions of lines* by algebraical symbols; he even uses (in No. 35 of his Memoir) such expressions as $\sqrt{2}\,(\cos 45° \pm \sin 45° \sqrt{-1})$ to denote two different and *directed diagonals* of a square: and there is the high authority of Peacock (Report, p. 228), for considering that the geometrical interpretation of the symbol $\sqrt{-1}$, as denoting *perpendicularity*, was "first formally maintained by Buée, though more than once suggested by other authors". In No. 43 of the Paper referred to, Buée constructs with much elegance, by a *bent line* AKE, or by an *inclined* line AE (where KE is a perpendicular, $= \frac{1}{2}a$, erected at the middle point K of a given line AB, or a), an *imaginary root* (x) of the quadratic equation, $x(a-x) = \frac{1}{2}a^2$, which had been proposed by Carnot (in p. 54 of the Géometrie de Position, Paris, 1804). But when he proceeds to explain (in No. 46) of his Paper) *in what sense* he regards the *two lines* AE and EB (or the two constructed *roots* of the quadratic) as having their *product* equal to the given value $\frac{1}{2}a^2$ or $\frac{1}{2}\,\overline{\mathrm{AB}}^2$, Buée *expressly limits* the signification of *such* a product to the result obtained by *multiplying the arithmetical values*, and *expressly excludes* the consideration of the *positions of the factor-lines* from his conception of their *multiplication*: whereas it seems to me to belong to the very *essence* of the method |36| of Argand and others, and generally to that system of geometrical interpretation whereon is based what Professor De Morgan has happily named *Double Algebra*, to *take account of those positions* (or directions), when *lines* are to be *multiplied* together. My *own* conception (as has been already hinted, and as will appear fully in the course of this work), of the *product of two directed lines in space* as a QUATERNION, is *altogether distinct*, both from the purely *arithmetical product* of numerical values of Buée, and from the *linear product* (or third coplanar line), in the method of Argand: yet I have thought it proper to submit the foregoing remarks, on the invention of this latter method, to the judgment of persons better versed than myself in scientific history. A few additional remarks and references on the subject will be found in a subsequent Note.

wide circle of readers: and thus, no doubt, by operations on right lines *in one plane*, the symbol $\sqrt{-1}$ received a perfectly clear interpretation, as denoting a *second unit-line, at right angles** to that line which had been selected to represent positive unity. But when it was proposed to *leave the plane*, and to construct a system which should have *some general analogy* to the known system

* Besides what has been already referred to, as having been done on this subject of the interpretation of the symbol $\sqrt{-1}$ by the Abbé Buée, it has been well remarked by Mr. Benjamin Gompertz, at page vi. of his very ingenious Tract on "The Principles and Applications of Imaginary Quantities, Book II., derived from a particular case of Functional Projections" (London, 1818), that the celebrated Dr. Wallis of Oxford, in his "Treatise of Algebra" (London, 1685), proposed to interpret the imaginary roots of a quadratic equation, by going *out of the line*, on which if real they should be measured. Thus Wallis (in his chapter lxvii.) observes:—"So that whereas in case of Negative Roots we are to say, the point B cannot be found, so as is supposed in AC Forward, but Backward it may in the same Line: we must here say, in case of a Negative Square, the point B cannot be found so as was supposed, in the Line AC; but Above that Line it may in the same Plain. This I have the more largely insisted on, because the Notion (I think) is new; and this, the plainest Declaration that at present I can think of, to explicate what we commonly call the *Imaginary Roots* of Quadratick Equations. For such are these." And again (in his following chapter lxviii., at page 269), Wallis proposes to construct thus the roots of the equation $aa \mp ba + æ = 0$:—"On AC$a = b$, bisected in C, erect a perpendicular CP $= \sqrt{æ}$. And taking PB $= \frac{1}{2}b$, make (on whether side you please of CP), PBC, a rectangled triangle. Whose right angle will therefore be at C or B, according as PB or PC is bigger; and accordingly, BC a sine or a tangent, (to the radius PB,) terminated in PC. The streight lines AB, Ba, are the two values of a. Both affirmative if (in the equation,) it be $-$ ba. Both negative, if $+$ ba. Which values be (what we call) *Real*, if the right angle be at C. But *Imaginary* if at B." These passages must always (I think) possess an historical interest, as exemplifying the manner in which, in the seventeenth century, one so eminent for his powers of *interpretation* of analytical expressions, as Dr. Wallis was, sought to apply those powers to the *geometrical construction* of the *imaginary roots* of an equation: and for the decision with which he held that such roots were quite *as cleary interpretable*, as "*what we call real*" values. His particular interpretation of those imaginary roots of a quadratic appears indeed to me to be inferior in elegance to that which was long afterwards proposed by Buée. But it may be noticed that, whether his point B were *on* or *off* the line ACa, Wallis seems (like Buée, and many other and more modern writers) to have regarded *that right line*, as being *in some sense a sum*, or at least *analogous to a sum*, of the *two successive lines* AB, Ba; which latter lines conduct, upon the whole, from the initial point A to the final point a; and construct according to him the two roots of the quadratic, whose algebraic sum is $= b$. Indeed, Wallis remarks (in the same page 269) that when those two roots are algebraically *imaginary*, or are geometrically constructed (according to him) by the help of a point B which is *above the line* ACa, then that straight line is *not equal to the aggregate* of AB + Ba; but this seems to be no more than guarding himself against being supposed to assert, that two sides of a triangle can be equal *in length* to the third. In chap. lxix., p. 272, he thus sums up:—"We find therefore, that in Equations, whether Lateral or Quadratick, which in the strict Sense, and first Prospect, appear Impossible; some mitigation may be allowed to make them Possible; and in such a mitigated interpretation they may yet be useful." For *lateral* equations (equations of the first degree), the *mitigation* here spoken of consists simply in the usual representation of *negative roots*, by lines drawn *backward* from a point, whereas they had been at first supposed to be drawn *forward*. For quadratic equations with *imaginary roots*, Wallis *mitigates* the problem, by substitutings a *bent line* ABa for that *straight line* ACa, which *constructs the given algebraical sum* (b) of the two roots of the equation, or *parts of the bent line*, AB, Ba. It is also to be noticed that he appears to have regarded the *algebraical semi-difference* of those two roots, AB, Ba, as being in all cases constructed by the *line* BC, drawn to the middle point C of the line Aa: which would again agree with many modern systems. Thus Wallis seems to have possessed, in 1685, *at least in germ* (for I do not pretend that he fully and consciously possessed them), some elements of the modern methods of *Addition* and *Subtraction* of directed lines. But on the equally essential point of *Multiplication* of directed lines in one plane, it does not appear that Wallis, any more than Buée (see the foregoing Note), had anticipated the method of Argand.

thus described, but should *extend to space*,* then difficulties of a new character arose, in the endeavour at surmounting which I was encouraged by the friend already mentioned (Mr. John T. Graves), who felt the wish, and formed the project, to surmount them in *some* way or other, as early, or perhaps earlier than myself.

⟦37⟧ A conjecture respecting such extension of the rule of multiplication of lines, from the plane to space, which long ago occurred to me (in 1831), may be stated briefly here, as an illustration of the general character of those old speculations. Let A denote a point assumed on the surface of a fixed sphere, described about the origin O of co-ordinates, with a radius equal to the unit of length; and let this point A be called the *unit-point*. Let also B and C be supposed to be two *factor-points*, on the same surface, representing the *directions* OA, OB, of the two *factor-lines* in space, of which lines it is required to perform, or to interpret, the multiplication; and so to determine, by some fixed rule to be assigned, the *product-point* D, or the direction of the *product-line*, OD. Then it appeared that the analogy to operations in the plane might be not ill observed, by conceiving D to be taken on the *circle* ABC; the *arcs*, AB, CD, of that (generally) *small circle* of the sphere being *equally long*, and *similarly measured*; so that the two *chords* AD, BC should be *parallel*: while the old rule of *multiplication of lengths* should be retained: and *addition of lines* be still interpreted as before. But in this system there were found to enter *radicals* and *fractions* into the expressions for the co-ordinates** of a product; and although the case of *squares of lines*, or products of equal factors, might be rendered *determinate* by agreeing to take the *great* circle AB, when the point C coincided with B, yet there seemed to be an essential *indetermination* in the construction of the *reciprocal* of a line: it being sufficient, according to the definition here considered, to take the chord BC parallel to the tangent plane to the sphere at the unit-point, in order to make the product point D coincide with that point A. There was also the great and (as I thought) fatal objection to this method of construction, that it did not preserve the *distributive principle* of multiplication; a *product of sums* not being equal, in it, to the *sum of the products*: and on the whole, I abandoned the conjecture.

⟦38⟧ Another construction, of a somewhat similar character, and liable to similar objections, for the product of two lines in space, occurred to me in 1835,

* At a much later period I learned that others had sought to accomplish some such extension to space, but in ways different from mine.

** The rectangular co-ordinates (or projections) of the two factor-lines and of the product-line being denoted by xyz, $x'y'z'$, $x''y''z''$, if we also write, for conciseness,

$$r = \sqrt{(x^2 + y^2 + z^2)},\ r' = \sqrt{(x'^2 + y'^2 + z'^2)},\ p = xx' + yy' + zz',$$

then the expressions which I found for $x''y''z''$ may be included briefly in the equations:

$$\frac{x'' - rr'}{rx' - r'x} = \frac{y''}{ry' - r'y} = \frac{z''}{rz' - r'z} = \frac{rx' - r'x}{p - rr'}.$$

and also independently to Mr. J.T. Graves in 1836, in which year he wrote to me on the subject. It may be briefly stated, by saying that instead of considering, as in the last-mentioned system, the *small* circle ABC, and drawing the *chord* AD, from unit-point to product-point, so as to be *parallel* to the chord BC from one factor-point to the other, it was now the *arc* AD of a *great* circle on the sphere, which was to be drawn so as to *bisect the arc* BC, of another great circle, and *be bisected* thereby. Or as Mr. Graves afterwards expressed to me the rule in question:—"Bisect the inclination of the factor-lines, and then double forward the angle between the linear unit and the bisecting line": the rule of multiplying *lengths* being understood to be still observed. Mr. Graves made several acute remarks on the consequences of this construction, and proposed a few supplementary *rules* to remove the *porismatic* character of some of them: but observed that, with these interpretations, the *square-root of the negative unit-line*, or the triplet $(-1, 0, 0)^{1/2}$, would still be indeterminate, and of the form $(0, \cos\theta, \sin\theta)$, where θ remained arbitrary: while cases might arise, in which the "minutest alteration" of a factor-line would make a "considerable change" in the position of the product-line: and this result he conceived to be, or to lead to, "a breach of the grand property of multiplication", respecting its operation on a *sum*. He left to me the investigation of the general expressions for the "constituent co-ordinates" of the resultant "triplet", or product-line, in terms of the constituents of the factors: and in fact I had already obtained such expressions, and had found them to involve radicals and fractions, and to violate the distributive principle, as in the system recently described [37]; with which indeed the one here mentioned had been perceived by me to have a very close analytical connexion.*

[39] Mr. J.T. Graves, however, communicated to me at the same time another method, which he said that he *preferred*, among all the modes that he had tried, "of representing lines in space, and of multiplying such lines together".

* With the notations recently employed, the expressions which I had found for the co-ordinates of the product, in the case or system [38], are included in the equations,

$$\frac{x'' + rr'}{rx' + r'x} = \frac{y''}{ry' + r'y} = \frac{z''}{rz' + r'z} = \frac{rx' + r'x}{p + rr'};$$

which only differ from those for the former case [37], by a change of sign in the radical r' (or r), which represents the length of a factor-line. The conditions for both systems are contained in these other equations,

$$xx'' + yy'' + zz'' = r^2x', x'x'' + y'y'' + z'z'' = r'^2x, x''^2 + y''^2 + z''^2 = r^2r'^2;$$

and the quadratic equation in x'', obtained by elimination of y'' and z'', resolves itself into two separate factors, each linear relatively to x'', namely,

$$(p - rr')(x'' - rr') - (rx' - r'x)^2 = 0,$$
$$(p + rr')(x'' + rr') - (rx' + r'x)^2 = 0.$$

The first corresponds to the system [37]; the second to the system [38].

This method consisted in considering such a line as a species of "compound couple", or as determined by *two couples*, one in the plane of xy, and the other perpendicular to that plane: it having been easily perceived that the rules proposed by me for the addition and multiplication ‖17‖ of *couples*, agreed in all respects with the previously known method ‖36‖, of representing the operations of the same names on *lines in one plane*. From this conception of *compound couples* Mr. Graves derived a "general rule for the multiplication of triplets", which I shall here transcribe,[†] only abridging the notation by writing ρ and ρ_1 to represent the radicals $\sqrt{(x^2+y^2)}$ and $\sqrt{(x_1{}^2+y_1{}^2)}$, or the projections of the factor-lines on the plane of xy: "$(x, y, z)\ (x_1, y_1, z_1) = (x_2, y_2, z_2)$, where

$$x_2 = (\rho\rho_1 - zz_1)\left(\frac{xx_1 - yy_1}{\rho\rho_1}\right),\ y_2 = (\rho\rho_1 - zz_1)\frac{xy_1 + yx_1}{\rho\rho_1},\ z_2 = z_1\rho + z\rho_1."$$

This particular system of expressions he does not seem to have developed farther, nor did it at the time attract much of my own attention: but I have thought it deserving of being put on record here, especially as, by a remarkable coincidence, it came to be independently and otherwise arrived at by another member of the same family, at a date later by ten years, and to be again communicated to me.* And perhaps I may be excused if I here leave the order of time, to give some short account of the train of thought by which his brother, the Rev. Charles Graves, appears to have been conducted, in 1846, to precisely the *same relations* between the constituents of three triplets.

‖40‖ Professor Graves employed a system of *two new imaginaries*, i and j, of which he conceived that i had the effect of causing a rotation (generally conical) through 90° round the axis of z, while j caused a line to revolve through an equal angle in its own vertical plane (that is, in the plane of the line and of z); and then he proceeded to *multiply* together the two triplets $x + iy + jz$, $x' + iy' + jz'$, by a peculiar process, and so to obtain a third triplet $x'' + iy'' + jz''$: the relations thus resulting, between the co-ordinates or constituents, being (as it turned out) identical with those which his brother had formerly found. These symbols i and j were *each a sort of fourth root of unity*: and the first, but *not* the second, had the property of operating on a *sum* by operating on each of its *parts* separately. Thus, as Professor Graves remarked, multiplication of triplets, on this plan, would *not* be a *distributive* operation, although it would be a *commutative* one. The method conducted him to an elegant exponential expression for a line in space, namely, $r\varepsilon^{il}\varepsilon^{j\lambda}$, where r was the *radius vector*, and l, λ might be called the *longitude* and *latitude* of the line, so that the co-ordinate projections were (some peculiar considerations being

[†] From Mr. Graves's Letter of August 8th, 1836.

* By the Rev. Charles Graves, Professor of Mathematics in the University of Dublin, in a letter of November 14th, 1846.

employed in order to justify these expressions of them, as connected with that of the line):

$$x = r \cos l \cos \lambda,\ y = r \sin l \cos \lambda,\ z = r \sin \lambda.$$

And then the rule for the *multiplication of two lines* came to be expressed by the very simple formula:

$$r\varepsilon^{il}\varepsilon^{j\lambda} . r'\varepsilon^{il'}\varepsilon^{j\lambda'} = rr'\varepsilon^{i(l+l')}\varepsilon^{j(\lambda+\lambda')};$$

the *lengths* being thus *multiplied* (as in the other systems above mentioned), but the *longitudes* and *latitudes* of the one line being respectively *added* to those of the other: which was in fact the rule expressed by Mr. J.T. Graves's co-ordinate formulæ ‖39‖.

‖41‖ It will not (I hope) be considered as claiming any merit to myself in this matter, but merely as recording an unpursued *guess*, which may assist to *illustrate* this whole inquiry, if I venture to mention here that the *first conjecture* respecting *geometrical triplets*, which I find noted among my papers (so long ago as 1830), was, that while *lines in space* might be *added* according to the same rule as in the plane, they might be *multiplied* by multiplying their lengths, and *adding* their polar angles. In the method ‖36‖, known to me then as that of Mr. Warren, if we write $x = r \cos \theta$, $y = r \sin \theta$, we have, for multiplication *within* the plane, equations which may be written thus, $r'' = rr'$, $\theta'' = \theta + \theta'$. It hence occurred to me, that if we employed for space these other known transformations of rectangular to polar co-ordinates,

$$x = r \cos \theta,\ y = r \sin \theta \cos \phi,\ z = r \sin \theta \sin \phi,$$

it might be natural to *define* multiplication of lines in space by the slightly extended but analogous formulæ,

$$r'' = rr',\ \theta'' = \theta + \theta',\ \phi'' = \phi + \phi':$$

which, however, conducted to *radicals*, as in the expression,

$$x'' = xx' - (y^2 + z^2)^{\frac{1}{2}} (y'^2 + z'^2)^{\frac{1}{2}},$$

whereas within the plane there were *rational* values for the rectangular co-ordinates of the product, namely (compare ‖17‖),

$$x'' = xx' - yy',\ y'' = xy' + yx'.$$

But this old (and uncommunicated) conjecture of mine, which was inconsistent with the distributive principle, though possessing some general *resemblance* to the lately mentioned results ‖39‖ ‖40‖ of Messrs. John and Charles Graves, cannot be considered to have been an *anticipation* of them. For while we all *agreed* in *adding* the *longitudes* of the two factors (in the sense lately mentioned), *they added latitudes* also; while I, less happily, had thought of *adding the colatitudes*, or the angular distances from a *line* (x), instead of those from a *plane* (xy). And this difference of plan produced a very important difference of results. Indeed the two systems are totally distinct, although there exists some sort of analogy between them.

‖42‖ I shall here mention one more system, which was communicated to me* in 1840, by the elder of those two brothers, and which involved a method of representing the usual imaginary quantities of algebra, *each by a corresponding unique point on the surface of a sphere*, described (as in ‖37‖) about the origin with a radius = 1: whence it appeared that the ordinary imaginary expression $r(\cos\theta + \sqrt{-1}\sin\theta)$ might be denoted by a *triplet* (x, y, z), under the *condition*, $x^2 + y^2 + z^2 = 1$: and that the *rules* thus obtained, for the multiplication of *such* triplets, might perhaps afford some *analogy*, suggesting rules† for the more *general* case, where the constituents x, y, z are wholly *independent* of each other. Mr. J.T. Graves's "mode of representing quantity spherically" was stated by him to me as follows:—

> All positive quantities r may be represented by points on an assumed semicircle, by taking the extremity of the arc $2\tan^{-1}r$ (counted from one end (A) of the semicircle) to represent r. Next let us consider our sphere as generated by the revolution of the semicircle‡ ABC round the axis AC (forwards or backwards, according to arbitrary convention). When the semicircle has moved through an angle θ, let the position of a point on its circumference denote $r(\cos\theta + \sqrt{-1}\sin\theta)$, if the same point in its original position denoted r.

I make a very easy transformation of this statement, when I present it thus:—Construct all quantities (so called), real and imaginary, according to the known method already described in ‖36‖, by drawing right lines from the assumed point (A) of the unit-sphere, in the tangent plane at that point; double all the lines so drawn, and treat the ends of the doubled lines as the stereographic projections of points upon the sphere. Infinity was thus represented, in the particular system of Mr. Graves here described, by the point diametrically opposite to A. And in this endeavour of mine, to furnish faithfully a record of every circumstance, which, even as remotely *suggesting* to a *friend* a train of thought, may have *indirectly* stimulated *myself*, I must not suppress the following acknowledgment of Mr. J.T. Graves:—"What led me to this was a passage in a letter from De Morgan,** in which be expressed a wish to be able to represent quantity *circularly*, in order to explain the passage from positive to negative through infinity."

‖43‖ The foregoing specimens may suffice to exemplify the attempts which were made, a considerable number of years ago, by Mr. Graves and by myself:

---

* In a letter of October 17th, 1840, from J.T. Graves, Esq.

† Mr. Graves appears not to have actually worked out such rules, at least I do not find that he communicated them to me. They would probably have been, on the plan described in ‖42‖, to have *multiplied* (as before) the *lengths*, and (as before) *added* the *longitudes*: but to have then *multiplied the tangents of the halves of the colatitudes* of the factors, in order to obtain the tangent of the half of the colatitude of the product.

‡ A figure, which it seems unnecessary here to reproduce, accompanied Mr. Graves's Letter.

** Augustus De Morgan, Esq., Professor of Mathematics in University College, London.

on the one hand, to *extend* to *space* that geometrical construction for the multiplication of *lines*, which was known to us from the work of Mr. Warren; and on the other hand, to render more entirely *definite* my conception of algebraical *triplets*. I will not here trouble my readers with any further account of the conjectures on those subjects which at various times occurred to him or me, before I was led to the quaternions, in a way which I shall presently explain. But I wish to mention first, that among the circumstances which assisted to prevent me from losing sight of the general subject, and from wholly abandoning the attempt to turn to some useful account those early speculations of mine, on triplets and on sets, was probably the publication of Professor De Morgan's first Paper on the Foundation of Algebra,† of which he sent me a copy in 1841. In that Paper, besides the discussion of other and more important topics, my Essay on Pure Time was noticed, in a free but friendly spirit; and the subject of triplets was alluded to, in such passages, for instance, as the following:—"But in this branch of logical algebra" (that referred to in paragraph ‖36‖ of the present Preface), "the lines must be all in one plane, or at least affected by only one modification of direction: the branch which shall apply to a line drawn in any direction from a point, or modified by two distinct directions, is yet to be found." . . . "An extension to geometry of three* dimensions is not practicable until we can assign two symbols, Ω and ω, such that $a + b\Omega + c\omega = a_1 + b_1\Omega + c_1\omega$ gives $a = a_1$, $b = b_1$, and $c = c_1$: and no *definite* symbol of ordinary algebra will fulfil this condition." My symbols $\times_2$, $\times_3$ (of 1834–5) had not then been published, nor otherwise exhibited to him; they were designed to fulfil precisely the foregoing conditions: but I was not myself satisfied with them, as not considering them "*definite*" enough (compare ‖29‖).

‖44‖ In the early numbers of the Cambridge Mathematical Journal, there appeared some ingenious and original Papers, by the late Mr. Gregory and by other able analysts, on the signs + and −, on the powers of +, on branches of curves in different planes, and on other connected subjects: but I hope that it will not be thought disrespectful if I confess that I do not remember their having had much influence on my own trains of thought. Perhaps I was not sufficiently prepared, or disposed, to look at algebra generally, and its

† In Vol. VII., Part II., of the Cambridge Philosophical Transactions.

* Professor De Morgan proposed at the same time a remarkable conjecture, which he may be considered to have afterwards illustrated and systematised, by his theory of *cube-roots* of negative unity, employed as *geometrical operators*, in his Paper on *Triple Algebra* (Camb. Phil. Trans., Vol. VIII., Part iii.); namely, that "an extension to three dimensions" might "require a solution of the equation $\phi^3 x = -x$." I much regret that my plan will not allow me to attempt the giving any further account, in this Preface, of that very original Paper of Professor De Morgan, the first suggestion of which he was pleased to attribute to the publication of my own remarks on Quaternions, in the Philosophical Magazine for July, 1844: and a similar expression of regret applies to the independent but somewhat later researches of Messrs. John and Charles Graves, in the same year, respecting other Triplet Systems, which involved cube-roots of *positive* unity, and of which some account has been preserved in the Proceedings of the Royal Irish Academy.

applications to geometry, from the same point of view, and was thereby prevented from studying those Papers with the requisite attention. At least, if anything in my own views shall be found to be inconsistent with those put forward in the Papers thus alluded to, I wish it to be considered as offered with every deference, and not in a controversial spirit. And if for the present I omit all further mention of them, it is partly because, without a closer study, I should fear to do them injustice: and partly because I make no pretensions to be here an *historian of science*, even in *one* department of mathematical speculation, or to give anything more than an account of the *progress of* my *own thoughts*, upon one class of subjects. For the same reasons, I pass over some other investigations having reference to the imaginary* symbol of algebra, which were not used as suggestions by myself, and proceed at once to the quaternions.

‖45‖ With such preparations as I have described, I resumed (in 1843) the endeavour to adapt the general conception of triplets to the multiplication of lines in space, resolving to *retain* the *distributive* principle, with which some formerly conjectured systems had been inconsistent, and at first supposing that I *could* preserve the *commutative* principle *also*, or the convertibility ‖24‖ ‖29‖ of the

---

* I am unwilling, however, to leave unmentioned here (although it did not happen to supply me with any suggestion), a remarkable use of the symbol $\sqrt{-1}$, which was made by the late Professor Mac Cullagh, of Dublin, whose great and original powers in mathematical and physical science must ever be remembered with admiration, and which he seems to have connected (in 1843) with investigations respecting the total reflexion of light. (See Proceedings of the R.I.A. for the date of January 13, 1845.) This use of imaginaries was founded on a theorem relative to the ellipse, which was expressed by him as follows, in a question proposed at the Examination for the Election of Junior Fellows in 1842 (see Dublin University Examination Papers for that year, published in 1843, p. lxxxiv.):—"Detur in spatio ellipsis, cujus centrum est origo co-ordinatarum. Puncta xyz, $x'y'z'$ in ellipsi sint termini diametrorum conjugatarum. Ostendendum est quantitates imaginarias

$$\frac{y + y'\sqrt{-1}}{x + x'\sqrt{-1}}, \quad \frac{z + z'\sqrt{-1}}{x + x'\sqrt{-1}},$$

constantes esse pro quolibet systemate diametrorum conjugatarum." ‖'Let there be given an ellipse in space whose centre is the origin of the coordinate system. Let the points xyz, $x'y'z'$ on the ellipse be the ends of the conjugate diameters. Show that the imaginary values

$$\frac{y + y'\sqrt{-1}}{x + x'\sqrt{-1}}, \quad \frac{z + z'\sqrt{-1}}{x + x'\sqrt{-1}},$$

are constant for any system of conjugate diameters.'‖ This elegant theorem of Professor Mac Cullagh may easily be proved, without employing any but the usual principles respecting the symbol $\sqrt{-1}$, by observing that the following expressions, for the six co-ordinates in question,

$$x = a \cos \upsilon + a' \sin \upsilon, \; y = b \cos \upsilon + b' \sin \upsilon, \; z = c \cos \upsilon + c' \sin \upsilon,$$

$$x' = a' \cos \upsilon - a \sin \upsilon, \; y' = b' \cos \upsilon - b \sin \upsilon, \; z' = c' \cos \upsilon - c \sin \upsilon,$$

give

$$\frac{x + x'\sqrt{-1}}{a + a'\sqrt{-1}} = \frac{y + y'\sqrt{-1}}{b + b'\sqrt{-1}} = \frac{z + z'\sqrt{-1}}{c + c'\sqrt{-1}} = \cos \upsilon - \sin \upsilon\sqrt{-1}.$$

factors as to their *order*. Instead of my old symbols $\times_1$, $\times_2$, $\times_3$ (see ⟦22⟧), I wrote more shortly 1, i, j; so that a numerical triplet took the form $x + iy + jz$, where I proposed to interpret x, y, z as three rectangular co-ordinates, and the triplet itself as denoting a line in space. From the analogy of couples, I assumed $i^2 = -1$; and tried the effect of assuming also $j^2 = -1$, which I interpreted as answering to a rotation through two right angles in the plane of xz, as $i^2 = -1$ had corresponded to such a rotation in the plane of xy. And because I at first supposed that ij and ji were to be *equal*, as in the ordinary calculations of algebra, the product of two triplets appeared to take the form,

$$(a + ib + jc)(x + iy + jz) = (ax - by - cz) + i(ay + bx) + j(az + cx) + ij(bz + cy):$$

but I did not at once see what to do with the *product* ij. The theory of triplets seemed to require that it should be *itself* a triplet, of the form,

$$ij = \alpha + i\beta + j\gamma,$$

the coefficients α, β, γ being some three constant numbers: but the question arose, how were those numbers to be determined, so as to adapt in the best way the resulting formula of multiplication to some *guiding geometrical analogies*.

⟦46⟧ To assist myself in applying such analogies, I considered the case where the co-ordinates b, c were *proportional* to y, z, so that the two factor-lines were in one common *plane*, containing the unit-line, or the axis of x. In that particular *case*, there was ready a *known* signification ⟦36⟧ for the product line, considered as the fourth proportional to the unit-line (assumed here on the last-mentioned axis), and to the two coplanar factor-lines. And I found, without difficulty, that the co-ordinate projections of such a fourth proportional were here,

$$ax - by - cz, \; ay + bx, \; az + cx,$$

that is to say, the coefficients of 1, i, j, in the recently written expression for the product of the two triplets, which had been supposed to represent the factor-lines. In fact, if we assume $y = \lambda b$, $z = \lambda c$, where λ is any coefficient, we have the two identical equations,

$$(ax \quad \lambda b^2 \quad \lambda c^2)^2 + (\lambda a + x)^2(b^2 + c^2) = (a^2 + b^2 + c^2)(x^2 + \lambda^2 b^2 + \lambda^2 c^2),$$

$$\tan^{-1}\frac{(\lambda a + x)(b^2 + c^2)^{\frac{1}{2}}}{ax - \lambda(b^2 + c^2)} = \tan^{-1}\frac{(b^2 + c^2)^{\frac{1}{2}}}{a} + \tan^{-1}\frac{\lambda(b^2 + c^2)^{\frac{1}{2}}}{x},$$

which express that the required geometrical conditions are satisfied. It was allowed then, in this *case of coplanarity*, or under the particular *condition*,

$$bz - cy = 0,$$

to treat the triplet,

$$(ax - by - cz) + i(ay + bx) + j(az + cx),$$

as denoting a *line* which might, consistently with known analogies, be regarded as the *product* of the two lines denoted by the two proposed triplets,

$$a + ib + jc, \text{ and } x + iy + jz.$$

And here the *fourth term,*

$$ij(bz + cy),$$

appeared to be simply *superfluous*: which induced me for a moment to fancy that perhaps the *product ij* was to be regarded as $= 0$. But I saw that this fourth term (or part) of the product was more immediately given, in the calculation, as the sum of the two following,

$$ib.jz,\ jc.iy;$$

and that this *sum* would vanish, under the present *condition* $bz = cy$, if we made what appeared to me a *less harsh* supposition, namely, the supposition (for which my old speculations on *sets* had prepared me) that

$$ij = -ji:$$

or that

$$ij = +k,\ ji = -k,$$

the value of the product k being still left undetermined.

⟦47⟧ In this manner, *without* now assuming $bz - cy = 0$, I had generally for the *product of two triplets*, the expression of *quadrinomial form,*

$$(a + ib + jc)(x + iy + jz) = (ax - by - cz) + i(ay + bx) + j(az + cx) + k(bz - cy);$$

and I saw that although the product of the sums of squares of the constituents of the two factors could not in general be decomposed into *three* squares of rational functions of them, yet it *could* be generally presented as the sum of *four* such squares, namely, the squares of the four coefficients of $1, i, j, k$, in the expression just deduced: for, without any relation being assumed between a, b, c, x, y, z, there was the identity,

$$(a^2 + b^2 + c^2)(x^2 + y^2 + z^2) = (ax - by - cz)^2 + (ay + bx)^2 + (az + cx)^2 + (bz - cy)^2.$$

This led me to conceive that perhaps instead of seeking to *confine* ourselves to *triplets*, such as $a + ib + jc$ or (a, b, c), we ought to regard these as only *imperfect forms of* QUATERNIONS, such as $a + ib + jc + kd$, or (a, b, c, d), the symbol k denoting *some new sort of unit operator*: and that thus my old conception of *sets* ⟦30⟧ might receive a new and useful application. But it was necessary, for operating *definitely* with such quaternions, to fix the value of the *square* k^2, of this new symbol k, and also the values of the *products*, ik, jk, ki, kj. It seemed natural, after assuming as above that $i^2 = j^2 = -1$, and that $ij = k$, $ji = -k$, to assume also that $ki = -ik = -i^2j = +j$, and $kj = -jk = j^2i = -i$. The assumption to be made respecting k^2 was less obvious; and I was for a while disposed to consider this square as equal to *positive* unity, because $i^2j^2 = +1$: but it appeared more convenient to suppose, in consistency with the foregoing expressions for the products of i, j, k, that

$$k^2 = ijij = -iijj = -i^2j^2 = -(-1)(-1) = -1.$$

⟦48⟧ Thus all the fundamental assumptions for the *multiplication of two quaternions* were completed, and were included in the formulæ,

$$i^2 = j^2 = k^2 = -1; \; ij = -ji = k; \; jk = -kj = i; \; ki = -ik = j:$$

which gave me the equation,

$$(a, b, c, d)(a', b', c', d') = (a'', b'', c'', d''),$$

or

$$(a + ib + jc + kd)(a' + ib' + jc' + kd') = a'' + ib'' + jc'' + kd'',$$

when and only when the following *four separate equations* were satisfied by the *constituents* of these three quaternions:

$$\begin{aligned} a'' &= aa' - bb' - cc' - dd', \\ b'' &= (ab' + ba') + (cd' - dc'), \\ c'' &= (ac' + ca') + (db' - bd'), \\ d'' &= (ad' + da') + (bc' - cb'). \end{aligned}$$

And I preceived on trial, for I was not acquainted with a theorem of Euler respecting *sums of four squares*, which might have enabled me to anticipate the result, that these expressions for a'', b'', c'', d'' had the following *modular property*:

$$a''^2 + b''^2 + c''^2 + d''^2 = (a^2 + b^2 + c^2 + d^2)(a'^2 + b'^2 + c'^2 + d'^2).$$

I saw also that if, instead of representing a line by a triplet of the form $x + iy + jz$, we should agree to represent it by this *other trinomial form*,

$$ix + jy + kz,$$

we should then be able to express the desired *product of two lines in space* by a QUATERNION, of which the constituents have very *simple geometrical significations*, namely, by the following,

$$(ix + jy + kz)(ix' + jy' + kz') = w'' + ix'' + jy'' + kz'',$$

where

$$w'' = -xx' - yy' - zz',$$
$$x'' = yz' - zy', y'' = zx' - xz', z'' = xy' - yx';$$

so that the part w'', independent of ijk, in this expression for the product, represents the *product of the lengths of the two factor-lines, multiplied by the cosine of the supplement of their inclination* to each other; and the remaining part $ix'' + jy'' + kz''$ of the same product of the two trinomials represents a *line*, which is in *length* the *product of the same two lengths, multiplied by the sine of the same inclination*, while in *direction* it is *perpendicular to the plane of the factor-lines*, and is such that the *rotation round the multiplier-line*, from the multiplicand-line towards the product-line (or towards the *line-part* of the whole quaternion product), has the *same right-handed* (or left-handed) *character*, as the rotation round the positive semiaxis of k (or of z), from the positive semiaxis of i (or of x), towards that of j (or of y).

‖49‖ When the conception, above described, had been so far unfolded and fixed in my mind, I felt that the *new instrument* for applying *calculation to geometry*, for which I had so long sought, was now, at least in part, attained. And although I had left several former conjectures respecting *triplets* for many years uncommunicated, except by name, even to friends, yet I at once proceeded to lay these results respecting *quaternions* before the Royal Irish Academy (at a Meeting of Council* in October, 1843, and at a General Meeting† shortly subsequent): introducing also a theory of their connexion with spherical trigonometry, some sketch of which appeared a few months later in London (in the Philosophical Magazine for July, 1844). On that *connexion of quaternions with spherical trigonometry*, and generally with *spherical geometry*, I need not at present dwell, since it is sufficiently explained in the concluding Lectures of this Volume: but it may be not improper that a brief account should here be given, of a not much later but hitherto unpublished speculation, of a character partly geometrical, but partly also metaphysical (or *à priori*), by which I sought to explain and confirm some results that might at first seem strange, among those to which my analysis had conducted me, respecting the *quadrinomial form*, and *non-commutative property*, of the *product* of two directed lines in space.

* The Minutes of Council of the R.I.A., for October 16th, 1843, record "Leave given to the President to read a paper on a new species of imaginary quantities, connected with a theory of quaternions". It may be necessary to state, in explanation, that the Chair of the Academy, which has since been so well filled by my friends, Drs. Lloyd and Robinson, was at that time occupied by me.

† At the Meeting of November 13th, 1843, as recorded in the "*Proceedings*" of that date, in which the fundamental formulæ and interpretations respecting the symbols *ijk* are given. Two letters on the subject, which have since been printed, were also written in October, 1843, to the friend so often mentioned in this Preface, Mr. J.T. Graves: and the chief results were also exhibited to his brother, the Rev. C. Graves, before the public communication of November, 1843. These circumstances (or some of them) have been stated elsewhere: but it seemed proper not to pass them over without some short notice here, as connected with the date of the invention and publication of the quaternions.

[50] Let, then, the PRODUCT of two co-initial lines, or of two vectors from a common origin, be conceived to be *something* which has QUANTITY, in the sense that it is doubled, tripled, &c., by doubling, tripling, &c., either factor; let it also be conceived to have in some sense, QUALITY, *analogous to direction*, which is in some way *definitely connected* with the directions of the two factor lines. In particular let us conceive, in order to preserve so far an analogy to *algebraic multiplication*, that its direction is in all respects *reversed*, when *either* of those directions is reversed; and therefore that it is *restored*, when *both* of them are reversed. On the other hand, for the sake of recognising what may be called the *symmetry of space*, let this *direction of the product*, so far as it can be constructed or represented by that of any *line in space*, be conceived as *not changing its relation to the system of those two factor directions*, when that system is in any manner *turned in space*: its own direction, *as a line*, being at the same time *turned with them*, as if it formed a part of one common and rigid system; and the *numerical element* of the same product (if it have any such) undergoing *no change* by such rotation. Let the product in question be conceived to be entirely *determined*, when the factors are determined; let it be made, if other conditions will allow, for the sake of general analogies, a *distributive function* of those two factors, summation of lines being performed by the same rules as composition of motions; and finally, if these various conditions can all be satisfied, and still leave anything undetermined, in the rules for *multiplication of lines*, let the indeterminateness be removed in such a way as to make these rules approach as much as possible to the other usual rules for the *multiplication of numbers* in algebra.

[51] The *square of a given line* must *not* be *any line* inclined to that given line; for, even if we chose any particular *angle* of inclination, there would be nothing to determine the *plane*, and thus the square would be *indeterminate*, unless we selected some one direction in space as *eminent*, which selection we are endeavouring to *avoid*. Nor can the square of a given line be a line in the *same* direction, nor in the direction *opposite*; for if *either* of these directions were selected, by a definition, then this definition would oblige us to consider the square as *reversed* in direction, when the line of which it is the square is reversed; whereas if the two factors of a product *both* change sign, the direction of the product is always (by what has been above agreed on) preserved, or rather *restored*. We must, therefore, consider the SQUARE OF A LINE as having *no direction in space*, and therefore as being *not* (properly) *itself a line*; but nothing hitherto prevents us from regarding the *square* as a NUMBER, which has always one determined *sign* (as yet unknown), and varies in the duplicate ratio of the length of the line to be squared. If, then, the length of a line α contain a times the unit of length, we are led to consider $\alpha\alpha$ or α^2 as a symbol equivalent to la^2, in which l is some numerical coefficient, positive or negative, as yet unknown, but constant for all lines in space, or having *one common value* for all. And, consequently, if α, β be *any two lines* in any *one common direction*, and having their *lengths* denoted by the *numbers* a and b, we are led to regard the product $\alpha\beta$ as equal to the number lab, l being the same *coefficient*

as before. But if the direction of β be exactly *opposite* to that of α, their lengths being still a and b, their product is then equal to the opposite number, $-lab$. The same general conclusions might perhaps have been more easily arrived at, if we had *begun* by considering the product of two equally long but *opposite* lines; for it might perhaps then have been even easier to see that, consistently with the *symmetry of space*, no *one* line rather than another could represent, even in part, the direction of the product.

⟦52⟧ Next, let us consider the product $\alpha\beta$ of *two mutually perpendicular lines*, α and β, of which each has its length equal to 1. Let α', β' be lines respectively equal in length to these, but respectively opposite in direction. Then $\alpha'\beta = -\alpha\beta = \alpha\beta'$; $\alpha'\beta' = \alpha\beta$. If the sought product $\alpha\beta$ were equal to any *number*, or even if it contained a number as a *part* of its expression, then, on our changing the multiplier α to its own opposite line α', this product or part ought *for one reason* (the symmetry of space) to remain *constant* (because the system of the factors would have been merely *turned in space*); and for another reason ($\alpha'\beta = -\alpha\beta$) the same product or part ought to *change sign* (because *one* factor would have been *reversed*): but this co-existence of opposite results would be absurd. We are led therefore to try whether the present condition (of *rectangularity of the two factors*) allows us to suppose the product $\alpha\beta$ to be a LINE.

⟦53⟧ Let γ be a third line, of which the length is unity, and which is at the positive side of β, with reference to α as an axis of rotation; right-handed (or left-handed) rotation having been previously selected as *positive*; let also γ' be the line opposite to γ. Then *any line* in space may be denoted by $m\alpha + n\beta + p\gamma$; we are therefore to try whether we can consistently suppose $\alpha\beta = m\alpha + n\beta + p\gamma$, m, n, p being some three numerical constants. If so, we should have (by the principle of the symmetry of space) $\alpha'\beta = m\alpha' + n\beta + p\gamma'$; and therefore (by a change of all the signs) $\alpha\beta = m\alpha + n\beta' + p\gamma$; therefore $n\beta' = n\beta$, and consequently $-n = n$, or finally $n = 0$. In like manner, since $\alpha\beta = -\alpha\beta' = -(m\alpha + n\beta' + p\gamma') = m\alpha' + n\beta + p\gamma$, we should have $m\alpha' = m\alpha$, and therefore $m = 0$. But there is no objection of *this* kind against supposing $\alpha\beta = p\gamma$, p being some numerical coefficient, constant for all pairs of rectangular lines in space: for the reversal of the direction of a factor has the effect of turning the system through two right angles round the other factor as an axis, and so reverses the direction of the product. And then if the lengths of these two lines, α, β, instead of being each $= 1$, are respectively a and b, their product $\alpha\beta$ will be $= pab\gamma$; that is, it will be a line perpendicular to both factors, with a length denoted by pab, and situated always to the positive or always to the negative side of the multiplicand line β, with respect to the multiplier line α as an axis of rotation, according as the constant number p is positive or negative.

⟦54⟧ So far, then, without having yet used any property of multiplication, algebraical or geometrical, beyond the three principles: 1st, that *no one direction in space is to be regarded as eminent above another*; 2nd, that *to multiply either factor by any number, positive or negative, multiplies the product by the same*; and 3rd, that *the product of two determined factors is itself determined*; we are led to assign interpretations: 1st, to the product of two *co-axal* vectors, or of two lines parallel to each other, or to one common axis; and 2nd, to the product of two *rectangular* vectors; which interpretations introduce only *two constant*, but as yet unknown, *numerical coefficients*, l and p, depending, however, partly on the assumed unit of length. And we see that for any two co-axal vectors, α, β, the equation $\alpha\beta - \beta\alpha = 0$ holds good; but that for any two rectangular vectors, $\alpha\beta + \beta\alpha = 0$. A *product of two rectangular lines* is, therefore, so far as the foregoing investigation leads us to conclude, *not a commutative function of them*.

⟦55⟧ Since then we are compelled, by considerations which appear more primary, to *give up the commutative property* of multiplication, as not holding *generally* for *lines*, let us at least try (as was proposed) whether we can retain the *distributive* property. If so, and if the multiplicand line β be the sum of two others, β_1 and β_2, of which one (β_1) is co-axal with the multiplier line α, while the other (β_2) is perpendicular thereto, we must interpret the product $\alpha\beta$ as equal to the *sum of the two partial products*, $\alpha\beta_1$ and $\alpha\beta_2$. But one of these is a number, and the other is a line; we are, therefore, led to consider a number as being under these circumstances *added* to a line, and as forming with it a certain *sum*, or *system*, denoted by $\alpha\beta_1 + \alpha\beta_2$, or more shortly by $\alpha\beta$. And such a *sum of line and number* may perhaps be called a GRAMMARITHM,* from the two Greek words, γραμμή, a line, and ἀριθμός, a number. A grammarithm is thus to be conceived as being entirely *determined*, when its *two parts* or elements are so; that is, when its *grammic* part is a known line, and its *arithmic* part is a known number. A change in *either* part is to be conceived as changing the grammarithm: thus, *an equation between two grammarithms includes generally two other equations*, one between two numbers, and another between two lines. Adopting this view of a grammarithm, and *defining* that $\alpha\beta = \alpha\beta_1 + \alpha\beta_2$, when $\beta = \beta_1 + \beta_2$, $\beta_1 \parallel \alpha$, $\beta_2 \perp \alpha$, the product of any determined multiplier line and any determined multiplicand line will be itself entirely determined, as soon as the unit of length and the numbers l and p shall have been chosen; and it remains to consider whether these numbers can now be so selected, as to make the rules of multiplication of *lines* approach more closely still to the rules of multiplication of *numbers*.

⟦56⟧ The *general distributive* principle will be found to give *no new condition*; and we have seen cause to *reject* the *commutative* principle or property, as *not generally* holding good in the present inquiry. It remains, then, to try whether we can determine or *connect* the two coefficients, l and p, so as to satisfy the *associative* principle, or to verify the formula,

---

* The word "grammarithm" was subsequently proposed in a communication to the Royal Irish Academy (see the Proceedings of July, 1846), as one which *might* replace the word "quaternion," at least in the geometrical view of the subject: but it did not appear that there would be anything gained by the systematic adoption of this change of expression, although the mere *suggestion* of another *name*, as not inapplicable, seemed to throw a little additional light on the whole theory.

$$\alpha\,.\,\beta\gamma = \alpha\beta\,.\,\gamma.$$

For this purpose we may first *distribute* the factors β, γ into others, $\beta_1\beta_2\gamma_1\gamma_2\gamma_3$ which shall be parallel or perpendicular to it and to each other; and then shall have to satisfy, if possible, *six* conditions, which may be reduced to the six following:

$$\alpha\,.\,\alpha\alpha = \alpha\alpha\,.\,\alpha;\ \alpha\,.\,\alpha\alpha' = \alpha\alpha\,.\,\alpha';\ \alpha\,.\,\alpha\alpha'' = \alpha\alpha\,.\,\alpha'';$$

$$\alpha\,.\,\alpha'\alpha = \alpha\alpha'\,.\,\alpha;\ \alpha\,.\,\alpha'\alpha' = \alpha\alpha'\,.\,\alpha';\ \alpha\,.\,\alpha'\alpha'' = \alpha\alpha'\,.\,\alpha'';$$

α, α', α'' being three rectangular unit-lines, so placed that the rotation round α from α' to α'' is positive. Then, but what has been already found, the following relations will hold good:

$$\alpha\alpha = \alpha'\alpha' = \alpha''\alpha'' = l;\ \alpha\alpha' = -\ \alpha'\alpha = p\alpha'';$$

$$\alpha\alpha'' = -\ \alpha''\alpha = -p\alpha';\quad \alpha'\alpha'' = -\alpha''\alpha' = +p\alpha;$$

and the six conditions to be satisfied become,

$$\alpha\,.\,l = l\,.\,\alpha;\ \alpha\,.\,p\alpha'' = l\,.\,\alpha';\ \alpha\,.\,-p\alpha' = l\,.\,\alpha'';$$

$$\alpha\,.\,-p\alpha'' = p\alpha''\,.\,\alpha;\ \alpha\,.\,l = p\alpha''\,.\,\alpha';\ \alpha\,.\,p\alpha = p\alpha''\,.\,\alpha''.$$

Of these the first suggests to us to treat an arithmic factor as *commutative* (as regards *order*) with a grammic one, or to treat the product "line into number" as equivalent to "number into line"; the fourth and sixth conditions afford no new information; and the second, third, and fifth become,

$$-p^2\alpha' = l\alpha';\ -p^2\alpha'' = l\alpha'';\ -p^2\alpha = l\alpha.$$

The *conditions of association* are therefore all satisfied by our assuming, with the present signification of the symbols,

$$\alpha l = l\alpha, \text{ and } l = -\ p^2;$$

and they cannot be satisfied otherwise. The constant l is, therefore, by those conditions, necessarily *negative*; and EVERY LINE in tridimensional space has its SQUARE (on this plan) equal to a NEGATIVE NUMBER: which is one of the most novel but essential elements of the whole quaternion theory. (Compare the recent paragraph ⟦48⟧; also art. 85, pages 81, 82, of the Lectures.) And that a *grammarithm* ⟦55⟧ may properly be called a *quaternion*, appears from the consideration that the *line*, which in it is *added* to a *number*, depends itself upon a *system of three numbers*, or may be represented by a *trinomial expression*, because it is always the *sum of three lines* (actual or null), which are parallel to three fixed directions (compare Lecture III.). The coefficient p remains still undetermined, and may be made equal to positive one, by a suitable choice of the unit of length, and the direction of positive rotation. In this way we shall have finally the very simple values,

$$p = +\ 1,\quad l = -\ 1;$$

and the *rules* for the *multiplication of lines in space* will then become entirely *definite*, and will *agree* in all respects with the relations ‖48‖, between the symbols *ijk*.

‖57‖ Another train of *à priori* reasoning, by which I early sought to confirm, or (if it had been necessary) to correct, the results expressed by those new symbols, was stated to the R.I. Academy* in (substantially) the following way. Admitting, for directed and *coplanar* lines, the conception ‖36‖ of *proportion*; and retaining the symbols *ijk*, or more fully, $+i$, $+j$, $+k$, to denote three rectangular unit-lines as above, while the three respectively opposite lines may be denoted by $-i$, $-j$, $-k$; but *not assuming* the knowledge of any laws respecting their *multiplication*, I sought to determine *what ought to be considered as the* FOURTH PROPORTIONAL, u, *to the three rectangular directions*† j, i, k, *consistently with that known conception* ‖36‖ *for directions within the plane*, and with some *general and guiding principles*, respecting *ratios* and *proportions*. These latter assumed principles (of a *regulative* rather than a *constitutive* kind) were simply the following: 1st, that ratios similar to the *same* ratio must be regarded as similar to *each other*; 2nd, that the respectively *inverse* ratios are also mutually similar; and 3rd, that ratios are similar, if they be *similarly compounded* of similar ratios: this similarity of *composition* being understood to include generally a sameness of *order*. It seemed to me that any proposed definitional‡ use of the word RATIO, which should be inconsistent with these

* See the Proceedings of November 11th, 1844.

† In the abstract published in the Proceedings, the words "South, West, Up" were used at first instead of the symbols i, j, k; and the sought fourth proportional to *jik*, which is here denoted by u, was called, provisionally, "Forward".

‡ As an example of the use of the first of these very simple principles, in serving to *exclude a definition* which might for a moment appear plausible, let us take the construction ‖38‖, and inquire whether (as that construction would suggest), we can *properly say* that *four directions* (or four diverging unit-lines), α, β, γ, δ, form generally a *proportion in space*, when the angles $\widehat{\alpha\delta}$, $\widehat{\beta\gamma}$, between the extremes and means have one *common bisector* (ε). If so, when the three directions α, β, γ became *rectangular*, we should have $\alpha : \beta :: \gamma : -\alpha$, and $\gamma : -\alpha :: \beta : -\gamma$; but we should have also, $\alpha : \beta :: \beta : -\alpha$, and *not* $\alpha : \beta :: \beta : -\gamma$; so that the two ratios, $\alpha : \beta$ and $\beta : -\gamma$, would be said to be similar to one *common* ratio ($\gamma : -\alpha$), without being similar to *each other*, if the foregoing construction for a *fourth proportional* were to be, by definition, adopted: and this objection *alone* would be held by me to be *decisive* against the introduction of such a *definition*; and therefore also against the adoption of the connected *rule* mentioned in ‖38‖, as having at one time occurred to a friend (J.T.G.) and to myself, for the multiplication of lines in space, even if there were *no other reasons* (as in fact there are), for the rejection of that rule. A similar objection applies, with equal decisiveness, against the rule mentioned in ‖37‖, as an earlier conjecture of my own. On the other hand, an analogous and equally simple argument may serve to *justify* the notation D − C = B − A, employed by me in the following Lectures, and elsewhere, to express that the two right lines AB and CD are *equally long* and *similarly directed*, against an objection made some years ago, in a perfectly candid spirit, by an able writer in the Philosophical Magazine (for June, 1849, p. 410); who thought that interpretation *more arbitrary* than it had appeared to me to be; and suggested that the *same notation* might as well have been employed to signify this *other conception*:—that the two equally long lines, AB, CD *met somewhere*, at a finite or infinite distance. I could not admit this extension; for it would lead to the conclusion that two lines AB, EF might be *equal* to the same *third* line CD, without being equal to *each other*: which would (in my opinion) be so great a violation of analogy, as to render the use of the *word* "EQUAL",

principles, would depart thereby *too widely* from known *analogies*, mathematical and metaphysical, and would involve an impropriety of *language*; while, on the other hand, it appeared that if these principles were attended to, and other analogies observed, it was permitted to extend the use of that word *ratio*, and the connected phrase *proportion*, not only from *quantity* to *direction, within one plane*, as had been done ‖36‖ by other writers,* but also from the

---

or of the *sign* =, with the interpretation referred to, an embarrassment instead of an assistance. But I do not feel that analogies are thus violated, by the simultaneous admission of the *two contrasted proportions* (see (3) (4) (5) of ‖57‖),

$$u : i :: j : k, \quad u : j :: i : -k;$$

for the elementary theorem called often "alternando," (ἐναλλὰξ λόγος, Euc. V. Def.13, and Prop. 16) is by its nature limited (in its original meaning) to the CASE where the *means* which change places are *homogeneous* with each other: whereas two *rectangular directions*, as here i and j, are in this whole theory regarded as being in some sense *heterogeneous*. They have at least no relation to each other, which can be represented by any *ratio*, such as EUCLID considers, of *magnitude to magnitude*; and therefore we have no right to *expect*, from analogy to old results, that *alternation* shall *generally* be allowed in a *proportion* involving such directions: although, *within* the plane, alternation is *found* to be admissible.

* Since the note to paragraph ‖36‖, pp. (31) (32), was in type, I have had an opportunity of re-consulting the fourth volume of the Annales de Mathématiques, and have found my recollections (agreeing indeed in the main with the formerly cited page 228 of Dr. Peacock's admirable *Report*), respecting the admitted priority of Argand, confirmed. Français, indeed (in 1813), published in those Annales (Tome IV., pp. 61, .. 71) a paper which contained a theory of "proportion de grandeur et de position", with a connected theory of multiplication (and also of addition) of lines in a given plane; but he expressly and honourably stated at the same time (p. 70), that he owed the substance of those new ideas to another person ("le fond de ces idées nouvelles ne m' appartient pas"): and on being soon afterwards shewn, through Gergonne, whose conduct in the whole matter deserves praise, a copy of Argand's earlier and printed Essay (Paris, 1806), Français most fully and distinctly recognised (p. 225) that the true author of the method was Argand ("il n'y a pas le moindre doute qu'on ne doive à M. Argand la premiere idée de représenter géométriquement les quantités imaginaires"). Nothing more lucid than Argand's own statements (see the same volume, pp. 136, 137, 138), as regards the *fundamental principles* of the theory of the *addition* and *multiplication* of coplanar lines, has since (so far as I know) appeared; not even in the writings of Professor De Morgan on Double Algebra, referred to in former notes. But Argand had not anticipated De Morgan's theory of Logometers; and was on the contrary disposed (pp. 144, .. 146) to regard the symbol $\sqrt{-1}^{\sqrt{-1}}$, notwithstanding Euler's well-known result, as denoting a *line* (KP), *perpendicular to the plane* of the lines 1 and $\sqrt{-1}$: and to consider it as offering an example of a quantity which was *irreducible to the form* $p + q\sqrt{-1}$, and was (according to him) as *heterogeneous* with respect to $\sqrt{-1}$, as the latter with respect to + 1 ("aussi hétérogéne" &c.). The word modulus ("module"), so well known by the important writings of M. Cauchy, occurs in a later paper by Argand, in the following volume of the Annales, as denoting the real quantity $\sqrt{p^2 + q^2}$. If I have seemed to dwell too much on the speculations of Argand (not all adopted by myself), it as been partly because (so far as I have observed) his merits as an original inventor have not yet been sufficiently recognised by mathematicians in these countries: and partly because *one of the two most essential links* (the other being *addition*) between Double Algebra and Quaternions, is ARGAND's main and *fundamental principle* respecting COPLANAR PROPORTION, expressed by him as follows (Annales, T. IV., pp. 136, 137):—"Si (fig. 2) *Ang.* AKB = *Ang.* A′K′B′, on a, abstraction faite des grandeurs absolues, KA: KB :: K′A′ : K′B′. C'est là le principe fondamental de la theorie dont nous avons essayé de poser les premières bases, dans l'écrit dont nous donnons ici un extrait" (namely, Argand's printed Essay of 1806, exhibited by Gergonne to Français, after the appearance of the first paper of the latter author on the subject in 1813). Argand continued thus (in p. 137): "Ce principe n'a rien au fond de plus étrange que celui sur lequel est fondée la conception du rapport géometrique entre deux lignes de signes différents, et il n'en est proprement qu' une généralisation": a remark in which I perfectly concur.

plane to *space*.* The supposed proportion,

$$j : i :: k : u, \tag{1}$$

gave thus, by inversion,

$$u : k :: i : j; \tag{2}$$

but also, in the planes of *ij*, *ik*, there were the two proportions,

$$i : j :: j : -i, \text{ and } k : i :: -i : k; \tag{3}$$

compounding therefore, on the one hand, the two ratios, $u{:}k$ and $k{:}i$, and, on the other hand, the two respectively similar ratios, $j : -i$, and $-i{:}k$, there resulted the new proportion,

$$u : i :: j : k; \tag{4}$$

which differed from the proportion (2) only by a *cyclical transposition* of the three directions *ijk*. For the same reason, we may make another cyclical change of the same sort, and may write

$$u : j :: k : i; \tag{5}$$

while, in this *cycle* of three rectangular directions, *ijk*, the *right-handed* (or left-handed) *character* of the *rotation*, round the first from the second to the third, is easily seen to be unaffected by such a transposition. Again compounding the two similar ratios (1) with these two others, which are evidently similar, whatever the unknown direction *u* may be,

---

* Although the observations in par. [57] relate rather to *proportions* than to *imaginaries*, yet the present may be a convenient occasion for remarking that Buée, and even Wallis, had speculated, before Argand and Français, on interpretations of the symbol $\sqrt{-1}$, which should extend to *space*: but that the *nearest approach* to an *anticipation of the quaternions*, or at least to an *anticipation of tiplets*, seems to me to have been made by Servois, in a passage of the lately cited volume of Gergonne's Annales, which appears curious and appropriate enough to be extracted here. Servois had been following up a hint of Gergonne, respecting the representation of ordinary imaginaries of the form $x + y\sqrt{-1}$ (x and y being whole numbers), by a *table of double argument* (p. 71); and thought (p. 235) that *such* a table might be regarded as only a *slice* (une tranche) of a table of TRIPLE argument, for representing *points* (*or lines*) *in* SPACE. He thus continued:—"Vous donneriez sans doute à chacune terme la forme *trinomiale*; mais quel coefficient aurait le troisième terme? Je ne le vois pas trop. L'analogie semblerait exiger que le trinôme fût de la forme, $p \cos\alpha + q \cos\beta + r \cos\gamma$, α, β, γ étant les angles d'une droite avec trois axes rectangulaires; et qu' on eût

$$(p\cos\alpha + q\cos\beta + r\cos\gamma)\,(p'\cos\alpha + q'\cos\beta + r'\cos\gamma) = \cos^2\alpha + \cos^2\beta + \cos^2\gamma = 1.$$

Les valeurs de p, q, r, p', q', r' qui satisferaient à cette condition seraient *absurdes*" ("quantités non-réelles," as he shortly afterwards calls them): "mais seraient-elles *imaginaires* réductibles à la forme génerale $A + B\sqrt{-1}$? Voila une question d'analise fort singulière, que je soumets à vos lumières." The six NON-REALS which Servois thus with remarkable sagacity *foresaw*, without being able to *determine* them, may now be identified with the then unknown symbols $+i$, $+j$, $+k$, $-i$, $-j$, $-k$, of the quaternion theory: at least, these latter symbols fulfil precisely the *condition* proposed by him, and furnish an *answer* to his "singular question". It may be proper to state that my own theory had been constructed and published for a long time, before the lately cited passage happened to meet my eye.

$$i : -i :: u : -u, \tag{6}$$

we find this other proportion,

$$j : -i :: k : -u; \tag{7}$$

and therefore, by (2) and (3),

$$u : k :: k : -u. \tag{8}$$

In like manner,

$$u : i :: i : -u, \text{ and } u : j :: j : -u; \tag{9}$$

and in any one of these proportions, any two terms, whether belonging to the same or to different ratios, may have their *signs* changed together. All these proportions, (2) . . (9), follow from the original supposition (1), by the general principles above stated, without the direction u being as yet any otherwise determined.

⟦58⟧ Suppose now that the two rectangular directions j and k are made to *turn together*, in their own plane, round i as an axis, till they take two new positions j_1 and k_1, which will therefore satisfy the proportion,

$$j : k :: j_1 : k_1. \tag{10}$$

We shall then have, by (4),

$$u : i :: j_1 : k_1; \tag{11}$$

and therefore, by a cyclical change of these three new rectangular directions,

$$u : j_1 :: k_1 : i :: l : i_1, \tag{12}$$

if l and i_1 be obtained from k_1 and i by any common rotation round j_1. Another cyclical change, combined with a rotation round the new line l, gives finally,

$$u : l :: i_1 : j_1 :: m : n; \tag{13}$$

where l, m, n may represent *any three rectangular directions whatever*, subject only to the condition that the *rotation* round l from m to n shall be of the *same character* as that round i from j to k. With this *condition*, therefore, the first assumed proportion (1) may be replaced by this *more general* one:

$$n : m :: l : u; \tag{14}$$

while for (8) and (9) may now be written, with the same signification of the symbols

$$u : l :: l : -u; \; u : m :: m : -u; \; u : n :: n : -u; \tag{15}$$

and because $n : m :: m : -n$, we have these other and not less general proportions,

$$m : -n :: l : u; \; m : n :: l : -u. \tag{16}$$

If, then, there be *any* such fourth proportional, u as has been above supposed, to the three *given* rectangular directions j, i, k, the *same* direction u, or the *opposite* direction $-u$, will also be, in the same sense, the fourth proportional to *any other three* rectangular directions, n, m, l, or m, n, l, according as the character of a certain rotation is *preserved* or *reversed*.

⟦59⟧ This remarkable result appeared to me to justify the regarding the directions here called $+u$ and $-u$ rather as *numerical* (or algebraical) than as *linear* (or geometrical) *units*; and to make it proper to denote them simply by the symbols $+1$ and -1; because their directions were seen to admit only of a certain *contrast* between themselves, but not of any *other* change: all that *geometrical variety*, which results from the conception of *tridimensional space*, having been found to *disappear*, as regarded them, in an investigation conducted as above. And in fact it is *not permitted*, on the foregoing principles, to *identify* the direction u with that of *any line* (l) *whatever*: for in that case the proportion (13) would give the result $l:l::m:n$, which must be regarded in this theory as an *absurd* one, the two terms of one ratio being *coincident* directions, while those of the other ratio are *rectangular*. But there is no objection of *this* sort against our supposing, as above, that

$$+u = +1, \quad -u = -1; \tag{17}$$

and *then* the *proportions*, derived from (13), (15),

$$1:l::m:n::n:-m; \quad 1:l::l:-1, \tag{18}$$

may be conveniently and concisely *expressed* by formulæ of *multiplication*, as follows:

$$lm = n; \quad ln = -m; \quad l^2 = -1. \tag{19}$$

⟦60⟧ In this way, then, or in one not essentially different, the fundamental formulæ ⟦48⟧ of the calculus of quaternions, as first exhibited to the R.I.A. in 1843, namely, the equations,

$$i^2 = -1, \; j^2 = -1, \; k^2 = -1, \tag{A}$$

$$ij = +k, \; jk = +i, \; ki = +j, \tag{B}$$

$$ji = -k, \; kj = -i, \; ik = -j, \tag{C}$$

were shewn (in 1844) to be consistent with *à priori* principles, and with considerations of a general nature; a *product* being *here* regarded as a FOURTH PROPORTIONAL, to a certain *extra-spatial** *unit*, and to *two directed factor-*

* It seemed (and still seems) to me natural to connect this *extra-spatial unit* with the conception ⟦3⟧ of TIME, regarded here merely as an *axis of continuous and uni-dimensional progression*. But whether we thus *consider jointly time and space*, or conceive generally *any system of four independent axes*, or scales of progression (u, i, j, k), I am disposed to infer from the above investigation the following LAW OF THE FOUR SCALES, as one which is at least consistent with analogy, and admissible as a *definitional extension* of the fundamental equations of quaternions:—"A formula

lines in space: whereas, in the investigation of paragraphs ⟦50⟧ to ⟦56⟧, it was viewed rather as a certain FUNCTION of those two factors, the *form* of which function was to be determined in the manner most consistent with some general and guiding analogies, and with the conception of the *symmetry of space*. But there was still *another view* of the whole subject, sketched not long afterwards in another communication to the R.I. Academy,[†] on which it is unnecessary to say more than a few words in this place, because it is, in substance, the view adopted in the following Lectures, and developed with some fulness in them: namely, that view according to which a QUATERNION is considered as the QUOTIENT of two directed lines in tridimensional space.

⟦61⟧ Of such a *geometrical quotient*,* b ÷ a, the fundamental property is in this theory conceived to be, that by *operating*, as a *multiplier* (or at least in a way *analogous* to multiplication), on the *divisor-line*, a, it *produces* (or generates) the *dividend-line*, b; and that thus it may be interpreted as satisfying the general and identical formula (compare ⟦9⟧):

$$(\mathrm{b} \div \mathrm{a}) \times \mathrm{a} = \mathrm{b}.$$

The *analogy to multiplication* consists partly in the operation being one which is performed at once on *length* and on *direction*, as in the ordinary multiplication of a line by a positive or negative number; or as is done in that known *generalization* ⟦36⟧ of such multiplication, for lines within one plane, which (for reasons assigned in notes to former paragraphs) ought (I think) to be called the *Method of Argand*: and partly in the circumstance that the new operation possesses, like that older one (from which, however, it is entirely *distinct*,[††] in many other and important respects), the *distributive* and *associative*,[‡] though

of *proportion between four independent and directed units* is to be considered as remaining true, when *any two* of them *change places* with each other (in the formula), provided that the *direction* (or *sign*) of *one* be *reversed*." Whatever may be thought of these abstract and semi-metaphysical *views*, the *formulæ* (A) (B) (C) of par. ⟦60⟧ are in any event a sufficient *basis* for the erection of a CALCULUS of quaternions.

[†] See the Proceedings of Feb. 10th, 1845.

* This view of a *geometrical quotient* was also developed to a certain extent, in an unfinished series of papers, which appeared a few years ago in the Cambridge and Dublin Mathematical Journal, under the head of *Symbolical Geometry*: a title adoped to mark that I had attempted, in the composition of that particular series, to allow a more prominent influence to the general *laws of symbolical language* than in some former papers of mine; and that to this extent I had on that occasion sought to imitate the *Symbolical Algebra* of Dr. Peacock, and to profit also by some of the remarks of Gregory and Ohm.

[††] Among these *distinctions* of method, it is important to bear in mind that *no one line* is taken, in my system, as representing the *direction* of *positive unity*: and that, on the contrary, *every vector-unit* is regarded as *one of the square roots of negative unity*. It is to be remarked, also, that the *product* of two inclined but non-rectangular vectors is considered in this theory as *not a line*, but a *quaternion*: all which will be found fully illustrated in the Lectures.

[‡] To this *associative* principle, or property of multiplication, I attach much importance, and have taken pains to shew, in the Fifth and Sixth Lectures, that it can be *geometrically proved* for quaternions, *independently* of the *distributive* principle, which may, however, in a different arrangement of the subject, be made to *precede* and *assist* the proof of the associative property, as shewn in the Seventh Lecture, and elsewhere. The *absence* of the associative princple appears to me to be an

not like it (generally) the *commutative* properties, of what is called *multiplication in algebra*;* at least when a few definitional formulæ (resembling those in par. ⟦9⟧) are established. And the *motive* (in this view) for calling such a *quotient* a QUATERNION, or the ground for connecting its conception with the NUMBER FOUR, is derived from the consideration that while the RELATIVE LENGTH of the two lines compared depends only on *one number*, expressing their RATIO (of the ordinary kind), their RELATIVE DIRECTION depends on a *system of three numbers*: *one* denoting the ANGLE (a $\wedge$ b) between the two lines, and the *two others* serving to determine the *aspect* of the PLANE of that angle, or the *direction* of the AXIS of the positive *rotation* in that plane, *from* the divisor-line (a) *to* the dividend-line (b).

⟦62⟧ For the unfolding of this general view,* and the deduction from it of many geometrical† and of some physical‡ consequences, I must refer to the

---

inconvenience in the *octaves* or octonomials of Messrs. J.T. Graves and Arthur Cayley (see Appendix B, p. 730): thus in the notation of the former we should indeed have, as in quaternions, $ij = k$, but *not generally* $i.j\omega = k\omega$, if ω represent an octave; for $i.jl = in = -o = -kl = -ij.l$.

* The expression "algebra", or "ordinary algebra", occurs several times in these Lectures, as denoting merely *that usual species of algebra*, in which the equation $ab = ba$ is treated as universally true, and not (of course) as implying any degree of disrespect to those many and eminent writers, who have not hitherto chosen to admit into their calculations such equations as $\alpha\beta = -\beta\alpha$, for the multiplication of two rectangulr lines, or for other and more abstract purposes. It is proper to state here, that a species of *non-commutative multiplication* for inclined lines (äussere Multiplikation) occurs in a very original and remarkable work by Prof. H. Grassmann (Ausdehnungslehre, Leipzig, 1844), which I did not meet with till after years had elapsed from the invention and communication of the quaternions: in which work I have also noticed (when too late to acknowledge it elsewhere) an employment of the symbol $\beta - \alpha$, to denote the *directed line* (Strecke), drawn from the point α to the point β. Nothwithstanding these, and perhaps some other coincidences of view, Prof. Grassmann's system and mine appear to be perfectly distinct and independent of each other, in their conceptions, methods, and results. At least, that the profound and philosophical author of the Ausdehnungslehre was not, at the time of its publication, in possession of the theory of the *quaternions*, which had in the preceding year (1843) been applied by me as a sort of organ or *calculus for spherical trigonometry*, seems clear from a passage of his Preface (Vorrede, p. xiv.), in which he states (under date of June 28th, 1844), that he had not then succeeded in *extending the use of imaginaries from the plane to space*; and generally that unsurmounted difficulties had opposed themselves to his attempts to construct, on his principles, a theory of *angles in space* (hingegen ist es nicht mehr möglich, vermittelst des Imaginären auch die Gesetze für den Raum abzuleiten. Auch stellen sich überhaupt der Betrachtung der Winkel im Raume Schwierigkeiten entgegen, zu deren allseitiger Lösung mir noch nicht hinreichende Musse geworden ist). The earlier treatise by Prof. A.F. Möbius (der barycentrische Calcul, Leipzig, 1827 ⟦*Möbius 1827*⟧), referred to in the same Preface by Grassmann, appears to be a work which likewise well deserves attention, for its conceptions, notations, and results; as does also another work of Möbius (Mechanik des Himmels, Leipzig, 1843 ⟦*Möbius 1843*⟧), elsewhere referred to in these Lectures (page 614).

* I may just hint here that the BIQUATERNIONS of Lect. VII. admit of being *geometrically interpreted* (comp. note to ⟦19⟧), by considering each as a *couple of quotients* $\left(\frac{\beta}{\alpha}, \frac{\gamma}{\alpha}\right)$ *constructed* by a TRIRADIAL (α, β, γ), and *multiplied* by a *commutative* factor of the form $\sqrt{-1}$ (compare ⟦16⟧), when the *line-couple* (β, γ) is changed to $(-\gamma, \beta)$, or when the *angle* $\widehat{\beta\gamma}$ is changed to an *adjacent angle*.

† Notwithstanding some references to works of M. Chasles, and other eminent foreign geometers, my acquaintance with their writings is far too imperfect to give me any confidence in the *novelty*

following *Lectures*; of which a considerable part has been drawn up in a more popular[§] style than this Preface: while the whole has been composed under the influence of a sincere desire to render the exposition of the subject as clear and elementary as possible. The prefixed *Table of Contents* (pp. ix. to lxxii.), though somewhat fuller than usual, will be found useful (it is hoped) not merely as an analytical *Index*, assisting a reader to *refer* easily to any part of the volume which he has once carefully read, but also as a general *abridgment* of the work, and in some places as a *commentary*.[‖] The *Diagrams* are numerous, and have been engraved* with care from my drawings: some of them may perhaps be thought to have been unnecessary, but it appeared better to err, if at all, on the side of clearness and fulness of illustration, especially in the early parts of a work based on a new mathematical conception, and designed to furnish, to those who may be disposed to employ it, a new mathematical organ. Whatever may be thought of the degree of success with which my exertions in this matter have been attended, it will be felt, at least, that they must have been arduous and persevering. My thanks are due, at this last stage, to the friends who have cheered me throughout by their continued sympathy; to the scientific contemporaries[†] who have at moments turned aside from their own original researches, to notice, and in some instances to extend, results or speculations of mine; to

---

of various theorems in the VIIth Lecture and Appendix (such as those respecting generations of the ellipsoid, and inscriptions of gauche polygons in surfaces of the second order), beyond what is derived from the opinion of a few geometrical friends.

[†] Some such *physical* applications were early suggested by Sir. J. Herschel.

[§] It had been designed that these Lectures should not go much more into detail than those which have been actually delivered on the subject by me, in successive years, in the Halls of this University; and the First Lecture, printed in 1848 (as the astronomical allusions at its commencement may indicate), was in fact delivered in that year, in very nearly the form in which it now appears. But it was soon found necessary to extend the plan of the composition: and it is evident that the subsequent Lectures, as printed, are too long, and that the last of them involves too much calculation, to have been delivered in their present form: thought something of the style of actual lecturing has been here and there retained. The real *divisions* of the work are not so much the *Lectures* themselves, as the shorter and more numerous *Articles*, to which accordingly the *references* have been chiefly made. An intermediate form of subdivision into *Sections* has however been used in drawing up the *Contents*, which the reader may adopt or not at his discretion, marking or leaving unmarked the margin of the Lectures accordingly. Some new terms and symbols have been unavoidably introduced into the work, but it is hoped that they will not be found embarrassing, or difficult to remember and apply.

[‖] For instance, as regards the formation of the Adeuteric Function (p. xliii.)

* By Mr. W. Oldham, whose fidelity and diligence are hereby acknowledged.

[†] In these countries, Messrs. Boole, Carmichael, Cayley, Cockle, De Morgan, Donkin, Charles and John Graves, Kirkman, O'Brien, Spottiswoode, Young, and perhaps others: some of whose researches or remarks on subjects connected with quaternions (such as the *triplets*, *tessarines*, *octaves*, and *pluquaternions*) have been elsewhere alluded to, but of which I much regret the impossibility of giving here a fuller account. As regards the theory of *algebraic keys* (clefs algébriques), lately proposed by one of the most eminent of continental analysts, as one that *includes* the quaternions (Comptes Rendus for Jan. 10, 1853, p. 75), it appears to me to be virtually *included* in that theory of SETS in algebra (explained in the present Praface), which was announced by me in 1835, and published in 1848 (Trans. R.I.A., Vol. XXI., Part II., p. 229, &c., the symbols $\times_r$ being in fact what M. Cauchy calls KEYS), as an *extension* of the theory of *couples* (and therefore also of imaginaries): of which SETS I have always considered the QUATERNIONS (in their *symbolical* aspect)

my academical superiors who have sanctioned, as a subject of public and repeated examination in this University, the theory to which this Volume relates, and have contributed to lighten, to an important extent, the pecuniary risk of its publication: but, above all, to that Great Being, who has graciously spared to me such a measure of health and energy as was required for bringing to a close this long and laborious undertaking.

WILLIAM ROWAN HAMILTON.

Observatory of T.C.D., June, 1853.

C. *FROM THE* CORRESPONDENCE OF HAMILTON WITH DE MORGAN (*GRAVES 1882–9*, Vol. 3)

For many years, Hamilton exchanged letters with Augustus De Morgan; selections from this scientific correspondence are reprinted in R.P. Graves's biography of Hamilton. The following excerpt, in which Hamilton and De Morgan puzzle over infinitesimals and the foundations of the calculus, provides a representative sample of the correspondence—of its intellectual flavour, and of the range of interests of Hamilton and De Morgan. The excerpt also shows the extent to which problems that arose in the days of Newton and Berkeley continued to bemuse mathematicians as late as the 1860s. The excerpt can usefully be compared with other selections dealing with the infinite in mathematics, notably with *Berkeley 1734*, *D'Alembert 1765a*, *Bolzano 1851*; with the early correspondence between Cantor and Dedekind; and with *Cantor 1883a*. The insertions in single brackets are those of R.P. Graves; references should be to the dates of the letters.

From Sir W.R. Hamilton to A. De Morgan

OBSERVATORY, *March* 8, 1862.

[After reporting progress made in the printing of the *Elements*, and having mentioned that the second part of the Third Book would treat of the Differential Calculus of Quaternions, he proceeds]:—As regards the Differential Calculus generally, I have lately been led to read again with care a good many works, your own included, at least in the parts bearing on *first principles*. When

---

to be merely a *particular* CASE. Before the publication of those *sets*, the closely connected conception of an "*algebra of the* n^{th} *character*" had occurred to Prof. De Morgan in 1844, avowedly as a suggestion from the quaternions. (Trans. Camb. Phil. Soc., Vol. VIII, Part III.)

your letter arrived this morning, I was deep in Berkeley's 'Defence of Freethinking in Mathematics'; the volume of his works, containing that Defence, &c., having just turned up. I think there is more than mere plausibility in the Bishop's criticisms on the remarks attached to the Second Lemma of the Second Book of the Principia; and that it is very difficult to understand the *logic* by which Newton proposes to prove, that the *momentum* (as he calls it) of the *rectangle* (or product) AB is equal to $aB + bA$, if the *momenta* of the *sides* (or factors) A and B be denoted by a and b. His mode of getting rid of ab appeared to me long ago (I must confess it) to involve so much of *artifice*, as to deserve to be called *sophistical*; although I should not like to say so publicly. He subtracts, you know $(A - \frac{1}{2}a)\ (B - \frac{1}{2}b)$ from $(A + \frac{1}{2}a)\ (B + \frac{1}{2}b)$; whereby, of course, ab disappears in the result. But *by what right*, or for *what reason*, other than to give an unreal air of *simplicity* to the calculation, does he *prepare* the *products* thus? Might it not be argued similarly that the difference,

$$(A - \tfrac{1}{2}a)^3 + (A - \tfrac{1}{2}a)^3 = 3aA^2 + \tfrac{1}{4}a^3,$$

was the moment of A^3; And is it not a sufficient *indication* that the mode of procedure adopted is *not* the *fit* one for the subject, that it quite *masks* the notion of a *limit*; or rather has the appearance of treating that notion as foreign and irrelevant, notwithstanding all that had been said so well before, in the First Section of the First Book?

Newton does not seem to have cared for being very consistent in his *philosophy*, if he could anyway get hold of *truth*, or what he considered to be such. In relation to *light*, he appears to have admitted *both rays and waves*, the latter being *excited* by the former. And I think that there is a real difficulty in reconciling the two parts of the Principia, above referred to. Indeed I have recently come to think that the '*momenta*,' or '*incrementa vel decrementa momentanea*,' of the Second Book (page 243 of the Third Edition), are just *infinitesimals in disguise*; although Sir Isaac did not like (it may be guessed) to employ the *word there*, after having so successfully avoided introducing the *thing*, or *thought*, in the First Section of the First Book. Even the '*motus*, *mutationes et fluxiones quantitatum*,' in page 244 of the edition cited, were (I think) *infinitesimal quantities*, *as there used*; although—or partly *because*—the great author *adds*, *immediately afterwards*, '*vel finitæ quævis quantitates velocitatibus hisce proportionales*'. Yet I have quoted this very passage, in the manuscript of my concluding chapter, as at least *suggesting finite fluxions*; and have taken care to express my obligation, for other suggestions, to that splendid First Section, above spoken of; especially for the suggestion of *magnified representations* of figures, the parts of which all *tend* to *vanish together*.

For my part, I stick to the *finite quantities*, suggested in that page 244, as (in some sense) *representing the velocities*, or as being *proportional* to them. My library is poor in fluxional books, but I notice that Thomas Simpson treats *fluxions* as *finite*. (I have only a rather late edition, London, 1823.) Thomas Simpson's conceptions appear to have been very clear and distinct, and I do not venture to *say* that the geometrical investigation which he gives, of the *fluxion of a rectangle*, avowedly supplied to him by a young but unnamed friend, is *insufficient* in itself, but it fails to *convince me*, perhaps because I was not early *accustomed* to fluxions. Certainly there is no *neglecting* of *ab*, or $\dot{x}\dot{y}$, as *small*; for in fact that *rectangle of the fluxions* is not *represented at all* in his Figure, which is here annexed [diagram inserted].... He conceives the *varying rectangle xy* to be the *sum of two mixtilinear triangles*, of which the *two separate fluxions are* $y\dot{x}$ and $x\dot{y}$. This is very ingenious, but I do not feel sure to what degree I could *rely* on it and build upon it any superstructure, if I were now coming, for the first time, as a *learner*, to the subject. However, I suppose that a pupil, if reasonably modest, or even prudent, will take, *for a while*, his teacher's statements upon *trust*; reserving to himself to *return* upon them, and to examine closely their truth and logic when he shall have acquired some degree of familiarity with the subject taught.

In a subsequent letter I shall perhaps be tempted to give you *my own proof*, resting on my own *definition* of *simultaneous differentials*, or *corresponding fluxions*, not only that the fluxion of a rectangle is as above, but *generally* that if $z = yx$ then $\dot{z} = y\dot{x} + \dot{y}x$, or $dz = y.dx + dy.x$, *exactly*; *even if xyz, and* $\dot{x}\dot{y}\dot{z}$, or *dx*, *dy*, *dz*, *be six finite quaternions*.

From A. De Morgan to Sir W.R. Hamilton

41, Chalcot Villas, Adelaide Road, N.W.,
March 11, 1862.

... Newton was *shuffling*. He had (1712) to back out of infinitesimals, in order to unleibnitize his system. See accompanying tract, which read while the subject is on your mind.

For myself, I am now fixed in the faith of the *subjective reality* of *infinitesimal quantity*. I intend to write on this subject when logic is off my mind, if that time should ever come. But *what* an infinitely small quantity is, I know no more than I know what a *straight line* is; but I know it *is*; and there I stop short. But I do not believe in *objectively realised* infinitesimals.

Newton's $A + \frac{1}{2}a$, &c., always reminded me of an observer I once knew, who took to differential calculus. He was puzzled, ϕt being the velocity at time t, to get a velocity with which to go cleverly through dt. After much meditation on ϕt at the beginning, and $\phi(t + dt)$ at the end, he had recourse to his old tools, and *took the mean*. This made his mind quite easy....

I hope Lady H. and all the young people are well. Has Miss Helen forgiven my algebra yet? It was not so atrocious as the trick I played a young lady not many days since. She wanted my autograph, and I sent her this to stick in her book

To Mr. *without* *Co.*

Pay *Miss* ______ ______, *or Order,*

Particular Attention.

A. De Morgan.

She said she would not put it in the book, but would keep it to present when she knew how to fill it up.

From Sir W.R. Hamilton to A. De Morgan

OBSERVATORY, *March* 14, 1862.

. . . I enclose a *leaf* of my new book (pp. 391, 392), which is not yet signed for press. . . . You will see that I have entered on the printing of my remarks on the Differential Calculus of Quaternions; and have touched, in passing, on Fluxions and on Newton. (I read your pamphlet yesterday.)

I wish that you would tell me whether you think that there is anything actually *wrong* in what I have said—though I dare say *you* would not have said it. At all events it is *true* subjectively, or *for myself*, since it has certainly been a study of the Principia, without much attempt to understand the *momenta*, which has gradually led me to adopt *Finite Differentials*, as in my *Lectures* I already did. I know indeed that Cauchy came to treat dx and dy as finite, in at least some of his writings; but Todhunter (whom, notwithstanding, I admire) has evidently a holy horror of treating them as in any way *separate* quantities. Hutton and Saunderson seem to be quite content to get at results, by neglecting some fluxions as indefinitely small; for example, in the fluxion of a rectangle; but T. Simpson is far from being so *content*, whether he is always *clear to me* or not. Thus I find that at page 150, of the edition cited in a former letter, he says:—'but hitherto no particular notice has been taken of *the method of increments*, or *indefinitely little parts*, used (and mistaken) by *many* for *that of fluxions*, &c.,' the *italics* being copied. About Berkeley, I am not unlikely to write again.

From A. De Morgan to Sir W.R. Hamilton

41, CHALCOT VILLAS, ADELAIDE ROAD, N.W.,
March 15, 1862.

The word "*ultimate*" [*Elements*, p. 392] sets you right as to Newton personally. I am familiar with *finite differentials*. I think Lacroix, *inter alios* has h his differential of x, and the *second term* of $\phi(x + h)$ the differential of ϕx. How he hocus-pocuses it into $\dfrac{d\phi x}{dx} h$ I forget. But it is only an evasion of the *infinitesimal question.* If you admit finite quantities which are, by a *finito-facient* infinite multiplier, as the *ultima* or *nascentia*, you admit the *ultimate ratio* to have been arrived at *before* vanishing of terms—unless indeed you bring the pure 0 into finitude by a multiplier. It is quite true, no doubt, that the *limit* of $\dfrac{m \times \Delta y}{m \times \Delta x}$ is $\dfrac{WV}{PV}$ [referring to a diagram]. And if, with easier conscience, you get at $\phi' x$ by the limit of $\dfrac{m\Delta y}{m\Delta x}$ than by that of $\dfrac{\Delta y}{\Delta x}$, why, no doubt, you have a right to do it. But you may just as well take

$$\frac{k}{h} = \text{Limit of} \frac{\Delta y}{\Delta x}$$

at once; and then call h and k differentials of x and y—which they *are not.*

But you have for you the authority of Legendre, who, when he has detected *in*commensurables, says, 'But if there be no common measure, we may imagine a quantity which *serves them as a common measure (qui leur sert de mesure commune).* This will do in mathematics; it would not be a culinary process. Next time your cook leaves, *imagine* a person who serves you as cook, and see if your meat be not too much underdone.

Take what plan you like, there is a road—a short cut—to limits, which sets you right.

For modes of nought let graceless zealots fight;
He can't be wrong whose *limits* are but right.

From Sir W.R. Hamilton to A. De Morgan

OBSERVATORY, *March* 18, 1862.

I should have some things to object to in your last note, but am in a mood for giving *you* some new opportunities of criticism, by applying my own *Definition* of Simultaneous Differentials to an Example. But first let me protest that I conceive that I have a right, without becoming *ipso facto* absurd, to *propose a definition for differentials*, provided that I keep within bounds of *convenience* and of *analogy*, and do not violate received and useful *rules* of calculation. When you say, 'But you may just as well take

$$\frac{k}{h} = \text{Limit of} \frac{\Delta y}{\Delta x}$$

at once; and then call h and k *differentials* of x and y—which they *are not*': I ask *on what grounds* do you *pronounce* so confidently that they are *not* differentials? Is it on the authority of Leibnitz? or from the etymology of the word? or do you appeal to general consent? The latter does not exist; and the former appear to me insufficient. I always thought it a trifling objection, which Lagrange made to the modern use of the word *limit*, in the beginning of the

Calcul des Fonctions, that the *ancients* had conceived a limit to be a thing which could only be *approached to*, but *never reached.* Let me, however, guard myself, so far as to protest that I am far from seeking to exclude the notion of *infinity* from mathematics; I only say that *my* differentials—and I am not singular therein—are *not* infinitesimals. I *use infinites*, great *or* small, in *getting at them*; but *they* are *finite.* As a little anecdote I may mention that I was asked, two or three years ago, by a clerical friend, to give him my opinion on a theological work of Dr. Mansel, especially as regarded the impossibility of a *Philosophy of the Infinite.* I replied, in substance, that 'Philosophy' was a high and sacred word—

> 'How charming is divine Philosophy!' &c.—

but that I was sure that there existed a *Science of the Infinite*, to wit, *Mathematical Science.*

Let me add, however, that *I* don't *adopt* the *definition*, that when $\frac{k}{h} =$ limit of $\frac{\Delta y}{\Delta x}$, then h and k are *differentials* of x and y; not that it does not seem to me quite competent to an elementary writer to do so, *if he pleases*, but because it *won't work* in *quaternions. I* must manage as well as I can, *without* assuming that $xy = yx$, or $xh = hx$, &c. Let then $y = x^2$, and let $\Delta x = h$: *my* calculus gives only, $\Delta y = xh + hx + h^2$, and *not*, generally, $= 2xh + h^2$. Consequently, my *quotient* of *differences* is

$$\frac{\Delta y}{\Delta x} = x + hxh^{-1} + h,$$

in which it is *by no means permitted* to change the *term*, hxh^{-1} to x, *in general.* Yet I *define* that the *part*, $xh + hx$, of the *difference* Δy, is the *differential* of y; and I *denote* it by dy, if h be denoted by dx; and this whether h be *small* or *large*; that is, in quaternions, whether what I call its *tensor* (answering to Argand's *modulus*) be a small or a large positive quantity. . . .

Your *opinion*, on any such subject, must have real *weight* with me, as with others; but I do not conceive that there exists *any authority*, by which the meaning of the *word 'differential'*, has yet been *so fixed*, as to make it *improper* in a new writer to propose a *new*, or at least a *modified*, meaning of it. It has been, perhaps happily, too much a subject of dispute for *that.* Even the word *fluxion* has not always been used in one sense; and whatever Newton meant by it (which seems to me doubtful) in that page 244 of the Principia, some English writers have certainly treated fluxions as *infinitesimals*, and justified the omission of $\dot{x}^2$, in the fluxion of the square, on the ground of its *insignificance*. . . .

I intended to *admit*, in a former letter, that Cauchy and *several* other French authors have considered that *differentials might be finite*, although apparently *preferring* to deal with *differential coefficients*, of which you see that I can make no use. . . .

From A. De Morgan to Sir W.R. Hamilton

41, CHALCOT VILLAS, ADELAIDE ROAD, N.W.,
March 23, 1862.

All your mathematics—all your use of a finite fraction $\frac{dy}{dx}$ — I can readily agree to. The sole question is the expediency of using the name *differential* for infinite multiple of a differential.

You ask on what grounds I say that h and k are *not* differentials. Simply, because hitherto they *are not*. The true question is, on what grounds do I dispute the expediency of *your making* them take the name. Is it the authority of Leibnitz? No. The authority of an inventor extends no further than this—his terms have a right to respectful consideration; but the community, *quem penes arbitrium*, finally accepts or rejects. Mathematical language has a fermentation always going on which throws up all that does not assimilate. My belief is, that if you call h and k differentials, the community, when quaternions become *publici juris*, will *uncall* them. Is it etymology? So far as this. Etymology should not be offended against without good reason shown. And especially, when the proposed change is other than *extension*. Is it general consent? Yes. You affirm that no such consent exists. If by general consent you mean logically universal consent, I agree. No one term has that consent. But I believe, as a matter of fact, that there is a vast preponderance of writers who mean by a differential—when they use the word apart from 'coefficient'—an infinitely small increment. This is a question of fact. If your reading denies the fact, I have nothing to say—mine affirms it.

But are you forced upon the word 'Differential'? They are—I mean h and k—the terms of Newton's *prime ratio*. I would rather use *primo-rationals* than differentials. But I dare say you may get a better word.

These terms need a shaking very much. I should like to call $\frac{dy}{dx}$ the 'x rate of y', in abbreviation of the rate at which y is changing relatively to that at which x is changing.

I think

$\frac{k}{h}$ prime ratio,

k the *prime* of y,

h the *prime* of x,

would give language that the common differential calculus might envy.

From Sir W.R. Hamilton to A. De Morgan

OBSERVATORY, *March* 25, 1862.

Thanks for your very interesting and partly controversial letter, which I received this morning, and which I hope to make soon the text, or at least the occasion, of another long letter from myself.

I trust that you have received your 'Mourey' [Mourey's Treatise on Negative Quantities, which had been lent by DeMorgan], minus a paper *wing*, which may yet be found; also the O and P of my *Elements*. The baker [who took his letters to the post] is at the door.

March 26.—You would be a most formidable opponent, in *any* controversy, but perhaps most of all in one which at all regards the history of the differential calculus. Yet I am somewhat surprised at your attributing a quasi-universality of usage to the interpretation of a differential as an *infinitesimal*, at least if you refer to *works* upon the *Differential Calculus*. I am quite aware that we are apt to *think* in *infinitesimals*, and that Laplace, for instance, continually *uses* them. You probably know *ten times* as many as I do, of such works; but really, at this moment, I cannot *remember one*, which *adopts* the interpretation in question; though *several*, perhaps *all*, *allude* to it, as the view of Leibnitz. *You mention it*, for instance, in pages 30 and 50, and, no doubt, in other places, of your Treatise on the Differential and Integral Calculus; although it seems clear that you *prefer*, as many others do, to use only *differential coefficients*; which will go a great way, I admit, but which, you know, *won't do at all* for *quaternions in general*, though of course you *could* not notice that circumstance when writing, and are *not bound* to take cognizance of it now.

[Hamilton then cites as on his side Lacroix (*Elementary Treatise*), and Peacock, whom he proves to have considered it in 1816 as an *established fact* that *Analysts meant*, by the *word* 'differential', the *first term* of the *development* of the *difference*; without the slightest hint that this term was to be *small*, or rather with the *express exclusion* of that condition. He adds, however, that Peacock was for excluding both infinitesimals and *limits*. 'In short, Peacock,' he says, 'was for relying on the method of the *Calcul des Fonctions*—which I greatly admired as a boy, but I believe that, when I was a young man, I had some small share in overthrowing (through $\varepsilon^{-x^{-2}}$, see note to your page 176)'. He next brings forward Cauchy (p. 18 of *Leçons sur le Calcul Différentiel*, Paris, 1829), making the statement, 'Quant à cette dernière quantité [dx], qui représente la différentielle de la variable indépendant, elle reste entièrement arbitraire; et on peut la supposer égale à une constante finie h, ou même la considérer comme une quantité infiniment petite'. He then speaks of Moigno as following Cauchy, and winds up as follows]:—

In short, instead of fearing to be thought *singular* in doing this last thing, I fear rather that I may be supposed to *pretend* to *originality* in a course in which so many have preceded me. But you know that the *only* originality I pretend to consists in my having *combined* the *conception* of the *finite differential* with that of the quaternion; and so *constructed* a *calculus* which, imperfect as it still is, admits already of many useful applications, and has become an *instrument of research* in both *geometrical* and *physical* questions.

If we were never to go beyond the *first order*, I think that *infinitesimals*, as they are very *natural*, might also perhaps content us. We can easily picture to ourselves little triangles, &c. But when we try to *imagine* an *infinitely small part* of an *infinitely small thing*, I think that we are apt to become *confused* in our conceptions; or at least we lose all that *facility* which the *first step* had *gained* for us. I know that it is possible to give *definitions* by which infinitesimals may have as respectable a *size* as any one can desire; but it is (as it appears to me) with *far greater violence to language* than any which any use of the word *differential* can require. '*Infinite*' is a good old English phrase; is it more (you can tell, which I can't) than about *fifty years* since '*Differential*' was naturalised in our language? You know, of course, that Euler (in his *Institutiones Calculi Differentialis**) held that differentials, *and fluxions too*, were rigorously zeros.

March 28.—You see that I am *in* for the *finite* differentials of quaternions; though, as you say, others may *uncall* them afterwards. After all, I am here only reproducing, in a more orderly manner, and I hope with greater clearness, a theory which was given in the *Lectures*.

From A. De Morgan to Sir W.R. Hamilton

41, Chalcot Villas, Adelaide Road, N.W.,
April 1, 1862.

I have received the Mourey. Never mind the cover. *Uno avulso non deficit alter*. I am thoroughly embusinessed for a few days—after which at you again.

[Here follows a long letter from Hamilton, dated April 5th, containing a proof by quaternions of *Taylor's Theorem*, which is an anticipation of his full treatment of it to be found in the *Elements*, pp. 423 . . . 432. The letter concludes with a P.S. in which he reverts, as follows, to the differential question]—

It was stupid of me, when writing lately, to forget that *you use*, within square brackets, the symbol dx to denote the *differential* of x *in Leibnitz's sense*, as signifying an *infinitely small difference*; but I think that the current now runs the other way *in elementary books*; and that the *differential of fx* is now generally stated, in such, to be *equal* to the *second term of the development* of $f(x + dx)$, where dx is an *arbitrary increment*; although *some*, as in the Notes to Lacroix, cited in a former letter, *define* dfx to *be* that second term; while others give an *independent definition* of the *derived function* $f'x$ as the *limit of a ratio*, &c., and *then* write, as a *new definition*, the formula $dfx = f'x . dx$.

* Ticini, 1787.—The first passage that I refer to occurs near the foot of page lxii. in the *Praefatio*. But there are others in the body of the book. Euler's only *infinitesimals* were quantities *absolutely null*; whose *ratios*, however, were to be investigated.

[De Morgan proceeds as follows in defence of his view:]

April 7 . . . Now for differentials. The following is my *resumé* of belief and of fact:—

1. Exclude fluxions. Fluxions were never anything but *velocities*. I believe Newton learnt the idea from the *intension* and *remission* of the schoolmen. You will soon receive a quotation from me on this point. The pure infinitesimals of Leibnitz mark all, I think, of the continental writers of note down to Lagrange, as Euler, Clairaut, D'Alembert, &c. Euler has a vagary about 0, but his *dx* is *not finite*.

2. Lagrange heads a small school on whom you rely. Lacroix and Cauchy are its chief members. Peacock, when a young man and a translator of Lacroix, spoke of this school as 'the Analysts.' Lagrange used infinitesimals in the *Méc. Anal.* and elsewhere. All the physical writers do it, I think. All the mathematicians do it in conversation.

Cauchy does it in several works, of which you have quoted one. In the last (Moigno) he relaxes and gives a choice. Moigno, p. 7. 'La différentielle sera d'ailleurs, en général, une quantité finie ou infiniment petite, suivant que l'accroissement $\Delta x = dx$ sera lui-même fini ou infiniment petit'. Looking over Cauchy's former works, I have no doubt the dog was thinking infinitesimals all the while he was making his poor young readers heave them up into finitude.

3. There is a reaction against Lagrange. Duhamel, Navier, Cournot, are pure infinitesimalists. Some of them say an infinitely small quantity is one which may be made as small *as you please*. This is an evasion; but they do not mean that *dx* is *finite*.

Lagrange dispensed with *dx*; Lacroix and Cauchy are your authorities. Against them I put all that I have stated.

The integral symbol $\int y dx$ will be utterly divested of analogy to summation upon a finite *dx*. And this would be a sore blow to application.

By-the-way, Poisson was a believer in the *reality* of infinitely small quantities—as I am. But it is late.

P.S.—The Differential Calculus was introduced into Cambridge in 1816 . . . I shall seriously propose the '*x* rate of *y*'.

From Sir W.R. Hamilton to A. De Morgan

OBSERVATORY, *April* 19, 1862.

. . . I admit that in the symbol $\int y dx$ we naturally treat *dx* as an infinitesimal. About this, however, at least as bearing on quaternions, I may have something more to say.

The Berkeleys which are here belong to the Earl of Dunraven, an old friend of mine, who lent them to me long ago, but wishes now to have them back again. Before I return them, is there anything that I could extract for you? In a pamphlet which you lately sent me, but which is not just now at my hand, I thought you seemed not to have seen the Bishop's 'Reasons for not replying,

&c.' No doubt he was occasionally waggish, as in those 'Reasons', in which I remember his remarking that perhaps 'some zealous fluxionist' might think it worth while to answer Mr. Walton. But on the whole, I think that Berkeley persuaded *himself* that he was in earnest against Fluxions, especially of orders higher than the first, as well as against matter.

I quite agree with you that it is an *evasion* to call things *infinitesimals*, which can only be *made* small.

From A. De Morgan to Sir W.R. Hamilton

41, Chalcot Villas, Adelaide Road, N.W.,
April 20, 1862.

. . . *Quoad* Berkeley, I have got him all. . . . I have no doubt Berkeley knew that the fluxions were sound enough. Remember that there had been three-quarters of a century of indivisibles, infinitesimals, &c., before Leibnitz published. Wallace alone wrote more of it than Newton and Leibnitz put together, but not of *calculus*, i.e. organised individual-into-species-collecting, rule-ending-in-method.

I sent you a Paper on Saturday. At the end you may form a guess where Newton first got the hint of fluxions. I do not mean from D'Oresme, but from some of the *intensionists*.

From Sir W.R. Hamilton to A. De Morgan

Observatory, *April* 22, 1862.

Thanks for your letter received to-day and your Paper (on Probabilities, &c.) yesterday. The concluding extracts and remarks are very curious.

I don't see why you should not succeed—if only, as you say, you could *provoke discussion*—in introducing your '*x*-rate of *y*', instead of what it is so tedious to write in the usual manner. But as I deal in quaternions, not with ultimate *ratios*, but with ultimate *laws*—not with rates of *variation*, but with *laws of growth*—or in more technical language, with *differential equations*, and *not* with *differential coefficients*, your new phraseology, good as it is, won't serve my purposes.

I hope you understand me as admitting that any one who chooses *may* introduce infinitesimals into quaternions, just as freely as into what may be called (by contrast) *algebra*. In fact I find that my friend, Dr. Salmon, in an Appendix which he has lately let me see, though it is not yet published, to what will doubtless be a very valuable volume on Geometry of Three Dimensions, just takes the bull by the horns, and says that I *have* determined (for instance) the radial direction of a normal to a sphere by infinitesimals; and of course I *do think* habitually of *small triangles*, &c., in geometrical and physical *applications*.

The only question is, *need we* regard *differentials* as other than *finite*? I know that Euler hold them to be *rigorously zeros*; but I observe that Mr. Price, in a book on 'Integral Calculus' (Oxford, 1854), which I thought that I had lost—it is called Vol. II. of a Treatise on 'Infinitesimal Calculus'—takes up infinitesimals in the boldest manner.

On the whole, I suppose you will be able to prove—or may *have* proved already—that you have the *weight* of modern authority on your side; but I still hold that the usage is not so *settled* as to make it *improper* for me to *define*, in my *own* book, as I am doing.

From A. De Morgan to Sir W.R. Hamilton

41, Chalchot villas, Adelaide Road, N.W.,
April 24, 1862.

Many thanks for Hart's Paper. I shall be very glad to keep it—as the author's autograph of the most elaborate and successful bit of rectilinearity I ever saw. The straight line is really a masterpiece of the nature of things. This reminds me of a joke I once heard: 'Sir!' said somebody, 'there is no straightforwardness about him; all he says is a parcel of fibs from one end to the other!' 'No straight-forwardness!' was the answer. 'Why! by your own account he lies evenly between his extreme points!'

My objection, as you know, is to your calling your magnified terms *differentials*. All the rest I admit. But as you give permission to people to take infinitesimals if they like, they will do it.

I think that people will not rouse themselves either to approve or disapprove of the '*x*-rate of *y*'.

I have nothing further to say about anything. I have just dispatched a big Paper on Logic to Cambridge, which I hope will be the last of a directly controversial character.

Here is a bit of Latin which was given to me a few days ago:—

> Ἀείδε είδυλλιον, θεα,
> Felis adest cum cithara,
> Vacca lunam transilit,
> Hoc jocoso motus visu
> Rumpitur catellus risu,
> Cum cochleari lanx abit.

‖The allusion is to the nursery rhyme by Edward Lear.‖

From Sir W.R. Hamilton to A. De Morgan

Observatory, *July* 2, 1862.

After a hitch about money, the impression of my Elements has been resumed, and I send you (to wind up) a proof of the sheet signed this morning.

It completes what I think useful to say, at present, about *differentials* of quaternions. You will see that I treat them (practically) as *infinitesimals*, when employed for purposes of *integration*. What follows is far more original, but I fancy that you would not care for it.

From the same to the same

OBSERVATORY, *September* 15, 1862.

Are you aware that it is possible with only *six* extractions of *square roots* (*two* being those required for chord and tangent of 36°) to approach so nearly to the cosine of the *seventh part* of four right angles, or to the positive root of the *cubic*,

$$2x^3 + x^2 - x = \tfrac{1}{4},$$

that the *sevenfold arc* resulting shall err (in defect) by only about *half a second* from the true and whole circumference? so that for *all practical purposes* a *regular heptagon* can be constructed by the right line and circle *alone*.

My part in the matter is merely the having assigned, this morning, *how much* the construction *errs*. The *rule itself* is attributed by Röber—and with great plausibility, as it seems to me at present—to the ancient Egyptians, who appear to have employed the heptagon in the architecture of some of their temples. The elder Röber was Professor of Architecture at Dresden about thirty years ago; and instead of claiming the construction as *his own*, professed (as I understand) to have divined it from a study of those ancient buildings. His son, Friedrich Röber, published it at Dresden in 1854, in a German Memoir on the Egyptian Temples, &c., which was sent me, with another on the Pyramids, from an unknown friend in Switzerland, through a son-in-law of the Archbishop of Dublin; so oddly do things come about.

When I began to examine the construction, I expected to find some *considerable* error; and was *astounded* to find that I could not, with Taylor's Logarithms, decide to my own satisfaction whether the error, assumed of course to exist, was in *excess*, or in *defect*. The consequence was that I threw *tables* entirely aside (except, you may say, the *multiplication table*, which pound and mil will make at last the *only* one for common life); performed all operations by *arithmetic* alone, extractions being the chief, yet scarcely the chief trouble, such long though contracted multiplications were to be employed, when I aimed at 15 decimal places; and finally arrived at results of which I have given you the general outline. . . .

Will you tell me how you would find, by Horner's Method—which I once *knew*, but have never really *practised*—the root x of the cubic above mentioned? I made it this morning by a clumsy process,

$$x = \cos\frac{2\pi}{7} = 0.62348\,98018\,5873\,45.$$

You wrote to me, a good many years ago, on the subject; but I do not know just now where that letter of yours is.

From A. De Morgan to Sir W.R. Hamilton

91, ADELAIDE ROAD, *September* 18, 1862.

$$1 + 3^2 + 3^4$$

I can believe anything of *six* square roots. But I cannot believe that the Egyptians employed them—though they may have hit on a method which requires six square roots to represent it arithmetically.

I send you every figure of my first attempt at the solution of the equations—done just as my pupils are made to do it. Horner himself did not hit the extreme of organisation. He and Young—his earliest expositor—both got the places to be thrown away in contraction, by mother-wit and look-out. The process I send you is purely mechanical, and done as a skeleton from beginning to end. You will find my scheme in an Appendix to my Arithmetic.

As to the rest, meaning everything in existence except Horner's Method, I have nothing particular to say. I have some remembrance of a good approximation to a heptagon in the common books of mensuration. But, after all, nothing is so short, even for most accurate work, as pure trial with the compasses. With *hair*-compasses, I would beat all the methods, in time and accuracy both. Do you know the *hair*-compasses? . . . I should like to see the formula of the six square roots.

September 23.—You have got the principle of Horner's Method . . . The table of Pitiscus is his corrected Rheticus—the Thesaurus—for which see my article on Tables. It is *rarissimus*. I have Cavendish's copy. The Duke of Devonshire sold all Cavendish's books—which looks odd. But all the C.'s hate their great chemical relation—and with good reason. He was a heartless brute. The original Rheticus, the Opus Palatinum, will do well enough; it is not quite so rare. Or Briggs' *Arithmetica Britannica*. Take notice that Horner's Method is not algebra. It is pure arithmetical organon—the true extension of division and the square-root. Whether Horner's immediate predecessor is a Sanscrit, or a Tartar, or an antediluvian, nobody knows.

I am surprised that the Egyptians should have got so close a method—and I admire Röber's ingenuity. But I am no way surprised at six square roots giving 15 places. . . . Seeing that my last figures differed from yours, I repeated the calculation, and find your figures are right *ab ovo usque ad mala* (*mala*—apples of discord, the last figures of a calculation). I also verified it by old Pitiscus as in accompanying Paper.

September 29. [Replying to Hamilton writing 'I should like to know exactly *what you claim*, in relation to Horner's Method. Is it more than the *arrangement* of the figures, which, however, I hold to be of *vast* importance in a long calculation?'] You evidently like calculation, in spite of sky-high quaternions, &c. If you had not the organ in your head, not all the quaternions could give it. I take it that you and I are the only persons who know $\frac{2\pi}{7}$ to 22 places.

As to what I claim, I know not *how much* it is; but I know *what* it is. I know that I bridged the chasm—how many arches it took I do not know—which separated Horner and his followers from seeing that his method *is* the natural and proper development of common arithmetic; so that division and the square-root are true cases of the whole method. Horner himself did not see his own fame; did not know that he was the first European who had gone beyond the nameless Brahmin, Tartar, antediluvian—whichever he was—who constructed the rule for the square-root. He had high notions, and looked upon his method—very justly—as a particular case of one which would solve, say $x^{\sin x} + (1 - x^2)^{\tan^{-1}x} = 1$, if there be a real solution, on which I do not pronounce.

My history of the problem is in the *Companion to the Almanac* for 1839. The article on *Involution* in the *English Cyclopædia* unites two articles of the *Penny Cyclopædia*. Part 16 of the Division of *Arts and Sciences* contained all the letter I; and I find I may say 'pars magna fui'. Witness: Incommensurables, Infinity, Involution, Integration (common and definite), Invention and Discovery, Irrational quantities (the only modem account of the 10th Book of Euclid), Interest.

P.S.—I know all about Horner but his *name*. He is *W.G.* Horner—but I never saw it stated what W and G stand for.

November 16. [Referring to a proposition suggested by a mode of tracing an hyperbola communicated by a Pennsylvanian visitor, he writes]:—I do not remember the *class* of propositions to which this belongs. I care no more for an isolated theorem now, in these things, than for 22 + 13 = 35, or any other case of addition. What is the connexion of all this with higher generalisations? And as I attend to it only once a year, when teaching comes on, I always write to Ireland for a wrinkle about conics or quarics (—I will not say qua*d*rics—quarrée, carré—I am not sure I will not stand up for *carrics* or *carics*) when I want one. It is 'Please remember the grotto—only once a year'—whether you understand the allusion or not.

I am excogitating the subject of infinity—with a view of writing upon it. There is a question I put to every thinking man who comes in my way—*I have various answers*. Consider space as objectively external—merely for convenience. Does the *totum spatium*—the universe of space—contain twice as many pints as quarts? Yes or no. No refinements about inconceivabilities, &c., &c. I think the idea of measuring out infinite space in quart measures a very grand one. For myself—after trying for years to evade it psychologically, ontologically, pseudologically, and amosgepotically—I find no rest for the sole of my foot except upon the assertion—which I now fully believe—that the universe *does* contain twice as many pints as quarts. That is, if worth 2*d*. a quart, it is worth 1*d*. a pint. You surely cannot deny this—and how can it be unless there be twice as many pints as quarts?

Amosgepotically—ἀμωσγεπως ‖*sic*‖—somehow or other. I have got this word to express my way of explaining things. If I am asked—as an omniscient philosopher—for the *why* of anything—and I say I don't know, I am despised—if I say it happens somehow or other, I am looked upon as a very common sort of thinker; but if I say it happens *amosgepotically*, I at once take rank with the high aristocracy.

From A. De Morgan to Sir W.R. Hamilton

91, ADELAIDE ROAD, *April* 13, 1863.

Are you gone to the moon, or to Java, or have you eloped with a mermaid? These are three of the possible questions, and are put as specimens; the whole list is too long. I should like to know among other things when the quaternions will be finished. . . .

I have been jotting down my ideas about infinity; that is quantitative infinity, not the absolute, or the unconditioned, or zero passing into being, or any of the Nicotian views of the subject. I want to test the question whether mathematicians do or do not actually admit an *infinitely small* quantity, by premises from which it may be actually squeezed, as therein contained. If I say that

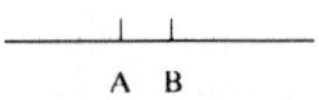

Is an infinitely small part of a line, I make it so, and I cannot be said to have *deduced* it. I may say, it shall be any finite line, however small, but I will not, by assumption, introduce any line except what is between $\frac{m}{n}$ and $\frac{m+1}{n}$ of the whole, where m and n are two finite numbers. Certainly a person who takes this ground may avoid being forced to acknowledge an infinitesimal.

But now, take him this way. He admits the concept, a straight line. He admits the concept, a point. He acknowledges that the straight line contains an infinite number of points; that is, more than any finite number which can be named. This is the common ground of all mathematicians. He then hands me over a *genus*, namely, that of which the individual is a point in a given straight line. From this genus I have a right to select a species by any difference I please. I select by this difference—'dividing the whole into two commensurable parts'—say each such point is a C. He grants me all the C's, because he gave them as individuals of an acknowledged genus. He will grant me that no part of the straight line has all its points among the C's; that is, he grants an *interval of length* between every two C's. I come at these intervals of length as necessary results of his own data and admissions. Now, all the C's being taken, that is, selected from his *genus*, what are these intervals? Are they finite? Certainly not; they are then infinitely small.

Here are points which approach without limit, and do not ultimately coincide. Or rather, without any subsequently ordained approach, here are points which do not coincide, but the distance between which is less than any finite quantity which can be named. Not laid down for that purpose and upon that hypothesis, nor increased *ad infinitum* by the proponent, but selected from a greater infinite acknowledged by all, and handed to him by common consent.

It appears to me that there is no getting out of the *concept*—concept without *image*—of an infinitesimal, as actually necessary to what all admit.

From Sir W.R. Hamilton to A. De Morgan

OBSERVATORY, *April* 16, 1863.

. . . I remember your proposing to me a question on the the *Cubature of Space*, to which I thought of replying by a question respecting the *Quadrature* of the *Infinitely Distant Surface*. Fine fun there will be for the Hamiltonians—I do not mean myself—when you come out with your *pints and quarts of infinity!*

But it appears to me a most legitimate and interesting subject of psychological and literary inquiry, what do *mathematicians mean* when they talk of *infinites and infinitesimals*?

Has it ever struck yourself—as it does me—that you *believe* more in infinites now than when you wrote your 'Differential and Integral Calculus'? I catch myself *believing* occassionally in the *line* at *infinity* in a *given plane*, and what is far more odd (yet useful), the *plane* and *circle at infinity* in *space*. Then I *look down on myself*!—but not on *you*.

From A. De Morgan to Sir W.R. Hamilton

91, ADELAIDE ROAD, *April* 26, 1862.

I do not believe in infinity *more* than I did in my 'Differential Calculus'; but I certainly had not so decided a belief, only a leaning. The fact—which I shall promulgate—is this. When I was a boy at Cambridge, I observed that the first thing every high undergraduate, and every mature B.A. did, was to settle definitely and irrevocably the true foundation of the Diff. Calc. I saw this, and made up my mind that they *ought* to wait, and that I *would* wait. All these years—say 38—I have pondered the subject, with a resolution never to commit myself on the wrong side of *fifty*. And now, having with a very unbiassed and very long meditation on the subject made myself sure that infinity is a subjective reality, I intend that it shall be so for the future. But I leave all other minds to stand on their own bases. I do not assume that it is a subjective reality in another subject.

But I do not say that two parallel straight lines meet at an infinite distance. They meet *never*. And I distinguish $\frac{1}{0}$ from all the lower infinites, by the same distinction which exists been 0 and the infinitesimals.

I am seriously of opinion that we should constantly flounder and get wrong in using *limits*, if we had not the light of infinity to guide us.

11
George Boole (1815–1864)

George Boole is today best remembered for his contributions to logic—as the man who, by breaking with the traditional, syllogistic methodology expounded in the logic textbooks, initiated the modern, mathematical study of logic. Bertrand Russell famously declared

> Pure mathematics was discovered by Boole, in a work which he called the *Laws of Thought* (1854). This work abounds in asseverations that it is not mathematical, the fact being that Boole was too modest to suppose his book the first ever written on mathematics. He was also mistaken in supposing that he was dealing with the laws of thought: the question how people actually think was quite irrelevant to him, and if his book had really contained the laws of thought, it was curious that no one should ever have thought in such a way before. His book was in fact concerned with formal logic, and this is the same thing as mathematics (*Russell 1901*, §2).[a]

But if Boole can be regarded as the initiator of a new era in logic, he should also be remembered in a less familiar role—as one of the great British algebraists of the nineteenth century, and as the culminating product of the approach to algebra pioneered by George Peacock and carried forward by Duncan Gregory and Augustus De Morgan.

Boole stands somewhat apart from his three algebraic predecessors. Unlike them, he was not educated at Cambridge. His father, John Boole, was a shoemaker in Lincoln—a remarkable man, by all accounts, with a strong amateur interest in mathematics, optics, and astronomy, and in the building of optical instruments. But John Boole did not have the social position or the financial means to send his son to University; and apart from a few years of elementary school, George Boole was entirely self-educated. As a child he was a linguistic prodigy, and taught himself Greek and Latin, German, French, and Italian—knowledge that was to enable him to read the works of the great Continental mathematicians, who as yet were little known in England. At the age of fifteen, in order to support his parents, he began work as a schoolteacher; for a time, still in his teens, he administered a school of his own. During this period,

[a] Russell's essay was written for a popular audience, and (as he notes) for an editor who asked him to make the essay 'as romantic as possible'. Russell's considered appraisal of Boole was more sober. For instance, in *Our knowledge of the external world*, Lecture II, he says of Boole: 'But in him and his successors, before Peano and Frege, the only thing really achieved, apart from certain details, was the invention of a mathematical symbolism for deducing consequences from the premises which the newer methods shared with Aristotle.'

encouraged by his father, Boole studied the great mathematicians, reading the *Principia*, Lacroix's *Calcul différentiel*, Lagrange's *Mécanique analytique*, Laplace's *Mécanique celeste*, Lagrange's *Calcul des fonctions*, and later the works of Jacobi and Poisson.

From about 1838 onwards, Boole was engaged in mathematical research. In 1839 he visited Duncan Gregory in Cambridge; Gregory spotted the talent of the young schoolteacher, and with his help and advice Boole began publishing articles in the *Cambridge Mathematical Journal* on differential equations, the calculus of variations, and the evaluation of integrals. Two of his articles on linear transformations, published in the *Cambridge Mathematical Journal* in 1842 and 1845 ⟦*Boole 1842a,b, 1845*⟧, inaugurated research into the theory of algebraic invariants, a topic which was immediately taken up by Eisenstein in 1844 and Cayley in 1845, and which was to occupy mathematicians of the stature of Sylvester, Gordan, and Hilbert for the remainder of the century. At about this time, Boole dropped several broad hints that he would like to come to Cambridge as a student, but was politely discouraged.

As an algebraist, Boole, from his earliest years as a creative mathematician, was paricularly adroit at 'operator methods'—that is, at treating the symbol d/dx for the operation of differentiation (or the difference operator D in the calculus of finite differences) as though it stood for a peculiar sort of number. (The method was also called the 'separation of symbols' because it abstracted the symbol d/dx from the symbol dy/dx.) By skilful formal manipulation of these operators (and of other operators that he introduced into the calculus of finite differences, like his Δ and E and π and ρ) Boole was able to solve various types of differential equation. (A general account of his 'symbolical methods' can be found in Chapters 16 and 17 of his *Treatise on differential equations* (*1859*). But the method itself dates from his work of the early 1840s.)

Boole's work on algebraic operators fits naturally with the view that algebra is the study of the lawlike manipulation of uninterpreted symbols. Indeed, Gregory anticipated Boole by several years on both points: he, too, studied the laws governing the iteration of differential operators and their multiplication by scalars; and he, too, developed a disinterpretational theory of algebra (*Gregory 1840*, reproduced above). Boole certainly knew of the syntactic approach to algebra developed by Peacock, Gregory, and De Morgan. However, the germ of the method of 'separation of symbols' is already evident in Boole's first paper, *Boole 1841*, which was inspired by his reading of Lagrange's *Mécanique analytique*, and was written in 1838, a year before his meeting with Gregory. So Boole may have hit on the central ideas independently.

Boole and Gregory exchanged letters discussing their research on 'symbolical algebra' in the years before Gregory's death at the age of thirty-one in 1844; and Gregory is cited along with Servois and De Morgan in Boole's most important paper on the application of operator methods to the study of differential equations, 'A general method in analysis' (*Boole 1844*). (In fact, it was Gregory who urged Boole to publish the paper in the *Transactions* of the Royal Society

(*Boole 1952*, p. 442).) This paper earned Boole the Royal Society's first Gold Medal for mathematics, and demonstrated in detail and in a dazzling variety of examples the power of operator methods. In an appendix dated 31 August 1844 to the published paper Boole added the following remarks:

Fearful of extending this paper beyond its due limits, I have abstained from introducing any researches not essential to the development of that general method in analysis which it was proposed to exhibit. It may however be remarked that the principles on which the method is founded have a much wider range. They may be applied to the solution of functional equations, to the theory of expansions and, to a certain extent, to the integration of non-linear differential equations. *The position which I am most anxious to establish is that any great advance in the higher analysis must be sought for by an increased attention to the laws of the combination of symbols. The value of this principle can scarcely be overrated: And I can only regret that in the absence of books, and circumstances unfavourable for mathematical investigation, I have not been able to do that justice to it in this essay which its importance demands.*

Boole's greatest accomplishment was to apply the new algebra and the mathematical techniques of operator theory to the study of deductive logic; the result was the first satisfactory algebraic calculus capable of formalizing the rules of traditional syllogistic logic. Boole performed this feat in *The mathematical analysis of logic*, a booklet which he wrote in a few weeks in 1847. His attention had been drawn to logic by the acrimonious public dispute then raging between Augustus De Morgan and Sir William Hamilton over the 'quantification of the predicate'. (This dispute is discussed above in the Introductory Note to the selections from De Morgan.) Although Boole and De Morgan corresponded regularly on scientific topics between 1842 and Boole's death in 1864—the ninety surviving letters have been published as *Boole and De Morgan 1982*—Boole's logical work was not significantly influenced by De Morgan, and there is even some evidence that the idea of an algebraic analysis of logic had occurred to him as early as 1832.[b] Moreover, in 1847 De Morgan was writing

[b] According to a memoir by his widow, Boole related to her that, while walking in a field in Doncaster as a boy of seventeen, he had first had the idea that algebraic formulae could express logical relations (*M. Boole 1878*, p. 326). The ultimate sources of this insight are a matter of conjecture. As a boy Boole had learned German grammar from a textbook by Becker which stressed the similarities between reasoning and syntax; and of course at the time he was deeply steeped in Continental algebra. Boole, a deeply religious man, seems to have regarded the Doncaster experience as a divine call to investigate mathematically the workings of the human mind; *pace* Russell, he always regarded his work in logic as a study of the 'laws of thought', and as having religious significance. But Boole never published an account of the philosophical underpinnings of his logic; some unfinished manuscripts on this topic are in the possession of the Library of the Royal Society, and were partially reproduced in *Boole 1952*.

The articles written shortly after Boole's death by Mary Everest Boole (the niece of Sir George Everest, for whom Mt Everest is named) are straightforward narratives, and probably reliable biographical guides. However, in later years Mary Boole also wrote extensively on the mystical and spiritualistic implications of Boole's logical discoveries, and her interpretations of his thought need to be taken with a considerable grain of salt. William Kneale gives the following account of the books she wrote: 'There are more than a dozen of them with such queer titles as *The Message of Psychic Science to Mothers and Nurses*, *Logic Taught by Love*, *Suggestions for Increasing* his own *Formal logic*, and having been publicly accused of plagiarism by Hamilton was perhaps understandably sensitive to sharing his ideas; in his letter to Boole of 31 May he said 'I would much rather not see your investigations till my own are quite finished; which they are not yet for I get something new every day. When my sheets are printed, I will ask for your publication: till then

Ethical Stability. The following passage (p. 29), taken at random from *The Philosophy and Fun of Algebra* (1909) will serve as a specimen of her later style: "The first Hebrew algebra is called Mosaism, from the name of Moses the Liberator, who was its great incarnator or Singular Solution. It ought hardly to be called an algebra: it is the master key of all algebras, the great central direction for all who wish to learn how to get into right relations to the Unknown, so that they can make algebras for themselves." She died in 1916 at the age of 84, and a little book in her honour was published as late as 1923 under the title *A Teacher of Brain Liberation*. This was by a devoted follower, Florence Daniels. It contains one gem: "Like the disciple of Pythagoras, for two years I sat at her feet without speaking (this is metaphorical, of course)"' (*Kneale 1948*, p. 155).

Despite these vagaries, Mary Boole seems to have been an intelligent and energetic woman, and a notable educational reformer and champion of women's rights; certainly she and George Boole produced one of the most remarkable families in the history of British science. They had five daughters. The oldest, Mary, married C.H. Hinton, a mathematician interested in four-dimensional geometry; while an instructor in mathematics at Princeton in the 1890s he invented an automatic baseball pitcher powered with gunpowder. (It was used by the Princeton team for batting practice, but found to be rather dangerous.) Their son, George Boole Hinton, was a distinguished metallurgist and botanist; his son, Howard Everest Hinton, FRS, a fervent Marxist, was one of the great entomologists of the twentieth century, publishing over 300 papers and a three-volume study of *The biology of insect eggs*. His cousin, Joan Hinton (also a great-grandchild of George Boole), was a theoretical physicist and Enrico Fermi's assistant at the University of Chicago; she worked with him on the Manhattan Project, and was present at the first atomic explosion in New Mexico. Horrified by the destruction at Hiroshima, she left the West for Communist China, where she devoted her energies to trying to build a new society and to the technological control of pollution.

The second Boole daughter, Margaret, married an artist, E.I. Taylor. Their son, Sir Geoffrey Taylor, FRS, was a mathematical physicist at the Cavendish Laboratory in Cambridge, and a scientist of the first rank—an expert on turbulent motion, meteorology, shock waves, and aeronautical dynamics. He published over 200 scientific papers, and like Joan Hinton was present at the Trinity explosion, where he calculated the effects of the blast. In 1933 he, like his grandfather, received the Royal Society's Gold Medal; in 1969 he received the Order of Merit.

The third daughter, Alicia, was the most mathematically talented of Boole's daughters, and a fertile geometer; in the 1930s she worked with H.S.M. Coxeter on four-dimensional polytopes, producing original contributions when she was in her seventies. Her son, Leonard Stott, a doctor, was an innovator in the treatment of tuberculosis; he also invented a portable X-ray machine, and developed a system of navigation based on spherical trigonometry.

The fourth daughter, Lucy, was a chemist and the first woman Professor of Chemistry at the Royal Free Hospital in London. She died unmarried in her early forties.

The youngest daughter, Ethel, had a romantic life as a novelist. In her early years she was intensely involved in Eastern European revolutionary politics; she lived in Russia for two years, and on returning to London worked with Eleanor Marx, Engels, and Bernard Shaw. In 1890 she met a Polish dissident who had recently escaped from imprisonment in Siberia; he arrived penniless in London with Ethel Boole's name and address on a scrap of paper. When they met, he recalled having glimpsed her from his prison window in Warsaw on Easter Sunday three years earlier, as she had stood in the square gazing at the Warsaw Citadel. They were married in 1891. They drifted apart, and Ethel became for a time the mistress of the spy Sydney Reilly, the original of Ian Fleming's James Bond; Reilly was said to have eleven passports and a different wife for each. After he abandoned her in Florence in 1895, Ethel wrote a novel based on Reilly's early adventures; *The gadfly* became a sensational best seller in the Soviet Union, selling five million copies, and a further million in China and Eastern Europe. Ethel Voynich eventually moved with her first husband to New York City, where she died in 1960 at the age of ninety-six. For further biographical details on Boole and his family, see *MacHale 1985*.

please not to send it.' According to tradition, Boole's *Mathematical analysis of logic* and De Morgan's *Formal logic* appeared in the bookstores on the same day.

The *Mathematical analysis of logic* was a milestone in the development of logic—the first work successfully to exploit the analogy between logical operations and arithmetical operations, and to display the whole of traditional logic as a species of algebra. Boole's project thus differed sharply from Frege's investigations in the *Begriffschrift*, which were rather intended to create a *lingua characteristica* and to explore the logical foundations of arithmetic—as it were, making arithmetic a branch of logic rather than logic a branch of algebra.

The insight that logical reasoning shares certain similarities with arithmetical calculation was not original to Boole: it had earlier occurred both to Leibniz and to Lambert. (Boole, however, was almost certainly unaware of their work.) Leibniz in particular had hoped to discover a *calculus ratiocinator*—a mechanical method for resolving arguments. He famously prophesied that, when his logical calculus had been perfected, philosophy would be a science as certain as mathematics. If a disagreement were to arise, the the philosophical disputants, instead of wrangling, would simply take pen and paper and say, *Calculemus*: 'Let us calculate' (*Leibniz 1875–90*, Vol. vii, p. 200.)

But although Leibniz attempted several times to exploit the formal analogies between multiplication (or addition) and logical conjunction, none of his systems was powerful enough to give an algebraic treatment even of the syllogism. The problem was not that Leibniz laboured too much under the influence of arithmetical analogies, that he was unwilling to abandon the familiar laws of arithmetic; for, like Boole, he realized that the laws of ordinary arithmetic would not hold for his logical calculus. (For example, he pointed out that representing conjunction of attributes by the sign of addition would result in the law, $A + A = A$). He was also aware that his calculus would admit of several equivalent interpretations—as a calculus of attributes, or of classes, or of shorter and longer line segments; and in this respect, he anticipated an important feature of the Peacock–Gregory–Boole view of algebra (see for example, *Leibniz 1875–90*, Vol. vii, pp. 236–47). The shortcomings of his system lie elsewhere. Under the influence of the metaphysical principle that the concept of the predicate must be contained within the concept of the subject, he thought of logic as a matter of adding and subtracting (or multiplying and dividing) the attributes of a particular subject, rather than as a matter of forming truth-functional combinations of propositions (or, equivalently, of Boole's selection operators); and this approach to logic proved to be sterile. For instance, in one of his systems Leibniz represented simple predicates by prime numbers; a complex predicate was then to be represented by the product of the numbers corresponding to its simple components (*Leibniz 1903*, pp. 57–77). In this system, some elementary logical laws can be shown to be analogous to elementary arithmetical laws. For example, if the predicate A is (integrally) divisible by B and if B is divisible by C, then A is divisible by C—which corresponds to the fact that, if all *A*s are *B*s, and all *B*s are *C*s, then all *A*s are *C*s. But Leibniz was never able to extend his system beyond these trivialities; indeed, he was unable, using representations of attributes by prime numbers, even to give a satisfactory algebraic formulation of negation or of disjunction.

Boole, in contrast, developed his logical system as a calculus of operators on classes, and was consequently able, for the first time, to give an algebraic representation of the whole of traditional Aristotelian logic; his algebraic methodology, his habit of regarding the symbols of algebra as standing for operations, was thus the crucial factor that enabled him to succeed where Leibniz had failed. Indeed, Boole opens the *Analysis* with a particularly clear and forceful statement of the Gregory–Boole philosophy of algebra:

> They who are acquainted with the present state of the theory of Symbolical Algebra, are aware, that the validity of the processes of analysis does not depend upon the interpretation of the symbols which are employed, but solely upon the laws of their combination. Every system of interpretation which does not affect the truth of the relations supposed, is equally admissible, and it is thus that the same process may, under one scheme of interpretation, represent the solution of a question on the properties of numbers, under another, that of a geometrical problem, and under a third, that of a problem in dynamics or optics.

In developing his system, Boole used the same methodology he had employed in studying differential operators—taking the attitude, implicit in *Gregory 1840* but never fully exploited in Gregory's mathematical research—that the 'laws of combination' for operators need not obey the familiar laws of arithmetic. By 1847 Boole was aware of Rowan Hamilton's quaternions; and although he never showed much interest in them (apart from a three-page 'Note on Quaternions' that he published in 1848) he knew that the quaternions did not satisfy the commutative law of multiplication. So he was psychologically well prepared to introduce an algebraic system in which the operators satisfied such unfamiliar laws as $x^2 = x$.

Boole, by modern standards, is exceptionally loose in describing the fundamental concepts of his system. He uses upper-case letters X, Y, Z, etc. to stand in effect for predicates; the class X is then the class of all individuals in the universe to which X applies. The fundamental objects in his system, represented by lower-case letters, are the 'elective symbols' x, y, z, etc.; the elective symbol x, if applied to a class or individual, yields all the Xs which the original class or individual contains. (Boole, like Dedekind and other mid-century logicians, did not possess a clear distinction between a singleton class and its sole element; hence the provision allowing the elective symbols to be applied to individuals.) The elective symbols are thus selection operators on classes, and can be compounded and manipulated in a lawlike manner much as differential operators are compounded and manipulated. Boole uses the symbol 1 to designate the universal class, so that $x(1)$ is the class X. In fact, he takes the operators in the *Mathematical analysis of logic* to be applied to the class 1, so that his resulting system is equivalent to a calculus of classes as well as of operators; later, in the chapter 'Of hypotheticals', Boole points out that the calculus applies to propositions as well. In *The laws of thought* he was to take the class-interpretation as fundamental; but the operator interpretation reveals more clearly the connection to his work in algebra.

Boole introduces the fundamental operations of multiplication and addition rather casually, without attempting any explicit definition: multiplication represents the composition of operators, so that xy is the selection operator that selects the Xs from the result of selecting the Ys from a given class; while addition (in contrast to most modern systems of Boolean algebra) represents the *exclusive* disjunction, so that $x + y$ selects those elements that are Xs or Ys but not both.

Boole's calculus contains a number of awkward features that had to be remedied by later writers. First, his use of exclusive disjunction leads to avoidable complications in the basic logical relationships. Peirce and Schröder pointed out that the Boolean system becomes tidier if one uses instead the inclusive disjunction, so that $x + x = x$; one then obtains De Morgan's laws as well as an elegant duality between addition and multiplication:

$$x(y + z) = (xy) + (xz)$$

$$x + (yz) = (x + y)(x + z).$$

Second, Boole's system does not handle existential propositions well. In order to express the proposition, 'Some Xs are Ys' he is forced to introduce a special set V with the property that it is non-empty and that it is to be indifferently interpreted as 'some Xs or some Ys'. (Unbeknownst to Boole, Leibniz had flirted with a similar idea years earlier (*Couturat 1901*, p. 357).) The proposition 'Some Xs are Ys' is then expressed in his system as '$v = xy$'. The elective symbol v corresponding to V must be handled with care, especially as regards negation; this is evident from the contortions Boole must go through on pp. 21–5 of the *Analysis*. Third, Boole's system contains no logic of relations, and is only capable of handling propositions that have been analysed into the subject–predicate form. This was a failing of all systems of logic before De Morgan's pioneering work on the logic of relations in *De Morgan 1864b*; the subject was to be explored in depth by Peirce and Schröder, with *Peirce 1870* making the first attempt to extend Boole's system to relational expressions. Fourth, and most importantly, Boole does not have the full-blown apparatus of quantification theory: his system is essentially limited to the syllogism, and was thus inadequate to the logical analysis of the foundations of mathematics that was to be pursued in the closing decades of the nineteenth century by Frege, Peano, Russell, Whitehead, and others.

In 1849, despite not possessing any university degree, Boole was appointed Professor of Mathematics at the newly-created Queen's College, Cork, Ireland. He continued to work on logic and on the theory of probability, publishing his results in a series of articles as well as in *The laws of thought* (*Boole 1854*). Boole called the *Mathematical analysis of logic* 'a hasty and (for this reason) regretted publication' (*Boole 1851*, p. 525). *The laws of thought* is indeed a more polished work; however, the *Mathematical analysis* already contains most of the logical ideas in *The laws of thought*, and in important respects is a clearer and more incisive presentation of Boole's system—in particular, it reveals more clearly than the later work the connection of Boole's logic to the algebraic ideas that had been ripening among British mathematicians in the early nineteenth century. The principal innovation in *The laws of thought* was an extension of Boole's analysis of logic to probabilistic reasoning—an innovation that seems to have been partly inspired by De Morgan's work on probability, and which was to influence the logical writings of C.S. Peirce, among others.

During his years at Cork Boole also wrote two influential textbooks, one on differential equations, the other on the calculus of finite differences (*Boole 1859* and *1860*); these works, which elaborate upon many of the mathematical discoveries in his early research articles, were in common use in England until the start of the First World War. A conscientious teacher, Boole died in 1864 from pneumonia which he contracted after insisting on walking through a rainstorm to deliver a lecture.

Boole's work illustrates the backtracking sort of progress that was made by British algebra during the nineteenth century, and shows that the development of the subject was not a steady, triumphant march towards the abstract algebras presented in, say, van der Waerden's *Moderne Algebra*. Let us briefly recapitulate the tangled main events in the last half-dozen selections. The new British approach to algebra began when Peacock, Babbage, and Herschel championed Leibniz's notation for the calculus over Newton's. Peacock some years later introduced the idea of a *symbolical algebra*. His aim in doing so was limited. It was, to justify the negative and the complex numbers and the techniques of the Continental analysts against the criticisms of Maseres, Frend, and others. Peacock did not describe his syntactic calculus in detail, and his 'principle of the permanence of equivalent forms' in effect guaranteed that symbolical algebra would always obey the familiar laws of arithmetic. Duncan Gregory refined Peacock's ideas, and contributed the important observation that the symbols of symbolical algebra need not stand for numbers, but can also be taken to stand for operations. Boole and De Morgan developed Gregory's insight in two different ways—Boole furnishing an algebraic analysis of logic (*1847*) in which the old laws of arithmetic are no longer valid, and in which the algebraic symbols stand neither for quantities nor for figures; and De Morgan (*1849b*) providing the first explicit description of a syntactic calculus.

But not all these achievements registered on the mathematicians of the time. Boole ignored De Morgan's formal system, and made no attempt to present Boolean algebra in a strict deductive framework; De Morgan, for his part, when he turned his attention to the logic of relations in *De Morgan 1864b*, largely ignored the accomplishments of *Boole 1847*.

Meanwhile, working independently in Ireland, Rowan Hamilton had discovered the representation of complex numbers by ordered 'couples' of real numbers, and then, in 1843, had discovered quaternions—an algebraic structure that does not satisfy the commutative law of arithmetic. But Hamilton explicitly rejected the syntactic theory of algebra developed by Peacock, Gregory, De Morgan, and Boole. He preferred instead to view algebra as 'the science of pure time'—a view which led him to mistrust algebraic systems in which the *associative* law does not hold. Boole and De Morgan in turn were aware of Hamilton's work on quaternions, but did not themselves pursue the subject in any detail. Oddly, none of these mathematicians placed any emphasis on *axioms*—an idea at least as ancient as Euclid, and one that would have fit naturally with their new theories of algebra. In short, the British algebraists of the second quarter of the nineteenth century discovered many of the *elements* of the modern conception of algebra, but did not weld them together into a unified whole.

In practice, British mathematics was not troubled by the unresolved difficulties in the foundations of algebra. The discoveries of Hamilton and Boole liberated algebraists from the old rules; and throughout the nineteenth century British mathematicians continued to discover and explore a bounty of new structures—matrices, vector spaces, linear algebras, groups. But they felt no need to dwell pedantically on distinctions between the manipulation of symbols and the manipulation of mathematical objects, or to present their algebras in a strict, axiomatic form, or to develop formal calculi.

An example from one of the great mid-century algebraists will illustrate this point. In one of his papers 'On the theory of groups' (*Cayley 1854*) Arthur Cayley introduces the concept of a group as follows: 'A set of symbols, 1, α, β, . . ., all of them different, and such that the product of any two of them (no matter in what order), or the product of any one of them into itself, belongs to the set, is said to be a *group*.' He adds in a footnote: 'The idea of a group as applied to permutations or substitutions is due to Galois, and the introduction of it may be considered as marking an epoch in the progress of the theory of algebraical equations.' In certain respects, Cayley's words are in accord with modern presentations of group theory. Since he speaks of a group as a set of *symbols* that are *applied to* permutations, he appears to have thought of groups abstractly. Moreover, he gives no further account of the reference of the symbols in his theory, nor does he discuss the nature of the abstract operation of multiplication. On the other hand, he is imprecise about the existence of a unit element and about the associativity of multiplication; indeed, it is unclear whether in his comments above he means to *define* groups or merely to *describe* them. Not that it much mattered: Cayley's group-theoretical researches did not depend on the distinction.

In the same way, such prescient ideas as De Morgan's lucid description of a formal system went unheeded by mathematicians in the nineteenth century—not because the idea was intrinsically difficult, but because De Morgan's level of rigour was not yet needed. The full-fledged concept of an abstract algebra—combining the ideas of a purely syntactic calculus, capable of multiple interpretations, not bound by the traditional rules of algebra, presented axiomatically, with explicit rules of inference, and capable of being studied metamathematically—emerged only at the end of the century, after mathematicians had acquired more experience with algebraic structures, non-Euclidean geometries, and axiomatizations of logic and number theory. These ideas were explicitly combined for the first time in Hilbert's early work on the foundations of geometry and of arithmetic (*Hilbert 1899*, *1900a*, *1900b*, and *1904*), where they were a necessary precondition for his investigations into the consistency and relative strengths of various axiom systems.

References to *Boole 1847* should be to the pagination of the original edition, as given in the margin. (This pagination is preserved in the modern reprints.)

A. THE MATHEMATICAL ANALYSIS OF LOGIC, BEING AN ESSAY TOWARDS A CALCULUS OF DEDUCTIVE REASONING (*BOOLE 1847*)

Ἐπικοινωνοῦσι δὲ πᾶσαι αἱ ἐπιστῆμαι ἀλλήλαις κατὰ τὰ κοινά. Κοινὰ δὲ λέγω, οἷς χρῶνται ὡς ἐκ τούτων ἀποδεικνύντες ἀλλ᾽ οὐ περὶ ὧν δεικνύουσιν, οὐδ᾽ ὃ δεικνύουσιν.

ARISTOTLE, *Anal. Post.*, lib. I. cap. XI.[a]

PREFACE

IN presenting this Work to public notice, I deem it not irrelevant to observe, that speculations similar to those which it records have, at different periods, occupied my thoughts. In the spring of the present year my attention was

[a] ['All sciences share with one another the use of common principles. By "common principles" I mean the things they use for the purpose of proof, not the subject-matter of their proofs, nor the proofs themselves.']

directed to the question then moved between Sir W. Hamilton and Professor De Morgan; and I was induced by the interest which it inspired, to resume the almost-forgotten thread of former inquiries. It appeared to me that, although Logic might be viewed with reference to the idea of quantity,* it had also another and a deeper system of relations. If it was lawful to regard it from *without*, as connecting itself through the medium of Number with the intuitions of Space and Time, it was lawful also to regard it from *within*, as based upon facts of another order which have their abode in the constitution of the Mind. The results of this view, and of the inquiries which it suggested, are embodied in the following Treatise.

It is not generally permitted to an Author to prescribe the mode in which his production shall be judged; but there are two conditions which I may venture to require of those who shall undertake to estimate the merits of this performance. The first is, that no preconceived notion of the impossibility of its objects shall be permitted to interfere with that candour and impartiality which the investigation of Truth demands; the second is, that their judgment of the
1|2 system as a whole shall not be founded either upon the examination of only |
a part of it, or upon the measure of its conformity with any received system, considered as a standard of reference from which appeal is denied. It is in the general theorems which occupy the latter chapters of this work,—results to which there is no existing counterpart,—that the claims of the method, as a Calculus of Deductive Reasoning, are most fully set forth.

What may be the final estimate of the value of the system, I have neither the wish nor the right to anticipate. The estimation of a theory is not simply determined by its truth. It also depends upon the importance of its subject, and the extent of its applications; beyond which something must still be left to the arbitrariness of human Opinion. If the utility of the application of Mathematical forms to the science of Logic were solely a question of Notation, I should be content to rest the defence of this attempt upon a principle which has been stated by an able living writer: "Whenever the nature of the subject permits the reasoning process to be without danger carried on mechanically, the language should be constructed on as mechanical principles as possible; while in the contrary case it should be so constructed, that there shall be the greatest possible obstacle to a mere mechanical use of it."** In one respect, the science of Logic differs from all others; the perfection of its method is chiefly valuable as an evidence of the speculative truth of its principles. To supersede the employment of common reason, or to subject it to the rigour of technical forms, would be the last desire of one who knows the value of that intellectual toil and warfare which imparts to the mind an athletic vigour, and teaches it
2|3 to contend with difficulties and to rely upon itself in emergencies. |

* See p. 42.
** Mill's *System of Logic, Ratiocinative and Inductive*, Vol. II. p. 292

MATHEMATICAL ANALYSIS OF LOGIC

INTRODUCTION

THEY who are acquainted with the present state of the theory of Symbolical Algebra, are aware, that the validity of the processes of analysis does not depend upon the interpretation of the symbols which are employed, but solely upon the laws of their combination. Every system of interpretation which does not affect the truth of the relations supposed, is equally admissible, and it is thus that the same process may, under one scheme of interpretation, represent the solution of a question on the properties of numbers, under another, that of a geometrical problem, and under a third, that of a problem of dynamics or optics. This principle is indeed of fundamental importance; and it may with safety be affirmed, that the recent advances of pure analysis have been much assisted by the influence which it has exerted in directing the current of investigation.

But the full recognition of the consequences of this important doctrine has been, in some measure, retarded by accidental circumstances. It has happened in every known form of analysis, that the elements to be determined have been conceived as measurable by comparison with some fixed standard. The predominant idea has been that of magnitude, or more strictly, of numerical ratio. The expression of magnitude, or | of operations upon magnitude, has been the 3|4
express object for which the symbols of Analysis have been invented, and for which their laws have been investigated. Thus the abstractions of the modern Analysis, not less than the ostensive diagrams of the ancient Geometry, have encouraged the notion, that Mathematics are essentially, as well as actually, the Science of Magnitude.

The consideration of that view which has already been stated, as embodying the true principle of the Algebra of Symbols, would, however, lead us to infer that this conclusion is by no means necessary. If every existing interpretation is shewn to involve the idea of magnitude, it is only by induction that we can assert that no other interpretation is possible. And it may be doubted whether our experience is sufficient to render such an induction legitimate. The history of pure Analysis is, it may be said, too recent to permit us to set limits to the extent of its applications. Should we grant to the inference a high degree of probability, we might still, and with reason, maintain the sufficiency of the definition to which the principle already stated would lead us. We might justly assign it as the definitive character of a true Calculus, that it is a method resting upon the employment of Symbols, whose laws of combination are known and general, and whose results admit of a consistent interpretation. That to the existing forms of Analysis a quantitative interpretation is assigned, is the result of the circumstances by which those forms were determined, and is not to be construed into a universal condition of Analysis. It is upon the foundation of this general principle, that I purpose to establish the Calculus of Logic, and that I claim for it a place among the acknowledged forms of Mathematical

Analysis, regardless that in its object and in its instruments it must at present stand alone.

That which renders Logic possible, is the existence in our minds of general notions,—our ability to conceive of a class, and to designate its individual
4|5 members by a common name. | The theory of Logic is thus intimately connected with that of Language. A successful attempt to express logical propositions by symbols, the laws of whose conbinations should be founded upon the laws of the mental processes which they represent, would, so far, be a step toward a philosophical language. But this is a view which we need not here follow into detail.* Assuming the notion of a class, we are able, from any conceivable collection of objects, to separate by a mental act, those which belong to the given class, and to contemplate them apart from the rest. Such, or a similar act of election, we may conceive to be repeated. The group of individuals left under consideration may be still further limited, by mentally selecting those among them which belong to some other recognised class, as well as to the one before contemplated. And this process may be repeated with other elements of distinction, until we arrive at an individual possessing all the distinctive characters which we have taken into account, and a member, at the same time, of every class which we have enumerated. It is in fact a method similar to this which we employ whenever, in common language, we accumulate descriptive epithets for the sake of more precise definition.

Now the several mental operations which in the above case we have supposed to be performed, are subject to peculiar laws. It is possible to assign relations among them, whether as respects the repetition of a given operation or the succession of different ones, or some other particular, which are never violated.
5|6 It is, for example, true that the result of two successive acts is | unaffected by the order in which they are performed ; and there are at least two other laws which will be pointed out in the proper place. These will perhaps to some appear so obvious as to be ranked among necessary truths, and so little important as to be undeserving of special notice. And probably they are noticed for the first time in this Essay. Yet it may with confidence be asserted, that if they were other than they are, the entire mechanism of reasoning, nay the very laws and constitution of the human intellect, would be vitally changed. A Logic might indeed exist, but it would no longer be the Logic we possess.

Such are the elementary laws upon the existence of which, and upon their capability of exact symbolical expression, the method of the following Essay

---

* This view is well expressed in one of Blanco White's Letters:—"Logic is for the most part a collection of technical rules founded on classification. The Syllogism is nothing but a result of the classification of things, which the mind naturally and necessarily forms, in forming a language. All abstract terms are classifications; or rather the labels of the classes which the mind has settled."—*Memoirs of the Rev. Joseph Blanco White*, vol. II, p. 163. See also, for a very lucid introduction, Dr. Latham's *First Outlines of Logic applied to Language*, Becker's *German Grammar*, &c. Extreme Nominalists make Logic entirely dependent upon language. For the opposite view, see Cudworth's *Eternal and Immutable Morality*, Book IV. Chap. III.

is founded; and it is presumed that the object which it seeks to attain will be thought to have been very fully accomplished. Every logical proposition, whether categorical or hypothetical, will be found to be capable of exact and rigorous expression, and not only will the laws of conversion and of syllogism be thence deducible, but the resolution of the most complex systems of propositions, the separation of any proposed element, and the expression of its value in terms of the remaining elements, with every subsidiary relation involved. Every process will represent deduction, every mathematical consequence will express a logical inference. The generality of the method will even permit us to express arbitrary operations of the intellect, and thus lead to the demonstration of general theorems in logic analogous, in no slight degree, to the general theorems of ordinary mathematics. No inconsiderable part of the pleasure which we derive from the application of analysis to the interpretation of external nature, arises from the conceptions which it enables us to form of the universality of the dominion of law. The general formulæ to which we are conducted seem to give to that element a visible presence, and the multitude of particular cases to which they apply, demonstrate the extent of its sway. Even the symmetry | 6|7
of their analytical expression may in no fanciful sense be deemed indicative of its harmony and its consistency. Now I do not presume to say to what extent the same sources of pleasure are opened in the following Essay. The measure of that extent may be left to the estimate of those who shall think the subject worthy of their study. But I may venture to assert that such occasions of intellectual gratification are not here wanting. The laws we have to examine are the laws of one of the most important of our mental faculties. The mathematics we have to construct are the mathematics of the human intellect. Nor are the form and character of the method, apart from all regard to its interpretation, undeserving of notice. There is even a remarkable exemplification, in its general theorems, of that species of excellence which consists in freedom from exception. And this is observed where, in the corresponding cases of the received mathematics, such a character is by no means apparent. The few who think that there is that in analysis which renders it deserving of attention for its own sake, may find it worth while to study it under a form in which every equation can be solved and every solution interpreted. Nor will it lessen the interest of this study to reflect that every peculiarity which they will notice in the form of the Calculus represents a corresponding feature in the constitution of their own minds.

It would be premature to speak of the value which this method may possess as an instrument of scientific investigation. I speak here with reference to the theory of reasoning, and to the principle of a true classification of the forms and cases of Logic considered as a Science.* The aim of these investigations

* "Strictly a Science"; also "an Art".—*Whately's Elements of Logic*. Indeed ought we not to regard all Art as applied Science; unless we are willing, with "the multitude", to consider Art as "guessing and aiming well"?—*Plato, Philebus*.

was in the first instance confined to the expression of the received logic, and
7|8 to the forms of the Aristotelian arrangement, | but it soon became apparent that restrictions were thus introduced, which were purely arbitrary and had no foundation in the nature of things. These were noted as they occurred, and will be discussed in the proper place. When it became necessary to consider the subject of hypothetical propositions (in which comparatively less has been done), and still more, when an interpretation was demanded for the general theorems of the Calculus, it was found to be imperative to dismiss all regard for precedent and authority, and to interrogate the method itself for an expression of the just limits of its application. Still, however, there was no special effort to arrive at novel results. But among those which at the time of their discovery appeared to be such, it may be proper to notice the following.

A logical proposition is, according to the method of this Essay, expressible by an equation the form of which determines the rules of conversion and of transformation, to which the given proposition is subject. Thus the law of what logicians term simple conversion, is determined by the fact, that the corresponding equations are symmetrical, that they are unaffected by a mutual change of place, in those symbols which correspond to the convertible classes. The received laws of conversion were thus determined, and afterwards another system, which is thought to be more elementary, and more general. See Chapter, *On the Conversion of Propositions.*

The premises of a syllogism being expressed by equations, the elimination of a common symbol between them leads to a third equation which expresses the conclusion, this conclusion being always the most general possible, whether Aristotelian or not. Among the cases in which no inference was possible, it was found, that there were two distinct forms of the final equation. It was a considerable time before the explanation of this fact was discovered, but it was at length seen to depend upon the presence or absence of a true medium of comparison between the premises. The distinction which is thought to be new is
8|9 illustrated in the Chapter, *On Syllogisms.* |

The nonexclusive character of the disjunctive conclusion of a hypothetical syllogism, is very clearly pointed out in the examples of this species of argument.

The class of logical problems illustrated in the chapter, *On the Solution of Elective Equations*, is conceived to be new: and it is believed that the method of that chapter affords the means of a perfect analysis of any conceivable system of propositions, an end toward which the rules for the conversion of a single categorical proposition are but the first step.

However, upon the originality of these or any of these views, I am conscious that I possess too slight an acquaintance with the literature of logical science, and especially with its older literature, to permit me to speak with confidence.

It may not be inappropriate, before concluding these observations, to offer a few remarks upon the general question of the use of symbolical language in the mathematics. Objections have lately been very strongly urged against this practice, on the ground, that by obviating the necessity of thought, and substituting a reference to general formulæ in the room of personal effort, it tends to weaken the reasoning faculties.

Now the question of the use of symbols may be considered in two distinct points of view. First, it may be considered with reference to the progress of scientific discovery, and secondly, with reference to its bearing upon the discipline of the intellect.

And with respect to the first view, it may be observed that as it is one fruit of an accomplished labour, that it sets us at liberty to engage in more arduous toils, so it is a necessary result of an advanced state of science, that we are permitted, and even called upon, to proceed to higher problems, than those which we before contemplated. The practical inference is obvious. If through the advancing power of scientific methods, we find that the pursuits on which we were once engaged, afford no longer a sufficiently ample field for intellectual effort, the remedy is, to proceed to higher inquiries, and, in new tracks, to seek for difficulties yet unsubdued. And such is, | indeed, the actual law of scientific 9|10
progress. We must be content, either to abandon the hope of further conquest, or to employ such aids of symbolical language, as are proper to the stage of progress, at which we have arrived. Nor need we fear to commit ourselves to such a course. We have not yet arrived so near to the boundaries of possible knowledge, as to suggest the apprehension, that scope will fail for the exercise of the inventive faculties.

In discussing the second, and scarcely less momentous question of the influence of the use of symbols upon the discipline of the intellect, an important distinction ought to be made. It is of most material consequence, whether those symbols are used with a full understanding of their meaning, with a perfect comprehension of that which renders their use lawful, and an ability to expand the abbreviated forms of reasoning which they induce, into their full syllogistic development; or whether they are mere unsuggestive characters, the use of which is suffered to rest upon authority.

The answer which must be given to the question proposed, will differ according as the one or the other of these suppositions is admitted. In the former case an intellectual discipline of a high order is provided, an exercise not only of reason, but of the faculty of generalization. In the latter case there is no mental discipline whatever. It were perhaps the best security against the danger of an unreasoning reliance upon symbols, on the one hand, and a neglect of their just claims on the other, that each subject of applied mathematics should be treated in the spirit of the methods which were known at the time when the application was made, but in the best form which those methods have assumed. The order of attainment in the individual mind would thus bear some relation to the actual order of scientific discovery, and the more abstract methods of the higher analysis would be offered to such minds only, as were prepared to receive them.

The relation in which this Essay stands at once to Logic and | to Mathematics, 10|11
may further justify some notice of the question which has lately been revived, as to the relative value of the two studies in a liberal education. One of the

chief objections which have been urged against the study of Mathematics in general, is but another form of that which has been already considered with respect to the use of symbols in particular. And it need not here be further dwelt upon, than to notice, that if it avails anything, it applies with an equal force against the study of Logic. The canonical forms of the Aristotelian syllogism are really symbolical; only the symbols are less perfect of their kind than those of mathematics. If they are employed to test the validity of an argument, they as truly supersede the exercise of reason, as does a reference to a formula of analysis. Whether men do, in the present day, make this use of the Aristotelian canons, except as a special illustration of the rules of Logic, may be doubted; yet it cannot be questioned that when the authority of Aristotle was dominant in the schools of Europe, such applications were habitually made. And our argument only requires the admission, that the case is possible.

But the question before us has been argued upon higher grounds. Regarding Logic as a branch of Philosophy, and defining Philosophy as the "science of a real existence", and "the research of causes", and assigning as its *main* business the investigation of the "why, (*τὸ διότι*)", while Mathematics display only the "that, (*τὸ ὅτὶ*)", Sir W. Hamilton has contended, not simply, that the superiority rests with the study of Logic, but that the study of Mathematics is at once dangerous and useless.* The pursuits of the mathematician "have not only not trained him to that acute scent, to that delicate, almost instinctive, tact which, in the twilight of probability, the search and discrimination of its finer facts demand; they have gone to cloud his vision, to indurate his touch, to all but the blazing light, the iron chain of demonstration, and left him out of the narrow confines of his science, to a passive *credulity* in any premises,
11 | 12 or to | an absolute *incredulity* in all". In support of these and of other charges, both argument and copious authority are adduced.** I shall not attempt a complete discussion of the topics which are suggested by these remarks. My object is not controversy, and the observations which follow are offered not in the spirit of antagonism, but in the hope of contributing to the formation of just views upon an important subject. Of Sir W. Hamilton it is impossible to speak otherwise than with that respect which is due to genius and learning.

Philosophy is then described as the *science of a real existence* and *the research of causes*. And that no doubt may rest upon the meaning of the word *cause*, it is further said, that philosophy "mainly investigates the *why*". These definitions are common among the ancient writers. Thus Seneca, one of Sir W. Hamilton's authorities, *Epistle* LXXXVIII., "The philosopher seeks and knows the *causes* of natural things, of which the mathematician searches out and com-

* *Edinburgh Review*, vol. LXII. p. 409, and *Letter to A. De Morgan, Esq.*

** The arguments are in general better than the authorities. Many writers quoted in condemnation of mathematics (Aristo, Seneca, Jerome, Augustine, Cornelius Agrippa, &c.) have borne a no less explicit testimony against other sciences, nor least of all, against that of logic. The treatise of the last named writer *De Vanitate Scientiarum*, must surely have been referred to by mistake.—*Vide* cap. CII.

putes the numbers and the measures." It may be remarked, in passing, that in whatever degree the belief has prevailed, that the business of philosophy is immediately with *causes*; in the same degree has every science whose object is the investigation of *laws*, been lightly esteemed. Thus the Epistle to which we have referred, bestows, by contrast with Philosophy, a separate condemnation on Music and Grammar, on Mathematics and Astronomy, although it is that of Mathematics only that Sir W. Hamilton has quoted.

Now we might take our stand upon the conviction of many thoughtful and reflective minds, that in the extent of the meaning above stated, Philosophy is impossible. The business of true Science, they conclude, is with laws and
phenomena. The nature of Being, the mode of the operation of Cause, the *why*, | 12|13
they hold to be beyond the reach of our intelligence. But we do not require the vantage-ground of this position; nor is it doubted that whether the aim of Philosophy is attainable or not, the desire which impels us to the attempt is an instinct of our higher nature. Let it be granted that the problem which has baffled the efforts of ages, is not a hopeless one; that the "science of a real existence", and "the research of causes", "that kernel" for which "Philosophy, is still militant", do not transcend the limits of the human intellect. I am then compelled to assert, that according to this view of the nature of Philosophy, *Logic forms no part of it*. On the principle of a true classification, we ought no longer to associate Logic and Metaphysics, but Logic and Mathematics.

Should any one after what has been said, entertain a doubt upon this point, I must refer him to the evidence which will be afforded in the following Essay. He will there see Logic resting like Geometry upon axiomatic truths, and its theorems constructed upon that general doctrine of symbols, which constitutes the foundation of the recognised Analysis. In the Logic of Aristotle he will be led to view a collection of the formulæ of the science, expressed by another, but, (it is thought) less perfect scheme of symbols. I feel bound to contend for the absolute exactness of this parallel. It is no escape from the conclusion to which it points to assert, that Logic not only constructs a science, but also inquires into the origin and the nature of its own principles,—a distinction which is denied to Mathematics. "It is wholly beyond the domain of mathematicians", it is said, "to inquire into the origin and nature of their principles."—*Review*, page 415. But upon what ground can such a distinction be maintained? What definition of the term Science will be found sufficiently arbitrary to allow such differences?

The application of this conclusion to the question before us is clear and decisive. The mental discipline which is afforded by the study of Logic, *as an*
exact science, is, in species, the same as that afforded by the study of Analysis. | 13|14

Is it then contended that either Logic or Mathematics can supply a perfect discipline to the Intellect? The most careful and unprejudiced examination of this question leads me to doubt whether such a position can be maintained. The exclusive claims of either must, I believe, be abandoned, nor can any others, partaking of a like exclusive character, be admitted in their room. It is an important observation, which has more than once been made, that it is one thing to arrive at correct premises, and another thing to deduce logical conclusions, and that the business of life depends more upon the former than upon the latter. The study of the exact sciences may teach us the one, and it may give us some general preparation of knowledge and of practice for the attainment of the other, but it is to the union of thought with action, in the field of Practical Logic, the arena of Human Life, that we are to look for its fuller and more perfect accomplishment.

I desire here to express my conviction, that with the advance of our knowledge of all true science, an ever-increasing harmony will be found to prevail among its separate branches. The view which leads to the rejection of one, ought, if consistent, to lead to the rejection of others. And indeed many of the authorities which have been quoted against the study of Mathematics, are even more explicit in their condemnation of Logic. "Natural science," says the Chian Aristo, "is above us, Logical science does not concern us." When such conclusions are founded (as they often are) upon a deep conviction of the preeminent value and importance of the study of Morals, we admit the premises, but must demur to the inference. For it has been well said by an ancient writer, that it is the "characteristic of the liberal sciences, not that they conduct us to Virtue, but that they prepare us for Virtue"; and Melancthon's sentiment, "abeunt studia in mores", has passed into a proverb. Moreover, there is a common ground upon which all sincere votaries of truth may meet, exchanging with each other the language of Flamsteed's appeal to Newton, "The works of the Eternal
14|15 Providence will be better understood through your labors and mine." |

FIRST PRINCIPLES

LET us employ the symbol 1, or unity, to represent the Universe, and let us understand it as comprehending every conceivable class of objects whether actually existing or not, it being premised that the same individual may be found in more than one class, inasmuch as it may possess more than one quality in common with other individuals. Let us employ the letters X, Y, Z, to represent the individual members of classes, X applying to every member of one class, as members of that particular class, and Y to every member of another class as members of such class, and so on, according to the received language of treatises on Logic.

Further let us conceive a class of symbols x, y, z, possessed of the following character.

The symbol x operating upon any subject comprehending individuals or classes, shall be supposed to select from that subject all the *X*s which it contains. In like manner the symbol y, operating upon any subject, shall be supposed to select from it all individuals of the class Y which are comprised in it, and so on.

When no subject is expressed, we shall suppose 1 (the Universe) to be the subject understood, so that we shall have

$$x = x(1),$$

the meaning of either term being the selection from the Universe of all the Xs which it contains, and the result of the operation | being in common language, 15|16 the class X, *i.e.* the class of which each member is an X.

From these premises it will follow, that the product xy will represent, in succession, the selection of the class Y, and the selection from the class Y of such individuals of the class X as are contained in it, the result being the class whose members are both Xs and Ys. And in like manner the product xyz will represent a compound operation of which the successive elements are the selection of the class Z, the selection from it of such individuals of the class Y as are contained in it, and the selection from the result thus obtained of all the individuals of the class X which it contains, the final result being the class common to X, Y, and Z.

From the nature of the operation which the symbols x, y, z, are conceived to represent, we shall designate them as elective symbols. An expression in which they are involved will be called an elective function, and an equation of which the members are elective functions, will be termed an elective equation.

It will not be necessary that we should here enter into the analysis of that mental operation which we have represented by the elective symbol. It is not an act of Abstraction according to the common acceptation of that term, because we never lose sight of the concrete, but it may probably be referred to an exercise of the faculties of Comparison and Attention. Our present concern is rather with the laws of combination and of succession, by which its results are governed, and of these it will suffice to notice the following.

1st. The result of an act of election is independent of the grouping or classification of the subject.

Thus it is indifferent whether from a group of objects considered as a whole, we select the class X, or whether we divide the group into two parts, select the Xs from them separately, and then connect the results in one aggregate conception.

We may express this law mathematically by the equation

$$x(u + v) = xu + xv,$$

| $u + v$ representing the undivided subject, and u and v the component parts of it. 16|17

2nd. It is indifferent in what order two successive acts of election are performed.

Whether from the class of animals we select sheep, and from the sheep those which are horned, or whether from the class of animals we select the horned, and from these such as are sheep, the result is unaffected. In either case we arrive at the class *horned sheep*.

The symbolical expression of this law is

$$xy = yx.$$

3rd. The result of a given act of election performed twice, or any number of times in succession, is the result of the same act performed once.

If from a group of objects we select the Xs, we obtain a class of which all the members are Xs. If we repeat the operation on this class no further change will ensue: in selecting the Xs we take the whole. Thus we have

$$xx = x,$$

or

$$x^3 = x;$$

and supposing the same operation to be *n* times performed, we have

$$x^n = x,$$

which is the mathematical expression of the law above stated.*

The laws we have established under the symbolical forms

$$x(u + v) = xu + xv \qquad (1),$$

$$xy = yx \qquad (2),$$

$$x^n = x \qquad (3),$$

17|18 | are sufficient for the basis of a Calculus. From the first of these, it appears that elective symbols are *distributive*, from the second that they are *commutative*; properties which they possess in common with symbols of *quantity*, and in virtue of which, all the processes of common algebra are applicable to the present system. The one and sufficient axiom involved in this application is that equivalent operations performed upon equivalent subjects produce equivalent results.**

The third law (3) we shall denominate the index law. It is peculiar to elective

---

* The office of the elective symbol x, is to select individuals comprehended in the class X. Let the class X be supposed to embrace the universe; then, whatever the class Y may be, we have

$$xy = y.$$

The office which x performs is now equivalent to the symbol $+$, in one at least of its interpretations, and the index law (3) gives

$$+^n = +,$$

which is the known property of that symbol.

** It is generally asserted by writers on Logic, that all reasoning ultimately depends on an application of the dictum of Aristotle, *de omni et nullo*. "Whatever is predicated universally of any class of things, may be predicated in like manner of any thing comprehended in that class." But it is agreed that this dictum is not immediately applicable in all cases, and that in a majority of instances, a certain previous process of reduction is necessary. What are the elements involved in that process of reduction? Clearly they are as much a part of general reasoning as the dictum itself.

Another mode of considering the subject resolves all reasoning into an application of one or other of the following canons, viz.

1. If two terms agree with one and the same third, they agree with each other.
2. If one term agrees, and another disagrees, with one and the same third, these two disagree with each other.

But the application of these canons depends on mental acts equivalent to those which are involved in the before-named process of reduction. We have to select individuals from classes, to convert propositions, &c., before we can avail ourselves of their guidance. Any account of the process of symbols, and will be found of great importance in enabling us to reduce our results to forms meet for interpretation.

From the circumstance that the processes of algebra may be applied to the present system, it is not to be inferred that the interpretation of an elective equation will be unaffected by such processes. The expression of a truth cannot be negatived by | a legitimate operation, but it may be limited. The equation $y = z$ 18|19
implies that the classes Y and Z are equivalent, member for member. Multiply it by a factor x, and we have

$$xy = xz,$$

which expresses that the individuals which are common to the classes X and Y are also common to X and Z, and *vice versâ*. This is a perfectly legitimate inference, but the fact which it declares is a less general one than was asserted in the original proposition. | 19|20

Of expression and interpretation

A Proposition is a sentence which either affirms or denies, as, All men are mortal, No creature is independent.

A Proposition has necessarily two terms, as *men*, *mortal*; the former of which, or the one spoken of, is called the subject; the latter, or that which is affirmed or denied of the subject, the predicate. These are connected together by the copula *is*, or *is not*, or by some other modification of the substantive verb.

The substantive verb is the only verb recognised in Logic; all others are resolvable by means of the verb *to be* and a participle or adjective, *e.g.* "The Romans conquered"; the word conquered is both copula and predicate, being equivalent to "were (copula) victorious" (predicate).

A Proposition must either be affirmative or negative, and must be also either universal or particular. Thus we reckon in all, four kinds of pure categorical Propositions.

1st. Universal-affirmative, usually represented by A,

Ex. All Xs are Ys.

2nd. Universal-negative, usually represented by E,

Ex. No Xs are Ys.

3rd. Particular-affirmative, usually represented by I,

Ex. Some Xs are Ys.

4th. Particular-negative, usually represented by O,*

Ex. Some Xs are not Ys.

---

reasoning is insufficient, which does not represent, as well the laws of the operation which the mind performs in that process, as the primary truths which it recognises and applies.

It is presumed that the laws in question are adequately represented by the fundamental equations of the present Calculus. The proof of this will be found in its capability of expressing propositions, and of exhibiting in the results of its processes, every result that may be arrived at by ordinary reasoning.

* The above is taken, with little variation, from the Treatises of Aldrich and Whately.

1. To express the class, not-X, that is, the class including all individuals that are not Xs.

The class X and the class not-X together make the Universe. But the Universe is 1, and the class X is determined by the symbol x, therefore the class not-X
20|21 will be determined by the symbol $1 - x$. |

Hence the office of the symbol $1 - x$ attached to a given subject will be, to select from it all the not-Xs which it contains.

And in like manner, as the product xy expresses the entire class whose members are both Xs and Ys, the symbol $y(1 - x)$ will represent the class whose members are Ys but not Xs, and the symbol $(1 - x)(1 - y)$ the entire class whose members are neither Xs nor Ys.

2. To express the Proposition, All Xs are Ys.

As all the Xs which exist are found in the class Y, it is obvious that to select out of the Universe all Ys, and from these to select all Xs, is the same as to select at once from the Universe all Xs.

Hence

$$xy = x,$$

or

$$x(1 - y) = 0, \quad (4).$$

3. To express the Proposition, No Xs are Ys.

To assert that no Xs are Ys, is the same as to assert that there are no terms common to the classes X and Y. Now all individuals common to those classes are represented by xy. Hence the Proposition that No Xs are Ys, is represented by the equation

$$xy = 0, \quad (5).$$

4. To express the Proposition, Some Xs are Ys.

If some Xs are Ys, there are some terms common to the classes X and Y. Let those terms constitute a separate class V, to which there shall correspond a separate elective symbol v, then

$$v = xy, \quad (6).$$

And as v includes all terms common to the classes X and Y, we can indifferently
21|22 interpret it, as Some Xs, or Some Ys. |

5. To express the Proposition, Some Xs are not Ys.

In the last equation write $1 - y$ for y, and we have

$$v = x(1 - y), \quad (7),$$

the interpretation of v being indifferently Some Xs or Some not-Ys.

The above equations involve the complete theory of categorical Propositions, and so far as respects the employment of analysis for the deduction of logical inferences, nothing more can be desired. But it may be satisfactory to notice some particular forms deducible from the third and fourth equations, and susceptible of similar application.

If we multiply the equation (6) by x, we have

$$vx = x^2y = xy \text{ by (3).}$$

Comparing with (6), we find

$$v = vx,$$

or

$$v(1 - x) = 0, \quad (8).$$

And multiplying (6) by y, and reducing in a similar manner, we have

$$v = vy,$$

or

$$v(1 - y) = 0, \quad (9).$$

Comparing (8) and (9),

$$vx = vy = v, \quad (10).$$

And further comparing (8) and (9) with (4), we have as the equivalent of this system of equations the Propositions

$$\left.\begin{array}{l}\text{All Vs are Xs}\\ \text{All Vs are Ys}\end{array}\right\}.$$

The system (10) might be used to replace (6), or the single equation

$$vx = vy, \quad (11),$$

might be used, assigning to vx the interpretation, Some Xs, and to vy the interpretation, Some Ys. But it will be observed that | this system does not express 22|23
quite so much as the single equation (6), from which it is derived. Both, indeed, express the Proposition, Some Xs are Ys, but the system (10) does not imply that the class V includes *all* the terms that are common to X and Y.

In like manner, from the equation (7) which expresses the Proposition Some Xs are not Ys, we may deduce the system

$$vx = v(1 - y) = v, \quad (12),$$

in which the interpretation of $v(1 - y)$ is Some not-Ys. Since in this case $vy = 0$, we must of course be careful not to interpret vy as Some Ys.

If we multiply the first equation of the system (12), viz.

$$vx = v(1 - y),$$

by y, we have

$$vxy = vy(1 - y);$$

$$\therefore\ vxy = 0, \quad (13),$$

which is a form that will occasionally present itself. It is not necessary to revert to the primitive equation in order to interpret this, for the condition that vx represents Some Xs, shews us by virtue of (5), that its import will be

Some Xs are not Ys,

the subject comprising *all* the Xs that are found in the class V.

Universally in these cases, difference of form implies a difference of interpretation with respect to the auxiliary symbol v, and each form is interpretable by itself.

Further, these differences do not introduce into the Calculus a needless perplexity. It will hereafter be seen that they give a precision and a definiteness to its conclusions, which could not otherwise be secured.

Finally, we may remark that all the equations by which particular truths are expressed, are deducible from any one general equation, expressing any one general Proposition, from which those particular Propositions are necessary
23|24 deductions. | This has been partially shewn already, but it is much more fully exemplified in the following scheme.

The general equation $$x = y,$$

implies that the classes X and Y are equivalent, member for member; that every individual belonging to the one, belongs to the other also. Multiply the equation by x, and we have

$$x^2 = xy;$$

$$\therefore\ x = xy,$$

which implies, by (4), that all Xs are Ys. Multiply the same equation by y, and we have in like manner

$$y = xy;$$

the import of which is, that all Ys are Xs. Take either of these equations, the latter for instance, and writing it under the form

$$(1 - x)y = 0,$$

we may regard it as an equation in which y, an unknown quantity, is sought to be expressed in terms of x. Now it will be shewn when we come to treat of the Solution of Elective Equations (and the result may here be verified by substitution) that the most general solution of this equation is

$$y = vx,$$

which implies that All Ys are Xs, and that Some Xs are Ys. Multiply by x, and we have

$$vy = vx,$$

which indifferently implies that some Ys are Xs and some Xs are Ys, being the particular form at which we before arrived.

For convenience of reference the above and some other results have been classified in the annexed Table, the first column of which contains propositions, the second equations, and the third the conditions of final interpretation. It is

to be observed, that the auxiliary equations which are given in this column are not independent: they are implied either in the equations of the second column, or in the condition for | the interpretation of v. But it has been thought better 24|25
to write them separately, for greater ease and convenience. And it is further to be borne in mind, that although three different forms are given for the expression of each of the *particular* propositions, everything is really included in the first form.

TABLE.

The class X	x	
The class not-X	$1 - x$	
All Xs are Ys All Ys are Xs	$x = y$	
All Xs are Ys	$x(1 - y) = 0$	
No Xs are Ys	$xy = 0$	
All Ys are Xs Some Xs are Ys	$y = vx$	$vx =$ some Xs $v(1 - x) = 0.$
No Ys are Xs Some not-Xs areYs	$y = v(1 - x)$	$v(1 - x) =$ some not-Xs $vx = 0.$
Some Xs are Ys	$v = xy$ or $vx = vy$ or $vx\ (1 - y) = 0$	$v =$ some Xs or some Ys $vx =$ some Xs, $vy =$ some Ys $v(1 - x) = 0$, $v(1 - y) = 0.$
Some Xs are not Ys	$v = x(1 - y)$ or $vx = v(1 - y)$ or $vxy = 0$	$v =$ some Xs, or some not-Ys $vx =$ some Xs, $v(1 - y) =$ some not-Ys $v(1 - x) = 0$, $vy = 0$. | 25|26

Of the conversion of propositions

A Proposition is said to be converted when its terms are tranposed; when nothing more is done, this is called simple conversion; *e.g.*

No virtuous man is a tyrant, *is converted into*
No tyrant is a virtuous man.

Logicians also recognise conversion *per accidens*, or by limitation, *e.g.*

All birds are animals, *is converted into*
Some animals are birds.

And conversion by *contraposition* or *negation*, as

Every poet is a man of genius, *converted into*
He who is not a man of genius is not a poet.

In one of these three ways every Proposition may be illatively converted, viz. E and I simply, A and O by negation, A and E by limitation.

The primary canonical forms already determined for the expression of Propositions, are

All Xs are Ys,	$x(1-y)=0,$	....A.
No Xs are Ys,	$xy=0,$	....E.
Some Xs are Ys,	$v=xy,$	....I.
Some Xs are not Ys,	$v=x(1-y)$	....O.

On examining these, we perceive that E and I are symmetrical with respect to x and y, so that x being changed into y, and y into x, the equations remain unchanged. Hence E and I may be interpreted into

No Ys are Xs,
Some Ys are Xs,

respectively. Thus we have the known rule of the Logicians, that particular
26|27 affirmative and universal negative Propositions admit of simple conversion. |

The equations A and O may be written in the forms

$$(1-y)\{1-(1-x)\}=0,$$
$$v=(1-y)\{1-(1-x)\}.$$

Now these are precisely the forms which we should have obtained if we had in those equations changed x into $1-y$, and y into $1-x$, which would have represented the changing in the original Propositions of the Xs into not-Ys, and the Ys into not-Xs, the resulting Propositions being

All not-Ys are not-Xs,
Some not-Ys are not not-Xs (a).

Or we may, by simply inverting the order of the factors in the second member of O, and writing it in the form

$$v=(1-y)x,$$

interpret it by I into

Some not-Ys are Xs,

which is really another form of (a). Hence follows the rule, that universal affirmative and particular negative Propositions admit of negative conversion, or, as it is also termed, conversion by contraposition.

The equations A and E, written in the forms

$$(1-y)x=0,$$
$$yx=0,$$

give on solution the respective forms

$$x=vy,$$

$$x = v(1 - y),$$

the correctness of which may be shewn by substituting these values of x in the equations to which they belong, and observing that those equations are satisfied quite independently of the nature of the symbol v. The first solution may be interpreted into

Some Ys are Xs,

and the second into

Some not-Ys are Xs. | 27|28

From which it appears that universal-affirmative, and universal-negative Propositions are convertible by limitation, or, as it has been termed, *per accidens*.

The above are the laws of Conversion recognized by Abp. Whately. Writers differ however as to the admissibility of negative conversion. The question depends on whether we will consent to use such terms as not-X, not-Y. Agreeing with those who think that such terms ought to be admitted, even although they change the *kind* of the Proposition, I am constrained to observe that the present classification of them is faulty and defective. Thus the conversion of No Xs are Ys, into All Ys are not-Xs, though perfectly legitimate, is not recognised in the above scheme. It may therefore be proper to examine the subject somewhat more fully.

Should we endeavour, from the system of equations we have obtained, to deduce the laws not only of the conversion, but also of the general transformation of propositions, we should be led to recognise the following distinct elements, each connected with a distinct mathematical process.

1st. The negation of a term, *i.e.* the changing of X into not-X, or not-X into X.

2nd. The translation of a Proposition from one *kind* to another, as if we should change

All Xs are Ys into Some Xs are Ys   A into I,

which would be lawful; or

All Xs are Ys into No Xs are Y.   A into E,

which would be unlawful.

3rd. The simple conversion of a Proposition.

The conditions in obedience to which these processes may lawfully be performed, may be deduced from the equations by which Propositions are expressed.

We have

All Xs are Ys ..... $x(1 - y) = 0$.   A,
No Xs are Ys .... $xy = 0$.   E. | 28|29

Write E in the form

$$x\{1 - (1 - y)\} = 0,$$

and it is interpretable by A into

All Xs are not-Ys,

so that we may change

No Xs are Ys into All Xs are not-Ys.

In like manner A interpreted by E gives

No Xs are not-Ys,

so that we may change

All Xs are Ys into No Xs are not-Ys.

From these cases we have the following Rule: A universal-affirmative Proposition is convertible into a universal-negative, and, *vice versâ*, by negation of the predicate.

Again, we have

Some Xs are Ys. . . . . . $v = xy$,
Some Xs are not Ys. . . . $v = x(1 - y)$.

These equations only differ from those last considered by the presence of the term v. The same reasoning therefore applies, and we have the Rule—

A particular-affirmative proposition is convertible into a particular-negative, and *vice versâ*, by negation of the predicate.

Assuming the universal Propositions

All Xs are Ys. . . . . . $x(1 - y) = 0$,
No Xs are Ys . . . . $xy = 0$.

Multiplying by v, we find

$$vx\ (1 - y) = 0,$$
$$vxy = 0,$$

which are interpretable into

Some Xs are Ys . . . . I,
29|30 Some Xs are not Ys . . . . O. |

Hence a universal-affirmative is convertible into a particular-affirmative, and a universal-negative into a particular-negative without negation of subject or predicate.

Combining the above with the already proved rule of simple conversion, we arrive at the following system of independent laws of transformation.

1st. An affirmative Proposition may be changed into its corresponding negative (A into E, or I into O), and *vice versâ*, by negation of the predicate.

2nd. A universal Proposition may be changed into its corresponding particular Proposition, (A into I, or E into O).

3rd. In a particular-affirmative, or universal-negative Proposition, the terms may be mutually converted.

Wherein negation of a term is the changing of X into not-X, and *vice versâ*, and is not to be understood as affecting the *kind* of the Proposition.

Every lawful transformation is reducible to the above rules. Thus we have

All Xs are Ys,	
No Xs are not-Ys	by 1st rule,
No not-Ys are Xs	by 3rd rule,
All not-Ys are not-Xs	by 1st rule,

which is an example of *negative conversion*. Again,

No Xs are Ys	
No Ys are Xs	3rd rule,
All Ys are not-Xs	1st rule,

which is the case already deduced. | 30|31

Of syllogisms

A Syllogism consists of three Propositions, the last of which, called the conclusion, is a logical consequence of the two former, called the premises; e.g.

Premises, { All Ys are Xs.
All Zs are Ys.

Conclusion, All Zs are Xs.

Every syllogism has three and only three terms, whereof that which is the subject of the conclusion is called the *minor* term, the predicate of the conclusion, the *major* term, and the remaining term common to both premises, the middle term. Thus, in the above formula, Z is the minor term, X the major term, Y the middle term.

The figure of a syllogism consists in the situation of the middle term with respect to the terms of the conclusion. The varieties of figure are exhibited in the annexed scheme.

1st Fig.	2nd Fig.	3rd Fig.	4th Fig.
YX	XY	YX	XY
ZY	ZY	YZ	YZ
ZX	ZX	ZX	ZX

When we designate the three propositions of a syllogism by their usual symbols (*A*, *E*, *I*, *O*), and in their usual order, we are said to determine the mood of the syllogism. Thus the syllogism given above, by way of illustration, belongs to the mood AAA in the first figure.

The moods of all syllogisms commonly received as valid, are represented by the vowels in the following mnemonic verses.

Fig. 1.—bArbArA, cElArEnt, dArII, fErIO que prioris.
Fig. 2.—cEsArE, cAmEstrEs, fEstInO, bArOkO, secundæ.
Fig. 3.—Tertia dArAptI, dIsAmIs, dAtIsI, fElAptOn,
bOkArdO, fErIsO, habet: quarta insuper addit.
Fig. 4.—brAmAntIp, cAmEnEs, dImArIs, fEsApO, frEsIsOn.

THE equation by which we express any Proposition concerning the classes X and Y, is an equation between the symbols x and y, and the equation by
31 | 32 which we express any | Proposition concerning the classes Y and Z, is an equation between the symbols y and z. If from two such equations we eliminate y, the result, if it do not vanish, will be an equation between x and z, and will be interpretable into a Proposition concerning the classes X and Z. And it will then constitute the third member, or Conclusion, of a Syllogism, of which the two given Propositions are the premises.

The result of the elimination of y from the equations

$$\begin{aligned} ay + b &= 0, \\ a'y + b' &= 0, \end{aligned} \qquad (14),$$

is the equation

$$ab' - a'b = 0, \qquad (15).$$

Now the equations of Propositions being of the first order with reference to each of the variables involved, all the cases of elimination which we shall have to consider, will be reducible to the above case, the constants a, b, a', b', being replaced by functions of x, z, and the auxiliary symbol v.

As to the choice of equations for the expression of our premises, the only restriction is, that the equations must not *both* be of the form $ay = 0$, for in such cases elimination would be impossible. When both equations are of this form, it is necessary to solve one of them, and it is indifferent which we choose for this purpose. If that which we select is of the form $xy = 0$, its solution is

$$y = v(1 - x), \quad (16),$$

if of the form $(1 - x)\, y = 0$, the solution will be

$$y = vx, \quad (17),$$

and these are the only cases which can arise. The reason of this exception will appear in the sequel.

For the sake of uniformity we shall, in the expression of particular propositions, confine ourselves to the forms

$vx = vy$, Some Xs are Ys,
32 | 33 $vx = v(1 - y)$, Some Xs are not Ys. |

These have a closer analogy with (16) and (17), than the other forms which might be used.

Between the forms about to be developed, and the Aristotelian canons, some points of difference will occasionally be observed, of which it may be proper to forewarn the reader.

To the right understanding of these it is proper to remark, that the essential structure of a Syllogism is, in some measure, arbitrary. Supposing the order of the premises to be fixed, and the distinction of the major and the minor term to be thereby determined, it is purely a matter of choice which of the two shall have precedence in the Conclusion. Logicians have settled this question in favour of the minor term, but it is clear, that this is a convention. Had it been agreed that the major term should have the first place in the conclusion, a logical scheme might have been constructed, less convenient in some cases than the existing one, but superior in others. What is lost in *barbara*, it would gain in *bramantip*. Convenience is *perhaps* in favour of the adopted arrangement,* but is is to be remembered that is *merely* an arrangement.

Now the method we shall exhibit, not having reference to one scheme of arrangement more than to another, will always give the more general conclusion, regard being paid only to its abstract lawfulness, considered as a result of pure reasoning. And therefore we shall sometimes have presented to us the spectacle of conclusions, which a logician would pronounce informal, but never of such as a reasoning being would account false.

The Aristotelian canons, however, beside restricting the *order* of the terms of a conclusion, limit their nature also;—and this limitation is of more consequence than the former. We may, by a change of figure, replace the particular conclusion | of *bramantip*, by the general conclusion of *barbara*; but we cannot 33|34
thus reduce to rule such inferences, as

Some not-Xs are not Ys.

Yet there are cases in which such inferences may lawfully be drawn, and in unrestricted argument they are of frequent occurrence. Now if an inference of this, or of any other kind, is lawful in itself, it will be exhibited in the results of our method.

We may by restricting the canon of interpretation confine our expressed results within the limits of the scholastic logic: but this would only be to restrict ourselves to the use of a part of the conclusions to which our analysis entitles us.

The classification we shall adopt will be purely mathematical, and we shall afterwards consider the logical arrangement to which it corresponds. It will be sufficient, for reference, to name the premises and the Figure in which they are found.

CLASS 1st.—Forms in which v does not enter.

Those which admit of an inference are AA, EA, Fig. 1; AE, EA, Fig. 2; AA, AE, Fig. 4.

Ex. AA, Fig. 1, and, by mutation of premises (change of order), AA, Fig. 4.

All Ys are Xs, $y(1-x)=0$, or $(1-x)y=0$,
All Zs are Ys, $z(1-y)=0$, or $zy-z=0$.

Eliminating y by (15) we have

* The contrary view was maintained by Hobbes. The question is very fairly discussed in Hallam's *Introduction to the Literature of Europe*, vol. III. p. 309. In the rhetorical use of Syllogism, the advantage appears to rest with the rejected form.

$$z(1-x)=0,$$
$$\therefore \text{ All Zs are Xs.}$$

A convenient mode of effecting the elimination, is to write the equation of the premises, so that y shall appear only as a factor of one member in the first equation, and only as a factor of the opposite member in the second equation,
34|35 and then to multiply the equations, omitting the y. This method we shall adopt. |

Ex. AE, Fig. 2, and, by mutation of premises, EA, Fig. 2.

$$\begin{array}{lll} \text{All Xs are Ys,} & x(1-y)=0, & \text{or } x=xy \\ \text{No Zs are Ys,} & zy=0, & \quad zy=0 \\ \hline & & \quad zx=0 \\ & & \therefore \text{ No Zs are Xs.} \end{array}$$

The only case in which there is no inference is AA, Fig. 2,

$$\begin{array}{lll} \text{All Xs are Ys,} & x(1-y)=0, & x=xy \\ \text{All Zs are Ys,} & z(1-y)=0, & zy=z \\ \hline & & xz=xz \\ & & \therefore 0=0. \end{array}$$

CLASS 2nd.—When v is introduced by the solution of an equation.

The lawful cases directly or indirectly* determinable by the Aristotelian Rules are AE, Fig. 1; AA, AE, EA, Fig. 3; EA, Fig. 4.

The lawful cases not so determinable, are EE, Fig. 1; EE, Fig. 2; EE, Fig. 3; EE, Fig. 4.

Ex. AE, Fig. 1, and, by mutation of premises, EA, Fig. 4.

$$\begin{array}{lll} \text{All Ys are Xs,} & y(1-x)=0, & y=vx \ (a) \\ \text{No Zs are Ys,} & zy=0, & 0=zy \\ \hline & & 0=vzx \\ & & \therefore \text{ Some Xs are not Zs.} \end{array}$$

The reason why we cannot interpret $vzx=0$ into Some Zs are not-Xs, is that by the very terms of the first equation (a) the interpretation of vx is fixed, as Some Xs; v is regarded as the representative of Some, only with reference to
35|36 the class X. |

For the reason of our employing a solution of one of the primitive equations, see the remarks on (16) and (17). Had we solved the second equation instead of the first, we should have had

---

* We say *directly* or *indirectly*, mutation or conversion of premises being in some instances required. Thus, AE (fig. 1) is resolvable by Fesapo (fig. 4), or by Ferio (fig. 1). Aristotle and his followers rejected the fourth figure as only a modification of the first, but this being a mere question of form, either scheme may be termed Aristotelian.

$$(1-x)y=0,$$
$$v(1-z)=y, \ (a),$$
$$v(1-z)(1-x)=0, \ (b),$$

$\therefore$ Some not-Zs are Xs.

Here it is to be observed, that the second equation (*a*) fixes the meaning of $v(1-z)$, as Some not-Zs. The full meaning of the result (*b*) is, that all the not-Zs which are found in the class Y are found in the class X, and it is evident that this could not have been expressed in any other way.

Ex. 2. AA, Fig. 3.

All Ys are Xs,	$y(1-x)=0,$	$y=vx$
All Ys are Zs,	$y(1-z)=0,$	$0=y(1-z)$
		$0=vx(1-z)$
		$\therefore$ Some Xs are Zs.

Had we solved the second equation, we should have had as our result, Some Zs are Xs. The form of the final equation particularizes what Xs or what Zs are referred to, and this remark is general.

The following, EE, Fig. 1, and, by mutation, EE, Fig. 4, is an example of a lawful case not determinable by the Aristotelian Rules.

No Ys are Xs,	$xy=0,$	$0=xy$
No Zs are Ys,	$zy=0,$	$y=v(1-z)$
		$0=v(1-z)x$
		$\therefore$ Some not-Zs are not Xs.

CLASS 3rd.—When v is met with in one of the equations, but not introduced by solution. | 36|37

The lawful cases determinable *directly* or *indirectly* by the Aristotelian Rules, are AI, EI, Fig. 1; AO, EI, OA, IE, Fig. 2; AI, AO, EI, EO, IA, IE, OA, OE, Fig. 3; IA, IE, Fig. 4.

Those not so determinable are OE, Fig. 1; EO, Fig. 4.

The cases in which no inference is possible, are AO, EO, IA, IE, OA, Fig. 1; AI, EO, IA, OE, Fig. 2; OA, OE, AI, EI, AO, Fig. 4.

Ex. 1. AI, Fig. 1, and, by mutation, IA, Fig. 4.

All Ys are Xs,	$y(1-x)=0$
Some Zs are Ys,	$vz=vy$
	$vz(1-x)=0$
	$\therefore$ Some Zs are Xs.

Ex. 2. AO, Fig. 2, and, by mutation, OA, Fig. 2.

All Xs are Ys,	$x(1-y)=0,$	$x = xy$
Some Zs are not Ys,	$vz = v(1-y),$	$vy = v(1-z)$

$$vx = vx(1-z)$$
$$vxz = 0$$

$\therefore$ Some Zs are not Xs.

The interpretation of vz as Some Zs, is implied, it will be observed, in the equation $vz = v(1-y)$ considered as representing the proposition Some Zs are not Ys.

The cases not determinable by the Aristotelian Rules are OE, Fig. 1, and, by mutation, EO, Fig. 4.

Some Ys are not Xs,	$xy = v(1-x)$
No Zs are Ys,	$0 = zy$

$$0 = v(1-x)z$$

$\therefore$ Some not-Xs are not Zs.

The equation of the first premiss here permits us to interpret $v(1-x)$,
37|38 but it does not enable us to interpret vz. |

Of cases in which no inference is possible, we take as examples—

AO, Fig. 1, and, by mutation, OA, Fig. 4,

All Ys are Xs,	$y(1-x)=0,$	$y(1-x)=0$
Some Zs are not Ys,	$vz = v(1-y)$ (*a*)	$v(1-z) = vy$

$$v(1-z)(1-x) = 0 \quad (b)$$
$$0 = 0$$

since the auxiliary equation in this case is $v(1-z) = 0$.

Practically it is not necessary to perform this reduction, but it is satisfactory to do so. The equation (*a*), it is seen, defines vz as Some Zs, but it does not define $v(1-z)$, so that we might stop at the result of elimination (*b*), and content ourselves with saying, that it is not interpretable into a relation between the classes X and Z.

Take as a second example AI, Fig. 2, and, by mutation, IA, Fig. 2,

All Xs are Ys,	$x(1-y)=0,$	$x = xy$
Some Zs are Ys,	$vz = vy,$	$vy = vz$

$$vx = vxz$$
$$v(1-z)x = 0$$
$$0 = 0,$$

the auxiliary equation in this case being $v(1-z) = 0$.

Indeed in every case in this class, in which no inference is possible, the result of elimination is reducible to the form $0 = 0$. Examples therefore need not be multiplied.

CLASS 4th.—When v enters into both equations.

No inference is possible in any case, but there exists a distinction among the unlawful cases which is peculiar to this class. The two divisions are,

1st. When the result of elimination is reducible by the auxiliary equations to the form $0 = 0$. The cases are II, OI, | Fig. 1; II, OO, Fig. 2; II, IO, OI, OO, 38|39
Fig. 3; II, IO, Fig. 4.

2nd. When the result of elimination is not reducible by the auxiliary equations to the form $0 = 0$.

The cases are IO, OO, Fig. 1; IO, OI, Fig. 2; OI, OO, Fig. 4.

Let us take as an example of the former case, II, Fig. 3.

$$\begin{array}{lll} \text{Some Xs are Ys,} & vx = vy, & vx = vy \\ \text{Some Zs are Ys,} & v'z = v'y, & v'y = v'z \\ \hline & & vv'x = vv'z \end{array}$$

Now the auxiliary equations $v(1 - x) = 0$, $v'(1 - z) = 0$,

$$\text{give } vx = v, \ v'z = v'.$$

Substituting we have

$$vv' = vv',$$
$$\therefore\ 0 = 0.$$

As an example of the latter case, let us take IO, Fig. 1,

$$\begin{array}{lll} \text{Some Ys are Xs,} & vy = vx, & vy = vx \\ \text{Some Zs are not Ys,} & v'z = v'(1 - y), & v'(1 - z) = v'y \\ \hline & & vv'(1 - z) = vv'x \end{array}$$

Now the auxiliary equations being $v(1 - x) = 0$, $v'(1 - z) = 0$, the above reduces to $vv' = 0$. It is to this form that all similar cases are reducible. Its interpretation is, that the classes v and v' have no common member, as is indeed evident.

The above classification is purely founded on mathematical distinctions. We shall now inquire what is the logical division to which it corresponds.

The lawful cases of the first class comprehend all those in which, from two universal premises, a universal conclusion may be drawn. We see that they include the premises of *barbara* and *celarent* in the first figure, of *cesare* and *camestres* in the second, and of *bramantip* and *camenes* in the fourth. | The 39|40
premises of *bramantip* are included, because they admit of an universal conclusion, although not in the same figure.

The lawful cases of the second class are those in which a particular conclusion only is deducible from two universal premises.

The lawful cases of the third class are those in which a conclusion is deducible from two premises, one of which is universal and the other particular.

The fourth class has no lawful cases.

Among the cases in which no inference of any kind is possible, we find six in the fourth class distinguishable from the others by the circumstance, that the result of elimination does not assume the form $0 = 0$. The cases are

$$\left\{\begin{array}{l}\text{Some Ys are Xs,}\\ \text{Some Zs are not Ys,}\end{array}\right\} \left\{\begin{array}{l}\text{Some Ys are not Xs,}\\ \text{Some Zs are not Ys,}\end{array}\right\} \left\{\begin{array}{l}\text{Some Xs are Ys,}\\ \text{Some Zs are not Ys,}\end{array}\right\}$$

and the three others which are obtained by mutation of premises.

It might be presumed that some logical peculiarity would be found to answer to the mathematical peculiarity which we have noticed, and in fact there exists a very remarkable one. If we examine each pair of premises in the above scheme, we shall find that there *is virtually* no middle term, i.e. *no medium of comparison*, in any of them. Thus, in the first example, the individuals spoken of in the first premiss are asserted to belong to the class Y, but those spoken of in the second premiss are *virtually* asserted to belong to the class not-Y: nor can we by any lawful transformation or conversion alter this state of things. The comparison will still be made with the class Y in one premiss, and with the class not-Y in the other.

Now in every cases beside the above six, there will be found a middle term,
40|41 either expressed or implied. I select two of the most difficult cases. |

In AO, Fig. 1, viz.

All Ys are Xs,
Some Zs are not Ys,

we have, by *negative conversion* of the first premiss,

All not-Xs are not-Ys,
Some Zs are not Ys,

and the middle term is now seen to be not-Y.

Again, in EO, Fig. 1,

No Ys are Xs,
Some Zs are not Ys,

a proved conversion of the first premiss (see *Conversion of Propositions*), gives

All Xs are not-Ys,
Some Zs are not-Ys,

and the middle term, the true medium of comparison, is plainly not-Y, although as the not-Ys in the one premiss *may be* different from those in the other, no conclusion can be drawn.

The mathematical condition in question, therefore,—the irreducibility of the final equation to the form $0 = 0$,—adequately represents the logical condition of there being no middle term, or common medium of comparison, in the given premises.

I am not aware that the distinction occasioned by the presence or absence of a middle term, in the strict sense here understood, has been noticed by logicians before. The distinction, though real and deserving attention, is indeed by no means an obvious one, and it would have been unnoticed in the present instance but for the peculiarity of its mathematical expression.

What appears to be novel in the above case is the proof of the existence of combinations of premises in which there | is absolutely no medium of compari- 41|42
son. When such a medium of comparison, or true middle term, does exist, the condition that its quantification in both premises together shall exceed its quantification as a single whole, has been ably and clearly shewn by Professor De Morgan to be necessary to lawful inference (*Cambridge Memoirs*, Vol. VIII. Part 3). And this is undoubtedly the true principle of the Syllogism, viewed from the standing-point of Arithmetic.

I have said that it would be possible to impose conditions of interpretation which should restrict the results of this calculus to the Aristotelian forms. Those conditions would be,

1st. That we should agree not to interpret the forms $v(1-x)$, $v(1-z)$.

2ndly. That we should agree to reject every interpretation in which the order of the terms should violate the Aristotelian rule.

Or, instead of the second condition, it might be agreed that, the conclusion being determined, the order of the premises should, if necessary, be changed, so as to make the syllogism formal.

From the *general* character of the system it is indeed plain, that it may be made to represent any conceivable scheme of logic, by imposing the conditions proper to the case contemplated.

We have found it, in a certain class of cases, to be necessary to replace the two equations expressive of universal Propositions, by their solutions; and it may be proper to remark, that it would have been allowable in all instances to have done this,* so that every case of the Syllogism, without | exception, might 42|43

* It may be satisfactory to illustrate this statement by an example. In *Barbara*, we should have

All Ys are Xs, $\quad y = vx$
All Zs are Ys, $\quad z = v'y$

$$z = vv'x$$

∴ All Zs are Xs. | 42|43

Or, we may multiply the resulting equation by $1 - x$, which gives

$$z(1-x) = 0,$$

whence the same conclusion, All Zs are Xs.

Some additional examples of the application of the system of equations in the text to the demonstration of general theorems, may not be inappropriate.

Let y be the term to be eliminated, and let x stand indifferently for either of the other symbols, then each of the equations of the premises of any given syllogism may be put in the form

$$ay + bx = 0, \quad (\alpha)$$

if the premiss is affirmative, and in the form

$$ay + b(1-x) = 0, \quad (\beta)$$

if it is negative, a and b being either constant, or of the form $\pm v$. To prove this in detail, let us examine each kind of proposition, making y successively subject and predicate.

A, All Ys are Xs, $y - vx = 0$, (γ),

All Xs are Ys, $x - vy = 0$, (δ),

E, No Ys are Xs, $xy = 0$,

No Xs are Ys, $y - v(1 - x) = 0$, (ε),

I, Some Xs are Ys,

Some Ys are Xs, $vx - vy = 0$, (ζ),

O, Some Ys are not Xs, $vy - v(1 - x) = 0$, (η),

Some Xs are not Ys, $vx = v(1 - y)$,

$\therefore\ vy - v(1 - x) = 0$, (θ).

The affirmative equations (γ), (δ) and (ζ), belong to (α), and the negative equations (ε), (η) and (θ), to (β). It is seen that the two last negative equations are alike, but there is a difference of interpretation. In the former

$$v(1 - x) = \text{Some not-Xs},$$

in the latter,

$$v(1 - x) = 0.$$

The utility of the two general forms of reference, (α) and (β), will appear from the following application.

1st. *A conclusion drawn from two affirmative propositions* is itself affirmative.

By (α) we have for the given propositions,

$$ay + bx = 0,$$

43|44 $$a'y + b'z = 0, \ |$$

and eliminating

$$ab'z - a'bx = 0,$$

which is of the form (α). Hence, if there is a conclusion, it is affirmative.

2nd. *A conclusion drawn from an affirmative and a negative proposition is negative.*

By (α) and (β), we have for the given propositions

$$ay + bx = 0,$$

$$a'y + b'(1 - z) = 0,$$

$$\therefore\ a'bx - ab'(1 - z) = 0,$$

which is of the form (β). Hence the conclusion, if there is one, is negative.

3rd. *A conclusion drawn from two negative premises will involve a negation, (not-X, not-Z) in both subject and predicate, and will therefore be inadmissible in the Aristotelian system, though just in itself.*

For the premises being

$$ay + b(1 - x) = 0,$$

$$a'y + b'(1 - z) = 0,$$

the conclusion will be

$$ab'(1 - z) - a'b(1 - x) = 0,$$

which is only interpretable into a proposition that has a negation in each term.

4th. *Taking into account those syllogisms only, in which the conclusion is the most general, that can be deduced from the premises,—if, in an Aristotelian syllogism, the minor premises be changed* have been treated by equations comprised in the general forms

$$
\begin{array}{lcr}
y = vx, & \text{or} & y - vx = 0.\ldots\text{A}, \\
y = v(1-x), & \text{or} & y + vx - v = 0.\ldots\text{E}, \\
vy = vx, & & vy - vx = 0.\ldots\text{I}, \\
vy = v(1-x), & & vy + vx - v = 0.\ldots\text{O}.\ |
\end{array}
$$ 43|44

Perhaps the system we have actually employed is better, as distinguishing the cases in which v only *may* be employed, | from those in which it *must*. But 44|45 for the demonstration of certain general properties of the Syllogism, the above system is, from its simplicity, and from the mutual analogy of its forms, very convenient. We shall apply it to the following theorem.*

---

in quality (from affirmative to negative or from negative to affirmative), whether it be changed in quantity or not, no conclusion will be deducible in the same figure.

An Aristotelian proposition does not admit a term of the form not-Z in the subject,—Now on changing the quantity of the minor proposition of a syllogism, we transfer it from the general form

$$ay + bz = 0,$$

to the general form

$$a'y + b'(1 - z) = 0,$$

see (α) and (β), or *vice versâ*. And therefore, in the equation of the conclusion, there will be a change from z to $1 - z$, or *vice versâ*. But this is equivalent to the change of Z into not-Z, or not-Z into Z. Now the subject of the original conclusion must have involved a Z and not a not-Z, therefore the subject of the new conclusion will involve a not-Z, and the conclusion will not be admissible in the Aristotelian forms, except by conversion, which would render necessary a change of Figure.

Now the conclusions of this calculus are always the most general that can be drawn, and therefore the above demonstration must not be supposed to extend to a syllogism, in which a particular conclusion is deduced, when a universal one is possible. This is the case with *bramantip* only, among the Aristotelian forms, and therefore the transformation of *bramantip* into *camenes*, and *vice versâ*, is the case of restriction contemplated in the preliminary statement of the theorem.| 44|45

5th. *If for the minor premiss of an Aristotelian syllogism, we substitute its contradictory, no conclusion is deducible in the same figure.*

It is here only necessary to examine the case of *bramantip*, all the others being determined by the last proposition.

On changing the minor of *bramantip* to its contradictory, we have AO, Fig. 4, and this admits of no legitimate inference.

Hence the theorem is true without exception. Many other general theorems may in like manner be proved.

* This elegant theorem was communicated by the Rev. Charles Graves, Fellow and Professor of Mathematics in Trinity College, Dublin, to whom the Author desires further to record his grateful acknowledgements for a very judicious examination of the former portion of this work, and for some new applications of the method. The following example of Reduction *ad impossibile* is among the number:

Reducend Mood,	All Xs are Ys,	$1 - y = v'(1 - x)$
Baroko	Some Zs are not Ys	$vz = v(1 - y)$
	Some Zs are not Xs	$vz = vv'(1 - x)$
Reduct Mood,	All Xs are Ys	$1 - y = v'(1 - x)$
Barbara	All Zs are Xs	$z(1 - x) = 0$
	All Zs are Ys	$z(1 - y) = 0.$

Given the three propositions of a Syllogism, prove that there is but one order in which they can be legitimately arranged, and determine that order.

All the forms above given for the expression of propositions, are particular cases of the general form,

45|46 $$a + bx + cy = 0. \mid$$

Assume then for the premises of the given syllogism, the equations

$$a + bx + cy = 0, \ (18),$$

$$a' + b'z + c'y = 0, \ (19),$$

then, eliminating y, we shall have for the conclusion

$$ac' - a'c + bc'x - b'cz = 0, \ (20).$$

Now taking this as one of our premises, and either of the original equations, suppose (18), as the other, if by elimination of a common term x, between them, we can obtain a result equivalent to the remaining premiss (19), it will appear that there are more than one order in which the Propositions may be lawfully written; but if otherwise, one arrangement only is lawful.

Effecting then the elimination, we have

$$bc(a' + b'z + c'y) = 0, \ (21),$$

which is equivalent to (19) multiplied by a factor bc. Now on examining the value of this factor in the equations A, E, I, O, we find it in each case to be v or $-v$. But it is evident, that if an equation expressing a given Proposition be multiplied by an extraneous factor, derived from another equation, its interpretation will either be limited or rendered impossible. Thus there will either be no result at all, or the result will be a *limitation* of the remaining Proposition.

If, however, one of the original equations were

$$x = y, \quad \text{or } x - y = 0,$$

the factor bc would be -1, and would *not* limit the interpretation of the other premiss. Hence if the first member of a syllogism should be understood to represent the double proposition All Xs are Ys, and All Ys are Xs, it would
46|47 be indifferent in what order the remaining Propositions were written.|

A more general form of the above investigation would be, to express the premises by the equations

The conclusion of the reduct mood is seen to be the contradictory of the suppressed minor premiss. Whence, &c. It may just be remarked that the mathematical test of contradictory propositions is, that on eliminating one elective symbol between their equations, the other elective symbol vanishes. The *ostensive* reduction of *Baroko* and *Bokardo* involves no difficulty.

Professor Graves suggests the employment of the equation $x = vy$ for the primary expression of the Proposition All Xs are Ys, and remarks, that on multiplying both members by $1 - y$, we obtain $x(1 - y) = 0$, the equation from which we set out in the text, and of which the previous one is a solution.

$$a + bx + cy + dxy = 0, \ (22),$$

$$a' + b'z + c'y + d'zy = 0, \ (23).$$

After the double elimination of y and x we should find

$$(bc - ad)\,(a' + b'z + c'y + d'zy) = 0;$$

and it would be seen that the factor $bc - ad$ must in every case either vanish or express a limitation of meaning.

The determination of the order of the Propositions is sufficiently obvious. | 47|48

OF HYPOTHETICALS

A hypothetical proposition is defined to be *two or more categoricals united by a copula* (or conjunction), and the different kinds of hypothetical Propositions are named from their respective conjunctions, viz. conditional (if), disjunctive (either, or), &c.

In conditionals, that categorical Proposition from which the other results is called the *antecedent*, that which results from it the *consequent*.

Of the conditional syllogism there are two, and only two formulæ.

1st. The constructive

If A is B, then C is D,
But A is B, therefore C is D.

2nd. The Destructive,

If A is B, then C is D,
But C is not D, therefore A is not B.

A dilemma is a complex conditional syllogism, with several antecedents in the major, and a disjunctive minor.

IF we examine either of the forms of conditional syllogism above given, we shall see that the validity of the argument does not depend upon any considerations which have reference to the terms A, B, C, D, considered as the representatives of individuals or of classes. We may, in fact, represent the Propositions A is B, C is D, by the arbitrary symbols X and Y respectively, and express our syllogisms in such forms as the following:

If X is true, then Y is true,
But X is true, therefore Y is true.

Thus, what we have to consider is not objects and classes of objects, but the truths of Propositions, namely, of those | elementary Propositions which are 48|49
embodied in the terms of our hypothetical premises.

To the symbols X, Y, Z, representative of Propositions, we may appropriate the elective symbols x, y, z, in the following sense.

The hypothetical Universe, 1, shall comprehend all conceivable case and conjunctures of circumstances.

The elective symbol x attached to any subject expressive of such cases shall select those cases in which the Proposition X is true, and similarly for Y and Z.

If we confine ourselves to the contemplation of a given proposition X, and hold in abeyance every other consideration, then two cases only are conceivable, viz. first that the given Proposition is true, and secondly that it is false.* As these cases together make up the Universe of the Proposition, and as the former is determined by the elective symbol x, the latter is determined by the symbol $1 - x$.

But if other considerations are admitted, each of these cases will be resolvable
49|50 into others, individually less extensive, the | number of which will depend upon the number of foreign considerations admitted. Thus if we associate the Propositions X and Y, the total number of conceivable cases will be found as exhibited in the following scheme.

	Cases.	Elective expressions.
1st	X true, Y true	xy
2nd	X true, Y false	$x(1 - y)$
3rd	X false, Y true	$(1 - x)y$
4th	X false, Y false	$(1 - x)(1 - y)$ (24).

If we add the elective expressions for the two first of the above cases the sum is x, which is the elective symbol appropriate to the more general case of X being true independently of any consideration of Y; and if we add the elective expressions in the two last cases together, the result is $1 - x$, which is the elective expression appropriate to the more general case of X being false.

Thus the extent of the hypothetical Universe does not at all depend upon the number of circumstances which are taken into account. And it is to be noted that however few or many those circumstances may be, the sum of the elective expressions representing every conceivable case will be unity. Thus let us consider the three Propositions, X, It rains, Y, It hails, Z, It freezes. The possible cases are the following:

* It was upon the obvious principle that a Proposition is either true of false, that the Stoics, applying it to assertions respecting future events, endeavoured to establish the doctrine of Fate. It has been replied to their argument, that it involves "an abuse of the word *true*, the precise meaning of which is id quod res *est*. An assertion respecting the future is neither true nor false."—*Copleston on Necessity and Predestination*, p. 36. Were the Stoic axiom, however, presented under the form, It is either certain that a given event will take place, or certain that it will not; the above reply would fail to meet the difficulty. The proper answer would be, that no merely verbal definition can settle the question, what is the actual course and constitution of Nature. When we affirm that it is either certain that an event will take place, or certain that it will not take place, we tacitly assume that the order of events is necessary, that the Future is but an evolution of the Present; so that the state of things which is, completely determines that which shall be. But this (at least as respects the conduct of moral agents) is the very question at issue. Exhibited under its proper form, the Stoic reasoning does not involve an abuse of terms, but a *petitio principii*.

It should be added, that enlightened advocates of the doctrine of Necessity in the present day, viewing the end as appointed only in and through the means, justly repudiate those practical ill consequences which are the reproach of Fatalism.

	Cases.	Elective expressions.
1st	It rains, hails, and freezes,	xyz
2nd	It rains and hails, but does not freeze	$xy(1-z)$
3rd	It rains and freezes, but does not hail	$xz\ (1-y)$
4th	It freezes and hails, but does not rain	$yz(1-x)$
5th	It rains, but neither hails nor freezes	$x(1-y)(1-z)$
6th	It hails, but neither rains nor freezes	$y(1-x)(1-z)$
7th	It freezes, but neither hails or rains	$z(1-x)(1-y)$
8th	It neither rains, hails, nor freezes	$(1-x)(1-y)(1-z)$
		$1 = \text{sum}$ | 50|51

Expression of Hypothetical Propositions

To express that a given Proposition X is true.

The symbol $1 - x$ selects those cases in which the Proposition X is false. But if the Proposition is true, there are no such cases in its hypothetical Universe, therefore

$$1 - x = 0,$$

or

$$x = 1, \quad (25).$$

To express that a given Proposition X is false.

The elective symbol x selects all those cases in which the Proposition is true, and therefore if the Proposition is false,

$$x = 0, \quad (26).$$

And in every case, having determined the elective expression appropriate to a given Proposition, we assert the truth of that Proposition by equating the elective expression to unity, and its falsehood by equating the same expression to 0.

To express that two Propositions, X and Y, are simultaneously true.

The elective symbol appropriate to this case is xy, therefore the equation sought is

$$xy = 1, \quad (27).$$

To express that two Propositions, X and Y, are simultaneously false.

The condition will obviously be

$$(1 - x)(1 - y) = 1,$$

or

$$x + y - xy = 0, \quad (28).$$

To express that either the Proposition X is true, or the Proposition Y is true.

To assert that either one or the other of two Propositions is true, is to assert that it is not true, that they are both false. Now the elective expression appropriate to their both being false is $(1 - x)\ (1 - y)$, therefore the equation required is

$$(1 - x)(1 - y) = 0,$$

51|52 or $x + y - xy = 1$, (29). |

And, by indirect considerations of this kind, may every disjunctive Proposition, however numerous its members, be expressed. But the following general Rule will usually be preferable.

RULE. *Consider what are those distinct and mutually exclusive cases of which it is implied in the statement of the given Proposition, that some one of them is true, and equate the sum of their elective expressions to unity. This will give the equation of the given Proposition.*

For the sum of the elective expressions for all distinct conceivable cases will be unity. Now all these cases being mutually exclusive, and it being asserted in the given Proposition that some one case out of a given set of them is true, it follows that all which are not included in that set are false, and that their elective expressions are severally equal to 0. Hence the sum of the elective expressions for the remaining cases, viz. those included in the given set, will be unity. Some one of those cases will therefore be true, and as they are mutually exclusive, it is impossible that more than one should be true. Whence the Rule in question.

And in the application of this Rule it is to be observed, that if the cases contemplated in the given disjunctive Proposition are not mutually exclusive, they must be resolved into an equivalent series of cases which are mutually exclusive.

Thus, if we take the Proposition of the preceding example, viz. Either X is true, or Y is true, and assume that the two members of this Proposition are not exclusive, insomuch that in the enumeration of possible cases, we must reckon that of the Propositions X and Y being both true, then the mutually exclusive cases which fill up the Universe of the Proposition, with their elective expressions, are

1st, X true and Y false, $x(1 - y)$,
2nd, Y true and X false, $y(1 - x)$,
52|53 3rd, X true and Y true, xy, |

and the sum of these elective expressions equated to unity gives

$$x + y - xy = 1. \quad (30),$$

as before. But if we suppose the members of the disjunctive Proposition to be exclusive, then the only cases to be considered are

1st, X true, Y false, $x(1 - y)$,
2nd, Y true, X false, $y(1 - x)$,

and the sum of these elective expressions equated to 0, gives

$$x - 2xy + y = 1, \quad (31).$$

The subjoined examples will further illustrate this method.

To express the Proposition, Either X is not true, or Y is not true, the members being exclusive.

The mutually exclusive cases are

1st, X not true, Y true, $y(1-x)$,
2nd, Y not true, X true, $x(1-y)$,

and the sum of these equated to unity gives

$$x - 2xy + y = 1, \quad (32),$$

which is the same as (31), and in fact the Propositions which they represent are equivalent.

To express the Proposition, Either X is not true, or Y is not true, the members not being exclusive.

To the cases contemplated in the last Example, we must add the following, viz.

X not true, Y not true, $(1-x)(1-y)$.

The sum of the elective expressions gives

$$x(1-y) + y(1-x) + (1-x)(1-y) = 1,$$

$$\text{or} \quad xy = 0, \quad (33).$$

To express the disjunctive Proposition, Either X is true, or Y is true, or Z is true, the members being exclusive. | 53|54

Here the mutually exclusive cases are

1st, X true, Y false, Z false, $x(1-y)(1-z)$,
2nd, Y true, Z false, X false, $y(1-z)(1-x)$,
3rd, Z true, X false, Y false, $z(1-x)(1-y)$,

and the sum of the elective expressions equated to 1, gives, upon reduction,

$$x + y + z - 2(xy + yz + zx) + 3xyz = 1, \quad (34).$$

The expression of the same Proposition, when the members are in no sense exclusive, will be

$$(1-x)(1-y)(1-z) = 0, \quad (35).$$

And it is easy to see that our method will apply to the expression of any similar Proposition, whose members are subject to any specified amount and character of exclusion.

To express the conditional Proposition, If X is true, Y is true.

Here it is implied that all the cases of X being true, are cases of Y being true. The former cases being determined by the elective symbol x, and the latter by y, we have, in virtue of (4),

$$x(1-y) = 0, \quad (36).$$

To express the conditional Proposition, If X be true, Y is not true.

The equation is obviously

$$xy = 0, \quad (37);$$

this is equivalent to (33), and in fact the disjunctive Proposition, Either X is not true, or Y is not true, and the conditional Proposition, If X is true, Y is not true, are equivalent.

To express that If X is not true, Y is not true.

In (36) write $1 - x$ for x, and $1 - y$ for y, we have

54|55 $$(1 - x)y = 0. \ |$$

The results which we have obtained admit of verification in many different ways. Let it suffice to take for more particular examination the equation

$$x - 2xy + y = 1, \quad (38),$$

which expresses the conditional Proposition, Either X is true, or Y is true, the members being in this case exclusive.

First, let the Proposition X be true, then $x = 1$, and substituting, we have

$$1 - 2y + y = 1, \quad \therefore \ -y = 0, \text{ or } y = 0,$$

which implies that Y is not true.

Secondly, let X be not true, then $x = 0$, and the equation gives

$$y = 1, \quad (39),$$

which implies that Y is true. In like manner we may proceed with the assumptions that Y is true, or that Y is false.

Again, in virtue of the property $x^2 = x$, $y^2 = y$, we may write the equation in the form

$$x^2 - 2xy + y^2 = 1,$$

and extracting the square root, we have

$$x - y = \pm 1, \quad (40),$$

and this represents the actual case; for, as when X is true or false, Y is respectively false or true, we have

$$x = 1 \text{ or } 0,$$

$$y = 0 \text{ or } 1,$$

$$\therefore \ x - y = 1 \text{ or } -1$$

There will be no difficulty in the analysis of other cases.

Examples of Hypothetical Syllogism

The treatment of every form of hypothetical Syllogism will consist in forming the equations of the premises, and eliminating the symbol or symbols which are
55|56 found in more than one of them. The result will express the conclusion. |

1st. Disjunctive Syllogism

Either X is true, or Y is true (exclusive), $x + y - 2xy = 1$
But X is true, $x = 1$

Therefore Y is not true, $\therefore\ y = 0.$

Either X is true, or Y is true (not exclusive), $x + y - xy = 1$
But X is not true, $x = 0$

Therefore Y is true, $\therefore\ y = 1.$

2nd. Constructive Conditional Syllogism

If X is true, Y is true, $x(1 - y) = 0$
But X is true, $x = 1$
Therefore Y is true, $\therefore\ 1 - y = 0$ or $y = 1.$

3rd. Destructive Conditional Syllogism

If X is true, Y is true, $x(1 - y) = 0$
But Y is not true, $y = 0$
Therefore X is not true, $\therefore\ x = 0.$

4th. Simple Constructive Dilemma, the minor premiss exclusive

If X is true, Y is true, $x(1 - y) = 0$, (41),
If Z is true, Y is true, $z(1 - y) = 0$, (42),
But Either X is true, or Z is true, $x + z - 2xz = 1$, (43).

From the equations (41), (42), (43), we have to eliminate x and z. In whatever way we effect this, the result is

$$y = 1;$$

whence it appears that the Proposition Y is true.

5th. Complex Constructive Dilemma, the minor premiss not exclusive

If X is true, Y is true, $x(1 - y) = 0,$
If W is true, Z is true, $w(1 - z) = 0,$
Either X is true, or W is true, $x + w - xw = 1.$

From these equations, eliminating x, we have

$$y + z - yz = 1, |$$ 56|57

which expresses the Conclusion, Either Y is true, or Z is true, the members being non-exclusive.

6th. Complex Destructive Dilemma, the minor premiss exclusive

If X is true, Y is true, $x(1 - y) = 0,$
If W is true, Z is true, $w(1 - z) = 0,$
Either Y is not true, or Z is not true, $y + z - 2yz = 1.$

From these equations we must eliminate y and z. The result is

$$xw = 0,$$

which expresses the Conclusion, Either X is not true, or Y is not true, the members *not being exclusive.*

7th. Complex Destructive Dilemma, the minor premiss not exclusive

If X is true, Y is true, $x(1 - y) = 0$
If W is true, Z is true, $w(1 - z) = 0$
Either Y is not true, or Z is not true, $yz = 0.$

On elimination of y and z, we have

$$xw = 0,$$

which indicates the same Conclusion as the previous example.

It appears from these and similar cases, that whether the members of the minor premiss of a Dilemma are exclusive or not, the members of the (disjunctive) Conclusion are never exclusive. This fact has perhaps escaped the notice of logicians.

The above are the principal forms of hypothetical Syllogism which logicians have recognised. It would be easy, however, to extend the list, especially by the blending of the disjunctive and the conditional character in the same Proposition, of which the following is an example.

If X is true, then either Y is true, or Z is true,
$x(1 - y - z + yz) = 0$
But Y is not true, $y = 0$
57|58 Therefore If X is true, Z is true, $\therefore\ x(1 - z) = 0.$ |

That which logicians term a *Causal* Proposition is properly a conditional Syllogism, the major premiss of which is suppressed.

The assertion that the Proposition X is true, *because* the Proposition Y is true, is equivalent to the assertion,

The Proposition Y is true,
Therefore the Proposition X is true;

and these are the minor premiss and conclusion of the conditional Syllogism,

If Y is true, X is true,
But Y is true,
Therefore X is true.

And thus causal Propositions are seen to be included in the applications of our general method.

Note, that there is a family of disjunctive and conditional Propositions, which do not, of right, belong to the class considered in this Chapter. Such are those in which the force of the disjunctive or conditional particle is expended upon the predicate of the Proposition, as if, speaking of the inhabitants of a particular island, we should say, that they are all *either Europeans or Asiatics*; meaning, that it is true of each individual, that he is either a European or an Asiatic. If we appropriate the elective symbol x to the inhabitants, y to Europeans, and z to Asiatics, then the equation of the above Proposition is

$$x = xy + xz, \text{ or } x(1 - y - z) = 0, \quad (a);$$

to which we might add the condition $yz = 0$, since no Europeans are Asiatics. The nature of the symbols x, y, z, indicates that the Proposition belongs to those which we have before designated as *Categorical*. Very different from the above is the Proposition, Either all the inhabitants are Europeans, or they are all Asiatics. Here the disjunctive particle separates Propositions. The case is that contemplated in (31) of the present Chapter; and the symbols by which it is expressed, | although subject to the same laws as those of (a), have a totally 58|59
different interpretation.*

The distinction is real and important. Every Proposition which language can express may be represented by elective symbols, and the laws of combination of those symbols are in all cases the same; but in one class of instances the symbols have reference to collections of objects, in the other, to the truths of constituent Propositions. | 59|60

Properties of elective functions

Since elective symbols combine according to the laws of quantity, we may, by Maclaurin's theorem, expand a given function $\phi(x)$, in ascending powers of x, known cases of failure excepted. Thus we have

$$\phi(x) = \phi(0) + \phi'(0)x + \frac{\phi''(0)}{1.2}x^2 + \&c., \ (44).$$

Now $x^2 = x$, $x^3 = x$, &c., whence

$$\phi(x) = \phi(0) + x\{\phi'(0) + \frac{\phi''(0)}{1.2} + \&c.\}, \ (45).$$

Now if in (44) we make $x = 1$, we have

$$\phi(1) = \phi(0) + \phi'(0) + \frac{\phi''(0)}{1.2} + \&c.,$$

whence

* Some writers, among whom is Dr. Latham (*First Outlines*), regard it as the exclusive office of a conjunction to connect *Propositions*, not *words*. In this view I am not able to agree. The Proposition, Every animal is *either* rational *or* irrational, cannot be resolved into, *Either* every animal is rational, *or* every animal is irrational. The former belongs to pure categoricals, the latter to hypotheticals. In *singular* Propositions, such conversions would seem to be allowable. This animal is *either* rational *or* irrational, is equivalent to, *Either* this animal is rational, *or* it is irrational. This peculiarity of *singular* Propositions would almost justify our ranking them, though truly universals, in a separate class, as Ramus and his followers did.

$$\phi'(0) + \frac{\phi''(0)}{1.2} + \frac{\phi'''(0)}{1.2.3} + \&c. = \phi(1) - \phi(0).$$

Substitute this value for the coefficient of x in the second member of (45), and we have*

60|61 $$\phi(x) = \phi(0) + \{\phi(1) - \phi(0)\}x, \quad (46), \mid$$

which we shall also employ under the form

$$\phi(x) = \phi(1)x + \phi(0)(1-x), \quad (47).$$

Every function of x, in which integer powers of that symbol are alone involved, is by this theorem reducible to the first order. The quantities $\phi(0)$, $\phi(1)$, we shall call the moduli of the function $\phi(x)$. They are of great importance in the theory of elective functions, as will appear from the succeeding Propositions.

PROP. 1. Any two functions $\phi(x)$, $\psi(x)$, are equivalent, whose corresponding moduli are equal.

This is a plain consequence of the last Proposition. For since

---

* Although this and the following theorems have only been proved for those forms of functions which are expansible by Maclaurin's theorem, they may be regarded as true for all forms whatever; this will appear from the applications. The reason seems to be that, as it is only through the one form of expansion that elective functions become interpretable, no conflicting interpretation is possible.

The development of $\phi(x)$ may also be determined thus. By the known formula for expansion in factorials,

$$\phi(x) = \phi(0) + \Delta\phi(0)x + \frac{\Delta^2\phi(0)}{1.2}x(x-1) + \&c.$$

Now x being an elective symbol, $x(x-1) = 0$, so that all the terms after the second, vanish. Also $\Delta\phi(0) = \phi(1) - \phi(0)$, whence

$$\phi(x) = \phi(0) + \{\phi(1) - \phi(0)\}x.$$

The mathematician may be interested in the remark, that this is not the only case in which an expansion stops at the second term. The expansions of the compound operative functions

$\phi\left(\frac{d}{dx} + x^{-1}\right)$ and $\phi\left\{x + \left(\frac{d}{dx}\right)^{-1}\right\}$ are, respectively,

$$\phi\left(\frac{d}{dx}\right) + \phi'\left(\frac{d}{dx}\right)x^{-1},$$

and

$$\phi(x) + \phi'(x)\left(\frac{d}{dx}\right)^{-1}.$$

See *Cambridge Mathematical Journal*, Vol. IV. p. 219.

$$\phi(x) = \phi(0) + \{\phi(1) - \phi(0)\}x,$$
$$\psi(x) = \psi(0) + \{\psi(1) - \psi(0)\}x,$$

it is evident that if $\phi(0) = \psi(0)$, $\phi(1) = \psi(1)$, the two expansions will be equivalent, and therefore the functions which they represent will be equivalent also.

The converse of this Proposition is equally true, viz.

If two functions are equivalent, their corresponding moduli are equal.

Among the most important applications of the above theorem, we may notice the following.

Suppose it required to determine for what forms of the function $\phi(x)$, the following equation is satisfied, viz.

$$\{\phi(x)\}^n = \phi(x). \mid$$ 61|62

Here we at once obtain for the expression of the conditions in question,

$$\{\phi(0)\}^n = \phi(0). \quad \{\phi(1)\}^n = \phi(1), \quad (48).$$

Again, suppose it required to determine the conditions under which the following equation is satisfied, viz.

$$\phi(x)\psi(x) = \chi(x),$$

The general theorem at once gives

$$\phi(0)\psi(0) = \chi(0). \quad \phi(1)\psi(1) = \chi(1), \quad (49).$$

This result may also be proved by substituting for $\phi(x)$, $\psi(x)$, $\chi(x)$, their expanded forms, and equating the coefficients of the resulting equation properly reduced.

All the above theorems may be extended to functions of more than one symbol. For, as different elective symbols combine with each other according to the same laws as symbols of quantity, we can first expand a given function with reference to any particular symbol which it contains, and then expand the result with reference to any other symbol, and so on in succession, the order of the expansions being quite indifferent.

Thus the given function being $\phi(xy)$ we have

$$\phi(xy) = \phi(x0) + \{\phi(x1) - \phi(x0)\}y,$$

and expanding the coefficients with reference to x, and reducing

$$\phi(xy) = \phi(00) + \{\phi(10) - \phi(00)\}x + \{\phi(01) - \phi(00)\}y$$
$$+ \{\phi(11) - \phi(10) - \phi(01) + \phi(00)\}xy, \quad (50),$$

to which we may give the elegant symmetrical form

$$\phi(xy) = \phi(00)(1-x)(1-y) + \phi(01)y(1-x)$$
$$+ \phi(10)x(1-y) + \phi(11)xy, \quad (51),$$

wherein we shall, in accordance with the language already employed, designate $\phi(00)$, $\phi(01)$, $\phi(10)$, $\phi(11)$, as the moduli of the function $\phi(xy)$.

By inspection of the above general form, it will appear that any functions
62|63 of two variables are equivalent, whose corresponding moduli are all equal. |

Thus the conditions upon which depends the satisfaction of the equation,

$$\{\phi(xy)\}^n = \phi(xy)$$

are seen to be

$$\begin{array}{ll} \{\phi(00)\}^n = \phi(00), & \{\phi(01)\}^n = \phi(01), \\ \{\phi(10)\}^n = \phi(10), & \{\phi(11)\}^n = \phi(11), \end{array} \qquad (52).$$

And the conditions upon which depends the satisfaction of the equation

$$\phi(xy)\psi(xy) = \chi(xy),$$

are

$$\begin{array}{ll} \phi(00)\psi(00) = \chi(00), & \phi(01)\psi(01) = \chi(01), \\ \phi(10)\psi(10) = \chi(10), & \phi(11)\psi(11) = \chi(11), \end{array} \qquad (53).$$

It is very easy to assign by induction from (47) and (51), the general form of an expanded elective function. It is evident that if the number of elective symbols is m, the number of the moduli will be 2^m, and that their separate values will be obtained by interchanging in every possible way the values 1 and 0 in the places of the elective symbols of the given function. The several terms of the expansion of which the moduli serve as coefficients, will then be formed by writing for each 1 that recurs under the functional sign, the elective symbol x, &c., which it represents, and for each 0 the corresponding $1 - x$, &c., and regarding these as factors, the product of which, multiplied by the modulus from which they are obtained, constitutes a term of the expansion.

Thus, if we represent the moduli of any elective function $\phi(xy\ldots)$ by a_1, $a_2, \ldots a_r$, the function itself, when expanded and arranged with reference to the moduli, will assume the form

$$\phi(xy) = a_1 t_1 + a_2 t_2 \ldots + a_r t_r, \quad (54),$$

in which $t_1 t_2 \ldots t_r$ are functions of $x, y \ldots$, resolved into factors of the forms $x, y, \ldots 1 - x, 1 - y, \ldots$ &c. These functions satisfy individually the index relations

$$t_1^n = t_1, \; t_2^n = t_2, \text{ \&c.} \quad (55),$$

and the further relations,

63|64 $$t_1 t_2 = 0 \ldots t_1 t_r = 0, \text{ \&c.} \quad (56),$$ |

the product of any two of them vanishing. This will at once be inferred from inspection of the particular forms (47) and (51). Thus in the latter we have for the values of t_1, t_2, &c., the forms

$$xy, \; x(1-y), \; (1-x)y, \; (1-x)\;(1-y);$$

and it is evident that these satisfy the index relation, and that their products all vanish. We shall designate $t_1 t_2 \ldots$ as the constituent functions of $\phi(xy)$, and we shall define the peculiarity of the vanishing of the binary products, by saying that those functions are *exclusive*. And indeed the classes which they represent are mutually exclusive.

The sum of all the constituents of an expanded function is unity. An elegant proof of this Proposition will be obtained by expanding 1 as a function of any proposed elective symbols. Thus if in (51) we assume $\phi(xy) = 1$, we have $\phi(11) = 1$,

$$\phi(10) = 1,\ \phi(01) = 1,\ \phi(00) = 1, \text{ and (51) gives}$$

$$1 = xy + x(1-y) + (1-x)y + (1-x)(1-y), \quad (57).$$

It is obvious indeed, that however numerous the symbols involved, all the moduli of unity are unity, whence the sum of the constituents is unity.

We are now prepared to enter upon the question of the general interpretation of elective equations. For this purpose we shall find the following Propositions of the greatest service.

PROP. 2. If the first member of the general equation $\phi(xy \ldots) = 0$, be expanded in a series of terms, each of which is of the form at, a being a modulus of the given function, then for every numerical modulus a which does not vanish, we shall have the equation

$$at = 0,$$

and the combined interpretations of these several equations will express the full significance of the original equation.

For, representing the equation under the form

$$a_1 t_1 + a_2 t_2 \ldots + a_r t_r = 0, \quad (58).$$

Multiplying by t_1, we have, by (56),

$$a_1 t_1 = 0, \quad (59), \mid$$ 64|65

whence if a_1 is a numerical constant which does not vanish,

$$t_1 = 0,$$

and similarly for all the moduli which do not vanish. And inasmuch as from these constituent equations we can form the given equation, their interpretations will together express its entire significance.

Thus if the given equation were

$$x - y = 0, \quad \text{Xs and Ys are identical}, \quad (60),$$

we should have $\phi(11) = 0$, $\phi(10) = 1$, $\phi(01) = -1$, $\phi(00) = 0$, so that the expansion (51) would assume the form

$$x(1-y) - y(1-x) = 0,$$

whence, by the above theorem,

$$x(1-y)=0, \qquad \text{All Xs are Ys,}$$

$$y(1-x)=0, \qquad \text{All Ys are Xs,}$$

results which are together equivalent to (60).

It may happen that the simultaneous satisfaction of equations thus deduced, may require that one or more of the elective symbols should vanish. This would only imply the nonexistence of a class: it may even happen that it may lead to a final result of the form

$$1=0,$$

which would indicate the nonexistence of the logical Universe. Such cases will only arise when we attempt to unite contradictory Propositions in a single equation. The manner in which the difficulty seems to be evaded in the result is characteristic.

It appears from this Proposition, that the differences in the interpretation of elective functions depend solely upon the number and position of the vanishing moduli. No change in the value of a modulus, but one which causes it to vanish, produces any change in the interpretation of the equation in which it is found. If among the infinite number of different values which we are thus permitted to give to the moduli which do not vanish in a proposed equation, any one value
65|66 should be | preferred, it is unity, for when the moduli of a function are all either 0 or 1, the function itself satisfies the condition

$$\{\phi(xy\ldots)\}^n=\phi(xy\ldots),$$

and this at once introduces symmetry into our Calculus, and provides us with fixed standards for reference.

PROP. 3. If $w=\phi(xy\ldots)$, $w, x, y, \ldots$ being elective symbols, and if the second member be completely expanded and arranged in a series of terms of the form at, we shall be permitted to equate separately to 0 every term in which the modulus a does not satisfy the condition

$$a^n=a,$$

and to leave for the value of w the sum of the remaining terms.

As the nature of the demonstration of this Proposition is quite unaffected by the number of the terms in the second member, we will for simplicity confine ourselves to the supposition of there being four, and suppose that the moduli of the two first only, satisfy the index law.

We have then

$$w=a_1t_1+a_2t_2+a_3t_3+a_4t_4, \quad (61),$$

with the relations

$$a_1^n=a_1,\ a_2^n=a_2,$$

in addition to the two sets of relations connecting t_1, t_2, t_3, t_4, in accordance with (55) and (56).

Squaring (61), we have

$$w = a_1 t_1 + a_2 t_2 + a_3^2 t_3 + a_4^2 t_4,$$

and subtracting (61) from this,

$$(a_3^2 - a_3)t_3 + (a_4^2 - a_4)t_4 = 0;$$

and it being an hypothesis, that the coefficients of these terms do not vanish, we have, by Prop. 2,

$$t_3 = 0, \ t_4 = 0, \quad (62),$$

whence (61) becomes

$$w = a_1 t_1 + a_2 t_2.$$

The utility of this Proposition will hereafter appear. | 66|67

PROP. 4. The functions $t_1 t_2 \ldots t_r$ being mutually exclusive, we shall always have

$$\psi(a_1 t_1 + a_2 t_2 \ldots + a_r t_r) = \psi(a_1)t_1 + \psi(a_2)t_2 \ldots + \psi(a_r)t_r, \quad (63),$$

whatever may be the values of $a_1 a_2 \ldots a_r$ or the form of ψ.

Let the function $a_1 t_1 + a_2 t_2 + \ldots + a_r t_r$ be represented by $\phi(xy \ldots)$, then the moduli $a_1 a_2 \ldots a_r$ will be given by the expressions

$$\phi(11 \ldots), \ \phi(10 \ldots), \ (\ldots)\phi(00 \ldots).$$

Also $\psi(a_1 t_1 + a_2 t_2 \ldots + a_r t_r) = \psi\{\phi(xy \ldots)\}$

$$= \psi\{\phi(11 \ldots)\}xy \ldots + \psi\{\phi(10)\}x(1-y) \ldots$$
$$+ \psi\{\phi(00)\}(1-x)(1-y) \ldots$$
$$= \psi(a_1)xy \ldots + \psi(a_2)x(1-y) \ldots + \psi(a_r)(1-x)(1-y) \ldots$$
$$= \psi(a_1)t_1 + \psi(a_2)t_2 \ldots + \psi(a_r)t_r, \quad (64).$$

It would not be difficult to extend the list of interesting properties, of which the above are examples. But those which we have noticed are sufficient for our present requirements. The following Proposition may serve as an illustration of their utility.

PROP. 5. Whatever process of reasoning we apply to a single given Proposition, the result will either be the same Proposition or a limitation of it.

Let us represent the equation of the given Proposition under its most general form,

$$a_1 t_1 + a_2 t_2 \ldots + a_r t_r = 0, \quad (65),$$

resolvable into as many equations of the form $t = 0$ as there are moduli which do not vanish.

Now the most general transformation of this equation is

$$\psi(a_1 t_1 + a_2 t_2 \ldots + a_r t_r) = \psi(0), \quad (66),$$

provided that we attribute to ψ a perfectly arbitrary character, allowing it even to involve new elective symbols, having *any proposed relation* to the original
67|68 ones. |

The development of (66) gives, by the last Proposition,

$$\psi(a_1)t_1 + \psi(a_2)t_2 \ldots + \psi(a_r)t_r = \psi(0).$$

To reduce this to the general form of reference, it is only necessary to observe that since

$$t_1 + t_2 \ldots + t_r = 1,$$

we may write for $\psi(0)$,

$$\psi(0)(t_1 + t_2 \ldots + t_r),$$

whence, on substitution and transposition,

$$\{\psi(a_1) - \psi(0)\}t_1 + \{\psi(a_2) - \psi(0)\}t_2 \ldots + \{\psi(a_r) - \psi(0)\}t_r = 0.$$

From which it appears, that if a be any modulus of the original equation, the corresponding modulus of the transformed equation will be

$$\psi(a) - \psi(0).$$

If $a = 0$, then $\psi(a) - \psi(0) = \psi(0) - \psi(0) = 0$, whence there are no *new terms* in the transformed equation, and therefore there are no *new Propositions* given by equating its constituent members to 0.

Again, since $\psi(a) - \psi(0)$ may vanish without a vanishing, terms may be wanting in the transformed equation which existed in the primitive. Thus some of the constituent truths of the original Proposition may entirely disappear from the interpretation of the final result.

Lastly, if $\psi(a) - \psi(0)$ do not vanish, it must either be a numerical constant, or it must involve new elective symbols. In the former case, the term in which it is found will give

$$t = 0,$$

which is one of the constituents of the original equation: in the latter case we shall have

$$\{\psi(a) - \psi(0)\}t = 0,$$

in which t has a limiting factor. The interpretation of this equation, therefore,
68|69 is a limitation of the interpretation of (65). |

The purport of the last investigation will be more apparent to the mathematician than to the logician. As from any mathematical equation an infinite number of others may be deduced, it seemed to be necessary to shew that when the original equation expresses a logical Proposition, every member of the derived series, even when obtained by expansion under a functional sign, admits of exact and consistent interpretation. | 69|70

Of the solution of elective equations

In whatever way an elective symbol, considered as unknown, may be involved in a proposed equation, it is possible to assign its complete value in terms of the remaining elective symbols considered as known. It is to be observed of such equations, that from the very nature of elective symbols, they are necessarily linear, and that their solutions have a very close analogy with those of linear differential equations, arbitrary elective symbols in the one, occupying the place of arbitrary constants in the other. The method of solution we shall in the first place illustrate by particular examples, and, afterwards, apply to the investigation of general theorems.

Given $(1-x)y=0$, (All Ys are Xs), to determine y in terms of x.

As y is a function of x, we may assume $y=vx+v'\ (1-x)$, (such being the expression of an arbitrary function of x), the moduli v and v' remaining to be determined. We have then

$$(1-x)\{vx+v'(1-x)\}=0,$$

or, on actual multiplication,

$$v'(1-x)=0:$$

that this may be generally true, without imposing any restriction upon x, we must assume $v'=0$, and there being no condition to limit v, we have

$$y=vx, \quad (67).$$

This is the complete solution of the equation. The condition that y is an elective symbol requires that v should be an elective | symbol also (since it must 70|71
satisfy the index law), its interpretation in other respects being arbitrary.

Similarly the solution of the equation, $xy=0$, is

$$y=v(1-x), \quad (68).$$

Given $(1-x)zy=0$, (All Ys which are Zs are Xs), to determine y.

As y is a function of x and z, we may assume

$$y=v(1-x)(1-z)+v'(1-x)z+v''x(1-z)+v'''zx.$$

And substituting, we get

$$v'(1-x)z=0,$$

whence $v'=0$. The complete solution is therefore

$$y=v(1-x)(1-z)+v''x(1-z)+v'''xz, \quad (69),$$

v', v'', v''', being arbitrary elective symbols, and the rigorous interpretation of this result is, that Every Y is *either* a not-X and not-Z, or an X and not-Z, or an X and Z.

It is deserving of note that the above equation may, in consequence of its linear form, be solved by adding the two particular solutions with reference to x and z; and replacing the arbitrary constants which each involves by an arbitrary function of the other symbol, the result is

$$y = x\phi(z) + (1 - z)\psi(x), \quad (70).$$

To shew that this solution is equivalent to the other, it is only necessary to substitute for the arbitrary functions $\phi(z)$, $\psi(x)$, their equivalents

$$wz + w'(1 - z) \text{ and } w''x + w'''(1 - x),$$

we get $y = wxz + (w' + w'')x(1 - z) + w'''(1 - x)(1 - z)$.

In consequence of the perfectly arbitrary character of w' and w'', we may replace their sum by a single symbol w', whence

$$y = wxz + w'x(1 - z) + w'''(1 - x)(1 - z),$$

71|72 which agrees with (69). |

The solution of the equation $wx(1 - y)z = 0$, expressed by arbitrary functions, is

$$z = (1 - w)\phi(xy) + (1 - x)\psi(wy) + y\chi(wx), \quad (71).$$

These instances may serve to shew the analogy which exists between the solutions of elective equations and those of the corresponding order of linear differential equations. Thus the expression of the integral of a partial differential equation, either by arbitrary functions or by a series with arbitrary coefficients, is in strict analogy with the case presented in the two last examples. To pursue this comparison further would minister to curiosity rather than to utility. We shall prefer to contemplate the problem of the solution of elective equations under its most general aspect, which is the object of the succeeding investigations.

To solve the general equation $\phi(xy) = 0$, with reference to y.

If we expand the given equation with reference to x and y, we have

$$\phi(00)(1 - x)(1 - y) + \phi(01)(1 - x)y + \phi(10)x(1 - y) + \phi(11)xy = 0, \quad (72),$$

the coefficients $\phi(00)$ &c. being numerical constants.

Now the general expression of y, as a function of x, is

$$y = vx + v'(1 - x),$$

v and v' being unknown symbols to be determined. Substituting this value in (72), we obtain a result which may be written in the following form,

$$[\phi(10) + \{\phi(11) - \phi(10)\}v]x + [\phi(00) + \{\phi(01) - \phi(00)\}v'](1 - x) = 0;$$

and in order that this equation may be satisfied without any way restricting the generality of x, we must have

$$\phi(10) + \{\phi(11) - \phi(10)\}v = 0,$$

$$\phi(00) + \{\phi(01) - \phi(00)\}v' = 0, \mid$$ 72|73

from which we deduce

$$v = \frac{\phi(10)}{\phi(10) - \phi(11)}, \qquad v' = \frac{\phi(00)}{\phi(01) - \phi(00)},$$

wherefore

$$y = \frac{\phi(10)}{\phi(10) - \phi(11)}x + \frac{\phi(00)}{\phi(00) - \phi(01)}(1 - x), \quad (73).$$

Had we expanded the original equation with respect to y only, we should have had

$$\phi(x0) + \{\phi(x1) - \phi(x0)\}y = 0;$$

but it might have startled those who are unaccustomed to the processes of Symbolical Algebra, had we from this equation deduced

$$y = \frac{\phi(x0)}{\phi(x0) - \phi(x1)},$$

because of the apparently meaningless character of the second member. Such a result would however have been perfectly lawful, and the expansion of the second member would have given us the solution above obtained. I shall in the following example employ this method, and shall only remark that those to whom it may appear doubtful, may verify its conclusions by the previous method.

To solve the general equation $\phi(xyz) = 0$, or in other words to determine the value of z as a function of x and y.

Expanding the given equation with reference to z, we have

$$\phi(xy0) + \{\phi(xy1) - \phi(xy0)\}.z = 0;$$

$$\therefore z = \frac{\phi(xy0)}{\phi(xy0) - \phi(xy1)} \ \ldots \ (74),$$

and expanding the second member as a function of x and y by aid of the general theorem, we have

$$z = \frac{\phi(110)}{\phi(110) - \phi(111)}xy + \frac{\phi(100)}{\phi(100) - \phi(101)}x(1 - y) +$$

$$+\frac{\phi(010)}{\phi(010)-\phi(011)}(1-x)y+\frac{\phi(000)}{\phi(000)-\phi(001)}$$

73|74 $$(1-x)(1-y) \ldots (75) \mid$$

and this is the complete solution required. By the same method we may resolve an equation involving any proposed number of elective symbols.

In the interpretation of any general solution of this nature, the following cases may present themselves.

The values of the moduli $\phi(00)$, $\phi(01)$, &c. being constant, one or more of the coefficients of the solution may assume the form $\frac{0}{0}$ or $\frac{1}{0}$. In the former case, the indefinite symbol $\frac{0}{0}$ must be replaced by an arbitrary elective symbol v. In the latter case, the term, which is multiplied by a factor $\frac{1}{0}$ (or by any numerical constant except 1), must be separately equated to 0, and will indicate the existence of a subsidiary Proposition. This is evident from (62).

Ex. Given $x(1-y)=0$, All Xs are Ys, to determine y as a function of x.

Let $\phi(xy)=x(1-y)$, then $\phi(10)=1$, $\phi(11)=0$, $\phi(01)=0$, $\phi(00)=0$; whence, by (73),

$$\begin{aligned} y &= \frac{1}{1-0}x+\frac{0}{0-0}(1-x) \\ &= x+\tfrac{0}{0}(1-x) \\ &= x+v(1-x), \quad (76), \end{aligned}$$

v being an arbitrary elective symbol. The interpretation of this result is that the class Y consists of the entire class X with an indefinite remainder of not-Xs. This remainder is indefinite in the highest sense, *i.e.* it may vary from 0 up to the entire class of not-Xs.

Ex. Given $x(1-z)+z=y$, (the class Y consists of the entire class Z, with such not-Zs as are Xs), to find Z.

Here $\phi(xyz)=x(1-z)-y+z$, whence we have the following set of values for the moduli,

$$\begin{array}{llll} \phi(110)=0 & \phi(111)=0, & \phi(100)=1, & \phi(101)=1, \\ \phi(010)=-1, & \phi(011)=0, & \phi(000)=0, & \phi(001)=1, \end{array}$$

and substituting these in the general formula (75), we have

74|75 $$z=\tfrac{0}{0}xy+\tfrac{1}{0}x(1-y)+(1-x)y, \quad (77), \mid$$

the infinite coefficient of the second term indicates the equation

$$x(1-y)=0, \text{ All Xs are Ys;}$$

and the indeterminate coefficient of the first term being replaced by v, an arbitrary elective symbol, we have

$$z=(1-x)y+vxy,$$

the interpretation of which is, that the class Z consists of all the Ys which are not Xs, and an *indefinite* remainder of Ys which are Xs. Of course this indefinite remainder may vanish. The two results we have obtained are logical inferences (not very obvious ones) from the original Propositions, and they give us all the information which it contains respecting the class Z, and its constituent elements.

Ex. Given $x = y(1 - z) + z(1 - y)$. The class X consists of all Ys which are not-Zs, and all Zs which are not-Ys: required the class Z.

We have

$$\phi(xyz) = x - y(1 - z) - z(1 - y),$$

$$\phi(110) = 0, \quad \phi(111) = 1, \quad \phi(100) = 1, \quad \phi(101) = 0,$$

$$\phi(010) = -1, \quad \phi(011) = 0, \quad \phi(000) = 0, \quad \phi(001) = -1;$$

whence, by substituting in (75),

$$z = x(1 - y) + y(1 - x), \quad (78),$$

the interpretation of which is, the class Z consists of all Xs which are not Ys, and of all Ys which are not Xs; an inference strictly logical.

Ex. Given $y\{1 - z(1 - x)\} = 0$, All Ys are Zs and not-Xs.

Proceeding as before to form the moduli, we have, on substitution in the general formulæ,

$$z = \tfrac{1}{0}xy + \tfrac{0}{0}x(1 - y) + y(1 - x) + \tfrac{0}{0}(1 - x)(1 - y),$$

$$\text{or } z = y(1 - x) + vx(1 - y) + v'(1 - x)(1 - y)$$

$$= y(1 - x) + (1 - y)\phi(x), \quad (79),$$

with the relation $xy = 0$:

from these it appears that No Ys are Xs, and that the class Z | consists of all 75|76
Ys which are not Xs, and of an indefinite remainder of not-Ys.

This method, in combination with Lagrange's method of indeterminate multipliers, may be very elegantly applied to the treatment of simultaneous equations. Our limits only permit us to offer a single example, but the subject is well deserving of further investigation.

Given the equations $x(1 - z) = 0$, $z(1 - y) = 0$, All Xs are Zs, All Zs are Ys, to determine the complete value of z with any subsidiary relations connecting x and y.

Adding the second equation multiplied by an indeterminate constant λ, to the first, we have

$$x(1 - z) + \lambda z(1 - y) = 0,$$

whence determining the moduli, and substituting in (75),

$$z = xy + \frac{1}{1-\lambda}x(1-y) + \tfrac{0}{0}(1-x)y, \quad (80),$$

from which we derive

$$z = xy + v(1-x)y,$$

with the subsidiary relation

$$x(1-y) = 0:$$

the former of these expresses that the class Z consists of all Xs that are Ys, with an indefinite remainder of not-Xs that are Ys; the latter, that All Xs are Ys, being in fact the conclusion of the syllogism of which the two given Propositions are the premises.

By assigning an appropriate meaning to our symbols, all the equations we have discussed would admit of interpretation in hypotheticals, but it may suffice to have considered them as examples of categoricals.

That peculiarity of elective symbols, in virtue of which every elective equation is reducible to a system of equations $t_1 = 0$, $t_2 = 0$, &c., so constituted, that all the binary products $t_1 t_2$, $t_1 t_3$, &c., vanish, represents a general doctrine in Logic with reference to the ultimate analysis of Propositions, of which it may
76|77 be desirable to offer some illustration. |

Any of these constituents t_1, t_2, &c. consists only of factors of the forms x, y, ... $1-w$, $1-z$, &c. In categoricals it therefore represents a compound class, *i.e.* a class defined by the presence of certain qualities, and by the absence of certain other qualities.

Each constituent equation $t_1 = 0$, &c. expresses a denial of the existence of some class so defined, and the different classes are mutually exclusive.

Thus all categorical Propositions are resolvable into a denial of the existence of certain compound classes, no member of one such class being a member of another.

The Proposition, All Xs are Ys, expressed by the equation $x\ (1-y) = 0$, is resolved into a denial of the existence of a class whose members are Xs and not-Ys.

The Proposition Some Xs are Ys, expressed by $v = xy$, is resolvable as follows. On expansion,

$$v - xy = vx(1-y) + vy(1-x) + v(1-x)(1-y) - xy(1-v);$$

$$\therefore vx(1-y) = 0,\ vy(1-x) = 0,\ v(1-x)(1-y) = 0,\ (1-v)xy = 0.$$

The three first imply that there is no class whose members belong to a certain unknown Some, and are 1st, Xs and not Ys; 2nd, Ys and not Xs; 3rd, not-Xs and not-Ys. The fourth implies that there is no class whose members are Xs and Ys without belonging to this unknown Some.

From the same analysis it appears that *all hypothetical Propositions may be resolved into denials of the coexistence of the truth or falsity of certain assertions.*

Thus the Proposition, If X is true, Y is true, is resolvable by its equation $x(1-y)=0$, into a denial that the truth of X and the falsity of Y coexist.

And the Proposition Either X is true, or Y is true, members exclusive, is resolvable into a denial, first, that X and Y are both true; secondly, that X and Y are both false.

But it may be asked, is not something more than a system of negations necessary to the constitution of an affirmative Proposition? is not a positive element required? Undoubtedly | there is need of one; and this positive element 77|78
is supplied in categoricals by the assumption (which may be regarded as a prerequisite of reasoning in such cases) that there *is* a Universe of conceptions, and that each individual it contains either belongs to a proposed class or does not belong to it; in hypotheticals, by the assumption (equally prerequisite) that, there is a Universe of conceivable cases, and that any given Proposition is either true or false. Indeed the question of the existence of conceptions (εἰ ἔστι) is preliminary to any statement of their qualities or relations (τί ἔστι).—*Aristotle, Anal. Post.* lib. II. cap. 2.

It would appear from the above, that Propositions may be regarded as resting at once upon a positive and upon a negative foundation. Nor is such a view either foreign to the sprit of Deductive Reasoning or inappropriate to its Method; the latter ever proceeding by limitations, while the former contemplates the particular as derived from the general.

Demonstration of the Method of Indeterminate Multipliers, as Applied to Simultaneous Elective Equations

To avoid needless complexity, it will be sufficient to consider the case of three equations involving three elective symbols, those equations being the most general of the kind. It will be seen that the case is marked by every feature affecting the character of the demonstration, which would present itself in the discussion of the more general problem in which the number of equations and the number of variables are both unlimited.

Let the given equations be

$$\phi(xyz)=0,\quad \psi(xyz)=0,\quad \chi(xyz)=0,\quad (1).$$

Multiplying the second and third of these by the arbitrary constants h and k, and adding to the first, we have

$$\phi(xyz)+h\psi(xyz)+k\chi(xyz)=0,\quad (2);\ |$$ 78|79

and we are to shew, that in solving this equation with reference to any variable z by the general theorem (75), we shall obtain not only the general value of z independent of h and k, but also any subsidiary relations which may exist between x and y independently of z.

If we represent the general equation (2) under the form $F(xyz)=0$, its solution may by (75) be written in the form

$$z = \frac{xy}{1-\dfrac{F(111)}{F(110)}} + \frac{x(1-y)}{1-\dfrac{F(101)}{F(100)}} + \frac{y(1-x)}{1-\dfrac{F(011)}{F(010)}} + \frac{(1-x)(1-y)}{1-\dfrac{F(001)}{F(000)}};$$

and we have seen, that any one of these four terms is to be equated to 0, whose modulus, which we may represent by M, does not satisfy the condition $M^n = M$, or, which is here the same thing, whose modulus has any other value than 0 or 1.

Consider the modulus (suppose M_1) of the first term, viz. $\dfrac{1}{1-\dfrac{F(111)}{F(110)}}$, and giving to the symbol F its full meaning, we have

$$M_1 = \frac{1}{1-\dfrac{\phi(111)+h\psi(111)+k\chi(111)}{\phi(110)+h\psi(110)+k\chi(110)}}.$$

It is evident that the condition $M_1^n = M_1$ cannot be satisfied unless the right-hand member be independent of h and k; and in order that this may be the case, we must have the function $\dfrac{\phi(111)+\mathrm{h}\psi(111)+\mathrm{k}\chi(111)}{\phi(110)+\mathrm{h}\psi(110)+\mathrm{k}\chi(110)}$ independent of h and k.

Assume then

$$\frac{\phi(111)+h\psi(111)+k\chi(111)}{\phi(110)+h\psi(110)+k\chi(110)} = c,$$

c being independent of h and k; we have, on clearing of fractions and equating coefficients,

$$\phi(111) = c\phi(110),\ \psi(111) = c\psi(110),\ \chi(111) = c\chi(110);$$

whence, eliminating c,

$$\frac{\phi(111)}{\phi(110)} = \frac{\psi(111)}{\psi(110)} = \frac{\chi(111)}{\chi(110)},$$

79|80 | being equivalent to the triple system

$$\left.\begin{array}{l} \phi(111)\psi(110) - \phi(110)\psi(111) = 0 \\ \psi(111)\chi(110) - \psi(110)\chi(111) = 0 \\ \chi(111)\phi(110) - \chi(110)\phi(111) = 0 \end{array}\right\} \quad (3);$$

and it appears that if any one of these equations is not satisfied, the modulus M_1 will not satisfy the condition $M_1^n = M_1$, whence the first term of the value of z must be equated to 0, and we shall have

$$xy = 0,$$

a relation between x and y independent of z.

Now if we expand in terms of z each pair of the primitive equations (1), we shall have

$$\phi(xy0) + \{\phi(xy1) - \phi(xy0)\}z = 0,$$

$$\psi(xy0) + \{\psi(xy1) - \psi(xy0)\}z = 0,$$

$$\chi(xy0) + \{\chi(xy1) - \chi(xy0)\}z = 0,$$

and successively eliminating z between each pair of these equations, we have

$$\phi(xy1)\psi(xy0) - \phi(xy0)\psi(xy1) = 0$$

$$\psi(xy1)\chi(xy0) - \psi(xy0)\chi(xy1) = 0,$$

$$\chi(xy1)\phi(xy0) - \chi(xy0)\phi(xy1) = 0,$$

which express all the relations between x and y that are formed by the elimination of z. Expanding these, and writing in full the first term, we have

$$\{\phi(111)\psi(110) - \phi(110)\psi(111)\}xy + \&c. = 0,$$

$$\{\psi(111)\chi(110) - \psi(110)\chi(111)\}xy + \&c. = 0,$$

$$\{\chi(111)\phi(110) - \chi(110)\phi(111)\}xy + \&c. = 0:$$

and it appears from Prop. 2. that if the coefficient of xy in any of these equations does not vanish, we shall have the equation

$$xy = 0;$$

but the coefficients in question are the same as the first members of the system (3), and the two sets of conditions exactly agree. Thus, as respects the first term of the expansion, the method of indeterminate coefficients leads to the same result as ordinary elimination; and it is obvious that from their similarity of form, the same reasoning will apply to all the other terms. | 80|81

Suppose, in the second place, that the conditions (3) are satisfied so that M_1 is independent of h and k. It will then indifferently assume the equivalent forms

$$M_1 = \frac{1}{1 - \dfrac{\phi(111)}{\phi(110)}} = \frac{1}{1 - \dfrac{\psi(111)}{\psi(110)}} = \frac{1}{1 - \dfrac{\chi(111)}{\chi(110)}}$$

These are the exact forms of the first modulus in the expanded values of z, deduced from the solution of the three primitive equations singly. If this common value of M_1 is 1 or $\frac{0}{0} = v$, the term will be retained in z; if any other constant value (except 0), we have a relation $xy = 0$, not given by elimination, but deducible from the primitive equations singly, and similarly for all the other terms. Thus in every case the expression of the subsidiary relations is a necessary accompaniment of the process of solution.

It is evident, upon consideration, that a similar proof will apply to the discussion of a system indefinite as to the number both of its symbols and of its equations.

Postscript

Some additional explanations and references which have occurred to me during the printing of this work are subjoined.

The remarks on the connexion between Logic and Language, p. 5, are scarcely sufficiently explicit. Both the one and the other I hold to depend very materially upon our ability to form general notions by the faculty of abstraction. Language is an instrument of Logic, but not an indispensable instrument.

To the remarks on Cause, p. 12, I desire to add the following: Considering Cause as an invariable antecedent in Nature (which is Brown's view), whether associated or not with the idea of Power, as suggested by Sir John Herschel, the knowledge of its existence is a knowledge which is properly expressed by the word *that* (τὸ ὅτὶ), not by *why* (τὸ διότὶ). It is very remarkable that the two greatest authorities in Logic, modern and ancient, agreeing in the latter interpretation, differ most widely in its application to Mathematics. Sir W.
81 | 82 Hamilton says that Mathematics | exhibit only the *that* (τὸ ὅτὶ): Aristotle says, The *why* belongs to mathematicians, for they have the demonstrations of Causes. *Anal. Post.* lib. I., cap. XIV. It must be added that Aristotle's view is consistent with the sense (albeit an erroneous one) which in various parts of his writings he virtually assigns to the word Cause, viz. an antecedent in Logic, a sense according to which the premises might be said to be the cause of the conclusion. This view appears to me to give even to his physical inquiries much of their peculiar character.

Upon reconsideration, I think that the view on p. 41, as to the presence or absence of a medium of comparison, would readily follow from Professor De Morgan's doctrine, and I therefore relinquish all claim to a discovery. The mode in which it appears in this treatise is, however, remarkable.

I have seen reason to change the opinion expressed in pp. 42, 43. The system of equations there given for the expression of Propositions in Syllogism is *always* preferable to the one before employed—first, in generality—secondly, in facility of interpretation.

In virtue of the principle, that a Proposition is either true or false, every elective symbol employed in the expression of hypotheticals admits only of the values 0 and 1, which are the only quantitative forms of an elective symbol. It is in fact possible, setting out from the theory of Probabilities (which is purely quantitative), to arrive at a system of methods and processes for the treatment of hypotheticals exactly similar to those which have been given. The two systems of elective symbols and of quantity osculate, if I may use the expression, in the points 0 and 1. It seems to me to be implied by this, that unconditional truth (categoricals) and probable truth meet together in the constitution of contingent truth (hypotheticals). The general doctrine of elective symbols and all the more characteristic applications are quite independent of any quantitative origin.

THE END.

12

James Joseph Sylvester (1814–1897)

Sylvester was educated at the University of London (later University College, London), at Trinity College, Dublin, and at St John's College, Cambridge. He had a chequered career. From 1838 to 1841 he was a colleague of De Morgan's at University College, London. From 1841 to 1843 he taught at the University of Virginia; he left after an incident in which he defended himself at swordpoint from assault by an insubordinate student. (His tribulations at the anti-Semitic University of Virginia are described in *Feuer 1988*.) During an eleven-year stint as a London lawyer (1844–55) he became acquainted with his principal mathematical collaborator, Arthur Cayley, who was also at the time a London lawyer. In 1855 he was appointed to the chair of mathematics at the Royal Military Academy, Woolwich. He resigned in 1870, and was for six years unemployed. From 1876 to 1883 he held the chair in mathematics at Johns Hopkins University. One of his colleagues at Johns Hopkins was Charles S. Peirce, whom he greatly respected, but with whom he quarrelled about the priority in an algebraic discovery. In 1883, at the age of 69, Oxford having lifted its prohibition against Jewish professors, Sylvester was appointed Savilian Professor of Geometry at Oxford, a post he held until 1894.

Previous selections from the British algebraists have traced the rise of a new, more abstract conception of algebra, and the accompanying widening of algebra to include the study, not just of the familiar real and natural numbers, but also of such subjects as quaternions and the forms of inference of deductive logic. Sylvester, one of the great algebraists of the century, played a major role in this development. Together with Cayley, he developed the theory of determinants and carried out deep investigations into the theory of invariants (a subject which had been inaugurated by George Boole). Calling himself the 'new Adam', he coined names for many of the new objects of algebra that were being discovered in lush abundance: combinants, reciprocants, concomitants, discriminants, zetaic multipliers, allotrious factors, and many others. But however much Sylvester participated in the trend to greater abstraction in algebra, he resisted the tendency to view mathematics as purely a matter of manipulating symbolic expressions and mechanically carrying out logical deductions from given premisses. The following speech, while illustrating the new conception of mathematics as occupied with the study of abstract forms, argues for the indispensability of observation and experimental reasoning in mathematics, and the great algebraist stresses the importance of geometric insight even for his own purely algebraic and arithmetical researches.[a] Similar warnings against an excessively formal and algebraic conception of mathematics were to surface again at the end of the century in *Klein 1895* and in Poincaré's criticisms of the writings of the mathematical logicians.

References should be to the paragraph numbers, which have been added in this reprinting.

A. PRESIDENTIAL ADDRESS TO SECTION 'A' OF THE BRITISH ASSOCIATION (*SYLVESTER 1869*)

LADIES AND GENTLEMEN,—

⟦1⟧ A few days ago I noticed in a shop window the photograph of a Royal mother and child, which seemed to me a very beautiful group; on scanning it more closely, I discovered that the faces were ordinary, or, at all events, not much above the average, and that the charm arose entirely from the natural action and expression of the mother stooping over and kissing her child which she held in her lap; and I remarked to myself that the homeliest features would become beautiful when lit up by the rays of the soul—like the sun "gilding pale streams with heavenly alchemy". By analogy, the thought struck me that if a man would speak naturally and as he felt on any subject of his predilection, he might hope to awaken a sympathetic interest in the minds of his hearers; and, in illustration of this, I remembered witnessing how the writer of a well-known article in the *Quarterly Review* so magnetized his audience at the Royal Institution by his evident enthusiasm that, when the lecture was over and the applause had subsided, some ladies came up to me and implored me to tell them what they should do to get up the Talmud; for that was what the lecture had been about.

⟦2⟧ Now, as I believe that even Mathematics are not much more repugnant than the Talmud to the common apprehension of mankind, and I really love my subject, I shall not quite despair of rousing and retaining your attention for a short time if I proceed to read (as, for greater assurance against breaking down, I shall beg your permission to do) from the pages I hold in my hand.

---

[a] Sylvester's speech led to a lengthy correspondence in *Nature* over the correct interpretation of Kant's doctrine of geometry. The correspondence was reprinted as an appendix to Sylvester's *The laws of verse; or, Principles of versification exemplified in metrical translations, together with an annotated reprint of the inaugural Presidential address to the Mathematical and physical section of the British association at Exeter* (*1870*), and also in *Sylvester 1908* (pp. 719–31). (*The laws of verse*, dedicated to Matthew Arnold, contains many of Sylvester's own poetic creations: like Boole and Rowan Hamilton, Sylvester had aspirations as a poet.)

⟦3⟧ It is not without a feeling of surprise and trepidation at my own temerity that I find myself in the position of one about to address this numerous and distinguished assembly. When informed that the Council of the British Association had it in contemplation to recommend me to the General Committee to fill the office of President of the Mathematical and Physical Section, the intimation was accompanied with the tranquilizing assurance that it would rest with myself to deliver or withhold an address as I might think fit, and that I should be only following in the footsteps of many of the most distinguished of my predecessors were I to resolve on the latter course.

⟦4⟧ Until the last few days I had made up my mind to avail myself of this option, by proceeding at once to the business before us without troubling you to listen to any address, swayed thereto partly by a consciousness of the very limited extent of my oratorical powers, partly by a disinclination, in the midst of various pressing private and official occupations, to undertake a kind of work new to one more used to thinking than to speaking (to making mathematics than to talking about them), and partly and more especially by a feeling of my inadequacy to satisfy the expectations that would be raised in the minds of those who had enjoyed the privilege of hearing or reading the allocution (which fills me with admiration and dismay) of my gifted predecessor, Dr Tyndall, a man in whom eloquence and philosophy seem to be inborn, whom Science and Poetry woo with an equal spell,* and whose ideas have a faculty of arranging themselves in forms of order and beauty as spontaneously and unfailingly as those crystalline solutions from which, in a striking passage of his address, he drew so vivid and instructive an illustration.

⟦5⟧ From this lotus-eater's dream of fancied security and repose I was rudely awakened by receiving from the Editor of an old-established journal in this city a note containing a polite but peremptory request that I should, at my earliest convenience, favour him with a "copy of the address I proposed to deliver at the forthcoming Meeting". To this invitation, my first impulse was to respond very much in the same way as did the "Needy knife-grinder" of the *Antijacobin*, when summoned to recount the story of his wrongs to his republi-

* So it is said of Jacobi, that he attracted the particular attention and friendship of Böckh, the director of the philological seminary at Berlin, by the zeal and talent he displayed for philology, and only at the end of two years' study at the University, and after a severe mental struggle, was able to make his final choice in favour of mathematics. The relation between these two sciences is not perhaps so remote as may at first sight appear, and indeed it has often struck me that metamorphosis runs like a golden thread through the most diverse branches of modern intellectual culture, and forms a natural link of connexion between subjects in their aims so unlike as Grammar, Ethnology, Rational Mythology, Chemistry, Botany, Comparative Anatomy, Physiology, Physics, Algebra, Music, all of which, under the modern point of view may be regarded as having morphology for their common centre. Even singing, I have been told, the advanced German theorists regard as being strictly a development of recitative, and infer therefrom that no essentially new melodic themes can be invented until a social cataclysm, or the civilization of some at present barbaric races, shall have created new necessities of expression and called into activity new forms of impassioned declamation.

can sympathizer, "Story, God bless you, I have none to tell, Sir!" "Address, Mr Editor, I have none to deliver."

⟦6⟧ I have found, however, that increase of appetite still grows with what it feeds on, that those who were present at the opening of the Section last year, and enjoyed my friend Dr Tyndall's melodious utterances, would consider themselves somewhat ill-treated if they were sent away quite empty on the present occasion, and that, failing an address, the Members would feel very much like the guests at a wedding-breakfast where no one was willing or able to propose the health of the bride and bridegroom.

⟦7⟧ Yielding, therefore, to these considerations and to the advice of some officially connected with the Association, to whose opinions I feel bound to defer, and unwilling also to countenauce by my example the too prevailing opinion that mathematical pursuits unfit a person for the discharge of the common duties of life and cut him off from the exercise of Man's highest prerogative, "discourse of reason and faculty of speech divine",—rather, I say, than favour the notion that we Algebraists (who regard each other as the flower and salt of the earth) are a set of mere calculating-machines endowed with organs of locomotion, or, at best, a sort of poor visionary dumb creatures only capable of communicating by signs and symbols with the outer world, I have resolved to take heart of grace and to say a few words, which I hope to render, if not interesting, at least intelligible, on a subject to which the larger part of my life has been devoted.

⟦8⟧ The President of the Association, Prof. Stokes, is so eminent alike as a mathematician and physicist, and so distinguished for accuracy and extent of erudition and research, that I felt assured I might safely assume he would, in his Address to the Association at large, take an exhaustive survey, and render a complete account of the recent progress and present condition and prospects of Mathematical and Physical Science. This consideration narrowed very much and brought almost to a point the ground available for me to occupy in this Section; and as I cannot but be aware that it is as a cultivator of pure mathematics (the subject in which my own researches have chiefly, though by no means exclusively lain*) that I have been placed in this Chair, I hope the Section will patiently bear with me in the observations I shall venture to make on the nature of that province of the human reason and its title to the esteem and veneration

* My first printed paper was on Fresnel's Optical Theory, published in the *Philosophical Magazine*; my latest contribution to the *Philosophical Transactions* is a memoir on the "Rotation of a Free Rigid Body". There is an old adage, "purus mathematicus, purus asinus". On the other hand, I once heard the great Richard Owen say, when we were opposite neighbours in Lincoln's-Inn Fields (doves nestling among hawks), that he would like to see *Homo Mathematicus* constituted into a distinct subclass, thereby suggesting to my mind sensation, perception, reflection, abstraction, as the successive stages or phases of protoplasm on its way to being made perfect in Mathematicised Man. Would it sound too presumptuous to speak of perception as a quintessence of sensation, language (that is, communicable thought) of perception, mathematic of language? We should then have four terms differentiating from inorganic matter and from each other the Vegetable, Animal, Rational, and Supersensual modes of existence.

with which through countless ages it has been and, so long as Man respects the intellectual part of his nature, must ever continue to be regarded.*

⟦9⟧ It is said of a great party leader and orator in the House of Lords that, when lately requested to make a speech at some religious or charitable (at all events a non-political) meeting, he declined to do so on the ground that he could not speak unless he saw an adversary before him—somebody to attack or reply to. In obedience to a somewhat similar combative instinct, I set to myself the task of considering certain recent utterances of a most distinguished member of this Association, one whom I no less respect for his honesty and public spirit than I admire for his genius and eloquence,† but from whose opinions on a subject which he has not studied I feel constrained to differ. Goethe has said—

"Verständige Leute kannst du irren sehn
In Sachen, nämlich, die sie nicht verstehn."

Understanding people you may see erring—in those things, to wit, which they do not understand.

⟦10⟧ I have no doubt that had my distinguished friend, the probable President-elect of the next Meeting of the Association, applied his uncommon powers of reasoning, induction, comparison, observation, and invention to the study of mathematical science, he would have become as great a mathematician as he is now a biologist; indeed he has given public evidence of his ability to grapple with the practical side of certain mathematical questions; but he has not made a study of mathematical science as such, and the eminence of his position and the weight justly attaching to his name render it only the more imperative that any assertions proceeding from such a quarter, which may appear to me erroneous, or so expressed as to be conducive to error, should not remain unchallenged or be passed over in silence.‡

⟦11⟧ He says "mathematical training is almost purely deductive. The mathematician starts with a few simple propositions, the proof of which is so obvious that they are called self-evident, and the rest of his work consists of subtle deductions from them. The teaching of languages, at any rate as ordinarily practised, is of the same general nature—authority and tradition furnish the data, and the mental operations are deductive." It would seem from the above

* Mr Spottiswoode favoured the Section, in his opening address, with a combined history of the program of Mathematics and Physics; Dr Tyndall's address was virtually on the limits of Physical Philosophy; the one here in print is an attempted faint adumbration of the nature of Mathematical Science in the abstract. What is wanting (like a fourth sphere resting on three others in contact) to build up the Ideal Pyramid is a discourse on the Relation of the two branches (Mathematics and Physics) to, their action and reaction upon, one another, a magnificent theme with which it is to be hoped some future President of Section A will crown the edifice and make the Tetralogy (symbolizable by $A + A'$, A, A', $A.A'$) complete.

† Although no great lecture-goer, I have heard three lectures in my life which have left a lasting impression as masterpieces on my memory—Clifford on Mind, Huxley on Chalk, Dumas on Faraday.

‡ In his *éloge* of Daubenton, Cuvier remarks, "Les savants jugent toujours comme le vulgaire les ouvrages qui ne sont pas de leur genre.".

somewhat singularly juxtaposed paragraphs that, according to Prof. Huxley, the business of the mathematical student is from a limited number of propositions (bottled up and labelled ready for future use) to deduce any required result by a process of the same general nature as a student of language employs in declining and conjugating his nouns and verbs—that to make out a mathematical proposition and to construe or parse a sentence are equivalent or identical mental operations. Such an opinion scarcely seems to need serious refutation. The passage is taken from an article in *Macmillan's Magazine* for June last, entitled "Scientific Education—Notes of an After-dinner Speech", and I cannot but think would have been couched in more guarded terms by my distinguished friend had his speech been made *before* dinner instead of *after*.

⟦12⟧ The notion that mathematical truth rests on the narrow basis of a limited number of elementary propositions from which all others are to be derived by a process of logical inference and verbal deduction, has been stated still more strongly and explicitly by the same eminent writer in an article of even date with the preceding in the *Fortnightly Review*, where we are told that "Mathematics is that study which knows nothing of observation, nothing of experiment, nothing of induction, nothing of causation." I think no statement could have been made more opposite to the undoubted facts of the case, that mathematical analysis is constantly invoking the aid of new principles, new ideas, and new methods, not capable of being defined by any form of words, but springing direct from the inherent powers and activity of the human mind, and from continually renewed introspection of that inner world of thought of which the phenomena are as varied and require as close attention to discern as those of the outer physical world (to which the inner one in each individual man may, I think, be conceived to stand in somewhat the same general relation of correspondence as a shadow to the object from which it is projected, or as the hollow palm of one hand to the closed fist which it grasps of the other), that it is unceasingly calling forth the faculties of observation and comparison, that one of its principal weapons is induction, that it has frequent recourse to experimental trial and verification, and that it affords a boundless scope for the exercise of the highest efforts of imagination and invention.

⟦13⟧ Lagrange, than whom no greater authority could be quoted, has expressed emphatically his belief in the importance to the mathematician of the faculty of observation; Gauss has called mathematics a science of the eye, and in conformity with this view always paid the most punctilious attention to preserve his text free from typographical errors; the ever to be lamented Riemann has written a thesis to show that the basis of our conception of space is purely empirical, and our knowledge of its laws the result of observation, that other kinds of space might be conceived to exist subject to laws different from those which govern the actual space in which we are immersed, and that there is no evidence of these laws extending to the ultimate infinitesimal elements of which space is composed. Like his master Gauss, Riemann refuses to accept Kant's doctrine of space and time being forms of intuition, and regards them as possessed of physical and objective reality. I may mention that Baron Sartorius von

Waltershausen (a member of this Association) in his biography of Gauss ("Gauss zu Gedächtniss"), published shortly after his death, relates that this great man used to say that he had laid aside several questions which he had treated analytically, and hoped to apply to them geometrical methods in a future state of existence, when his conceptions of space should have become amplified and extended; for as we can conceive beings (like infinitely attenuated bookworms* in an infinitely thin sheet of paper) which possess only the notion of space of two dimensions, so we may imagine beings capable of realising space of four or a greater number of dimensions†. Our Cayley, the central luminary, the Darwin of the English school of mathematicians, started and elaborated at an early age, and with happy consequences, the same bold hypothesis.

‖14‖ Most, if not all, of the great ideas of modern mathematics have had their origin in observation. Take, for instance, the arithmetical theory of forms, of which the foundation was laid in the diophantine theorems of Fermat, left without proof by their author, which resisted all the efforts of the myriad-minded Euler to reduce to demonstration, and only yielded up their cause of being when turned over in the blowpipe flame of Gauss's transcendent genius; or the doctrine of double periodicity, which resulted from the observation by Jacobi of a purely analytical fact of transformation; or Legendre's law of reciprocity; or Sturm's theorem about the roots of equations, which, as he informed me with his own lips, stared him in the face in the midst of some mechanical investigations connected with the motion of compound pendulums; or Huyghens' method of continued fractions, characterized by Lagrange as one of the principal discoveries of "that great mathematician, and to which he appears to have been led by the construction of his Planetary Automaton"; or the New Algebra, speaking of which one of my predecessors (Mr Spottiswoode) has said, not without just reason and authority, from this Chair, "that it reaches out and indissolubly connects itself each year with fresh branches of mathematics, that the theory of equations has almost become new through it, algebraic geometry transfigured in its light, that the calculus of variations, molecular physics, and mechanics" (he might, if speaking at the present moment, go on to add the theory of elasticity and the highest developments of the integral calculus) "have

* I have read or been told that eye of observer has never lighted on these depredators, living or dead. Nature has gifted me with eyes of exceptional microscopic power, and I can speak with some assurance of having repeatedly seen the creature wriggling on the learned page. On approaching it with breath or finger-nail it stiffens out into the semblance of a streak of dirt, and so eludes detection.

† It is well known to those who have gone into these views that the laws of motion accepted as a fact suffice to prove in a general way that the space we live in is a flat or level space (a "homaloid"), our existence therein being assimilable to the life of the bookworm in an *unrumpled page*: but what if the page should be undergoing a process of gradual bending into a curved form? Mr W.K. Clifford has indulged in some remarkable speculations as to the possibility of our being able to infer, from certain unexplained phenomena of light and magnetism, the fact of our level space of three dimensions being in the act of undergoing in space of four dimensions (space as inconceivable to us as our space to the supposititious bookworm) a distortion analogous to the rumpling of the page to which that creature's powers of direct perception have been postulated to be limited.

all felt its influence."

‖15‖ Now this gigantic outcome of modern analytical thought, itself, too, only the precursor and progenitor of a future still more heaven-reaching theory, which will comprise a complete study of the interoperation, the actions and reactions, of algebraic forms (Analytical Morphology in its absolute sense), how did this originate? In the accidental observation by Eisenstein, some score or more years ago, of a single invariant (the Quadrinvariant of a Binary Quartic) which he met with in the course of certain researches just as accidentally and unexpectedly as M. Du Chaillu might meet a Gorilla in the country of the Fantees, or any one of us in London a White Polar Bear escaped from the Zoological Gardens. Fortunately he pounced down upon his prey and preserved it for the contemplation and study of future mathematicians. It occupies only part of a page in his collected posthumous works. This single result of observation (as well entitled to be so called as the discovery of Globigerinæ in chalk or of the Confoco-ellipsoidal structure of the shells of the Foraminifera), which remained unproductive in the hands of its distinguished author, has served to set in motion a train of thought and to propagate an impulse which have led to a complete revolution in the whole aspect of modern analysis, and whose consequences will continue to be felt until Mathematics are forgotten and British Associations meet no more.

‖16‖ I might go on, were it necessary, piling instance upon instance to prove the paramount importance of the faculty of observation to the process of mathematical discovery.* Were it not unbecoming to dilate on one's personal experience, I could tell a story of almost romantic interest about my own latest researches in a field where Geometry, Algebra, and the Theory of Numbers melt in a surprising manner into one another, like sunset tints or the colours of the dying dolphin, "the last still loveliest" (a sketch of which has just appeared in the *Proceedings of the London Mathematical Society* **), which would very strikingly illustrate how much observation, divination, induction, experimental trial, and verification, causation, too (if that means, as I suppose it must, mounting from phenomena to their reasons or causes of being), have to do with the work of the mathematician. In the face of these facts, which every analyst in this room or out of it can vouch for out of his own knowledge and personal experience, how can it be maintained, in the words of Professor Huxley, who, in this instance, is speaking of the sciences as they are in themselves and without any reference to scholastic discipline, that Mathematics "is that study which knows nothing of observation, nothing of induction, nothing of experiment, nothing of causation"?

---

* Newton's Rule was to all appearance, and according to the more received opinion, obtained inductively by its author. My own reduction of Euler's problem of the Virgins (or rather one slightly more general than this) to the form of a question (or, to speak more exactly, a set of questions) in simple partitions was, strange to say, first obtained by myself inductively, the result communicated to Prof. Cayley, and proved subsequently by each of us independently, and by perfectly distinct methods.

** Under the title of "Outline Trace of the Theory of Reducible Cyclodes".

⟦17⟧ I, of course, am not so absurd as to maintain that the habit of observation of external nature will be best or in any degree cultivated by the study of mathematics, at all events as that study is at present conducted; and no one can desire more earnestly than myself to see natural and experimental science introduced into our schools as a primary and indispensable branch of education: I think that that study and mathematical culture should go on hand in hand together, and that they would greatly influence each other for their mutual good. I should rejoice to see mathematics taught with that life and animation which the presence and example of her young and buoyant sister could not fail to impart, short roads preferred to long ones, Euclid honourably shelved or buried "deeper than did ever plummet sound" out of the schoolboy's reach, morphology introduced into the elements of Algebra—projection, correlation, and motion accepted as aids to geometry—the mind of the student quickened and elevated and his faith awakened by early initiation into the ruling ideas of polarity, continuity, infinity, and familiarization with the doctrine of the imaginary and inconceivable.

⟦18⟧ It is this living interest in the subject which is so wanting in our traditional and mediaeval modes of teaching. In France, Germany, and Italy, everywhere where I have been on the Continent, mind acts direct on mind in a manner unknown to the frozen formality of our academic institutions; schools of thought and centres of real intellectual cooperation exist; the relation of master and pupil is acknowledged as a spiritual and a lifelong tie, connecting successive generations of great thinkers with each other in an unbroken chain, just in the same way as we read, in the catalogue of our French Exhibition, or of the Salon at Paris, of this man or that being the pupil of one great painter or sculptor and the master of another. When followed out in this spirit, there is no study in the world which brings into more harmonious action all the faculties of the mind than the one of which I stand here as the humble representative, there is none other which prepares so many agreeable surprises for its followers, more wonderful than the changes in the transformation-scene of a pantomime, or, like this, seems to raise them, by successive steps of initiation, to higher and higher states of conscious intellectual being.

⟦19⟧ This accounts, I believe, for the extraordinary longevity of all the greatest masters of the Analytical art, the Dii Majores of the mathematical Pantheon. Leibnitz lived to the age of 70; Euler to 76; Lagrange to 77; Laplace to 78; Gauss to 78; Plato, the supposed inventor of the conic sections, who made mathematics his study and delight, who called them the handles or aids to philosophy, the medicine of the soul, and is said never to have let a day go by without inventing some new theorems, lived to 82; Newton, the crown and glory of his race, to 85; Archimedes, the nearest akin, probably, to Newton in genius, was 75, and might have lived on to be 100, for aught we can guess to the contrary, when he was slain by the impatient and ill-mannered sergeant, sent to bring him before the Roman general, in the full vigour of his faculties, and in the very act of working out a problem; Pythagoras, in whose school, I believe, the word mathematician (used, however, in a somewhat wider than its present sense) originated, the second founder of geometry, the inventor of the matchless theorem which goes by his name, the precognizer of the undoubtedly mis-called Copernican theory, the discoverer of the regular solids and the musical canon, who stands at the very apex of this pyramid of fame, (if we may credit the tradition) after spending 22 years studying in Egypt, and 12 in Babylon, opened school when 56 or 57 years old in Magna Græcia, married a young wife when past 60, and died, carrying on his work with energy unspent to the last, at the age of 99. The mathematician lives long and lives young; the wings of his soul do not early drop off, nor do its pores become clogged with the earthy particles blown from the dusty highways of vulgar life.

⟦20⟧ Some people have been found to regard all mathematics, after the 47th proposition of Euclid, as a sort of morbid secretion, to be compared only with the pearl said to be generated in the diseased oyster, or, as I have heard it described, "une excroissance maladive de l'esprit humain". Others find its justification, its "raison d'être", in its being either the torch-bearer leading the way, or the handmaiden holding up the train of Physical Science; and a very clever writer in a recent magazine article, expresses his doubts whether it is, in itself, a more serious pursuit, or more worthy of interesting an intellectual human being, than the study of chess problems or Chinese puzzles. What is it to us, they say, if the three angles of a triangle are equal to two right angles, or if every even number is, or may be, the sum of two primes, or if every equation of an odd degree must have a real root. How dull, stale, flat, and unprofitable are such and such like announcements! Much more interesting to read an account of a marriage in high life, or the details of an international boat-race. But this is like judging of architecture from being shown some of the brick and mortar, or even a quarried stone of a public building, or of painting from the colours mixed on the palette, or of music by listening to the thin and screechy sounds produced by a bow passed haphazard over the strings of a violin. The world of ideas which it discloses or illuminates, the contemplation of divine beauty and order which it induces, the harmonious connexion of its parts, the infinite hierarchy and absolute evidence of the truths with which it is concerned, these, and such like, are the surest grounds of the title of mathematics to human regard, and would remain unimpeached and unimpaired were the plan of the universe unrolled like a map at our feet, and the mind of man qualified to take in the whole scheme of creation at a glance.

⟦21⟧ In conformity with general usage, I have used the word mathematics in the plural; but I think it would be desirable that this form of word should be reserved for the applications of the science, and that we should use mathematic in the singular number to denote the science itself, in the same way as we speak of logic, rhetoric, or (own sister to algebra*) music. Time was when

* I have elsewhere (in my 'Trilogy' published in the *Philosophical Transactions*) referred to the close connexion between these two cultures, not merely as having Arithmetic for their common parent, but as similar in their habits and affections. I have called "Music the Algebra of sense, Algebra the Music of the reason; Music the dream, Algebra the waking life,—the soul of each the same!"

all the parts of the subject were dissevered, when algebra, geometry, and arithmetic either lived apart or kept up cold relations of acquaintance confined to occasional calls upon one another; but that is now at an end; they are drawn together and are constantly becoming more and more intimately related and connected by a thousand fresh ties, and we may confidently look forward to a time when they shall form but one body with one soul. Geometry formerly was the chief borrower from arithmetic and algebra, but it has since repaid its obligations with abundant usury; and if I were asked to name, in one word, the pole-star round which the mathematical firmament revolves, the central idea which pervades as a hidden spirit the whole corpus of mathematical doctrine, I should point to Continuity as contained in our notions of space, and say, it is this, it is this! Space is the *Grand Continuum* from which, as from an inexhaustible reservoir, all the fertilizing ideas of modern analysis are derived; and as Brindley, the engineer, once allowed before a parliamentary committee that, in his opinion, rivers were made to feed navigable canals, I feel almost tempted to say that one principal reason for the existence of space, or at least one principal function which it discharges, is that of feeding mathematical invention. Everybody knows what a wonderful influence geometry has exercised in the hands of Cauchy, Puiseux, Riemann, and his followers Clebsch, Gordan, and others, over the very form and presentment of the modern calculus, and how it has come to pass that the tracing of curves, which was once to be regarded as a puerile amusement, or at best useful only to the architect or decorator, is now entitied to take rank as a high philosophical exercise, inasmuch as every new curve or surface, or other circumscription of space is capable of being regarded as the embodiment of some specific organized system of continuity.*

⟦22⟧ The early study of Euclid made me a hater of Geometry, which I hope may plead my excuse if I have shocked the opinions of any in this room (and I know there are some who rank Euclid as second in sacredness to the Bible alone, and as one of the advanced outposts of the British Constitution) by the tone in which I have previously alluded to it as a school-book; and yet, in spite of this repugnance, which had become a second nature in me, whenever I went far enough into any mathematical question, I found I touched, at last, a geometrical bottom: so it was, I may instance, in the purely arithmetical theory of partitions; so, again, in one of my more recent studies, the purely algebraical question of the invariantive criteria of the nature of the roots of an equation of the fifth degree: the first inquiry landed me in a new theory of polyhedra; the latter found its perfect and only possible complete solution in the construction of a surface of the ninth order and the subdivision of its infinite content into three distinct natural regions.

---

* M. Camille Jordan's application of Dr Salmon's Eikosi-heptagram to Abelian functions is one of the most recent instances of this reverse action of geometry on analysis. Mr Crofton's admirable apparatus of a reticulation with infinitely fine meshes rotated successively through indefinitely small angles, which he applies to obtaining whole families of definite integrals, is another equally striking example of the same phenomenon.

[23] Having thus expressed myself at much greater length than I originally intended on the subject, which, as standing first on the muster-roll of the Association, and as having been so recently and repeatedly arraigned before the bar of public opinion, is entitled to be heard in its defence (if anywhere) in this place,—having endeavoured to show what it is not, what it is, and what it is probably destined to become, I feel that I must enough and more than enough have trespassed on your forbearance, and shall proceed with the regular business of the Meeting.

[24] Before calling upon the authors of the papers contained in the varied bill of intellectual fare which I see before me, I hope to be pardoned if I direct attention to the importance of practising brevity and condensation in the delivery of communications to the Section, not merely as a saving of valuable time, but in order that what is said may be more easily followed and listened to with greater pleasure and advantage. I believe that immense good may be done by the oral interchange and discussion of ideas which takes place in the Sections; but for this to be possible, details and long descriptions should be reserved for printing and reading, and only the general outlines and broad statements of facts, methods, observations, or inventions brought before us here, such as can be easily followed by persons having a fair average acquaintance with the several subjects treated upon. I understand the rule to be that, with the exception of the author of any paper who may answer questions and reply at the end of the discussion, no member is to address the Section more than once on the same subject, or occupy more than a quarter of an hour in speaking.

[25] In order to get through the business set down in each day's paper, it may sometimes be necessary for me to bring a discussion to an earlier close than might otherwise be desirable, and for that purpose to request the authors of papers, and those who speak upon them, to be brief in their addresses. I have known most able investigators at these Meetings, and especially in this Section, gradually part company with their audience, and at last become so involved in digressions as to lose entirely the thread of their discourse, and seem to forget, like men waking out of sleep, where they were or what they were talking about. In such cases I shall venture to give a gentle pull to the string of the kite before it soars right away out of sight into the region of the clouds. I now call upon Dr Magnus to read his paper and recount to the Section his wondrous story on the Emission, Absorption, and Reflection of Obscure Heat.*

POSTSCRIPT.—The remarks on the use of experimental methods in mathematical investigation led to Dr Jacobi, the eminent physicist of St Petersburg, who was present at the delivery of the address, favouring me with the annexed anecdote relative to his illustrious brother C.G.J. Jacobi.

* Curiously enough, and as if symptomatic of the genial warmth of the proceedings in which seven sages from distant lands (Jacobi, Magnus, Newton, Janssen, Morren, Lyman, Neumayer) took frequent part, the opening and concluding papers (each of surpassing interest, and a letting-out of mighty waters) were on Obscure Heat, by Prof. Magnus, and on Stellar Heat, by Mr Huggins.

"En causant un jour avec mon frère défunt sur la necessité de contrôler par des expériences réitérées toute observation, même si elle confirme l'hypothèse, il me raconta avoir découvert un jour une loi très-remarquable de la théorie des nombres, dont il ne douta guère qu'elle fût générale. Cependant par un excès de précaution ou plutôt pour faire le superflu, il voulut substituer un chiffre quelconque réel aux termes généraux, chiffre qu'il choisit au hasard ou, peut-être, par une espèce de divination, car en effet ce chiffre mit sa formule en défaut; tout autre chiffre qu'il essaya en confirma la généralité. Plus tard il réussit à prouver que le chiffre choisi par lui par hasard, appartenait à un système de chiffres qui faisait la seule exception à la règle.

"Ce fait curieux m'est resté dans la mémoire, mais comme il s'est passé il y a plus d'une trentaine d'années, je ne rappelle plus des détails.

"M.H. JACOBI".

"EXETER, 24. *Août*, 1869".

13
William Kingdon Clifford (1845–1879)

Clifford, the son of a justice of the peace, was educated in mathematics at Trinity College, Cambridge. He took his degree in 1869; in the same year, at the age of twenty-four, he was elected professor of applied mathematics at University College, London. For a few years he had a glittering career as a mathematician and as a public lecturer on scientific subjects; but in his early thirties his health began to fail, and he died of tuberculosis in Madeira at thirty-three. Four volumes of his writings were collected by his friends and published after his death.

As a mathematician, Clifford worked on projective geometry, non-Euclidean geometry, quaternions, and Riemann surfaces; his efforts to generalize Hamilton's quaternions led him to the discovery of Clifford algebras. Most of his papers on these subjects are collected in *Clifford 1882*. Clifford was furthermore responsible for calling the attention of British mathematicians to Riemann's paper 'On the hypotheses which lie at the foundations of geometry'; his translation of this paper (which he published in *Nature* in 1873) is reproduced below (*Riemann 1868*).

Clifford is most widely remembered for his popular writings on scientific subjects—in particular, for *The common sense of the exact sciences* (*1885*), and for his many public lectures (collected in *Clifford 1901*). One of these lectures is reproduced below; in addition to exhibiting Clifford's sparkling style, it shows him struggling to rethink the relationship between mathematics and the empirical sciences, and in particular to reconcile the insights of Riemann, Darwin, and Kant. In this respect, this lecture is characteristic, both for himself and for the closing decades of the century; in particular we shall find that many of the themes broached by Clifford in *1872* will reappear in the selections from Helmholtz.

A. ON THE SPACE THEORY OF MATTER (*CLIFFORD 1876*)

(*Abstract.*)

RIEMANN has shewn that as there are different kinds of lines and surfaces, so there are different kinds of space of three dimensions; and that we can only find out by experience to which of these kinds the space in which we live belongs. In particular, the axioms of plane geometry are true within the limits of experiment on the surface of a sheet of paper, and yet we know that the sheet is really covered with a number of small ridges and furrows, upon which (the total curvature not being zero) these axioms are not true. Similarly, he says although the axioms of solid geometry are true within the limits of experiment for finite portions of our space, yet we have no reason to conclude that they are true for very small portions; and if any help can be got thereby for the explanation of physical phenomena, we may have reason to conclude that they are not true for very small portions of space.

I wish here to indicate a manner in which these speculations may be applied to the investigation of physical phenomena. I hold in fact

(1) That small portions of space *are* in fact of a nature analogous to little hills on a surface which is on the average flat; namely, that the ordinary laws of geometry are not valid in them.

(2) That this property of being curved or distorted is continually being passed on from one portion of space to another after the manner of a wave.

(3) That this variation of the curvature of space is what really happens in that phenomenon which we call the *motion of matter*, whether ponderable or etherial.

(4) That in the physical world nothing else takes place but this variation, subject (possibly) to the law of continuity.

I am endeavouring in a general way to explain the laws of double refraction on this hypothesis, but have not yet arrived at any results sufficiently decisive to be communicated.

B. ON THE AIMS AND INSTRUMENTS OF SCIENTIFIC THOUGHT (*CLIFFORD 1872*)

The following lecture was delivered to the meeting of the British Association at Brighton on 19 August 1872. References to *Clifford 1872* should be to the paragraph numbers, which have been added in this reprinting.

⟦1⟧ It may have occurred (and very naturally too) to such as have had the curiosity to read the title of this lecture, that it must necessarily be a very dry and difficult subject; interesting to very few, intelligible to still fewer, and, above all, utterly incapable of adequate treatment within the limits of a discourse like this. It is quite true that a complete setting-forth of my subject would require a comprehensive treatise on logic, with incidental discussion of the main questions of metaphysics; that it would deal with ideas demanding close study for their apprehension, and investigations requiring a peculiar taste to relish them. It is not my intention now to present you with such a treatise.

[2] The British Association, like the world in general, contains three classes of persons. In the first place, it contains scientific thinkers; that is to say, persons whose thoughts have very frequently the characters which I shall presently describe. Secondly, it contains persons who are engaged in work upon what are called scientific subjects, but who in general do not, and are not expected to, think about these subjects in a scientific manner. Lastly, it contains persons who suppose that their work and their thoughts are unscientific, but who would like to know something about the business of the other two classes aforesaid. Now, to any one who belonging to one of these classes considers either of the other two, it will be apparent that there is a certain gulf between him and them; that he does not quite understand them, nor they him; and that an opportunity for sympathy and comradeship is lost through this want of understanding. It is this gulf that I desire to bridge over, to the best of my power. That the scientific thinker may consider his business in relation to the great life of mankind; that the noble army of practical workers may recognise their fellowship with the outer world, and the spirit which must guide both; that this so-called outer world may see in the work of science only the putting in evidence of all that is excellent in its own work,—may feel that the kingdom of science is within it: these are the objects of the present discourse. And they compel me to choose such portions of my vast subject as shall be intelligible to all, while they ought at least to command an interest universal, personal, and profound.

[3] In the first place, then, what is meant by scientific thought? You may have heard some of it expressed in the various Sections this morning. You have probably also heard expressed in the same places a great deal of unscientific thought; notwithstanding that it was about mechanical energy, or about hydrocarbons, or about eocene deposits, or about malacopterygii. For scientific thought does not mean thought about scientific subjects with long names. There are no scientific subjects. The subject of science is the human universe; that is to say, everything that is, or has been, or may be related to man. Let us then, taking several topics in succession, endeavour to make out in what cases thought about them is scientific, and in what cases not.

[4] Ancient astronomers observed that the relative motions of the sun and moon recurred all over again in the same order about every nineteen years. They were thus enabled to predict the time at which eclipses would take place. A calculator at one of our observatories can do a great deal more than this. Like them, he makes use of past experience to predict the future; but he knows of a great number of other cycles besides that one of the nineteen years, and takes account of all of them; and he can tell about the solar eclipse of six years hence exactly when it will be visible, and how much of the sun's surface will be covered at each place, and, to a second, at what time of day it will begin and finish there. This prediction involves technical skill of the highest order; but it does not involve scientific thought, as any astronomer will tell you.

[5] By such calculations the places of the planet Uranus at different times

of the year had been predicted and set down. The predictions were not fulfilled. Then arose Adams, and from these errors in the prediction he calculated the place of an entirely new planet, that had never yet been suspected; and you all know how the new planet was actually found in that place. Now this prediction does involve scientific thought, as any one who has studied it will tell you.

⟦6⟧ Here then are two cases of thought about the same subject, both predicting events by the application of previous experience, yet we say one is *technical* and the other *scientific*.

⟦7⟧ Now let us take an example from the building of bridges and roofs. When an opening is to be spanned over by a material construction, which must bear a certain weight without bending enough to injure itself, there are two forms in which this construction can be made, the arch and the chain. Every part of an arch is compressed or pushed by the other parts; every part of a chain is in a state of tension, or is pulled by the other parts. In many cases these forms are united. A girder consists of two main pieces or booms, of which the upper one acts as an arch and is compressed, while the lower one acts as a chain and is pulled; and this is true even when both the pieces are quite straight. They are enabled to act in this way by being tied together, or braced, as it is called, by cross pieces, which you must often have seen. Now suppose that any good practical engineer makes a bridge or roof upon some approved pattern which has been made before. He designs the size and shape of it to suit the opening which has to be spanned; selects his material according to the locality; assigns the strength which must be given to the several parts of the struture according to the load which it will have to bear. There is a great deal of thought in the making of this design, whose success is predicted by the application of previous experience; it requires technical skill of a very high order; but it is not scientific thought. On the other hand, Mr. Fleeming Jenkin[1] designs a roof consisting of two arches braced together, instead of an arch and a chain braced together; and although this form is quite different from any known structure, yet before it is built he assigns with accuracy the amount of material that must be put into every part of the structure in order to make it bear the required load, and this prediction may be trusted with perfect security. What is the natural comment on this? Why, that Mr. Fleeming Jenkin is a scientific engineer.

⟦8⟧ Now it seems to me that the difference between scientific and merely technical thought, not only in these but in all other instances which I have considered, is just this: Both of them make use of experience to direct human action; but while technical thought or skill enables a man to deal with the same circumstances that he has met with before, scientific thought enables him to deal with different circumstances that he has never met with before. But how can experience of one thing enable us to deal with another quite different thing? To answer this question we shall have to consider more closely the nature of scientific thought.

⟦9⟧ Let us take another example. You know that if you make a dot on a piece of paper, and then hold a piece of Iceland spar over it, you will see not

[1] *On Braced Arches and Suspension Bridges*. Edinburgh: Neill, 1870.

one dot but two. A mineralogist, by measuring the angles of a crystal, can tell you whether or no it possesses this property without looking through it. He requires no scientific thought to do that. But Sir William Rowan Hamilton, the late Astronomer-Royal of Ireland, knowing these facts and also the explanation of them which Fresnel had given, thought about the subject, and he predicted that by looking through certain crystals in a particular direction we should see not two dots but a continuous circle. Mr. Lloyd made the experiment, and saw the circle, a result which had never been even suspected. This has always been considered one of the most signal instances of scientific thought in the domain of physics. It is most distinctly an application of experience gained under certain circumstances to entirely different circumstances.

⟦10⟧ Now suppose that the night before coming down to Brighton you had dreamed of a railway accident caused by the engine getting frightened at a flock of sheep and jumping suddenly back over all the carriages; the result of which was that your head was unfortunately cut off, so that you had to put it in your hat-box and take it back home to be mended. There are, I fear, many persons even at this day, who would tell you that after such a dream it was unwise to travel by railway to Brighton. This is a proposal that you should take experience gained while you are asleep, when you have no common sense,—experience about a phantom railway, and apply it to guide you when you are awake and have common sense, in your dealings with a real railway. And yet this proposal is not dictated by scientific thought.

⟦11⟧ Now let us take the great example of Biology. I pass over the process of classification, which itself requires a great deal of scientific thought; in particular when a naturalist who has studied and monographed a fauna or a flora rather than a family is able at once to pick out the distinguishing characters required for the subdivision of an order quite new to him. Suppose that we possess all this minute and comprehensive knowledge of plants and animals and intermediate organisms, their affinities and differences, their structures and functions;—a vast body of experience, collected by incalculable labour and devotion. Then comes Mr. Herbert Spencer: he takes that experience of life which is not human, which is apparently stationary, going on in exactly the same way from year to year, and he applies that to tell us how to deal with the changing characters of human nature and human society. How is it that experience of this sort, vast as it is, can guide us in a matter so different from itself? How does scientific thought, applied to the development of a kangaroo foetus or the movement of the sap in exogens, make prediction possible for the first time in that most important of all sciences, the relations of man with man?

⟦12⟧ In the dark or unscientific ages men had another way of applying experience to altered circumstances. They believed, for example, that the plant called Jew's ear, which does bear a certain resemblance to the human ear, was a useful cure for diseases of that organ. This doctrine of "signatures", as it was called, exercised an enormous influence on the medicine of the time. I need hardly tell you that it is hopelessly unscientific; yet it agrees with those other examples that we have been considering in this particular; that it applies experience about the shape of a plant—which is one circumstance connected with it

—to dealings with its medicinal properties, which are other and different circumstances. Again, suppose that you had been frightened by a thunder-storm on land, or your heart had failed you in a storm at sea; if any one then told you that in consequence of this you should always cultivate an unpleasant sensation in the pit of your stomach, till you took delight in it, that you should regulate your sane and sober life by the sensations of a moment of unreasoning terror: this advice would not be an example of scientific thought. Yet it would be an application of past experience to new and different circumstances.

‖13‖ But you will already have observed what is the additional clause that we must add to our definition in order to describe scientific thought and that only. The step between experience about animals and dealings with changing humanity is the law of evolution. The step from errors in the calculated places of Uranus to the existence of Neptune is the law of gravitation. The step from the observed behaviour of crystals to conical refraction is made up of laws of light and geometry. The step from old bridges to new ones is the laws of elasticity and the strength of materials.

‖14‖ The step, then, from past experience to new circumstances must be made in accordance with an observed uniformity in the order of events. This uniformity has held good in the past in certain places; if it should also hold good in the future and in other places, then, being combined with our experience of the past, it enables us to predict the future, and to know what is going on elsewhere; so that we are able to regulate our conduct in accordance with this knowledge.

‖15‖ The aim of scientific thought, then, is to apply past experience to new circumstances; the instrument is an observed uniformity in the course of events. By the use of this instrument it gives us information transcending our experience, it enables us to infer things that we have not seen from things that we have seen; and the evidence for the truth of that information depends on our supposing that the uniformity holds good beyond our experience. I now want to consider this uniformity a little more closely; to show how the character of scientific thought and the force of its inferences depend upon the character of the uniformity of Nature. I cannot of course tell you all that is known of this character without writing an encyclopaedia; but I shall confine myself to two points of it about which it seems to me that just now there is something to be said. I want to find out what we mean when we say that the uniformity of Nature is *exact*; and what we mean when we say that it is *reasonable*.

‖16‖ When a student is first introduced to those sciences which have come under the dominion of mathematics, a new and wonderful aspect of Nature bursts upon his view. He has been accustomed to regard things as essentially more or less vague. All the facts that he has hitherto known have been expressed qualitatively, with a little allowance for error on either side. Things which are let go fall to the ground. A very observant man may know also that they fall faster as they go along. But our student is shown that, after falling for one second in a vacuum, a body is going at the rate of thirty-two feet per second, that after falling for two seconds it is going twice as fast, after going two and a half seconds two and a half times as fast. If he makes the experiment, and finds a single inch per second too much or too little in the rate, one of two things must have happened: either the law of falling bodies has been wrongly stated, or the experiment is not accurate—there is some mistake. He finds reason to think that the latter is always the case; the more carefully he goes to work, the more of the error turns out to belong to the experiment. Again, he may know that water consists of two gases, oxygen and hydrogen, combined; but he now learns that two pints of steam at a temperature of 150° Centigrade will always make two pints of hydrogen and one pint of oxygen at the same temperature, all of them being pressed as much as the atmosphere is pressed. If he makes the experiment and gets rather more or less than a pint of oxygen, is the law disproved? No; the steam was impure, or there was some mistake. Myriads of analyses attest the law of combining volumes; the more carefully they are made, the more nearly they coincide with it. The aspects of the faces of a crystal are connected together by a geometrical law, by which, four of them being given, the rest can be found. The place of a planet at a given time is calculated by the law of gravitation; if it is half a second wrong, the fault is in the instrument, the observer, the clock, or the law; now, the more observations are made, the more of this fault is brought home to the instrument, the observer, and the clock. It is no wonder, then, that our student, contemplating these and many like instances, should be led to say, "I have been short-sighted; but I have now put on the spectacles of science which Nature had prepared for my eyes; I see that things have definite outlines, that the world is ruled by exact and rigid mathematical laws; *καὶ σύ, θεός, γεωμετρεῖς*."[a] It is our business to consider whether he is right in so concluding. Is the uniformity of Nature absolutely exact, or only more exact than our experiments?

‖17‖ At this point we have to make a very important distinction. There are two ways in which a law may be inaccurate. The first way is exemplified by that law of Galileo which I mentioned just now: that a body falling *in vacuo* acquires equal increase in velocity in equal times. No matter how many feet per second it is going, after an interval of a second it will be going thirty-two *more* feet per second. We now know that this rate of increase is not exactly the same at different heights, that it depends upon the distance of the body from the centre of the earth; so that the law is only approximate; instead of the increase of velocity being exactly *equal* in equal times, it itself increases very slowly as the body falls. We know also that this variation of the law from the truth is *too small to be perceived* by direct observation on the change of velocity. But suppose we have invented means for observing this, and have verified that the increase of velocity is inversely as the squared distance from the earth's centre. Still the law is not accurate; for the earth does not attract accurately towards her centre, and the direction of attraction is continually varying with the motion of the sea; the body will not even fall in a straight line. The sun and the planets, too, especially the moon, will produce deviations; yet the sum of all these errors

[a] 'Thou too, O God, art a geometrician'.

will escape our new process of observation, by being a great deal smaller than the necessary errors of that observation. But when these again have been allowed for, there is still the influence of the stars. In this case, however, we only give up one exact law for another. It may still be held that if the effect of every particle of matter in the universe on the falling body were calculated according to the law of gravitation, the body would move exactly as this calculation required. And if it were objected that the body must be slightly magnetic or diamagnetic, while there are magnets not an infinite way off; that a very minute repulsion, even at sensible distances, accompanies the attraction; it might be replied that these phenomena are themselves subject to exact laws, and that when *all* the laws have been taken into account, the actual motion will exactly correspond with the calculated motion.

⟦18⟧ I suppose there is hardly a physical student (unless he has specially considered the matter) who would not at once assent to the statement I have just made; that if we knew all about it, Nature would be found universally subject to exact numerical laws. But let us just consider for another moment what this means.

⟦19⟧ The word "exact" has a practical and a theoretical meaning. When a grocer weighs you out a certain quantity of sugar very carefully, and says it is exactly a pound, he means that the difference between the mass of the sugar and that of the pound weight he employs is too small to be detected by his scales. If a chemist had made a special investigation, wishing to be as accurate as he could, and told you this was exactly a pound of sugar, he would mean that the mass of the sugar differed from that of a certain standard piece of platinum by a quantity too small to be detected by *his* means of weighing, which are a thousandfold more accurate than the grocer's. But what would a mathematician mean, if he made the same statement? He would mean this. Suppose the mass of the standard pound to be represented by a length, say a foot, measured on a certain line; so that half a pound would be represented by six inches, and so on. And let the difference between the mass of the sugar and that of the standard pound be drawn upon the same line to the same scale. Then, if that difference were magnified an infinite number of times, it would still be invisible. This is the theoretical meaning of exactness; the practical meaning is only very close approximation; *how* close, depends upon the circumstances. The knowledge then of an exact law in the theoretical sense would be equivalent to an infinite observation. I do not say that such knowledge is impossible to man; but I do say that it would be absolutely different in kind from any knowledge that we possess at present.

⟦20⟧ I shall be told, no doubt, that we do possess a great deal of knowledge of this kind, in the form of geometry and mechanics; and that it is just the example of these sciences that has led men to look for exactness in other quarters. If this had been said to me in the last century, I should not have known what to reply. But it happens that about the beginning of the present century the foundations of geometry were criticised independently by two mathematicians,

Lobatschewsky[1] and the immortal Gauss;[2] whose results have been extended and generalised more recently by Riemann[3] and Helmholtz.[4] And the conclusion to which these investigations lead is that, although the assumptions which were very properly made by the ancient geometers are practically exact—that is to say, more exact than experiment can be—for such finite things as we have to deal with, and such portions of space as we can reach; yet the truth of them for very much larger things, or very much smaller things, or parts of space which are at present beyond our reach, is a matter to be decided by experiment, when its powers are considerably increased. I want to make as clear as possible the real state of this question at present, because it is often supposed to be a question of words or metaphysics, whereas it is a very distinct and simple question of fact. I am supposed to know then that the three angles of a rectilinear triangle are exactly equal to two right angles. Now suppose that three points are taken in space, distant from one another as far as the Sun is from *α* Centauri, and that the shortest distances between these points are drawn so as to form a triangle. And suppose the angles of this triangle to be very accurately measured and added together; this can at present be done so accurately that the error shall certainly be less than one minute, less therefore than the five-thousandth part of a right angle. Then I do not know that this sum would differ at all from two right angles; but also I do not know that the difference would be less than ten degrees, or the ninth part of a right angle.[5] And I have reasons for not knowing.

⟦21⟧ This example is exceedingly important as showing the connection between exactness and universality. It is found that the deviation if it exists must be nearly proportional to the area of the triangle. So that the error in the case of a triangle whose sides are a mile long would be obtained by dividing that in the case I have just been considering by four hundred quadrillions; the result must be a quantity inconceivably small, which no experiment could detect. But between this inconceivably small error and no error at all, there is fixed an enormous gulf; the gulf between practical and theoretical exactness, and, what is even more important, the gulf between what is practically universal and what is theoretically universal. I say that a law is practically universal which is more exact than experiment for all cases that might be got at by such experiments as we can make. We assume this kind of universality, and we find that it pays us to assume it. But a law would be theoretically universal if it were true of all cases whatever; and this is what we do not know of any law at all.

---

[1] *Geometrische Untersuchungen zur Theorie der Parallellinien*, Berlin, 1840. Translated by Hoüel. Gauthier-Villars, 1866.

[2] Letter to Schumacher, Nov. 28, 1846 (refers to 1792).

[3] *Ueber die Hypothesen welche der Geometrie zu Grunde liegen*. Göttingen, Abhandl., 1866–67. Translated by Hoüel in *Annali di Matematica*, Milan, vol. iii.

[4] *The Axioms of Geometry*, Academy, vol. i, p. 128 (a popular exposition).

[5] Assuming that parallax observations prove the deviation less than half a second for a triangle whose vertex is at the star and base a diameter of the earth's orbit.

⟦22⟧ I said there were two ways in which a law might be inexact. There is a law of gases which asserts that when you compress a perfect gas the pressure of the gas increases exactly in the proportion in which the volume diminishes. Exactly; that is to say, the law is more accurate than the experiment, and experiments are corrected by means of the law. But it so happens that this law has been explained; we know precisely what it is that happens when a gas is compressed. We know that a gas consists of a vast number of separate molecules, rushing about in all directions with all manner of velocities, but so that the mean velocity of the molecules of air in this room, for example, is about twenty miles a minute. The pressure of the gas on any surface with which it is in contact is nothing more than the impact of these small particles upon it. On any surface large enough to be seen there are millions of these impacts in a second. If the space in which the gas is confined be diminished, the average rate at which the impacts take place will be increased in the same proportion; and because of the enormous number of them, the actual rate is always exceedingly close to the average. But the law is one of statistics; its accurracy depends on the enormous numbers involved; and so, from the nature of the case, its exactness cannot be theoretical or absolute.

⟦23⟧ Nearly all the laws of gases have received these statistical explanations; electric and magnetic attraction and repulsion have been treated in a similar manner; and an hypothesis of this sort has been suggested even for the law of gravity. On the other hand the manner in which the molecules of a gas interfere with each other proves that they repel one another inversely as the fifth power of the distance; so that we here find at the basis of a statistical explanation a law which has the form of theoretical exactness. Which of these forms is to win? It seems to me again that we do not know, and that the recognition of our ignorance is the surest way to get rid of it.

⟦24⟧ The world in general has made just the remark that I have attributed to a fresh student of the applied sciences. As the discoveries of Galileo, Kepler, Newton, Dalton, Cavendish, Gauss, displayed ever new phenomena following mathematical laws, the theoretical exactness of the physical universe was taken for granted. Now, when people are hopelessly ignorant of a thing, they quarrel about the source of their knowledge. Accordingly many maintained that we know these exact laws by intuition. These said always one true thing, that we did not know them from experience. Others said that they were really given in the facts, and adopted ingenious ways of hiding the gulf between the two. Others again deduced from transcendental considerations sometimes the laws themselves, and sometimes what through imperfect information they supposed to be the laws. But more serious consequences arose when these conceptions derived from Physics were carried over into the field of Biology. Sharp lines of division were made between kingdoms and classes and orders; an animal was described as a miracle to the vegetable world; specific differences which are practically permanent within the range of history were regarded as permanent through all time; a sharp line was drawn between organic and inorganic matter. Further investigation, however, has shown that accuracy had been prematurely attributed to the science, and has filled up all the gulfs and gaps that hasty observers had invented. The animal and vegetable kingdoms have a debateable ground between them, occupied by beings that have the characters of both and yet belong distinctly to neither. Classes and orders shade into one another all along their common boundary. Specific differences turn out to be the work of time. The line dividing organic matter from inorganic, if drawn to-day, must be moved to-morrow to another place; and the chemist will tell you that the distinction has now no place in his science except in a technical sense for the convenience of studying carbon compounds by themselves. In Geology the same tendency gave birth to the doctrine of distinct periods, marked out by the character of the strata deposited in them all over the sea; a doctrine than which, perhaps, no ancient cosmogony has been further from the truth, or done more harm to the progress of science. Refuted many years ago by Mr. Herbert Spencer,[1] it has now fairly yielded to an attack from all sides at once, and may be left in peace.

‖25‖ When then we say that the uniformity which we observe in the course of events is exact and universal, we mean no more than this: that we are able to state general rules which are far more exact than direct experiment, and which apply to all cases that we are at present likely to come across. It is important to notice, however, the effect of such exactness as we observe upon the nature of inference. When a telegram arrived stating that Dr. Livingstone had been found by Mr. Stanley, what was the process by which you inferred the finding of Dr. Livingstone from the appearance of the telegram? You assumed over and over again the existence of uniformity in nature. That the newspapers had behaved as they generally do in regard to telegraphic messages; that the clerks had followed the known laws of the action of clerks; that electricity had behaved in the cable exactly as it behaves in the laboratory; that the actions of Mr. Stanley were related to his motives by the same uniformities that affect the actions of other men; that Dr. Livingstone's handwriting conformed to the curious rule by which an ordinary man's handwriting may be recognised as having persistent characteristics even at different periods of his life. But you had a right to be much more sure about some of these inferences than about others. The law of electricity was known with practical exactness, and the conclusions derived from it were the surest things of all. The law about the handwriting, belonging to a portion of physiology which is unconnected with consciousness, was known with less, but still with considerable accuracy. But the laws of human action in which consciousness is concerned are still so far from being completely analysed and reduced to an exact form that the inferences which you made by their help were felt to have only a provisional force. It is possible that by and by, when psychology has made enormous advances and become an exact science, we may be able to give to testimony the sort of weight which we give to the inferences of physical science. It will then be

[1] 'Illogical Geology', in *Essays*, vol. i. Originally published in 1859.

possible to conceive a case which will show how completely the whole process of inference depends on our assumption of uniformity. Suppose that testimony, having reached the ideal force I have imagined, were to assert that a certain river runs uphill. You could infer nothing at all. The arm of inference would be paralysed, and the sword of truth broken in its grasp; and reason could only sit down and wait until recovery restored her limb, and further experience gave her new weapons.

⟦26⟧ I want in the next place to consider what we mean when we say that the uniformity which we have observed in the course of events is *reasonable* as well as exact.

⟦27⟧ No doubt the first form of this idea was suggested by the marvellous adaptation of certain natural structures to special functions. The first impression of those who studied comparative anatomy was that every part of the animal frame was fitted with extraordinary completeness for the work that it had to do. I say extraordinary, because at the time the most familiar examples of this adaptation were manufactures produced by human ingenuity; and the completeness and minuteness of natural adaptations were seen to be far in advance of these. The mechanism of limbs and joints was seen to be adapted, far better than any existing ironwork, to those motions and combinations of motion which were most useful to the particular organisms. The beautiful and complicated apparatus of sensation caught up indications from the surrounding medium, sorted them, analysed them, and transmitted the results to the brain in a manner with which, at the time I am speaking of, no artificial contrivance could compete. Hence the belief grew amongst physiologists that every structure which they found must have its function and subserve some useful purpose; a belief which was not without its foundation in fact, and which certainly (as Dr. Whewell remarks) has done admirable service in promoting the growth of physiology. Like all beliefs found successful in one subject, it was carried over into another, of which a notable example is given in the speculations of Count Rumford about the physical properties of water. Pure water attains its greatest density at a temperature of about $39\frac{1}{2}°$ Fahrenheit; it expands and becomes lighter whether it is cooled or heated, so as to alter that temperature. Hence it was concluded that water in this state must be at the bottom of the sea, and that by such means the sea was kept from freezing all through; as it was supposed must happen if the greatest density had been that of ice. Here then was a substance whose properties were eminently adapted to secure an end essential to the maintenance of life upon the earth. In short, men came to the conclusion that the order of nature was reasonable in the sense that everything was adapted to some good end.

⟦28⟧ Further consideration, however, has led men out of that conclusion in two different ways. First, it was seen that the facts of the case had been wrongly stated. Cases were found of wonderfully complicated structures that served no purpose at all; like the teeth of that whale of which you heard in Section D the other day, or of the Dugong, which has a horny palate covering them all up and used instead of them; like the eyes of the unborn mole, that are never used, though perfect as those of a mouse until the skull opening closes up, cutting them off from the brain, when they dry up and become incapable of use; like the outsides of your own ears, which are absolutely of no use to you. And when human contrivances were more advanced it became clear that the natural adaptations were subject to criticism. The eye, regarded as an optical instrument of human manufacture, was thus described by Helmholtz—the physiologist who learned physics for the sake of his physiology, and mathematics for the sake of his physics, and is now in the first rank of all three. He said, "If an optician sent me that as an instrument, I should send it back to him with grave reproaches for the carelessness of his work, and demand the return of my money."

⟦29⟧ The extensions of the doctrine into Physics were found to be still more at fault. That remarkable property of pure water, which was to have kept the sea from freezing, does not belong to salt water, of which the sea itself is composed. It was found, in fact, that the idea of a reasonable adaptation of means to ends, useful as it had been in its proper sphere, could yet not be called universal, or applied to the order of nature as a whole.

⟦30⟧ Secondly, this idea has given way because it has been superseded by a higher and more general idea of what is reasonable, which has the advantage of being applicable to a large portion of physical phenomena besides. Both the adaptation and the non-adaptation which occur in organic structures have been *explained*. The scientific thought of Dr. Darwin, of Mr. Herbert Spencer, and of Mr. Wallace, has described that hitherto unknown process of adaptation as consisting of perfectly well-known and familiar processes. There are two kinds of these: the direct processes, in which the physical changes required to produce a structure are worked out by the very actions for which that structure becomes adapted—as the backbone or notochord has been modified from generation to generation by the bendings which it has undergone; and the indirect processes included under the head of Natural Selection—the reproduction of children slightly different from their parents, and the survival of those which are best fitted to hold their own in the struggle for existence. Naturalists might give you some idea of the rate at which we are getting explanations of the evolution of all parts of animals and plants—the growth of the skeleton, of the nervous system and its mind, of leaf and flower. But what then do we mean by *explanation*?

⟦31⟧ We were considering just now an explanation of a law of gases—the law according to which pressure increases in the same proportion in which volume diminishes. The explanation consisted in supposing that a gas is made up of a vast number of minute particles always flying about and striking against one another, and then showing that the rate of impact of such a crowd of particles on the sides of the vessel containing them would vary exactly as the pressure is found to vary. Suppose the vessel to have parallel sides, and that there is only one particle rushing backwards and forwards between them; then it is clear that if we bring the sides together to half the distance, the particle will hit each of them twice as often, or the pressure will be doubled. Now it turns out that this would be just as true for millions of particles as for one, and when they are flying in all directions instead of only in one direction and its opposite. Observe now; it is a perfectly well-known and familiar thing that a body should strike against an opposing surface and bound off again; and it is a mere everyday occurrence that what has only half so far to go should be back in half the time; but that pressure should be strictly proportional to density is a comparatively strange, unfamiliar phenomenon. The explanation describes the unknown and unfamiliar as being made up of the known and the familiar; and this, it seems to me, is the true meaning of explanation.[1]

⟦32⟧ Here is another instance. If small pieces of camphor are dropped into water, they will begin to spin round and swim about in a most marvellous way. Mr. Tomlinson gave, I believe, the explanation of this. We must observe, to begin with, that every liquid has a skin which holds it; you can see that to be true in the case of a drop, which looks as if it were held in a bag. But the tension of this skin is greater in some liquids than in others; and it is greater in camphor and water than in pure water. When the camphor is dropped into water it begins to dissolve and get surrounded with camphor and water instead of water. If the fragment of camphor were exactly symmetrical, nothing more would happen; the tension would be greater in its immediate neighbourhood, but no motion would follow. The camphor, however, is irregular in shape; it dissolves more on one side than the other; and consequently gets pulled about, because the tension of the skin is greater where the camphor is most dissolved. Now it is probable that this is not nearly so satisfactory an explanation to you as it was to me when I was first told of it; and for this reason. By that time I was already perfectly familiar with the notion of a skin upon the surface of liquids, and I had been taught by means of it to work out problems in capillarity. The explanation was therefore a description of the unknown phenomenon which I did not know how to deal with as made up of known phenomena which I did know how to deal with. But to many of you possibly the liquid skin may seem quite as strange and unaccountable as the motion of camphor on water.

⟦33⟧ And this brings me to consider the source of the pleasure we derive from an explanation. By known and familiar I mean that which we know how to deal with, either by action in the ordinary sense, or by active thought. When therefore that which we do not know how to deal with is described as made up of things that we do know how to deal with, we have that sense of increased power which is the basis of all higher pleasures. Of course we may afterwards by association come to take pleasure in explanation for its own sake. Are we then to say that the observed order of events is reasonable, in the sense that all of it admits of explanation? That a process may be capable of explanation, it must break up into simpler constituents which are already familiar

---

[1] This view differs from those of Mr. J.S. Mill and Mr. Herbert Spencer in requiring every explanation to contain an addition to our knowledge about the thing explained. Both these writers regard subsumption under a general law as a species of explanation. See also Ferrier's *Remains*, vol. ii, p. 436.

to us. Now, first, the process may itself be simple, and not break up; secondly, it may break up into elements which are as unfamiliar and impracticable as the original process.

⟦34⟧ It is an explanation of the moon's motion to say that she is a falling body, only she is going so fast and is so far off that she falls quite round to the other side of the earth, instead of hitting it; and so goes on for ever. But it is no explanation to say that a body falls because of gravitation. That means that the motion of the body may be resolved into a motion of every one of its particles towards every one of the particles of the earth, with an acceleration inversely as the square of the distance between them. But this attraction of two particles must always, I think, be less familiar than the original falling body, however early the children of the future begin to read their Newton. Can the attraction itself be explained? Le Sage said that there is an everlasting hail of innumerable small ether-particles from all sides, and that the two material particles shield each other from this and so get pushed together. This is an explanation; it may or may not be a true one. The attraction may be an ultimate simple fact; or it may be made up of simpler facts utterly unlike anything that we know at present; and in either of these cases there is no explanation. We have no right to conclude, then, that the order of events is always capable of being explained.

⟦35⟧ There is yet another way in which it is said that Nature is reasonable; namely, inasmuch as every effect has a cause. What do we mean by this?

⟦36⟧ In asking this question, we have entered upon an appalling task. The word represented by "cause" has sixty-four meanings in Plato and forty-eight in Aristotle. These were men who liked to know as near as might be what they meant; but how many meanings it has had in the writings of the myriads of people who have not tried to know what they meant by it will, I hope, never be counted. It would not only be the height of presumption in me to attempt to fix the meaning of a word which has been used by so grave authority in so many and various senses; but it would seem a thankless task to do that once more which has been done so often at sundry times and in divers manners before. And yet without this we cannot determine what we mean by saying that the order of nature is reasonable. I shall evade the difficulty by telling you Mr. Grote's opinion.[1] You come to a scarecrow and ask, what is the cause of this? You find that a man made it to frighten the birds. You go away and say to yourself, "Everything resembles this scarecrow. Everything has a purpose." And from that day the word "cause" means for you what Aristotle meant by "final cause". Or you go into a hairdresser's shop, and wonder what turns the wheel to which the rotatory brush is attached. On investigating other parts of the premises, you find a man working away at a handle. Then you go away and say, "Everything is like that wheel. If I investigated enough, I should always find a man at a handle". And the man at the handle, or whatever

[1] Plato, vol. ii. (Phaedo).

corresponds to him, is from henceforth known to you as "cause".

⟦37⟧ And so generally. When you have made out any sequence of events to your entire satisfaction, so that you know all about it, the laws involved being so familiar that you seem to see how the beginning must have been followed by the end, then you apply that as a simile to all other events whatever, and your idea of cause is determined by it. Only when a case arises, as it always must, to which the simile will not apply, you do not confess to yourself that it was only a simile and need not apply to everything, but you say, "The cause of that event is a mystery which must remain for ever unknown to me". On equally just grounds the nervous system of my umbrella is a mystery which must remain for ever unknown to me. My umbrella has no nervous system; and the event to which your simile did not apply has no cause in your sense of the word. When we say then that every effect has a cause, we mean that every event is connected with something in a way that might make somebody call that the cause of it. But I, at least, have never yet seen any single meaning of the word that could be fairly applied to the *whole* order of nature.

⟦38⟧ From this remark I cannot even except an attempt recently made by Mr. Bain to give the word a universal meaning, though I desire to speak of that attempt with the greatest respect. Mr. Bain[1] wishes to make the word "cause" hang on in some way to what we call the law of energy; but though I speak with great diffidence I do think a careful consideration will show that the introduction of this word "cause" can only bring confusion into a matter which is distinct and clear enough to those who have taken the trouble to understand what energy means. It would be impossible to explain that this evening; but I may mention that "energy" is a technical term out of mathematical physics, which requires of most men a good deal of careful study to understand it accurately.

⟦39⟧ Let us pass on to consider, with all the reverence which it demands, another opinion held by great numbers of the philosophers who have lived in the Brightening Ages of Europe; the opinion that at the basis of the natural order there is something which we can know to be *unreasonable*, to evade the processes of human thought. The opinion is set forth first by Kant, so far as I know, in the form of his famous doctrine of the antinomies or contradictions, a later form[2] of which I will endeavour to explain to you. It is said, then, that space must either be infinite or have a boundary. Now you cannot conceive infinite space; and you cannot conceive that there should be any end to it. Here, then, are two things, one of which must be true, while each of them is inconceivable; so that our thoughts about space are hedged in, as it were, by a contradiction. Again, it is said that matter must either be infinitely divisible, or must consist of small particles incapable of further division. Now you cannot conceive a piece of matter divided into an infinite number of parts, while, on the other hand, you cannot conceive a piece of matter, however small, which

---

[1] *Inductive Logic*, chap. iv.

[2] That of Mr. Herbert Spencer, *First Principles*. I believe Kant himself would have admitted that the antinomies do not exist for the empiricist.

absolutely cannot be divided into two pieces; for, however great the forces are which join the parts of it together, you can imagine stronger forces able to tear it in pieces. Here, again, there are two statements, one of which must be true, while each of them is separately inconceivable; so that our thoughts about matter also are hedged in by a contradiction. There are several other cases of the same thing, but I have selected these two as instructive examples. And the conclusion to which philosophers were led by the contemplation of them was that on every side, when we approach the limits of existence, a contradiction must stare us in the face. The doctrine has been developed and extended by the great successors of Kant; and this unreasonable, or unknowable, which is also called the absolute and the unconditioned, has been set forth in various ways as that which we know to be the true basis of all things. As I said before, I approach this doctrine with all the reverence which should be felt for that which has guided the thoughts of so many of the wisest of mankind. Nevertheless I shall endeavour to show that in these cases of supposed contradiction there is always something which we do not know now, but of which we cannot be sure that we shall be ignorant next year. The doctrine is an attempt to found a positive statement upon this ignorance, which can hardly be regarded as justifiable. Spinoza said, "A free man thinks of nothing so little as of death"; it seems to me we may parallel this maxim in the case of thought, and say, "A wise man only remembers his ignorance in order to destroy it". A boundary is that which divides two adjacent portions of space. The question, then, "Has space (in general) a boundary?" involves a contradiction in terms, and is, therefore, unmeaning. But the question, "Does space contain a finite number of cubic miles, or an infinite number?" is a perfectly intelligible and reasonable question which remains to be answered by experiment.[1] The surface of the sea would still contain a finite number of square miles, if there were no land to bound it. Whether or no the space in which we live is of this nature remains to be seen. If its extent is finite, we may quite possibly be able to assign that extent next year; if, on the other hand, it has no end, it is true that the knowledge of that fact would be quite different from any knowledge we at present possess, but we have no right to say that such knowledge is impossible. Either the question will be settled once for all, or the extent of space will be shown to be greater than a quantity which will increase from year to year with the improvement of our sources of knowledge. Either alternative is perfectly conceivable, and there is no contradiction. Observe especially that the supposed contradiction arises from the assumption of theoretical exactness in the laws of geometry. The other case that I mentioned has a very similar origin. The idea of a piece of matter the parts of which are held together by forces, and are capable of being torn asunder by greater forces, is entirely derived from the large pieces of matter which we have to deal with. We do not know whether this idea applies in any sense even to the *molecules* of gases; still less can we

[1] The very important distinction between *unboundedness* and *infinite extent* is made by Riemann, *loc. cit.*

apply it to the *atoms* of which they are composed. The word force is used of two phenomena: the pressure, which when two bodies are in contact connects the motion of each with the position of the other; and attraction or repulsion,—that is to say, a change of velocity in one body depending on the position of some other body which is not in contact with it. We do not know that there is anything corresponding to either of these phenomena in the case of a molecule. A meaning can, however, be given to the question of the divisibility of matter in this way. We may ask if there is any piece of matter so small that its properties as matter depend upon its remaining all in one piece. This question is reasonable; but we cannot answer it at present, though we are not at all sure that we shall be equally ignorant next year. If there is no such piece of matter, no such limit to the division which shall leave it matter, the knowledge of that fact would be different from any of our present knowledge; but we have no right to say that it is impossible. If, on the other hand, there *is* a limit, it is quite possible that we may have measured it by the time the Association meets at Bradford. Again, when we are told that the infinite extent of space, for example, is something that we cannot conceive at present, we may reply that this is only natural, since our experience has never yet supplied us with the means of conceiving such things. But then we cannot be sure that the facts will not make us learn to conceive them; in which case they will cease to be inconceivable. In fact, the putting of limits to human conception must always involve the assumption that our previous experience is universally valid in a theoretical sense; an assumption which we have already seen reason to reject. Now you will see that our consideration of this opinion has led us to the true sense of the assertion that the Order of Nature is reasonable. If you will allow me to define a reasonable question as one which is asked in terms of ideas justified by previous experience, without itself contradicting that experience, then we may say, as the result of our investigation, that to every reasonable question there is an intelligible answer which either we or posterity may know.

‖40‖ We have, then, come somehow to the following conclusions. By scientific thought we mean the application of past experience to new circumstances by means of an observed order of events. By saying that this order of events is exact we mean that it is exact enough to correct experiments by, but we do not mean that it is theoretically or absolutely exact, because we do not know. The process of inference we found to be in itself an assumption of uniformity, and we found that, as the known exactness of the uniformity became greater, the stringency of the inference increased. By saying that the order of events is reasonable we do not mean that everything has a purpose, or that everything can be explained, or that everything has a cause; for neither of these is true. But we mean that to every reasonable question there is an intelligible answer, which either we or posterity may know *by the exercise of scientific thought*.

‖41‖ For I specially wish you not to go away with the idea that the exercise of scientific thought is properly confined to the subjects from which my illustrations have been chiefly drawn to-night. When the Roman jurists applied their experience of Roman citizens to dealings between citizens and aliens, showing by the difference of their actions that they regarded the circumstances as essentially different, they laid the foundations of that great structure which has guided the social progress of Europe. That procedure was an instance of strictly scientific thought. When a poet finds that he has to move a strange new world which his predecessors have not moved; when, nevertheless, he catches fire from their flashes, arms from their armoury, sustentation from their footprints, the procedure by which he applies old experience to new circumstances is nothing greater or less than scientific thought. When the moralist, studying the conditions of society and the ideas of right and wrong which have come down to us from a time when war was the normal condition of man and success in war the only chance of survival, evolves from them the conditions and ideas which must accompany a time of peace, when the comradeship of equals is the condition of national success; the process by which he does this is scientific thought and nothing else. Remember, then, that it is the guide of action; that the truth which it arrives at is not that which we can ideally contemplate without error, but that which we may act upon without fear; and you cannot fail to see that scientific thought is not an accompaniment or condition of human progress, but human progress itself. And for this reason the question what its characters are, of which I have so inadequately endeavoured to give you some glimpse, is the question of all questions for the human race.

14
Arthur Cayley (1821–1895)

Cayley was educated at Trinity College, Cambridge, which elected him to a Fellowship in 1842. He resigned after three years to enter Lincoln's Inn; from 1849 to 1863 he practised law in London. There he met and collaborated with another legal algebraist, James Joseph Sylvester; the two worked closely on the theory of invariants. During this time, Cayley produced some three hundred research papers. In 1863 he returned to academia, and from then until his death he was the Sadlerian Professor of Pure Mathematics at Cambridge.

His 1883 Presidential address to the British Association for the Advancement of Science gives a masterly survey of the developments in nineteenth-century mathematics as they appeared to one of its leading practitioners. Cayley touches on many of the topics discussed in earlier selections: empiricism in mathematics; the geometric representation of complex numbers; the status of non-Euclidean geometries and their relevance to the study of empirical space; Riemann surfaces; *n*-dimensional geometries; Hamilton's view of algebra as the 'science of pure time'; developments in number-theory; the contributions of Boole and the Peirces to algebra and logic.

Cayley's own contributions to this history were considerable. His total mathematical output of nearly a thousand papers fills thirteen large volumes. They range over algebraic geometry, the theory of invariants, the study of *n*-dimensional spaces (which he began to study in the 1840s, long before Riemann), the algebra of matrices, and the theory of groups. Klein's classification of non-Euclidean geometries built directly on Cayley's work in projective geometry.

The first four paragraphs of Cayley's Address have been omitted. (They contain obituaries of members of the Association.) References to *Cayley 1883* should be to the page numbers of the reprinting in volume nine of Cayley's *Collected mathematical papers* (*Cayley 1889–98*); they are given in the margins.

A. PRESIDENTIAL ADDRESS TO THE BRITISH ASSOCIATION, SEPTEMBER 1883 (*CAYLEY 1883*)

I wish to speak to you to-night upon Mathematics. I am quite aware of the difficulty arising from the abstract nature of my subject; and if, as I fear, many or some of you, recalling the Presidential Addresses at former meetings—for instance, the *résumé* and survey which we had at York of the progress, during the half century of the lifetime of the Association, of a whole circle of sciences—Biology, Palæontology, Geology, Astronomy, Chemistry—so much more familiar to you, and in which there was so much to tell of the fairy-tales of science; or at Southampton, the discourse of my friend who has in such kind terms introduced me to you, on the wondrous practical applications of science to electric lighting, telegraphy, the St Gothard Tunnel and the Suez Canal, gun-cotton, and a host of other purposes, and with the grand concluding speculation on the conservation of solar energy: if, I say, recalling these or any earlier Addresses, you should wish that you were now about to have, from a different President, a discourse on a different subject, I can very well sympathise with you in the feeling.

But be this as it may, I think it is more respectful to you that I should speak to you upon and do my best to interest you in the subject which has occupied me, and in which I am myself most interested. And in another point of view, I think it is right that the Address of a President should be on his own subject, and that different subjects should be thus brought in turn before the meetings. So much the worse, it may be, for a particular meeting; but the meeting is the individual, which on evolution principles must be sacrificed for the development of the race.

Mathematics connect themselves on the one side with common life and the physical sciences; on the other side with philosophy, in regard to our notions of space and time, and in the questions which have arisen as to the universality and necessity of the truths of mathematics, and the foundation of our knowledge of them. I would remark here that the connexion (if it exists) of arithmetic and algebra with the notion of time is far less obvious than that of geometry with the notion of space.

As to the former side, I am not making before you a defence of mathematics, but if I were I should desire to do it—in such manner as in the *Republic* Socrates was required to defend justice, quite irrespectively of the worldly advantages
which | may accompany a life of virtue and justice, and to show that, indepen- 430|
dently of all these, justice was a thing desirable in itself and for its own sake— 431
not by speaking to you of the utility of mathematics in any of the questions of common life or of physical science. Still less would I speak of this utility before, I trust, a friendly audience, interested or willing to appreciate an interest in mathematics in itself and for its own sake. I would, on the contrary, rather consider the obligations of mathematics to these different subjects as the sources of mathematical theories now as remote from them, and in as different a region of thought—for instance. geometry from the measurement of land, or the Theory of Numbers from arithmetic—as a river at its mouth is from its mountain source.

On the other side, the general opinion has been and is that it is indeed by experience that we arrive at the mathematics, but that experience is not their proper foundation: the mind itself contributes something. This is involved in the Platonic theory of reminiscence; looking at two things, trees or stones or anything else, which seem to us more or less equal, we arrive at the idea of equality: but we must have had this idea of equality before the time when first seeing the two things we were led to regard them as coming up more or less perfectly to this idea of equality; and the like as regards our idea of the beautiful, and in other cases.

The same view is expressed in the answer of Leibnitz, the *nisi intellectus, ipse*, to the scholastic dictum, *nihil in intellectu quod non prius in sensu*: there is nothing in the intellect which was not first in sensation, except (said Leibnitz) the intellect itself. And so again in the *Critick of Pure Reason*, Kant's view is that while there is no doubt but that all our cognition begins with experience, we are nevertheless in possession of cognitions *a priori*, independent, not of this or that experience, but absolutely so of all experience, and in particular that the axioms of mathematics furnish an example of such cognitions *a priori*. Kant holds further that space is no empirical conception which has been derived from external experiences, but that in order that sensations may be referred to something external, the representation of space must already lie at the foundation; and that the external experience is itself first only possible by this representation of space. And in like manner time is no empirical conception which can be deduced from an experience, but it is a necessary representation lying at the foundation of all intuitions.

And so in regard to mathematics, Sir W.R. Hamilton, in an Introductory Lecture on Astronomy (1836), observes:

> These purely mathematical sciences of algebra and geometry are sciences of the pure reason, deriving no weight and no assistance from experiment, and isolated or at least isolable from all outward and accidental phenomena. The idea of order with its subordinate ideas of number and figure, we must not indeed call innate ideas, if that phrase be defined to imply that all men must possess them with equal clearness and fulness: they are, however, ideas which seem to be so far born with us that the possession of them in any conceivable degree is only the development of our original powers, the unfolding of our proper humanity.

The general question of the ideas of space and time, the axioms and defini-
tions of geometry, the axioms relating to number, and the nature of mathe-
431| matical reasoning, are | fully and ably discussed in Whewell's *Philosophy of*
432 *the Inductive Sciences* (1840), which may be regared as containing an exposition
of the whole theory.

But it is maintained by John Stuart Mill that the truths of mathematics, in particular those of geometry, rest on experience; and as regards geometry, the same view is on very different grounds maintained by the mathematician Riemann.

It is not so easy as at first sight it appears to make out how far the views taken by Mill in his *System of Logic Ratiocinative and Inductive* (9th ed. 1879) are absolutely contradictory to those which have been spoken of; they profess to be so; there are most definite assertions (supported by argument), for instance, p. 263:—

> It remains to enquire what is the ground of our belief in axioms, what is the evidence on which they rest. I answer, they are experimental truths, generalisations from

experience. The proposition 'Two straight lines cannot enclose a space', or, in other words, two straight lines which have once met cannot meet again, is an induction from the evidence of our senses.

But I cannot help considering a previous argument (p. 259) as very materially modifying this absolute contradiction. After enquiring "Why are mathematics by almost all philosophers . . . considered to be independent of the evidence of experience and observation, and characterised as systems of necessary truth?" Mill proceeds (I quote the whole passage) as follows:—

The answer I conceive to be that this character of necessity ascribed to the truths of mathematics, and even (with some reservations to be hereafter made) the peculiar certainty ascribed to them, is a delusion, in order to sustain which it is necessary to suppose that those truths relate to and express the properties of purely imaginary objects. It is acknowledged that the conclusions of geometry are derived partly at least from the so-called definitions, and that these definitions are assumed to be correct representations, as far as they go, of the objects with which geometry is conversant. Now, we have pointed out that, from a definition as such, no proposition unless it be one concerning the meaning of a word can ever follow, and that what apparently follows from a definition, follows in reality from an implied assumption that there exits a real thing conformable thereto. This assumption in the case of the definitions of geometry is not strictly true: there exist no real things exactly conformable to the definition. There exist no real points without magnitude, no lines without breadth, nor perfectly straight, no circles with all their radii exactly equal, nor squares with all their angles perfectly right. It will be said that the assumption does not extend to the actual but only to the possible existence of such things. I answer that according to every test we have of possibility they are not even possible. Their existence, so far as we can form any judgment, would seem to be inconsistent with the physical constitution of our planet at least, if not of the universal ⟦*sic*⟧. To get rid of this difficulty and at the same time to save the credit of the supposed system of necessary truth, it is customary to say that the points, lines, circles and squares which are the subject of geometry exist in our conceptions merely and are part of our minds; which minds by working on their own materials construct an *a priori* science, the evidence of which is purely mental and has nothing to do with outward experience. By howsoever high authority this doctrine has been sanctioned, it appears to me psychologically incorrect. The points, lines and squares which anyone has
in his mind are (as I apprehend) simply copies | of the points, lines and squares which 432|
he has known in his experience. Our idea of a point I apprehend to be simply our idea 433
of the *minimum visibile*, the small portion of surface which we can see. We can reason about a line as if it had no breadth, because we have a power which we can exercise over the operations of our minds: the power, when a perception is present to our senses or a conception to our intellects, of *attending* to a part only of that perception or conception instead of the whole. But we cannot *conceive* a line without breadth: we can form no mental picture of such a line; all the lines which we have in our mind are lines possessing breadth. If anyone doubt this, we may refer him to his own experience. I much question if anyone who fancies that he can conceive of a mathematical line thinks so from the evidence of his own consciousness. I suspect it is rather because he supposes that, unless such a perception be possible, mathematics could not exist as a science: a supposition which there will be no difficulty in showing to be groundless.

I think it may be at once conceded that the truths of geometry are truths precisely because they relate to and express the properties of what Mill calls "purely imaginary objects"; that these objects do not exist in Mill's sense, that they do not exist in nature, may also be granted; that they are "not even possible", if this means not possible in an existing nature, may also be granted. That we cannot "conceive" them depends on the meaning which we attach to the word conceive. I would myself say that the purely imaginary objects are the only realities, the *ὄντως ὄντα* ⟦'the realities that really exist'⟧, in regard to which the corresponding physical objects are as the shadows in the cave; and it is only by means of them that we are able to deny the existence of a corresponding physical object; if there is no conception of straightness, then it is meaningless to deny the existence of a perfectly straight line.

But at any rate the objects of geometrical truth are the so-called imaginary objects of Mill, and the truths of geometry are only true, and *a fortiori* are only necessarily true, in regard to these so-called imaginary objects; and these objects, points, lines, circles, &c., in the mathematical sense of the terms, have a likeness to and are represented more or less imperfectly, and from a geometer's point of view no matter how imperfectly, by corresponnding physical points, lines, circles, &c. I shall have to return to geometry, and will then speak of Riemann, but I will first refer to another passage of the Logic.

Speaking of the truths of arithmetic, Mill says (p. 297) that even here there is one hypothetical element: "In all propositions concerning numbers a condition is implied without which none of them would be true, and that condition is an assumption which may be false. The condition is that 1 = 1: that all the numbers are numbers of the same or of equal units." Here at least the assumption may be absolutely true; one shilling = one shilling in purchasing power, although they may not be absolutely of the same weight and fineness: but it is hardly necessary; one coin + one coin = two coins, even if the one be a shilling and the other a half-crown. In fact, whatever difficulty be raisable as to geometry, it seems to me that no similar difficulty applies to arithmetic; mathematician or not, we have each of us, in its most abstract form, the idea of a number; we can each of us appreciate the truth of a proposition in regard to numbers; and we cannot but see that a truth in regard to numbers is something
433| different in kind from | an experimental truth generalised from experience.
434 Compare, for instance, the proposition that the sun, having already risen so many times, will rise to-morrow, and the next day, and the day after that, and so on; and the proposition that even and odd numbers succeed each other alternately *ad infinitum*: the latter at least seems to have the characters of universality and necessity. Or again, suppose a proposition observed to hold good for a long series of numbers, one thousand numbers, two thousand numbers, as the case may be: this is not only no proof, but it is absolutely no evidence, that the proposition is a true proposition, holding good for all numbers whatever; there are in the Theory of Numbers very remarkable instances of propositions observed to hold good for very long series of numbers and which are nevertheless untrue.

I pass in review certain mathematical theories.

In arithmetic and algebra, or say in analysis, the numbers or magnitudes which we represent by symbols are in the first instance ordinary (that is, positive) numbers or magnitudes. We have also in analysis and in analytical geometry *negative* magnitudes; there has been in regard to these plenty of philosophical discussion, and I might refer to Kant's paper, *Ueber die negativen Grössen in die Weltweisheit* (1763), but the notion of a negative magnitude has become quite a familiar one, and has extended itself into common phraseology. I may remark that it is used in a very refined manner in bookkeeping by double entry.

But it is far otherwise with the notion which is really the fundamental one (and I cannot too strongly emphasise the assertion) underlying and pervading the whole of modern analysis and geometry, that of imaginary magnitude in analysis and of imaginary space (or space as a *locus in quo* of imaginary points and figures) in geometry: I use in each case the word imaginary as including real. This has not been, so far as I am aware, a subject of philosophical discussion or enquiry. As regards the older metaphysical writers this would be quite accounted for by saying that they knew nothing, and were not bound to know anything, about it; but at present, and, considering the prominent position which the notion occupies—say even that the conclusion were that the notion belongs to mere technical mathematics, or has reference to nonentities in regard to which no science is possible, still it seems to me that (as a subject of philosophical discussion) the notion ought not to be thus ignored; it should at least be shown that there is a right to ignore it.

Although in logical order I should perhaps now speak of the notion just referred to, it will be convenient to speak first of some other quasi-geometrical notions; those of more-than-three-dimensional space, and of non-Euclidian two- and three-dimensional space, and also of the generalised notion of distance. It is in connexion with these that Riemann considered that our notion of space is founded on experience, or rather that it is only by experience that we know that our space is Euclidian space.

It is well known that Euclid's twelfth axiom, even in Playfair's form of it, has been considered as needing demonstration; and that Lobatschewsky constructed a perfectly consistent theory, wherein this axiom was assumed not to hold good, or say a system of non-Euclidian plane geometry. There is a like
system of non-Euclidian | solid geometry. My own view is that Euclid's twelfth 434|
axiom in Playfair's form of it does not need demonstration, but is part of our 435
notion of space, of the physical space of our experience—the space, that is, which we become acquainted with by experience, but which is the representation lying at the foundation of all external experience. Riemann's view before referred to may I think be said to be that, having *in intellectu* a more general notion of space (in fact a notion of non-Euclidian space), we learn by experience that space (the physical space of our experience) is, if not exacty, at least to the highest degree of approximation, Euclidian space.

But suppose the physical space of our experience to be thus only approximately Euclidian space, what is the consequence which follows? *Not* that the propositions of geometry are only approximately true, but that they remain absolutely true in regard to that Euclidian space which has been so long regarded as being the physical space of our experience.

It is interesting to consider two different ways in which, without any modification at all of our notion of space, we can arrive at a system of non-Euclidian (plane or two-dimensional) geometry; and the doing so will, I think, throw some light on the whole question.

First, imagine the earth a perfectly smooth sphere; understand by a plane the surface of the earth, and by a line the apparently straight line (in fact, an arc of great circle) drawn on the surface; what experience would in the first instance teach would be Euclidian geometry; there would be intersecting lines which produced a few miles or so would seem to go on diverging: and apparently parallel lines which would exhibit no tendency to approach each other; and the inhabitants might very well conceive that they had by experience established the axiom that two straight lines cannot enclose a space, and the axiom as to parallel lines. A more extended experience and more accurate measurements would teach them that the axioms were each of them false; and that any two lines if produced far enough each way, would meet in two points: they would in fact arrive at a spherical geometry, accurately representing the properties of the two-dimensional space of their experience. But their original Euclidian geometry would not the less be a true system: only it would apply to an ideal space, not the space of their experience.

Secondly consider an ordinary, indefinitely extended plane; and let us modify only the notion of distance. We measure distance, say, by a yard measure or a foot rule, anything which is short enough to make the fractions of it of no consequence (in mathematical language, by an infinitesimal element of length); imagine, then, the length of this rule constantly changing (as it might do by an alteration of temperature), but under the condition that its actual length shall depend only on its situation on the plane and on its direction: viz. if for a given situation and direction it has a certain length, then whenever it comes back to the same situation and direction it must have the same length. The distance along a given straight or curved line between any two points could then be measured in the ordinary manner with this rule, and would have a perfectly determinate value: it could be measured over and over again, and would always
435| be the same; but of course it would be the distance, not in the ordinary | accep-
436 tation of the term, but in quite a different acceptation. Or in a somewhat different way: if the rate of progress from a given point in a given direction be conceived as depending only on the configuration of the ground, and the distance along a given path between any two points thereof be measured by the time required for traversing it, then in this way also the distance would have a perfectly determinate value; but it would be a distance, not in the ordinary acceptation of the term, but in quite a different acceptation. And corresponding to the new notion of distance we should have a new non-Euclidian system of plane geometry; all theorems involving the notion of distance would be altered.

We may proceed further. Suppose that as the rule moves away from a fixed central point of the plane it becomes shorter and shorter; if this shortening takes place with sufficient rapidity, it may very well be that a distance which in the ordinary sense of the word is finite will in the new sense be infinite; no number of repetitions of the length of the ever-shortening rule will be sufficient to cover it. There will be surrounding the central point a certain finite area such that (in the new acceptation of the term distance) each point of the boundary thereof will be at an infinite distance from the central point; the points outside this area you cannot by any means arrive at with your rule; they will form a *terra incognita*, or rather an unknowable land: in mathematical language, an imaginary or impossible space: and the plane space of the theory will be that within the finite area—that is, it will be finite instead of infinite.

We thus with a proper law of shortening arrive at a system of non-Euclidian geometry which is essentially that of Lobatschewsky. But in so obtaining it we put out of sight its relation to spherical geometry: the three geometries (spherical, Euclidian, and Lobatschewsky's) should be regarded as members of a system: viz. they are the geometries of a plane (two-dimensional) space of constant positive curvature, zero curvature, and constant negative curvature respectively; or again, they are the plane geometries corresponding to three different notions of distance; in this point of view they are Klein's elliptic, parabolic, and hyperbolic geometries respectively.

Next as regards solid geometry: we can by a modification of the notion of distance (such as has just been explained in regard to Lobatschewsky's system) pass from our present system to a non-Euclidian system; for the other mode of passing to a non-Euclidian system, it would be necessary to regard our space as a flat three-dimensional space existing in a space of four dimensions (i.e., as the analogue of a plane existing in ordinary space); and to substitute for such flat three-dimensional space a curved three-dimensional space, say of constant positive or negative curvature. In regarding the physical space of our experience as possibly non-Euclidian, Riemann's idea seems to be that of modifying the notion of distance, not that of treating it as a locus in four-dimensional space.

I have just come to speak of four-dimensional space. What meaning do we attach to it? Or can we attach to it any meaning? It may be at once admitted that we cannot conceive of a fourth dimension of space; that space as we conceive of it, and the physical space of our experience, are alike three-dimensional; but we can, I think, conceive of space as being two- or even one-dimensional;
we can imagine rational | beings living in a one-dimensional space (a line) or in 436|
a two-dimensional space (a surface), and conceiving of space accordingly, and 437
to whom, therefore, a two-dimensional space, or (as the case may be) a three-dimensional space would be as inconceivable as a four-dimensional space is to us. And very curious speculative questions arise. Suppose the one-dimensional space a right line, and that it afterwards becomes a curved line: would there be any indication of the change? Or, if originally a curved line, would there be anything to suggest to them that it was not a right line? Probably not, for a one-dimensional geometry hardly exists. But let the space be two-dimensional, and imagine it originally a plane, and afterwards bent or converted into a curved space (converted, that is, into some form of developable surface): or imagine it originally a developable or curved surface. In the former case there should be an indication of the change, for the geometry originally applicable to the

space of their experience (our own Euclidian geometry) would cease to be applicable; but the change could not be apprehended by them as a bending or deformation of the plane, for this would imply the notion of a three-dimensional space in which this bending or deformation could take place. In the latter case their geometry would be that appropriate to the developable or curved surface which is their space: viz. this would be their Euclidian geometry: would they ever have arrived at our own more simple system? But take the case where the two-dimensional space is a plane, and imagine the beings of such a space familiar with our own Euclidian plane geometry; if, a third dimension being still inconceivable by them, they were by their geometry or otherwise led to the notion of it, there would be nothing to prevent them from forming a science such as our own science of three-dimensional geometry.

Evidently all the foregoing questions present themselves in regard to ourselves, and to three-dimensional space as we conceive of it, and as the physical space of our experience. And I need hardly say that the first step is the difficulty, and that granting a fourth dimension we may assume as many more dimensions as we please. But whatever answer be given to them, we have, as a branch of mathematics, potentially, if not actually, an analytical geometry of n-dimensional space. I shall have to speak again upon this.

Coming now to the fundamental notion already referred to, that of imaginary magnitude in analysis and imaginary space in geometry: I connect this with two great discoveries in mathematics made in the first half of the seventeenth century, Harriot's representation of an equation in the form $f(x) = 0$, and the consequent notion of the roots of an equation as derived from the linear factors of $f(x)$, (Harriot, 1560–1621: his *Algebra*, published after his death, has the date 1631), and Descartes' method of coordinates, as given in the *Géométrie*, forming a short supplement to his *Traité de la Méthode, etc.*, (Leyden, 1637).

Taking the coefficients of an equation to be real magnitudes, it at once follows from Harriot's form of an equation that an equation of the order n ought to have n roots. But it is by no means true that there are always n real roots. In particular, an equation of the second order, or quadric equation, may have no real root; but if we assume the existence of a root i of the quadric equa-
437| tion $x^2 + 1 = 0$, then the | other root is $= -i$; and it is easily seen that every
438 quadric equation (with real coefficients as before) has two roots, $a \pm bi$, where a and b are real magnitudes. We are thus led to the conception of an imaginary magnitude, $a + bi$, where a and b are real magnitudes, each susceptible of any positive or negative value, zero included. The general theorem is that, taking the coefficients of the equation to be imaginary magnitudes, then an equation of the order n has always n roots, each of them an imaginary magnitude, and it thus appears that the foregoing form $a + bi$ of imaginary magnitude is the only one that presents itself. Such imaginary magnitudes may be added or multiplied together or dealt with in any manner; the result is always a like imaginary magnitude. They are thus the magnitudes which are considered in analysis, and analysis is the science of such magnitudes. Observe the leading character that the imaginary magnitude $a + bi$ is a magnitude composed of

the two real magnitudes a and b (in the case $b = 0$ it is the real magnitude a, and in the case $a = 0$ it is the pure imaginary magnitude bi). The idea is that of considering, in place of real magnitudes, these imaginary or complex magnitudes $a + bi$.

In the Cartesian geometry a curve is determined by means of the equation existing between the coordinates (x, y) of any point thereof. In the case of a right line, this equation is linear; in the case of a circle, or more generally of a conic, the equation is of the second order; and generally, when the equation is of the order n, the curve which it represents is said to be a curve of the order n. In the case of two given curves, there are thus two equations satisfied by the coordinates (x, y) of the several points of intersection, and these give rise to an equation of a certain order for the coordinate x or y of a point of intersection. In the case of a straight line and a circle, this is a quadric equation; it has two roots, real or imaginary. There are thus two values, say of x, and to each of these corresponds a single value of y. There are therefore two points of intersection—viz. a straight line and a circle intersect *always* in two points, real or imaginary. It is in this way that we are led analytically to the notion of imaginary points in geometry. The conclusion as to the two points of intersection cannot be contradicted by experience: take a sheet of paper and draw on it the straight line and circle, and try. But you might say, or at least be strongly tempted to say, that it is meaningless. The question of course arises, What is the meaning of an imaginary point? and further, In what manner can the notion be arrived at geometrically?

There is a well-known construction in perspective for drawing lines through the intersection of two lines, which are so nearly parallel as not to meet within the limits of the sheet of paper. You have two given lines which do not meet, and you draw a third line, which, when the lines are all of them produced, is found to pass through the intersection of the given lines. If instead of lines we have two circular arcs not meeting each other, then we can, by means of these arcs, construct a line; and if on completing the circles it is found that the circles intersect each other in two real points, then it will be found that the line passes through these two points: if the circles appear not to intersect, then the line will appear not to interset either of the circles. But the geometrical construction being in each case the same, we say that in the second case also the line passes through the two intersections of the circles. | 438|

Of course it may be said in reply that the conclusion is a very natural one, 439
provided we assume the existence of imaginary points; and that, this assumption not being made, then, if the circles do not intersect, it is meaningless to assert that the line passes through their points of intersection. The difficulty is not got over by the analytical method before referred to, for this introduces difficulties of its own: is there in a plane a point the coordinates of which have given imaginary values? As a matter of fact, we do consider in plane geometry imaginary points introduced into the theory analytically or geometrically as above.

The like considerations apply to solid geometry, and we thus arrive at the notion of imaginary space as a *locus in quo* of imaginary points and figures.

I have used the word imaginary rather than complex, and I repeat that the word has been used as including real. But, this once understood, the word becomes in many cases superfluous, and the use of it would even be misleading. Thus, "a problem has so many solutions": this means, so many imaginary (including real) solutions. But if it were said that the problem had "so many imaginary solutions", the word "imaginary" would here be understood to be used in opposition to real. I give this explanation the better to point out how wide the application of the notion of the imaginary is—viz. (unless expressly or by implication excluded), it is a notion implied and presupposed in all the conclusions of modern analysis and geometry. It is, as I have said, the fundamental notion underlying and pervading the whole of these branches of mathematical science.

I shall speak later on of the great extension which is thereby given to geometry, but I wish now to consider the effect as regards the theory of a function. In the original point of view, and for the original purposes, a function, algebraic or transcendental, such as $\sqrt{x}$, $\sin x$, or $\log x$, was considered as known, when the value was known for every real value (positive or negative) of the argument; or if for any such values the value of the function became imaginary, then it was enough to know that for such values of the argument there was no real value of the function. But now this is not enough, and to know the function means to know its value—of course, in general, an imaginary value $X + iY$,—for every imaginary value $x + iy$ whatever of the argument.

And this leads naturally to the question of the geometrical representation of an imaginary variable. We represent the imaginary variable $x + iy$ by means of a point in a plane, the coordinates of which are (x, y). This idea, due to Gauss, dates from about the year 1831. We thus picture to ourselves the succession of values of the imaginary variable $x + iy$ by means of the motion of the representative point: for instance, the succession of values corresponding to the motion of the point along a closed curve to its original position. The value $X + iY$ of the function can of course be represented by means of a point (taken for greater convenience in a different plane), the coordinates of which are X, Y.

We may consider in general two points, moving each in its own plane, so that the position of one of them determines the position of the other, and conse-
439| quently | the motion of the one determines the motion of the other: for instance,
440 the two points may be the tracing-point and the pencil of a pentagraph. You may with the first point draw any figure you please, there will be a corresponding figure drawn by the second point: for a good pentagraph, a copy on a different scale (it may be); for a badly-adjusted pentagraph, a distorted copy: but the one figure will always be a sort of copy of the first, so that to each point of the one figure there will correspond a point of the other figure.

In the case above referred to, where one point represents the value $x + iy$ of the imaginary variable and the other the value $X + iY$ of some function $\phi\,(x + iy)$ of that variable, there is a remarkable relation between the two figures: this is the relation of orthomorphic projection, the same which presents itself between a portion of the earth's surface, and the representation thereof by a map on the stereographic projection or on Mercator's projection—viz. any indefinitely small area of the one figure is represented in the other figure by an indefinitely small area of the same shape. There will possibly be for different parts of the figure great variations of scale, but the shape will be unaltered; if for the one area the boundary is a circle, then for the other area the boundary will be a circle; if for one it is an equilateral triangle, then for the other it will be an equilateral triangle.

I have for simplicity assumed that to each point of either figure there corresponds one, and only one, point of the other figure; but the general case is that to each point of either figure there corresponds a determinate number of points in the other figure; and we have thence arising new and very complicated relations which I must just refer to. Suppose that to each point of the first figure there correspond in the second figure two points: say one of them is a red point, the other a blue point; so that, speaking roughly, the second figure consists of two copies of the first figure, a red copy and a blue copy, the one superimposed on the other. But the difficulty is that the two copies cannot be kept distinct from each other. If we consider in the first figure a closed curve of any kind—say, for shortness, an oval—this will be in the second figure represented in some cases by a red oval and a blue oval, but in other cases by an oval half red and half blue; or, what comes to the same thing, if in the first figure we consider a point which moves continuously in any manner, at last returning to its original position, and attempt to follow the corresponding points in the second figure, then it may very well happen that, for the corresponding point of either colour, there will be abrupt changes of position, or say jumps, from one position to another; so that, to obtain in the second figure a continuous path, we must at intervals allow the point to change from red to blue, or from blue to red. There are in the first figure certain critical points called branch-points (*Verzweigungspunkte*), and a system of lines connecting these, by means of which the colours in the second figure are determined; but it is not possible for me to go further into the theory at present. The notion of colour has of course been introduced only for facility of expression; it may be proper to add that in speaking of the two figures I have been following Briot and Bouquet rather than Riemann, whose representation of the function of an imaginary variable is a different one.

I have been speaking of an imaginary variable $(x + iy)$, and of a function $\phi\,(x + iy) = X + iY$ of that variable, but the theory may equally well be stated
in | regard to a plane curve: in fact, the $x + iy$ and the $X + iY$ are two imaginary 440|
variables connected by an equation; say their values are u and v, connected by 441
an equation $F(u, v) = 0$; then, regarding u, v as the coordinates of a point *in plano*, this will be a point on the curve represented by the equation. The curve, in the widest sense of the expression, is the whole series of points, real or imaginary, the coordinates of which satisfy the equation, and these are exhibited by the foregoing corresponding figures in two planes, but in the ordinary sense the curve is the series of real points, with coordinates u, v, which satisfy the equation.

In geometry it is the curve, whether defined by means of its equation, or in any other manner, which is the subject for contemplation and study. But we also use the curve as a representation of its equation—that is, of the relation existing between two magnitudes x, y, which are taken as the coordinates of a point on the curve. Such employment of a curve for all sorts of purposes—the fluctuations of the barometer, the Cambridge boat races, or the Funds—is familiar to most of you. It is in like manner convenient in analysis, for exhibiting the relations between any three magnitudes x, y, z, to regard them as the coordinates of a point in space; and, on the like ground, we should at least wish to regard any four or more magnitudes as the coordinates of a point in space of a corresponding number of dimensions. Starting with the hypothesis of such a space, and of points therein each determined by means of its coordinates, it is found possible to establish a system of n-dimensional geometry analogous in every respect to our two- and three-dimensional geometries, and to a very considerable extent serving to exhibit the relations of the variables. To quote from my memoir "On Abstract Geometry" (1869), [413]:

> The science presents itself in two ways: as a legitimate extension of the ordinary two- and three-dimensional geometries, and as a need in these geometries and in analysis generally. In fact, whenever we are concerned with quantities connected in any manner, and which are considered as variable or determinable, then the nature of the connexion between the quantities is frequently rendered more intelligible by regarding them (if two or three in number) as the coordinates of a point in a plane or in space. For more than three quantities there is, from the greater complexity of the case, the greater need of such a representation; but this can only be obtained by means of the notion of a space of the proper dimensionality; and to use such representation we require a corresponding geometry. An important instance in plane geometry has already presented itself in the question of the number of curves which satisfy given conditions; the conditions imply relations between the coefficients in the equation of the curve; and for the better understanding of these relations it was expedient to consider the coefficients as the coordinates of a point in a space of the proper dimensionality.

It is to be borne in mind that the space, whatever its dimensionality may be, must always be regarded as an imaginary or complex space such as the two- or three-dimensional space of ordinary geometry; the advantages of the representation would otherwise altogether fail to be obtained.

I have spoken throughout of Cartesian coordinates; instead of these, it is in plane geometry not unusual to employ trilinear coordinates, and these may be regarded as absolutely undetermined in their magnitude—viz. we may take x,
441| y, z to be, not equal, | but only proportional to the distances of a point from
442 three given lines; the ratios of the coordinates (x, y, z) determine the point; and so in one-dimensional geometry, we may have a point determined by the ratio of its two coordinates x, y, these coordinates being proportional to the distances of the point from two fixed points; and generally in n-dimensional geometry a point will be determined by the ratios of the $(n + 1)$ coordinates $(x, y, z, \ldots)$. The corresponding analytical change is in the expression of the original magnitudes as fractions with a common denominator; we thus, in place of rational and integral non-homogeneous functions of the original variables,

introduce rational and integral homogeneous functions (quantics) of the next succeeding number of variables—viz. we have binary quantics corresponding to one-dimensional geometry, ternary to two-dimensional geometry, and so on.

It is a digression, but I wish to speak of the representation of points or figures in space upon a plane. In perspective, we represent a point in space by means of the intersection with the plane of the picture (suppose a pane of glass) of the line drawn from the point to the eye, and doing this for each point of the object we obtain a representation or picture of the object. But such representation is an imperfect one, as not determining the object: we cannot by means of the picture alone find out the form of the object; in fact, for a given point of the picture the corresponding point of the object is not a determinate point, but it is a point anywhere in the line joining the eye with the point of the picture. To determine the object we need two pictures, such as we have in a plan and elevation, or, what is the same thing, in a representation on the system of Monge's descriptive geometry. But it is theoretically more simple to consider two projections on the same plane, with different positions of the eye: the point in space is here represented on the plane by means of two points which are such that the line joining them passes through a fixed point of the plane (this point is in fact the intersection with the plane of the picture of the line joining the two positions of the eye); the figure in space is thus represented on the plane by two figures, which are such that the lines joining corresponding points of the two figures pass always through the fixed point. And such two figures completely replace the figure in space; we can by means of them perform on the plane any constructions which could be performed on the figure in space, and employ them in the demonstration of properties relating to such figure. A curious extension has recently been made: two figures in space such that the lines joining corresponding points pass through a fixed point have been regarded by the Italian geometer Veronese as representations of a figure in four-dimensional space, and have been used for the demonstration of properties of such figure.

I referred to the connexion of Mathematics with the notions of space and time, but I have hardly spoken of time. It is, I believe, usually considered that the notion of number is derived from that of time; thus Whewell in the work referred to, p. xx, says number is a modification of the conception of repetition, which belongs to that of *time*. I cannot recognise that this is so: it seems to me that we have (independently, I should say, of space or time, and in any case not more depending on time than on space) the notion of plurality; we think
of, say, the letters *a*, *b*, *c*, &c., and thence in the case | of a finite set—for 442|
instance *a*, *b*, *c*, *d*, *e*—we arrive at the notion of number; coordinating them 443
one by one with any other set of things, or, suppose, with the words first, second, &c., we find that the last of them goes with the word fifth, and we say that the number of things is = five: the notion of cardinal number would thus appear to be derived from that of ordinal number.

Questions of combination and arrangement present themselves, and it might be possible from the mere notion of plurality to develope a branch of mathematical science: this, however, would apparently be of a very limited extent, and it is difficult *not* to introduce into it the notion of number; in fact, in the case of a finite set of things, to avoid asking the question, How many? If we do this, we have a large enough subject, including the partition of numbers, which Sylvester has called Tactic.

From the notion thus arrived at of an integer number, we pass to that of a fractional number, and we see how by means of these the ratio of any two concrete magnitudes of the same kind can be expressed, not with absolute accuracy, but with any degree of accuracy we please: for instance, a length is so many feet, tenths of a foot, hundredths, thousandths, &c.; subdivide as you please, *non constat* that the length can be expressed accurately, we have in fact incommensurables; as to the part which these play in the Theory of Numbers, I shall have to speak presently: for the moment I am only concerned with them in so far as they show that we cannot from the notion of number pass to that which is required in analysis, the notion of an abstract (real and positive) magnitude susceptible of continuous variation. The difficulty is got over by a Postulate. We consider an abstract (real and positive) magnitude, and regard it as susceptible of continuous variation, without in anywise concerning ourselves about the actual expression of the magnitude by a numerical fraction or otherwise.

There is an interesting paper by Sir W.R. Hamilton, "Theory of Conjugate Functions, or Algebraical Couples: with a preliminary and elementary Essay on Algebra as the Science of Pure Time", 1833–35 (*Trans. R.I. Acad.* t. XVII.), in which, as appears by the title, he purposes to show that algebra is the science of pure time. He states there, in the General Introductory Remarks, his conclusions: first, that the notion of time is connected with existing algebra; second, that this notion or intuition of time may be unfolded into an independent pure science; and, third, that the science of pure time thus unfolded is coextensive and identical with algebra, so far as algebra itself is a science; and to sustain his first conclusion he remarks that

> the history of algebraic science shows that the most remarkable discoveries in it have been made either expressly through the notion of *time*, or through the closely connected (and in some sort coincident) notion of continuous progression. It is the genius of algebra to consider what it reasons upon as *flowing*, as it was the genius of geometry to consider what it reasoned on as *fixed*. . . . And generally the revolution which Newton made in the higher parts of both pure and applied algebra was founded mainly on the notion of *fluxion*, which involves the notion of *time*.

Hamilton uses the term algebra in a very wide sense, but whatever else he includes under it, he includes all that in contradistinction to the Differential Calculus would be called algebra. Using the word in this restricted sense, I cannot myself recognise the connexion of algebra with the notion of time:
443| granting that the notion of continuous progression presents itself, and is of |
444 importance, I do not see that it is in anywise the fundamental notion of the science. And still less can I appreciate the manner in which the author connects with the notion of time his algebraical couple, or imaginary magnitude $a + bi$ ($a + b\sqrt{-1}$, as written in the memoir).

I would go further: the notion of continuous variation is a very fundamental one, made a foundation in the Calculus of Fluxions (if not always so in the Differential Calculus) and presenting itself or implied throughout in mathematics: and it may be said that a change of any kind takes place only in time; it seems to me, however, that the changes which we consider in mathematics are for the most part considered quite irrespectively of time.

It appears to me that we do not have in Mathematics the notion of time until we bring it there: and that even in kinematics (the science of motion) we have very little to do with it; the motion is a hypothetical one; if the system be regarded as actually moving, the rate of motion is altogether undetermined and immaterial. The relative rates of motion of the different points of the system are nothing else than the ratios of purely geometrical quantities, the indefinitely short distances simultaneously described, or which might be simultaneously described, by these points respectively. But whether the notion of time does or does not sooner enter into mathematics, we at any rate have the notion in Mechanics, and along with it several other new notions.

Regarding Mechanics as divided into Statics and Dynamics, we have in dynamics the notion of time, and in connexion with it that of velocity: we have in statics and dynamics the notion of force; and also a notion which in its most general form I would call that of corpus: viz. this may be, the material point or particle, the flexible inextensible string or surface, or the rigid body, of ordinary mechanics; the incompressible perfect fluid of hydrostatics and hydrodynamics; the ether of any undulatory theory; or any other imaginable corpus; for instance, one really deserving of consideration in any general treatise of mechanics is a developable or skew surface with absolutely rigid generating lines, but which can be bent about these generating lines, so that the element of surface between two consecutive lines rotates as a whole about one of them. We have besides, in dynamics necessarily, the notion of mass or inertia.

We seem to be thus passing out of pure mathematics into physical science; but it is difficult to draw the line of separation, or to say of large portions of the *Principia*, and the *Mécanique céleste*, or of the whole of the *Mécanique analytique*, that they are not pure mathematics. It may be contended that we first come to physics when we attempt to make out the character of the corpus as it exists in nature. I do not at present speak of any physical theories which cannot be brought under the foregoing conception of mechanics.

I must return to the Theory of Numbers; the fundamental idea is here integer number: in the first instance positive integer number, but which may be extended to include negative integer number and zero. We have the notion of a product, and that of a prime number, which is not a product of other numbers; and thence also that of a number as the product of a determinate system of prime
factors. We have here the | elements of a theory in many respects analogous 444|
to algebra: an equation is to be solved—that is, we have to find the integer 445
values (if any) which satisfy the equation; and so in other cases: the congruence notation, although of the very highest importance, does not affect the character of the theory.

But as already noticed we have incommensurables, and the consideration of these gives rise to a new universe of theory. We may take into consideration any surd number such as $\sqrt{2}$, and so consider numbers of the form $a + b\sqrt{2}$, (a and b any positive or negative integer numbers not excluding zero); calling these integer numbers, every problem which before presented itself in regard to integer numbers in the original and ordinary sense of the word presents itself equally in regard to integer numbers in this new sense of the word; of course all definitions must be altered accordingly: an ordinary integer, which is in the ordinary sense of the word a prime number, may very well be the product of two integers of the form $a + b\sqrt{2}$, and consequently not a prime number in the new sense of the word. Among the incommensurables which can be thus introduced into the Theory of Numbers (and which was in fact *first* so introduced) we have the imaginary i of ordinary analysis : viz. we may consider numbers $a + bi$ (a and b ordinary positive or negative integers, not excluding zero), and, calling these integer numbers, establish in regard to them a theory analogous to that which exists for ordinary real integers. The point which I wish to bring out is that the imaginary i does not in the Theory of Numbers occupy a unique position, such as it does in analysis and geometry; it is in the Theory of Numbers one out of an indefinite multitude of incommensurables.

I said that I would speak to you, not of the utility of mathematics in any of the questions of common life or of physical science, but rather of the obligations of mathematics to these different subjects. The consideration which thus presents itself is in a great measure that of the history of the development of the different branches of mathematical science in connexion with the older physical sciences, Astronomy and Mechanics: the mathematical theory is in the first instance suggested by some question of common life or of physical science, is pursued and studied quite independently thereof, and perhaps after a long interval comes in contact with it, or with quite a different question. Geometry and algebra must, I think, be considered as each of them originating in connexion with objects or questions of common life—geometry, notwithstanding its name, hardly in the measurement of land, but rather from the contemplation of such forms as the straight line, the circle, the ball, the top (or sugar-loaf): the Greek geometers appropriated for the geometrical forms corresponding to the last two of these, the words *σφαῖρα* and *κῶνος*, our sphere and cone, and they extended the word cone to mean the complete figure obtained by producing the straight lines of the surface both ways indefinitely. And so algebra would seem to have arisen from the sort of easy puzzles in regard to numbers which may be made, either in the picturesque forms of the Bija-Ganita with its maiden with the beautiful locks, and its swarms of bees amid the fragrant blossoms, and the one queen-bee left humming around the lotus flower; or in the more prosaic form in which a student has presented to him in a modern text-book
445| a problem leading to a simple equation.|

446 The Greek geometry may be regarded as beginning with Plato (B.C. 430–347): the notions of geometrical analysis, loci, and the conic sections are attributed to him, and there are in his Dialogues many very interesting allusions to mathe-

matical questions: in particular the passage in the *Theætetus*, where he affirms the incommensurability of the sides of certain squares. But the earliest extant writings are those of Euclid (B.C. 285): there is hardly anything in mathematics more beautiful than his wondrous fifth book; and he has also in the seventh, eighth, ninth and tenth books fully and ably developed the first principles of the Theory of Numbers, including the theory of incommensurables. We have next Apollonius. (about B.C. 247), and Archimedes (B.C. 287–212), both geometers of the highest merit, and the latter of them the founder of the science of statics (including therein hydrostatics): his dictum about the lever, his "Εὕρηκα" and the story of the defence of Syracuse, are well known. Following these we have a worthy series of names, including the astronomers Hipparchus (B.C. 150) and Ptolemy (A.D. 125), and ending, say, with Pappus (A.D. 400), but continued by their Arabian commentators, and the Italian and other European geometers of the sixteenth century and later, who pursued the Greek geometry.

The Greek arithmetic was, from the want of a proper notation, singularly cumbrous and difficult; and it was for astronomical purposes superseded by the sexagesimal arithmetic, attributed to Ptolemy, but probably known before his time. The use of the present so-called Arabic figures became general among Arabian writers on arithmetic and astronomy about the middle of the tenth century, but was not introduced into Europe until about two centuries later. Algebra among the Greeks is represented almost exclusively by the treatise of Diophantus (A.D. 150), in fact a work on the Theory of Numbers containing questions relating to square and cube numbers, and other properties of numbers, with their solutions; this has no historical connexion with the later algebra, introduced into Italy from the East by Leonardi Bonacci of Pisa (A.D. 1202–1208) and successfully cultivated in the fifteenth and sixteenth centuries by Lucas Paciolus, or de Burgo, Tartaglia, Cardan, and Ferrari. Later on, we have Vieta (1540–1603), Harriot, already referred to, Wallis, and others.

Astronomy is of course intimately connected with geometry; the most simple facts of observation of the heavenly bodies can only be *stated* in geometrical language: for instance, that the stars describe circles about the pole-star, or that the different positions of the sun among the fixed stars in the course of the year form a circle. For astronomical calculations it was found necessary to determine the arc of a circle by means of its chord: the notion is as old as Hipparchus, a work of whom is referred to as consisting of twelve books on the chords of circular arcs; we have (A.D. 125) Ptolemy's *Almagest*, the first book of which contains a table of arcs and chords with the method of construction; and among other theorems on the subject he gives there the theorem afterwards inserted in Euclid (Book VI. Prop. D) relating to the rectangle contained by the diagonals of a quadrilateral inscribed in a circle. The Arabians made the improvement of using in place of the chord of an arc the sine, or half chord, of double the arc; and so brought the theory into the form in which it is used in modern trigonometry: the before-mentioned theorem of Ptolemy, or rather a particular
case of it, translated into the notation of sines, gives the expression for the sine 446|
of the sum | of two arcs in terms of the sines and cosines of the component arcs; 447

and it is thus the fundamental theorem on the subject. We have in the fifteenth and sixteenth centuries a series of mathematicians who with wonderful enthusiasm and perseverance calculated tables of the trigonometrical or circular functions, Purbach, Müller or Regiomontanus, Copernicus, Reinhold, Maurolycus, Vieta, and many others; the tabulations of the functions tangent and secant are due to Reinhold and Maurolycus respectively.

Logarithms were invented, not exclusively with reference to the calculation of trigonometrical tables, but in order to facilitate numerical calculations generally; the invention is due to John Napier of Merchiston, who died in 1618 at 67 years of age; the notion was based upon refined mathematical reasoning on the comparison of the spaces described by two points, the one moving with a uniform velocity, the other with a velocity varying according to a given law. It is to be observed that Napier's logarithms were nearly but not exactly those which are now called (sometimes Napierian, but more usually) hyperbolic logarithms–those to the base e; and that the change to the base 10 (the great step by which the invention was perfected for the object in view) was indicated by Napier but actually made by Henry Briggs, afterwards Savilian Professor at Oxford (d. 1630). But it is the hyperbolic logarithm which is mathematically important. The direct function e^x or $\exp. x$, which has for its inverse the hyperbolic logarithm, presented itself, but not in a prominent way. Tables were calculated of the logarithms of numbers, and of those of the trigonometrical functions.

The circular functions and the logarithm were thus invented each for a practical purpose, separately and without any proper connexion with each other. The functions are connected through the theory of imaginaries and form together a group of the utmost importance throughout mathematics: but this is mathematical theory; the obligation of mathematics is for the discovery of the functions.

Forms of spirals presented themselves in Greek architecture, and the curves were considered mathematically by Archimedes; the Greek geometers invented some other curves, more or less interesting, but recondite enough in their origin. A curve which might have presented itself to anybody, that described by a point in the circumference of a rolling carriage-wheel, was first noticed by Mersenne in 1615, and is the curve afterwards considered by Roberval, Pascal, and others under the name of the Roulette, otherwise the Cycloid. Pascal (1623–1662) wrote at the age of seventeen his *Essais pour les Coniques* in seven short pages, full of new views on these curves, and in which he gives, in a paragraph of eight lines, his theorem of the inscribed hexagon.

Kepler (1571–1630) by his empirical determination of the laws of planetary motion, brought into connexion with astronomy one of the forms of conic, the ellipse, and established a foundation for the theory of gravitation. Contemporary with him for most of his life, we have Galileo (1564–1642), the founder of the science of dynamics; and closely following upon Galileo we have Isaac Newton (1643–1727): the *Philosophiæ naturalis Principia Mathematica* known as the *Principia* was first published in 1687.

The physical, statical, or dynamical questions which presented themselves before the publication of the *Principia* were of no particular mathematical dif-

ficulty, but it | is quite otherwise with the crowd of interesting questions arising 447|
out of the theory of gravitation, and which, in becoming the subject of mathe- 448
matical investigation, have contributed very much to the advance of mathematics. We have the problem of two bodies, or what is the same thing, that of the motion of a particle about a fixed centre of force, for any law of force; we have also the (mathematically very interesting) problem of the motion of a body attracted to two or more fixed centres of force; then, next preceding that of the actual solar system—the problem of three bodies; this has ever been and is far beyond the power of mathematics, and it is in the lunar and planetary theories replaced by what is mathematically a different problem, that of the motion of a body under the action of a principal central force and a disturbing force: or (in one mode of treatment) by the problem of disturbed elliptic motion. I would remark that we have here an instance in which an astronomical fact, the observed slow variation of the orbit of a planet, has directly suggested a mathematical method, applied to other dynamical problems, and which is the basis of very extensive modern investigations in regard to systems of differential equations. Again, immediately arising out of the theory of gravitation, we have the problem of finding the attraction of a solid body of any given form upon a particle, solved by Newton in the case of a homogeneous sphere, but which is far more difficult in the next succeeding cases of the spheroid of revolution (very ably treated by Maclaurin) and of the ellipsoid of three unequal axes: there is perhaps no problem of mathematics which has been treated by as great a variety of methods, or has given rise to so much interesting investigation as this last problem of the attraction of an ellipsoid upon an interior or exterior point. It was a dynamical problem, that of vibrating strings, by which Lagrange was led to the theory of the representation of a function as the sum of a series of multiple sines and cosines; and connected with this we have the expansions in terms of Legendre's functions P_n, suggested to him by the question just referred to of the attraction of an ellipsoid; the subsequent investigations of Laplace on the attractions of bodies differing slightly from the sphere led to the functions of two variables called Laplace's functions. I have been speaking of ellipsoids, but the general theory is that of attractions, which has become a very wide branch of modern mathematics; associated with it we have in particular the names of Gauss, Lejeune-Dirichlet, and Green; and I must not omit to mention that the theory is now one relating to n-dimensional space. Another great problem of celestial mechanics, that of the motion of the earth about its centre of gravity, in the most simple case, that of a body not acted upon by any forces, is a very interesting one in the mathematical point of view.

I may mention a few other instances where a practical or physical question has connected itself with the development of mathematical theory. I have spoken of two map projections—the stereographic, dating from Ptolemy; and Mercator's projection, invented by Edward Wright about the year 1600: each of these, as a particular case of the orthomorphic projection, belongs to the theory of the geometrical representation of an imaginary variable. I have spoken also of perspective, and of the representation of solid figures employed in Monge's

descriptive geometry. Monge, it is well known, is the author of the geometrical
448| theory of the curvature of surfaces and of curves of | curvature: he was led
449 to this theory by a problem of earthwork; from a given area, covered with earth of uniform thickness, to carry the earth and distribute it over an equal given area, with the least amount of cartage. For the solution of the corresponding problem in solid geometry he had to consider the intersecting normals of a surface, and so arrived at the curves of curvature. (See his "Mémoire sur les Déblais et les Remblais", *Mém. de l'Acad.*, 1781.) The normals of a surface are, again, a particular case of a doubly infinite system of lines, and are so connected with the modern theories of congruences and complexes.

The undulatory theory of light led to Fresnel's wave-surface, a surface of the fourth order, by far the most interesting one which had then presented itself. A geometrical property of this surface, that of having tangent planes each touching it along a plane curve (in fact, a circle), gave to Sir W.R. Hamilton the theory of conical refraction. The wave-surface is now regarded in geometry as a particular case of Kummer's quartic surface, with sixteen conical points and sixteen singular tangent planes.

My imperfect acquaintance as well with the mathematics as the physics prevents me from speaking of the benefits which the theory of Partial Differential Equations has received from the hydrodynamical theory of vortex motion, and from the great physical theories of heat, electricity, magnetism, and energy.

It is difficult to give an idea of the vast extent of modern mathematics. This word "extent" is not the right one: I mean extent crowded with beautiful detail—not an extent of mere uniformity such as an objectless plain, but of a tract of beautiful country seen at first in the distance, but which will bear to be rambled through and studied in every detail of hillside and valley, stream, rock, wood, and flower. But, as for anything else, so for a mathematical theory—beauty can be perceived, but not explained. As for mere extent, I can perhaps best illustrate this by speaking of the dates at which some of the great extensions have been made in several branches of mathematical science.

As regards geometry, I have already spoken of the invention of the Cartesian coordinates (1637). This gave to geometers the whole series of geometric curves of higher order than the conic sections: curves of the third order, or cubic curves; curves of the fourth order, or quartic curves; and so on indefinitely. The first fruits of it were Newton's *Enumeratio linearum tertii ordinis*, and the extremely interesting investigations of Maclaurin as to corresponding points on a cubic curve. This was at once enough to show that the new theory of cubic curves was a theory quite as beautiful and far more extensive than that of conics. And I must here refer to Euler's remark in the paper "Sur une contradiction apparente dans la théorie des courbes planes" (Berlin Memoirs, 1748), in regard to the nine points of intersection of two cubic curves (viz. that when eight of the points are given the ninth point is thereby completely determined): this is not only a fundamental theorem in cubic curves (including in itself Pascal's theorem of the hexagon inscribed in a conic), but it introduces into plane geometry a new notion—that of the point-system, or system of the points of intersection of two curves. | 449|

A theory derived from the conic, that of polar reciprocals, led to the general 450
notion of geometrical duality—viz. that in plane geometry the point and the line are correlative figures; and founded on this we have Plücker's great work, the *Theorie der algebraischen Curven* (Bonn, 1839), in which he establishes the relation which exists between the order and class of a curve and the number of its different point- and line-singularities (Plücker's six equations). It thus appears that the true division of curves is not a division according to order only, but according to order and class, and that the curves of a given order and class are again to be divided into families according to their singularities: this is not a mere subdivision, but is really a widening of the field of investigation; each such family of curves is in itself a subject as wide as the totality of the curves of a given order might previously have appeared.

We *unite* families by considering together the curves of a given *Geschlecht*, or deficiency; and in reference to what I shall have to say on the Abelian functions, I must speak of this notion introduced into geometry by Riemann in the memoir "Theorie der Abel'schen Functionen," *Crelle*, t. LIV. (1857). For a curve of a given order, reckoning cusps as double points, the deficiency is equal to the greatest number $\frac{1}{2}(n-1)(n-2)$ of the double points which a curve of that order can have, less the number of double points which the curve actually has. Thus a conic, a cubic with one double point, a quartic with three double points, &c., are all curves of the deficiency 0; the general cubic is a curve, and the most simple curve, of the deficiency 1; the general quartic is a curve of deficiency 3; and so on. The deficiency is usually represented by the letter p. Riemann considers the general question of the rational transformation of a plane curve: viz. here the coordinates, assumed to be homogeneous or trilinear, are replaced by any rational and integral functions, homogeneous of the same degree in the new coordinates; the transformed curve is in general a curve of a different order, with its own system of double points; but the deficiency p remains unaltered; and it is on this ground that he unites together and regards as a single class the whole system of curves of a given deficiency p. It must not be supposed that all such curves admit of rational transformation the one into the other: there is the further theorem that any curve of the class depends, in the case of a cubic, upon one parameter, but for $p > 1$ upon $3p-3$ parameters, each such parameter being unaltered by the rational transformation; it is thus only the curves having the same one parameter, or $3p-3$ parameters, which can be rationally transformed the one into the other.

Solid geometry is a far wider subject: there are more theories, and each of them is of greater extent. The ratio is not that of the numbers of the dimensions of the space considered, or, what is the same thing, of the elementary figures—point and line in the one case; point, line and plane in the other case—belonging to these spaces respectively, but it is a very much higher one. For it is very inadequate to say that in plane geometry we have the curve, and in solid geometry the curve and surface: a more complete statement is required for the comparison. In plane geometry we have the curve, which may be regarded as a singly

infinite system of points, and also as a singly infinite system of lines. In solid geometry we have, first, that which under one aspect is the curve, and under
450| another aspect the developable, and which may be | regarded as a singly infinite
451 system of points, of lines, or of planes; secondly, the surface, which may be regarded as a doubly infinite system of points or of planes, and also as a special triply infinite system of lines (viz. the tangent-lines of the surface are a special complex): as distinct particular cases of the former figure, we have the plane curve and the cone; and as a particular case of the latter figure, the ruled surface or singly infinite system of lines; we have *besides* the congruence, or doubly infinite system of lines, and the complex, or triply infinite system of lines. But, even if in solid geometry we attend only to the curve and the surface, there are crowds of theories which have scarcely any analogues in plane geometry. The relation of a curve to the various surfaces which can be drawn through it, or of a surface to the various curves that can be drawn upon it, is different in kind from that which in plane geometry most nearly corresponds to it, the relation of a system of points to the curves through them, or of a curve to the points upon it. In particular, there is nothing in plane geometry corresponding to the theory of the curves of curvature of a surface. To the single theorem of plane geometry, a right line is the shortest distance between two points, there correspond in solid geometry two extensive and difficult theories—that of the geodesic lines upon a given surface, and that of the surface of minimum area for any given boundary. Again, in solid geometry we have the interesting and difficult question of the representation of a curve by means of equations; it is not every curve, but only a curve which is the complete intersection of two surfaces, which can be properly represented by two equations $(x, y, z, w)^m = 0$, $(x, y, z, w)^n = 0$, in quadriplanar coordinates; and in regard to this question, which may also be regarded as that of the classification of curves in space, we have quite recently three elaborate memoirs by Nöther, Halphen, and Valentiner respectively.

In n-dimensional geometry, only isolated questions have been considered. The field is simply too wide; the comparison with each other of the two cases of plane geometry and solid geometry is enough to show how the complexity and difficulty of the theory would increase with each successive dimension.

In Transcendental Analysis, or the Theory of Functions, we have all that has been done in the present century with regard to the general theory of the function of an imaginary variable by Gauss, Cauchy, Puiseux, Briot, Bouquet, Liouville, Riemann, Fuchs, Weierstrass, and others. The fundamental idea of the geometrical representation of an imaginary variable $x + iy$, by means of the point having x, y for its coordinates, belongs, as I mentioned, to Gauss; of this I have already spoken at some length. The notion has been applied to differential equations; in the modern point of view, the problem in regard to a given differential equation is, not so much to reduce the differential equation to quadratures, as to determine from it directly the course of the integrals for all positions of the point representing the independent variable: in particular, the differential equation of the second order leading to the hypergeometric series $F(\alpha, \beta, \gamma, x)$ has been treated in this manner, with the most interesting results;

the function so determined for all values of the parameters (α, β, γ) is thus becoming a known function. I would here also refer to the new notion in this part of analysis introduced by Weierstrass—that of the one-valued integer func-
tion, as defined by an | infinite series of ascending powers, convergent for all 451|
finite values, real or imaginary, of the variable x or $1/x - c$, and so having the 452
one essential singular point $x = \infty$ or $x = c$, as the case may be: the memoir is published in the Berlin *Abhandlungen*, 1876.

But it is not only general theory: I have to speak of the various special functions to which the theory has been applied, or say the various known functions.

For a long time the only known transcendental functions were the circular functions sine, cosine, &c.; the logarithm—i.e. for analytical purposes the hyperbolic logarithm to the base e; and, as implied therein, the exponential function e^x. More completely stated, the group comprises the direct circular functions sin, cos, &c.; the inverse circular functions $\sin^{-1}$ or arc sin, &c.; the exponential function, exp.; and the inverse exponential, or logarithmic, function, log.

Passing over the very important Eulerian integral of the second kind or gamma-function, the theory of which has quite recently given rise to some very interesting developments—and omitting to mention at all various functions of minor importance,—we come (1811–1829) to the very wide groups, the elliptic functions and the single theta-functions. I give the interval of date so as to include Legendre's two systematic works, the *Exercices de Calcul Intégral* (1811–1816) and the *Théorie des Fonctions Elliptiques* (1825–1828); also Jacobi's *Fundamenta nova theoriæ Functionum Ellipticarum* (1829), calling to mind that many of Jacobi's results were obtained simultaneously by Abel. I remark that Legendre started from the consideration of the integrals depending on a radical $\sqrt{X}$, the square root of a rational and integral quartic function of a variable x; for this he substituted a radical $\Delta\varphi = \sqrt{1 - k^2\sin^2\varphi}$, and he arrived at his three kinds of elliptic integrals $F\varphi$, $E\varphi$, $\Pi\varphi$, depending on the argument or amplitude φ, the modulus k, and also the last of them on a parameter n; the function F is properly an inverse function, and in place of it Abel and Jacobi each of them introduced the direct functions corresponding to the circular functions sine and cosine, Abel's functions called by him φ, f, F, and Jacobi's functions sinam, cosam, Δam, or as they are also written sn, cn, dn. Jacobi, moreover, in the development of his theory of transformation obtained a multitude of formulæ containing q, a transcendental function of the modulus defined by the equation $q = e^{-\pi K'/K}$, and he was also led by it to consider the two new functions H, Θ, which (taken each separately with two different arguments) are in fact the four functions called elsewhere by him Θ_1, Θ_2, Θ_3, Θ_4; these are the so-called theta-functions, or, when the distinction is necessary, the single theta-functions. Finally, Jacobi using the transformation $\sin\varphi = \text{sinam}\ u$, expressed Legendre's integrals of the second and third kinds as integrals depending on the new variable u, denoting them by means of the letters Z, Π, and connecting them with his own functions H and Θ: and the elliptic functions sn, cn, dn are expressed with these, or say with Θ_1, Θ_2, Θ_3, Θ_4, as fractions having a common denominator.

It may be convenient to mention that Hermite in 1858, introducing into the theory in place of q the new variable ω connected with it by the equation $q = e^{i\pi\omega}$ (so that ω is in fact $= iK'/K$), was led to consider the three functions $\varphi\omega$, $\psi\omega$, $\chi\omega$, which denote respectively the values of $\sqrt[4]{k}$, $\sqrt[4]{k'}$ and $\sqrt[12]{kk'}$
452| regarded as functions of ω. | A theta-function, putting the argument $= 0$, and
453 then regarding it as a function of ω, is what Professor Smith in a valuable memoir, left incomplete by his death, calls an omega-function, and the three functions $\varphi\omega$, $\psi\omega$, $\chi\omega$ are his modular functions.

The proper elliptic functions sn, cn, dn form a system very analogous to the circular functions sine and cosine (say they are a sine and two separate cosines), having a like addition-theorem, viz. the form of this theorem is that the sn, cn and dn of $x + y$ are each of them expressible rationally in terms of the sn, cn and dn of x and of the sn, cn and dn of y; and, in fact, reducing itself to the system of the circular functions in the particular case $k = 0$. But there is the important difference of form that the expressions for the sn, cn and dn of $x + y$ are fractional functions having a common denominator: this is a reason for regarding these functions as the ratios of four functions A, B, C, D, the absolute magnitudes of which are and remain indeterminate (the functions sn, cn, dn are in fact quotients $[\Theta_1, \Theta_2, \Theta_3] \div \Theta_4$ of the four theta-functions, but this is a further result in nowise deducible from the addition-equations, and which is intended to be for the moment disregarded; the remark has reference to what is said hereafter as to the Abelian functions). But there is in regard to the functions sn, cn, dn (what has no analogue for the circular functions), the whole theory of transformation of any order n prime or composite, and, as parts thereof, the whole theory of the modular and multiplier equations; and this theory of transformation spreads itself out in various directions, in geometry, in the Theory of Equations, and in the Theory of Numbers. Leaving the theta-functions out of consideration, the theory of the proper elliptic functions sn, cn, dn is at once seen to be a very wide one.

I assign to the Abelian functions the date 1826–1832. Abel gave what is called his theorem in various forms, but in its most general form in the *Mémoire sur une propriété générale d'une classe très-étendue de Fonctions Transcendantes* (1826), presented to the French Academy of Sciences, and crowned by them after the author's death, in the following year. This is in form a theorem of the integral calculus, relating to integrals depending on an irrational function y determined as a function of x by any algebraical equation $F(x, y) = 0$ whatever: the theorem being that a sum of any number of such integrals is expressible by means of the sum of a determinate number p of like integrals, this number p depending on the form of the equation $F(x, y) = 0$ which determines the irrational y (to fix the ideas, remark that considering this equation as representing a curve, then p is really the deficiency of the curve; but as already mentioned, the notion of deficiency dates only from 1857): thus in applying the theorem to the case where y is the square root of a function of the fourth order, we have in effect Legendre's theorem for elliptic integrals $F\varphi + F\psi$ expressed by

means of a single integral $F\mu$, and not a theorem applying in form to the elliptic functions sn, cn, dn. To be intelligible I must recall that the integrals belonging to the case where y is the square root of a rational and integral function of an order exceeding four are (in distinction from the general case) termed hyperelliptic integrals: viz. if the order be 5 or 6, then these are of the class $p = 2$; if the order be 7 or 8, then they are of the class $p = 3$, and so on; the *general* Abelian integral of the class $p = 2$ is a hyperelliptic integral: but if $p = 3$, or any greater value, | then the hyperelliptic integrals are only a particular case of 453|
the Abelian integrals of the same class. The further step was made by Jacobi 454
in the short but very important memoir "Considerationes generales de transcendentibus Abelianis," *Crelle*, t. IX. (1832): viz. he there shows for the hyperelliptic integrals of any class (but the conclusion may be stated generally) that the direct functions to which Abel's theorem has reference are not functions of a single variable, such as the elliptic sn, cn, or dn, but functions of p variables. Thus, in the case $p = 2$, which Jacobi specially considers, it is shown that Abel's theorem has reference to two functions $\lambda(u, v)$, $\lambda_1(u, v)$ each of two variables, and gives in effect an addition-theorem for the expression of the functions $\lambda(u + u', v + v')$, $\lambda_1(u + u', v + v')$ algebraically in terms of the functions $\lambda(u, v)$, $\lambda_1(u, v)$, $\lambda(u' v')$, $\lambda_1(u', v')$.

It is important to remark that Abel's theorem does not directly give, nor does Jacobi assert that it gives, the addition-theorem in a perfect form. Take the case $p = 1$: the result from the theorem is that we have a function $\lambda(u)$, which is such that $\lambda(u + v)$ can be expressed algebraically in terms of $\lambda(u)$ and $\lambda(v)$. This is of course perfectly correct, $\text{sn}(u+v)$ is expressible algebraically in terms of sn u, sn v, but the expression involves the radicals $\sqrt{1 - \text{sn}^2 u}$, $\sqrt{1 - k^2\text{sn}^2 u}$, $\sqrt{1 - \text{sn}^2 v}$, $\sqrt{1 - \text{k}^2\text{sn}^2 v}$; but it does not give the three functions sn, cn, dn, or in anywise amount to the statement that the sn, cn and dn u of $u+v$ are expressible rationally in terms of the sn, cn and dn of u and of v. In the case $p = 1$, the right number of functions, each of one variable, is 3, but the three functions sn, cn and dn are properly considered as the ratios of 4 functions; and so, in general, the right number of functions, each of p variables, is $4^p - 1$, and these may be considered as the ratios of 4^p functions. But notwithstanding this last remark, it may be considered that the notion of the Abelian functions of p variables is established, and the addition-theorem for these functions in effect given by the memoirs (Abel 1826, Jacobi 1832) last referred to.

We have next for the case $p = 2$, which is hyperelliptic, the two extremely valuable memoirs, Göpel, "Theoria transcendentium Abelianarum primi ordinis adumbratio læva", *Crelle*, t. XXXV. (1847), and Rosenhain, "Mémoire sur les fonctions de deux variables et à quatre périodes qui sont les inverses des intégrales ultra-elliptiques de la première classe" (1846), Paris, *Mém. Savans Étrang*. t. XI. (1851), each of them establishing on the analogy of the single theta-functions the corresponding functions of two variables, or double theta-functions, and in connexion with them the theory of the Abelian functions of two variables. It may be remarked that in order of simplicity the theta-functions certainly precede the Abelian functions.

Passing over some memoirs by Weierstrass which refer to the general hyperelliptic integrals, p any value whatever, we come to Riemann, who died 1866, at the age of forty: collected edition of his works, Leipzig, 1876. His great memoir on the Abelian and theta-functions is the memoir already incidentally referred to, "Theorie der Abel'schen Functionen", *Crelle*, t. LIV. (1857); but intimately connected therewith we have his Inaugural Dissertation (Göttingen, 1851), *Grundlagen für eine allgemeine Theorie der Functionen einer veränderlichen complexen Grösse*: his treatment of the problem of the Abelian functions, and establishment for the purpose of this theory of the multiple theta-
454| functions, are alike founded on his general principles of the | theory of the
455 functions of a variable complex magnitude $x + iy$, and it is this which would have to be gone into for any explanation of his method of dealing with the problem.

Riemann, starting with the integrals of the most general form, and considering the inverse functions corresponding to these integrals—that is, the Abelian functions of p variables—defines a theta-function of p variables, or p-tuple theta-function, as the sum of a p-tuply infinite series of exponentials, the general term of course depending on the p variables; and he shows that the Abelian functions are algebraically connected with theta-functions of the proper arguments. The theory is presented in the broadest form; in particular as regards the theta-functions, the 4^p functions are not even referred to, and there is no development as to the form of the algebraic relations between the two sets of functions.

In the Theory of Equations, the beginning of the century may be regarded as an epoch. Immediately preceding it, we have Lagrange's *Traité des Équations Numériques* (1st ed. 1798), the notes to which exhibit the then position of the theory. Immediately following it, the great work by Gauss, the *Disquisitiones Arithmeticæ* (1801), in which he establishes the theory for the case of a prime exponent n, of the binomial equation $x^n - 1 = 0$: throwing out the factor $x - 1$, the equation becomes an equation of the order $n - 1$, and this is decomposed into equations the orders of which are the prime factors of $n - 1$. In particular, Gauss was thereby led to the remarkable geometrical result that it was possible to construct geometrically—that is, with only the ruler and compass—the regular polygons of 17 sides and 257 sides respectively. We have then (1826–1829) Abel, who, besides his demonstration of the impossibility of the solution of a quintic equation by radicals, and his very important researches on the general question of the algebraic solution of equations, established the theory of the class of equations since called Abelian equations. He applied his methods to the problem of the division of the elliptic functions, to (what is a distinct question) the division of the complete functions, and to the very interesting special case of the lemniscate. But the theory of algebraic solutions in its most complete form was established by Galois (born 1811, killed in a duel 1832), who for this purpose introduced the notion of a group of substitutions; and to him also are due some most valuable results in relation to another set of equations presenting themselves in the theory of elliptic functions—viz. the modular equations. In 1835 we have Jerrard's transformation of the general quintic equation. In 1870 an elaborate work, Jordan's *Traité des Substitutions et des équations algébriques*: a mere inspection of the table of contents of this would serve to illustrate my proposition as to the great extension of this branch of mathematics.

The Theory of Numbers was, at the beginning of the century, represented by Legendre's *Théorie des Nombres* (1st ed. 1798), shortly followed by Gauss' *Disquisitiones Arithmeticæ* (1801). This work by Gauss is, throughout, a theory of ordinary real numbers. It establishes the notion of a congruence; gives a proof of the theorem of reciprocity in regard to quadratic residues; and contains a very complete theory of binary quadratic forms $(a, b, c)\ (x, y)^2$, of negative
and positive determinant, including | the theory, there first given, of the composi- 455|
tion of such forms. It gives also the commencement of a like theory of ternary 456
quadratic forms. It contains also the theory already referred to, but which has since influenced in so remarkable a manner the whole theory of numbers—the theory of the solution of the binomial equation $x^n - 1 = 0$: it is, in fact, the roots or periods of roots derived from these equations which form the incommensurables, or unities, of the complex theories which have been chiefly worked at; thus, the i of ordinary analysis presents itself as a root of the equation $x^4 - 1 = 0$. It was Gauss himself who, for the development of a real theory—that of biquadratic residues—found it necessary to use complex numbers of the before-mentioned form, $a + bi$ (a and b positive or negative real integers, including zero), and the theory of these numbers was studied and cultivated by Lejeune-Dirichlet. We have thus a new theory of these complex numbers, side by side with the former theory of real numbers: everything in the real theory reproducing itself, prime numbers, congruences, theories of residues, reciprocity, quadratic forms, &c., but with greater variety and complexity, and increased difficulty of demonstration. But instead of the equation $x^4 - 1 = 0$, we may take the equation $x^3 - 1 = 0$: we have here the complex numbers $a + b\rho$ composed with an imaginary cube root of unity, the theory specially considered by Eisenstein: again a new theory, corresponding to but different from that of the numbers $a + bi$. The general case of any prime value of the exponent n, and with periods of roots, which here present themselves instead of single roots, was first considered by Kummer: viz. if $n - 1 = ef$, and $\eta_1, \eta_2, \ldots, \eta_e$ are the e periods, each of them a sum of f roots, of the equation $x^n - 1 = 0$, then the complex numbers considered are the numbers of the form $a_1\eta_1 + a_2\eta_2 + \ldots + a_e\eta_e$ ($a_1, a_2, \ldots, a_e$ positive or negative ordinary integers, including zero): f may be $= 1$, and the theory for the periods thus includes that for the single roots.

We have thus a new and very general theory, including within itself that of the complex numbers $a + bi$ and $a + b\rho$. But a new phenomenon presents itself; for these special forms the properties in regard to prime numbers corresponded precisely with those for real numbers; a non-prime number was in one way only a product of prime factors; the power of a prime number has only factors which are lower powers of the same prime number: for instance, if p be a prime number, then, excluding the obvious decomposition $p.p^2$, we cannot have

p^3 = a product of two factors A, B. In the general case this is not so, but the exception first presents itself for the number 23; in the theory of the numbers composed with the 23rd roots of unity, we have prime numbers p, such that $p^3 = AB$. To restore the theorem, it is necessary to establish the notion of ideal numbers; a prime number p is by definition not the product of two actual numbers, but in the example just referred to the number p is the product of two ideal numbers having for their cubes the two actual numbers A, B, respectively, and we thus have $p^3 = AB$. It is, I think, in this way that we most easily get some notion of the meaning of an ideal number, but the mode of treatment (in Kummer's great memoir, "Ueber die Zerlegung der aus Wurzeln der Einheit gebildeten complexen Zahlen in ihre Primfactoren," *Crelle*, t. XXXV. 1847) is a much more refined one; an ideal number, without ever being isolated, is made to manifest itself in the properties of the prime number of which it is a factor,
456| and without reference to the | theorem afterwards arrived at, that there is always
457 some power of the ideal number which is an actual number. In the still later developments of the Theory of Numbers by Dedekind, the units, or incommensurables, are the roots of any irreducible equation having for its coefficients ordinary integer numbers, and with the coefficient unity for the highest power of x. The question arises, What is the analogue of a whole number? thus, for the very simple case of the equation $x^2 + 3 = 0$, we have as a whole number the apparently fractional form $\frac{1}{2}(1 + i\sqrt{3})$ which is the imaginary cube root of unity, the ρ of Eisenstein's theory. We have, moreover, the (as far as appears) wholly distinct complex theory of the numbers composed with the congruence-imaginaries of Galois: viz. these are imaginary numbers assumed to satisfy a congruence which is not satisfied by any real number; for instance, the congruence $x^2 - 2 = 0 \pmod 5$ has no real root, but we assume an imaginary root i, the other root is then $= -i$, and we then consider the system of complex numbers $a + bi \pmod 5$, viz. we have thus the 5^2 numbers obtained by giving to each of the numbers a, b, the values 0, 1, 2, 3, 4, successively. And so in general, the consideration of an irreducible congruence $F(x) = 0 \pmod p$ of the order n, to any prime modulus p, gives rise to an imaginary congruence root i, and to complex numbers of the form $a + bi + ci^2 + \ldots + ki^{n-1}$, where $a, b, k, \ldots$ &c., are ordinary integers each $= 0, 1, 2, \ldots, p-1$.

As regards the theory of forms, we have in the ordinary theory, in addition to the binary and ternary quadratic forms, which have been very thoroughly studied, the quaternary and higher quadratic forms (to these last belong, as very particular cases, the theories of the representation of a number as a sum of four, five or more squares), and also binary cubic and quartic forms, and ternary cubic forms, in regard to all of which something has been done; the binary quadratic forms have been studied in the theory of the complex numbers $a + bi$.

A seemingly isolated question in the Theory of Numbers, the demonstration of Fermat's theorem of the impossibility for any exponent λ, greater than 3, of the equation $x^\lambda + y^\lambda = z^\lambda$, has given rise to investigations of very great interest and difficulty.

Outside of ordinary mathematics, we have some theories which must be refer-

red to: algebraical, geometrical, logical. It is, as in many other cases, difficult to draw the line; we do in ordinary mathematics use symbols not denoting quantities, which we nevertheless combine in the way of addition and multiplication, $a + b$, and ab, and which may be such as not to obey the commutative law $ab = ba$: in particular, this is or may be so in regard to symbols of operation; and it could hardly be said that any development whatever of the theory of such symbols of operation did not belong to ordinary algebra. But I do separate from ordinary mathematics the system of multiple algebra or linear associative algebra, developed in the valuable memoir by the late Benjamin Peirce, *Linear Associative Algebra* (1870, reprinted 1881 in the *American Journal of Mathematics*, vol. IV., with notes and addenda by his son, C.S. Peirce); we here consider symbols A, B, &c. which are linear functions of a determinate number of letters or units i, j, k, l, &c., with coefficients which are ordinary analytical magnitudes, real or imaginary, viz. the coefficients are in general of the form
$x + iy$, where | i is the before-mentioned imaginary or $\sqrt{-1}$ of ordinary analysis. 457|
The letters i, j, &c., are such that every binary combination i^2, ij, ji, &c., (the 458
ij being in general not $= ji$), is equal to a linear function of the letters, but under the restriction of satisfying the associative law: viz. for each combination of three letters $ij.k$ is $= i.jk$, so that there is a determinate and unique product of three or more letters; or, what is the same thing, the laws of combination of the units i, j, k, are defined by a multiplication table giving the values of i^2, ij, ji, &c.: the original units may be replaced by linear functions of these units, so as to give rise, for the units finally adopted, to a multiplication table of the most simple form; and it is very remarkable, how frequently in these simplified forms we have nilpotent or idempotent symbols ($i^2 = 0$, or $i^2 = i$, as the case may be), and symbols i, j, such that $ij = ji = 0$; and consequently how simple are the forms of the multiplication tables which define the several systems respectively.

I have spoken of this multiple algebra before referring to various geometrical theories of earlier date, because I consider it as the general analytical basis, and the true basis, of these theories. I do not realise to myself directly the notions of the addition or multiplication of two lines, areas, rotations, forces, or other geometrical, kinematical, or mechanical entities; and I would formulate a general theory as follows: consider any such entity as determined by the proper number of parameters a, b, c (for instance, in the case of a finite line given in magnitude and position, these might be the length, the coordinates of one end, and the direction-cosines of the line considered as drawn from this end); and represent it by or connect it with the linear function $ai + bj + ck +$ &c., formed with these parameters as coefficients, and with a given set of units, i, j, k, &c. Conversely, any such linear function represents an entity of the kind in question. Two given entities are represented by two linear functions; the sum of these is a like linear function representing an entity of the same kind, which may be regarded as the sum of the two entities; and the product of them (taken in a determined order, when the order is material) is an entity of the same kind, which may be regarded as the product (in the same order) of the two entities.

We thus establish by definition the notion of the sum of the two entities, and that of the product (in a determinate order, when the order is material) of the two entities. The value of the theory in regard to any kind of entity would of course depend on the choice of a system of units, $i, j, k, \ldots$, with such laws of combination as would give a geometrical or kinematical or mechanical significance to the notions of the sum and product as thus defined.

Among the geometrical theories referred to, we have a theory (that of Argand, Warren, and Peacock) of imaginaries in plane geometry; Sir W.R. Hamilton's very valuable and important theory of Quaternions; the theories developed in Grassmmann's *Ausdehnungslehre*, 1841 and 1862; Clifford's theory of Biquaternions; and recent extensions of Grassmann's theory to non-Euclidian space, by Mr Homersham Cox. These different theories have of course been developed, not in anywise from the point of view in which I have been considering them, but from the points of view of their several authors respectively.

The literal symbols x, y, &c., used in Boole's *Laws of Thought* (1854) to represent things as subjects of our conceptions, are symbols obeying the laws
458| of algebraic | combination (the distributive, commutative, and associative laws)
459 but which are such that for any one of them, say x, we have $x - x^2 = 0$, this
equation not implying (as in ordinary algebra it would do) either $x = 0$ or else $x = 1$. In the latter part of the work relating to the Theory of Probabilities, there is a difficulty in making out the precise meaning of the symbols; and the remarkable theory there developed has, it seems to me, passed out of notice, without having been properly discussed. A paper by the same author, "Of Propositions numerically definite" (*Camb. Phil. Trans.* 1869), is also on the borderland of logic and mathematics. It would be out of place to consider other systems of mathematical logic, but I will just mention that Mr C.S. Peirce in his "Algebra of Logic", *American Math. Journal*, vol. III., establishes a notation for relative terms, and that these present themselves in connexion with the systems of units of the linear associative algebra.

Connected with logic, but primarily mathematical and of the highest importance, we have Schubert's *Abzählende Geometrie* (1878). The general question is, How many curves or other figures are there which satisfy given conditions? for example, How many conics are there which touch each of five given conics? The class of questions in regard to the conic was first considered by Chasles, and we have his beautiful theory of the characteristics μ, ν, of the conics which satisfy four given conditions; questions relating to cubics and quartics were afterwards considered by Maillard and Zeuthen; and in the work just referred to the theory has become a very wide one. The noticeable point is that the symbols used by Schubert are in the first instance, not numbers, but mere logical symbols: for example, a letter g denotes the condition that a line shall cut a given line; g^2 that it shall cut each of two given lines; and so in other cases; and these logical symbols are combined together by algebraical laws: they first acquire a numerical signification when the number of conditions becomes equal to the number of parameters upon which the figure in question depends.

In all that I have last said in regard to theories outside of ordinary mathematics,

I have been still speaking on the text of the vast extent of modern mathematics. In conclusion I would say that mathematics have steadily advanced from the time of the Greek geometers. Nothing is lost or wasted; the achievements of Euclid, Archimedes, and Apollonius are as admirable now as they were in their own days. Descartes' method of coordinates is a possession for ever. But mathematics have never been cultivated more zealously and diligently, or with greater succcss, than in this century—in the last half of it, or at the present time: the advances made have been enormous, the actual field is boundless, the future full of hope. In regard to pure mathematics we may most confidently say:—

Yet I doubt not through the ages one increasing purpose runs,
And the thoughts of men are widened with the process of the suns.

15
Charles Sanders Peirce (1839–1914)

Near the end of his strange and beleaguered life, shortly after the death of his friend William James, C.S. Peirce despairingly wrote: 'Who could be of a nature so different from his as I? He so concrete, so living; I a mere table of contents, so abstract, a very snarl of twine' (*Peirce 1931–58*, Vol. vi, p. 131.)

Anybody who has struggled to understand Peirce's thought—or who has turned to Peirce from tidier thinkers like James or Frege—will feel at once the justice of his self-appraisal. Always trying to squeeze his thoughts into a system, he never managed to make them fit. They keep brimming over, refusing to stay within their assigned boundaries. The insights and suggestions and projects come tumbling out of him in an exuberant cascade: logic machines, an indeterminate physics, a differential calculus based on infinitesimals, pungent observations on scientific method, and a steady stream of articles on logic and the foundations of mathematics.

Most of the great mathematical logicians of the nineteenth century—De Morgan and Boole in particular—came to logic from a previous interest in algebra: they wrote as mathematicians, not as philosophers. But with Peirce the motivation is more complex. The philosophical roots of his thought are deeper, the range of his scientific interests is wider, and the applications of his logic are more extensive, than in any of his contemporaries. He was perhaps the greatest logician of the nineteenth century; but he was also a chemist, astronomer, geodesist, and mathematician; an authority on scholastic logic; a philosopher of language and of scientific method; a historian of medieval and renaissance science; and a remarkably fertile thinker who saw all of these topics as interwoven with the mathematical logic he did so much to develop.

But despite the sparkle of his writing and the magnitude of his technical accomplishments and the praise that has come his way from the likes of John Dewey and Karl Popper, C.I. Lewis and A.J. Ayer, Peirce remains largely ignored in modern philosophy. Certainly he has not had the influence of Frege or Russell or Wittgenstein. His name is generally mispronounced (it rhymes, inappropriately, with *terse*), and he is best known for coining a label—*pragmatism*—he later disowned. Even his discovery of quantification theory is generally passed over in silence.

This neglect of Peirce has at least three sources. The first is the inherent difficulty of his thought. He is a subtle thinker, and his range of reference is enormous. His scientific work roams over both the theoretical and the experimental natural sciences; he read fluently French and German, Latin and Greek, and plunged deeply into the history of science and of philosophy. Much of what he wrote was never intended for publication and shows him jotting down ideas in response to his reading—groping to understand the consequences of a recent scientific discovery, or arguing with some obscure logician of the fourteenth century.

The second is the oddity and dissonance of his doctrines—both the doctrines themselves, and the idiosyncratic way in which they are combined. On one page he will insist on the importance of logic, of mathematical rigour, of scientific philosophy, and will proclaim his robust contempt for the reasoning powers of 'seminarians and literary types'. And on the next he will offer a cosmological metaphysics of 'evolutionary love', express his admiration for Hegel, and attempt to reconcile science and religion. He contains something to repulse almost everybody, from positivists and pragmatists to Wittgensteinians, post-modernists, and philosophers of ordinary language. The real or perceived quirkiness of many of his views has kept him from being wholeheartedly adopted by later philosophers, and Peirce, when he has been noticed at all, has generally been relegated to the tame status of a brilliant forerunner—a perceptive and independent thinker who glimpsed the doctrines of Popper or Reichenbach or some other later philosopher, but who was frequently muddled, and who had a tendency to lapse into deplorable error.

Third is the state of his writings. Peirce left a voluminous but chaotic record of his thought—published articles, logical and scientific notebooks, drafts of articles, drafts of books. This material would easily fill a hundred large volumes. In the 1930s the overwhelming task of editing this material was entrusted by Harvard University to a pair of scholars just out of graduate school. They coped as best they could, working in haste and producing six volumes of Peirce's so-called *Collected Papers* in four years. The *Papers* were arranged by subject-matter rather than chronologically, making it difficult to retrace Peirce's footsteps—an important *desideratum*, since Peirce was constantly revising his theories and exploring new ideas. Paragraphs excerpted from one manuscript were grafted on to paragraphs written years later; continuous manuscripts were broken up and the fragments published in separate volumes. In consequence, Peirce emerged as a hopelessly confused thinker—even more a 'snarl of twine' than he had seemed to himself. Not surprisingly, subsequent philosophers have preferred to praise Peirce for being a brilliant though erratic forerunner of themselves than to struggle through his artificially tangled writings.

I

Charles Sanders Peirce was born in Cambridge, Massachusetts in 1839; like Bertrand Russell and Ludwig Wittgenstein he grew up amid the local aristocracy, a fact which may explain something of his independent intellectual temperament. The Peirce family had been established in Boston for two hundred years, the descendants of a John Pers who emigrated from Norwich in 1637; Charles Peirce's grandfather was the Harvard University librarian; and his father, Benjamin Peirce (1809–80), was for forty years Perkins Professor of Astronomy and Mathematics at Harvard, and was the leading American scientist of the day—a pioneer in linear algebra; a distinguished astronomer; and an early defender of the theory of continental drift and of Darwin's theory of evolution. Benjamin Peirce was also President of the American Academy for the Advancement of Science, a founder of the National Academy of Sciences, and Superintendent of the US Coast Survey (which, at the time, was the pre-eminent American institution for experimental science). His house was the meeting-place for many of the scientific organizations in Boston—the Cambridge Scientific Club, the Cambridge Astronomical Society, and the Mathematical Club—as well as for such literary figures as Emerson, Longfellow, Lowell, Henry James, Sr, and Oliver Wendell Holmes, Sr. Not without reason, Peirce often declared in later life that he had been 'reared in a laboratory'. His older brother James Mills Peirce became a mathematician, and eventually succeeded to the father's Harvard professorship; a younger brother went into the diplomatic service, and in time was appointed minister to Norway.

A few weeks after his twelfth birthday, Charles Peirce chanced upon his older brother's copy of Whately's *Elements of logic* (*1832*); absorbed, he spent the next several days mastering its contents. The experience (he later wrote) transformed his intellectual outlook so completely that he was never able to regard any of his subsequent scientific work as anything but an exercise in logic.

Benjamin Peirce soon recognized his son's intellectual gifts, and supervised his early study of mathematics and physics. He also held him up as one who was destined to be an even greater scientist than the father. Objectively speaking, the prediction was correct; but it imposed a heavy burden on Charles, who felt called upon to surpass the accomplishments of a father who was widely regarded as the greatest scientist America had yet produced. Peirce did not, like John Stuart Mill, suffer actual breakdown; but his personality developed in unfortunate ways. By all accounts, he was a difficult man to deal with—prickly, ill at ease with other people, intellectually vain, overbearing in his criticisms. He was also wholeheartedly devoted to his research, capable of prodigious hard work, willing to change his mind, and inclined to jeer loudly at ideas that struck him as silly. All these traits were later to cause him difficulty in his dealings with colleagues and superiors.

When Peirce entered Harvard in 1855 he had already developed a strong interest in philosophy, and in addition to his official studies of chemistry, mathematics, and physics he immersed himself in the works of the great philosophers:

> The first strictly philosophical books that I read were of the classical German schools; and I became so deeply imbued with many of their ways of thinking that I have never been able to disabuse myself of them. Yet my attitude was always that of a dweller in a laboratory, eager only to learn what I did not yet know, and not that of philosophers bred in theological seminaries, whose ruling impulse is to teach what they hold to be infallibly true. I devoted two hours a day to the study of Kant's *Critic of the Pure Reason* for more than three years, until I almost knew the whole book by heart, and had critically examined every section of it (*Peirce 1931–58*, Vol. i, p. ix).

Peirce was not, however, a slavish follower of Kant. In particular, he early realized that there were serious problems with Kant's reliance on traditional logic. Kant had believed logic to be a 'closed and completed' body of doctrine, substantially perfected by Aristotle; the intervening centuries, he thought, had added only minor refinements (see, for example, *Kant 1787*, pp. viii–ix). The 'metaphysical deduction of the categories' in the first *Critique* (in which Kant derives his table of the categories from the basic principles of logic) was explicitly dependent on the assumption that Aristotelian logic was complete. But Peirce was aware that Boole and De Morgan had advanced the subject beyond what was known to Aristotle and Kant, and he saw in this fact the opportunity for a philosophical advance. One of the initial motives for Peirce's study of logic was his desire to amend Kant by building a new bridge between metaphysics and logic, and over the next years he made several attempts to deduce from the new logic a revised table of the categories. (For a representative attempt see his 'On a new list of categories', *Peirce 1868*.)

After his graduation from Harvard in 1859, Peirce studied zoological classification for six months with Louis Agassiz; Agassiz gave him the task of sorting fossil brachiopods. (Peirce's interest in schemes of classification dates from this experience, and in later life he was to make repeated attempts to classify the dependencies of the sciences on each other—a task which he regarded as of a piece with his work in logic.) Peirce next entered the Lawrence Scientific School, where he received his graduate training in chemistry, graduating *summa cum laude* in 1863. He worked as a research chemist for several years, and in 1869 published his most significant contribution to chemistry, the note on 'The pairing of the elements' (*Peirce 1869*) in which he partially anticipated Mendeleev's discovery of the periodic table of the elements. For decades afterwards Peirce listed his profession as 'chemist'.

II

The bulk of Peirce's energies during the 1860s were directed, not towards chemistry, but towards philosophy and logic. In 1867 he published his first logic papers: 'On an improvement in Boole's calculus of logic', 'On the natural classification of arguments', and 'Upon the logic of mathematics'. These papers simplify Boole's system and correct several mistakes; they also explore extensions of syllogistic logic, and connections between logic, arithmetic, and algebra. At about the same time, Peirce took his first important step in philosophy, one that unites him with the mainstream of analytical philosophy. Ever since Descartes, philosophers had made epistemology the central philosophical discipline. The foundations of modern science, it was assumed, were to be explored by examining the mental functioning and capacities for knowledge of the individual consciousness. But Peirce rejected both the individualism and the psychologism of the Cartesian approach. Without prohibiting psychological inquiries altogether, he argued that they should be founded on an objective, interpersonal study of the principles of scientific reasoning—that is, on logic, if 'logic' is understood in a sufficiently broad sense.

Peirce was aware that in taking this step he was in a certain measure returning to the standpoint of the Schoolmen, to the philosophical logic of the Middle Ages; indeed, he argued that the scholastic methodology, which represented the search for truth as a collective enterprise of fallible human beings, was closer to the spirit of modern science than the Cartesian picture of an individual ego erecting its knowledge upon certain and indubitable foundations. Accordingly in the mid-1860s Peirce undertook a thorough study of medieval logic and metaphysics, which he read as diligently as he had Kant. (Indeed, while still in his twenties he owned a larger library of medieval philosophy than did Harvard University; his collection of scholastic treatises was eventually left to Harvard, and still forms an important part of its collection.) He concentrated particularly on the medieval logicians—John of Salisbury, Robert of Lincoln, Roger Bacon, William of Occam, Petrus Hispanus, and above all Duns Scotus, who became as great an influence on his thought as Kant or Aristotle:

> The works of Duns Scotus have strongly influenced me. If his logic and metaphysics, not slavishly worshiped, but torn away from its medievalism, be adapted to modern culture, under continual wholesome reminders of nominalistic criticisms, I am convinced it will go far toward supplying the philosophy which is best to harmonize with physical science (*Peirce 1931–58*, Vol. i, p. ix).

Peirce summed up his break with Cartesianism in a series of papers on epistemology, logic, and scientific method. (All these papers are reprinted in *Peirce 1982–*, Vol. ii.) The papers argue the centrality of logic and the philosophy of language to philosophy; they explore the relationships between truth, reality, cognition, and the active, investigating scientific community whose theories gradually converge on a settled view of the world. Peirce argued in the *sic et non* style of the scholastics, but despite the medieval influences his doctrines themselves have a decidedly modern ring. They were already formulated by Peirce in 1868, and lay at the heart of his philosophy long before they were laboriously rediscovered by analytical philosophy. For instance: (1) 'We cannot begin with complete doubt. We must begin with all the prejudices which we actually have when we enter upon the study of philosophy'. (2) 'We have no power of Introspection, but all knowledge of the internal world is derived by hypothetical reasoning from our knowledge of external facts'. (3) 'The science of thought in its intellectual significance is one and the same thing with the science of the laws of signs'. (4) The ultimate test of certainty is not to be found in the *individual*, but in the *community* of inquirers. (5) Philosophy ought 'to trust rather to the multitude and variety of its arguments than to the conclusiveness of any one. Its reasoning should not form a chain which is no stronger than its weakest link, but a cable whose fibers may be ever so slender, provided they are sufficiently numerous and intimately connected' (*Peirce 1982–*, Vol. ii, pp. 212–13; Vol. iii, p. 83).

III

The links forged at this early stage in Peirce's development between philosophy, formal logic, and the study of scientific reasoning were to be a guiding influence not only on his thought but also on his professional life, and during the next thirty years he simultaneously pursued several overlapping careers. In broad overview, his trajectory was as follows. For thirty years (1861–91) he was employed as a cartographer, astronomer, and physicist by the US Coast Survey. During part of this period (1869–78) he was also employed as an astronomer by the Harvard Observatory; and for a further five years (1879–84) he was a lecturer in logic at Johns Hopkins. For the last twenty-three years of his life (1891–1914) he was unemployed and lived as a recluse in Milford, Pennsylvania. Throughout the entire span of fifty-three years he wrote on philosophy and formal logic, particularly during three periods: the early 1870s, his years at Johns Hopkins, and the long period of his retirement.

Peirce during the 1870s made a number of significant contributions to the natural sciences. During his years at the Harvard Observatory he carried out pioneering studies of the spectrum of the aurora borealis, the chemical composition of the sun, and the size and shape of the Milky Way Galaxy. His *Photometric researches* appeared in 1878, and established him as the leading American astronomer of his generation. In 1872 the Coast Survey made him responsible for ascertaining the shape of the earth by measuring the variability of its gravitational field. This assignment called for pendulum experiments more precise than had previously been performed. Peirce obtained the necessary precision by two innovations: he designed a new type of pendulum whose stand did not flex to the same extent as previous stands; and he increased the accuracy of his measurements by stating the length of the meter in terms of the wavelength of light—the first time this had been done. At the same time, Peirce was working for the Survey on problems in mathematical cartography. By applying elliptic integrals to the theory of conformal map projections he was able to produce his 'quincuncial' maps of the surface of the earth, which are still in use by the Coast and Geodetic Survey for charting international air routes.

IV

Peirce's involvement in the natural sciences left a deep imprint on his philosophy, and from the early 1870s we find in his thought a new emphasis on the active, experimenting scientist, who was 'to carry his mind into his laboratory, and to make of his alembics and cucurbits instruments of thought, giving a new conception of reasoning, as something which was to be done with one's eyes open, by manipulating real things instead of words and fancies'. Peirce's work in the sciences of measurement had persuaded him that absolute accuracy in the physical sciences was not only unattainable, but 'irresistibly comical', and the core of his thought during this period was an effort to develop this insight, to scrutinize the consequences of regarding truth as a *limit* successively approached by increasingly refined and accurate scientific investigations.

In the early 1870s, Peirce and a number of friends—including William James (Peirce's junior by three years) and the legal scholar Oliver Wendell Holmes, Jr.—founded the 'Metaphysical Club' in Cambridge. In the meetings of this club, Peirce presented his complex network of views on logic, philosophy, and scientific method, coining the name 'pragmatism' for his position. The precipitate from these meetings he later published in an influential series of articles on 'Illustrations of the logic of science' that appeared in 1877–8. (*Peirce 1982–*, Vol. iii, pp. 241–374.) In his old age, Peirce described his philosophy as:

> the attempt of a physicist to make such conjecture as to the constitution of the universe as the methods of science may permit, with the aid of all that has been done by previous philosophers. I shall support my propositions by such arguments as I can. Demonstrative proof is not to be thought of. The demonstrations of the metaphysicians are all moonshine. The best that can be done is to supply a hypothesis, not devoid of all likelihood, in the general line of growth of scientific ideas, and capable of being verified or refuted by future observers (*Peirce 1931–58*, Vol. i, p. x).

William James was later to make the name 'pragmatism' famous, giving credit to Peirce for the terminology. But the pragmatism presented in James's famous lectures was at best a pale oversimplification of Peirce's, deprived of its logical and scientific core, overlaid with psychological elements, and offered to the world as a key that would unlock every door. Peirce disapproved of James's tendency to over-dramatize the importance of his own doctrine. 'You and Schiller carry pragmatism too far for me,' Peirce wrote to James on 7 March 1904. 'I don't want to exaggerate it but keep it within the bounds to which the evidences of it are limited. . . . I also want to say that pragmatism after all solves no real problem. It only shows that supposed problems are not real problems. But when one comes to such questions as immortality, the nature of the connection of mind and matter (further than that mind acts on matter not like a *cause* but like a *law*) we are left completely in the dark. The effect of pragmatism here is simply to open our minds to receiving evidence, not to furnish evidence.' In 1905 the ever-tetchy Peirce made his disagreement public. Deploring the fact that the word had fallen into 'literary clutches', Peirce changed the terminology: 'The writer, finding his bantling "pragmatism" so promoted, feels that it is time to kiss his child good-by and relinquish it to its higher destiny; while to serve the precise purpose of expressing the original definition, he begs to announce the birth of the word "pragmaticism", which is ugly enough to be safe from kidnappers' (*Peirce 1905*).

The backbone of Peirce's entire system of philosophy was formal logic, which he regarded as the theory of sound scientific reasoning, whether deductive or probabilistic. Thus it was that he could declare that 'each chief step in science has been a lesson in logic'—or that his 1883 lectures in logic at Johns Hopkins began with a discussion of Boolean algebra, the logic of relatives, and the quantifiers; turned to the theory of probability and induction; and ended with an examination of Kepler's *De motibus stellae Martis* ('the greatest piece of inductive reasoning ever produced') (*Peirce 1982–*, Vol. iv, pp. 476, 382). Even in his early years he insisted that logic must encompass more than the theory of valid inference:

> As long as the logician contents himself with tracing out the forms of propositions and arguments, his science is one of the most exact and satisfactory. It may be confused; it can hardly be erroneous. But logic cannot stop here. It is bound, by its very nature, to push its research into the manner of reality itself, and in doing so can no longer confine its attention to mere forms of language but must inevitably consider how and what we think (*Peirce 1982–*, Vol. ii, p. 165).

Like Kant and the Schoolmen he saw metaphysics as being closely allied to logic; and indeed his conception of philosophy as based on a public, non-psychologistic study of scientific reasoning was bound to tie logic, natural science, and philosophy into a tight knot. No one of these activities had priority in Peirce's thinking. His work in estimating the margin of error in the measurement of physical phenomena suggested questions about probabilistic reasoning; his pragmatism and fallibilism were tied to the idea of ever-closer approximations to scientific truth; his evolutionary conception of the growth of knowledge caused him to investigate the history of science and philosophy; his logical and scientific research reacted on his philosophy—on his conception of scientific method, language, truth, logic, reality; and his work in philosophy and the history of science in turn suggested new lines of research in logic and the natural sciences.

This is an important point, and distinguishes Peirce from those logicians who thought of logic, in Russell's phrase, as 'those general statements which can be made concerning everything without mentioning any one thing or predicate or relation' (*Russell 1914*). Peirce emphatically rejected this conception, detecting in it not only the seeds of paradox but an excessve generality that would obliterate the differences between the styles of reasoning in the several sciences. Not for him the project of providing a logical foundation for all of knowledge:

> the main advantages which we have to expect from logical studies are rather, first, clear disentanglements of reasoning which is felt to be cogent without our precisely knowing wherein the elenchus lies—such, for instance, as the reasoning of elementary geometry; and, second, broad and philosophical *aperçus* covering several sciences, by which we are made to see how the methods used in one science may be made to apply to another (*Peirce 1982–*, Vol. iv, p. 239).

This conception of logic and his deep knowledge of the history of science had various advantages. It led Peirce to consider probabilistic and counterfactual and modal reasoning, and also gave him a balanced sense of the benefits to be expected from the new logic. The new logic, he said, could be expected to improve our theoretical understanding of the reasoning employed by mathematicians and scientists; but there was no more reason to believe that it would itself be useful *within* the sciences than that a theory of muscular action would be a useful tool for winning foot-races. On this crucial point regarding the likely contribution of logic to mathematics and philosophy, Peirce displayed a greater sobriety and maturity than many of his contemporaries, whose exaggerated claims for the new logic he dismissed as 'infantine'.

Indeed, he warned that formal logic could be a positive nuisance—a hammer that would make every problem resemble a nail. 'If I am asked,' he wrote,

> whether the study of logic really makes a man reason better, I am obliged to confess that in most cases it has the directly opposite result. It makes a captious reasoner, who appears to himself and others to reason in a superior manner, who is consecutive in his thoughts, self-conscious, free from contradictions, but whose thought is not nearly so good as that of the perfectly untutored person for the purpose of finding out what is true. The average uneducated woman has a mind far better adapted to that purpose than the average graduate of Oxford; and the reason seems to be that the latter has been sophisticated with logic,—not directly from treatises but from conversation and reading. The young man who makes a course in logic with a feeling of satisfaction at having gained something, has by that token certainly got nothing but a mental morbid diathesis of which he is unlikely ever to be cured. It is the object of this little book, so to inoculate the student with an innocent form of the malady that he may be exempt from any malignant seizure (*Peirce 1982–*, Vol. iv, p. 401).

V

In 1879 Peirce received his first and only academic appointment, a part-time lectureship in logic at Johns Hopkins University. While at Johns Hopkins, he was closely associated with the mathematics department, and in particular with its chairman, the algebraist J.J. Sylvester; among his students in the philosophy department were John Dewey and Thorstein Veblen. (Dewey, however, stopped coming to Peirce's lectures after one semester, finding the mathematics too difficult.)

During this time Peirce made several fundamental contributions to logic. He stated the procedure for reducing formulae of the sentential calculus to conjunctive and disjunctive normal form; carried out some early investigations into the logical foundations of the concept of number; demonstrated that the Boolean sentential calculus can be obtained from the single connective of joint denial ('neither p nor q'); and introduced truth-values, and used them to provide a decision procedure for the sentential calculus. In 1885, independently of Frege but six years after the *Begriffschrift*, he produced a system of quantification theory with an elegant and flexible notation that was in use well into the twentieth century. He discussed the rules for transforming a quantified formula into prenex normal form, distinguished first-order logic from second-order, stated the modern second-order definition of identity, and gave the definition of a finite set as one which cannot be put into a one-to-one correspondence with any proper subset. Peirce also during his years at Johns Hopkins continued to refine his pragmatism; made two long trips to Europe to pursue his pendulum experiments; wrote a handful of papers on quaternions and linear algebra; and published one of the first papers in America on experimental psychology (*Peirce and Jastrow 1884*).

By the middle of the 1880s Peirce had more than surpassed his father's accomplishments. He had acquired a formidable international reputation in astronomy, geodesy, and logic, and was elected to numerous domestic and foreign learned societies; his brilliance was acknowledged on all sides, and he seems to have been an inspiring teacher. Nevertheless, in 1884 the university authorities at Johns Hopkins went to some trouble to remove Peirce from his lectureship, a step which involved dismissing the entire philosophy department and re-hiring everybody except Peirce. The reasons for this action are not wholly clear. But Peirce clearly inspired dislike and jealousy among his colleagues; and the scandal caused by his divorce and remarriage in 1883 probably provided the stimulus or the opportunity to remove him.

Soon after his dismissal from Hopkins, Peirce quarrelled with the new director of the Coast Survey, and found himself again dismissed, thereby ending his career as a natural scientist. Unable to find work, Peirce, having inherited a small bequest from his father, withdrew to a farmhouse in Milford, Pennsylvania, where he continued to pursue his studies in logic and metaphysics; he was forty-nine, and for the rest of his life would have no steady employment. Almost everything Peirce wrote in these years he regarded as being connected to his discoveries in logic. He wrote voluminously on pragmatism, scientific method, and the philosophy of language. He attempted to derive a system of metaphysics from his logic of relatives. He explored the connections between logic, probability, and inductive reasoning. He delved into medieval astronomy, non-Euclidean geometry, the four-colour conjecture, Cantor's transfinite numbers, infinitesimals, and the structure of the continuum. He sketched the first known design for an electric switching circuit that could perform logical and arithmetical operations (*Fisch 1980*, p. 272). He developed a pictorial manner of exploring logical relationships—the system of 'logical graphs'. He worked on modal logic, probabilistic logic, inductive logics, and logics with more than two truth-values. Most of this work was unpublished during Peirce's lifetime; the manuscripts were bequeathed to Harvard University, and extracts from them were published in the *Collected papers* in the 1930s.

During this long last period of his life Peirce lived in extreme poverty, and indeed only the charity of friends—above all, the steadfast William James—kept him from actual starvation; Peirce's letters are interspersed with observations on the difficulty of writing when the temperature in his house was below freezing. He became a recluse, going for weeks without human contact. As he himself realized, this way of life had an effect on his writing, which became increasingly strange and introverted; and amid the logical and scientific writings there are outbursts against rival positions, a good deal of self-praise, and rambling speculations about the metaphysics of 'thirdness' and 'evolutionary love'. Most of this material (which James justifiably described as 'flashes of brilliant light relieved against Cimmerian darkness') has never been published, and will fill some twenty volumes of the thirty-volume edition of Peirce that is now under way. If the volumes that have hitherto appeared are any guide, the simple expedient of putting Peirce's writings into chronological order and publishing the manuscripts entire should make it easier to follow the train of his reasoning, and diminish the amount of Cimmerian darkness that surrounds his thought.

It would be futile to attempt a summary of Peirce's tangled thought, or to pretend that everything he wrote deserves equally close study. Much of his *œuvre* is of merely historical interest. As a natural scientist he was a distinguished figure, but hardly a Helmholtz or a Maxwell, let alone an Einstein; many of his discoveries were partial anticipations of work that was done better by others. He never welded his views into a system, and his writing is frequently slapdash.

On the other hand, Peirce had a wider range of interests, a deeper knowledge of the history of philosophy, a greater command of the scientific culture of his day than any philosopher since Leibniz; and his tangledness, his 'snarl of twine' quality is as often as not a consequence of his immense erudition and his refusal to be content with oversimplifications. 'You see here,' Peirce said in some lectures on medieval logic, 'how differently Scotus and Ockham regard the same question, how much more simple and lucid Ockham's view is, and how much more certain Scotus's complex theory is to take into account all the facts than Ockham's simple one.' If Peirce lacks polish and literary grace, he offers instead great breadth and a sense of complexity—an astonishing talent for generating questions, for seeing the interconnections between disparate fields, for pointing out the weaknesses in his own theories. His writings are teeming with incisive observations on logic, methodology, and the history of philosophy: a rich mine of insights and suggestions, the residue of a life spent shuttling as few have done between logic, philosophy, and the natural sciences. Readers who study him with the patience he demands will not go home empty-handed.

A. *FROM* LINEAR ASSOCIATIVE ALGEBRA (*BENJAMIN PEIRCE 1870*)

This selection comprises the introductory material from C.S. Peirce's 1881 reprinting of his father's memoir on linear associative algebra. Benjamin Peirce was one of the earliest champions of William Rowan Hamilton's quaternions, and was already lecturing on them to his class at Harvard in 1848. His work on linear associative algebra summarizes all the algebras of hypercomplex numbers that were known at the time, quaternions included; it was the most original piece of mathematics yet to come out of the United States—elegant in its presentation, and one of the first mathematical memoirs to treat of *algebras* in the plural, explicitly recognizing that there were a great variety of algebraic systems, rather than a single, all-encompassing 'algebra'. The portion reprinted here contains the introductory material, in which Peirce states his conception of mathematics and some of the fundamental definitions for his algebraic investigations.

References to *B. Peirce 1870* should be to the section numbers, which appear in the original text.

---

1. Mathematics is the science which draws necessary conclusions.

This definition of mathematics is wider than that which is ordinarily given, and by which its range is limited to quantitative research. The ordinary definition, like those of other sciences, is objective; whereas this is subjective. Recent investigations, of which quaternions is the most noteworthy instance, make it manifest that the old definition is too restricted. The sphere of mathematics is here extended, in accordance with the derivation of its name, to all demonstrative research, so as to include all knowledge strictly capable of dogmatic teaching. Mathematics is not the discoverer of laws, for it is not induction; neither is it the framer of theories, for it is not hypothesis; but it is the judge over both, and it is the arbiter to which each must refer its claims; and neither law can rule nor theory explain without the sanction of mathematics. It deduces from a law all its consequences, and develops them into the suitable form for comparison with observation, and thereby measures the strength of the argument from observation in favor of a proposed law or of a proposed form of application of a law.

Mathematics, under this definition, belongs to every enquiry, moral as well as physical. Even the rules of logic, by which it is rigidly bound, could not be deduced without its aid. The laws of argument admit of simple statement, but they must be curiously transposed before they can be applied to the living speech and verified by observation. In its pure and simple form the syllogism cannot be directly compared with all experience, or it would not have required an Aristotle to discover it. It must be transmuted into all the possible shapes in which reasoning loves to clothe itself. The transmutation is the mathematical process in the establishment of the law. Of some sciences, it is so large a portion that they have been quite abandoned to the mathematician,—which may not have been altogether to the advantage of philosophy. Such is the case with geometry and analytic mechanics. But in many other sciences, as in all those of mental philosophy and most of the branches of natural history, the deductions are so immediate and of such simple construction, that it is of no practical use to separate the mathematical portion and subject it to isolated discussion.

2. The branches of mathematics are as various as the sciences to which they belong, and each subject of physical enquiry has its appropriate mathematics. In every form of material manifestation, there is a corresponding form of human thought, so that the human mind is as wide in its range of thought as

the physical universe in which it thinks. The two are wonderfully matched. But where there is a great diversity of physical appearance, there is often a close resemblance in the processes of deduction. It is important, therefore, to separate the intellectual work from the external form. Symbols must be adopted which may serve for the embodiment of forms of argument, without being trammeled by the conditions of external representation or special interpretation. The words of common language are usually unfit for this purpose, so that other symbols must be adopted, and mathematics treated by such symbols is called *algebra*. Algebra, then, is formal mathematics.

3. All relations are either qualitative or quantitative. Qualitative relations can be considered by themselves without regard to quantity. The algebra of such enquiries may be called logical algebra, of which a fine example is given by Boole.

Quantitative relations may also be considered by themselves without regard to quality. They belong to arithmetic, and the corresponding algebra is the common or arithmetical algebra.

In all other algebras both relations must be combined, and the algebra must conform to the character of the relations.

4. The symbols of an algebra, with the laws of combination, constitute its *language*; the methods of using the symbols in the drawing of inferences is its *art*; and their interpretation is its *scientific application*. This three-fold analysis of algebra is adopted from President Hill, of Harvard University, and is made the basis of a division into books.

BOOK I*

THE LANGUAGE OF ALGEBRA

5. The language of algebra has its alphabet, vocabulary, and grammar.

6. The symbols of algebra are of two kinds: one class represent its fundamental conceptions and may be called its *letters*, and the other represent the relations or modes of combination of the letters and are called *the signs*.

7. The *alphabet* of an algebra consists of its letters; the *vocabulary* defines its signs and the elementary combinations of its letters; and the *grammar* gives the rules of composition by which the letters and signs are united into a complete and consistent system.

The Alphabet

8. Algebras may be distinguished from each other by the number of their independent fundamental conceptions, or of the letters of their alphabet. Thus an algebra which has only one letter in its alphabet is a *single* algebra; one which has two letters is a *double* algebra; one of three letters a *triple* algebra; one of four letters a *quadruple* algebra, and so on.

---

* Only this book was ever written. [C.S.P.]

This artificial division of the algebras is cold and uninstructive like the artificial Linnean system of botany. But it is useful in a preliminary investigation of algebras, until a sufficient variety is obtained to afford the material for a natural classification.

Each fundamental conception may be called a *unit*; and thus each unit has its corresponding letter, and the two words, unit and letter, may often be used indiscriminately in place of each other, when it cannot cause confusion.

9. The present investigation, not usually extending beyond the sextuple algebra, limits the demand of the algebra for the most part to six letters; and the six letters, i, j, k, l, m and n, will be restricted to this use except in special cases.

10. *For any given letter another may be substituted*, provided a new letter represents a combination of the original letters of which the replaced letter is a necessary component.

For example, any combination of two letters, which is entirely dependent for its value upon both of its components, such as their sum, difference, or product, may be substituted for either of them.

This *principle of the substitution of letters* is radically important, and is a leading element of originality in the present investigation; and without it, such an investigation would have been impossible. It enables the geometer to analyse an algebra, reduce it to its simplest and characteristic forms, and compare it with other algebras. It involves in its principle a corresponding substitution of *units* of which it is in reality the formal representative.

There is, however, no danger in working with the symbols, irrespective of the ideas attached to them, and the consideration of the change of the original conceptions may be safely reserved for the *book of interpretation*.

11. In making the substitution of letters, the original letter will be preserved with the distinction of a subscript number.

Thus, for the letter i there may successively be substituted i_1, i_2, i_3, etc. In the final forms, the subscript numbers can be omitted, and they may be omitted at any period of the investigation, when it will not produce confusion.

It will be practically found that these subscript numbers need scarcely ever be written. They pass through the mind, as a sure ideal protection from erroneous substitution, but disappear from the writing with the same facility with which those evanescent chemical compounds, which are essential to the theory of transformation, escape the eye of the observer.

12. A *pure* algebra is one in which every letter is connected by some indissoluble relation with every other letter.

13. When the letters of an algebra can be separated into two groups, which are mutually independent, it is a *mixed algebra*. It is mixed even when there are letters common to the two groups, provided those which are not common to the two groups are mutually independent. Were an algebra employed for the simultaneous discussion of distinct classes of phenomena, such as those of sound and light, and were the peculiar units of each class to have their appropriate letters, but were there no recognized dependence of the phenomena upon

each other, so that the phenomena of each class might have been submitted to independent research, the one algebra would be actually a mixture of two algebras, one appropriate to sound, the other to light.

It may be farther observed that when, in such a case as this, the component algebras are identical in form, they are reduced to the case of one algebra with two diverse interpretations.

The Vocabulary

14. Letters which are not appropriated to the alphabet of the algebra* may be used in any convenient sense. But it is well to employ *the small letters* for expressions of common algebra, and *the capital letters* for those of the algebra under discussion.

There must, however, be exceptions to this notation; thus the letter D will denote the derivative of an expression to which it is applied, and Σ the summation of cognate expressions, and other exceptions will be mentioned as they occur. Greek letters will generally be reserved for angular and functional notation.

15. The three symbols $\mathfrak{J}$, ∂, and $\mathfrak{S}$ will be adopted with the signification

$$\mathfrak{J} = \sqrt{-1}$$

$$\partial = \text{the ratio of circumference to diameter of circle} = 3.1415926536$$

$$\mathfrak{S} = \text{the base of Naperian logarithms} = 2.7182818285,$$

which gives the mysterious formula

$$\mathfrak{J}^{-\mathfrak{J}} = \sqrt{\mathfrak{S}^{\partial}} = 4.810477381.$$

16. All the signs of common algebra will be adopted; but any signification will be permitted them which is not inconsistent with their use in common algebra; so that, if by any process an expression to which they refer is reduced to one of common algebra, they must resume their ordinary signification.

17. The sign =, which is called that of equality, is used in its ordinary sense to denote that the two expressions which it separates are the same whole, although they represent different combinations of parts.

18. The signs > and < which are those of inequality, and denote “more than” or “less than” in quantity, will be used to denote the relations of a whole to its part, so that the symbol which denotes the part shall be at the vertex of the angle, and that which denotes the whole at its opening. This involves the proposition that the smaller of the quantities is included in the class expressed by the larger. Thus

$$B < A \text{ or } A > B$$

denotes that A is a whole of which B is a part, so that all B is A.†

---

* See §9.

† The formula in the text implies, also, that some A is not B. [C.S.P.]

If the usual algebra had originated in qualitative, instead of quantitative, investigations, the use of the symbols might easily have been reversed; for it seems that all conceptions involved in A must also be involved in B, so that B is more than A in the sense that it involves more ideas.

The combined expression

$$B > C < A$$

denotes that there are quantities expressed by C which belong to the class A and also to the class B. It implies, therefore, that some B is A and that some A is B.* The intermediate C might be omitted if this were the only proposition intended to be expressed, and we might write

$$B > < A.$$

In like manner the combined expression

$$B < C > A$$

denotes that there is a class which includes both A and B,† which proposition might be written

$$B < > A.$$

19. A vertical mark drawn through either of the preceding signs reverses its signification. Thus

$$A \neq B$$

denotes that B and A are essentially different wholes;

$$A \not> B \text{ or } B \not< A$$

denotes that all B is not A,‡ so that if they have only quantitative relations, they must bear to each other the relation of

$$A = B \text{ or } A < B.$$

20. The sign $+$ is called *plus* in common algebra and denotes *addition*. It may be retained with the same name, and the process which it indicates may be called addition. In the simplest cases it expresses a mere mixture, in which the elements preserve their mutual independence. If the elements cannot be mixed without mutual action and a consequent change of constitution, the mere union is still expressed by the sign of addition, although some other symbol is required to express the character of the mixture as a peculiar compound having properties different from its elements. It is obvious from the simplicity of the

* This, of course, supposes that C does not vanish. [C.S.P.]

† The universe will be such a class unless A or B is the universe. [C.S.P.]

‡ The general interpretation is rather that either A and B are identical or that some B is not A. [C.S.P.]

union recognized in this sign, that the order of the admixture of the elements cannot affect it; so that it may be assumed that

$$A + B = B + A$$

and

$$(A + B) + C = A + (B + C) = A + B + C.$$

21. The sign − is called *minus* in common algebra, and denotes *subtraction*. Retaining the same name, the process is to be regarded as the reverse of addition; so that if an expression is first added and then subtracted, or the reverse, it disappears from the result; or, in algebraic phrase, it is *canceled*. This gives the equations

$$A + B - B = A - B + B = A$$

and

$$B - B = 0.$$

The sign minus is called the negative sign in ordinary algebra, and any term preceded by it may be united with it, and the combination may be called a *negative term*. This use will be adopted into all the algebras, with the provision that the derivation of the word negative must not transmit its interpretation.

22. The sign × may be adopted from ordinary algebra with the name of the sign of *multiplication*, but without reference to the meaning of the process. The result of multiplication is to be called the *product*. The terms which are combined by the sign of multiplication may be called *factors*; the factor which precedes the sign being distinguished as the *multiplier*, and that which follows it being the *multiplicand*. The words multiplier, multiplicand, and product, may also be conveniently replaced by the terms adopted by Hamilton, of *facient*, *faciend*, and *factum*. Thus the equation of the product is

multiplier × multiplicand = product; *or* facient × faciend = factum.

When letters are used, the sign of multiplication can be *omitted* as in ordinary algebra.

23. When an expression used as a factor in certain combinations gives a product which vanishes, it may be called in those combinations a *nilfactor*. Where as the multiplier it produces vanishing products it is *nilfacient*, but where it is the multiplicand of such a product it is *nilfaciend*.

24. When an expression used as a factor in certain combinations overpowers the other factors and is itself the product, it may be called an *idemfactor*. When in the production of such a result it is the multiplier, it is *idemfacient*, but when it is the multiplicand it is *idemfaciend*.

25. When an expression raised to the square or any higher power vanishes, it may be called *nilpotent*; but when, raised to a square or higher power, it gives itself as the result, it may be called *idempotent*.

The defining equation of nilpotent and idempotent expressions are respectively $A^n = 0$, and $A^n = A$; but with reference to idempotent expressions, it will always be assumed that they are of the form

$$A^2 = A,$$

unless it be otherwise distinctly stated.

26. *Division* is the reverse of multiplication, by which its results are verified. It is the process for obtaining one of the factors of a given product when the other factor is given. It is important to distinguish the position of the given factor, whether it is facient or faciend. This can be readily indicated by combining the sign of multiplication, and placing it before or after the given factor just as it stands in the product. Thus when the multiplier is the given factor, the correct equation of division is

$$\text{quotient} = \frac{\text{dividend}}{\text{divisor } \times}$$

and the equation of verification is

$$\text{divisor} \times \text{quotient} = \text{dividend}.$$

But when the multiplicand is the given factor, the equation of division is

$$\text{quotient} = \frac{\text{dividend}}{\times \text{ divisor}}$$

and the equation of verification is

$$\text{quotient} \times \text{divisor} = \text{dividend}.$$

27. Exponents may be introduced just as in ordinary algebra, and they may even be permitted to assume the forms of the algebra under discussion. There seems to be no necessary restriction to giving them even a wider range and introducing into one algebra the exponents from another. Other signs will be defined when they are needed.

The definition of the fundamental operations is an essential part of the vocabulary, but as it is subject to the rules of grammar which may be adopted, it must be reserved for special investigation in the different algebras.

The Grammar

28. Quantity enters as a form of thought into every inference. It is always implied in the syllogism. It may not, however, be the direct object of inquiry; so that there may be logical and chemical algebras into which it only enters accidentally, agreeably to §1. But where it is recognized, it should be received in its most general form and in all its variety. The algebra is otherwise unnecessarily restricted, and cannot enjoy the benefit of the most fruitful forms of philosophical discussion. But while it is thus introduced as a part of the formal algebra, it is *subject to every degree and kind of limitation in its interpretation.*

The free introduction of quantity into an algebra does not even involve the reception of its unit as one of the independent units of the algebra. But it is probable that without such a unit, no algebra is adapted to useful investigation. It is so admitted into quaternions, and its admission seems to have misled some philosophers into the opinion that quaternions is a triple and not a quadruple algebra. This will be the more evident from the form in which quaternions first present themselves in the present investigation, and in which the unit of quantity is not distinctly recognizable without a transmutation of the form.*

29. The introduction of quantity into an algebra naturally carries with it, not only the notation of ordinary algebra, but likewise many of the rules to which it is subject. Thus, when a quantity is a factor of a product, it has the same influence whether it be facient or faciend, so that with the notation of §14, there is the equation

$$Aa = aA,$$

and in such a product, the quantity a may be called the *coefficient.*

In like manner, terms which only differ in their coefficients, may be added by adding their coefficients; thus,

$$(a \pm b)A = aA \pm bA = Aa \pm Ab = A(a \pm b).$$

30. The exceeding simplicity of the conception of an equation involves the identity of the equations

$$A = B \text{ and } B = A$$

and the substitution of B for A in every expression, so that

$$MA \pm C = MB \pm C,$$

or that, *the members of an equation may be mutually transposed or simultaneously increased or decreased or multiplied or divided by equal expressions.*

31. How far the principle of §16 limits the extent within which the ordinary symbols may be used, cannot easily be decided. But it suggests limitations which may be adopted during the present discussion, and leave an ample field for curious investigation.

The distributive principle of multiplication may be adopted; namely, the principle that the product of an algebraic sum of factors into or by a common factor, is equal to the corresponding algebraic sum of the individual products of the various factors into or by the common factor; and it is expressed by the equations

* Hamilton's total exclusion of the imaginary of ordinary algebra from the calculus as well as from the interpretation of quaternions will not probably be accepted in the future development of this algebra. It evinces the resources of his genius that he was able to accomplish his investigations under these trammels. But like the restrictions of the ancient geometry, they are inconsistent with the generalizations and broad philosophy of modern science. With the restoration of the ordinary imaginary, quaternions becomes Hamilton's biquaternions. From this point of view, all the algebras of this research would be called bi-algebras. But with the ordinary imaginary is involved a vast power of research, and the distinction of names should correspond; and the algebra which loses it should have its restricted nature indicated by such a name as that of a *semi-algebra*.

$$(A \pm B)C = AB \pm BC.$$

$$C(A \pm B) = CA \pm CB.$$

32. *The associative principle of multiplication* may be adopted; namely, that the product of successive multiplications is not affected by the order in which the multiplications are performed, provided there is no change in the relative position of the factors; and it is expressed by the equations

$$ABC = (AB)C = A(BC).$$

This is quite an important limitation, and the algebras which are subject to it will be called *associative*.

33. The principle that the value of a product is not affected by the relative position of the factors is called *the commutative principle*, and is expressed by the equation

$$AB = BA.$$

This principle is *not* adopted in the present investigation.

34. An algebra in which every expression is reducible to the form of an algebraic sum of terms, each of which consists of a single *letter* with a quantitative coefficient, is called a *linear algebra*.* Such are all the algebras of the present investigation.

35. Wherever there is a limited number of independent conceptions, a linear algebra may be adopted. For a combination which was not reducible to such an algebraic sum as those of linear algebra, would be to that extent independent of the original conceptions, and would be an independent conception additional to those which were assumed to constitute the elements of the algebra.

36. An algebra in which there can be complete interchange of its independent units, without changing the formulae of combination, is a *completely symmetrical algebra*; and one in which there may be a partial interchange of its units is *partially symmetrical*. But the term symmetrical should not be applied, unless the interchange is more extensive than that involved in the distributive and commutative principles. An algebra in which the interchange is effected in a certain order which returns into itself is a cyclic algebra.

Thus, quaternions is a cyclic algebra, because in any of its fundamental equations, such as

$$i^2 = -1$$

$$ij = -ji = k$$

$$ijk = -1$$

there can be an interchange of the letters in the order i, j, k, i, each letter being

* In the various algebras of De Morgan's "Triple Algebra," the distributive, associative and commutative principles were all adopted, and they were all linear. [De Morgan's algebras are "semi-algebras." See Cambridge Phil. Trans., viii, 241.] [C.S.P.]

changed into that which follows it. The double algebra in which

$$i^2 = i, \quad ij = i$$

$$j^2 = j, \quad ji = j$$

is cyclic because the letters are interchangeable in the order i, j, i. But neither of these algebras is commutative.

37. When an algebra can be reduced to a form in which all the letters are expressed as powers of some one of them, it may be called a *potential algebra*. If the powers are all squares, it may be called *quadratic*; if they are cubes, it may be called *cubic*; and similarly in other cases.

B. NOTES ON BENJAMIN PEIRCE'S *LINEAR ASSOCIATIVE ALGEBRA* (*PEIRCE 1976*)

The following comments by C.S. Peirce on the previous selection were first published in *Peirce 1976* (Vol. iii, pp. 526–8).

§1. This definition of mathematics was entirely novel in 1870, when it was first put forth. Since then all important writers on the foundations of mathematics have been led to similar definitions, no doubt more or less influenced, directly or indirectly, by Peirce. After broad and deep study of the question, I am definitively convinced that the definition here given is better than any of the modifications of it. The chief points to be considered are as follows:

1st, Is not the definition too broad because mathematics is limited to deductions from a certain kind of hypotheses; namely, from such as are perfectly definite and somewhat complicated?

2nd, Is not the definition too broad, because mathematical reasoning is schematic while philosophical deductions deal with pure concepts?

3rd, Is not the definition too broad in that all sciences draw necessary conclusions? Should we not say that mathematics is the logic of necessary inference?

4th, Is not the definition too narrow in that some of the highest achievements of mathematics have consisted in the formation of mathematical hypotheses?

5th, Is not the definition too narrow in that induction, hypotheses, and every variety of reasoning is used in mathematics?

6th, Is it not, after all, true that when the conception of quantity is sufficiently generalized, all mathematics relates to quantity and is distinguished from other sciences by this?

The first question is suggested in the Article *Mathematics* in the 9th Edition of the *Encyclopaedia Britannica*, and elsewhere. But this question really is whether an improved definition is forthcoming; and it does not appear that anything more than a vague distinction between much and little is given, or can be given, to support an objection in this sense. No doubt, when hypotheses are simple, as vague hypotheses from which anything can be deduced are apt to be, people work out their consequences for themselves, while in cases where they are too complex for them to handle they consult mathematicians. But no more precise distinction has been offered; and the utility of it, if it should be offered, is very doubtful.

The second question is asked by nobody. But blustering intuitionalists, as ignorant of logic as they are arrogant, still maintain the opinion that there are conceptual deductions of a radically different nature from mathematical deductions. The only tenable distinction is that the conceptual deductions are loose and superficial reasonings that are inconclusive, while those of the mathematicians are accepted by every understanding that grasps them. My thirty years' study of the logic of relations would have brought to light any distinction that could have existed between mathematical reasonings and other deductions. But I only find distinctions of degree, especially the distinction that mathematicians seldom reason inconclusively and metaphysicians seldom conclusively.

The third question is amply justified by the fact that Dedekind holds mathematics to be a branch of logic. At the time my father was writing this book, I was writing my paper on the logic of relations that was published in the 9th volume of the *Memoirs of the American Academy*. There was no collaboration, but there were frequent conversations on the allied subjects, especially about the algebra. The only way in which I think that anything I said influenced anything in my father's book (except that it was partly on my urgent prayer that he undertook the research) was that when at one time he seemed inclined to the opinion which Dedekind long afterward embraced, I argued strenuously against it, and thus he came to take the middle ground of his definition. In truth, no two things could be more directly opposite than the cast of mind of the logician and that of the mathematician. It is almost inconceivable that a man should be great in both ways. Leibniz came the nearest to it. He was, indeed, a great logician, for all his nominalism, which clung to him like the coat of Nessus, but which he more and more surmounted as the idea of continuity gained strength in his mind. But his mathematical power, though far from being mean, was inferior to that of either of the two Bernouillis. The mathematician's interest in a reasoning is as a means of solving problems—both a problem that has come up and possible problems like it that have not yet come up. His whole endeavor is to find short cuts to the solution. The logician, on the other hand, is interested in picking a method to pieces and in finding what its essential ingredients are. He cares little how these may be put together to form an effective method, and still less for the solution of any particular problem. In short, logic is the theory of all reasoning, while mathematics is the practice of a particular kind of reasoning. Mathematics might be called an art instead of a science were it not that the last achievement that it has in view is an achievement of knowing.

The fourth question is one which has exercised me considerably. But I have now obtained a satisfactory solution of it. It is true that there is great exercise of intellect in framing a mathematical hypothesis like that of the theory of functions. But as a purely arbitrary hypothesis, there is no element of cognition in it, and consequently it has nothing to do with the nature of mathematics *as a science*.

The fifth question is raised by a remark of Sylvester. But the reply is that all those other kinds of reasoning are in mathematics merely ancillary and provisional. Neither Sylvester nor any mathematician is satisfied until they have been swept away and replaced by demonstrations.

The sixth question is suggested by modern studies of the nature of quantity. The reply to it is that it is so far from being possible to refute B. Peirce's definition by any reply to this question that no justification of the old definition can be based on such reply except so far as that reply justifies the new definition.

§3. This distinction is merely a distinction between a system of quantity in which there are only two values and a system of quantity in which are more than two values.

§12, 13. Whether or not mixed algebras ought to be excluded is a question of what one's aim may be. Certainly, some mixed algebras are very interesting. It will be found that, after all, no inconsiderable percentage of my father's algebras are mixed, according to the definition here given.

§28. The argument in favor of allowing the coefficients of the algebra to be imaginary is certainly inconclusive. Every logician must admit that imaginary quantity has two dimensions and therefore should be recognized as double. To say that where quantity is introduced "it should be received in its most general form and in all its variety", if sound would forbid our separating an imaginary into a modulus and argument. It is evident that every form of algebra which will result from making the coefficients imaginary would coincide with some algebra that would result from restricting them to being real; but it is not evident that the converse is true. Therefore, nothing can be gained but much may be lost by allowing the coefficients to be imaginary.

§41. It may be observed that no argument is introduced to prove the advantage of thus taking the question of whether there is an idempotent expression or not as the first question to be asked concerning an algebra; and the fact that there is not shows the difference between the mathematical and the scientifically logical type of mind.

C. ON THE LOGIC OF NUMBER (*PEIRCE 1881*)

In 1870 Peirce published his 'Description of a notation for the logic of relatives' (*Peirce 1870*). This was the first and most influential of his series of papers on the logic of relations; Peirce in later life seems to have regarded it as his most important contribution to logic.[a] Characteristically for Peirce, it had roots in philosophy, in logic, and in mathematics. Peirce had become interested in relations as a consequence both of his reading of the Schoolmen and of his investigations into the new table of the categories; and he knew, at the latest from reading *De Morgan 1864b*, that the Aristotelian syllogism is incapable of handling relative terms. In addition, Benjamin Peirce had generalized the work of the British algebraists on complex numbers and quaternions, thereby producing a bounty of algebraical structures with unusual multiplication tables. (His monograph on linear associative algebras also appeared in 1870; Charles Peirce was to publish a second, annotated edition of his father's work, *B. Peirce 1881*.) It seemed to Peirce that these new algebras could be applied to logic much as Boole had successfully applied operator methods in his *Mathematical analysis of logic* (*Boole 1847*, reproduced above). Commenting on De Morgan's *On the syllogism, No. IV, and on the logic of relations* (*De Morgan 1864b*), Peirce wrote:

> This system still leaves something to be desired. Moreover, Boole's logical algebra has such singular beauty, so far as it goes, that it is interesting to inquire whether it cannot be extended over the whole realm of formal logic, instead of being restricted to that simplest and least useful part of the subject, the logic of absolute terms, which, when he wrote, was the only formal logic known (*Peirce 1870*, §1; reprinted in *Peirce 1982–*, Vol. ii, p. 359).

In his *1870*, Peirce studied the composition of relations with each other and with class-terms. For instance, if 'w' signifies 'woman' and '*l*' signifies 'lover of . . .' and '*s*' signifies 'servant of . . .', then in Peirce's system the composition '*sl*w' denotes the servant of the lover of a woman. Treating this operation of composition as *logical multiplication*, Peirce introduced a complicated *logical addition*: '$l + s$' signifies 'lover of every non-servant of . . .'. He then worked out the principal laws for the resulting abstract algebraic system, and showed in the process that many of the linear associative algebras in his father's memoir *B. Peirce 1870* could be defined in terms of what he called 'elementary relatives'. (In his *1875* Peirce was able to prove that *all* linear associative algebras can be so expressed.) For all its merits, Peirce's *1870* is frequently obscure, and the notation is exceedingly awkward. For instance, Peirce uses the notation l_0 to designate 'lover of itself', and l_∞ to designate 'lover of something'; worse, his paper is filled with logarithms, exponentiations, and power-series expansions whose logical significance is difficult to fathom. In retrospect, it is clear that his paper demanded the theory of quantification; but it was nevertheless the first important attempt to extend Boole's logic to the logic of relations.[b]

In the summer of 1870, Peirce travelled to Europe to observe the solar eclipse

[a] See, for instance, his remarks in *Peirce 1983*, Vol. i, p. 143.

[b] For further details on *Peirce 1870*, see *Martin 1978*; for the subsequent development of the logic of relations, see *Tarski 1941*.

for the Coast Survey. *En route* to Sicily he passed through England, where he met De Morgan, Jevons, and Clifford; he presented them with copies of *Peirce 1870*. This article, despite its shortcomings, made a stir among the logicians of Europe. (Clifford, indeed, was later to declare that 'Charles Peirce ... is the greatest living logician, and the second man since Aristotle who has added to the subject something material, the other man being George Boole, author of *The laws of thought*' (*Peirce 1982–*, Vol. ii, p. xxx).)

During his years at Johns Hopkins, Peirce worked principally on logic, to which he made several fundamental contributions. His *1880a* states the procedure for reducing formulae of the sentential calculus to conjunctive and disjunctive normal form. His unpublished *1880b* demonstrated that the sentential calculus can be obtained from the single connective of joint denial ('neither *p* nor *q*'). (Henry Sheffer's independent discovery of this fact appeared in *Sheffer 1913*. Peirce's manuscript was first published in the *Collected papers* in 1933.) In his *1881*, 'On the logic of number', Peirce examined the foundations of arithmetic, analysing the natural numbers in terms of discrete, linearly ordered sets without a maximum element. He gave informal recursive definitions of addition and multiplication,[c] proved that both operations were associative and commutative, and proved the distributive principle. This paper was one of the earliest attempts to explore the logical foundations of the concept of number, and it contains some of the central ideas in Dedekind's influential memoir, *Was sind und was sollen die Zahlen?* (*Dedekind 1888*). But despite Peirce's exaggerated claims (*Peirce 1931–58*, Vol. iv, p. 268) that Dedekind's book 'proves no difficult theorem that I had not proved or published years before', his paper does not have the mathematical power of Dedekind's analysis. Missing from Peirce's paper are the rudiments of set theory (*Dedekind 1888*, §1), the general theorem on recursive definitions (§9), the isomorphism theorem (§10), and the Peano axioms (§10). And despite the title of his article and some suggestive remarks in the first paragraphs, Peirce (in contrast to Frege and Dedekind) did not attempt to derive mathematics from logic (or from logic and set theory).[d] In consequence, many of the number-theoretical accomplishments of the logicists eluded his grasp.

References to *Peirce 1881* should be to the section headings.

---

[c] Hermann Grassmann had already given a similar analysis in his *Lehrbuch der Arithmetik* (*H. Grassmann 1861*). Peirce appears not to have known of this work.

[d] Indeed, Peirce often asserted that logic is grounded in mathematics, rather than *vice versa*. For example (*Peirce 1931–58*, Vol. i, p. 112):

> It might, indeed, very easily be supposed that even pure mathematics itself would have need of one department of philosophy; that is to say, of logic. Yet a little reflection would show, what the history of science confirms, that that is not true. Logic will, indeed, like every other science, have its mathematical parts. There will be a mathematical logic just as there is a mathematical physics and a mathematical economics. If there is any part of logic of which mathematics stands in need—logic being a science of fact and mathematics only a science of the consequences of hypotheses—it can only be that very part of logic which consists merely in an application of mathematics, so that the appeal will be, not of mathematics to a prior science of logic, but of mathematics to mathematics.

See also *Peirce 1902*, reproduced below.

Nobody can doubt the elementary propositions concerning number: those that are not at first sight manifestly true are rendered so by the usual demonstrations. But although we see they *are* true, we do not so easily see precisely *why* they are true; so that a renowned English logician has entertained a doubt as to whether they were true in all parts of the universe. The object of this paper is to show that they are strictly syllogistic consequences from a few primary propositions. The question of the logical origin of these latter, which I here regard as definitions, would require a separate discussion. In my proofs I am obliged to make use of the logic of relatives, in which the forms of inference are not, in a narrow sense, reducible to ordinary syllogism. They are, however, of that same nature, being merely syllogisms in which the objects spoken of are pairs or triplets. Their validity depends upon no conditions other than those of the validity of simple syllogism, unless it be that they suppose the existence of singulars, while syllogism does not.

The selection of propositions which I have proved will, I trust, be sufficient to show that all others might be proved with like methods.

Let r be any relative term, so that one thing may be said to be r of another, and the latter r'd by the former. If in a certain system of objects, whatever is r of an r of anything is itself r of that thing, then r is said to be a transitive relative in that system. (Such relatives as "lover of everything loved by—" are transitive relatives.) In a system in which r is transitive, let the q's of anything include that thing itself, and also every r of it which is not r'd by it. Then q may be called a fundamental relative of quantity; its properties being, first, that it is transitive; second, that everything in the system is q of itself, and, third, that nothing is both q of and q'd by anything except itself. The objects of a system having a fundamental relation of quantity are called quantities, and the system is called a system of quantity.

A system in which quantities may be q's of or q'd by the same quantity without being either q's of or q'd by each other is called multiple;* a system in which of every two quantities one is a q of the other is termed simple.

Simple Quantity

In a simple system every quantity is either "as great as" or "as small as" every other; whatever is as great as something as great as a third is itself as great as that third, and no quantity is at once as great as and as small as anything except itself.

A system of simple quantity is either continuous, discrete, or mixed. A continuous system is one in which every quantity greater than another is also greater than some intermediate quantity greater than that other. A discrete system is

* For example, in the ordinary algebra of imaginaries two quantities may both result from the addition of quantities of the form $a^2 + b^2 i$ to the same quantity without either being in this relation to the other.

one in which every quantity greater than another is next greater than some quantity (that is, greater than without being greater than something greater than). A mixed system is one in which some quantities greater than others are next greater than some quantities, while some are continuously greater than some quantities.

Discrete Quantity

A simple system of discrete quantity is either limited, semi-limited, or unlimited. A limited system is one which has an absolute maximum and an absolute minimum quantity; a semi-limited system has one (generally considered a minimum) without the other; an unlimited has neither.

A simple, discrete, system, unlimited in the direction of increase or decrement, is in that direction either infinite or super-infinite. An infinite system is one in which any quantity greater than x can be reached from x by successive steps to the next greater (or less) quantity than the one already arrived at. In other words, an infinite, discrete, simple, system is one in which, if the quantity next greater than an attained quantity is itself attained, then any quantity greater than an attained quantity is attained; and by the class of attained quantities is meant any class whatever which satisfies these conditions. So that we may say that an infinite class is one in which if it is true that every quantity next greater than a quantity of a given class itself belongs to that class, then it is true that every quantity greater that a quantity of that class belongs to that class. Let the class of numbers in question be the numbers of which a certain proposition holds true. Then, an infinite system may be defined as one in which from the fact that a certain proposition, if true of any number, is true of the next greater, it may be inferred that that proposition if true of any number is true of every greater.

Of a super-infinite system this proposition, in its numerous forms, is untrue.

Semi-infinite Quantity

We now proceed to study the fundamental propositions of semi-infinite, discrete, and simple quantity, which is ordinary number.

Definitions

The minimum number is called one.

By $x+y$ is meant, in case $x=1$, the number next greater than y; and in other cases, the number next greater than $x' + y$, where x' is the number next smaller than x.

By $x \times y$ is meant, in case $x=1$, the number y, and in other cases $y + x'y$, where x' is the number next smaller than x.

It may be remarked that the symbols $+$ and $\times$ are triple relatives, their two correlates being placed one before and the other after the symbols themselves.

Theorems

The proof in each case will consist in showing, 1st, that the proposition is true of the number one, and 2d, that if true of the number n it is true of the number $1 + n$, next larger than n. The different transformations of each expression will be ranged under one another in one column, with the indications of the principles of transformation in another column.

1. To prove the associative principle of addition, that

$$(x + y) + z = x + (y + z)$$

whatever numbers x, y, and z, may be. First it is true for $x = 1$; for $(1 + y) + z = 1 + (y + z)$ by the definition of addition, 2d clause. Second, if true for $x = n$, it is true for $x = 1 + n$; that is, if $(n + y) + z = n + (y + z)$ then $((1 + n) + y) + z = (1 + n) + (y + z)$. For

$((1 + n) + y) + z$

$= (1 + (n + y)) + z$ by the definition of addition:

$= 1 + ((n + y) + z)$ by the definition of addition:

$= 1 + (n + (y + z))$ by hypothesis:

$= (1 + n) + (y + z)$ by the definition of addition.

2. To prove the commutative principle of addition that

$$x + y = y + x$$

whatever numbers x and y may be. First, it is true for $x = 1$ and $y = 1$, being in that case an explicit identity. Second, if true for $x = n$ and $y = 1$, it is true for $x = 1 + n$ and $y = 1$. That is, if $n + 1 = 1 + n$, then $(1 + n) + 1 = 1 + (1 + n)$. For $(1 + n) + 1$

$= 1 + (n + 1)$ by the associative principle:

$= 1 + (1 + n)$ by hypothesis.

We have thus proved that, whatever number x may be, $x + 1 = 1 + x$, or that $x + y = y + x$ for $y = 1$. It is now to be shown that if this be true for $y = n$, it is true for $y = 1 + n$; that is, that if $x + n = n + x$, then $x + (1 + n) = (1 + n) + x$. Now,

$x + (1 + n)$

$= (x + 1) + n$ by the associative principle:

$= (1 + x) + n$ as just seen:

$= 1 + (x + n)$ by the definition of addition:

$= 1 + (n + x)$ by hypothesis:

$= (1 + n) + x$ by the definition of addition.

Thus the proof is complete.

3. To prove the distributive principle, first clause. The distributive principle consists of two propositions:

$$\text{1st,} \quad (x + y)z = xz + yz$$

$$\text{2d,} \quad x(y + z) = xy + xz.$$

We now undertake to prove the first of these. First, it is true for $x = 1$. For

$$\begin{aligned} &(1 + y)z \\ &= z + yz \quad \text{by the definition of multiplication:} \\ &= 1.z + yz \quad \text{by the definition of multiplication.} \end{aligned}$$

Second, if true for $x = n$, it is true for $x = 1 + n$; that is, if $(n + y)z = nz + yz$, then $((1 + n) + y)z = (1 + n)z + yz$. For

$$\begin{aligned} &((1 + n) + y)z \\ &= (1 + (n + y))z \quad \text{by the definition of addition:} \\ &= z + (n + y)z \quad \text{by the definition of multiplication:} \\ &= z + (nz + yz) \quad \text{by hypothesis:} \\ &= (z + nz) + yz \quad \text{by the associative principle of addition:} \\ &= (1 + n)z + yz \quad \text{by the definition of multiplication.} \end{aligned}$$

4. To prove the second proposition of the distributive principle, that

$$x(y + z) = xy + xz.$$

First, it is true for $x = 1$; for

$$\begin{aligned} &1\,(y + z) \\ &= y + z \quad \text{by the definition of multiplication:} \\ &= 1y + 1z \quad \text{by the definition of multiplication.} \end{aligned}$$

Second, if true for $x = n$, it is true for $x = 1 + n$; that is, if $n(y + z) = ny + nz$, then $(1 + n)(y + z) = (1 + n)y + (1 + n)z$. For

$$\begin{aligned} &(1 + n)(y + z) \\ &= (y + z) + n(y + z) \quad \text{by the definition of multiplication:} \\ &= (y + z) + (ny + nz) \quad \text{by hypothesis:} \\ &= (y + ny) + (z + nz) \quad \text{by the principles of addition:} \\ &= (1 + n)y + (1 + n)z \quad \text{by the definition of multiplication.} \end{aligned}$$

5. To prove the associative principle of multiplication; that is, that

$$(xy)z = x(yz),$$

whatever numbers x, y, and z, may be. First, it is true for $x = 1$, for

$$\begin{aligned}
&(1y)z \\
&= yz \quad \text{by the definition of multiplication:} \\
&= 1.yz \quad \text{by the definition of multiplication.}
\end{aligned}$$

Second, if true for $x = n$, it is true for $x = 1 + n$; that is, if $(ny)z = n(yz)$, then $((1+n)y)z = (1+n)(yz)$. For

$$\begin{aligned}
&((1 + n)y)z \\
&= (y + ny)z \quad \text{by the definition of multiplication:} \\
&= yz + (ny)z \quad \text{by the distributive principle:} \\
&= yz + n(yz) \quad \text{by hypothesis:} \\
&= (1 + n)(yz) \quad \text{by the definition of multiplication.}
\end{aligned}$$

6. To prove the commutative principle of multiplication; that

$$xy = yx,$$

whatever numbers x and y may be. In the first place, we prove that it is true for $y = 1$. For this purpose, we first show that it is true for $y = 1$, $x = 1$; and then that if true for $y = 1$, $x = n$, it is true for $y = 1$, $x = 1 + n$. For $y = 1$ and $x = 1$, it is an explicit identity. We have now to show that if $n1 = 1n$ then $(1 + n)1 = 1(1 + n)$. Now,

$$\begin{aligned}
&(1 + n)1 \\
&= 1 + n1 \quad \text{by the definition of multiplication:} \\
&= 1 + 1n \quad \text{by hypothesis:} \\
&= 1 + n \quad \text{by the definition of multiplication:} \\
&= 1(1 + n) \quad \text{by the definition of multiplication.}
\end{aligned}$$

Having thus shown the commutative principle to be true for $y = 1$, we proceed to prove that if it is true for $y = n$, it is true for $y = 1 + n$; that is, if $xn = nx$, then $x(1 + n) = (1 + n)x$. For

$$\begin{aligned}
&(1 + n)x \\
&= x + nx \quad \text{by the definition of multiplication:} \\
&= x + xn \quad \text{by hypothesis:} \\
&= 1x + xn \quad \text{by the definition of multiplication:} \\
&= x1 + xn \quad \text{as already seen:} \\
&= x(1 + n) \quad \text{by the distributive principle.}
\end{aligned}$$

Discrete Simple Quantity Infinite in Both Directions

A system of number infinite in both directions has no minimum, but a certain quantity is called *one*, and the numbers as great as this constitute a partial system of semi-infinite number, of which this one is a minimum. The definitions of addition and multiplication require no change, except that the *one* therein is to be understood in the new sense.

To extend the proofs of the principles of addition and multiplication to unlimited number, it is necessary to show that if true for any number $(1 + n)$ they are also true for the next smaller number n. For this purpose we can use the same transformations as in the second clauses of the former proof; only we shall have to make use of the following lemma.

If $x + y = x + z$, then $y = z$ whatever numbers x, y, and z, may be. First this is true in case $x = 1$, for then y and z are both next smaller than the same number. Therefore, neither is smaller than the other, otherwise it would not be next smaller to $1 + y = 1 + z$. But in a simple system, of any two different numbers one is smaller. Hence, y and z are equal. Second, if the proposition is true for $x = n$, it is true for $x = 1 + n$. For if $(1 + n) + y = (1 + n) + z$, then by the definition of addition $1 + (n + y) = 1 + (n + z)$; whence it would follow that $n + y = n + z$, and, by hypothesis, that $y = z$. Third, if the proposition is true for $x = 1 + n$, it is true for $x = n$. For if $n + y = n + z$, then $1 + n + y = 1 + n + z$, because the system is simple. The proposition has thus been proved to be true of 1, of every greater and of every smaller number, and therefore to be universally true.

An inspection of the above proofs of the principles of addition and multiplication for semi-infinite number will show that they are readily extended to doubly infinite number by means of the proposition just proved.

The number next smaller than one is called naught, 0. This definition in symbolic form is $1+0=1$. To prove that $x+0=x$, let x' be the number next smaller than x. Then,

$$
\begin{aligned}
& x + 0 \\
&= (1 + x') + 0 && \text{by the definition of } x' \\
&= (1 + 0) + x' && \text{by the principles of addition:} \\
&= 1 + x' && \text{by the definition of naught:} \\
&= x && \text{by the definition of } x'.
\end{aligned}
$$

To prove that $x0 = 0$. First, in case $x = 1$, the proposition holds by the definition of multiplication. Next, if true for $x = n$, it is true for $x = 1 + n$. For

$$
\begin{aligned}
& (1 + n)0 \\
&= 1.0 + n.0 && \text{by the distributive principle:} \\
&= 1.0 + 0 && \text{by hypothesis:}
\end{aligned}
$$

$= 1.0$ by the last theorem:

$= 0$ as above.

Third, the proposition, if true for $x = 1 + n$ is true for $x = n$. For, changing the order of the transformations,

$$1.0 + 0 = 1.0 = 0 = (1 + n)0 = 1.0 + n.0$$

Then by the above lemma, $n.0 = 0$, so that the proposition is proved.

A number which added to another gives naught is called the negative of the latter. To prove that every number greater than naught has a negative. First, the number next smaller than naught is the negative of one; for, by the definition of addition, one plus this number is naught. Second, if any number n has a negative, then the number next greater than n has for its negative the number next smaller than the negative of n. For let m be the number next smaller than the negative of n. Then $n + (1 + m) = 0$.

But $n + (1 + m)$

$= (n + 1) + m$ by the associative principle of addition:

$= (1 + n) + m$ by the commutative principle of addition.

So that $(1 + n) + m = 0$. *Q. E. D.* Hence, every number greater than 0 has a negative, and naught is its own negative.

To prove that $(-x)y = -(xy)$. We have

$0 = x + (-x)$ by the definition of the negative:

$0 = 0y = (x + (-x))y$ by the last proposition but one:

$0 = xy + (-x)y$ by the distributive principle:

$-(xy) = (-x)y$ by the definition of the negative.

The negative of the negative of a number is that number. For $x + (-x) = 0$. Whence by the definition of the negative $x = -(-x)$.

Limited Discrete Simple Quantity

Let such a relative term, c, that whatever is a c of anything is the only c of that thing, and is a c of that thing only, be called a relative of simple correspondence. In the notation of the logic of relatives,

$$c\breve{c} \prec 1, \ \breve{c}c \prec 1.$$

If every object, s, of a class is in any such relation c, with a number of a semi-infinite discrete simple system, and if, further, every number smaller than a number c'd by an s is itself c'd by an s, then the numbers c'd by the s's are said to count them, and the system of correspondence is called a count. In logical notation, putting g for as great as, and n for a positive integral number,

$$s \prec \check{c}n \quad \breve{g}cs \prec cs.$$

If in any count there is a maximum counting number, the count is said to be finite, and that number is called the number of the count. Let $[s]$ denote the number of a count of the s's, then

$$[s] \prec cs \quad gcs \prec \overline{[s]}$$

The relative "identical with" satisfies the definition of a relative of simple correspondence, and the definition of a count is satisfied by putting "identical with" for c, and "positive integral number as small as x" for s. In this mode of counting, the number of numbers as small as x is x.

Suppose that in any count the number of numbers as small as the minimum number, one, is found to be n. Then, by the definition of a count, every number as small as n counts a number as small as one. But, by the definition of one there is only one number as small as one. Hence, by the definition of single correspondence, no other number than one counts one. Hence, by the definition of one, no other number than one counts any number as small as one. Hence, by the definition of the count, one is, in every count, the number of numbers as small as one.

If the number of numbers as small as x is in some count y, then the number of numbers as small as y is in some count x. For if the definition of a simple correspondence is satisfied by the relative c, it is equally satisfied by the relative c'd by.

Since the number of numbers as small as x is in some count y, we have, c being some relative of simple correspondence,

1st. Every number as small as x is c'd by a number.

2d. Every number as small as a number that is c of a number as small as x is itself c of a number as small as x.

3d. The number y is c of a number as small as x.

4th. Whatever is not as great as a number that is c of a number as small as x is not y.

Now let c_1 be the converse of c. Then the converse of c_1 is c; whence, since c satisfies the definition of a relative of simple correspondence, so also does c_1. By the 3d proposition above, every number as small as y is as small as a number that is c of a number as small as x. Whence, by the 2d proposition, every number as small as y is c of a number as small as x; and it follows that every number as small as y is c_1'd by a number. It follows further that every number c_1 of a number as small as y is c_1 of something c_1'd by (that is, c_1 being a relative of simple correspondence, is identical with) some number as small as x. Also, "as small as" being a transitive relative, every number as small as a number c of a number as small as y is as small as x. Now by the 4th proposition y is as great as any number that is c of a number as small as x, so that what is not as small as y is not c of a number as small as x; whence whatever number is c'd by a number not as small as y is not a number as small as x. But by the 2d proposition every number as small as x not c'd by a number not as small as y is c'd by a number as small as y. Hence, every number as small as x is c'd by a number as small as y. Hence, every number as small as a number c_1 of a number as small as y is c_1 of a number as small as y. Moreover, since we have shown that every number as small as x is c_1 of a number as small as y, the same is true of x itself. Moreover, since we have seen that whatever is c_1 of a number as small as y is as small as x, it follows that whatever is not as great as a number c_1 of a number as small as y is not as great as a number as small as x; *i.e.* ("as great as" being a transitive relative) is not as great as x, and consequently is not x. We have now shown—

1st, that every number as small as y is c_1'd by a number;

2d, that every number as small as a number that is c_1 of a number as small as y is itself c_1 of a number as small as y;

3d, that the number x is c_1 of a number as small as y; and

4th, that whatever is not as great as a number that is c_1 of a number as small as y is not x.

These four propositions taken together satisfy the definition of the number of numbers as small as y counting up to x.

Hence, since the number of numbers as small as one cannot in any count be greater than one, it follows that the number of numbers as small as any number greater than one cannot in any count be one.

Suppose that there is a count in which the number of numbers as small as $1 + m$ is found to be $1 + n$, since we have just seen that it cannot be 1. In this count, let m' be the number which is c of $1 + n$, and let n' be the number which is c'd by $1 + m$. Let us now consider a relative, e, which differs from c only in excluding the relation of m' to $1 + n$ as well as the relation of $1 + m$ to n' and in including the relation of m' to n'. Then e will be a relative of single correspondence; for c is so, and no exclusion of relations from a single correspondence affects this character, while the inclusion of the relation of m' to n' leaves m' the only e of n' and an e of n' only. Moreover, every number as small as m is e of a number, since every number except $1 + m$ that is c of anything is e of something, and every number except $1 + m$ that is as small as $1 + m$ is as small as m. Also, every number as small as a number e'd by a number is itself e'd by a number; for every number c'd is e'd except $1 + m$, and this is greater than any number e'd. It follows that e is the basis of a mode of counting by which the numbers as small as m count up to n. Thus we have shown that if in any way $1 + m$ counts up to $1 + n$, then in some way m counts up to n. But we have already seen that for $x = 1$ the number of numbers as small as x can in no way count up to other than x. Whence it follows that the same is true whatever the value of x.

If every S is a P, and if the P's are a finite lot counting up to a number as

---

NOTE.—It may be remarked that when we reason that a certain proposition, if false of any number, is false of some smaller number, and since there is no number (in a semi-limited system) smaller than every number, the proposition must be true, our reasoning is a mere logical transformation of the reasoning that a proposition, if true for n, is true for $1 + n$, and that it is true for 1.

small as the number of S's, then every P is an S. For if, in counting the P's, we begin with the S's (which are a part of them), and having counted all the S's arrive at the number n, there will remain over no P's not S's. For if there were any, the number of P's would count up to more than n. From this we deduce the validity of the following mode of inference:

Every Texan kills a Texan,
Nobody is killed by but one person,
Hence, every Texan is killed by a Texan,

supposing Texans to be a finite lot. For, by the first premise, every Texan killed by a Texan is a Texan killer of a Texan. By the second premise, the Texans killed by Texans are as many as the Texan killers of Texans. Whence we conclude that every Texan killer of a Texan is a Texan killed by a Texan, or, by the first premise, every Texan is killed by a Texan. This mode of reasoning is frequent in the theory of numbers.

D. ON THE ALGEBRA OF LOGIC: A CONTRIBUTION TO THE PHILOSOPHY OF NOTATION (*PEIRCE 1885*)

Peirce's next two papers, the brief note *1883* and the longer article, 'On the algebra of logic' (*1885*), presented his discovery of the quantifiers. Gottlob Frege, in his *Begriffschrift* (*1879*), had already made the same discovery (and had carried the analysis of number further than Peirce was to do); and Peirce's student O.H. Mitchell (*1883*) had, under Peirce's guidance, in effect developed a system of monadic quantification theory. But these discoveries had little impact at the time. It was Peirce's *1885* that successfully launched upon the world the theory of quantification *via* the three volumes of Schröder's *Vorlesungen über die Algebra der Logik* (*1890*, *1891*, *1895*). (These *Vorlesungen* were for some years the standard reference work in mathematical logic, and were largely based on Peirce's discoveries.)

'On the algebra of logic' is noteworthy for other reasons as well. It begins with an important passage (§2) on the propositional calculus, containing the first explicit use of two truth-values.[a] Peirce then describes a decision procedure for the truth of any formula of the sentential calculus: '[T]o find whether a formula is necessarily true substitute **f** and **v** for the letters and see whether

[a] Truth-values and truth-tables have their roots in the work of George Boole (*1854*, pp. 72–6). They are implicit in the work of Venn and Jevons (see the discussion in *Lewis 1918*, pp. 74 and 175 ff.). Truth-tables are also implicit in §5 of the *Begriffschrift*, although Frege did not introduce 'The True' and 'The False' until his *1891*. For further references on this topic, see *Post 1921*, reproduced in *van Heijenoort 1967*, pp. 264–83.

it can be supposed false by any such assignment of values.' He also gives a lucid defence of material implication, and shows how to define negation in terms of implication and a special symbol α for absurdity. Next (§3) Peirce treats first-order quantification theory. He coins the term 'quantifier' (probably derived from Sir William Hamilton's terminology of 'quantifying the predicate'); the propositional matrix of a quantified formula he calls its 'Boolian'. He uses the symbols Σ and Π to represent the existential and universal quantifiers. This felicitous notation—like his use of the Boolean sentential connectives—was a major advantage of his system, and enabled Peirce to discuss the rules for transforming a quantified formula into prenex normal form.[b] Peirce next (§4) proceeds to second-intentional logic. (Following the Schoolmen, he clearly distinguishes first-intentional logic from second-intentional.) He states the modern second-order definition of identity, avoiding Leibniz's confusion of use and mention.[c] The paper closes with his definition of a finite set as one which cannot be put into a one-to-one correspondence with any proper subset. (Dedekind's later independent definition of an infinite collection in his *1888* is equivalent to Peirce's.)

References to *Peirce 1885* should be to the section numbers, which appear in the original text.

I.—Three kinds of signs

Any character or proposition either concerns one subject, two subjects, or a plurality of subjects. For example, one particle has mass, two particles attract one another, a particle revolves about the line joining two others. A fact concerning two subjects is a dual character or relation; but a relation which is a mere combination of two independent facts concerning the two subjects may be called *degenerate*, just as two lines are called a degenerate conic. In like manner a plural character or conjoint relation is to be called degenerate if it is a mere compound of dual characters.

A sign is in a conjoint relation to the thing denoted and to the mind. If this triple relation is not of a degenerate species, the sign is related to its object only

[b] The Peircean notation was standard in the work of the Polish set-theoretic logicians of the 1920s and 1930s: see the papers of Kuratowski or Sierpinski, or any volume of *Fundamenta mathematicae* from that period. Löwenheim and Skolem continued to use the Peirce–Schröder notation well into the twentieth century; and as late as his '*Einkleidung der Mathematik in Schröderschen Relativkalkül*' (*Löwenheim 1940*), Löwenheim was urging the superiority of the Peirce–Schröder notation to that of Peano and Russell.

[c] Leibniz's definition of identity was as follows: 'Those things are the same of which one can be substituted for the other *salva veritate*'—'Eadem sunt quorum unum potest substitui alteri salva veritate' (*Leibniz 1875–90*, Vol. vii, pp. 228, 236). Quine observes that Aristotle and Aquinas had already given a similar definition (*Quine 1960*, p. 116).

in consequence of a mental association, and depends upon a habit. Such signs are always abstract and general, because habits are general rules to which the organism has become subjected. They are, for the most part, conventional or arbitrary. They include all general words, the main body of speech, and any mode of conveying a judgment. For the sake of brevity I will call them *tokens*.

But if the triple relation between the sign, its object, and the mind, is degenerate, then of the three pairs

sign object
sign mind
object mind

two at least are in dual relations which constitute the triple relation. One of the connected pairs must consist of the sign and its object, for if the sign were not related to its object except by the mind thinking of them separately, it would not fulfil the function of a sign at all. Supposing, then, the relation of the sign to its object does not lie in a mental association, there must be a direct dual relation of the sign to its object independent of the mind using the sign. In the second of the three cases just spoken of, this dual relation is not degenerate, and the sign signifies its object solely by virtue of being really connected with it. Of this nature are all natural signs and physical symptoms. I call such a sign an *index*, a pointing finger being the type of the class.

The index asserts nothing; it only says "There!" It takes hold of our eyes, as it were, and forcibly directs them to a particular object, and there it stops. Demonstrative and relative pronouns are nearly pure indices, because they denote things without describing them; so are the letters on a geometrical diagram, and the subscript numbers which in algebra distinguish one value from another without saying what those values are.

The third case is where the dual relation between the sign and its object is degenerate and consists in a mere resemblance between them. I call a sign which stands for something merely because it resembles it, an *icon*. Icons are so completely substituted for their objects as hardly to be distinguished from them. Such are the diagrams of geometry. A diagram, indeed, so far as it has a general signification, is not a pure icon; but in the middle part of our reasonings we forget that abstractness in great measure, and the diagram is for us the very thing. So in contemplating a painting, there is a moment when we lose the consciousness that it is not the thing, the distinction of the real and the copy disappears, and it is for the moment a pure dream,—not any particular existence, and yet not general. At that moment we are contemplating an *icon*.

I have taken pains to make my distinction* of icons, indices, and tokens clear, in order to enunciate this proposition: in a perfect system of logical notation signs of these several kinds must all be employed. Without tokens there would be no generality in the statements, for they are the only general signs; and generality

* See *Proceedings American Academy of Arts and Sciences*, Vol. VII, p. 294, May 14, 1867.

is essential to reasoning. Take, for example, the circles by which Euler represents the relations of terms. They well fulfil the function of icons, but their want of generality and their incompetence to express propositions must have been felt by everybody who has used them. Mr. Venn has, therefore, been led to add shading to them; and this shading is a conventional sign of the nature of a token. In algebra, the letters, both quantitative and functional, are of this nature. But tokens alone do not state what is the subject of discourse; and this can, in fact, not be described in general terms; it can only be indicated. The actual world cannot be distinguished from a world of imagination by any description. Hence the need of pronouns and indices, and the more complicated the subject the greater the need of them. The introduction of indices into the algebra of logic is the greatest merit of Mr. Mitchell's system.* He writes F_1 to mean that the proposition F is true of every object in the universe, and F_u to mean that the same is true of some object. This distinction can only be made in some such way as this. Indices are also required to show in what manner other signs are connected together. With these two kinds of signs alone any proposition can be expressed; but it cannot be reasoned upon, for reasoning consists in the observation that where certain relations subsist certain others are found, and it accordingly requires the exhibition of the relations reasoned with in an icon. It has long been a puzzle how it could be that, on the one hand, mathematics is purely deductive in its nature, and draws its conclusions apodictically, while on the other hand, it presents as rich and apparently unending a series of surprising discoveries as any observational science. Various have been the attempts to solve the paradox by breaking down one or other of these assertions, but without success. The truth, however, appears to be that all deductive reasoning, even simple syllogism, involves an element of observation; namely, deduction consists in constructing an icon or diagram the relations of whose parts shall present a complete analogy with those of the parts of the object of reasoning, of experimenting upon this image in the imagination, and of observing the result so as to discover unnoticed and hidden relations among the parts. For instance, take the syllogistic formula,

$$\begin{array}{rcl} \text{All } M & \text{is} & P \\ S & \text{is} & M \\ \therefore S & \text{is} & P. \end{array}$$

This is really a diagram of the relations of S, M, and P. The fact that the middle term occurs in the two premises is actually exhibited, and this must be done or the notation will be of no value. As for algebra, the very idea of the art is that it presents formulae which can be manipulated, and that by observing the effects of such manipulation we find properties not to be otherwise discerned. In such manipulation, we are guided by previous discoveries which are embodied in general formulae. These are patterns which we have the right to imitate in our procedure, and are the *icons par excellence* of algebra. The letters of applied

* *Studies in Logic*, by members of the Johns Hopkins University. Boston: Little & Brown, 1883.

algebra are usually tokens, but the x, y, z, etc. of a general formula, such as

$$(x + y)z = xz + yz,$$

are blanks to be filled up with tokens, they are indices of tokens. Such a formula might, it is true, be replaced by an abstractly stated rule (say that multiplication is distributive); but no application could be made of such an abstract statement without translating it into a sensible image.

In this paper, I purpose to develope an algebra adequate to the treatment of all problems of deductive logic, showing as I proceed what kinds of signs have necessarily to be employed at each stage of the development. I shall thus attain three objects. The first is the extension of the power of logical algebra over the whole of its proper realm. The second is the illustration of principles which underlie all algebraic notation. The third is the enuneration of the essentially different kinds of necessary inference; for when the notation which suffices for exhibiting one inference is found inadequate for explaining another, it is clear that the latter involves an inferential element not present to the former. Accordingly, the procedure contemplated should result in a list of categories of reasoning, the interest of which is not dependent upon the algebraic way of considering the subject. I shall not be able to perfect the algebra sufficiently to give facile methods of reaching logical conclusions: I can only give a method by which any legitimate conclusion may be reached and any fallacious one avoided. But I cannot doubt that others, if they will take up the subject, will succeed in giving the notation a form in which it will be highly useful in mathematical work. I even hope that what I have done may prove a first step toward the resolution of one of the main problems of logic, that of producing a method for the discovery of methods in mathematics.

II.—NON-RELATIVE LOGIC

According to ordinary logic, a proposition is either true or false, and no further distinction is recognized. This is the descriptive conception, as the geometers say; the metric conception would be that every proposition is more or less false, and that the question is one of amount. At present we adopt the former view.

Let propositions be represented by quantities. Let **v** and **f** be two constant values, and let the value of the quantity representing a proposition be **v** if the proposition is true and be **f** if the proposition is false. Thus, x being a proposition, the fact that x is either true or false is written

$$(x - \mathbf{f})(\mathbf{v} - x) = 0.$$

So

$$(x - \mathbf{f})(\mathbf{v} - y) = 0$$

will mean that either x is false or y is true. This may be said to be the same as 'if x is true, y is true'. A hypothetical proposition, generally, is not confined to stating what actually happens, but states what is invariably true throughout a universe of possibility. The present proposition is, however, limited to that one individual state of things, the Actual.

We are, thus, already in possession of a logical notation, capable of working syllogism. Thus, take the premises, 'if x is true, y is true', and 'if y is true, z is true'. These are written

$$(x - \mathbf{f})(\mathbf{v} - y) = 0$$
$$(y - \mathbf{f})(\mathbf{v} - z) = 0.$$

Multiply the first by $(\mathbf{v} - z)$ and the second by $(x - \mathbf{f})$ and add. We get

$$(x - \mathbf{f})(\mathbf{v} - \mathbf{f})(\mathbf{v} - z) = 0,$$

or dividing by $\mathbf{v} - \mathbf{f}$, which cannot be 0,

$$(x - \mathbf{f})(\mathbf{v} - z) = 0;$$

and this states the syllogistic conclusion, "if x is true, z is true".

But this notation shows a blemish in that it expresses propositions in two distinct ways, in the form of quantities, and in the form of equations; and the quantities are of two kinds, namely those which must be either equal to $\mathbf{f}$ or to $\mathbf{v}$, and those which are equated to *zero*. To remedy this, let us discard the use of equations, and perform no operations which can give rise to any values other than $\mathbf{f}$ and $\mathbf{v}$.

Of operations upon a simple variable, we shall need but one. For there are but two things that can be said about a single proposition, by itself; that it is true and that it is false,

$$x = \mathbf{v} \text{ and } x = \mathbf{f}.$$

The first equation is expressed by x itself, the second by any function, ϕ, of x, fulfilling the conditions

$$\phi\mathbf{v} = \mathbf{f} \quad \phi\mathbf{f} = \mathbf{v}.$$

The simplest solution of these equations is

$$\phi x = \mathbf{f} + \mathbf{v} - x.$$

A product of n factors of the two forms $(x - \mathbf{f})$ and $(\mathbf{v} - y)$, if not zero equals $(\mathbf{v} - \mathbf{f})^n$. Write P for the product. Then $\mathbf{v} - \dfrac{P}{(\mathbf{v} - \mathbf{f})^{n-1}}$ is the simplest function of the variables which becomes $\mathbf{v}$ when the product vanishes and $\mathbf{f}$ when it does not. By this means any proposition relating to a single individual can be expressed.

If we wish to use algebraical signs with their usual significations, the meanings of the operations will entirely depend upon those of $\mathbf{f}$ and $\mathbf{v}$. Boole chose $\mathbf{v} = 1$, $\mathbf{f} = 0$. This choice gives the following forms:

$$\mathbf{f} + \mathbf{v} - x = 1 - x$$

which is best written $\bar{x}$.

$$\mathbf{v} - \frac{(x - \mathbf{f})(\mathbf{v} - y)}{\mathbf{v} - \mathbf{f}} = 1 - x + xy = \overline{x\bar{y}}$$

$$\mathbf{v} - \frac{(\mathbf{v} - x)(\mathbf{v} - y)}{\mathbf{v} - \mathbf{f}} = x + y - xy$$

$$\mathbf{v} - \frac{(\mathbf{v} - x)(\mathbf{v} - y)(\mathbf{v} - z)}{(\mathbf{v} - \mathbf{f})^2} = x + y + z - xy - xz - yz + xyz$$

$$\mathbf{v} - \frac{(x - \mathbf{f})(y - \mathbf{f})}{\mathbf{v} - \mathbf{f}} = 1 - xy = \overline{xy}.$$

It appears to me that if the strict Boolian system is used, the sign + ought to be altogether discarded. Boole and his adherent, Mr. Venn (whom I never disagree with without finding his remarks profitable), prefer to write $x + \bar{x}y$ in place of $\overline{x\bar{y}}$. I confess I do not see the advantage of this, for the distributive principle holds equally well when written

$$\overline{\bar{x}\bar{y}}z = \overline{\overline{xz}\,\overline{yz}}$$

$$\overline{\overline{xy}\bar{z}} = \overline{\bar{x}\bar{z}}.\overline{\bar{y}\bar{z}}.$$

The choice of $\mathbf{v} = 1$, $\mathbf{f} = 0$, is agreeable to the received measurement of probabililies. But there is no need, and many times no advantage, in measuring probabilities in this way. I presume that Boole, in the formation of his algebra, at first considered the letters as denoting propositions or events. As he presents the subject, they are class-names; but it is not necessary so to regard them. Take, for example, the equation

$$t = n + hf,$$

which might mean that the body of taxpayers is composed of all the natives, together with householding foreigners. We might reach the signification by either of the following systems of notation, which indeed differ grammatically rather than logically.

Sign	Signification 1st System	Signification 2nd System
t	Taxpayer	He is a Taxpayer
n	Native	He is a Native
h	Householder	He is a Householder
f	Foreigner	He is a Foreigner

There is no *index* to show who the "He" of the second system is, but that makes no difference. To say that he is a taxpayer is equivalent to saying that he is a native or is a householder and a foreigner. In this point of view, the constants

1 and 0 are simply the probabilities, to one who knows, of what is true and what is false; and thus unity is conferred upon the whole system.

For my part, I prefer for the present not to assign determinate values to $\mathbf{f}$ and $\mathbf{v}$, nor to identify the logical operations with any special arithmetical ones, leaving myself free to do so hereafter in the manner which may be found most convenient. Besides, the whole system of importing arithmetic into the subject is artificial, and modern Boolians do not use it. The algebra of logic should be self-developed, and arithmetic should spring out of logic instead of reverting to it. Going back to the beginning, let the writing of a letter by itself mean that a certain proposition is true. This letter is a *token*. There is a general understanding that the actual state of things or some other is referred to. This understanding must have been established by means of an *index*, and to some extent dispenses with the need of other indices. The denial of a proposition will be made by writing a line over it.

I have elsewhere shown that the fundamental and primary mode of relation between two propositions is that which we have expressed by the form

$$\mathbf{v} - \frac{(x - \mathbf{f})\,(\mathbf{v} - y)}{\mathbf{v} - \mathbf{f}}.$$

We shall write this

$$x \prec y,$$

which is also equivalent to

$$(x - \mathbf{f})\,(\mathbf{v} - y) = 0.$$

It is stated above that this means "if x is true, y is true". But this meaning is greatly modified by the circumstance that only the actual state of things is referred to.

To make the matter clear, it will be well to begin by defining the meaning of a hypothetical proposition, in general. What the usages of language may be does not concern us; language has its meaning modified in technical logical formulae as in other special kinds of discourse. The question is what is the sense which is most usefully attached to the hypothetical proposition in logic? Now, the peculiarity of the hypothetical proposition is that it goes out beyond the actual state of things and declares what *would* happen were things other than they are or may be. The utility of this is that it puts us in possession of a rule, say that "if A is true, B is true", such that should we hereafter learn something of which we are now ignorant, namely that A is true, then, by virtue of this rule, we shall find that we know something else, namely, that B is true. There can be no doubt that the Possible, in its primary meaning, is that which may be true for aught we know, that whose falsity we do not know. The purpose is subserved, then, if, throughout the whole range of possibility, in every state of things in which A is true, B is true too. The hypothetical proposition may therefore be falsified by a single state of things, but only by one in which A is true while B is false. States of things in which A is false, as well as those in which B is true, cannot falsify it. If, then, B is a proposition true in every case throughout the whole range of possibility, the hypothetical proposition, taken in its logical sense, ought to be regarded as true, whatever may be the usage of ordinary speech. If, on the other hand, A is in no case true, throughout the range of possibility, it is a matter of indifference whether the hypothetical be understood to be true or not, since it is useless. But it will be more simple to class it among true propositions, because the cases in which the antecedent is false do not, in any other case, falsify a hypothetical. This, at any rate, is the meaning which I shall attach to the hypothetical proposition in general, in this paper.

The range of possibility is in one case taken wider, in another narrower; in the present case it is limited to the actual state of things. Here, therefore, the proposition

$$a \prec b$$

is true if a is false or if b is true, but is false if a is true while b is false. But though we limit ourselves to the actual state of things, yet when we find that a formula of this sort is true by logical necessity, it becomes applicable to any single state of things throughout the range of logical possibility. For example, we shall see that from $x \overline{\prec} y$ we can infer $z \prec x$. This does not mean that because in the actual state of things x is true and y false, therefore in every state of things either z is false or x true; but it does mean that in whatever state of things we find x true and y false, in that state of things either z is false or x is true. In that sense, it is not limited to the actual state of things, but extends to any single state of things.

The *first icon* of algebra is contained in the formula of identity

$$x \prec x.$$

This formula does not of itself justify any transformation, any inference. It only justifies our continuing to hold what we have held (though we may, for instance, forget how we were originally justified in holding it).

The *second icon* is contained in the rule that the several antecedents of a *consequentia* may be transposed; that is, that from

$$x \prec (y \prec z)$$

we can pass to

$$y \prec (x \prec z).$$

This is stated in the formula

$$\{x \prec (y \prec z)\} \prec \{y \prec (x \prec z)\}.$$

Because this is the case, the brackets may be omitted, and we may write

$$y \prec x \prec z.$$

By the formula of identity

$$(x \prec y) \prec (x \prec y);$$

and transposing the antecedents

$$x \prec \{(x \prec y) \prec y\}$$

or, omitting the unnecessary brackets

$$x \prec (x \prec y) \prec y.$$

This is the same as to say that if in any state of things x is true, and if the proposition "if x, then y" is true, then in that state of things y is true. This is the *modus ponens* of hypothetical inference, and is the most rudimentary form of reasoning.

To say that $(x \prec x)$ is generally true is to say that it is so in every state of things, say in that in which y is true; so that we may write

$$y \prec (x \prec x),$$

and then, by transposition of antecedents,

$$x \prec (y \prec x),$$

or from x we may infer $y \prec x$.

The *third icon* is involved in the principle of the transitiveness of the copula, which is stated in the formula

$$(x \prec y) \prec (y \prec z) \prec x \prec z.$$

According to this, if in any case y follows from x and z from y, then z follows from x. This is the principle of the syllogism in *Barbara*.

We have already seen that from x follows $y \prec x$. Hence, by the transitiveness of the copula, if from $y \prec x$ follows z, then from x follows z, or from

$$(y \prec x) \prec z$$

follows

$$x \prec z,$$

or

$$\{(y \prec x) \prec z\} \prec x \prec z.$$

The original notation $x \prec y$ served without modification to express the pure formula of identity. An enlargement of the conception of the notation so as to make the terms themselves complex was required to express the principle of the transposition of antecedents; and this new *icon* brought out new propositions. The third *icon* introduces the image of a chain of consequence. We must now again enlarge the notation so as to introduce negation. We have already seen that if a is true, we can write $x \prec a$, whatever x may be. Let b be such that we can write $b \prec x$ whatever x may be. Then b is false. We have here a *fourth icon*, which gives a new sense to several formulæ. Thus the principle of the interchange of antecedents is that from

$$x \prec (y \prec z)$$

we can infer

$$y \prec (x \prec z).$$

Since z is any proposition we please, this is as much as to say that if from the truth of x the falsity of y follows, then from the truth of y the falsity of x follows.

Again the formula

$$x \prec \{(x \prec y) \prec y\}$$

is seen to mean that from x we can infer that anything we please follows from that things ⟦*sic*⟧ following from x, and *a fortiori* from everything following from x. This is, therefore, to say that from x follows the falsity of the denial of x; which is the principle of contradiction.

Again the formula of the transitiveness of the copula, or

$$\{x \prec y\} \prec \{(y \prec z) \prec (x \prec z)\}$$

is seen to justify the inference

$$x \prec y$$

$$\therefore\ \bar{y} \prec \bar{x}.$$

The same formula justifies the *modus tollens*,

$$x \prec y$$

$$\bar{y}$$

$$\therefore\ \bar{x}.$$

So the formula

$$\{(y \prec x) \prec z\} \prec (x \prec z)$$

shows that from the falsity of $y \prec x$ the falsity of x may be inferred.

All the traditional moods of syllogism can easily be reduced to *Barbara* by this method.

A *fifth icon* is required for the principle of excluded middle and other propositions connected with it. One of the simplest formulæ of this kind is

$$\{(x \prec y) \prec x\} \prec x.$$

This is hardly axiomatical. That it is true appears as follows. It can only be false by the final consequent x being false while its antecedent $(x \prec y) \prec x$ is true. If this is true, either its consequent, x, is true, when the whole formula would be true, or its antecedent $x \prec y$ is false. But in the last case the antecedent of $x \prec y$, that is x, must be true.*

From the formula just given, we at once get

$$\{(x \prec y) \prec \alpha\} \prec x,$$

where the α is used in such a sense that $(x \prec y) \prec \alpha$ means that from $(x \prec y)$ every proposition follows. With that understanding, the formula states the principle of excluded middle, that from the falsity of the denial of x follows the truth of x.

The logical algebra thus far developed contains signs of the following kinds:

1st, Tokens; signs of simple propositions, as t for 'He is a taxpayer', etc.

2d, The single operative sign $\prec$; also of the nature of a token.

3d, The juxtaposition of the letters to the right and left of the operative sign. This juxtaposition fulfils the function of an index, in indicating the connections of the tokens.

4th, The parentheses, subserving the same purpose.

5th, The letters α, β, etc. which are indices of no matter what tokens, used for expressing negation.

6th, The indices of tokens, x, y, z, etc. used in the general formulæ.

7th, The general formulæ themselves, which are *icons*, or exemplars of algebraic proceedings.

8th, The fourth *icon* which affords a second interpretation of the general formulæ.

We might dispense with the fifth and eighth species of signs—the devices by

---

* It is interesting to observe that this reasoning is dilemmatic. In fact, the dilemma involves the fifth icon. The dilemma was only introduced into logic from rhetoric by the humanists of the *renaissance*; and at that time logic was studied with so little accuracy that the peculiar nature of this mode of reasoning escaped notice. I was thus led to suppose that the whole non-relative logic was derivable from the principles of the ancient syllogistic, and this error is involved in Chapter II of my paper in the third volume of this Journal. My friend, Professor Schröder, detected the mistake and showed that the distributive formulæ

$$(x + y)z \prec xz + yz$$

$$(x + z)(y + z) \prec xy + z$$

could not be deduced from syllogistic principles. I had myself independently discovered and virtually stated the same thing. (*Studies in Logic*, p. 189). There is some disagreement as to the definition of the dilemma (see Keynes's excellent *Formal Logic*, p. 241); but the most useful definition would be a syllogism depending on the above distribution formulæ. The distribution formulæ

$$xz + yz \prec (x + y)z$$

$$xy + z \prec (x + z)(y + z)$$

are strictly syllogistic. DeMorgan's added moods are virtually dilemmatic, depending on the principle of excluded middle.

which we express negation—by adopting a second operational sign $\overline{\prec}$, such that

$$x \overline{\prec} y$$

should mean that $x = \mathbf{v}$, $y = \mathbf{f}$. With this we should require new indices of connections, and new general formulae. Possibly this might be the preferable notation. We should thus have two operational signs but no sign of negation. The forms of Boolian algebra hitherto used, have either two operational signs and a special sign of negation, or three operational signs. One of the operational signs is in that case superfluous. Thus, in the usual notation we have

$$\overline{x + y} = \bar{x}\bar{y}$$

$$\bar{x} + \bar{y} = \overline{xy}$$

showing two modes of writing the same fact. The apparent balance between the two sets of theorems exhibited so strikingly by Schröder, arises entirely from this double way of writing everything. But while the ordinary system is not so analytically fitted to its purpose as that here set forth, the character of superfluity here, as in many other cases in algebra, brings with it great facility in working.

The general formulæ given above are not convenient in practice. We may dispense with them altogether, as well as with one of the indices of tokens used in them, by the use of the following rules. A proposition of the form

$$x \prec y$$

is true if $x = \mathbf{f}$ or $y = \mathbf{v}$. It is only false if $y = \mathbf{f}$ and $x = \mathbf{v}$. A proposition written in the form

$$x \overline{\prec} y$$

is true if $x = \mathbf{v}$ and $y = \mathbf{f}$, and is false if either $x = \mathbf{f}$ or $y = \mathbf{v}$. Accordingly, to find whether a formula is necessarily true substitute **f** and **v** for the letters and see whether it can be supposed false by any such assignment of values. Take, for example, the formula

$$(x \prec y) \prec \{(y \prec z) \prec (x \prec z)\}.$$

To make this false we must take

$$(x \prec y) = \mathbf{v}$$

$$\{(y \prec z) \prec (x \prec z)\} = \mathbf{f}.$$

The last gives $(y \prec z) = \mathbf{v}$, $(x \prec z) = \mathbf{f}$, $x = \mathbf{v}$, $z = \mathbf{f}$. Substituting these values in

$$(x \prec y) = \mathbf{v} \quad (y \prec z) = \mathbf{v}$$

we have $$(\mathbf{v} \prec y) = \mathbf{v} \quad (y \prec \mathbf{f}) = \mathbf{v},$$

which cannot be satisfied together.

As another example, required the conclusion from the following premises. Any one I might marry would be either beautiful or plain; any one whom I might marry would be a woman; any beautiful woman would be an ineligible wife; any plain woman would be an ineligible wife. Let

m be any one whom I might marry,
b, beautiful,
p, plain,
w, woman,
i, ineligible.

Then the premises are

$$\begin{aligned} &m \prec (b \prec \mathbf{f}) \prec p, \\ &m \prec w, \\ &w \prec b \prec i, \\ &w \prec p \prec i. \end{aligned}$$

Let x be the conclusion. Then

$$[m \prec (b \prec \mathbf{f}) \prec p] \prec (m \prec w) \prec (w \prec b \prec i) \\ \prec (w \prec p \prec i) \prec x$$

is necessarily true. Now if we suppose $m = \mathbf{v}$, the proposition can only be made false by putting $w = \mathbf{v}$ and either b or $p = \mathbf{v}$. In this case the proposition can only be made false by putting $i = \mathbf{v}$. If therefore, x can only be made $\mathbf{f}$ by putting $m = \mathbf{v}$, $i = \mathbf{f}$, that is if $x = (m \prec i)$ the proposition is necessarily true.

In this method, we introduce the two special tokens of second intention $\mathbf{f}$ and $\mathbf{v}$, we retain two indices of tokens x and y, and we have a somewhat complex *icon*, with a special prescription for its use.

A better method may be found as follows. We have seen that

$$x \prec (y \prec z)$$

may be conveniently written

$$x \prec y \prec z;$$

while

$$(x \prec y) \prec z$$

ought to retain the parenthesis. Let us extend this rule, so as to be more general, and hold it necessary *always* to include the antecedent in parenthesis. Thus, let us write

$$(x) \prec y$$

instead of $x \prec y$. If now, we merely change the external appearance of two signs; namely, if we use the vinculum instead of the parenthesis, and the sign $+$ in place of $\prec$, we shall have

$$x \prec y \quad \text{written} \quad \bar{x} + y$$

$$x \prec y \prec z \quad \text{"} \quad \bar{x} + \bar{y} + z$$

$$(x \prec y) \prec z \quad \text{"} \quad \overline{\bar{x}+y}+z, \text{ etc.}$$

We may further write for $x \mathrel{\overline{\prec}} y$, $\overline{\bar{x}+y}$ implying that $\bar{x}+y$ is an antecedent for whatever consequent may be taken, and the vinculum becomes identified with the sign of negation. We may also use the sign of multiplication as an abbreviation, putting

$$xy = \overline{\bar{x} + \bar{y}} = \overline{x \prec \bar{y}}.$$

This subjects addition and multiplication to all the rules of ordinary algebra, and also to the following:

$$y + x\bar{x} = y \quad y(x + \bar{x}) = y$$

$$x + \bar{x} = \mathbf{v} \quad \bar{x}x = \mathbf{f}$$

$$xy + z = (x + z)(y + z).$$

To any proposition we have a right to add any expression at pleasure; also to strike out any factor of any term. The expressions for different propositions separately known may be multiplied together. These are substantially Mr Mitchell's rules of procedure. Thus the premises of *Barbara* are

$$\bar{x} + y \text{ and } \bar{y} + z.$$

Multiplying these, we get $(\bar{x} + y)(\bar{y} + z) = \bar{x}\bar{y} + yz$.
Dropping $\bar{y}$ and y we reach the conclusion $\bar{x} + z$.

III.—First-intentional logic of relations

The algebra of Boole affords a language by which anything may be expressed which can be said without speaking of more than one individual at a time. It is true that it can assert that certain characters belong to a whole class, but only such characters as belong to each individual separately. The logic of relatives considers statements involving two and more individuals at once. Indices are here required. Taking, first, a degenerate form of relation, we may write $x_i y_j$ to signify that x is true of the individual i while y is true of the individual j. If z be a relative character z_{ij} will signify that i is in that relation to j. In this way we can express relations of considerable complexity. Thus, if

$$\begin{matrix} 1, & 2, & 3, \\ 4, & 5, & 6, \\ 7, & 8, & 9, \end{matrix}$$

are points in a plane, and l_{123} signifies that 1, 2, and 3 lie on one line, a well-known proposition of geometry may be written

$$l_{159} \prec l_{267} \prec l_{348} \prec l_{147} \prec l_{258} \prec l_{369} \prec l_{123} \prec l_{456} \prec l_{789}.$$

In this notation is involved a *sixth icon*.

We now come to the distinction of *some* and *all*, a distinction which is precisely on a par with that between truth and falsehood; that is, it is descriptive, not metrical.

All attempts to introduce this distinction into the Boolian algebra were more or less complete failures until Mr. Mitchell showed how it was to be effected. His method really consists in making the whole expression of the proposition consist of two parts, a pure Boolian expression referring to an individual and a Quantifying part saying what individual this is. Thus, if k means 'he is a king', and h, 'he is happy', the Boolian

$$(\bar{k} + h)$$

means that the individual spoken of is either not a king or is happy. Now, applying the quantification, we may write

$$\text{Any } (\bar{k} + h)$$

to mean that this is true of any individual in the (limited) universe, or

$$\text{Some } (\bar{k} + h)$$

to mean that an individual exists who is either not a king or is happy. So

$$\text{Some } (kh)$$

means some king is happy, and

$$\text{Any } (kh)$$

means every individual is both a king and happy. The rules for the use of this notation are obvious. The two propositions

$$\text{Any } (x) \quad \text{Any } (y)$$

are equivalent to

$$\text{Any } (xy).$$

From the two propositions

$$\text{Any } (x) \quad \text{Some } (y)$$

we may infer

$$\text{Some } (xy).*$$

* I will just remark, quite out of order, that the quantification may be made numerical; thus producing the numerically definite inferences of DeMorgan and Boole. Suppose at least $\frac{2}{3}$ of the company have white neckties and at least $\frac{3}{4}$ have dress coats. Let w mean 'he has a white necktie', and d 'he has a dress coat'. Then, the two propositions are

$$\tfrac{2}{3}\,(w) \text{ and } \tfrac{3}{4}\,(d).$$

Mr. Mitchell has also a very interesting and instructive extension of his notation for *some* and *all*, to a two-dimensional universe, that is, to the logic of relatives. Here, in order to render the notation as iconical as possible we may use Σ for *some*, suggesting a sum, and Π for *all*, suggesting a product. Thus $\Sigma_i x_i$ means that x is true of some one of the individuals denoted by i or

$$\Sigma_i x_i = x_i + x_j + x_k + \text{etc.}$$

In the same way, $\Pi_i x_i$ means that x is true of all these individuals, or

$$\Pi_i x_i = x_i x_j x_k, \text{etc.}$$

If x is a simple relation, $\Pi_i \Pi_j x_{ij}$ means that every i is in this relation to every j, $\Sigma_i \Pi_j x_{ij}$ that some one i is in this relation to every j, $\Pi_j \Sigma_i x_{ij}$ that to every j some i or other is in this relation, $\Sigma_i \Sigma_j x_{ij}$ that some i is in this relation to some j. It is to be remarked that $\Sigma_i x_i$ and $\Pi_i x_i$ are only *similar* to a sum and a product; they are not strictly of that nature, because the individuals of the universe may be innumerable.

At this point, the reader would perhaps not otherwise easily get so good a conception of the notation as by a little practice in translating from ordinary language into this system and back again. Let l_{ij} mean that i is a lover of j, and b_{ij} that i is a benefactor of j. Then

$$\Pi_i \Sigma_j l_{ij} b_{ij}$$

means that everything is at once a lover and a benefactor of something; and

$$\Pi_i \Sigma_j l_{ij} b_{ji}$$

that everything is a lover of a benefactor of itself.

$$\Sigma_i \Sigma_k \Pi_j (l_{ij} + b_{jk})$$

means that there are two persons, one of whom loves everything except benefac-

---

These are to be multiplied together. But we must remember that xy is a mere abbreviation for $\overline{\bar{x} + \bar{y}}$, and must therefore write

$$\overline{\overline{\tfrac{2}{3}\, w} + \overline{\tfrac{3}{4}\, d}}.$$

Now $\overline{\frac{2}{3}\, w}$ is the denial of $\frac{2}{3}\, w$, and this denial may be written $(>\frac{1}{3})\bar{w}$, or more than $\frac{1}{3}$ of the universe (the company) have not white neckties. So $\overline{\frac{3}{4} d} = (>\frac{1}{4})\bar{d}$. The combined premises thus become

$$\overline{(>\tfrac{1}{3})\bar{w} + (>\tfrac{1}{4})\bar{d}}.$$

Now $(>\frac{1}{3})\bar{w} + (>\frac{1}{4})\bar{d}$ gives May be $(\frac{1}{3} + \frac{1}{4})(\bar{w} + \bar{d})$.

Thus we have

$$\overline{\text{May be } (\tfrac{7}{12})(\bar{w} + \bar{d})},$$

and this is

$$(\text{At least } \tfrac{5}{12})(\overline{\bar{w} + \bar{d}}),$$

which is the conclusion.

tors of the other (whether he loves any of these or not is not stated). Let g_i mean that i is a griffin, and c_i that i is a chimera, then

$$\Sigma_i \Pi_j (g_i l_{ij} + \bar{c}_j)$$

means that if there be any chimeras there is some griffin that loves them all; while

$$\Sigma_i \Pi_j g_i (l_{ij} + \bar{c}_j)$$

means that there is a griffin and he loves every chimera that exists (if any exist). On the other hand,

$$\Pi_j \Sigma_i g_i (l_{ij} + \bar{c}_j)$$

means that griffins exist (one, at least), and that one or other of them loves each chimera that may exist; and

$$\Pi_j \Sigma_i (g_i l_{ij} + \bar{c}_j)$$

means that each chimera (if there is any) is loved by some griffin or other.

Let us express: every part of the world is either sometimes visited with cholera, and at others with small-pox (without cholera), or never with yellow fever and the plague together. Let

c_{ij}	mean the place i has cholera at the time j.				
s_{ij}	"	"	"	small-pox	" "
y_{ij}	"	"	"	yellow fever	" "
p_{ij}	"	"	"	plague	" "

Then we write

$$\Pi_i \Sigma_j \Sigma_k \Pi_l (c_{ij} \bar{c}_{ik} s_{ik} + \bar{y}_{il} + \bar{p}_{il}).$$

Let us express this: one or other of two theories must be admitted, 1st, that no man is at any time unselfish or free, and some men are always hypocritical, and at every time some men are friendly to men to whom they are at other times inimical, or 2d, at each moment all men are alike either angels or fiends. Let

u_{ij}	mean the man i is unselfish at the time j,				
f_{ij}	"	"	"	free	" "
h_{ij}	"	"	"	hypocritical	" "
a_{ij}	"	"	"	an angel	" "
d_{ij}	"	"	"	a fiend	" "
p_{ijk}	"	"	"	friendly	" " to the man k,
e_{ijk}	the man i is an enemy at the time j to the man k;				
1_{jm}	the two objects j and m are identical.				

Then the proposition is

$$\Pi_i \Sigma_h \Pi_j \Sigma_k \Sigma_l \Sigma_m \Pi_n \Pi_p \Pi_q (\bar{u}_{ij} \bar{f}_{ij} h_{hj} p_{kjl} e_{kml} \bar{1}_{jm} + a_{pn} + d_{qn}).$$

We have now to consider the procedure in working with this calculus. It is far from being true that the only problem of deduction is to draw a conclusion from given premises. On the contrary, it is fully as important to have a method for ascertaining what premises will yield a given conclusion. There are besides other problems of transformation, where a certain system of facts is given, and it is required to describe this in other terms of a definite kind. Such, for example, is the problem of the 15 young ladies, and others relating to synthemes. I shall, however, content myself here with showing how, when a set of premises are given, they can be united and certain letters eliminated. Of the various methods which might be pursued, I shall here give the one which seems to me the most useful on the whole.

1st. The different premises having been written with distinct indices (the same index not used in two propositions) are written together, and all the Π's and Σ's are to be brought to the left. This can evidently be done, for

$$\Pi_i x_i . \Pi_j x_j = \Pi_i \Pi_j x_i x_j$$

$$\Sigma_i x_i . \Pi_j x_j = \Sigma_i \Pi_j x_i x_j$$

$$\Sigma_i x_i . \Sigma_j x_j = \Sigma_i \Sigma_j x_i x_j .$$

2d. Without deranging the order of the indices of any one premise, the Π's and Σ's belonging to different premises may be moved relatively to one another, and as far as possible the Σ's should be carried to the left of the Π's. We have

$$\Pi_i \Pi_j x_{ij} = \Pi_j \Pi_i x_{ij}$$

$$\Sigma_i \Sigma_j x_{ij} = \Sigma_j \Sigma_i x_{ij}$$

and also

$$\Sigma_i \Pi_j x_i y_j = \Pi_j \Sigma_i x_i y_j .$$

But this formula does not hold when the *i* and *j* are not separated. We do have, however,

$$\Sigma_i \Pi_j x_{ij} \prec \Pi_j \Sigma_i x_{ij} .$$

It will, therefore, be well to begin by putting the Σ's to the left, as far as possible, because at a later stage of the work they can be carried to the right but not to the left. For example, if the operators of the two premises are $\Pi_i \Sigma_j \Pi_k$ and $\Sigma_x \Pi_y \Sigma_z$, we can unite them in either of the two orders

$$\Sigma_x \Pi_y \Sigma_z \Pi_i \Sigma_j \Pi_k$$

$$\Sigma_x \Pi_i \Sigma_j \Pi_y \Sigma_z \Pi_k ,$$

and shall usually obtain different conclusions accordingly. There will often be room for skill in choosing the most suitable arrangement.

3d. It is next sometimes desirable to manipulate the Boolian part of the expression, and the letters to be eliminated can, if desired, be eliminated now. For this purpose they are replaced by relations of second intention, such as "other than", etc. If, for example, we find anywhere in the expression

$$a_{ijk}\bar{a}_{xyz},$$

this may evidently be replaceable by

$$(n_{ix} + n_{jy} + n_{kz})$$

where, as usual, *n* means not or other than. This third step of the process is frequently quite indispensable, and embraces a variety of processes; but in ordinary cases it may be altogether dispensed with.

4th. The next step, which will also not commonly be needed, consists in making the indices refer to the same collections of objects, so far as this is useful. If the quantifying part, or Quantifier, contains Σ_x, and we wish to replace the x by a new index i, not already in the Quantifier, and such that every x is an i, we can do so at once by simply multiplying every letter of the Boolian having x as an index by x_i. Thus, if we have "some woman is an angel" written in the form $\Sigma_w a_w$ we may replace this by $\Sigma_i(a_i w_i)$. It will be more often useful to replace the index of a Π by a wider one; and this will be done by adding $\bar{x}_i$ to every letter having x as an index. Thus, if we have "all dogs are animals, and all animals are vertebrates" written thus

$$\Pi_d \alpha_d \Pi_a v_a,$$

where a and α alike mean animal, it will be found convenient to replace the last index by i, standing for any object, and to write the proposition

$$\Pi_i(\bar{\alpha}_i + v_i).$$

5th. The next step consists in multiplying the whole Boolian part, by the modification of itself produced by substituting for the index of any Π any other index standing to the left of it in the Quantifier. Thus, for

$$\Sigma_i \Pi_j l_{ij},$$

we can write

$$\Sigma_i \Pi_j l_{ij} l_{ii}.$$

6th. The next step consists in the re-manipulation of the Boolian part, consisting, 1st, in adding to any part any term we like; 2d, in dropping from any part any factor we like, and 3d, in observing that

$$x\bar{x} = \mathbf{f}, \qquad x + \bar{x} = \mathbf{v},$$

so that

$$x\bar{x}y + z = z, \ (x + \bar{x} + y)z = z.$$

7th. Π's and Σ's in the Quantifier whose indices no longer appear in the Boolian are dropped.

The fifth step will, in practice, be combined with part of the sixth and seventh. Thus, from $\Sigma_i \Pi_j l_{ij}$ we shall at once proceed to $\Sigma_i l_{ii}$ if we like.

The following examples will be sufficient.

From the premises $\Sigma_i a_i b_i$ and $\Pi_j(\bar{b}_j + c_j)$, eliminate b. We first write

$$\Sigma_i \Pi_j a_i b_i (\bar{b}_j + c_j).$$

The distributive process gives

$$\Sigma_i \Pi_j a_i (b_i \bar{b}_j + b_i c_j).$$

But *we always have a right to drop a factor or insert an additive term.* We thus get

$$\Sigma_i \Pi_j a_i (b_i \bar{b}_j + c_j).$$

By the third process, we can, if we like, insert n_{ij} for $b_i \bar{b}_j$. In either case, we identify j with i and get the conclusion

$$\Sigma_i a_i c_i.$$

Given the premises

$$\Sigma_h \Pi_i \Sigma_j \Pi_k (\alpha_{hik} + s_{jk} l_{ji})$$

$$\Sigma_u \Sigma_v \Pi_x \Pi_y (\varepsilon_{uyx} + \bar{s}_{yv} b_{vx}).$$

Required to eliminate s. The combined premise is

$$\Sigma_u \Sigma_v \Sigma_h \Pi_i \Sigma_j \Pi_x \Pi_k \Pi_y (\alpha_{hik} + s_{jk} l_{ji}) (\varepsilon_{uyx} + \bar{s}_{yv} b_{vx}).$$

Identify k with v and y with j, and we get

$$\Sigma_u \Sigma_v \Sigma_h \Pi_i \Sigma_j \Pi_x (\alpha_{hiv} + s_{jv} l_{ji}) (\varepsilon_{ujx} + \bar{s}_{jv} b_{vx}).$$

The Boolian part then reduces, so that the conclusion is

$$\Sigma_u \Sigma_v \Sigma_h \Pi_i \Sigma_j \Pi_x (\alpha_{hiv} \varepsilon_{ujx} + \alpha_{hiv} b_{vx} + \varepsilon_{ujx} l_{ji}).$$

IV.—SECOND-INTENTIONAL LOGIC

Let us now consider the logic of terms taken in collective senses. Our notation, so far as we have developed it, does not show us even how to express that two indices, i and j, denote one and the same thing. We may adopt a special token of second intention, say 1, to express identity, and may write 1_{ij}. But this relation of identity has peculiar properties. The first is that if i and j are identical, whatever is true of i is true of j. This may be written

$$\Pi_i \Pi_j \{ \bar{1}_{ij} + \bar{x}_i + x_j \}.$$

The use of the general index of a token, x, here, shows that the formula is iconical. The other property is that if everything which is true of i is true of j, then i and j are identical. This is most naturally written as follows: Let the token, q, signify the relation of a quality, character, fact, or predicate to its subject. Then the property we desire to express is

$$\Pi_i \Pi_j \Sigma_k (1_{ij} + \bar{q}_{ki} q_{kj}).$$

And identity is defined thus

$$1_{ij} = \Pi_k (q_{ki} q_{kj} + \bar{q}_{ki} \bar{q}_{kj}).$$

That is, to say that things are identical is to say that every predicate is true of both or false of both. It may seem circuitous to introduce the idea of a quality to express identity; but that impression will be modified by reflecting that $q_{ki}q_{kj}$ merely means that i and j are both within the class or collection k. If we please, we can dispense with the token q, by using the index of a token and by referring to this in the Quantifier just as subjacent indices are referred to. That is to say, we may write

$$1_{ij} = \Pi_x(x_i x_j + \bar{x}_i \bar{x}_j).$$

The properties of the token q must now be examined. These may all be summed up in this, that taking any individuals i_1, i_2, i_3, etc., and any individuals, j_i, j_2, j_3, etc., there is a collection, class, or predicate embracing all the i's and excluding all the j's except such as are identical with some one of the i's. This might be written

$$(\Pi_\alpha \Pi_{i_\alpha})(\Pi_\beta \Pi_{j_\beta})\Sigma_k(\Pi_\alpha \Sigma_{i'_\alpha})\Pi_l q_{ki_\alpha}(\bar{q}_{kj_\beta} + q_{li'_\alpha}q_{lj_\beta} + \bar{q}_{li'_\alpha}\bar{q}_{lj_\beta}),$$

where the i's and the i''s are the same lot of objects. This notation presents indices of indices. The $\Pi_\alpha \Pi_{i_\alpha}$ shows that we are to take any collection whatever of i's, and then any i of that collection. We are then to do the same with the j's. We can then find a quality k such that the i taken has it, and also such that the j taken wants it unless we can find an i that is identical with the j taken. The necessity of some kind of notation of this description in treating of classes collectively appears from this consideration: that in such discourse we are neither speaking of a single individual (as in the non-relative logic) nor of a small number of individuals considered each for itself, but of a whole class, perhaps an infinity of individuals. This suggests a relative term with an indefinite series of indices as $x_{ijkl\ldots}$. Such a relative will, however, in most, if not in all cases, be of a degenerate kind and is consequently expressible as above. But it seems preferable to attempt a partial decomposition of this definition. In the first place, any individual may be considered as a class. This is written,

$$\Pi_i \Sigma_k \Pi_j q_{ki}(\bar{q}_{kj} + 1_{ij}).$$

This is the *ninth icon*. Next, given any class, there is another which includes all the former excludes and excludes all the former includes. That is,

$$\Pi_l \Sigma_k \Pi_i (q_{li}\bar{q}_{ki} + \bar{q}_{li}q_{ki}).$$

This is the *tenth icon*.Next, given any two classes, there is a third which includes all that either includes and excludes all that both exclude. That is

$$\Pi_l \Pi_m \Sigma_k \Pi_i (q_{li}q_{ki} + q_{mi}q_{ki} + \bar{q}_{li}\bar{q}_{mi}\bar{q}_{ki}).$$

This is the *eleventh icon*. Next, given any two classes, there is a class which includes the whole of the first and any one individual of the second which there may be not included in the first and nothing else. That is,

$$\Pi_l \Pi_m \Pi_i \Sigma_k \Pi_j \{q_{li} + \bar{q}_{mi} + q_{ki}(q_{kj} + \bar{q}_{lj})\}.$$

This is the *twelfth icon.*

To show the manner in which these formulæ are applied let us suppose we have given that everything is either true of i or false of j. We write

$$\Pi_k(q_{ki} + \bar{q}_{kj}).$$

The tenth icon gives

$$\Pi_l\Sigma_k(q_{li}\bar{q}_{ki} + \bar{q}_{li}q_{ki})\,(q_{lj}\bar{q}_{kj} + \bar{q}_{lj}q_{kj}).$$

Multiplication of these two formulæ gives

$$\Pi_l\Sigma_k(q_{ki}\bar{q}_{li} + q_{lj}\bar{q}_{kj}),$$

or, dropping the terms in k

$$\Pi_l(\bar{q}_{li} + q_{lj}).$$

Mutliplying this with the original datum and identifying l with k, we have

$$\Pi_k\,(q_{ki}q_{kj} + \bar{q}_{ki}\bar{q}_{kj}).$$

No doubt, a much more direct method of procedure could be found.

Just as q signifies the relation of predicate to subject, so we need another token, which may be written r, to signify the conjoint relation of a simple relation, its relate and its correlate. That is, $r_{j_\alpha i}$ is to mean that i is in the relation α to j. Of course, there will be a series of properties of r similar to those of q. But it is singular that the uses of the two tokens are quite different. Namely, the chief use of r is to enable us to express that the number of one class is at least as great as that of another. This may be done in a variety of different ways. Thus, we may write that for every a there is a b, in the first place, thus:

$$\Sigma_a\Pi_i\Sigma_j\Pi_h\{\bar{a}_i + b_jr_{j_\alpha i}(\bar{r}_{j_\alpha h} + \bar{a}_h + 1_{ih})\}.$$

But, by an icon analogous to the eleventh, we have

$$\Pi_\alpha\Pi_\beta\Sigma_\gamma\Pi_u\Pi_v(r_{u\alpha v}r_{u\gamma v} + r_{u\beta v}r_{u\gamma v} + \bar{r}_{u\alpha v}\bar{r}_{u\beta v}\bar{r}_{u\gamma v}).$$

From this, by means of an icon analogous to the *tenth*, we get the general formula

$$\Pi_\alpha\Pi_\beta\Sigma_\gamma\Pi_u\Pi_v\{r_{u\alpha v}r_{u\beta v}r_{u\gamma v} + \bar{r}_{u\gamma v}(\bar{r}_{u\alpha v} + \bar{r}_{u\beta v})\}.$$

For $r_{u\beta v}$ substitute a_u and multiply by the formula the last but two. Then, identifying u with h and v with j, we have

$$\Sigma_a\Pi_i\Sigma_h\Pi_h\{\bar{a}_i + b_jr_{jai}(\bar{r}_{jah} + 1_{ih})\}$$

a somewhat simpler expression. However, the best way to express such a proposition is to make use of the letter c as a token of a one-to-one correspondence. That is to say, c will be defined by the three formulæ,

$$\Pi_a\Pi_u\Pi_v\Pi_w(\bar{c}_a + \bar{r}_{uav} + \bar{r}_{uaw} + 1_{vw})$$
$$\Pi_a\Pi_u\Pi_v\Pi_w(\bar{c}_a + \bar{r}_{uaw} + r_{vaw} + 1_{uv})$$
$$\Pi_a\Sigma_u\Sigma_v\Sigma_w(c_a + r_{uav}r_{uaw}\,\bar{1}_{vw} + r_{uaw}r_{vaw}\,\bar{1}_{uv}).$$

Making use of this token, we may write the proposition we have been considering in the form

$$\Sigma_a \Pi_i \Sigma_j c_a(\bar{a}_i + b_j r_{j\alpha i}).$$

In an appendix to his memoir on the logic of relatives, DeMorgan enriched the science of logic with a new kind of inference, the syllogism of transposed quantity. DeMorgan was one of the best logicians that ever lived and unquestionably the father of the logic of relatives. Owing, however, to the imperfection of his theory of relatives, the new form, as he enunciated it, was a down-right paralogism, one of the premises being omitted. But this being supplied, the form furnishes a good test of the efficacy of a logical notation. The following is one of DeMorgan's examples:

Some X is Y,
For every X there is something neither Y nor Z;
Hence, something is neither X nor Z.

The first premise is simply

$$\Sigma_a x_a y_a.$$

The second may be written

$$\Sigma_a \Pi_i \Sigma_j c_a(\bar{x}_i + r_{jai}\bar{y}_j\bar{z}_j).$$

From these two premises, little can be inferred. To get the above conclusion it is necessary to add that the class of X's is a finite collection; were this not necessary the following reasoning would hold good (the limited universe consisting of numbers); for it precisely conforms to DeMorgan's scheme.

Some odd number is prime;
Every odd number has its square, which is neither prime nor even;
Hence, some number is neither odd nor even.*

Now, to say that a lot of objects is finite, is the same as to say that if we pass through the class from one to another we shall necessarily come round to one of those individuals already passed; that is, if every one of the lot is in any one-to-one relation to one of the lot, then to every one of the lot some one is in this same relation. This is written thus:

$$\Pi_\beta \Pi_u \Sigma_v \Sigma_s \Pi_t \{\bar{c}_\beta + \bar{x}_u + x_v r_{u\beta v} + x_s(\bar{x}_t + \bar{r}_{t\beta s})\}$$

Uniting this with the two premises and the second clause of the definition of c, we have

* Another of DeMorgan's examples is this: "Suppose a person, on reviewing his purchases for the day, finds, by his counterchecks, that he has certainly drawn as many checks on his banker (and maybe more) as he has made purchases. But he knows that he paid some of his purchases in money, or otherwise than by checks. He infers then that he has drawn checks for something else except that day's purchases. He infers rightly enough". Suppose, however, that what happened was this: He bought something and drew a check for it; but instead of paying with the check, he paid cash. He then made another purchase for the same amount, and drew another check. Instead, however, of paying with that check, he paid with the one previously drawn. And thus he continued without cessation, or *ad infinitum*. Plainly the premises remain true, yet the conclusion is false.

$$\Sigma_a\Sigma_\alpha\Pi_\beta\Pi_u\Sigma_v\Sigma_s\Pi_i\Sigma_j\Pi_t\Pi_\gamma\Pi_e\Pi_f\Pi_g x_a y_a c_\alpha(\bar{x}_i + r_{j\alpha i}\bar{y}_j\bar{z}_j)$$
$$\{\bar{c}_\beta + \bar{x}_u + x_v r_{u\beta v} + x_s(\bar{x}_t + \bar{r}_{t\beta s})\}\,(\bar{c}_\gamma + \bar{r}_{e\gamma g} + \bar{r}_{f\gamma w} + 1_{ef}).$$

We now substitute α for β and for γ, a for u and for e, j for t and for f, v for g. The factor in i is to be repeated, putting first s and then v for i. The Boolian part thus reduces to

$$(\bar{x}_s + r_{j\alpha s}\bar{y}_j\bar{z}_j)c_\alpha x_a y_a r_{a\alpha v} x_v r_{j\alpha v}\bar{y}_j\bar{z}_j 1_{aj} + r_{j\alpha s}\bar{y}_j\bar{z}_j x_s\bar{x}_j(\bar{x}_v + r_{j\alpha v}\bar{y}_j\bar{z}_j)$$
$$(\bar{r}_{a\alpha v} + \bar{r}_{j\alpha v} + 1_{aj}),$$

which, by the omission of factors, becomes

$$y_a\bar{y}_j 1_{aj} + \bar{x}_j\bar{z}_j.$$

Thus we have the conclusion

$$\Sigma_j\bar{x}_j\bar{z}_j.$$

It is plain that by a more iconical and less logically analytical notation this procedure might be much abridged. How minutely analytical the present system is, appears when we reflect that every substitution of indices of which nine were used in obtaining the last conclusion is a distinct act of inference. The annulling of $(y_a\bar{y}_j 1_{aj})$ makes ten inferential steps between the premises and conclusion of the syllogism of transposed quantity.

E. THE LOGIC OF MATHEMATICS IN RELATION TO EDUCATION (*PEIRCE 1898*)

The following article appeared in the *Educational review*; despite the concluding declaration that the series was to be continued, no more articles appeared.

References to *Peirce 1898* should be to the paragraph numbers, which have been added in this edition.

---

§1 Of mathematics in general

⟦1⟧ In order to understand what number is, it is necessary first to acquaint ourselves with the nature of the business of mathematics in which number is employed.

⟦2⟧ I wish I knew with certainty the precise origin of the definition of mathematics as the science of quantity. It certainly cannot be Greek, because the Greeks were advanced in projective geometry, whose problems are such as these: whether or not four points obtained in a given way lie in one plane; whether or not four planes have a point in common; whether or not two rays (or unlimited straight lines) intersect, and the like—problems which have nothing to do with quantity, as such. Aristotle names, as the subjects of mathematical study, quantity and continuity. But though he never gives a formal definition of mathematics, he makes quite clear, in more than a dozen places, his view that mathematics ought not to be defined by the things which it studies but by its peculiar mode and degree of abstractness. Precisely what he conceives this to be it would require me to go too far into the technicalities of his philosophy to explain; and I do not suppose anybody would to-day regard the details of his opinion as important for my purpose. Geometry, arithmetic, astronomy, and music were, in the Roman schools of the fifth century[1] and earlier, recognized as the four branches of mathematics. And we find Boethius (A.D. 500) defining them as the arts which relate, not to quantity, but to *quantities*, or *quanta*. What this would seem to imply is, that mathematics is the foundation of the minutely exact sciences; but really it is not worth our while, for the present purpose, to ascertain what the schoolmasters of that degenerate age conceived mathematics to be.

⟦3⟧ In modern times projective geometry was, until the middle of this century, almost forgotten, the extraordinary book of Desargues[2] having been completely lost until, in 1845, Chasles came across a MS copy of it: and, especially before imaginaries became very prominent, the definition of mathematics as the science of quantity suited well enough such mathematics as existed in the seventeenth and eighteenth centuries.

⟦4⟧ Kant, in the *Critique of pure reason* (Methodology, chapter 1, section I), distinctly rejects the definition of mathematics as the science of quantity. What really distinguishes mathematics, according to him, is not the subject of which it treats, but its method, which consists in studying constructions, or diagrams. That such is its method is unquestionably correct; for, even in algebra, the great purpose which the symbolism subserves is to bring a skeleton representation of the relations concerned in the problem before the mind's eye in a schematic shape, which can be studied much as a geometrical figure is studied.

⟦5⟧ But Rowan Hamilton and De Morgan, having a superficial acquaintance with Kant, were just enough influenced by the *Critique* to be led, when they found reason for rejecting the definition as the science of quantity, to conclude that mathematics was the science of pure time and pure space. Notwithstanding the profound deference which every mathematician must pay to Hamilton's opinions and my own admiration for De Morgan, I must say that it is rare to meet with a careful definition of a science so extremely objectionable as this. If Hamilton and De Morgan had attentively read what Kant himself has to say about number, in the first chapter of the *Analytic of principles* and elsewhere, they would have seen that it has no more to do with time and space than has every conception. Hamilton's intention probably was, by means of this definition, to throw a slur upon the introduction of imaginaries into geometry, as a false science; but what De Morgan, who was a student of multiple algebra,

[1] Davidson, *Aristotle and the ancient educational ideals*. Appendix: The Seven Liberal Arts. (New York: Charles Scribner's Sons.)

[2] Brouillon, *Projet d'une atteinte aux événemens des rencontres du cône avec son plan*, 1639.

and whose own formal logic is plainly mathematical, could have had in view, it is hard to comprehend, unless he wished to oppose Boole's theory of logic. Not only do mathematicians study hypotheses which, both in truth and according to the Kantian epistemology, no otherwise relate to time and space than do all hypotheses whatsoever, but we now all clearly see, since the non-Euclidean geometry has become familiar to us, that there *is* a real science of space and a real science of time, and that these sciences are positive and experiential—branches of physics, and so not mathematical except in the sense in which thermotics and electricity are mathematical; that is, as calling in the aid of mathematics. But the gravest objection of all to the definition is that is altogether ignores the veritable characteristics of this science, as they were pointed out by Aristotle and by Kant.

⟦6⟧ Of late decades philosophical mathematicians have come to a pretty just understanding of the nature of their own pursuit. I do not know that anybody struck the true note before Benjamin Pierce ⟦sic⟧, who, in 1870,[3] declared mathematics to be "the science which draws necessary conclusions", adding that it must be defined "subjectively" and not "objectively". A view substantially in accord with his, though needlessly complicated, is given in the article Mathematics, in the ninth edition of the *Encyclopædia Britannica*. The author, Professor George Chrystal, holds that the essence of mathematics lies in its making pure hypotheses, and in the character of the hypotheses which it makes. What the mathematicians mean by a "hypothesis" is a proposition imagined to be strictly true of an ideal state of things. In this sense, it is only about hypotheses that necessary reasoning has any application; for, in regard to the real world, we have no right to presume that any given intelligible proposition is true in absolute strictness. On the other hand, probable reasoning deals with the ordinary course of experience; now, nothing like *a course of experience* exists for ideal hypotheses. Hence to say that mathematics busies itself in drawing necessary conclusions, and to say that it busies itself with hypotheses, are two statements which the logician perceives come to the same thing.

⟦7⟧ A simple way of arriving at a true conception of the mathematician's business is to consider what service it is which he is called in to render in the course of any scientific or other inquiry. Mathematics has always been more or less a trade. An engineer, or a business company (say, an insurance company), or a buyer (say, of land), or a physicist, finds it suits his purpose to ascertain what the necessary consequences of possible facts would be; but the facts are so complicated that he cannot deal with them in his usual way. He calls upon a mathematician and states the question. Now the mathematician does not conceive it to be any part of his duty to verify the facts stated. He accepts them absolutely without question. He does not in the least care whether they are correct or not. He finds, however, in almost every case that the statement has one inconvenience, and in many cases that it has a second. The first inconve-

[3] In his *Linear associative algebra*.

nience is that, though the statement may not at first sound very complicated, yet, when it is accurately analyzed, it is found to imply so intricate a condition of things that it far surpasses the power of the mathematician to say with exactitude what its consequence would be. At the same time, it frequently happens that the facts, as stated, are insufficient to answer the question that is put. Accordingly, the first business of the mathematician, often a most difficult task, is to frame another simpler but quite fictitious problem (supplemented, perhaps, by some supposition), which shall be within his powers, while at the same time it is sufficiently like the problem set before him to answer, well or ill, as a substitute for it.[4] This substituted problem differs also from that which was first set before the mathematician in another respect: namely, that it is highly abstract. All features that have no bearing upon the relations of the premises to the conclusion are effaced and obliterated. The skeletonization or diagrammatization of the problem serves more purposes than one; but its principal purpose is to strip the significant relations of all disguise. Only one kind of concrete clothing is permitted—namely, such as, whether from habit or from the constitution of the mind, has become so familiar that it decidedly aids in tracing the consequences of the hypothesis. Thus, the mathematician does two very different things: namely, he first frames a pure hypothesis stripped of all features which do not concern the drawing of consequences from it, and this he does without inquiring or caring whether it agrees with the actual facts or not; and, secondly, he proceeds to draw necessary consequences from that hypothesis.

⟦8⟧ Kant is entirely right in saying that, in drawing those consequences, the mathematician uses what, in geometry, is called a "construction", or in general a diagram, or visual array of characters or lines. Such a construction is formed according to a precept furnished by the hypothesis. Being formed, the construction is submitted to the scrutiny of observation, and new relations are discovered among its parts, not stated in the precept by which it was formed, and are found, by a little mental experimentation, to be such that they will always be present in such a construction. Thus, the necessary reasoning of mathematics is performed by means of observation and experiment, and its necessary character is due simply to the circumstance that the subject of this observation and experiment is a diagram of our own creation, the conditions of whose being we know all about.

⟦9⟧ But Kant, owing to the slight development which formal logic had received in his time, and especially owing to his total ignorance of the logic of relatives, which throws a brilliant light upon the whole of logic, fell into error in supposing that mathematical and philosophical necessary reasoning are distinguished by the circumstance that the former uses constructions. This is not true. All necessary reasoning whatsoever proceeds by constructions; and the only difference between mathematical and philosophical necessary deductions is that the latter are so excessively simple that the construction attracts no attention

[4] See this well put in Thomson and Tait's *Natural philosophy*, §447.

and is overlooked. The construction exists in the simplest syllogism in Barbara. Why do the logicians like to state a syllogism by writing the major premise on one line and the minor below it, with letters substituted for the subject and predicates? It is merely because the reasoner has to notice that relation between the parts of those premises which such a diagram brings into prominence. If the reasoner makes use of syllogistic in drawing his conclusion, he has such a diagram or construction in his mind's eye, and observes the result of eliminating the middle term. If, however, he trusts to his unaided reason, he still uses some kind of a diagram which is familiar to him personally. The true difference between the necessary logic of philosophy and mathematics is merely one of degree. It is that, in mathematics, the reasoning is frightfully intricate, while the elementary conceptions are of the last degree of familiarity; in contrast to philosophy, where the reasonings are as simple as they can be, while the elementary conceptions are abstruse and hard to get clearly apprehended. But there is another much deeper line of demarcation between the two sciences. It is that mathematics studies nothing but pure hypotheses, and is the only science which never inquiries what the actual facts are; while philosophy, although it uses no microscopes or other apparatus of special observation, is really an experimental science, resting on that experience which is common to us all; so that its principal reasonings are not mathematically necessary at all, but are only necessary in the sense that all the world knows beyond all doubt those truths of experience upon which philosophy is founded. This is why the mathematician holds the reasoning of the metaphysician in supreme contempt, while he himself, when he ventures into philosophy, is apt to reason fantastically and not solidly, because he does not recognize that he is upon ground where elaborate deduction is of no more avail than it is in chemistry or biology.

‖10‖ I have thus set forth what I believe to be the prevalent opinion of philosophical mathematicians concerning the nature of their science. It will be found to be significant for the question of number. But were I to drop this branch of the subject without saying one word more, my criticism of the old definition, "mathematics is the science of quantity", would not be quite just. It must be admitted that quantity is useful in almost every branch of mathematics. Jevons wrote a book entitled *Pure logic, the science of quality*, which expounded, with a certain modification, the logical algebra of Boole. But it is a mistake to regard that algebra as one in which there is no system of quantity. As Boole rightly holds, there is a quadratic equation which is fundamental in it. The meaning of that equation may be expressed as follows: Every proposition has one or other of two *values*, being either *true* (which gives it one value) or *false* (which gives it the other). So stated, we see that the algebra of Boole is nothing but the algebra of that system of quantities which has but two values—the simplest conceivable system of quantity. The widow of the great Boole has lately written a little book[5] in which she points out that, in solving a mathematical problem,

[5] *The Mathematical psychology of Boole and Gratry.*

we usually introduce some part or element into the construction which, when it has served our purpose, is removed. Of that nature is a scale of quantity, together with the apparatus by which it is transported unchanged from one part of the diagram to another, for the purpose of comparing those two parts. Something of this general description seems to be indispensable in mathematics. Take, for example, the Theorem of Pappus concerning ten rays in a plane. The demonstration of it which is now usual, that of von Staudt, introduces a third dimension; and the utility of that arises from the fact that a ray, or unlimited straight line, being the intersection of two planes, these planes show us exactly where the ray runs, while, as long as we confine ourselves to the consideration of a single plane, we have no easy method of describing precisely what the course of the ray is. Now this is not precisely a system of quantity; but it is closely analogous to such a system, and that it serves precisely the same purpose will appear when we remember that that same theorem can easily (though not *so* easily) be demonstrated by means of the barycentric calculus. Although, then, it is not true that all mathematics is a science of quantity, yet it is true that all mathematics makes use of a scaffolding altogether *analogous* to a system of quantity; and quantity itself has more or less utility in every branch of mathematics which has as yet developed into any large theory.

⟦11⟧ I have only to add that the hypotheses of mathematics may be divided into those *general hypotheses* which are adhered to throughout a whole branch of mathematics, and the *particular hypotheses* which are peculiar to different special problems.

CHARLES S. PEIRCE

MILFORD, PA

(To be continued)

F. *FROM* THE SIMPLEST MATHEMATICS (*PEIRCE 1902*)

The following selection is the introductory section of Chapter 3 of the manuscript of the 'Minute logic'; it bears the date 'January–February 1902'. It was first published in *Peirce 1931–58* (Vol. iv, pp. 189–204); the footnotes in double square brackets are taken from that edition. References to *Peirce 1902* should be to the paragraph numbers, which were added in that printing.

---

§1. The essence of mathematics

⟦227⟧ In this chapter, I propose to consider certain extremely simple branches of mathematics which, owing to their utility in logic, have to be treated in considerable detail, although to the mathematician they are hardly worth consideration. In Chapter 4,[a] I shall take up those branches of mathematics upon which the interest of mathematicians is centred, but shall do no more than make a rapid examination of their logical procedure. In Chapter 5,[a] I shall treat formal logic by the aid of mathematics. There can really be little logical matter in these chapters; but they seem to me to be quite indispensable preliminaries to the study of logic.

⟦228⟧ It does not seem to me that mathematics depends in any way upon logic. It reasons, of course. But if the mathematician ever hesitates or errs in his reasoning, logic cannot come to his aid. He would be far more liable to commit similar as well as other errors there. On the contrary, I am persuaded that logic cannot possibly attain the solution of its problems without great use of mathematics. Indeed all formal logic is merely mathematics applied to logic.

⟦229⟧ It was Benjamin Peirce, whose son I boast myself, that in 1870 first defined mathematics as "the science which draws necessary conclusions". This was a hard saying at the time; but today, students of the philosophy of mathematics generally acknowledge its substantial correctness.

⟦230⟧ The common definition, among such people as ordinary schoolmasters, still is that mathematics is the science of quantity. As this is inevitably understood in English, it seems to be a misunderstanding of a definition which may be very old,[1] the original meaning being that mathematics is the science of *quantities*, that is, forms possessing quantity. We perceive that Euclid was aware that a large branch of geometry had nothing to do with measurement (unless as an aid in demonstrating); and, therefore, a Greek geometer of his age (early in the third century B.C.) or later could not define mathematics as the science of that which the abstract noun quantity expresses. A line, however, was classed as a quantity, or *quantum*, by Aristotle[b] and his followers; so that even perspective (which deals wholly with intersections and projections, not at all with lengths) could be said to be a science of quantities, "quantity" being taken in the concrete sense. That this was what was originally meant by the definition "Mathematics is the science of quantity", is sufficiently shown by the circumstance that those writers who first enunciate it, about A.D. 500, that is Ammonius Hermiæ[c] and Boëthius,[d] make astronomy and music branches of

[a] ⟦These chapters were not written.⟧

[1] From what is said by Proclus Diadochus, A.D. 485 ⟦*Commentarii in Primum Euclidis Elementorum Librum*, Prologi pars prior, c. 12⟧, it would seem that the Pythagoreans understood mathematics to be the answer to the two questions "how many?" and "how much?"

[b] ⟦*Metaphysica*, 1020a, 14–20.⟧

[c] ⟦In *Porphyrii Isogogen sine* v *voces*, p. 5v., 1.11 *et seq.*⟧

[d] ⟦*de institutione Arithmetica*, L.I, c. 1.⟧

mathematics; and it is confirmed by the reasons they give for doing so.[2] Even Philo of Alexandria (100 B.C.), who defines mathematics as the science of ideas furnished by sensation and reflection in respect to their necessary consequences, since he includes under mathematics, besides its more essential parts, the theory of numbers and geometry, also the practical arithmetic of the Greeks, geodesy, mechanics, optics (or projective geometry), music, and astronomy, must be said to take the word 'mathematics' in a different sense from ours. That Aristotle did not regard mathematics as the science of quantity, in the modern abstract sense, is evidenced in various ways. The subjects of mathematics are, according to him, the how much and the continuous. (See *Metaph*. K iii 1061 a 33). He referred the continuous to his category of *quantum*; and therefore he did make *quantum*, in a broad sense, the one object of mathematics.

⟦231⟧ Plato, in the Sixth book of the *Republic*,[3] holds that the essential characteristic of mathematics lies in the peculiar kind and degree of its abstraction, greater than that of physics but less than that of what we now call philosophy; and Aristotle[e] follows his master in this definition. It has ever since been the habit of metaphysicians to extol their own reasonings and conclusions as vastly more abstract and scientific than those of mathematics. It certainly would seem that problems about God, Freedom, and Immortality are more exalted than, for example, the question how many hours, minutes, and seconds would elapse before two couriers travelling under assumed conditions will come together; although I do not know that this has been proved. But that the methods of thought of the metaphysicians are, as a matter of historical fact, in any aspect, not far inferior to those of mathematics is simply an infatuation. One singular consequence of the notion which prevailed during the greater part of the history of philosophy, that metaphysical reasoning ought to be similar to that of mathematics, only more so, has been that sundry mathematicians have thought themselves, as mathematicians, qualified to discuss philosophy; and no worse metaphysics than theirs is to be found.

⟦232⟧ Kant regarded mathematical propositions as synthetical judgements *a priori*; wherein there is this much truth, that they are not, for the most part, what he called analytical judgements; that is, the predicate is not, in the sense he intended, contained in the definition of the subject. But if the propositions of arithmetic, for example, are true cognitions, or even forms of cognition, this circumstance is quite aside from their mathematical truth. For all modern mathematicians agree with Plato and Aristotle that mathematics deals exclusively with hypothetical states of things, and asserts no matter of fact whatever; and further, that it is thus alone that the necessity of its conclusions is to be explained.[4] This is the true essence of mathematics; and my father's definition

---

[2] I regret I have not the passage of Ammonius to which I refer. It is probably one of the excerpts given by Brandis. My MS. note states that he gives reasons showing this to be his meaning.

[3] 510C to the end; but in the *Laws* his notion is improved.

[e] ⟦See *Metaphysica*, $1025^{b}1$–$1026^{a}33$; $1060^{b}31$–$1061^{b}34$.⟧

[4] A view which J.S. Mill (*Logic* II, V, §2) rather comically calls "the important doctrine of Dugald Stewart".

is in so far correct that it is impossible to reason necessarily concerning anything else than a pure hypothesis. Of course, I do not mean that if such pure hypothesis happened to be true of an actual state of things, the reasoning would thereby cease to be necessary. Only, it never would be known apodictically to be true of an actual state of things. Suppose a state of things of a perfectly definite, general description. That is, there must be no room for doubt as to whether anything, itself determinate, would or would not come under that description. And suppose, further, that this description refers to nothing occult—nothing that cannot be summoned up fully into the imagination. Assume, then, a range of possibilities equally definite and equally subject to the imagination; so that, so far as the given description of the supposed state of things is general, the different ways in which it might be made determinate could never introduce doubtful or occult features. The assumption, for example, must not refer to any matter of fact. For questions of fact are not within the purview of the imagination. Nor must it be such that, for example, it could lead us to ask whether the vowel *OO* can be imagined to be sounded on as high a pitch as the vowel *EE*. Perhaps it would have to be restricted to pure spatial, temporal, and logical relations. Be that as it may, the question whether in such a state of things, a certain other similarly definite state of things, equally a matter of the imagination, could or could not, in the assumed range of possibility, ever occur, would be one in reference to which one of the two answers, *Yes* and *No*, would be true, but never both. But all pertinent facts would be within the beck and call of the imagination; and consequently nothing but the operation of thought would be necessary to render the true answer. Nor, supposing the answer to cover the whole range of possibility assumed, could this be rendered otherwise than by reasoning that would be apodictic, general, and exact. No knowledge of what actually is, no *positive* knowledge, as we say, could result. On the other hand, to assert that any source of information that is restricted to actual facts could afford us a necessary knowledge, that is, knowledge relating to a whole general range of possibility, would be a flat contradiction in terms.

‖233‖ Mathematics is the study of what is true of hypothetical states of things. That is its essence and definition. Everything in it, therefore, beyond the first precepts for the construction of the hypotheses, has to be of the nature of apodictic inference. No doubt, we may reason imperfectly and jump at a conclusion; still, the conclusion so guessed at is, after all, that in a certain supposed state of things something would necessarily be true. Conversely, too, every apodictic inference is, strictly speaking, mathematics. But mathematics, as a serious science, has, over and above its essential character of being hypothetical, an accidental characteristic peculiarity—a *proprium*, as the Aristotelians used to say—which is of the greatest logical interest. Namely, while all the "philosophers" follow Aristotle in holding no demonstration to be thoroughly satisfactory except what they call a "direct" demonstration, or a "demonstration why"—by which they mean a demonstration which employs only general concepts and concludes nothing but what would be an item of a definition if all its terms were themselves distinctly defined—the mathematicians, on the contrary, entertain a contempt for that style of reasoning, and glory in what the philosophers stigmatize as "mere" indirect demonstrations, or "demonstrations that". Those propositions which can be deduced from others by reasoning of the kind that the philosophers extol are set down by mathematicians as "corollaries". That is to say, they are like those geometrical truths which Euclid did not deem worthy of particular mention, and which his editors inserted with a garland, or corolla, against each in the margin, implying perhaps that it was to them that such honor as might attach to these insignificant remarks was due. In the theorems, or at least in all the major theorems, a different kind of reasoning is demanded. Here, it will not do to confine oneself to general terms. It is necessary to set down, or to imagine, some individual and definite schema, or diagram—in geometry, a figure composed of lines with letters attached; in algebra an array of letters of which some are repeated. This schema is constructed so as to conform to a hypothesis set forth in general terms in the thesis of the theorem. Pains are taken so to construct it that there would be something closely similar in every possible state of things to which the hypothetical description in the thesis would be applicable, and furthermore to construct it so that it shall have no other characters which could influence the reasoning. How it can be that, although the reasoning is based upon the study of an individual schema, it is nevertheless necessary, that is, applicable, to all possible cases, is one of the questions we shall have to consider. Just now, I wish to point out that after the schema has been constructed according to the precept virtually contained in the thesis, the assertion of the theorem is not evidently true, even for the individual schema; nor will any amount of hard thinking of the philosophers' corollarial kind ever render it evident. Thinking in general terms is not enough. It is necessary that something should be DONE. In geometry, subsidiary lines are drawn. In algebra permissible transformations are made. Thereupon, the faculty of observation is called into play. Some relation between the parts of the schema is remarked. But would this relation subsist in every possible case? Mere corollarial reasoning will sometimes assure us of this. But, generally speaking, it may be necessary to draw distinct schemata to represent alternative possibilities. Theorematic reasoning invariably depends upon experimentation with individual schemata. We shall find that, in the last analysis, the same thing is true of the corollarial reasoning, too; even the Aristotelian "demonstration why". Only in this case, the very words serve as schemata. Accordingly, we may say that corollarial, or "philosophical" reasoning is reasoning with words; while theorematic, or mathematical reasoning proper, is reasoning with specially constructed schemata.

‖234‖ Another characteristic of mathematical thought is the extraordinary use it makes of abstractions. Abstractions have been a favorite butt of ridicule in modern times. Now it is very easy to laugh at the old physician who is represented as answering the question, why opium puts people to sleep, by saying that it is because it has a dormative virtue. It is an answer that no doubt carries vagueness to its last extreme. Yet, invented as the story was to show how little meaning there might be in an abstraction, nevertheless the physician's answer does contain a truth that modern philosophy has generally denied: it does assert that there really is in opium *something* which explains its always putting people to sleep. This has, I say, been denied by modern philosophers generally. Not, of course, explicitly; but when they say that the different events of people going to sleep after taking opium have really nothing in common, but only that the mind classes them together—and this is what they virtually do say in denying the reality of generals—they do implicitly deny that there is any true explanation of opium's generally putting people to sleep.

⟦235⟧ Look through the modern logical treatises, and you will find that they almost all fall into one or other of two errors, as I hold them to be; that of setting aside the doctrine of abstraction (in the sense in which an abstract noun marks an abstraction) as a grammatical topic with which the logician need not particularly concern himself; and that of confounding abstraction, in this sense, with that operation of the mind by which we pay attention to one feature of a percept to the disregard of others. The two things are entirely disconnected. The most ordinary fact of perception, such as "it is light", involves *precisive* abstraction, or *prescission*. But *hypostatic* abstraction, the abstraction which transforms "it is light" into "there is light here", which is the sense which I shall commonly attach to the word abstraction (since *prescission* will do for precisive abstraction) is a very special mode of thought. It consists in taking a feature of a percept or percepts (after it has already been prescinded from the other elements of the percept), so as to take propositional form in a judgement (indeed, it may operate upon any judgement whatsoever), and in conceiving this fact to consist in the relation between the subject of that judgement and another subject, which has a mode of being that merely consists in the truth of propositions of which the corresponding concrete term is the predicate. Thus, we transform the proposition, "honey is sweet", into "honey possesses sweetness". "Sweetness" might be called a fictitious thing, in one sense. But since the mode of being attributed to it *consists* in no more than the fact that some things are sweet, and it is not pretended, or imagined, that it has any other mode of being, there is, after all, no fiction. The only profession made is that we consider the fact of honey being sweet under the form of a relation; and so we really can. I have selected sweetness as an instance of one of the least useful of abstractions. Yet even this is convenient. It facilitates such thoughts as that the sweetness of honey is particularly cloying; that the sweetness of honey is something like the sweetness of a honeymoon; etc. Abstractions are particularly congenial to mathematics. Everyday life first, for example, found the need of that class of abstractions which we call *collections*. Instead of saying that some human beings are males and all the rest females, it was found convenient to say that *mankind* consists of the male *part* and the female *part*. The same thought makes classes of collections, such as pairs, leashes, quatrains, hands, weeks, dozens, baker's dozens, sonnets, scores, quires, hundreds, long hundreds, gross, reams, thousands, myriads, lacs, millions, milliards, milliasses, etc. These have suggested a great branch of mathematics.[5] Again, a point moves: it is by abstraction that the geometer says that it "describes a line". This line, though an abstraction, itself moves; and this is regarded as generating a surface; and so on. So likewise, when the analyst treats operations as themselves subjects of operations, a method whose utility will not be denied, this is another instance of abstraction. Maxwell's notion of a tension exercised upon lines of electrical force, transverse to them, is somewhat similar. These examples exhibit the great rolling billows of abstraction in the ocean of mathematical thought; but when we come to a minute examination of it, we shall find, in every department, incessant ripples of the same form of thought, of which the examples I have mentioned give no hint.

⟦236⟧ Another characteristic of mathematical thought is that it can have no success where it cannot generalize. One cannot, for example, deny that chess is mathematics, after a fashion; but, owing to the exceptions which everywhere confront the mathematician in this field—such as the limits of the board; the single steps of king, knight, and pawn; the finite number of squares; the peculiar mode of capture by pawns; the queening of pawns; castling—there results a mathematics whose wings are effectually clipped, and which can only run along the ground. Hence it is that a mathematician often finds what a chess-player might call a gambit to his advantage; exchanging a smaller problem that involves exceptions for a larger one free from them. Thus, rather than suppose that parallel lines, unlike all other pairs of straight lines in a plane, never meet, he supposes that they intersect at infinity. Rather than suppose that some equations have roots while others have not, he supplements real quantity by the infinitely greater realm of imaginary quantity. He tells us with ease how many inflexions a plane curve of any description has; but if we ask how many of these are real, and how many merely fictional, he is unable to say. He is perplexed by three-dimensional space, because not all pairs of straight lines intersect, and finds it to his advantage to use quaternions which represent a sort of four-fold continuum, in order to avoid the exception. It is because exceptions so hamper the mathematician that almost all the relations with which he chooses to deal are of the nature of correspondences; that is to say, such relations that for every relate there is the same number of correlates, and for every correlate the same number of relates.

⟦237⟧ Among the minor, yet striking characteristics of mathematics, may be mentioned the fleshless and skeletal build of its propositions; the peculiar difficulty, complication, and stress of its reasonings; the perfect exactitude of its results; their broad universality; their practical infallibility. It is easy to speak with precision upon a general theme. Only, one must commonly surrender all ambition to be certain. It is equally easy to be certain. One has only to be suffi-

[5] Of course, the moment a collection is recognized as an abstraction we have to admit that even a percept is an abstraction or represents an abstraction, if matter has parts. It therefore becomes difficult to maintain that all abstractions are fictions.

ciently vague. It is not so difficult to be pretty precise and fairly certain at once about a very narrow subject. But to reunite, like mathematics, perfect exactitude and practical infallibility with unrestricted universality, is remarkable. But it is not hard to see that all these characters of mathematics are inevitable consequences of its being the study of hypothetical truth.

‖238‖ It is difficult to decide between the two definitions of mathematics; the one by its method, that of drawing necessary conclusions; the other by its aim and subject matter, as the study of hypothetical states of things. The former makes or seems to make the deduction of the consequences of hypotheses the sole business of the mathematician as such. But it cannot be denied that immense genius has been exercised in the mere framing of such general hypotheses as the field of imaginary quantity and the allied idea of Riemann's surface, in imagining non-Euclidian measurement, ideal numbers, the perfect liquid. Even the framing of the particular hypotheses of special problems almost always calls for good judgement and knowledge, and sometimes for great intellectual power, as in the case of Boole's logical algebra. Shall we exclude this work from the domain of mathematics? Perhaps the answer should be that, in the first place, whatever exercise of intellect may be called for in applying mathematics to a question not propounded in mathematical form ‖it‖ is certainly not pure mathematical thought; and in the second place, that the mere creation of a hypothesis may be a grand work of poietic genius, but cannot be said to be scientific, inasmuch as that which it produces is neither true nor false, and therefore is not knowledge. This reply suggests the further remark that if mathematics is the study of purely imaginary states of things, poets must be great mathematicians, especially that class of poets who write novels of intricate and enigmatical plots. Even the reply, which is obvious, that by *studying* imaginary states of things we mean *studying* what is true of them, perhaps does not fully meet the objection. The article *Mathematics* in the ninth edition of the *Encyclopædia Britannica* makes mathematics consist in the study of a particular sort of hypotheses, namely, those that are exact, etc., as there set forth at some length. The article is well worthy of consideration.

‖239‖ The philosophical mathematician, Dr. Richard Dedekind, holds mathematics to be a branch of logic. This would not result from my father's definition, which runs, not that mathematics is the science of *drawing* necessary conclusions—which would be deductive logic—but that it is the science which *draws* necessary conclusions. It is evident, and I know as a fact, that he had this distinction in view. At the time when he thought out this definition, he, a mathematician, and I, a logician, held daily discussions about a large subject which interested us both; and he was struck, as I was, with the contrary nature of his interest and mine in the same propositions. The logician does not care particularly about this or that hypothesis or its consequences, except so far as these things may throw a light upon the nature of reasoning. The mathematician is intensely interested in efficient methods of reasoning, with a view to their possible extension to new problems; but he does not, *quâ* mathematician, trouble himself minutley to dissect those parts of this method whose correctness is a matter of course. The different aspects which the algebra of logic will assume for the two men is instructive in this respect. The mathematician asks what value this algebra has as a calculus. Can it be applied to unravelling a complicated question? Will it, at one stroke, produce a remote consequence? The logician does not wish the algebra to have that character. On the contrary, the greater number of distinct logical steps, into which the algebra breaks up an inference, will for him constitute a superiority of it over another which moves more swiftly to its conclusions. He demands that the algebra shall analyze a reasoning into its last elementary steps. Thus, that which is a merit in a logical algebra for one of these students is a demerit in the eyes of the other. The one studies the science of drawing conclusions, the other the science which draws necessary conclusions.

‖240‖ But, indeed, the difference between the two sciences is far more than that between two points of view. Mathematics is purely hypothetical: it produces nothing but conditional propositions. Logic, on the contrary, is categorical in its assertions. True, it is not merely, or even mainly, a mere discovery of what really is, like metaphysics. It is a normative science. It thus has a strongly mathematical character, at least in its methodeutic division; for here it analyzes the problem of how, with given means, a required end is to be pursued. This is, at most, to say that it has to call in the aid of mathematics; that it has a mathematical branch. But so much may be said of every science. There is a mathematical logic, just as there is a mathematical optics and a mathematical economics. Mathematical logic is formal logic. Formal logic, however developed, is mathematics. Formal logic, however, is by no means the whole of logic, or even its principal part. It is hardly to be reckoned as a part of logic proper. Logic has to define its aim; and in doing so is even more dependent upon ethics, or the philosophy of aims, by far, than it is, in the methodeutic branch, upon mathematics. We shall soon come to understand how a student of ethics might well be tempted to make his science a branch of logic; as, indeed, it pretty nearly was in the mind of Socrates. But this would be no truer a view than the other. Logic depends upon mathematics; still more intimately upon ethics; but its proper concern is with truths beyond the purview of either.

‖241‖ There are two characters of mathematics which have not yet been mentioned, because they are not exclusive characteristics of it. One of these, which need not detain us, is that mathematics is distinguished from all other sciences except only ethics, in standing in no need of ethics. Every other science, even logic—logic, especially—is in its early stages in danger of evaporating into airy nothingness, degenerating, as the Germans say, into an anachrioid film, spun from the stuff that dreams are made of. There is no such danger for pure mathematics; for that is precisely what mathematics ought to be.

‖242‖ The other character—and of particular interest it is to us just now—is that mathematics, along with ethics and logic alone of the sciences, has no need of any appeal to logic. No doubt, some reader may exclaim in dissent to this, on first hearing it said. Mathematics, they may say, is preëminently a science of reasoning. So it is; preëminently a science that reasons. But just as it is not necessary, in order to talk, to understand the theory of the formation of vowel sounds, so it is not necessary, in order to reason, to be in possession of the theory of reasoning. Otherwise, plainly, the science of logic could never be developed. The contrary objection would have more excuse, that no science stands in need of logic, since our natural power of reason is enough. Make of logic what the majority of treatises in the past have made of it, and a very common class of English and French books still make of it—that is to say, mainly formal logic, and that formal logic represented as an art of reasoning—and in my opinion this objection is more than sound, for such logic is a great hindrance to right reasoning. It would, however, be aside from our present purpose to examine this objection minutely. I will content myself with saying that undoubtedly our natural power of reasoning is enough, in the same sense that it is enough, in order to obtain a wireless transatlantic telegraph, that men should be born. That is to say, it is bound to come sooner or later. But that does not make research into the nature of electricity needless for gaining such a telegraph. So likewise if the study of electricity had been pursued resolutely, even if no special attention had ever been paid to mathematics, the requisite mathematical ideas would surely have been evolved. Faraday, indeed, did evolve them without any acquaintance with mathematics. Still it would be far more economical to postpone electrical researches, to study mathematics by itself, and then to apply it to electricity, which was Maxwell's way. In this same manner, the various logical difficulties which arise in the course of every science except mathematics, ethics, and logic, will, no doubt, get worked out after a time, even though no special study of logic be made. But it would be far more economical to make first a systematic study of logic. If anybody should ask what are these logical difficulties which arise in all the sciences, he must have read the history of science very irreflectively. What was the famous controversy concerning the measure of force but a logical difficulty? What was the controversy between the uniformitarians and the catastrophists but a question of whether or not a given conclusion followed from acknowledged premisses? This will fully appear in the course of our studies in the present work.[f]

⟦243⟧ But it may be asked whether mathematics, ethics, and logic have not encountered similar difficulties. Are the doctrines of logic at all settled? Is the history of ethics anything but a history of controversy? Have no logical errors been committed by mathematicians? To that I reply, first, as to logic; that not only have the rank and file of writers on the subject been, as an eminent psychiatrist, Maudsley, declares, men of arrested brain-development, and not only have they generally lacked the most essential qualification for the study, namely mathematical training, but the main reason why logic is unsettled is that thirteen different opinions are current as to the true aim of the science. Now this is not a logical difficulty but an ethical difficulty; for ethics is the science of aims. Secondly, it is true that pure ethics has been, and always must be, a theatre

[f] ⟦This point is not discussed in the 'Minute Logic'.⟧

of discussion, for the reason that its study consists in the gradual development of a distinct recognition of a satisfactory aim. It is a science of subtleties, no doubt; but it is not logic, but the development of the ideal, which really creates and resolves the problems of ethics. Thirdly, in mathematics errors of reasoning have occurred, nay, have passed unchallenged for thousands of years. This, however, was simply because they escaped notice. Never, in the whole history of the science, has a question whether a given conclusion followed *mathematically* from given premisses, when once started, failed to receive a speedy and unanimous reply. Very few have been even the apparent exceptions; and those few have been due to the fact that it is only within the last half century that mathematicians have come to have a perfectly clear recognition of what is mathematical soil and what foreign to mathematics. Perhaps the nearest approximation to an exception was the dispute about the use of divergent series. Here neither party was in possession of sufficient pure mathematical reasons covering the whole ground; and such reasons as they had were not only of an extra-mathematical kind, but were used to support more or less vague positions. It appeared then, as we all know now, that divergent series are of the utmost utility.[6]

Struck by this circumstance, and making an inference, of which it is sufficient to say that it was not mathematical, many of the old mathematicians pushed the use of divergent series beyond all reason. This was a case of mathematicians disputing about the validity of a kind of inference that is not mathematical. No doubt, a sound logic (such as has not hitherto been developed) would have shown clearly that that non-mathematical inference was not a sound one. But this is, I believe, the only instance in which any large party in the mathematical world ever proposed to rely, in mathematics, upon unmathematical reasoning. My proposition is that true mathematical reasoning is so much more evident than it is possible to render any doctrine of logic proper—without just such reasoning—that an appeal in mathematics to logic could only embroil a situation. On the contrary, such difficulties as may arise concerning necessary reasoning have to be solved by the logician by reducing them to questions of mathematics. Upon those mathematical dicta, as we shall come clearly to see,

---

[6] It would not be fair, however, to suppose that every reader will know this. Of course, there are many series so extravagantly divergent that no use at all can be made of them. But even when a series is divergent from the very start, some use might commonly be made of it, if the same information could not otherwise be obtained more easily. The reason is—or rather, one reason is—that most series, even when divergent, approximate at last somewhat to geometrical series, at least, for a considerable succession of terms. The series $\log(1 + x) = x - \frac{1}{2}x^2 + \frac{1}{3}x^3 - \frac{1}{4}x^4 +$, etc., is one that would not be judiciously employed in order to find the natural logarithm of 3, which is 1.0986, its successive terms being $2 - 2 + 8/3 - 4 + 32/5 - 32/3 +$, etc. Still, employing the common device of substituting for the last two terms that are to be used, say M and N, the expression $M/(1 - N/M)$, the succession of the first six values is 0.667, 1.143, 1.067, 1.128, 1.067, which do show some approximation to the value. The mean of the last two, which any professional computer would use (supposing him to use this series, at all) would be 1.098, which is not very wrong. Of course, the computer would practically use the series $\log 3 = 1 + 1/12 + 1/80 + 1/448 +$, etc., of which the terms written give the correct values to four places, if they are properly used.

the logician has ultimately to repose.

⟦244⟧ So a double motive induces me to devote some preliminary chapters to mathematics. For, in the first place, in studying the theory of reasoning, we are concerned to acquaint ourselves with the methods of that prior science of which acts of reasoning form the staple. In the second place, logic, like any other science, has its mathematical department, and of that, a large portion, at any rate, may with entire convenience be studied as soon as we take up the study of logic, without any propedeutic. That portion is what goes by the name of Formal Logic.[7] It so happens that the special kind of mathematics needed for formal logic, which, therefore, we need to study in detail, as we need not study other branches of mathematics, is so excessively simple as neither to have much mathematical interest, nor to display the peculiarities of mathematical reasoning. I shall, therefore, devote the present chapter—a very dull one, I am sorry to say, it must be—to this kind of mathematics. Chapter 4 will treat of the more truly mathematical mathematics; and Chapter 5 will apply the results of the present chapter to the study of Formal Logic.

[7] "Formal Logic" is also used, by Germans chiefly, to mean that sect of Logic which makes Formal Logic pretty much the whole of Logic.

References

The following References contain only works referred to in Volume I. A full Bibliography, incorporating all references from both volumes in this collection, can be found at the end of Volume II. The principles of dating are explained in the Introduction. Works marked with an asterisk are contained, either in whole or in part, in the present collection. To make the Bibliography as useful for historical reference as possible, an effort was made to supply full names of authors whenever they could be determined. Reprints or translations are listed only when they were considered likely to be of assistance to the reader. In addition to the titles listed here, the reader will find additional specialist bibliographical information in *Benacerraf and Putnam 1983, M. Cantor 1894–1908, Hallett 1984, van Heijenoort 1967, Hofmann 1963, Moore 1982*, and *Stäckel and Engel 1895*, as well as in the collected works of the major mathematicians represented in this collection.

Aristotle

1984 *The complete works of Aristotle*, (ed. Jonathan Barnes). Princeton University Press.

Baltzer, Richard

1866–7 *Die Elemente der Mathematik*. Hirzel, Leipzig.

Baron, Margaret Elanor

1969 *The origins of the infinitesimal calculus*. Pergamon, Oxford.

Barrow, Isaac

1670 *Lectiones geometricae*. In *The mathematical works of Isaac Barrow* (ed. William Whewell, Cambridge University Press, 1860).

Baumann, Julius

1868–9 *Die Lehre von Raum, Zeit und Mathematik*, 2 vols. Reimer, Berlin.

1908 Dedekind und Bolzano. *Annalen der Naturphilosophie*, 7, 444–9.

Baumgarten, Alexander Gottlieb

1739 *Metaphysica*. Hemmerde, Halle.

Baxter, Andrew

c.1730 *An enquiry into the nature of the human soul*. J. Bettenham, London.

Becker, Oskar (ed.)
1964 *Grundlagen der Mathematik in geschichtlicher Entwicklung* (2nd extended edn, 1974). Suhrkamp, Frankfurt.

Benacerraf, Paul and Putnam, Hilary (eds.)
1964 *Philosophy of mathematics: Selected readings.* Prentice-Hall, Englewood Cliffs, NJ.
1983 Second edition of *Benacerraf and Putnam 1964.*

Berkeley, George
1707 *Arithmetica et miscellanea mathematica.* Churchill, London and Pepyat, Dublin.
1707–8* *Philosophical commentaries.* (First published in *Berkeley 1871*; reprinted in *Berkeley 1948–57*, Vol. 1, pp. 7–139, and partially reprinted in this collection, Vol. 1, pp. 13–16.)
1710* *A treatise concerning the principles of human knowledge.* Rhames, Dublin. (2nd edn: Jacob Tonson, London, 1734. Partially reprinted in this collection, Vol. 1, pp. 21–37.)
1721* *De motu.* Jacob Tonson, London. (Reprinted, with a translation by the editors, in *Berkeley 1948–57*, Vol. 4, pp. 11–52. Translation reprinted in this collection, Vol. 1, pp. 37–54.)
1732 *Alciphron: or, the minute philosopher. In seven dialogues. Containing an apology for the Christian religion, against those who are called free-thinkers.* J. Tonson, London and G. Risk, Dublin.
1734* *The analyst, or, A discourse addressed to an infidel mathematician.* Jacob Tonson, London. (Reprinted in *Berkeley 1948–57*, Vol. 4, pp. 65–102, and in this collection, Vol. 1, pp. 60–92.)
1735a *A defence of free-thinking in mathematics.* Rhames, Dublin. (Reprinted in *Berkeley 1948–57*, Vol. 4, pp. 109–41.)
1735b *Reasons for not replying to Mr. Walton's* Full answer. (Reprinted in *Berkeley 1948–57*, Vol. 4, pp. 147–56.)
1871 *The works of George Berkeley, D.D., formerly Bishop of Cloyne: including many of his writings hitherto unpublished*, 4 vols, (ed. Alexander Campbell Fraser). Clarendon Press, Oxford.
1901* Of infinites. (First printed in 1901 in *Hermathena*, Vol. 11 (ed. Swift Payne Johnston). Reprinted in *Berkeley 1948–57*, Vol. 4, pp. 235–8, and in this collection, Vol. 1, pp. 16–19. Probably written in 1707.)
1948–57 *The works of George Berkeley Bishop of Cloyne*, 9 vols, (ed. Arthur Aston Luce and Thomas Edmund Jessop). Thomas Nelson, London.

Bernoulli, Jacob (= Jacques = James) (1654–1705)
1744 *Opera*, 2 vols. G. Cramer, Geneva.

Bernoulli, John (= Johann = Jean) (1667–1748)
1742 *Opera omnia.* Bousquet, Lausanne and Geneva.

Bertrand, Joseph
1849 *Traité d'arithmétique.* Hachette, Paris.

Birkhoff, Garrett (ed.)
1973a *A source book in classical analysis.* Harvard University Press, Cambridge, Mass.

Bolzano, Bernard

1804* *Betrachtungen über einige Gegenstände der Elementargeometrie.* Karl Barth, Prague. (English translation by Steven Russ, this collection, Vol. 1, pp. 172–4.)

1810* *Beiträge zu einer begründeteren Darstellung der Mathematik.* Caspar Widtmann, Prague. (English translation by Steven Russ, this collection, Vol. 1, pp. 174–224.)

1816 *Der binomische Lehrsatz, und als Folgerung aus ihm der polynomische und die Reihen, die zur Berechnung der Logarithmen und Exponentialgrössen dienen, genauer als bisher erwiesen.* C.W. Enders, Prague.

1817a* *Rein analytischer Beweis des Lehrsatzes, dass zwischen je zwei Werten, die ein entgegengesetztes Resultat gewähren, wenigstens eine reelle Wurzel der Gleichung liege.* Gottlieb Haase, Prague. (English translation by Steven Russ, this collection, Vol. 1, pp. 225–48.)

1817b *Die drey Probleme der Rectification, der Complanation und der Cubierung, ohne Betrachtung des unendlich Kleinen, ohne die Annahme des Archimedes und ohne irgend eine nicht streng erweisliche Voraussetzung gelöst; zugleich als Probe einer gänzlichen Umgestaltung der Raumwissenschaft allen Mathematikern zur Prüfung vorgelegt.* Gotthelf Kummer, Leipzig.

1836 *Lebensbeschreibung des Dr. B. Bolzano mit einigen seiner ungedrückten Aufsätze.* Sulzbach.

1837 *Dr. Bolzanos Wissenschaftslehre. Versuch einer ausführlichen und grössenteils neuen Darstellung der Logik, mit steter Rücksicht auf deren bisherige Bearbeiter.* Seidel, Sulzbach.

1842 *Versuch einer objektiven Begründung der Lehre von der Zusammensetzung der Kräfte.* Kronberger and Rziwnas, Prague.

1843 *Versuch einer objektiven Begründung der Lehre von den drei Dimensionen des Raumes.* Kronberger and Rziwnas, Prague.

1849 *Über die Einteilung der schönen Künste. Eine aesthetische Abhandlung.* Gottlieb Haase, Prague.

1851 *Dr. Bernard Bolzanos Paradoxien des Unendlichen, herausgegeben aus dem schriftlichen Nachlasse des Verfassers von Dr. Fr. Prihonsky.* Reclam, Leipzig.

1930 *Funktionenlehre* (ed. Karel Rychlik). Königliche Böhmische Gesellschaft der Wissenschaften, Prague. (Written 1834.)

1950* *Paradoxes of the infinite.* Routledge and Kegan Paul, London. (Translation by Donald A. Steele of *Bolzano 1851.* Partially reprinted in this collection, Vol. 1, pp. 249–92.)

1969– *Bernard Bolzano Gesamtausgabe* (ed. Eduard Winter *et alii*). Fromann, Stuttgart.

1972 *Theory of science* (ed. and trans. Rolf George). Blackwell, Oxford. (Selections from *Bolzano 1837.*)

See Russ, Steven.

Bonola, Roberto

1955 *Non-Euclidean geometry, A critical and historical study of its developments* (trans. H.S. Carslaw). Dover, New York.

Boole, George

1841 On certain theorems in the calculus of variations. *Cambridge Mathematical Journal*, **2**, 97–102.

1842a Exposition of a general theory of linear transformations, Part I. *Cambridge Mathematical Journal*, **3**, 1-20.

1842b Exposition of a general theory of linear transformations, Part II. *Cambridge Mathematical Journal*, **3**, 106-19.

1844 On a general method in analysis. *Philosophical Transactions of the Royal Society of London*, **134**, 225-82.

1845 Notes on linear transformations. *Cambridge Mathematical Journal*, **4**, 167-71.

1847* *The mathematical analysis of logic*. Macmillan, Barclay, and Macmillan, Cambridge. (Reprinted Blackwell, Oxford, 1948, and in this collection, Vol. 1, pp. 451-509.)

1848 The calculus of logic. *Cambridge and Dublin Mathematical Journal*, May 1848. (Reprinted in *Boole 1952*, pp. 125-40.)

1851 On the theory of probabilities, and in particular on Mitchell's problem of the distribution of the fixed stars. *Philosophical Magazine*, **1**, 521-30.

1854 *An investigation of the laws of thought, on which are founded the mathematical theories of logic and probabilities*. Macmillan, Cambridge.

1859 *A treatise on differential equations*. Macmillan, Cambridge.

1860 *A treatise on the calculus of finite differences*. Macmillan, Cambridge.

1952 *Studies in logic and probability* (ed. Rush Rhees). Watts, London; and Open Court, La Salle.

Boole, George and De Morgan, Augustus

1982 *The Boole-De Morgan correspondence, 1842-1864* (ed. G.C. Smith). Oxford University Press.

Boole, Mary Everest

1878 The home side of a scientific mind. *The University Magazine*, **1**, 105-14; 173-83; 326-36; 454-60.

Bourbaki, Nicolas (or Nicholas, or Nicolaus)

1969 *Eléments d'histoire des mathématiques* (2nd edn). Hermann, Paris.

Boyer, Carl B.

1939 *The concepts of the calculus*. Columbia University Press, New York. (Reprinted 1949 and 1959 with the title, *The history of the calculus and its conceptual development*. Dover, New York.)

Cajori, Florian

1919 *A history of the conceptions of limits and fluxions in Great Britain from Newton to Woodhouse*. Open Court, Chicago.

1924 *A history of mathematics*. Macmillan, New York.

1925 Indivisibles and 'ghosts of departed quantities' in the history of mathematics. *Scientia*, **37**, 303-6.

1928 *A history of mathematical notations, Vol. 1: Notations in elementary mathematics*. Open Court, La Salle.

Cantor, Georg

1874* Über eine Eigenschaft des Inbegriffs aller reellen algebraischen Zahlen. *Journal für die reine und angewandte Mathematik*, **77**, 258-62. (Reprinted in *Cantor 1932*, pp. 115-18. Translated by William Ewald, this collection, Vol. 2, pp. 839-43.)

1883d* *Grundlagen einer allgemeinen Mannigfaltigkeitslehre. Ein mathematisch-philosophischer Versuch in der Lehre des Unendlichen.*

Teubner, Leipzig. (Separate printing of *Cantor 1883b*, with an additional preface and footnotes not reproduced in *Cantor 1932*. Translated by William Ewald, this collection, Vol. 2, pp. 878–920.)

Cantor, Moritz
1894–1908 *Vorlesungen über die Geschichte der Mathematik*, 4 vols, (2nd edn). Teubner, Leipzig.

Carnot, Lazare Nicolas Marguerite
1813 *Réflexions sur la métaphysique du calcul infinitésimal* (2nd edn). Courcier, Paris. (1st edn: 1787, Duprat, Paris).

Cauchy, Augustin-Louis
1821 *Cours d'analyse algébrique*. Imprimerie Royale, Paris. (Reprinted in *Cauchy 1882–1958*, Vol. 3 (1897).)
1882–1958 *Œuvres complètes d'Augustin Cauchy*, 12 + 15 vols. Gauthier-Villars, Paris.

Cayley, Arthur
1854 On the theory of groups. *Philosophical Magazine*, 7, 40–7. (Reprinted in *Cayley 1889–98*, Vol. 2, pp. 123–30.)
1857 Note on the recent progress of theoretical dynamics, *Reports of the British Association, 1857*, 1–42. (Reprinted in *Cayley 1889–98*, Vol. 3, pp. 156–204.)
1883* Presidential address to the British Association, September 1883. *Report of the British Association for the Advancement of Science*, (1883), 3–37. (Reprinted in *Cayley 1889–98*, Vol. 9, pp. 429–59, and in this collection, Vol. 1, pp. 542–73.)
1889–98 *Collected mathematical papers*, 13 vols. Cambridge University Press.

Clairault, Alexis Claude
1797 *Elémens d'algèbre*, 2 vols, (5th edn). Duprat, Paris.

Clifford, William Kingdon
1872* On the aims and instruments of scientific thought. (Lecture delivered before the members of the British Association, Brighton, 19 August 1872; first published in *Clifford 1901*, Vol. 1, pp. 139–80. Reprinted in this collection, Vol. 1, pp. 524–41.)
1873* On the hypotheses which lie at the bases of geometry *Nature*, 7, 14–17, 36, 37, 183–4. (Translation of *Riemann 1868*. Reprinted in *Clifford 1882*, pp. 55–71, and in this collection, Vol. 2, pp. 652–61.)
1876* On the space theory of matter. *Proceedings of the Cambridge Philosophical Society*, 2, 157–8. (Read 21 Feb. 1870; reprinted in *Clifford 1882*, pp. 21–2, and in this collection, Vol. 1, pp. 523–4.)
1882 *Mathematical papers*. Macmillan, London.
1885 *The common sense of the exact sciences*. Kegan, Paul, London. (Reprinted, with an introduction by James R. Newman, Knopf, New York, 1946.)
1901 *Lectures and essays*, 2 vols, (ed. Leslie Stephen and Sir Frederick Pollock). Macmillan, London.

Couturat, Louis
1901 *La logique de Leibniz d'après des documents inédits*. Alcan, Paris.

Crowe, Michael J.
1967 *A history of vector analysis: the evolution of the idea of a vectorial system*. University of Notre Dame Press, Indiana.

D'Alembert, Jean le Rond
1751 *Discours préliminaire de l'Encyclopédie*. In *Diderot* et alii *1751–72*, Vol. 1, pp. i-xlv.
1754* Différentiel. In *Diderot* et alii *1751–72*, Vol. 4. (English translation by William Ewald and Dirk J. Struik, this collection, Vol. 1, pp. 123–8.)
1765a* Infini. In *Diderot* et alii *1751–72*, Vol. 8. (English translation by William Ewald, this collection, Vol. 1, pp. 128–30.)
1765b* Limite. In *Diderot* et alii *1751–72*, Vol. 9. (English translation by William Ewald, this collection, Vol. 1, pp. 130–1.)

Dauben, Joseph W.
1982 Peirce's place in mathematics. *Historia Mathematica*, **9**, 311–25.

Dedekind, Richard
1872* *Stetigkeit und irrationale Zahlen*. Vieweg, Braunschweig. (Translated by Wooster W. Beman as 'Continuity and irrational numbers' in *Dedekind 1901*. Reprinted, with corrections by William Ewald, this collection, Vol. 2, pp. 765–79.)
1877* Sur la théorie des nombres entiers algébriques. *Bulletin des sciences mathématiques et astronomiques*, **11**, 278–88. (Reprinted in *Dedekind 1930–2*, Vol. 3, pp. 262–98. Translated by David Reed, this collection, Vol. 2, pp. 779–87.)
1888* *Was sind und was sollen die Zahlen*? Vieweg, Braunschweig. (Translated by Wooster W. Beman as 'The Nature and Meaning of Numbers' in *Dedekind 1901*. Reprinted, with corrections, by William Ewald, this collection, Vol. 2, pp. 787–833.
1901 *Essays on the theory of numbers* (trans. Wooster W. Beman). Open Court, Chicago, (Reprinted Dover, New York, 1963.)

De Morgan, Augustus
1831 *On the study and difficulties of mathematics*. Society for the Diffusion of Useful Knowledge, London.
1836 *A treatise of the calculus of functions*. Encyclopedia Metropolitana, London.
1837 *The elements of algebra preliminary to the differential calculus and fit for the higher classes of schools in which the principles of arithmetic are taught*. Taylor and Walton, London.
1839 *First notions of logic, preparatory to the study of geometry*. Taylor and Walton, London. (Reprinted as Chapter 1 of *De Morgan 1847*.)
1842a* On the foundation of algebra. *Transactions of the Cambridge Philosophical Society*, **7**, 173–87. (Reprinted in this collection, Vol. 1, pp. 336–48.) (Read 9 Dec 1839.)
1842b On the foundation of algebra, No. II. *Transactions of the Cambridge Philosophical Society*, **7**, 287–300. (Read 29 Nov. 1841.)
1846 On the syllogism, No. I. On the structure of the syllogism, and on the application of the theory of probabilities to questions of argument and authority. *Transactions of the Cambridge Philosophical Society*, **8**, 379–408. (Read 9 Nov. 1846.)

1847 *Formal logic: or, The calculus of inference, necessary and probable.* Taylor and Walton, London.

1849a On the foundation of algebra, Nos. III and IV. *Transactions of the Cambridge Philosophical Society*, **8**, 139–42 and 241–54. (Read 27 Nov. 1843 and 28 Oct. 1844.)

1849b* *Trigonometry and double algebra.* Taylor, Walton, and Maberly, London. (Partially reprinted in this collection, Vol. 1, pp. 349–61.)

1850 On the syllogism, No. II. On the symbols of logic, the theory of the syllogism, and in particular of the copula. *Transactions of the Cambridge Philosophical Society*, **9**, 79–127. (Read 25 Feb. 1850.)

1852 On the early history of infinitesimals in England. *Philosophical Magazine* (series 4), **4**, 321–30.

1860a *Syllabus of a proposed system of logic.* Walton and Maberly, London.

1860b Logic. Entry in the *English Cyclopædia*, Vol. 5. (Reprinted, with omissions, in *De Morgan 1966*, pp. 247–70). Broadbury and Evans, London.

1864a On the syllogism, No. III, and on logic in general. *Transactions of the Cambridge Philosophical Society*, **10**, 173–230. (Read 8 Feb. 1858.)

1864b On the syllogism, No. IV, and on the logic of relations. *Transactions of the Cambridge Philosophical Society*, **10**, 331–58. (Read 23 Apr. 1860).

1864c On the syllogism, No. V. and on various points of the onymatic system. *Transactions of the Cambridge Philosophical Society*, **10**, 428–87. (Read 4 May 1863.)

1872 *A budget of paradoxes.* Longmans, Green, London.

1966 On the syllogism and other logical writings (ed. Peter Heath). Yale University Press, New Haven.

See Boole, George and De Morgan, Augustus.

Diderot, Denis *et alii.* (eds.)

1751–72 *Encyclopédie, ou Dictionnaire raisonné des sciences, des arts et des métiers, par une société de gens de lettres*, 28 vols. Samuel Faulche, Neufchastel [Neuchâtel].

Dieudonné, Jean A. (ed.)

1978 *Abrégé d'histoire des mathématiques, 1700–1900*, 2 vols. Hermann, Paris.

Dirichlet, Peter Gustav Lejeune

1863 *Vorlesungen über Zahlentheorie* (ed. Richard Dedekind). Vieweg, Braunschweig. (Subsequent editions in 1871, 1879, 1894. These contain important 'Supplements' written by Dedekind, of which portions are translated by William Ewald in this collection, Vol. 2, pp. 762–5 and 833–4.)

Edwards, Charles Henry

1979 *The historical development of the calculus.* Springer Verlag, New York.

Encyklopädie

1898–1935 *Encyklopädie der mathematischen Wissenschaften, mit Einschluss ihrer Anwendungen.* Herausgegeben im Auftrag der Akademien der Wissenschaften zu Göttingen, Leipzig, München, und Wien. Teubner, Leipzig. (Numerous volumes; appeared in parts.)

Engel, Friedrich
See Stäckel, Paul and Engel, Friedrich.

Euclid
1896 *Data, cum commentario Marini et scholiis antiquis* (ed. H. Mengel). Leipzig.

1925 *The thirteen books of the elements*, 2 vols, (2nd edn; trans. and ed. Thomas L. Heath, with extensive commentary). Cambridge University Press. (Reprinted Dover, New York, 1956.)

Feuer, Lewis S.
1988 Sylvester in Virginia. *The Mathematical Intelligencer*, **9**, 13–19.

Fisch, Max F.
1980 The range of Peirce's relevance. *The Monist*, **63**, 269–76.

Fischer, Ernst Gottfried
1808 *Untersuchung über den eigentlichen Sinn der Höheren Analysis nebst einer idealen Übersicht der Mathematik und Naturkunde nach ihrem ganzen Umfange*. S.F. Weiss, Berlin.

Fontenelle, Bernard
1727 *Elémens de la géométrie de l'infini*, Suite des Mémoires de l'Académie royale des sciences. De l'Imprimerie royale, Paris.

Frege, Gottlob
1879 *Begriffschrift, eine der arithmetischen nachgebildete Formelsprache des reinen Denkens*. Nebert, Halle. (Translated by Stefan Bauer-Mengelberg in *van Heijenoort 1967*, pp. 1–82.)
1884 *Die Grundlagen der Arithmetik, eine logisch-mathematische Untersuchung über den Begriff der Zahl*, Koebner, Breslau. (Translated by John Langshaw Austin as *The foundations of arithmetic, A logico-mathematical enquiry into the concept of number*, Basil Blackwell, Oxford, 1950; rev. 2nd edn, 1953.)
1891 *Function und Begriff*. H. Pohle, Jena.
1972 *Conceptual notation and related articles* (ed. and trans. Terrell Bynum). Clarendon Press, Oxford. (Translation of *Frege 1879*.)

Frend, William
1796–9 *The principles of algebra; or, the true theory of equations established on mathematical demonstration*, 2 vols. J. Davis, London.

Friedman, Michael
1992 *Kant and the exact sciences*. Harvard University Press, Cambridge, Mass.

Gauss, Carl Friedrich
1801 *Disquisitiones arithmeticae*. Fleischer, Leipzig.
1816* Review of J.C. Schwab, *Commentatio in primum elementorum Euclidis librum, qua veritatem geometriae principiis ontologicis niti evincitur, omnesque propositiones, axiomatum geometricorum loco habitae, demonstrantur* (Stuttgart, 1814), and of Matthias Metternich, *Vollständige Theorie der Parallel-Linien* (Mainz, 1815). *Göttingische gelehrte Anzeigen*, 20 April 1816. (Reprinted in *Stäckel and Engel 1895*, pp. 220–3. English translation by William Ewald, this collection, Vol. 1, pp. 299–300.)

1822* Review of Carl Reinhard Müller's *Theorie der Parallelen* (Marburg, 1822). *Göttingische gelehrte Anzeigen*, 28 October 1822. (Reprinted in *Stäckel and Engel 1895*, pp. 223–6. English translation by William Ewald, this collection, Vol. 1, p. 300.)

1827 Disquisitiones generales circa superficies curvas. *Commentationes societatis regiae scientiarum Gottingensis recentiores*, Vol. 6. (Reprinted in *Gauss 1863–1929*, Vol. 4, (1873), pp. 217–58.)

1828 Theoria residuorum biquadraticorum. Commentatio prima. *Commentationes societatis regiae scientiarum Gottingensis recentiores*, Vol. 6. (Presented 5 April 1825. Reprinted in *Gauss 1863–1929*, Vol. 2, pp. 65–92.)

1831* Anzeige der Theoria residuorum biquadraticorum, Commentatio secunda. *Göttingische gelehrte Anzeigen*, 23 April 1831. (Reprinted in *Gauss 1863–1929*, Vol. 2, (1876), pp. 169–78. English translation by William Ewald, this collection, Vol. 1, pp. 306–13.)

1832 Theoria residuorum biquadraticorum. Commentatio secunda. *Commentationes societatis regiae scientiarum Gottingensis recentiores*, Vol. 7. (Presented 15 Apr. 1831. Reprinted in *Gauss 1863–1929*, Vol. 2, pp. 93–148.)

1863–1929 *Werke*, 12 vols. Königliche Gesellschaft der Wissenschaften, Göttingen, Leipzig, and Berlin.

1929* Zur Metaphysik der Mathematik. *Gauss 1863–1929*, Vol. 12, pp. 57–61 (1929). (English translation by William Ewald, this collection, Vol. 1, pp. 293–6.)

Gergonne, Joseph-Diez

1818 Essai sur la théorie des définitions. *Annales de mathématiques pures et appliquées*, **9**, 1–35. (The *Annales* were later renamed the *Journal de mathématiques pures et appliquées*.)

Gibson, George Alexander

1899 Berkeley's *Analyst* and its critics: An episode in the development of the doctrine of limits. *Bibliotheca Mathematica*, NS, **13**, 65–70.

Gillispie, Charles Coulston *et alii* (eds.)

1970–6 *Dictionary of scientific biography*, 14 vols. Scribner, New York.

Grassmann, Hermann Günther

1844 *Die lineale Ausdehnungslehre, ein neuer Zweig der Mathematik*. Wigand, Leipzig. (2nd edn: 1878).

1861 *Lehrbuch der Arithmetik und Trigonometrie für höhere Lehranstalten*. Enslin, Berlin.

1865 Volume 2 of *Grassmann 1861*.

Grassmann, Robert

1872 *Die Formenlehre oder Mathematik in strenger Formelentwicklung*. R. Grassmann, Stettin.

Grattan-Guinness, Ivor

1972b Bolzano, Cauchy, and the 'new analysis' of the nineteenth century. *Archive for History of Exact Sciences*, **6**, 372–400.

1980 (ed.) *From the calculus to set theory: 1630–1910. An introductory history*. Duckworth, London.

Graves, Robert Perceval
1882–9 *Life of Sir William Rowan Hamilton*, 3 vols. Longmans, London.

Gregory, Duncan Farquharson
1840* On the real nature of symbolical algebra. *Transactions of the Royal Society of Edinburgh*, 14, 208–16. (Read 7 May 1838. Reprinted in this collection, Vol. 1, pp. 323–30.)
1841 *Examples of the processes of the differential and integral calculus*. J. Deighton, Cambridge.
1845 *A treatise on the application of analysis to solid geometry* (ed. William Walton). J. Deighton, Cambridge.
1865 *The mathematical writings of Duncan Farquharson Gregory* (ed. William Walton). Deighton Bell, Cambridge.

Hallett, Michael
1984 *Cantorian set theory and limitation of size*. Clarendon Press, Oxford.

Hamilton, William
1852 *Discussions on philosophy and literature, education and university reform*. Longmans, London. (3rd edn: 1866.)

Hamilton, William Rowan
1834 On a general method in dynamics. *Philosophical Transactions of the Royal Society*, part 2 for 1834, 247-308. (Reprinted in *Hamilton 1931-67*, Vol. 2, pp. 103-61.)
1837* Theory of conjugate functions, or algebraic couples: with a preliminary and elementary essay on algebra as the science of pure time. *Transactions of the Royal Irish Academy*, **17**, 293-422. (Reprinted in *Hamilton 1931-67*, Vol. 3, pp. 3-96, and partially reprinted in this Collection, Vol. 1, pp. 369–75.) (Read 4 Nov. 1833 and 1 June 1835.)
1853* *Lectures on quaternions*. Hodges and Smith, Dublin. (Preface reprinted in Hamilton 1931-67, Vol. 3, pp. 117–58, and in this collection, Vol. 1, pp. 375–425.)
1931-67 *The mathematical papers of Sir William Rowan Hamilton*, 3 vols, (ed. H. Halberstam and R. E. Ingram). Cambridge University Press.

Hankel, Hermann
1867 *Vorlesungen über die complexen Zahlen und ihren Functionen*. Voss, Leipzig.

Hankins, Thomas L.
1980 *Sir William Rowan Hamilton*. Johns Hopkins University Press, Baltimore and London.

Hausen, Christian
1734 *Elementa matheseos*. Marcheana, Leipzig.

Heath, Thomas L.
1921 *A history of Greek mathematics*, 2 vols. Clarendon Press, Oxford.

Helmholtz, Hermann von
1876* The origin and meaning of geometrical axioms. *Mind*, **1**, 301-21. (German text in *Helmholtz 1865–76*, Vol. 3. Reprinted, with notes and

commentary by Paul Hertz, in *Helmholtz 1921*; Hertz edition and notes translated, with title 'On the origin and significance of the axioms of geometry', by Malcolm F. Lowe in *Helmholtz 1977*.) Reprinted in this collection, Vol. 2, pp. 663–85.

Hilbert, David

1899 *Grundlagen der Geometrie*. Teubner, Leipzig. (Translated by E.J. Townsend, Open Court, Chicago, 1902.) (Later editions published in 1903, 1909, 1913, 1922, 1923, and 1930; these contain supplements reprinting articles on the foundations of mathematics that do not appear in the *Gesammelte Abhandlungen*.)

1900a* Über den Zahlbegriff. *Jahresbericht der Deutschen Mathematiker-Vereinigung*, **8**, 180–94. (Reprinted in *Hilbert 1909*, pp. 256–62; *1913*, pp. 237–42; *1922b*, pp. 237–42; *1923b*, pp. 237–42; *1930a*, pp. 241–6. Translated by William Ewald, this collection, Vol. 2, pp. 1089–95.)

1900b* Mathematische Probleme. Vortrag, gehalten auf dem internationalen Mathematiker-Kongress zu Paris. 1900. *Nachrichten der Königlichen Gesellschaft der Wissenschaften zu Göttingen*, 253–97. (Reprinted, with additions, in *Archiv der Mathematik und Physik* (ser. 3), **1**, (1901), 44–63, 213–37. Translated by Mary Winston Newson in *Bulletin of the American Mathematical Society* (ser. 2), **8** (1902), 437–79; partially reprinted in this collection, Vol. 2, pp. 1096–1105.)

1904 Über die Grundlagen der Logik und der Arithmetik. In *Verhandlungen des dritten internationalen Mathematiker-Kongresses in Heidelberg vom 8. bis 13. August 1904*, pp. 174–85. Teubner, Leipzig, 1905. (Reprinted in *Hilbert 1909*, pp. 263–79; *1913*, pp. 243–58; *1922b*, pp. 243–58; *1923b*, pp. 243–58; *1930a*, pp. 247–61. Translated by Beverly Woodward in *van Heijenoort 1967*, pp. 129–38.)

1922a* Neubegründung der Mathematik. Erste Mitteilung. *Abhandlungen aus dem mathematischen Seminar der Hamburgischen Universität*, **1**, 157–77. (Reprinted in *Hilbert 1932–5*, Vol. 3, pp. 157–77. Translated by William Ewald, this collection, Vol. 2, pp. 1115–34.)

Hobbes, Thomas

1845 *The Latin works of Thomas Hobbes*, 5 vols, (ed. William Molesworth). Green and Longman, London. (*De principiis et ratiocinatione geometrarum* is in Vol. 4, pp. 385–465.)

Hofmann, Joseph Ehrenfried

1963 *Geschichte der Mathematik*, 3 vols. De Gruyter, Berlin.

Jesseph, Douglas M.

1993 *Berkeley's philosophy of mathematics*. University of Chicago Press.

Jevons, William Stanley

1879 *The principles of science, a treatise on logic and scientific method*, (3rd edn). Macmillan, London.

Kant, Immanuel

1747* *Gedanken von der wahren Schätzung der lebendingen Kräften und Beurteilung der Beweise, derer sich Herr von Leibniz und andere Mechaniker in dieser Streitsache bedient haben, nebst einigen* *vorhergehenden Betrachtungen, welche die Kraft der Körper überhaupt betreffen.* (Written and submitted 1746. Partial English translation by William Ewald, this collection, Vol. 1, pp. 133–4.)

1755 *Allgemeine Naturgeschichte und Theorie des Himmels, oder Versuch von der Verfassung und dem mechanischen Ursprunge des ganzen Weltgebäudes nach Newtonischen Grundsätzen abgehandelt.* Johann Friedrich Petersen, Königsberg and Leipzig.

1781 *Critik der reinen Vernunft.* Hartknoch, Riga.

1787 *Kritik der reinen Vernunft*, Second edition of *Kant 1781*. Hartknoch, Riga.

1902–23 *Kants Werke.* Preussische Akademie der Wissenschaften, Berlin.

1933 *Critique of pure reason.* Macmillan, London. (Translation of *Kant 1781* and *1787* by Norman Kemp Smith.)

1972 *Briefwechsel* (ed. Otto Schöndörfer). Meiner, Hamburg.

Kästner, Abraham Gotthelf

1758 *Anfangsgründe der Arithmetik, Geometrie, ebenen und sphärischen Trigonometrie und Perspektiv.* Vandenhoeck, Königsberg.

1794 *Anfangsgründe der Analysis endlicher Grössen*, (3rd edn). Vandenhoeck and Ruprecht, Göttingen. (1st edn: 1760.)

Kennedy, Hubert

1979 James Mills Peirce and the cult of quaternions. *Historia Mathematica*, **6**, 423–9.

Klein, Felix

1895 Über Arithmetisierung der Mathematik. *Nachrichten der Königlichen Gesellschaft der Wissenschaften zu Göttingen*, Geschäftliche Mitteilungen 1895, Heft 2. (Reprinted in *Klein 1921–3*, Vol. 2, pp. 232–40.)

1926–7 *Vorlesungen über die Entwicklung der Mathematik im 19. Jahrhundert*, 2 vols, (ed. Richard Courant, Otto Neugebauer, and Stephen Cohn-Vossen). Springer Verlag, Berlin.

Kline, Morris

1972 *Mathematical thought from ancient to modern times.* Oxford University Press, New York.

Klügel, Georg Simon

1763 *Conatuum praecipuorum theoriam parallelarum demonstrandi recensio.* F.A. Rosenbusch, Göttingen.

1803–31 *Mathematisches Wörterbuch*, 5 vols. Schwickert, Leipzig.

Kneale, William

1948 Boole and the revival of logic. *Mind*, **57**, 149–75.

Kneale, William and Kneale, Martha

1962 *The development of logic.* Clarendon Press, Oxford.

Lacroix, Silvestre François

*c.*1805 *Elémens d'algèbre*, (5th edn). Courcier, Paris. (24 editions published altogether; approximate date for this edition.)

1811 *Anfangsgründe der Algebra*. Mainz. (German translation, by one Metternich, of the seventh edition of *Lacroix 1805*.)

See Peacock, George.

Lagrange, Joseph-Louis

1797 *Théorie des fonctions analytiques contenant les principes du calcul différentiel, dégagés de toute consideration d'infiniment petits, d'évanouissans, de limites et de fluxions, et réduits à l'analyse algébrique des quantités finies*. Imprimerie de la République, Prairial, An 5 (=1797), Paris.

1806 *Leçons sur le calcul des fonctions. Nouvelle edition*. Courcier, Paris. (1st edn: 1804.)

1808 *Traité de la résolution des équations numériques de tous les degrés*, (2nd edn). Courcier, Paris. (1st edn: Duprat, Paris, 1798.)

Lambert, Johann Heinrich

1761 *Cosmologische Briefe über die Einrichtung des Weltbaues*. E. Kletts, Augsburg.

1764 *Neues Organon, oder Gedanken über die Erforschung und Bezeichnung des Wahres und dessen Unterscheidung vom Irrthum und Schein*, 2 vols. Wendler, Leipzig.

1771 *Anlage zur Architectonic, oder Theorie des Einfachen und des Ersten in der philosophischen und mathematischen Erkenntnis*, 2 vols. Hartknoch, Riga.

1781–84 *Deutscher gelehrter Briefwechsel*, 4 vols, (ed. Joh. Bernoulli). Bernoulli, Berlin.

1786* Theorie der Parallellinien. *Magazin für reine und angewandte Mathematik für 1786*, 137–64, 325–58. (Reprinted in *Stäckel and Engel 1895*, pp. 152–208. Partial English translation by William Ewald, this collection, Vol. 1, pp. 158–67.) (Written in 1766.)

1946–8 *Opera mathematica*, 2 vols. Orell Füssli, Zurich.

1967– *Gesammelte philosophische Werke* (ed. Hans Werner Arndt). Olms, Hildesheim. (10 volumes planned.)

Legendre, Adrien Marie

1794 *Eléments de géométrie* F. Didot, Paris. (12th edn: 1823.)

1833 Réflexions sur différentes manières de démontrer la théorie des parallèles ou le théorèm sur la somme des trois angles du triangle. *Mémoires de l'Académie royale des sciences*, **12**, 367–410.

Leibniz, Gottfried Wilhelm

1875–90 *Die philosophischen Schriften von Gottfried Wilhelm Leibniz*, 7 vols, (ed. Carl J. Gerhardt). Weidmann, Berlin. (Reprinted 1960–1, Georg Olms, Hildesheim.)

1849–63 *Mathematische Schriften*, 7 vols, (ed. Carl J. Gerhardt). Asher, Berlin (Vols 1–2) and Schmidt, Halle (Vols 3–7).

1903 *Opuscules et fragments inédits de Leibniz* (ed. Louis Couturat). Alcan, Paris. (Reprinted 1961, Georg Olms, Hildesheim.)

Lewis, Clarence Irving

1918 *A survey of symbolic logic*. University of California, Berkeley. (Reprinted, omitting Chapters V and VI, Dover, New York, 1960.)

L'Hospital, Guillaume François Antoine de
1696 *Analyse des infiniment petits*. Imprimerie Royale, Paris.

Lipschitz, Rudolf Otto Sigismund
1877–80 *Lehrbuch der Analysis*, 2 vols. Cohen, Bonn.

Lobatchevsky, Nikolai Ivanovich
1840 *Geometrische Untersuchungen zur Theorie der Parallellinien*. G. Funcke, Berlin.

Löwenheim, Leopold
1940 Einkleidung der Mathematik in Schröderschen Relativkalkül. *The Journal of Symbolic Logic*, **5**, 1–15.

Luce, Arthur Aston
1949 *The life of George Berkeley, Bishop of Cloyne*. Nelson, London.

Ludlam, William
1809 *The rudiments of mathematics: designed for the use of students at the universities*, (5th edn). John Evans, London.

MacHale, Desmond
1985 *George Boole: His life and work*. Boole Press, Dublin.

MacLaurin, Colin
1742* *A treatise of fluxions*, 2 vols. Ruddimans, Edinburgh. (Partially reprinted in this collection, Vol. 1, pp. 95–122.)
1748 *An account of Sir Isaac Newton's philosophical discoveries*. Murdoch, London.

Manning, Thomas
1796–8 *An introduction to arithmetic and algebra*, 2 vols. B. Flower, Cambridge.

Mansion, Paul
1908 Gauss contra Kant sur la géometrie non euclidienne. In *Bericht über den 3. internationalen Kongress für Philosphie, Heidelberg 1908* (ed. Theodor Elsenhans). Winter, Heidelberg.

Martin, Gottfried
1972 *Arithmetik und Kombinatorik bei Kant*. De Gruyter, Berlin.

Martin, Richard M.
1976 Some comments on De Morgan, Peirce, and the logic of relations. *Transactions of the Charles S. Peirce Society*, **12**, 223–30.
1978 Of servants, benefactors, and lovers: Peirce's algebra of relatives of 1870. *Journal of Philosophical Logic*, **7**, 27–48.

Maseres, Francis
1758 *A dissertation on the use of the negation sign in algebra*. S. Richardson, London. (An appendix contains 'Mr. Machin's quadrature of the circle'.)

Mercator, Nicolaus
*c.*1678 *Euclidis elementa geometriae.* London. (Approximate date.)

Michael, Emily
1976 Peirce's earliest contact with scholastic logic. *Transactions of the Charles S. Peirce Society*, **12**, 46–55.

Michelsen, Johann Andreas
1789 *Gedanken über den gegenwärtigen Zustand der Mathematik und die Art der Vollkommenheit und Brauchbarkeit derselben zu vergrössern.* S.F. Hesse, Berlin.
1790 *Beiträge zur Beförderung des Studiums der Mathematik.* Berlin. (This work is cited by Bolzano; the editor has been unable to locate a copy.)

Minnigerode, B.
1871 Bemerkungen über irrationale Zahlen. *Mathematische Annalen*, **4**, 497–8.

Mitchell, Oscar Howard
1881 Some theorems in numbers. *American Journal of Mathematics*, **4**, 25–38.
1883 On a new algebra of logic. In *Studies in logic by members of the Johns Hopkins University* (ed. Charles S. Peirce). Little, Brown, Boston.

Möbius, August Ferdinand
1827 *Der barycentrische Calcul.* Barth, Leipzig.
1843 *Die Elemente der Mechanik des Himmels.* Weidmann, Leipzig.

Moore, Gregory H.
1982 *Zermelo's axiom of choice: Its origins, development, and influence.* Springer Verlag, New York.

Newton, Isaac
1686 *Philosophiae naturalis principia mathematica.* S. Pepys, London.
1700 *Tractatus de quadratura curvarum.* Paris. (Exact date and location uncertain; multiple later editions, e.g. London 1704 and 1706.)
1726 Third edition of *Newton 1686.*
1934 *Sir Isaac Newton's mathematical principles of natural philosophy and his system of the world*, (trans. Andrew Motte, 1729; rev. trans. Florian Cajori, 1934). University of California, Berkeley.
1972 *Philosophiae naturalis principia mathematica*, Vol. 1, (ed. Alexander Koyré and I. Bernard Cohen). Cambridge University Press. (A reprinting, with variant readings, of *Newton 1726.*)

North, J.D.
1976 Sylvester, James Joseph. In *Gillispie* et alii *1970–6*, Vol. 13, pp. 216–22.

Nový, L.
1973 *Origins of modern algebra.* Academia, Prague

Pappus of Alexandria
1588 *Mathematicae collectiones* (ed. Federigo Commandino). Pisa and Venice. (Multiple editions between 1588 and 1660.)

Paris Logic Group
1987 *Logic colloquium '85* (ed. Anita B. Feferman and Solomon Feferman). North-Holland, Amsterdam.

Pasch, Moritz
1882a *Einleitung in der differential- und integral-Rechnung*. Teubner, Leipzig.
1882b *Vorlesungen über neuere Geometrie*. Teubner, Leipzig. (2nd edn: 1912.)
1927 *Mathematik am Ursprung. Gesammelte Abhandlungen über Grundfragen der Mathematik*. Meiner, Leipzig.

Peacock, George
1816 *A treatise on the differential and integral calculus. By Silvestre François Lacroix*, (trans. Charles Babbage, John Frederick William Herschel, and George Peacock). J. Smith, Cambridge.
1820 *A collection of examples of the application of the differential and integral calculus*. J. Smith, Cambridge.
1830 *A treatise on algebra*. Deighton, Cambridge.
1833 Report on the recent progress and present state of certain branches of analysis, *British Association for the Advancement of Science, Report 3*, 185–352. (Reprinted in 1834 as *Report on certain branches of analysis*. R. Taylor, London.)
1842–5 *Treatise on algebra*, 2 vols. (2nd edn; 1st edn is *Peacock 1830*). Deighton, Cambridge. (Vol. 1 (1842) entitled *Arithmetical algebra*. Vol. 2 (1845) entitled *Symbolical algebra*.)

Peano, Giuseppe
1894 *Notations du logique mathématique: Introduction au formulaire de mathématique*. Charles Guadagnini, Turin.
1895–1908 *Formulaire de mathématiques*, 5 vols. Bocca, Turin.
1958 *Opere scelte*, Vol. 2. Editione Cremonese, Rome.
1973 *Selected works of Giuseppe Peano* (ed. and trans. Hubert C. Kennedy). University of Toronto Press.

Peirce, Benjamin
1870* *Linear associative algebra*. Washington, DC. (Partially reprinted in this collection, Vol. 1, pp. 584–94.) (Lithograph.)
1881 ——. *American Journal of Mathematics*, **4**, 97–229. (Reprinting of *B. Peirce 1870*, with notes and addenda by Charles Sanders Peirce.)

Peirce, Charles Sanders
1868 On a new list of categories. *Proceedings of the American Academy of Arts and Sciences*, **7**, 287–98. (Reprinted in *Peirce 1931–58*, Vol. 1, pp. 287–99; and in *Peirce 1982–*, Vol. 2, pp. 49–58.)
1869 The pairing of the elements. *Chemical News*, **4**, 339–40. (Reprinted in *Peirce 1982–*, Vol. 2, pp. 282–3.)
1870 Description of a notation for the logic of relatives, resulting from an amplification of the conceptions of Boole's calculus of logic. *Memoirs of the American Academy of Arts and Sciences*, NS, **9**, 317–78. (Reprinted in *Peirce 1931–58*, Vol. 3, pp. 27–98; and in *Peirce 1982–*, Vol. 2, pp. 359–429.)
1871 Review of Fraser's *The works of George Berkeley*, *North American Review*, **113**, 449–72. (Reprinted in *Peirce 1982–*, Vol. 2, pp. 462–87.)

1875 On the application of logical analysis to multiple algebra. *Proceedings of the American Academy of Arts and Sciences*, NS, **2**, 392–4. (Reprinted in *Peirce 1982–*, Vol. 3, pp. 177–9.)

1878 *Photometric researches: Made in the years 1872–1875*. Wilhelm Engelmann, Leipzig. (Reprinted in *Peirce 1982–*, Vol. 3, pp. 382–493.)

1880a On the algebra of logic. *American Journal of Mathematics*, **3**, 15–57. (Reprinted in *Peirce 1931–58*, Vol. 3, pp. 104–57; and in *Pierce 1982–*, Vol. 4, pp. 163–209.)

1880b On a Boolean algebra with one constant. (First published in 1933 in *Peirce 1931–58*, Vol. 4, pp. 13–18; title supplied by the editors. Also in *Pierce 1982–*, Vol. 4, pp. 218–21.)

1881* On the logic of number. *The American Journal of Mathematics*, **4**, 85–95. (Reprinted in *Peirce 1931–58*, Vol. 3, pp. 158–70; in *Pierce 1982–*, Vol. 4, pp. 299–309; and in this collection, Vol. 1, pp. 598–608.)

1883 A theory of probable inference. (With two Notes; Note B entitled 'The logic of relatives'.) In *Studies in logic by members of the Johns Hopkins University* (ed. Charles S. Peirce). Little, Brown, Boston. (Reprinted in *Peirce 1983*, and in *Pierce 1982–*, Vol. 4, pp. 408–466; Note B reprinted in *Peirce 1931–58*, Vol. 3, pp. 195–209.)

1885* On the algebra of logic: A contribution to the philosophy of notation. *The American Journal of Mathematics*, **7**, 180–202. (Reprinted in *Peirce 1931–58*, Vol. 3, pp. 210–38; in *Peirce 1982–*, Vol. 5, pp. 162–90; and in this collection, Vol. 1, pp. 608–32.)

1898* The logic of mathematics in relation to education. *Educational Review*, 1898, 209–16. (Reprinted in *Peirce 1931–58*, Vol. 3, pp. 346–59, and in this collection, Vol. 1, pp. 632–7.)

1902* The simplest mathematics. First published in 1933 in *Peirce 1931–58*, Vol. 4, pp. 189–262. (Reprinted in this collection, Vol. 1, pp. 637–48.)

1905 What pragmatism is. *The Monist*, **15**, 161–81.

1931-58 *Collected papers of Charles Sanders Peirce*, 8 vols, (ed. Charles Hartshorne, Paul Weiss, and Arthur Burks). Harvard University Press, Cambridge, Mass.

1976 *The new elements of mathematics*, 4 vols, (ed. Carolyn Eisele). Mouton, The Hague.

1982– *Writings of Charles Sanders Peirce: A chronological edition*, (ed. Max H. Fisch *et alii*). Indiana University Press, Bloomington.

1983 *Studies in logic by members of the Johns Hopkins University*, (ed. C.S. Peirce; with an introduction by Max H. Fisch). Benjamins, Amsterdam.

1985 *Historical perspectives on Peirce's logic of science*, 2 vols, (ed. Carolyn Eisele). Mouton, Berlin.

Peirce, Charles Sanders and Jastrow, Charles

1884 On small differences in sensation. *Memoirs of the National Academy of Sciences*, Vol. 3 (5th memoir), Part 1, 75–83.

Perry, Ralph Barton

1935 *The thought and character of William James*. Little, Brown, Boston

Platner, Ernst

1793–1800 *Philosophische Aphorismen*, (2nd edn). Schwickert, Leipzig.

Popper, Karl
1953–4 Berkeley as a precursor of Mach. *British Journal for the Philosophy of Science*, **4**, 26–36.

Post, Emil Leon
1921 Introduction to a general theory of elementary propositions. *American Journal of Mathematics*, **43**, 163–85. (Reprinted in *van Heijenoort 1967*, pp. 264–83.)

Proclus
1970 *A commentary on the first book of Euclid's 'Elements'*, (translated with introduction and notes by Glenn R. Morrow). Princeton University Press.

Quine, Willard van Orman
1960 *Word and object*. MIT Press, Cambridge, Mass.
1963 *Set theory and its logic*. Harvard University Press, Cambridge, Mass.

Raphson, Joseph
1690 *Analysis aequationum universalis*. Abel Swalle, London.
1697 *De spatio reali, seu ente infinito conamen mathematico-metaphysicum*. Thomas Braddyll, London.
1715 *The history of fluxions, shewing in a compendious manner the first rise of, and various improvements made in, that incomparable method*. W. Pearson, London. (Also published in Latin as *Historia fluxionum*.)

Riccardi, Pietro
1887–93 *Saggio di una bibliografia Euclidea*, 4 parts. Istituto delle Scienze ed Arti Liberali, *Memorie*, Ser. 4, Vols 8 and 9; Ser. 5, Vol. 1. The Institute, Bologna.

Riemann, Bernhard
1868* Über die Hypothesen, welche der Geometrie zu Grunde liegen. *Abhandlungen der Königlichen Gesellschaft der Wissenschaften zu Göttingen*, **13**, 133–52. (Read as a *Probevorlesung* in Göttingen, 10 June 1854. Translation by William Kingdon Clifford reprinted in this collection, Vol. 2, pp. 652–61.

Rösling, Christian Lebrecht
1805 *Grundlehren von den Formen, Differenzen, Differentialen und Integralien der Functionen*. J.S. Palm, Erlangen.

Russ, Steven B.
1980* A translation of Bolzano's paper on the intermediate value theorem. *Historia Mathematica*, **7**, 156–85. (Revised version reprinted in this collection, Vol. 1, pp. 225–48.)

Russell, Bertrand Arthur William
1901 Recent work in the philosophy of mathematics. *The International Monthly*, **4**, 83–101. (Reprinted under the title 'Mathematics and the metaphysicians' as Ch. 5 of *Mysticism and logic and other essays*, Longmans, Green, New York, 1918.)

1907 The regressive method of discovering the principles of mathematics. (Read before the Cambridge Mathematical Club on 9 March 1907; first printed in *Russell 1973*, pp. 272–83.)

1914 On scientific method in philosophy. Herbert Spencer Lecture, Oxford. (Reprinted in *Mysticism and logic and other essays*, Longmans, Green, New York, 1918.)

1937 Second edition of Russell 1903. Allen and Unwin, London.

1973 *Essays in analysis* (ed. Douglas Lackey). Braziller, New York.

Rychlik, Karel (ed.)

1962 *Theorie der reellen Zahlen in Bolzanos handschriftlichem Nachlass*. Verlag der Tschechoslowakischen Akademie der Wissenschaften, Prague.

Saccheri, Girolamo

1733 *Euclides ab omni naevo vindicatus*. (Translated by George Bruce Halsted in *American Mathematical Monthly*, **1-5**, 1894-8. Reprinted, Open Court, Chicago, 1920; and Chelsea, New York, 1970. German translation in *Stäckel and Engel 1895*, pp. 41–136.)

Schröder, Ernst

1873 *Lehrbuch der Arithmetik und Algebra*. Teubner, Leipzig.

1880 Review of *Frege 1879*. *Zeitschrift für Mathematik und Physik*, **25**, 81–94. (Translated by Terrell Bynum in *Frege 1972*.)

1890 *Vorlesungen über die Algebra der Logik (exakte Logik)*, Vol 1. Teubner, Leipzig. (Reprinted in *Schröder 1966*.)

1891 —— Vol. 2, part 1. (Reprinted in *Schröder 1966*.)

1895 —— Vol. 3, *Algebra und Logik der Relative*, part 1. (Reprinted in Schröder 1966.)

1966 *Vorlesungen über die Algebra der Logik*. Chelsea, New York.

Schrödinger, Erwin

1945 The Hamilton postage stamp: An announcement by the Irish minister of posts and telegraphs. In *A collection of papers in memory of Sir William Rowan Hamilton* (ed. David Eugene Smith), Scripta mathematica studies, no. 2. Scripta mathematica, New York.

Schultz, Johann

1784 *Entdeckte Theorie der Parallelen nebst einer Untersuchung über den Ursprung ihrer bisherigen Schwierigkeit*. Kanter, Königsberg.

1785 *Erläuterung über des Herrn Professor Kant Critik der reinen Vernunft*. Dengel, Königsberg. (2nd edn: Frankfurt and Leipzig, 1791.) [Same title can be found under the spelling Johann Schulze.] (Modernized reprint, (ed. R.C. Hafferberg), Rassmann, Jena and Leipzig, 1897.)

1788 *Versuch einer genauen Theorie des Unendlichen*. Hartung, Königsberg and Leipzig.

1789-92 *Prüfung der Kantischen Kritik der reinen Vernunft*, 2 vols. Hartung, Königsberg.

1790 *Anfangsgründe der reinen Mathesis*. Königsberg.

1797 *Kurzer Lehrbegriff der Mathematik*. Nicolovius, Königsberg.

Schweickart (or Schweikart), Ferdinand Karl
1807 *Theorie der Parallellinien, nebst dem Vorschlag ihrer Verbannung aus der Geometrie*. Jena.

Scott, Joseph F.
1938 *The mathematical work of John Wallis*. Oxford University Press.
1973 MacLaurin, Colin. In *Gillispie* et alii *1970–6*, Vol. 7, pp. 609–12.

Scriba, Christoph
1973 Lambert, Johann. In *Gillispie* et alii *1970–6*, Vol. 7.

Servois, François-Joseph
1814a Sur la théorie des quantités imaginaires. Lettre de M. Servois. *Annales de mathématiques pures et appliquées* (later renamed *Journal de mathématiques pures et appliquées*), **4**, 228–35. (The letter is dated 23 Nov. 1813.)
1814b Essai sur un nouveau mode d'exposition des principes du calcul différentiel. *Annales de mathématiques pures et appliquées* (later renamed *Journal de mathématiques pures et appliquées*), **5**, 93–140.

Sheffer, Henry M.
1913 A set of five independent postulates for Boolean algebras, with application to logical constants. *Transactions of the American Mathematical Society*, **14**, 481–8.

Sigwart, Christoph
1904 *Logik*, (3rd edn). Mohr, Tübingen.

Stäckel, Paul and Engel, Friedrich (eds.)
1895 *Die Theorie der Parallellinien von Euklid bis auf Gauss*. Teubner, Leipzig.
1898 *Urkunden zur Geschichte der nichteuklidischen Geometrien*, 2 vols. Teubner, Leipzig.

Stammler, Gerhard
1922 *Berkeleys Philosophie der Mathematik*. Reuther and Reichard, Berlin.

Steck, Max
1970 *Bibliographia Lambertiana*, (2nd edn). Olms, Hildesheim.

Stolz, Otto
1881 B. Bolzano's Bedeutung in der Geschichte der Infinitesimalrechnung. *Mathematische Annalen*, **18**, 255–79.

Struik, Dirk J. (ed.)
1969 *A source book in mathematics, 1200–1800*. Harvard University Press, Cambridge, Mass.

Sylvester, James Joseph
1869* Presidential address to Section 'A' of the British Association. *Exeter British Association Report*, 1869, 1–9. (Reprinted in *Sylvester 1908*, pp. 650–61, and in this collection, Vol. 1, pp. 511–22.)

1870 *Laws of verse; or, Principles of versification exemplified in metrical translations, together with an annotated reprint of the inaugural Presidential address to the Mathematical and Physical Section of the British Association at Exeter*. Longmans, London.

1908 *The collected mathematical papers of James Joseph Sylvester*, Vol. 2. Cambridge University Press.

Synge, John L.

1937 *Geometrical optics: an introduction to Hamilton's method*. Cambridge University Press.

Tarski, Alfred

1941 On the calculus of relations. *Journal of Symbolic Logic*, **6**, 73–89.

Taurinus, Franz Adolph

1826 Geometriae prima elementa. Cologne.

Turnbull, Herbert Western

1947 Colin MacLaurin. *American Mathematical Monthly*, **54**, 318–22.

1951 *Bi-centenary of the death of Colin MacLaurin*. University Press, Aberdeen.

van Heijenoort, Jean (ed.)

1967 *From Frege to Gödel: A source book in mathematical logic, 1879–1931*. Harvard University Press, Cambridge, Mass. (Second printing 1971.)

van Rootselaar, Bob

1962–6 Bolzano's theory of real numbers. *Archive for History of Exact Sciences*, **2**, 168–80.

1976 Zermelo, Ernst Friedrich Ferdinand. In *Gillispie* et alii *1970–6*, Vol. 14, pp. 613–16.

Vilant, Nicholas

1798 *The elements of mathematical analysis, abridged, for the use of students*. F.W. Wingrave, London.

von Waltershausen, W. Sartorius

1856 Metaphysik der Geometrie. In *Gauss 1863–1929*, Vol. 8, pp. 267–8 (1900). (This article is an excerpt from von Waltershausen's biographical recollections of Gauss in the memorial volume, *Gauss zum Gedächtnis*, pp. 80–1, Leipzig. The title 'Metaphysik der Geometrie' was supplied by the editors of Gauss's *Werke*.)

Vorländer, Karl

1977 *Immanuel Kant: der Mann und das Werk*, 2 vols, (2nd edn). Meiner, Hamburg.

Wallis, John

1685 *A treatise of algebra, both historical and practical*. J. Playford, London.

1693 *De postulato quinto et definitione quinta lib. 6. Euclidis disceptatio geometrica*. (Lecture on Euclid delivered in Oxford on the evening of 11 July 1663; reprinted in *Opera mathematica*, Vol. 2. German translation in *Stäckel and Engel 1895*, pp. 17–30.)

1695–9 *Opera mathematica*, 2 vols. Oxford University Press. (Vol. 1 contains the *Arithmetica infinitarum. Opera* reprinted 1968, Georg Olms, Hildesheim.)

Warda, Arthur
1922 *Immanuel Kants Bücher*. Breslauer, Berlin.

Warnock, Geoffrey J.
1982 *Berkeley*, (2nd edn). Basil Blackwell, Oxford. (1st edn: 1953.)

Warren, John
1828 *A treatise on the geometrical representation of the square roots of negative quantities*. J. Smith, Cambridge.

Weyl, Hermann
1944b David Hilbert and his mathematical work. *Bulletin of the American Mathematical Society*, **50**, 612–54. (Reprinted in *Weyl 1968*, pp. 130–72.)
1968 *Gesammelte Abhandlungen*, 4 vols, (ed. Komaravolu Chandrasekharan). Springer Verlag, Berlin.

Whately, Richard
1832 *Elements of logic*. William Jackson, New York.

Zeller, Eduard
1875 *Geschichte der deutschen Philosophie seit Leibniz*, (2nd edn). Oldenbourg, Munich.

Zermelo, Ernst
1930* Über Grenzzahlen und Mengenbereiche: Neue Untersuchungen über die Grundlagen der Mengenlehre. *Fundamenta Mathematicae*, **16**, 29–47. (Translated by Michael Hallett in this collection, Vol. 2, pp. 1219–33.)

Index to Volume 1

The following index is to the present volume only. The index at the conclusion of Volume 2 includes references to both volumes in this collection. An attempt has been made to supply full names; when this was not possible, or when the attribution of a name is based on conjecture, the uncertain information is enclosed in square brackets.

mmanuel Kant's *Critique of* ... int
of the modern philosophy of ... great
nainstream mathematician to ... t
ad their origin in the works of Kant. This two-volume work provides an
overview of this important era of foundational research through a carefully
chosen selection of articles. They provide insight into the foundations of each of
the main branches of mathematics—algebra, geometry, number theory, analysis,
ogic, and set theory. A scholarly commentary supplies historical background and
explains how the various mathematical and philosophical themes are interlinked.

Classic works by Bolzano, Riemann, Dedekind, and Poincaré are reproduced in
reliable translations and many selections from writers such as Gauss, Cantor,
Kronecker, and Zermelo are translated here for the first time. The collection is
an invaluable resource for anyone wishing to gain an understanding of the
foundation of modern mathematics.

From Reviews:

'... well prepared and paves the way to a profound study of the concerned subject.'
E. Knobloch, Zentralblatt für Mathematik

'... the book is a wide-ranging and substantial contribution to the accessibility of writings on the foundation of various branches of mathematics.'
I. Grattan-Guinness, History and Philosophy of Logic

OXFORD
UNIVERSITY PRESS

www.oup.com